Lecture Notes in Electrical Engineering

Volume 472

"Lecture Notes in Electrical Engineering (LNEE)" is a book series which reports the latest research and developments in Electrical Engineering, namely:

- Communication, Networks, and Information Theory
- Computer Engineering
- Signal, Image, Speech and Information Processing
- Circuits and Systems
- Bioengineering

LNEE publishes authored monographs and contributed volumes which present cutting edge research information as well as new perspectives on classical fields, while maintaining Springer's high standards of academic excellence. Also considered for publication are lecture materials, proceedings, and other related materials of exceptionally high quality and interest. The subject matter should be original and timely, reporting the latest research and developments in all areas of electrical engineering.

The audience for the books in LNEE consists of advanced level students, researchers, and industry professionals working at the forefront of their fields. Much like Springer's other Lecture Notes series, LNEE will be distributed through Springer's print and electronic publishing channels.

More information about this series at http://www.springer.com/series/7818

Vijay Janyani · Manish Tiwari
Ghanshyam Singh · Paolo Minzioni
Editors

Optical and Wireless Technologies

Proceedings of OWT 2017

Springer

Editors
Vijay Janyani
Department of Electronics and Communication Engineering
Malaviya National Institute of Technology Jaipur
Jaipur, Rajasthan
India

Ghanshyam Singh
Department of Electronics and Communication Engineering
Malaviya National Institute of Technology Jaipur
Jaipur, Rajasthan
India

Manish Tiwari
Department of Electronics and Communication Engineering
Manipal University Jaipur
Jaipur, Rajasthan
India

Paolo Minzioni
Department of Electrical, Computer, and Biomedical Engineering
University of Pavia
Pavia
Italy

ISSN 1876-1100 ISSN 1876-1119 (electronic)
Lecture Notes in Electrical Engineering
ISBN 978-981-10-7394-6 ISBN 978-981-10-7395-3 (eBook)
https://doi.org/10.1007/978-981-10-7395-3

Library of Congress Control Number: 2017958733

Printed on acid-free paper

This Springer imprint is published by Springer Nature
The registered company is Springer Nature Singapore Pte Ltd.
The registered company address is: 152 Beach Road, #21-01/04 Gateway East, Singapore 189721, Singapore

Preface

Optical and wireless technologies are advancing at an accelerating rate recently. The traditional approaches to providing high data rates to the masses are transforming and expanding in a way that is beyond our imagination. The challenges in providing uninterrupted data and broadband communications have not changed. Our mission as a technical community is to understand these challenges and find ways to mitigate them. This includes the development and management of appropriate channels, novel devices, new protocols, efficient networks, and their integration. Keeping in view the amalgamation of these issues, the proceedings of First International Conference on Optical and Wireless Technologies (OWT 2017) is being presented herewith.

The conference (OWT 2017) was held in the campus of Malaviya National Institute of Technology Jaipur during July 1–2, 2017. A total of 175 participants including the invited speakers, contributing authors, and attendees participated in this conference. The participants were explored to a broad range of topics critical to our society and industry in the related areas. The conference provided an opportunity to exchange ideas among global leaders and experts from academia and industry on topics such as optical materials, optical signal processing and networking, photonic communication systems and networks, all-optical systems, microwave photonics, optical devices for optical communications, nonlinear optics, nanophotonics, software-defined and cognitive radio, signal processing for wireless communications, antenna systems, spectrum management and regulatory issues, vehicular communications, wireless sensor networks, machine-to-machine communications, cellular-Wi-fi integration.

Apart from a high-quality contributed paper presented by delegates from all over the country and abroad, the conference participants also witnessed informative demonstrations and technical sessions from the industry as well as invited talks from renowned experts aimed at advances in these areas. Overall response to the conference was quite encouraging. A large number of papers were received. After a rigorous editorial and review process, 80 papers were invited for presentation during the conference. Among the presented papers, 77 papers were selected for inclusion in the conference proceedings. We are confident that the papers presented in this proceedings shall provide a platform for young as well as experienced professionals to generate new ideas and networking opportunities.

The editorial team members would like to extend gratitude and sincere thanks to all the contributed authors, reviewers, panelists, local organizing committee members, and the session chairs for paying attention to the quality of the publication. We are thankful to our sponsors for generously supporting this event and institutional partner (Malaviya National Institute of Technology Jaipur) for providing all the necessary support, encouragement, and infrastructure in this beautiful campus. Finally, we pay the highest regard to the Irisworld Science & Technology Education and Research (IRISWORLD), a "Not-for-Profit" society from Jaipur, for extending their support for financial management of the OWT 2017.

Best wishes from

Jaipur, India — Dr. Vijay Janyani
Jaipur, India — Prof. Manish Tiwari
Jaipur, India — Dr. Ghanshyam Singh
Pavia, Italy — Dr. Paolo Minzioni

Acknowledgements

The editors wish to extend their heartfelt acknowledgment to all contributing authors, esteemed reviewers for their timely response, members of various organizing committee, and production staff whose diligent work put shape to the OWT 2017 proceedings. We especially thank our dedicated reviewers for their volunteering efforts to check the manuscript thoroughly to maintain the technical quality and for their useful suggestions.

We also pay our best regards to the faculty members from institutional partners (Malaviya National Institute of Technology Jaipur and Manipal University Jaipur) for extending their enormous assistance during the conference-related assignments, especially to Mr. Dinesh Kumar Yadav and Mr. Lokesh Kumar from Manipal University Jaipur; Sanjeev Yadav, Ashok Kumar, and Jitendra Deegwal from Government Engineering College Ajmer; Mr. Nidhish Tiwari from JNIT Jaipur; Mr. Ramesh Dewanda, Executive Member, Irisworld Society Jaipur; and Sh. Narendra Godara from Malaviya National Institute of Technology Jaipur.

We also acknowledge the financial support received from our esteemed sponsors such as Viavi; Agmatel; JV Micronics; Keysight Technologies; SriDutt Technologies Private Limited, Bangalore; Photon Design; Rohde & Schwarz India Pvt. Ltd.; Fiber Optic Services; Eigen Technologies; IETE Rajasthan Centre Jaipur; ni logic Pvt. Ltd., Pune; and OSA student chapter of Malaviya National Institute of Technology Jaipur. Finally, we extend our sincere thanks to Springer for agreeing to be our publishing partner. Specially, the efforts made by Swati Meherishi, Executive Editor, and Aparajita Singh, Senior Editorial Assistant, from the Applied Sciences and Engineering group at Springer, are highly appreciated.

Dr. Vijay Janyani
Prof. Manish Tiwari
Dr. Ghanshyam Singh
Dr. Paolo Minzioni

Acknowledgements

Organizing Committee

Chief Patron

Prof. Udaykumar R. Yaragatti, Director, Malaviya National Institute of Technology Jaipur

Patrons

Prof. Deepak Bhatnagar, Director, CCT, University of Rajasthan, Jaipur
Mr. B. Singh, Director, ETDC Jaipur

General Chairs

Dr. Vijay Janyani, Malaviya National Institute of Technology Jaipur
Dr. Ghanshyam Singh, Malaviya National Institute of Technology Jaipur
Dr. Manish Tiwari, Manipal University Jaipur

Conveners

Mr. Dinesh Yadav, Manipal University Jaipur
Dr. Jitendra Deegwal, Government Engineering College Ajmer

Organizing Secretaries

Mr. Nidhish Tiwari, JNIT Jaipur
Dr. Vipin Pal, Manipal University Jaipur

IETE Oversight Committee

Dr. M. M. Sharma, FIETE
Prof. S. K. Bhatnagar, FIETE
Mr. K. M. Bajaj, FIETE

Sponsorship Committee

Dr. Lokesh Tharani, RTU Kota
Dr. Sanjeev Jain, EC Bikaner
Mr. Sanjeev Yadav, Government Women Engineering College Ajmer
Mr. Ashok Kumar, Government Women Engineering College Ajmer
Mr. Narendra Kr. Godara, Malaviya National Institute of Technology Jaipur
Mr. Chetan Selwal, Government Women Engineering College Ajmer

Registration Committee

Dr. Narendra Kumar Yadav, JECRC University, Jaipur
Dr. Arun Dev D. Dwivedi, Poornima University, Jaipur
Dr. Jankiballabh Sharma, RTU Kota
Mr. Sandeep Vyas, VIT Jaipur
Mr. Upendra Choudhary, Malaviya National Institute of Technology Jaipur
Ms. Kanchan Sengar, JECRC University, Jaipur

Publication Committee

Dr. Ravi Kumar Maddila, Malaviya National Institute of Technology Jaipur
Dr. Seema Varma, Banasthali University, Tonk
Dr. Dhirendra Mathur, RTU Kota
Dr. Rekha Mehra, Government Engineering College Ajmer
Mr. Lokesh Sharma, Manipal University Jaipur
Ms. Dimple Bansal, SGVU Jaipur

Hospitality Committee

Dr. Vivekanand Tiwari, Manipal University Jaipur
Dr. Ashok Sirohi, RCEW Jaipur
Mr. K. R. Yadav, Khaitan Polytechnic College, Jaipur
Mr. Mukesh Kumar Gupta, Malaviya National Institute of Technology Jaipur
Mr. Sourabh Sahu, Malaviya National Institute of Technology Jaipur
Mr. Saurabh Maheshwari, Government Women Engineering College Ajmer

International Advisory Committee

Prof. Hiroyuki Tsuda, Keio University, Japan
Prof. Ali Gharsallah, University of Tunis E. M., Tunisia
Prof. Buryy Oleh Anatolievyach, LPNU, Ukraine
Prof. Ajoy Kar, HW University, Edinburgh, UK
Dr. Suchandan Pal, CEERI, Pilani, India
Prof. Kolin Poul, IIT Delhi, India
Prof. Konstantin Kozadaev, BSU, Minsk, Belarus
Dr. Lotfi Osman, University of Carthage, Tunisia
Prof. Mário F. S. Ferreira, University of Aveiro, Portugal
Dr. Miklos Veres, HAS, Budapest, Hungary
Prof. Sergii Ubizskii, LPNU, Ukraine
Prof. Yuri Shpolyanskiy, University of Saint Petersburg, Russia

Technical Program Committee

Dr. Akshay Kr. Rathore, Concordia University, Canada
Prof. Takasumi Tanabe, Keio University, Japan
Prof. Toshiharu Saiki, Keio University, Japan
Dr. Bishnu Prasad Gautam, WAKHOK, Japan
Dr. Reza Abdi-Ghaleh, University of Bonab, Iran
Dr. Kalpana Dhaka, IIT Guwahati, India
Dr. Upena D. Dalal, SVNIT, Surat
Prof. Rajeev Gupta, RTU Kota
Prof. Mithiliesh Kumar, RTU Kota
Dr. Manish Mathew, CEERI, Pilani, India
Dr. Preetam Kumar, IIT Patna
Dr. C. Periasamy, Malaviya National Institute of Technology Jaipur
Dr. Seema Verma, Banasthali University, Tonk
Dr. Sanjeev Kumar Metya, NIT, Arunachal Pradesh
Dr. Narendra Kumar Yadav, JECRC University, Jaipur
Dr. Bramha P. Pandey, GLA University, Mathura
Dr. Sanyog Rawat, Manipal University, Jaipur
Dr. Lokesh Tharani, RTU Kota
Dr. Girish Parmar, RTU Kota
Dr. Anil Yadav, Amity University Gurgaon
Dr. Dinesh Goyal, SGVU Jaipur
Dr. Ashok Sirohi, RCEW Jaipur

Our Reviewers

Mr. Surendra Agarwal, NIT Kurukshetra, email: skagarwal5@rediffmail.com
Dr. Mushtaq Ahmed, Malaviya National Institute of Technology Jaipur, India, email: mahmed.cse@mnit.ac.in
Dr. Parvez Alvi, Banasthali University, email: drpaalvi@gmail.com
Dr. Mohd. Samar Ansari, Malaviya National Institute of Technology Jaipur, email: msansari.ece@mnit.ac.in
Dr. Ashwini Arya, KAIST, Republic of Korea, email: ashwiniarya.iitr@gmail.com
Mr. Ramesh Battula, Malaviya National Institute of Technology Jaipur, email: ramsbattula@gmail.com
Dr. Sanjay Dabhole, Sant Gajanan Maharaj Rural Polytechnic, email: sankop1217@gmail.com
Dr. Upena Dalal, Sardar Vallabhbhai National Institute of Technology, Surat, email: upena_dalal@yahoo.com
Dr. Jitendra Deegwal, Government Engineering College Ajmer, Rajasthan, email: jitendradeegwal@gmail.com
Dr. Kalpana Dhaka, Indian Institute of Technology Guwahati, email: dhaka.kalpana@gmail.com
Dr. Tarun Dubey, Manipal University Jaipur, email: tarunkumar.dubey@jaipur.manipal.edu
Dr. Umesh Dwivedi, Amity University Rajasthan, email: umeshkudwivedi@gmail.com
Mr. Kuldeep Goswami, Government Women Engineering College Ajmer, email: kuldeep4career@gmail.com
Dr. Dinesh Goyal, Suresh Gyan Vihar University, email: dinesh8dg@gmail.com
Mr. Mukesh Gupta, Malaviya National Institute of Technology Jaipur, email: mkgupta06@gmail.com
Dr. Nikhil Deep Gupta, Malaviya National Institute of Technology Jaipur, email: 2012rec9530@mnit.ac.in
Dr. Tawfik Ismail, NILES, Cairo University, email: tawfik@niles.edu.eg
Dr. Shruti Jain, JUIT, Waknaghat, email: shruti.jain@juit.ac.in
Dr. Nagesh Janrao, Technical Education, email: janrao.nl@gmail.com
Dr. Alok Joshi, JIIT, Noida, email: 20.alok@gmail.com

Dr. Amit Joshi, Malviya National Institute of Technology,
email: amjoshi.ece@mnit.ac.in
Dr. Rajesh Khanna, Thapar University, email: rkhanna@thapar.edu
Dr. Vijaya kumar Krishnasamy, Malaviya National Institute of Technology Jaipur,
email: vijayk.ee@mnit.ac.in
Dr. Arjun Kumar, Intel, email: akdec.iitr@gmail.com
Mr. Ashok Kumar, Government Women Engineering College Ajmer, India,
email: kumarashoksaini@gmail.com
Dr. Preetam Kumar, Indian Institute of Technology Patna, email: pkumar@iitp.ac.in
Dr. Ravi Kumar Maddila, Malaviya National Institute of Technology Jaipur,
email: rkmaddila.ece@mnit.ac.in
Mr. Saurabh Maheshwari, Government Women Engineering College Ajmer,
email: dr.msaurabh@gmail.com
Dr. Manish Mathew, CEERI, Pilani, email: manish.mathew@gmail.com
Ms. Monika Mathur, Rajasthan Technical University,
email: monikamathur16@gmail.com
Mr. Sudarshan Maurya, Government Mahila Engineering College,
email: maurya.sudarshan@gmail.com
Dr. Arka Prokash Mazumdar, Malaviya National Institute of Technology Jaipur,
email: apmazumdar.cse@mnit.ac.in
Dr. Rekha Mehra, Government Engineering College Ajmer,
email: mehrarekha710@gmail.com
Dr. Sanjeev Metya, National Institute of Technology Arunachal Pradesh,
email: smetya@ieee.org
Dr. Satyasai Nanda, Malaviya National Institute of Technology Jaipur,
email: nanda.satyasai@gmail.com
Prof. Lotfi Osman, Higher School of Communication of Tunis, University of Carthage, email: lotfi.osman@supcom.tn
Dr. Vipin Pal, Manipal University Jaipur, email: vipinrwr@yahoo.com
Dr. Trilochan Panigrahi, National Institute of Technology Goa,
email: tpanigrahi80@gmail.com
Dr. Girish Parmar, Rajasthan Technical University,
email: girish_parmar2002@yahoo.com
Dr. C. Periasamy, Malaviya National Institute of Technology Jaipur,
email: cpsamy.ece@mnit.ac.in
Dr. Emmanuel Pilli, Malaviya National Institute of Technology Jaipur,
email: espilli.cse@mnit.ac.in
Mr. Pravin Prajapati, Gujarat Technological University,
email: ec.pravin.prajapati@adit.ac.in
Dr. Sanyog Rawat, Manipal University, email: sanyog.rawat@jaipur.manipal.edu
Prof. Kanad Ray, Amity University Rajasthan, email: kanadray00@gmail.com
Dr. Chitrakant Sahu, Malaviya National Institute of Technology Jaipur,
email: chitrakant.ece@mnit.ac.in
Dr. Preeta Sharan, The Oxford College of Engineering, Bangalore,
email: sharanpreeta@gmail.com

Dr. Jankiballabh Sharma, Rajasthan Technical University, email: jbsharma@rtu.ac.in
Ms. Neeru Sharma, Jaypee University of Information Technology, email: neeru.sharma@juit.ac.in
Dr. Ritu Sharma, Malaviya National Institute of Technology Jaipur, email: rsharma.ece@mnit.ac.in
Dr. Sumit Srivastava, Manipal University Jaipur, email: sumit.310879@gmail.com
Dr. Fazal Talukdar, National Institute of Technology Silchar, email: fatalukdar@gmail.com
Dr. MeenakshiTripathi, Malaviya National Institute of Technology Jaipur, email: mtripathi.cse@mnit.ac.in
Dr. Yogesh Trivedi, Nirma University, email: yogesh.trivedi@nirmauni.ac.in
Prof. Hiroyuki Tsuda, Keio University, email: tsuda@elec.keio.ac.jp
Prof. Manisha Upadhyay, Nirma University, email: manisha.upadhyay@nirmauni.ac.in
Dr. Karan Verma, Universiti Teknologi Petronas, email: vermakara@gmail.com
Dr. Pankaj Verma, National Institute of Technology Kurukshetra, email: pankaj@nitkkr.ac.in
Mr. Santosh Vipparthi, Malaviya National Institute of Technology Jaipur, email: skvipparthi@mnit.ac.in
Dr. Rajesh Vishwakarma, JUET Noida, email: rajesh.vishwakarma@juet.ac.in
Dr. Sandeep Vyas, Malaviya National Institute of Technology Jaipur, email: vyas.sandeep@vitej.ac.in
Mr. Dinesh Yadav, Manipal University Jaipur, email: dinesh.yadav@jaipur.manipal.edu
Dr. Narendra Yadav, JECRC University, email: narensinghyadav@yahoo.com
Mr. Sanjeev Yadav, Government Women Engineering College Ajmer, email: sanjeev.yadav.in@ieee.org
Dr. Rajveer Yaduvanshi, AIACTR, email: yaduvanshirs007@gmail.com
Dr. Vijay Janyani, Malaviya National Institute of Technology Jaipur, email: vjanyani.ece@mnit.ac.in
Dr. Ghanshyam Singh, Malaviya National Institute of Technology Jaipur, email: gsingh.ece@mnit.ac.in
Dr. Manish Tiwari, Manipal University Jaipur, email: mt.jaipur@gmail.com

Contents

Invited Speakers

Dr. Paolo Minzioni, University of Pavia, Italy (OSA Travelling Lecturer: On invitation by OSA Student Chapter of Malaviya National Institute of Technology Jaipur).

Profile Summary: Dr. Paolo Minzioni completed his Ph.D. degree in Electronics from University of Pavia, Italy, in 2006 and works in the Department of Electrical, Computer, and Biomedical Engineering, University of Pavia. From October to December 2016, he worked as a Visiting Professor at Tufts University, Medford, MA, USA. He is co-investigator of the European Project FABULOUS and also responsible for the University of Pavia of the European Project NISTAS (development of contactless measurement techniques for the evaluation of pulse-wave velocity in the carotid artery). He has been involved in different national research projects, including European projects ATLAS and NOBEL. He was author of different invited seminars on integrated optical devices for biomedical applications and transmission systems in Italy and abroad: CNIT, Pisa, Italy; Aston University, Birmingham, UK; UCI, Irvine, USA; UCLA, Los Angeles, USA; Nankai University, Tianjin, China; Peking University, Beijing, China; Chinese Academy of Sciences, Beijing, China. He has contributed more than 70 research articles to international conferences and published over 50 scientific papers in peer-reviewed journals. He is co-inventor of six international patent applications, regarding optical transmission systems four applications) and optical tweezers for single-particle trapping and manipulation (two applications). He is a guest editor of the special issues on "Silicon Photonics Components and Applications" published by MDPI Applied Sciences; "Cell manipulation and analysis in microfluidic chips" published by De Gruyter Optofluidics, Microfluidics and Nanofluidics; and "Roadmap Paper on Optofluidics" published by Institute of Physics, Journal of Optics. He is also the Member of the Editorial Board of Scientific Reports, Nature Publishing Group, and the Editorial Board of Applied Sciences, MDPI.

Dr. Sigang Yang, Tsinghua University, Beijing, China (SPIE Visiting Lecturer: On invitation by SPIE Student Chapter of Manipal University Jaipur).

Profile Summary: Dr. Sigang Yang received his Ph.D. degree in Physical Electronics from the Department of Electronics Engineering, Tsinghua University, in February 2008. He was a Postdoctoral Fellow in the Department of Electrical and Electronic Engineering at the University of Hong Kong, from August 2008 to July 2010. He joined as faculty at Tsinghua University in August 2010. He is now an Associate Professor in the Department of Electronic Engineering, Tsinghua University. His research fields include computational electromagnetism, photonic crystal fibers, fiber nonlinearities, fiber-based optical parametric amplifications (FOPAs) and oscillators (FOPOs), and terahertz wave and frequency comb. He was the recipient of the Best Student Paper Award in 2004 Asia-Pacific Optical Communication Conference (APOC) and the SPIE Educational Scholarship in Optics & Photonics and 51st SPIE Annual Meeting, San Diego, California, in 2006. His researches, such as broadband dispersion-compensation photonic crystal fiber (PCF) and slow light based on PCF, have been featured by Photonics Spectra both in 2006 and 2008, respectively. He is also the reviewer for Optics Letters, Journal of the Optical Society of America B, Optics Express, Applied Optics, and PIER & JEMWA. He is a member of the Optical Society of America (OSA).

Dr. Tawfik Ismail, Cairo University, Egypt.

Profile Summary: Dr. Tawfik Ismail received his B.Sc. degree (with honors) from the Department of Electronics and Communication, Faculty of Engineering, Cairo University, Giza, Egypt, in 2001. He joined as research assistant in the Department of Engineering Application of Laser (EAL), National Institute of Laser Enhanced Science (NILES), Cairo University, in 2003. He received his M.Sc. and Ph.D. degrees in the Department of EAL in 2009 and 2013, respectively. He is currently an Assistant Professor in the Department of Engineering Applications of Laser, NILES, Cairo University, Egypt. His areas of research interest include optical and wireless networks, IoT, biomedical telemetry, and optoelectronics. He is currently the Co-PI of two successful research projects funded by the National Telecom Regularity Authority (NTRA) and Information Technology Industry Development Agency (ITIDA). These are collaborative projects between University of Cairo, Egypt, and University of Toronto, Canada.

Dr. Hossam A. Selmy, Cairo University, Egypt.

Profile Summary: Dr. Hossam A. Selmy obtained his B.Sc. degree in Electronics and Communication Engineering (ECE) from the Faculty of Engineering, Cairo University, Egypt in 2001; M.Sc. degree in Optical Communication and Networking, from ECE Department, Faculty of Engineering, Egypt, in 2007; and Ph.D. degree in Optical Communication and Networking from Egypt–Japan University for Science and Technology, Egypt, in 2013. Presently, he is Assistant Professor in the Department of Engineering Applications of Laser, NILES, Cairo University, Egypt. He also worked as Research Associate Professor in the Department of Optical Communication, Egypt–Japan University for Science and Technology, Egypt. He was co-investigator (Co-PI) for a research project "Cooperative Resource Allocation for Free Space Optical Communication" funded by STDF, Egypt, during July 2014–June 2015 and presently engaged with another research project "Modeling and implementation of free space optical transceiver for next generation optical communication applications" under the India–Egypt Agreement on Science and Technology Cooperation as Co-PI (2016–2018). His areas of current research interest include optical communication engineering: optical modulation, optical devices, free space optical communication.

Dr. Manish Mathew, CEERI, Pilani.

Profile Summary: Manish Mathew received his M.Tech. degree from Shri. G.S. Institute of Technology and Science, Indore, India, in 2003 and Ph.D. degree from the University of Tokyo, Japan, in 2014. He joined as a Scientist at the Council of Scientific and Industrial Research—Central Electronics Engineering Research Institute (CSIR-CEERI), Pilani, India, in 2005. In 2011, he was awarded a MEXT Scholarship to pursue graduate work in Japan. Currently, he is working on the growth, characterization, and fabrication of III-nitride devices. He has 6 international patents and more than 25 publications in the field of III-N light-emitting diodes (LEDs).

Dr. Debasish Pal, CEERI, Pilani.

Profile Summary: Debasish Pal was born in Burdwan district of West Bengal. He did his schooling at Hindustan Cables Higher Secondary School. He joined as Junior scientist in CSIR-CEERI, Pilani, in 2005. Currently, he is working as a senior scientist in the same institute. He has completed his B.E. degree in Electronics and Communication with distinction from Bangalore University. He obtained his M.Tech. in Microwaves from the University of Burdwan with the first rank (for which he was awarded University Gold Medal) in 2005. He obtained his Ph.D. degree from IIT Roorkee in 2016 in the fields of antenna and metamaterials. Currently, he is working in the field of klystron. He is involved in the development of various klystrons which are sponsored by Bhaba Atomic Research Centre, Mumbai, Defence Research and Development Organization, Institute of Plasma Research, Gandhinagar. He is also involved in the development of terahertz devices. His areas of research interest are metamaterial antenna, photonic band gap structures, defected ground structure antenna, fractal antenna, application of biological optimizing techniques such as PSO, BFO, and ANN on antenna and microwave passive components.

Dr. Ayan Kumar Bandyopadhyay, CEERI, Pilani.

Profile Summary: Ayan Kumar Bandyopadhyay is currently working as Principal Scientist at the Central Electronics Engineering Research Institute, Pilani, India. He has completed his M.Sc. degree in Physics and M.Tech. degree in Microwaves from Burdwan University, Burdwan, West Bengal, in 1998 and 2001, respectively. He has obtained his Ph.D. degree with distinction from Otto-von-Guericke University, Germany, and worked as a Postdoctoral Fellow in the European Synchrotron Radiation Facility (ESRF), France. During his postdoctoral research, he has contributed to the design of the higher-order-mode free radio frequency-accelerating cavities of the ESRF. He was a Visiting Fellow in the Department of Electrical Engineering, Perugia, Italy, and at the Deutsche Elektronen Synchrotron (DESY), Hamburg, Germany. His areas of current research interest include design and development of microwave components, systems and high-power microwave tubes, especially klystrons.

Editors and Contributors

About the Editors

Dr. Vijay Janyani obtained bachelor's and master's degrees in Electronics and Communication Engineering from Malaviya Regional Engineering College, Jaipur (now Malaviya National Institute of Technology Jaipur), and Ph.D. degree from the George Green Institute (GGIEMR) of Nottingham, UK, under Commonwealth Scholarship Plan in 2005. He is the recipient of various awards and honors such as the Derrick Kirk Prize of University of Nottingham, Commonwealth Scholarship UK, and AICTE Career Award. He has completed various national and international government-funded major research projects funded by UKIERI, DRDO, DST, etc., in collaboration with UK, Japan, Egypt, Ukraine, etc. He has been a Visiting Faculty at AIT Bangkok and UoTEM Tunisia. His areas of current research interest include optical communication, optoelectronics and photonics, numerical modeling, nonlinear optics, RF and microwaves, optical networks, and solar energy.

Dr. Manish Tiwari received his Ph.D. degree in ECE in the field of Photonics from Malaviya National Institute of Technology Jaipur. Presently, he is Professor in the Department of ECE, Manipal University Jaipur. He has been Visiting Researcher at City University, London, under UKIERI project in Microstructured Optical Fibers during 2010 and 2011; and Tsinghua University, Beijing, during 2016. He has given talks in PolyU, Hong Kong; KMUT, Bangkok; Kasetsart University, Bangkok; City University, London; and several UKIERI workshops. He has also served on panel of experts in various workshops by CSTT, MHRD, Government of India. His areas of current research interest include micro-/nanostructure photonic devices, nonlinear optics, and photonic crystal fibers.

Dr. Ghanshyam Singh received his bachelor's degree in ECE from National Institute of Technology Silchar and master's and Ph.D. degrees in ECE from Malaviya National Institute of Technology Jaipur. He has been Visiting Scholar/Visiting Professor in the Department of Physics, Herriot-Watt University, Edinburgh, UK; at the Institute of Photonics, University of Eastern Finland, Joensuu campus, Finland (CIMO Fellowship, Government of Finland); and in the Department of EEE, Keio University, Hiyoshi Campus, Japan. He is engaged with joint research projects with partner researchers from Keio University, Japan; University of Vienna, Austria; University of Cairo, Egypt; and LNPU, Lviv, Ukraine. His areas of current research interest include antenna engineering, micro- and nanostructured photonic devices, and networks and nonlinear characteristics of photonic crystal fibers.

Dr. Paolo Minzioni completed his Ph.D. degree in Electronics from the University of Pavia, Italy, in 2006, and works in the Department of Electrical, Computer, and Biomedical Engineering, University of Pavia. From October to December 2016, he worked as a Visiting Professor at Tufts University, Medford, MA, USA. He is co-investigator of the European projects FABULOUS and NISTAS. He has been involved in different national research projects, including European projects ATLAS and NOBEL. He has contributed more than 70 research articles to international conferences and published over 50 scientific papers in peer-reviewed journals. He is co-inventor of six international patent applications.

Contributors

E. R. Aarathy Department of Physics, Amrita University, Kollam, Kerala, India; Tata Institute of Fundamental Research, Mumbai, India

Mahesh P. Abegaonkar Centre for Applied Research in Electronics (CARE), Indian Institute of Technology (IIT) Delhi, Hauz Khas, New Delhi, India

Mais F. Abid Electrical Technical Engineering College, Baghdad, Iraq

Rupesh Acharya Poornima College of Engineering, Rajasthan Technical University, Jaipur, India

R. Adeline Mellita Department of Electronics Engineering, Indian Institute of Information Technology, Design and Manufacturing, Kancheepuram, Chennai, India

S. K. Agarwal Government Women Engineering College Ajmer, Ajmer, Rajasthan, India

Sachin Agrawal Discipline of Electronics & Communication Engineering, Indian Institute of Information Technology, Design & Manufacturing Jabalpur, Jabalpur, India

P. A. Alvi Department of Physics, Banasthali University, Rajasthan, India

Shivani Ameta Department of ECE, Government Women Engineering College, Ajmer, Rajasthan, India

S. G. Anjum Department of Electronics, F/o Engineering and Technology, Aligarh Muslim University, Aligarh, UP, India

Payal Awwal Computer Science, Gweca, Ajmer, India

Rubal Bansal Computer Science, Gweca, Ajmer, India

Vani Bansal National Institute of Technology, Kurukshetra, Haryana, India

Pourush Bara Electronics and Communication Engineering, Manipal University Jaipur, Jaipur, Rajasthan, India

Ramesh Babu Battula Department of CSE, Malaviya National Institute of Technology, Jaipur, India

Sangama Bhadouria Pranveer Singh Institute of Technology, Kanpur, India

Garima Bhardwaj Department of Electronics, Banasthali University, Rajasthan, India

Deepak Bhatnagar Microwave Lab, Department of Physics, University of Rajasthan, Jaipur, Rajasthan, India

Udhav Bhosle Department of Electronics and Telecommunication Engineering, Rajiv Gandhi Institute of Technology, University of Mumbai, Mumbai, India

Vipra Bohara Electronics Communication, Kautilya Institute of Technology and Engineering, Jaipur, India

Oleh Buryy Institute of Telecommunications, Radioelectronics and Electronic Engineering, Lviv Polytechnic National University, Lviv, Ukraine

Rekha Chahar Government Women Engineering College Ajmer, Ajmer, India

Ritesh Chandra Department of ECE, Malaviya National Institute of Technology, Jaipur, Rajasthan, India

S. Chandra Department of Electronics Engineering, Indian Institute of Technology (Indian School of Mines), Dhanbad, India

D. S. Chandu Department of Electronics Engineering, Indian Institute of Information Technology, Design and Manufacturing, Kancheepuram, Chennai, India

Ashish Chaudhary Government Engineering College, Ajmer, India

Santosh Chaudhary Department of Mathematics, Malaviya National Institute of Technology Jaipur, Jaipur, Rajasthan, India

Ashok Chauhan Optoelectronic Devices Group, Council of Scientific and Industrial Research-Central Electronics and Engineering Research Institute (CSIR-CEERI), Pilani, Rajasthan, India

Naresh Pratap Singh Chauhan Government Engineering College, Ajmer, India

Suchismita Chinara Department of Computer Science and Engineering, National Institute of Technology, Rourkela, India

Dungar Nath Chouhan Department of Electronics and Communication Engineering, Manipal University Jaipur, Jaipur, India

V. Civin Department of Electronics and Communication, Rajiv Gandhi Institute of Technology, Kottayam, Kerala, India

Nandanwar Chetan Damodar Department of Computer Science and Engineering, National Institute of Technology, Rourkela, India

Utpal Das Center for Lasers and Photonics, IIT, Kanpur, UP, India

Sanjay Deshmukh Department of Electronics and Telecommunication Engineering, Rajiv Gandhi Institute of Technology, University of Mumbai, Mumbai, India; Ramrao Adik Institute of Technology, Nerul, Navi Mumbai, India

Abhinav Deshpande Department of Electronics and Communication Engineering, Malviya National Institute of Technology, Jaipur, India

Julie Devi Department of Electronics and Communication Engineering, Malaviya National Institute of Technology Jaipur, Jaipur, Rajasthan, India

Sugak Dmytro Institute of Telecommunications, Radioelectronics and Electronic Engineering, Lviv Polytechnic National University, Lviv, Ukraine

Prateek Dolas Department of Electronics and Communication Engineering, Indian Institute of Technology Roorkee, Roorkee, India

Tarun Kumar Dubey Department of Electronics and Communication Engineering, Manipal University Jaipur, Jaipur, India

Manu Elappila Department of Computer Science and Engineering, National Institute of Techonology, Rourkela, India

Essar Farooq Department of Electronics and Communication Engineering, Shri Mata Vaishno Devi University, Kakryal, Katra, Jammu & Kashmir, India

Kumkum Garg Manipal University Jaipur, Dahmi Kalan, Jaipur, Rajasthan, India

Monika Garg Government Women Engineering College Ajmer, Ajmer, India

Shweta Garg Government Women Engineering College Ajmer, Ajmer, India

Amar Ghar Center for Lasers and Photonics, IIT, Kanpur, UP, India

Zaineb Gharsallah National Engineering School of Tunis Communications Systems Laboratory (SysCom), University of Tunis El Manar, Ariana, Tunisia; Department of Electronics & Communication, Malaviya National Institute of Technology, Jaipur, India

D. Ghosh Department of Electronics and Communication Engineering, Indian Institute of Technology Roorkee, Roorkee, India

Ashish Kumar Ghunawat Department of Electronics and Communication Engineering, Malviya National Institute of Technology, Jaipur, India

Suparna Goswami Department of ECE, NERIST, Nirjuli, Arunachal Pradesh, India

Mukta Goyal Department of Computer Science and Information Technology, Jaypee Institute of Information Technology, Noida, India

Ashok Gundumalla Discipline of Electronics & Communication Engineering, Indian Institute of Information Technology, Design & Manufacturing Jabalpur, Jabalpur, India

Nikhil Deep Gupta Electronics and Communication Department, Malaviya National Institute of Technology, Jaipur, Rajasthan, India

Ravi Gupta Government Engineering College Bharatpur, Bharatpur, Rajasthan, India

S. C. Gupta IIT Delhi, New Delhi, India

Sachin Kumar Gupta Department of Electronics and Communication Engineering, Shri Mata Vaishno Devi University, Kakryal, Katra, Jammu & Kashmir, India

M. A. Hassan Department of Mechanical Engineering, Birla Institute of Technology, Patna Campus, Patna, India

Pawan Kumar Inaniya Department of ECE, Government Women Engineering College, Ajmer, Rajasthan, India

Anjali Jain Department of Electronics and Communication, MNIT Jaipur, JLN Marg, Jaipur, Rajasthan, India

Jyoti Jain Department of Electronics & Communication Engineering, Rajasthan College of Engineering for Women, Rajasthan, India

P. C. Jain EE Department, School of Engineering, Shiv Nadar University, G. Noida, UP, India

Ruchika Jain Swami Keshvanand Institute of Technology, Management & Gramothan, Jaipur, India

Sanjeev Jain Department of Electronics and Communication Engineering, Government Engineering College, Bikaner, India

Shubhi Jain Swami Keshvanand Institute of Technology, Management & Gramothan, Jaipur, India

Alka Jakhar Electronics and Communication Department, Malaviya National Institute of Technology, Jaipur, Rajasthan, India

Anil Kumar Jangir Department of Electronics and Communication Engineering, Malaviya National Institute of Technology, Jaipur, India

Vijay Janyani Department of Electronics and Communication Engineering, Malaviya National Institute of Technology, Jaipur, India

Priya Jyotiyana Department of Computer Science and Engineering, Government Women Engineering College, Ajmer, India

Pallavi Kaliyar SKIT, College of Engineering, Jaipur, India

Shruti Kalra Department of ECE, Jaipur Engineering College & Research Centre, Jaipur, India

Kamaljeet Singh Kalsi Department of Computer Science, GGITS, Jabalpur, MP, India

Vinay Kanungo SKIT, Jaipur, India

S. S. Karthikeyan Department of Electronics Engineering, Indian Institute of Information Technology, Design and Manufacturing, Kancheepuram, Chennai, India

Amritveer Kaur Department of Electronics and Communication Engineering, Malaviya National Institute of Technology Jaipur, Jaipur, Rajasthan, India

Ushaben Keshwala Electronics and Communication Engineering Department, Amity School of Engineering, Amity University, Noida, Uttar Pradesh, India

Keyur G. Khatsuriya Shantilal Shah Engineering College, Bhavnagar, Gujarat, India

Shiban K. Koul Centre for Applied Research in Electronics (CARE), Indian Institute of Technology (IIT) Delhi, Hauz Khas, New Delhi, India

Ashok Kumar USICT, GGSIPU, Dwarka, India

Durgesh Kumar Poornima College of Engineering, Rajasthan Technical University, Jaipur, India

Harsh Kumar Department of Electronics and Communication Engineering, Malaviya National Institute of Technology, Jaipur, India

Manish Kumar Thukral Department of Electronics Engineering, IIT(ISM), Dhanbad, Jharkhand, India

N. S. Kumar Tata Institute of Fundamental Research, Mumbai, India

Sanjay Kumar Department of Electronics and Communication Engineering, Malaviya National Institute of Technology, Jaipur, India

V. Kumar Department of Electronics Engineering, Indian Institute of Technology (Indian School of Mines), Dhanbad, India

Guddan Kumari Department of Electronics and Communication Engineering, Government Women Engineering College, Ajmer, India

Neha Kumari Department of Electronics and Communication Engineering, Birla Institute of Technology, Patna Campus, Patna, India

Richa Kumari Department of Computer Science and Engineering, Government Mahila Engineering College, Ajmer, Rajasthan, India

Kumarswamy RVCE, Bangalore, India

Bhanu Priya Kumawat Government Women Engineering College Ajmer, Ajmer, India

Ishwar Ram Kumawat Department of Electronics and Communication Engineering, Malviya National Institute of Technology, Jaipur, Rajasthan, India

Soma Kumawat MNIT, Jaipur, Rajasthan, India

Chhagan Lal University of Padova, Padova, Italy

Fateh Lal Lohar Department of Electronics and Communication Engineering, Malviya National Institute of Technology, Jaipur, India

Ravi Kumar Maddila Department of Electronics and Communication Engineering, Malviya National Institute of Technology, Jaipur, Rajasthan, India

Saurabh Maheshwari Computer Science, Gweca, Ajmer, India

Mandvi Department of Electronics and Communication Engineering, Malviya National Institute of Technology, Jaipur, India

Baria Dipikaben Manharbhai Department of Electronics and Communication Engineering, Malaviya National Institute of Technology, Jaipur, India

Kishor G. Maradia EC Department, GEC, Gandhinagar, Gujarat, India

Natalia Martynyuk Semiconductor Electronics Department, Lviv Polytechnic National University, Lviv, Ukraine

Manish Mathew Optoelectronic Devices Group, Council of Scientific and Industrial Research-Central Electronics and Engineering Research Institute (CSIR-CEERI), Pilani, Rajasthan, India

Garima Mathur Poornima College of Engineering, Rajasthan Technical University, Jaipur, India

Rohit Mathur Department of Electronics and Communication Engineering, Manipal University Jaipur, Jaipur, India

Sudarshan Maurya Government Mahila Engineering College, Ajmer, Rajasthan, India

Santosh Meena Government Women Engineering College Ajmer, Ajmer, India

S. Meghana EE Department, School of Engineering, Shiv Nadar University, G. Noida, UP, India

Ajaz Hussain Mir Department of Electronics and Communication Engineering, National Institute of Technology Srinagar, Srinagar, India

Mahmood F. Mosleh Electrical Technical Engineering College, Baghdad, Iraq

Saba Mushtaq Department of Electronics and Communication Engineering, National Institute of Technology Srinagar, Srinagar, India

Kalpana Naidu Department of ECE, Indian Institute of Information Technology-Kota, Kota, India

Monia Najjar National Engineering School of Tunis Communications Systems Laboratory (SysCom), University of Tunis El Manar, Ariana, Tunisia; Higher Institute of Computer Science of El Manar (ISI), University of Tunis El Manar, Ariana, Tunisia

Satyasai Jagannath Nanda Department of Electronics and Communication Engineering, Malviya National Institute of Technology, Jaipur, Rajasthan, India

Vijayakumar Narayanan Department of Electronics and Communication, Government Engineering College, Thiruvananthapuram, Kerala, India

Harshal Nigam Swami Keshvanand Institute of Technology, Management & Gramothan, Jaipur, India

Kumari Nikita Department of Electronics and Communication, MNIT Jaipur, JLN Marg, Jaipur, Rajasthan, India

Darakshanda Noor Government Women Engineering College Ajmer, Ajmer, India

Anuj Kumar Ojha Department of Electrical and Electronics Engineering, BITS Pilani, Pilani Campus, Pilani, Rajasthan, India

Vipin Pal Manipal University, Jaipur, India

P. K. Panigrahi Center for Lasers and Photonics, IIT, Kanpur, UP, India

Trilochan Panigrahi Department of E.C.E, National Institute of Technology Goa, Farmagudi, Goa, India

Manoj Singh Parihar Discipline of Electronics & Communication Engineering, Indian Institute of Information Technology, Design & Manufacturing Jabalpur, Jabalpur, India

Bhupendra M. Parmar GTU, Ahmedabad, Gujarat, India

Anurag Patro Department of Computer Science and Engineering, National Institute of Techonology, Rourkela, India

R. Pradeep Fiber Optics Laboratory, Department of ECE, College of Engineering Trivandrum, Thiruvananthapuram, India

K. B. Pramod MCT's RGIT, Mumbai, India; JAIN University, Bangalore, India

A. V. Praveen Kumar Department of Electrical and Electronics Engineering, BITS Pilani, Pilani Campus, Pilani, Rajasthan, India

Alisha Priya Department of Electronics Engineering, IIT(ISM), Dhanbad, Jharkhand, India

Sanjeev Kumar Raghuwanshi Department of Electronics Engineering, IIT(ISM), Dhanbad, Jharkhand, India

Preeti Rai Department of Computer Science, GGITS, Jabalpur, MP, India

Rajesh Kumar Raj Government Engineering College, Ajmer, India

K. Rajalakshmi Department of Computer Science and Information Technology, Jaypee Institute of Information Technology, Noida, India

Mahavirsingh Rajpurohit MCT's RGIT, Mumbai, India

Abhishek Ranjan OFC and Photonics Laboratory, Department of Electronics and Communication Engineering, National Institute of Technology, Patna, Bihar, India

Rakesh Ranjan OFC and Photonics Laboratory, Department of Electronics and Communication Engineering, National Institute of Technology, Patna, Bihar, India

Arpit Rawankar Tata Institute of Fundamental Research, Mumbai, India

Sanyog Rawat Electronics and Communication Engineering Department, Manipal University, Jaipur, India

Kanad Ray Amity School of Applied Science, Amity University, Noida, Rajasthan, India

I. V. Reshma Department of Electronics and Communication, College of Engineering Trivandrum, Trivandrum, Kerala, India

Rakesh Roshan Pranveer Singh Institute of Technology, Kanpur, India

Anupam Sahu Department of Electronics and Communication Engineering, Madan Mohan Malaviya University of Technology, Gorakhpur, Uttar Pradesh, India

Jyoti Saini Government Women Engineering College Ajmer, Ajmer, Rajasthan, India

Makni Sana National Engineering School of Tunis Communications Systems Laboratory (SysCom), University of Tunis El Manar, Ariana, Tunisia

Sandeep Sancheti Department of ECE, Manipal University Jaipur, Jaipur, Rajasthan, India

K. Sandhya Department of Physics, Banasthali University, Rajasthan, India

Sanyogita Center for Lasers and Photonics, IIT, Kanpur, UP, India

Kishor G. Sawarkar MCT's RGIT, Mumbai, India

Chetan Selwal Department of Electronics and Communication Engineering, Government Women Engineering College, Ajmer, India

Ubizskii Serhij Institute of Telecommunications, Radioelectronics and Electronic Engineering, Lviv Polytechnic National University, Lviv, Ukraine

Shahiruddin Department of Electronics and Communication Engineering, Birla Institute of Technology, Patna Campus, Patna, India

Shalini Department of Electronics and Communication Engineering, Birla Institute of Technology, Patna, India

Vivek Shandil National Institute of Technology, Kurukshetra, Haryana, India

Abha Sharma Department of Electronics & Communication Engineering, Rajasthan College of Engineering for Women, Rajasthan, India

Arvind Sharma Department of ECE, Government Women Engineering College, Ajmer, Rajasthan, India

Devi Prasad Sharma Manipal University Jaipur, Jaipur, India

Indra Bhooshan Sharma Department of Electronics and Communication Engineering, Malviya National Institute of Technology, Jaipur, India

Lokesh Sharma Manipal University Jaipur, Jaipur, India

M. M. Sharma Department of Electronics and Communication Engineering, Malviya National Institute of Technology, Jaipur, India

Mahendra Mohan Sharma Department of Electronics and Communication Engineering, Malaviya National Institute of Technology, Jaipur, Rajasthan, India

Pankaj Kumar Sharma Department of Computer Science and Engineering, Government Mahila Engineering College, Ajmer, Rajasthan, India

Ritu Sharma Department of Electronics and Communication Engineering, Malaviya National Institute of Technology Jaipur, Jaipur, Rajasthan, India

Roopkishor Sharma Government Engineering College, Ajmer, India

Varshali Sharma Department of Electronics and Communication Engineering, Manipal University Jaipur, Jaipur, Rajasthan, India

Karma Sonam Sherpa Sikkim Manipal Institute of Technology, Majhitar, Sikkim, India

M. Shree Prasad Department of E.C.E, National Institute of Technology Goa, Farmagudi, Goa, India

M. J. Siddiqui Department of Electronics, F/o Engineering and Technology, Aligarh Muslim University, Aligarh, UP, India

Priyanka Sidhar Deenbandhu Chotu Ram University of Science & Technology, Murthal, Haryana, India

Poonam Singal Deenbandhu Chotu Ram University of Science & Technology, Murthal, Haryana, India

A. Dinamani Singh Department of ECE, NERIST, Nirjuli, Arunachal Pradesh, India

Dharmendra Kumar Singh OFC and Photonics Laboratory, Department of Electronics and Communication Engineering, National Institute of Technology, Patna, Bihar, India

Ghanshyam Singh Department of Electronics and Communication Engineering, Malaviya National Institute of Technology, Jaipur, India

Kuldip Singh Optoelectronic Devices Group, Council of Scientific and Industrial Research-Central Electronics and Engineering Research Institute (CSIR-CEERI), Pilani, Rajasthan, India

Shefali Singla Deenbandhu Chotu Ram University of Science & Technology, Murthal, Haryana, India

Surya Srivastava National Institute of Technology, Kurukshetra, Haryana, India

Dmytro Sugak Semiconductor Electronics Department, Lviv Polytechnic National University, Lviv, Ukraine; Scientific Research Company 'Carat', Lviv, Ukraine

Kriti Suneja MNIT, Jaipur, India

Ihor Syvorotka Scientific Research Company 'Carat', Lviv, Ukraine

Prateek Thakral National Institute of Technology, Kurukshetra, Haryana, India

Manish Kumar Thukral Manipal University Jaipur, Dahmi Kalan, Jaipur, Rajasthan, India

Manish Tiwari Department of Electronics and Communication Engineering, Manipal University Jaipur, Jaipur, India

Naveen Kumar Tiwari Cognizant Technology Solutions, Noida, Uttar Pradesh, India

Vivekanand Tiwari Department of Electronics and Communication Engineering, Manipal University Jaipur, Jaipur, Rajasthan, India

Sandeep Toshniwal Electronics Communication, Kautilya Institute of Technology and Engineering, Jaipur, India

Meenakshi Tripathi Department of CSE, Malaviya National Institute of Technology, Jaipur, India

Kushal Tuckley R&D, AGV System, Mumbai, India

Serhij Ubizskii Semiconductor Electronics Department, Lviv Polytechnic National University, Lviv, Ukraine

Vaishali Department of ECE, Manipal University Jaipur, Jaipur, Rajasthan, India

Tunisha Varshney Department of Computer Science and Engineering, Central University of Rajasthan, Ajmer, India

Karan Verma Department of Computer Science and Engineering, Central University of Rajasthan, Ajmer, India

N. Vijayakumar GEC Barton Hill, Thiruvananthapuram, Kerala, India

Sandeep Vyas Department of ECE, Vivekanand Institute of Technology, Jaipur, India

Anu Wilson Fiber Optics Laboratory, Department of ECE, College of Engineering Trivandrum, Thiruvananthapuram, India

Uliana Yakhnevych Semiconductor Electronics Department, Lviv Polytechnic National University, Lviv, Ukraine

Anju Yadav Manipal University, Jaipur, India

Dinesh Yadav Electronics and Communication Engineering, Manipal University Jaipur, Jaipur, Rajasthan, India

Sandeep Kumar Yadav Government Women Engineering College Ajmer, Ajmer, India

Sanjeev Yadav Government Women Engineering College Ajmer, Ajmer, India

Nisha Yadav Department of Physics, Banasthali University, Rajasthan, India

R. P. Yadav MNIT, Jaipur, India

Rajveer S. Yaduvanshi USICT, GGSIPU, Dwarka, India

Yogita UIET, Panjab University, Chandigarh, India

Joseph Zacharias Department of Electronics and Communication, Rajiv Gandhi Institute of Technology, Kottayam, Kerala, India

Shalini Department of Electronics and Communication Engineering, Birla Institute of Technology, Patna, India

Analysis of OFDM-MIMO with BPSK Modulation and Different Antenna Configurations Using Alamouti STBC

Sanjay Deshmukh and Udhav Bhosle

Abstract Orthogonal frequency division multiplexing (OFDM) is a multicarrier modulation scheme widely used in 4G. It is effective for reducing the effects of multipath delay in wireless communication. For OFDM system, total delay spread (Td) is less than symbol period (T); it can be combined with antenna arrays at transmitter and receiver (MIMO) to increase diversity gain called as MIMO-OFDM system. MIMO-OFDM improves system capacity on time-variant and frequency-selective channels. The quality of service provided by a system greatly depends upon the correct selection of modulation scheme. In this paper, analysis and simulation of OFDM-MIMO system with different digital modulation schemes like BPSK, 8-PSK, QPSK, and QAM is performed using Alamouti space–time block codes (STBC). STBC uses spatial and time dimensions for error control unlike traditional methods that uses only time dimension. A comparison study shows that for a given data rate and channel conditions, OFDM-MIMO using STBC with BPSK modulation is better compared with other digital modulation techniques in terms of error performance. OFDM-MIMO using STBC with BPSK modulation is further analyzed and simulated for various antenna configurations. It is observed that as numbers of antennas at transmitter and receiver side are increased, the system diversity increases and gives better performance in terms of BER.

Keywords Bit error rate (BER) · Transmit diversity · Orthogonal space–time block codes (OSTBC) · Orthogonal frequency division multiplexing (OFDM) Multi-input multi-output (MIMO) system

S. Deshmukh (✉) · U. Bhosle
Department of Electronics and Telecommunication Engineering,
Rajiv Gandhi Institute of Technology, University of Mumbai, Mumbai, India
e-mail: deshmukhsd25@gmail.com

S. Deshmukh
Ramrao Adik Institute of Technology, Nerul, Navi Mumbai, India

V. Janyani et al. (eds.), *Optical and Wireless Technologies*, Lecture Notes in Electrical Engineering 472, https://doi.org/10.1007/978-981-10-7395-3_1

1 Introduction

In wireless environment, signal propagates from transmitter to receiver via direct, reflected, and scattered paths collectively referred as multipath. Multipath propagation leads to severe selective fading and inter-symbol interference (ISI) of the signal. Orthogonal frequency division multiplexing (OFDM) is used to mitigate the effects of multipath propagation. In OFDM, the high-speed data stream is divided into narrowband data streams that are transmitted simultaneously over a number of subcarriers. As a result, the symbol duration is several times longer than a single carrier system with same symbol rate. When symbol duration is greater than delay spread, OFDM offers inter-symbol interference (ISI) free transmission. The system with multiple antennas at transmitter and receiver is combined with OFDM to improve reliability of signal and spectral efficiency [1].

Lot of research has been carried in the area of combining OFDM with MIMO systems. MIMO system converts wideband frequency-selective MIMO channels into many flat MIMO channels [2]. Cuibo Yu and Daoben Li [1] proposed a technique to improve spectral efficiency and link reliability using multiple antennas at both ends of wireless communication system. Author in [3] analyzes the error performance of MIMO systems in frequency-selective Rayleigh fading channels. Wireless communication using multiple antennas and channel coding is presented in [4]. Authors in [5] propose closed form bit error probability expressions for OFDM system for frequency-selective Rayleigh and Ricean channels. Channel estimation algorithm to simulate an Alamouti coded OFDM system with two transmit and two receive antennas is proposed in [2]. Author in [6] analyzes the impact of spatial correlation on the BER performance of OSTBC-MIMO systems. Markus Rupp et al [7] propose space–time coding schemes for multiple transmit and receive antenna systems to reduce system complexity. Inter-symbol interference (ISI) reduction approach for BPSK- and QPSK-modulated MIMO system is proposed in [8]. Authors in [9] show significant increase in system capacity using spatial diversity. The system with two transmit antennas and one receive antenna scheme provides the same diversity order as maximal-ratio receiver combining (MRRC) with one transmit antenna, and two receive antennas are reported in [10]. Work in this paper focuses on simulating and analyzing OFDM–MIMO using different types of digital modulation techniques as the quality of service provided by a system greatly depends upon the correct selection of modulation scheme. The system with best modulation is further analyzed and simulated for different antenna configurations. The rest of the paper is organized as follows. Section 2 explains the details of system model. Section 3 discusses the system analysis. Section 4 discusses the results, and conclusion is reported in Sect. 5.

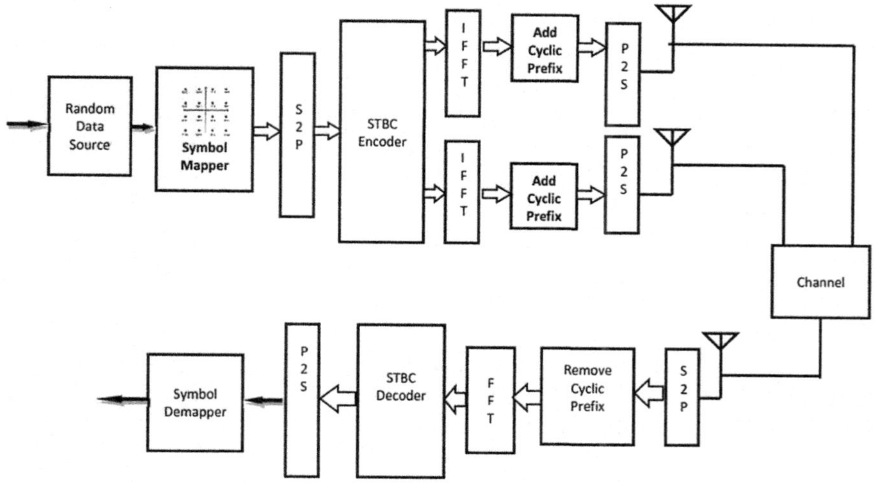

Fig. 1 OFDM–MIMO system for various modulation schemes using Tx-Rx diversity (Alamouti STBC)

2 System Model

The MIMO-OFDM is modeled using OSTBC technique uses spatial diversity gain to attain higher data rates. The binary information sequence of data rate 1 Kb/s is generated by a random sequence generator. The data sequence is modulated using various phase shift keying (PSK) techniques. The output is further encoded using an OSTBC for transmission over the Rayleigh fading channel. OSTBC is used for multiple antenna transmission. OSTBC because of MIMO structure uses time as well as space dimensions for coding. MIMO channel model simulates the Rayleigh fading MIMO channel. For an IFFT transformation, a cyclic prefix (CP) is added to the signal and transmitted through the channel. The model adds white Gaussian noise. At the receiver, the CP is removed and signal is reconstructed using FFT. Block diagram of proposed system is shown in Fig. 1.

3 The System Analysis

Consider $r \times t$ MIMO system that transmits t symbols on T_i transmit antennas and R_i receive antennas. The t-dimensional transmit vector is given as $[x_1 x_2, \ldots, x_{T_i}]^T$, and receive vector is $[y_1 y_2, \ldots, y_{Ri}]^T$ Therefore, the system model is expressed as [11]

$$y = Hx + w \tag{1}$$

$$\begin{bmatrix} y_1 \\ y_2 \\ \vdots \\ y_{R_i} \end{bmatrix} = H_{r\times t} \begin{bmatrix} x_1 \\ x_2 \\ \vdots \\ x_{T_i} \end{bmatrix} + \begin{bmatrix} w_1 \\ w_2 \\ \vdots \\ w_{R_i} \end{bmatrix} \tag{2}$$

where $r \times t$ channel matrix H is given as, $H = \begin{bmatrix} h_{11} & h_{12} & \cdots & h_{1t} \\ h_{21} & h_{22} & \ldots & h_{2t} \\ \vdots & \vdots & \ddots & \vdots \\ h_{r1} & h_{r2} & \cdots & h_{rt} \end{bmatrix}$

Let x_1 and x_2 denote the symbols that are transmitted during 1st time instant. Consider MISO system with two transmit and one receiving antenna. The received symbol during the 1st time instant is given as [11]

$$y_1 = \begin{bmatrix} h_1 & h_2 \end{bmatrix} \begin{bmatrix} x_1 \\ x_2 \end{bmatrix} + w_1 \tag{3}$$

where h_1 is channel coefficient between Tx antenna 1 and Rx antenna and h_2 is channel coefficient between Tx antenna 2 and Rx antenna.

Let $-x_2^*, x_1^*$ be transmitted from transmit antenna 1, 2, respectively, during 2nd time instant [11].

$$y_2 = \begin{bmatrix} h_1 & h_2 \end{bmatrix} \begin{bmatrix} -x_2^* \\ x_1^* \end{bmatrix} + w_2 \tag{4}$$

Combining Eqs. (3) and (4)

$$\begin{bmatrix} y_1 \\ y_2^* \end{bmatrix} = \begin{bmatrix} h_1 & h_2 \\ h_2^* & -h_1^* \end{bmatrix} \begin{bmatrix} x_1 \\ x_2 \end{bmatrix} + \begin{bmatrix} w_1 \\ w_2^* \end{bmatrix} \tag{5}$$

w_2^* is zero mean Gaussian noise with variance σ_n^2.

Alamouti code is a space–time block code. Symbols convert Alamouti coded system into a 2 × 2 MIMO system with channel matrix $\begin{bmatrix} h_1 & h_2 \\ h_2^* & -h_1^* \end{bmatrix}$,

The columns c_1, c_2 of effective channel matrix H are $c_1 = \begin{bmatrix} h_1 \\ h_2^* \end{bmatrix} c_2 = \begin{bmatrix} h_2 \\ -h_1^* \end{bmatrix}$

Here

$$c_1^H c_2 = \left(\begin{bmatrix} h_1 \\ h_2^* \end{bmatrix} \right)^H \times \begin{bmatrix} h_2 \\ -h_1^* \end{bmatrix} = \begin{bmatrix} h_1^* & h_2 \end{bmatrix} \begin{bmatrix} h_2 \\ -h_1^* \end{bmatrix} = h_1^* h_2 - h_2 h_1^* = 0$$

This property is used to design receiver beam forming column C_1

Let $\tilde{y} = c_1^H y = c_1^H(c_1x_1 + c_2x_2 + w)$ Since $c_1^H c_2 = 0$
Then

$$\tilde{y} = c_1{}^2 x_1 + c_1^H w$$

Using Eq. (5), Alamouti system model can be equivalently represented as

$$y = c_1X_1 + c_2X_2 + W \tag{6}$$

Noise power for Alamouti code is

$$\mathrm{E}\left\{|\tilde{w}|^2\right\} = \sigma^2 c_1^H c_1 = \sigma^2 \| c_1 \|^2 \tag{7}$$

SNR at the output of beam former for Alamouti code is [11]

$$\mathrm{SNR} = \frac{\|c_1\|^4 \mathrm{E}\left\{|x_1|^2\right\}}{\mathrm{E}\left\{|\tilde{w}|^2\right\}} = \frac{\|c_1\|^4 \frac{p}{2}}{\sigma^2 \|c_1\|^2} \tag{8}$$

where p is total transmit power assigned to both transmit antennas, so $\mathrm{E}\left\{|x_1|^2\right\} = \frac{p}{2}$

$$\mathrm{SNR} = \frac{1}{2} \|c_1\|^2 \frac{P}{\sigma^2} \tag{9}$$

We have

$$c_1 = \begin{bmatrix} h_1 \\ h_2^* \end{bmatrix} = \|c_1\|^2 = |h_1|^2 + |h_2^*|^2 = |h_1|^2 + |h_2|^2 = \| h^2 \|$$

$$SNR = \frac{1}{2} \| h_1{}^2 \| \frac{P}{\sigma^2} \tag{10}$$

which shows that Alamouti code achieves a diversity order of 2 without channel state information (CSI).

4 Results

OFDM-MIMO system shown in Fig. 1 is implemented with antenna configurations 2 Tx- 2 Rx using Alamouti space–time block coding. The system is simulated for BER performance using different digital modulation techniques like BPSK, 8-PSK, QPSK, QAM under same channel conditions with different values of E_b/N_0. The plot of E_b/N_0 against BER is obtained for system using different digital modulation

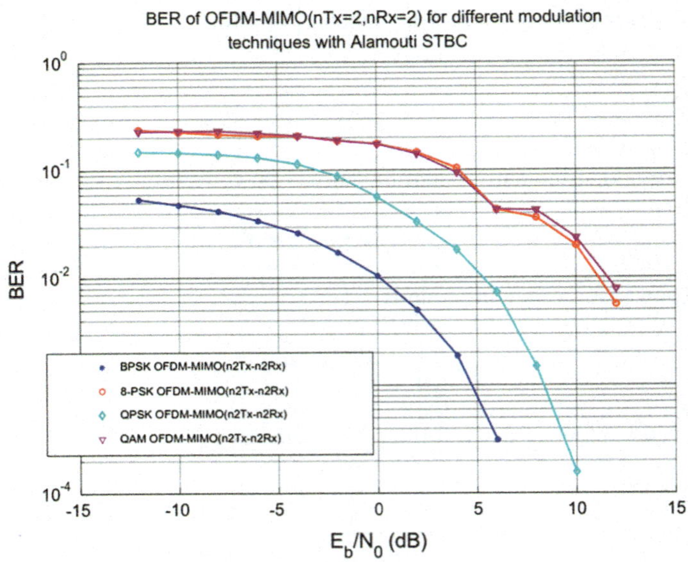

Fig. 2 BER of OFDM-MIMO system for different modulation techniques with Alamouti STBC

Table 1 List of Simulation parameters

System parameters	Specifications
Modulation	BPSK, 8-PSK, QPSK, and QAM
Channel model	Rayleigh, AWGN
No. of Transmitting and receiving antennas	(2 Tx × 1 Rx), (2 Tx × 2 Rx) (3 Tx × 1 Rx) and (3 Tx × 2 Rx)
FFT length	256

techniques as shown in Fig. 2. **The system with best modulation is further analyzed and simulated for different antenna configurations**. Parameters used for simulation are listed in Table 1.

It is observed that performance of OFDM-MIMO system improves with increase in E_b/N_0 for all modulation schemes. It is further noted that performance of system with BPSK modulation is consistent and better as compared to other band-pass modulation and demodulation techniques.

A comparison study is as shown in Table 2. The BER with QPSK system approaches to 00 at E_b/N_0 equal 12 db, whereas system with 8- PSK and QAM needs still higher values of E_b/N_0 than 12 db. Above results show that system shows better results with BPSK as compared to other modulation techniques considered. **The system with best modulation is further analyzed and simulated for different antenna configurations** (Fig. 3).

As seen from Table 3, it is observed that for E_b/N_0 equal to −12 db, the BER for OFDM-MISO BPSK system (2Tx & 1Rx) is 0.0707. If receiver antennas are increased to 2, BER is reduced to 0.0542. It is found that for the same system (2Tx

Table 2 Comparison of BER vs E_b/N_0 of OFDM–MIMO system for various modulation schemes using TX-RX diversity (Alamouti STBC)

Eb/No	BPSK	8PSK	QPSK	QAM
−12	0.0542	0.2387	0.1493	0.2300
−10	0.0481	0.2274	0.1467	0.2317
−8	0.0422	0.2170	0.1406	0.2317
−6	0.0344	0.2092	0.1319	0.2213
−4	0.0264	0.2083	0.1145	0.2083
−2	0.0173	0.1883	0.0885	0.1892
0	**0.0105**	**0.1762**	**0.0570**	**0.1753**
2	0.0050	0.1475	0.0335	0.1423
4	0.0018	0.1050	0.0184	0.0954
6	3.15e-4	0.0434	0.0073	0.0434
8	0.0	0.0364	0.0014	0.0428
10	0.0	0.0201	1.54e-4	0.0236
12	0.0	0.0056	0.0	0.0078

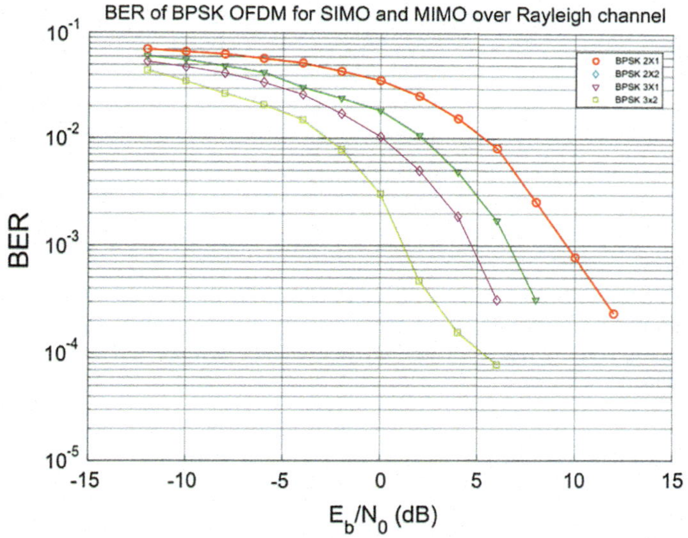

Fig. 3 BER of OFDM-MIMO for BPSK modulation techniques with various antenna configurations

& 2Rx), BER approaches to 0 at E_b/N_0 equal to 8 db and above. For system (2Tx & 1Rx), BER is 0.0026 at E_b/N_0 equal to 8 db. Further BER performance of (3Tx & 1Rx) MISO-OFDM system with Alamouti STBC is analyzed with different values of E_b/N_0 ranging from −12 to 12 db. It is observed that for E_b/N_0 equal to −12 db, the BER for system (3Tx & 1Rx) is 00607. If receiver antennas are

Table 3 Comparison of BER vs Eb/No of OFDM–MIMO system with BPSK for various antenna configurations

Eb/No	2Tx, 1Rx	2Tx, 2Rx	3Tx, 1Rx	3Tx, 2Rx
−12	0.0707	0.0542	0.0607	0.0448
−10	0.0672	0.0481	0.0564	0.0353
−8	0.0638	0.0422	0.0490	0.0269
−6	0.0581	0.0344	0.0428	0.0210
−4	0.0529	0.0264	0.0309	0.0151
−2	0.0442	0.0173	0.0243	0.0079
0	**0.0361**	**0.0105**	**0.0185**	**0.0030**
2	0.0256	0.0050	0.0108	4.73×10^{-4}
4	0.0156	0.0018	0.0048	1.57×10^{-4}
6	0.0082	3.1×10^{-4}	0.0017	7.89×10^{-4}
8	0.0026	**0.0**	3.5×10^{-4}	0.0
10	7.8×10^{-4}	**0.0**	**0.0**	**0.0**
12	2.3×10^{-4}	**0.0**	**0.0**	**0.0**

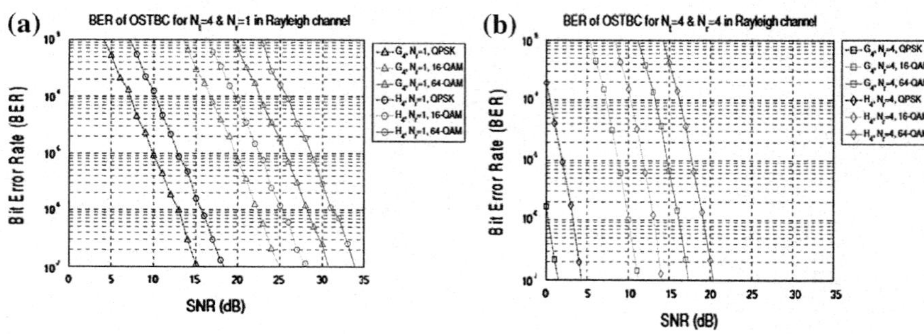

Fig. 4 BER of OFDM-MIMO for higher M-ary phase modulation techniques with **a** four transmit and one receive (Tx = 4, Rx = 1) antenna; **b** with four transmit and four receive (Tx = 4, Rx = 4) antenna

increased to 2, BER is reduced to 0.0448. It is observed from Table 3 that the BER of system (3Tx & 2Rx) approaches to 0 at E_b/N_0 equal to 8 db, whereas for MISO-OFDM (3Tx & 1Rx), it is equal to 3.5×10^{-4} at E_b/N_0 equal to 8db. It is clear from above observations that numbers of receiving antennas are increased; then system gives a better performance in terms of BER.

From Fig. 4a, b, it can be observed that higher M-ary phase modulation techniques like QPSK and QAM provide better results if more receiver antennas are used. This is because as the number of receiver antennas increases, the diversity increases which will give better performance.

5 Conclusion

This paper simulates and analyzes the performance of OFDM-MIMO using different types of digital modulation. OFDM-MIMO using STBC with BPSK modulation is better than other digital modulation techniques in terms of error performance. For higher order of PSK modulation schemes, higher values of E_b/N_0 are needed to achieve a suitable BER as compared to BPSK. The system with best modulation is further analyzed and simulated for different antenna configurations. It is observed that as the number of receiving antennas increases, system diversity increases. Higher diversity of the system gives a better performance in terms of BER. Above systems do not spread the data sequence. Future scope includes testing the performance of system with increased processing gain by spreading the data by suitable spreading factor.

References

1. Yu C, Li D (2003) A study of multi-input multi-output (MIMO) channel model and its impact on the capacity. In: Proceedings of ICCT 2003, pp 1461–1463
2. Hoseinzade M, Mohamed-pour K, Andargoli SMH, Gharanjik A (2010) Decision feedback channel estimation for alamouti coded OFDM-MIMO systems. In: 2010 Wireless telecommunications symposium (WTS), April 2010, pp 1–8
3. Liu H (2003) Error performance of MIMO systems in frequency elective rayleigh fading channels. In: IEEE GLOBECOM 2003, pp 2104–2108
4. Brink ST (2006) Realtek semiconductor corp, "coding over space and time for wireless systems". In: IEEE wireless communications magazine, August 2006, pp 18–30
5. Tan P, Beaulieu NC(2006) Bit error probability analysis of OFDM systems in the presence of channel estimation error over rayleigh and ricean fading channels. In: IEEE ICC proceedings-2006, pp 5172–5179
6. Kim M (2006) Exact BER analysis of OSTBCs in spatially correlated MIMO channels. IEEE Trans Commun 54(8), 1365–1373
7. Rupp M, Mecklenbr¨auker CF (2003) Improving transmission by MIMO channelstructuring. IEEE 2003, pp 3066–3070
8. Mesleh R, Haas H, Lee Y, Yun S (2005) Interchannel interference avoidance in MIMO transmission by exploiting spatial information. In: 2005 IEEE 16th International Symposium on Personal, Indoor and Mobile Radio Communications, pp 141–145
9. Winters JH, Salz J, Gitlin RD (1994) The impact of antenna diversity on the capacity of wireless communication systems. IEEE Trans Commun 42(21314):1740–1751
10. Alamouti SM (1998) A simple transmit diversity technique for wireless communications. IEEE J Sel Areas Commun 16(8):1451–1458
11. Jagannatham AK (2016) Principles of modern wireless communication systems. In: Theory and practice. McGraw Hill Education Private Ltd, pp 245–247

Performance Analysis of Atmospheric Conditions Over Terrestrial Free-Space Optical Communication

Vaishali and Sandeep Sancheti

Abstract In this investigation, relation between bit error rate (BER) versus atmospheric effect present in the channel has been analyzed for free-space optical communication (FSO) system. This paper deals with the performance evaluation of BER with respect to different atmospheric conditions. A terrestrial FSO link is designed that offers high data rate optical communication system. Absorption, scattering, and turbulence are the three major adverse effects present in the atmosphere that cause serious degradation in the performance of BER that may lead to communication link inoperable. During this analysis, a relation between minimum BER, optical power of laser, and link distance would be observed.

Keywords Free-space optical communication · Optical wireless communication Scintillation · Line of sight · Laser

1 Introduction

The optical wireless communication (OWC) is the other name of optical communication in which ultraviolet (UV) light, infrared (IR), or unguided visible light is used to carry information. It has been classified into two major groups. One is indoor OWC which is also known as wireless infrared communication, and another is outdoor OWC. FSO is also known as outdoor optical wireless communication (OWC) or free-space photonics (FSP) system which is defined as a clear line-of-sight (LOS) technology that uses invisible infrared light to carry information from one place to another [1]. In case of terrestrial FSO communication, optical LASER beam is used to carry information operating at the wavelength from 850 to 1550 nm through the atmosphere in order to achieve high data rate and large

Vaishali (✉) · S. Sancheti
Department of ECE, Manipal University Jaipur, Jaipur, Rajasthan, India
e-mail: vvaishali.ak@gmail.com

S. Sancheti
e-mail: sandeepsancheti@gmail.com

V. Janyani et al. (eds.), *Optical and Wireless Technologies*, Lecture Notes in Electrical Engineering 472, https://doi.org/10.1007/978-981-10-7395-3_2

bandwidth [2]. Currently, the commercial available terrestrial FSO system offers the data rate ranges from 100 Mbps to 2.5 Gbps. Although, in laboratories, 160 Gbps data rate is achieved by researchers [1].

In case of short-range FSO link, optical power of laser should be very moderate because a significant fraction of the sent power can be incident on the receiver (e.g., a photodiode) [3–5]. In a particular case where 1550 nm spectral region is considered, laser safety issues regarding safety of eye are not required [6]. However, it is less efficient than optical fiber because of the availability of services is smaller than with a cable due to which the link may be altered either by atmospheric influences like heavy fog, snow, rain or strong wind, or by flying objects such as birds or planes. Due to this, FSO is less robust than other wireless technologies, but it has a higher transmission capacity, is secured against electromagnetic interference, and does not affected by electro-smog [7]. Also, it does not lead to interference between data links which are at different points; so it does not require a license for operating purpose, and it is far more remarkable in terms of data security, since it is more difficult to intercept due to the presence of tightly collimated laser beam than a radio link and uses of schemes of quantum cryptography [2].

2 Free-Space Optical Communication System

A free-space optical link system is shown in Fig. 1, where continuous wave laser is taken as an optical source which is operated at 1550 nm wavelength that means the operable frequency required is in terahertz, i.e., 193.1 THz. On transmitter side, a continuous wave source, i.e., a semiconductor laser is used for achieving high output power by emitting light through stimulated emission in place of spontaneous emission [8–13]. Laser diodes are directly modulated so that the current is directly applied to the device to control light output. The main element of an optical receiver is a photodetector, which uses photoelectric effect. The primary photodetectors are typically a semiconductor-based device made from indium gallium arsenide (InGaAs) [14].

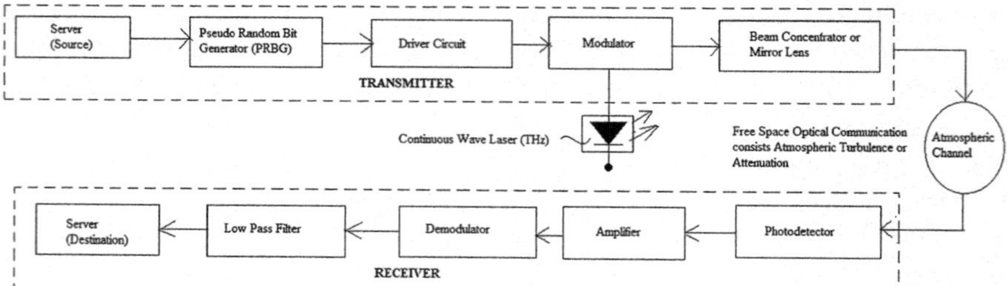

Fig. 1 Block diagram of free-space optical communication link

2.1 System Design

As shown in Fig. 1 that on the transmitting part, it consists a pseudo-random bit generator (PRBS) that generates bits in non-return-to-zero (NRZ) form having data rate of 10 Gbps [15]. In NRZ pulse generator, bit sequences are converted into electrical signals and after that deliver to the Mach–Zehnder interferometer. In Mach–Zehnder interferometer, these bit sequences in electrical form are combined with the laser output which is an optical signal which results in intensity modulation [13]. Resulting optical signal is carried by FSO channel where attenuation level is set according to the distance between source and destination. In FSO channel, attenuation is considered in the form of geometrical attenuation [16].

At the receiving end, optical attenuator is used for reduction of power level which is obtained through FSO channel. This power reduction is also done by various processes like absorption, diffusion, dispersion, reflection. Signal is received by the photodetector where the values of responsivity and dark current are observed. It generates an electrical signal that is to be delivered in low-pass Butterworth filter. On the basis of Gaussian algorithm for transmission of short sequences of bits, BER analyzer is used to generate pattern and also estimate the bit error rate between the sequences. BER analyzer is used as a measuring tool that gives results in the form of BER, Q-Factor, eye diagram, etc. [17]. RF spectrum analyzer is attached with the output so that it can calculate the magnitude of power or other factors with frequency within the full frequency cycle of the instrument. Parameters that are considered for system design and fixed during simulations have been detailed in Table 1.

Table 1 System parameters for designing FSO link

S. No.	Parameters	Values
1.	Type of laser	Continuous wave
2.	Optical power	(0.01–30) W
3.	Attenuation in FSO channel	(178–478) dB/km
4.	Transmitter aperture diameter	2.5 cm
5.	Receiver aperture diameter	8 cm
6.	Beam divergence	2 mrad
7.	Responsivity	1 A/W
8.	Dark current	10 nA
9.	Attenuation of optical attenuator	(0–5) dB
10.	Link distance	100 m

2.2 Free-Space Path Loss

Free-space path loss (FSPL) which is directly related with Friis transmission equation [18] is an important parameter to calculate the level of signal strength in case of RF signal. It occurs in case of line-of-sight path in free-space system where there is no obstacle that might cause the disturbance in form of reflection or refraction. These calculations are purely based on channels or path's specifications. Friis transmission equation depicts the power received when the transmitted source is situated at some distance. It also states that the path loss is huge at higher frequencies. Free-space path loss (FSPL) is the main cause of weak the signal strength of electromagnetic wave. The equation for FSPL is

$$FSPL = \left(\frac{4\pi d}{\lambda}\right)^2 = \left(\frac{4\pi df}{c}\right)^2 \tag{1}$$

where λ is the signal wavelength; f shows the signal frequency; d is the distance from the transmitter, and c is the speed of light.

2.3 Calculation of Atmospheric Losses

Terrestrial free-space optical link is immensely affected by weather conditions which may be varied from normal or harsh. Several case studies have been simulated which are used for identification and estimation of attenuation level of atmospheric turbulence. For transmission of optical signal through atmosphere, the system would be modeled by the help of Beer Lambert's law [19] that can be stated as:

$$\tau(\gamma, L) = \frac{P_t(\gamma, 0)}{P_r(\gamma, L)} = \exp(-\gamma(\lambda)L) \tag{2}$$

where $\tau(\gamma, L)$ is the transmittance of the atmosphere; $P_t(\gamma, 0)$ represents the emitted power from the transmitter; $P_r(\gamma, L)$ shows the received power after a distance of propagation; L and $\gamma(\lambda)$ are the atmospheric attenuation coefficient (km^{-1})

For assessing the total link attenuation, it is important to calculate the value of atmospheric attenuation level that can be found by Kruse model [20]. Kruse model is used to estimate the scattering attenuation coefficient, $\beta_a(\lambda)$ [19]. The Kruse model relates the atmospheric attenuation coefficient to the meteorological visibility of the atmosphere and the optical wavelength, λ. It can be expressed as:

$$\gamma(\lambda) = \beta_a(\lambda) = \frac{10\log\left(\frac{1}{th}\right)}{V(\text{km})} * \left(\frac{\lambda(\text{nm})}{\lambda_0(\text{nm})}\right)^{-q} (\text{dB/km}) \tag{3}$$

Table 2 International codes of visibility range for different weather conditions [22]

S. no.	Weather conditions	Visibility range (in Km)
1.	Thick fog	0.2
2.	Moderate fog	0.5
3.	Light fog	0.770–1
4.	Thin fog/heavy rainfall (25 mm/hr)	1.9–2
5.	Haze/medium rain (12.5 mm/hr)	2.8–4
6.	Clear weather/drizzle (0.25 mm/hr)	18–20
7.	Very clear weather	23–50

where th is the optical threshold (2 or 5%). 2% optical threshold is used in optical communication systems, and 5% threshold is used in airports to measure the runway visual range; V denotes the visibility of the atmosphere in km; λ_o represents wavelength corresponding to the maximum spectrum of the solar band (500 nm), and q is the parameter relating to the particle size distribution of the atmosphere. Table 2 illustrates about the international codes of visibility range in kilometers for different atmospheric conditions. The data in Table 2 are required to calculate the atmospheric attenuation in adverse condition, e.g., in case of thick fog, the value of α (attenuation) is 373.78 dB/Km at the wavelength of 1550 nm while for very clear weather it is just 178.33 dB/Km.

3 Analytical and Simulation Results

In the analytical process, different atmospheric conditions have been considered. By using mathematical expressions mentioned in previous section, attenuation level of short link FSO system can be evaluated in a computationally optimized form. Tables 1 and 2 show the specifications of transmitter, channel, and receiver that are to be considered for investigation.

3.1 Analysis of Atmospheric Attenuation

Performance of FSO link is highly affected by atmosphere and physical characteristics of installation location. These atmospheric characteristics are extremely dependable upon the weather conditions that cause fading. These faded links in FSO system can cause performance degradation in the form of increased bit error rate (BER) and transmission delays. Table 3 shows all the considered weather conditions in the form of snow, fog, rainfall, drizzle, clear weather, etc. Their respective attenuation levels (from α_1 to α_8) in terms of dB/km are calculated with respect to wavelengths that vary from 500 to 1500 nm.

Table 3 Attenuation level of atmosphere in different weather conditions

Wavelength (nm)	Weather conditions	λ_1	λ_2	λ_3	λ_4	λ_5	λ_6
Attenuation (dB/km)		500	850	1060	1250	1310	1550
α_1	Snow	478	473.4	471.48	470.05	469.64	468.18
α_2	Thick fog	390.13	379	377.08	375.65	375.24	373.78
α_3	Moderate fog	268.85	261.64	259.72	258.29	257.88	256.42
α_4	Light fog	228.43	214.87	209.66	206	204.99	201.48
α_5	Thin fog/heavy rainfall (25 mm/hr)	208.21	198.08	194.17	191.43	190.67	188.05
α_6	Haze/medium rain (12.5 mm/hr)	198.11	190.74	187.82	185.74	185.16	183.12
α_7	Clear weather/drizzle (0.25 mm/hr)	190.22	184.51	182.32	180.72	180.28	178.69
α_8	Very clear weather	188.91	183.79	181.76	180.26	179.84	178.33

It is noted that for efficient analysis, the link distance and bit rate to transfer information through channel for all weather conditions are taken as a constant, i.e., 100 m and 750 Mbps, respectively. Figure 2 clearly illustrates that attenuation is inversely proportional to the wavelength. So, for better prospects and optimized FSO link, 1550 nm wavelength provides better proficiency as compared with 850 nm in terms of atmospheric attenuation. Figure 2 also depicts about the margin of variation between the attenuations that is to be generated between lowest and highest points of wavelength for infrared band. So, attenuation taken as α1 for snow is 468.18 dB/km, and attenuation taken as α_8 for clear weather is 178.33 dB/km at 1550 nm wavelength.

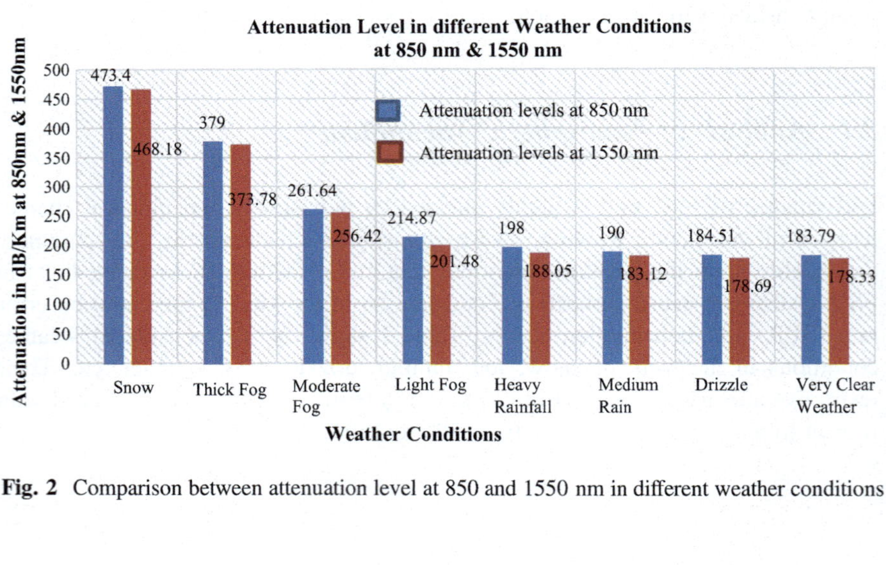

Fig. 2 Comparison between attenuation level at 850 and 1550 nm in different weather conditions

Table 4 Values of BER for respective attenuation levels

S. No.	Weather conditions	λ_1	λ_2	λ_3	Optical power
		500 nm	1250 nm	1550 nm	
1.	Snow	1.584×10^{-8}	1.592×10^{-11}	2.136×10^{-12}	30 W
2.	Thick fog	1.133×10^{-5}	2.632×10^{-9}	5.497×10^{-10}	3 W
3.	Moderate fog	2.324×10^{-6}	3.484×10^{-9}	5.497×10^{-10}	200 mw
4.	Light fog	3.878×10^{-9}	2.813×10^{-22}	6.549×10^{-27}	100 mW
5.	Thin Fog/heavy rainfall (25 mm/hr)	1.708×10^{-6}	6.115×10^{-12}	1.229×10^{-13}	50 mW
6.	Haze/medium rain (12.5 mm/hr)	3.017×10^{-9}	5.883×10^{-15}	1.257×10^{-16}	50 mW
7.	Clear weather/drizzle (0.25 mm/hr)	3.63×10^{-12}	2.486×10^{-15}	6.274×10^{-20}	50 mW
8.	Very clear weather	3.526×10^{-13}	1.108×10^{-18}	3.16×10^{-20}	50 mW

3.2 *Calculation of Bit Error Rate (BER)*

Numerical analysis of attenuation has been performed, and on the basis of outcomes, FSO system is simulated to find out the result in the form of BER. Attenuation level of channel is an essential specification that is required to obtain the output in terms of bit error rate (BER). BER shows the probability of making bit decision error that is required to be maintained in between 10^{-8} and 10^{-13} [21] for free-space optical link. Simulation results are shown in Table 4 that illustrates the BER levels which are taken at 500, 1250, and 1550 nm for comparative purpose. So, there is a comparison between the values of BER that are obtained in both infrared and visible conditions. Table 4 also highlights that in last four cases (very clear weather, drizzle, haze, and heavy rainfall), all the assumed parameters and input conditions are same. It is necessary to find out the methods which are helpful in maintaining the attenuation level in spite of different weather conditions. For the same source, channel, and receiver parameters, BER for last four cases drastically shows improvement from 10^{-13} to 10^{-20} for 1550 nm wavelength which is shown in Table 4. So, it may be concluded that BER could be improved by improving the channel's parameter.

3.3 *Minimum Required Value of Optical Power*

It becomes necessary to calculate the minimum required value of optical power for the source. These values are achieved by the help of simulation process. Figure 3 illustrates the intensity of optical power that would be required during link setup under different weather conditions. From Table 4, it is concluded that for thin fog,

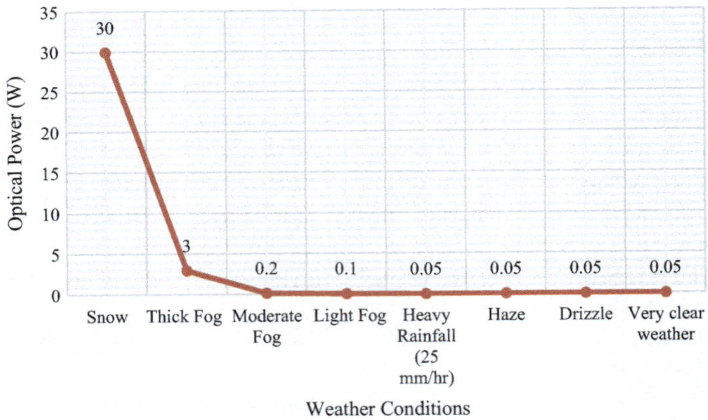

Fig. 3 Minimum required value of optical power with respect to weather conditions

haze, drizzle, and very clear weather conditions, system can be operable at the same intensity of source, i.e., 50 mW. But for other cases, where there is to obtain a required level of performance, intensity of optical source is exponentially increased, i.e., shown in Fig. 3.

For obtaining consistent values of BER, minimum required value of optical power is 30 W during the season of snow that is much higher than the limit of maximum permissible exposure (MPE). Same problem occurs with thick fog condition in which the minimum optical power required is 3 W.

3.4 *Synthesis of Atmospheric Conditions*

As discussed earlier that for achieving the optimum performance of free-space optical link, it is required to maintain the value of BER in adverse weather conditions too. For designing the ideal system that is operable at any atmospheric situation, consider the value of BER is chosen at 10^{-12} [21]. Therefore, it is required to estimate the value of optical power that is needed for consistent performance at the wavelength of 1550 nm.

In case of frosty conditions like in the range of Leh-Ladakh, to attain the required value of BER, optical power of the source needed 30 W. While in case of heavy rainfall that is very common in coastal areas of India, the desired value of optical power is 65 mW. Considering the third condition which is for desert regions where the climatic conditions are quite dry needed just 37 mW of optical power to maintain the ideal value.

4 Conclusions

In the immediate future, the major challenge that the researchers face is to design a trustworthy and pervasive FSO link. In this paper, extensive work on theoretical analysis and simulation studies has been performed to investigate about the link impairments occur in FSO channel which are imposed by combine effect of atmospheric turbulence and attenuation. So, the effects of atmospheric conditions such as snow, fog, haze, mist, rainfall have been examined in detail.

References

1. Khalighi MA, Uysal M (2014) survey on free space optical communication: a communication theory perspective. IEEE Commun Surv Tutor 16(4). IEEE Press
2. Zhou D, LoPresti PG, Refai HH (2011) Enlargement of beam coverage in FSO mobile network. J Lightwave Technol 29(10). IEEE Press
3. Sharma D, Khan SA, Singh S (2015) Literature survey and issue on free space optical communication system. Int J Eng Res Technol 4(02), IJERT
4. Henniger H, Wilfert O (2010) An introduction to free space optical communication. Radioengineering 19(2). IEEE Press
5. Kedar D, Arnon S (2004) Urban optical wireless communication network: the main challenges and possible solutions. IEEE Opt Commun. IEEE Press
6. Anguita JA, Cisternas JE (2010) Experimental evaluation of transmitter and receiver diversity in a terrestrial FSO link, In: IEEE globecom workshop on optical wireless communications
7. Borah DK, Boucouvalas AC, Davis CC, Hranilovic S, Yiannopoulos K (2012) A review of communication-oriented optical wireless systems. J Wirel Commun Netw. IEEE Press
8. Zhu X, Kahn JM (2002) Free-space optical communication through atmospheric turbulence channels. IEEE Trans Commun 50(8). IEEE Press
9. Soni G, Malhotra JS (2012) Impact of beam divergence on the performance of free space optical system. Int J Sci Res Publ 2(2), IJSRP
10. Hulea M, Ghassemlooy Z, Rajbhandari S, Tang X (2014) Compensating for optical beam scattering and wandering in FSO communications. J Lightwave Technol 32(7). IEEE Press
11. Kora AD, Hontinfinde R, Ouattara T (2015) Free space optics attenuation model for visibilities ranging from 9 to 12 Km. Elsevier Sci Direct Proced Comput Sci 56, Science Direct
12. Yang L, Gao X, Alouini M-S (2014) Performance analysis of free-space optical communication systems with multiuser diversity over atmospheric turbulence channels. IEEE Photonics J 6(2). IEEE Press
13. Usman M, Yang H-C, Alouini M-S (2014) Practical switching-based hybrid FSO/RF transmission and its performance analysis. IEEE Photonics J 6(5). IEEE Press
14. Bekkali A, Ben Naila C, Kazaura K, Wakamori K, Matsumoto M (2010) Transmission analysis of OFDM-based wireless services over turbulent radio-on-FSO links modelled by gamma–gamma distribution. IEEE Photonics J 2(3). IEEE Press
15. Chan VWS (2006) Free-space optical communications. J Lightwave Technol 24(12). IEEE Press
16. Garg N, Kumar S (2013) Design of free space optical communication link with mach-zehnder optical modulator for long distance. In: 4th ICCCNT. IEEE Press

17. Reinhardt C, Kuga Y, Jaruwatanadilok S, Ishimaru A (2009) Improving bit-error-rate performance of the free-space optical communications system with channel estimation based on radiative transfer theory. IEEE J Sel Areas Commun 27(9). IEEE Press
18. Zhu B, Cheng J, Wu L (2015) A distance-dependent free-space optical cooperative communication system. IEEE Commun Lett 19(6)
19. Awan MS, Csurgai Horwath L, Muhammad SS, Leitgeb E, Nadeem F, Khan MS (2009) Characterization of fog and snow attenuations for free-space optical propagation. J Commun 4 (8). IEEE Press
20. Ali MAA (2014) Analysis of data rate for free space optical communications system. Int J Electron Commun Technol 5(Spl-1), IJECT
21. Navidpour SM, Uysal M, Kavehrad M (2007) BER performance of free space optical transmission with spatial diversity. IEEE Trans Wirel Commun 6(8). IEEE Press
22. Sahbudin RKZ, Chun TTK, Anas SBA, Hitam S, Mokhtar M (2015) SAC-OCDMA over hybrid FTTx free space optical communication networks. J Telecommun Inf Technol

Design and Analysis of Spiral Photonic Crystal Fiber for Ultra-Flattened Dispersion for C+L+U

Julie Devi, Amritveer Kaur, Ritu Sharma, Varshali Sharma and Santosh Chaudhary

Abstract A design of two-dimensional photonic crystal fiber with circular air hole, and it analyzes for large fattened dispersion for diameter 0.6 um in the wavelength range 1.55–1.78 μm. A step-by-step design optimization is demonstrated and maximum optimized value of total dispersion for diameter of 0.6 μm at 1.25 μm is 3.65 ps/nm.km. The PCF has spiral layout. There are five rings around the air core. Three layouts of spiral PCF are designed for different diameter of air hole and compared with layout having decagonal structure using finite element method. The optimized design can be used as dispersion control and as well as dispersion shifted photonic crystal fiber.

Keywords Photonic crystal fiber · Flattened dispersion · Finite element method Scattering boundary condition

1 Introduction

Photonic crystal fiber has created significant attention in the field of optical fiber communication in recent year [1]. Index-guiding photonic crystal fiber is a single material optical fiber consisting of microscopic array of air channel running down the entire fiber length that provides the confinement and guidance of light in the Centre core [2, 3]. In comparison with single-mode fiber/photonic-crystal fiber

J. Devi · A. Kaur · R. Sharma (✉)
Department of Electronics and Communication Engineering, Malaviya National Institute of Technology Jaipur, Jaipur, Rajasthan 302017, India
e-mail: rsharma.ece@gmail.com

V. Sharma
Department of Electronics and Communication Engineering, Manipal University Jaipur, Jaipur, Rajasthan 302017, India

S. Chaudhary
Department of Mathematics, Malaviya National Institute of Technology Jaipur, Jaipur, Rajasthan 302017, India

V. Janyani et al. (eds.), *Optical and Wireless Technologies*, Lecture Notes in Electrical Engineering 472, https://doi.org/10.1007/978-981-10-7395-3_3

provides design freedom, which lead to flexible tailoring of various guiding properties such as dispersion, nonlinearity, and birefringence. The control of chromatic dispersion is very important aspect in optical fiber communication system [4–7]. To achieve this, flattened dispersion fiber or dispersion compensating fiber has been incorporated as a part of communication system. Variety of design reported in the literature achieves ultra-flattened dispersion or large negative dispersion. Maximum value of average dispersion −393 ps/nm-km with flattened dispersion profile using variable diameter of air hole is reported in Ref. [2]. Reference [8] has analyzed the dispersion properties with variable diameter of inner air hole and reports minimum dispersion of 0.8 ps/nm-km and dispersion slop −0.7 ps/km $\cdot$ nm^2 at 1060 nm. Reference [9] reported the effect of varying the parameter of PCF geometry, namely the inner ring air hole diameter and distance from center in terms of dispersion, nonlinearity and got zero dispersion at 1.55 um and has ultra-flattened dispersion at wavelength in the C and L band.

In this paper, we proposed a design of photonic crystal fiber having spiral layout giving ultra-flattened dispersion over a wide range 1.55–1.78 μm covering C+L+U band. The proposed design is optimized for different diameter of air hole and giving lower dispersion of 3.65 ps/nm.km at wavelength 1.25 μm for diameter 0.6 um. The proposed design also compares with decagonal fiber structure and reported lower dispersion than decagonal photonic crystal fiber. All layouts are designed and analyzed using Comsol (5.1) with scattering boundary condition.

2 Proposed PCF Structure and Theory

The structure is designed using Comsol 5.1 version based on finite element method. Finite element method (FEM) is a numerical method for solving a differential or integral equation. The method essentially consists of assuming the piecewise continuous function for the solution and obtaining the parameters of the functions in a manner that reduces the error in the solution. The designed structure of photonic crystal fiber with circular air hole is shown in Fig. 1. The structure has a spiral lattice with circular air hole. The structure is designed using silica having refractive index 1.45. The angular displacement between two adjacent air hole is 180/N where N is the number of air hole in core region and it depends on the size of the core and diameter of air hole [10–12]. The proposed design has five rings and distance between two circular air hole (pitch) is 2 μm. Radius of circular rings (1, 2, 3, 4, 5) are (2, 3.8, 5.6, 7.4, 9.2) μm, respectively, and radius of core region is 2 μm.

A full vector finite element method with scattering boundary condition is used for mode analysis, which provide effective mode index that is further used to calculate waveguide dispersion. Effective mode index for fundamental mode is described as

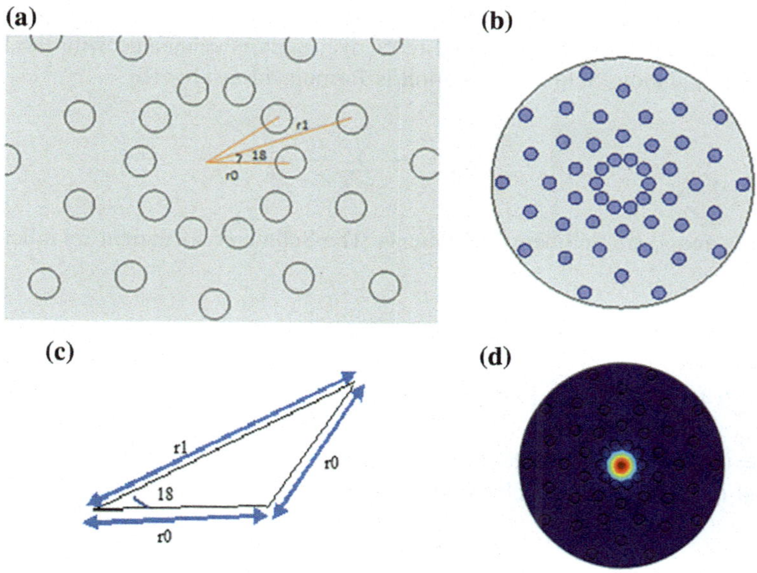

Fig. 1 Design of spiral PCF: **a** geometric interpretation of spiral PCF. **b** Geometry of spiral PCF. **c** Lattice of spiral PCF. **d** Mode profile of spiral PCF at 1.55 μm

$$\beta = \frac{2\pi}{\lambda} \times n_{eff} \tag{1}$$

where β is propagation constant and is operating wavelength [13].

Chromatic dispersion in PCF can be adjusted by changing the geometry of PCF and size of air hole. We can analyze the minimum dispersion and ultra-flattened dispersion by optimizing the geometry of PCF [6–8]. All light sources used for optical fiber transmission emit light not at single wavelength but a band of spectral width $\Delta\lambda$, distributed around . The speed at which light travel through fiber depend on its wavelength and design of fiber thus some wavelength of band of which the pulse is comprised may be delayed compared with other, leading to pulse spreading with time after traversing a significant length of fiber called chromatic dispersion. Chromatic dispersion is the combination of two form of dispersion: (1) material dispersion (2) waveguide dispersion. The waveguide dispersion D of a PCF is easily calculated from the effective index of the fundamental mode n_{eff} versus the wavelength using

$$D_w = -\frac{\lambda}{C} \times \frac{d^2_{neff}}{d\lambda^2} \tag{2}$$

where is operating wavelength and c is speed of light.

Material refractive index itself depends on the value of wavelength. The wavelength dependency of material refractive index is calculated with the help of Sellmeier equation. Sellmeier equation is formulated as [14–16]

$$n_{eff}^2 = 1 + \sum_{i=1}^{3} \frac{a_i \lambda^2}{\lambda^2 - \lambda_i^2} \tag{3}$$

where a_i and λ_i are Sellmeier coefficients. The Sellmeier coefficient for silica given as

$a_1 = 0.6961663$ $\lambda_1 = 0.0684043$

$a_2 = 0.4079426$ $\lambda_2 = 0.1162414$

$a_3 = 0.8974794$ $\lambda_3 = 9.896161$
And total dispersion or chromatic dispersion is given as

$$D = D_w + D_m \tag{4}$$

2.1 *Effective Area*

A measure of area which photonic crystal fiber mode effectively covers in transverse direction. Effective area can be calculated by following relation

$$A_{eff} = \frac{\left(\int\int |E_t^2| dxdy\right)^2}{\int\int |E_t^4| dxdy} \mu\text{m}^2 \tag{5}$$

where E represents the electric field amplitude.

Nonlinearity is calculated from effective area, and it's directly impact on power density of light injected into photonic crystal fiber, i.e., smaller the effective area larger the power density, which lead higher nonlinear effect, nonlinear coefficient calculated by the given formula, i.e., [13]

$$\gamma = \frac{2\pi n_2}{\lambda A_{eff}} \text{W}^{-1}\text{m}^{-1} \tag{6}$$

where n_2 is nonlinear coefficient and is operating wavelength. Value of n_2 for silica is taken 3.0×10^{-20} m^2/w [13].

3 Result and Discussion

Design parameter

Material used	silica.
Type of air hole	circular.
Type of structure	spiral
Type of lattice	isosceles triangle

Pitch of air hole = 2 μm

Diameter of air hole = 1 μm

Proposed structure of PCF is shown in Fig. 1a, b.

The plot of n_{eff} versus wavelength is shown in Fig. 2, the double derivative of $n_{eff}$is calculated using Matlab code, and finally, the waveguide dispersion is calculated using Eq 2. The material dispersion is calculated using Sellemeir Eq. (3), and the result is added with waveguide dispersion to find the total (chromatic) dispersion given by Eq. (4) and shown in Fig. 3.

The wavelength is varied from 0.2 to 1.7 μm during the analysis. The value of n_{eff} is obtained from Comsol 5.1 for different diameter of PCF. It can be seen from the graph that the value of refractive index decreases as value of wavelength increases as per the Eq. (1). We also observe from the graph that the value of refractive index increases as we decrease the value of diameter of air hole.

Figure 3 shows the variation of total (chromatic) dispersion with wavelength as per the Eqs. (2) and (3). From Fig. 3, we can observe that if we reduce the value of diameter of air hole, dispersion also decreases. We also observe that diameter of 0.6 μm gives flat dispersion 19.35 ps/nm-km in a range of 1.55–1.78 μm, so we can use this range for flattened dispersion fiber.

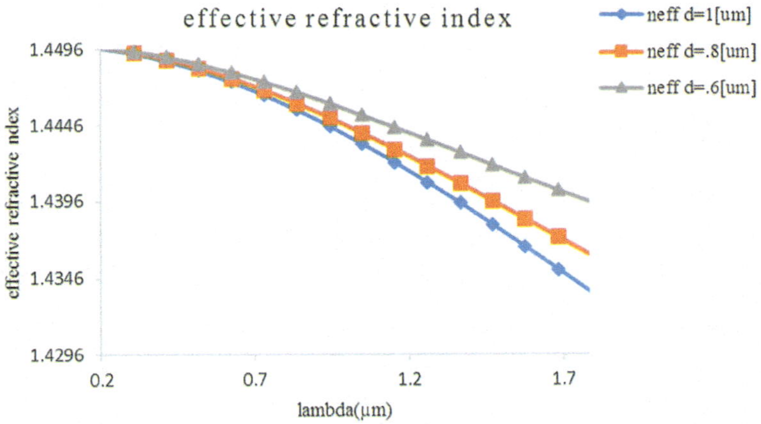

Fig. 2 Effective refractive versus wavelength plot for different diameter of air hole

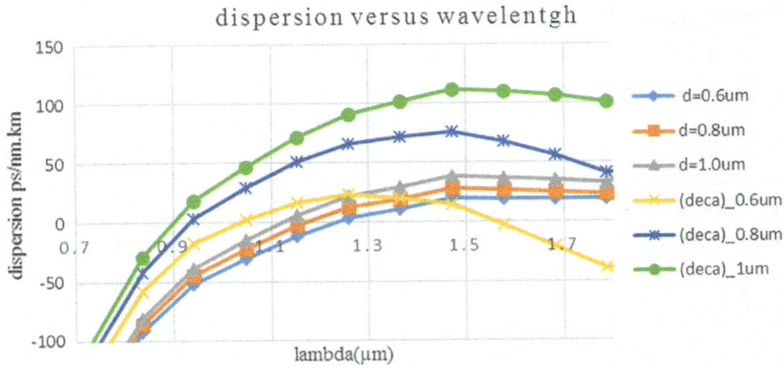

Fig. 3 Plot of total dispersion versus wavelength for different diameter of air hole

From Fig. 3, it can be observed that as we reduce the value of diameter of air hole and zero-dispersion wavelength gets shifted toward right-hand side. The proposed structure is also used as zero-dispersion wavelength-shifted fiber.

In Fig. 4a, the effect of diameter of air hole on effective area and nonlinearity is analyzed. Effective area increases nonlinearly with the wavelength. We can alter the value of effective area by optimizing the geometry and examine large nonlinearity for different parameter. The value of effective area is obtained using Eq. (5) by using Comsol 5.1. The effective area of spiral PCF is smaller than conventional optical fiber. This is an advantage of photonic crystal fiber over the conventional fiber.

Figure 4b shows the variation of nonlinearity with the wavelength. We examine high nonlinearity for silica for spiral PCF. Nonlinearity coefficient of silica is calculated as n_2 $(3.0 \times \frac{10^{-20}\text{m}^2}{w})$. By using n_2 and effective area which can be calculated using Eq. (5), it can be analyzed from the graph that nonlinearity increases with diameter.

4 Conclusion

A novel equiangular spiral photonic crystal fiber in silica has been designed and simulated for large flattened dispersion in the range of 1.55–1.78 μm for diameter 0.6 μm. These fibers can be used as dispersion control fiber. In this paper, we compare our design with decagonal photonic crystal fiber and analyze the lower dispersion in spiral PCF than decagonal PCF. We also observe for wavelength 1.1 and 1.56 μm the magnitude of dispersion is very close to zero for decagonal structure. As the diameter of spiral PCF is decreased, dispersion also decreases, and zero-dispersion wavelength gets shifted toward right-hand side. Hence, this optimized design can be used as dispersion control and dispersion shifted photonic crystal fiber.

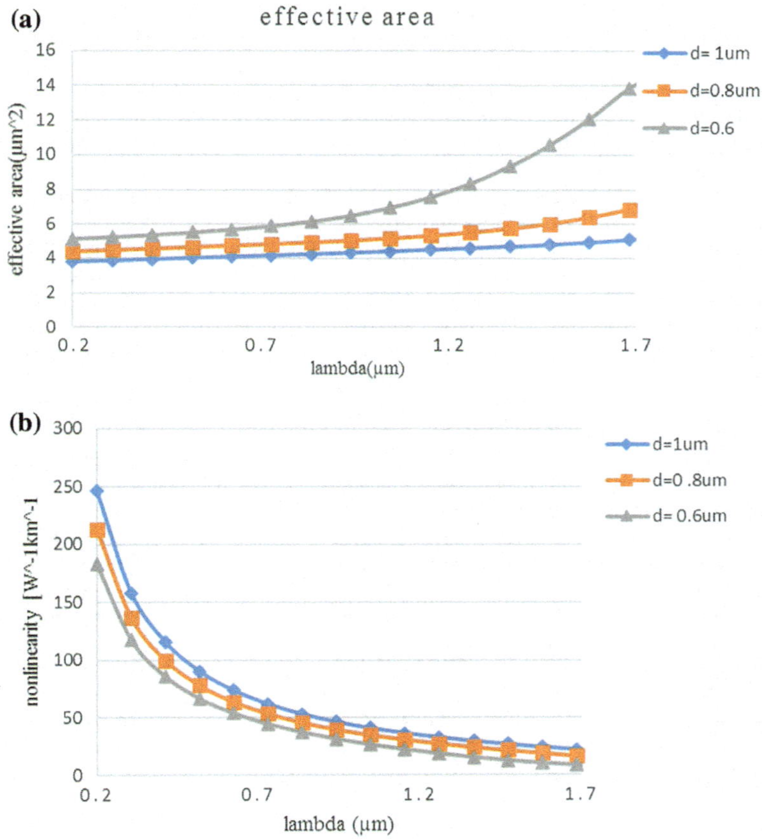

Fig. 4 **a** Effective area versus wavelength plot for wavelength 0.2 –1.7 μm. **b** Wavelength versus nonlinearity (W^{-1} km^{-1}) for wavelength 0.2–1.7 μm

Acknowledgements The authors wish to acknowledge DST Rajasthan, India for funding this project.

References

1. Knight JC, Arriaga J, Birks TA (2000) Anomalous dispersion in photonic crystal fiber. IEEE Photonics Technol Lett 12(7)
2. Islam MA, Alam MS (2012) Design of a polarization-maintaining equiangular spiral photonic crystal fiber for residual dispersion compensation over E+S+C+L+U wavelength bands. IEEE Photonics Technol Lett 24(11)
3. Liao J, Sun J, Du M, Qin Y (2014) Highly nonlinear dispersion-flattened slotted spiral photonic crystal fibers. IEEE Photonics Technol Lett 26(4)
4. Soref R (2006) The past, present, and future of silicon photonics. IEEE J Sel Top Quantum Electron 12(6)

5. Hasan MS, Habib MS, Habib MS, Abdur Razzak SM (2014) Design of hybrid photonic crystal fiber: polarization and dispersion properties. Photonics Nanostruct Fundam Appl 12:205–211
6. Revathi S, Inabathini S, Sandeep R (2015) Soft glass spiral photonic crystal fiber for large nonlinearity and high birefringence. Opt Appl xlv(1)
7. Sharma R, Janyani V, Bhatnagar SK (2011) Low chromatic dispersion and high birefringence investigated in elliptical air hole photonic crystal fiber. J Mod Opt 59:205–212
8. Agrawal A, Azabi YO, Rahman BMA (2013) Stacking the equiangular spiral. Photonics Technol Lett 25(3):291–294. https://doi.org/10.1109/LPT.2012.2236309
9. Heiadarialamdarloo J, Carvalho MI, Nogueira RN, Jesus Teixeira AL (2016) Highly nonlinear, dispersion flattened equiangular spiral photonic crystal fiber in telecommunication window. J Laser Opt Photonics https://doi.org/10.4172/2469-410X.1000
10. Mondal J, Rahaman MS (2015) Design of highly birefringent dispersion compensating spiral photonic crystal fiber. In: 2nd International Conference on Electrical Engineering and Information & Communication Technology (ICEEICT)
11. Liao J, Sun J, Du M, Qin Y (2014) Highly nonlinear dispersion-flattened slotted spiral photonic crystal fibers. IEEE Photonics Technol Lett 26(4)
12. Agrawal A, Kejalakshmy N, Rahman BMA, Grattan KTV (2009) Soft glass Equiangular Spiral Photonic crystal fiber for Supercontinuum generation. IEEE Photonics Technol Lett 21 (22):1722–1724. https://doi.org/10.1109/lpt.2009.2032523
13. Samiul Habib M, Selim Habib M, Imran Hasan M, Razzak SMA (2013) Highly nonlinear polarization maintaining two zero dispersion spiral photonic crystal fiber using artificial defects. Opt Fiber Technol 19:539–542
14. Hao Rui, Sun Guifang (2015) Design of photonic crystal fiber with large negative dispersion and high nonlinearity. Optic 126:3353–3356
15. Sharma R, Janyani V, Bhatnagar SK (2011) Improved single mode property in elliptical air hole photonic crystal fiber. J Mod Opt 58:604–610
16. Sharma V, Sharma R (2016) Design of hybrid photonic crystal fiber with elliptical and circular air holes analysed for large flattened dispersion and birefriengence. J Nanophotonics 10/8

A New Technique to Construct Zero Cross-Correlation Code for SAC-OCDMA

Soma Kumawat and Ravi Kumar Maddila

Abstract Spectral amplitude coding-optical code division multiple access (SAC-OCDMA) systems reduce the multiple access interference (MAI) but they are affected by phase induced intensity noise (PIIN). Zero cross-correlation code (ZCCC) is the code with the property of zero cross-correlation. Due to its cross-correlation property, it is not affected by PIIN. ZCCC uses the direct detection (DD) technique to detect signals. DD technique reduces the complexity of receiver and uses fewer components. A new technique is proposed to construct ZCCC without mapping in 3 steps. The proposed ZCCC require less FBG at encoder and decoder compare to ZCCC. It has 1 s (wavelengths) in form of combination of all weights.

Keywords SAC-OCDMA · ZCCC · Cross-correlation · Direct detection · *BER*

1 Introduction

Optical code division multiple access (OCDMA) technique allows the different users to access the entire channel asynchronously [1]. There are different techniques which give unique codes for different users. Spectral amplitude coding-optical code division multiple access (SAC-OCDMA) provide codes for different users in spectral domain [2, 3]. It reduces the multiple access interference (MAI) by using ideal in-phase cross-correlation property of code and detection technique [4].

Various codes are reported for SAC-OCDMA systems. These are modified quadratic congruence code and modified frequency hopping code [5], random diagonal code which uses code segment and data segment to generate the code in which data segment zero cross-correlation [6], DW code [7] is designed for a weight of 2 only, modified double weight (MDW) code for even weight only [7], and enhanced double weight (EDW) code for odd weight only [7] and many more. Due to their cross-correlation property, these codes are affected by phase induced intensity noise (PIIN). PIIN is completely reduced by zero cross-correlation code (ZCCC).

S. Kumawat · R. K. Maddila (✉)
MNIT, Jaipur, Rajasthan, India
e-mail: 2013rec9531@mnit.ac.in

V. Janyani et al. (eds.), *Optical and Wireless Technologies*, Lecture Notes in Electrical Engineering 472, https://doi.org/10.1007/978-981-10-7395-3_4

These codes have zero cross-correlation (ZCC) property. ZCCC uses Direct Detection (DD) technique for its detection. DD reduces the complexity of receiver along with fewer components.

ZCCC is reported in [8]. Paper [9] explains the ZCCC code construction along with LED spectrum slicing. They are using mapping technique to provide codes for higher number of users. Repetition of basic matrix to construct the code for higher number of users is mapping technique. As the number of users are increased, the increment of code length is not constant [10]. That is, due to mapping technique [10], multi-diagonal code [11] and modified zero cross-correlation code [12] are reported with ZCC property. Multi-diagonal code [11] first computes the diagonal matrices depending on weight and number of users. After computing, these matrices are combined to make multi-diagonal code. Modified zero cross-correlation code [12] first defined unit code matrix according to given number of users, which are repeated depending on weight. Paper describes an easy way to construct ZCCC without mapping in 3 steps. Any number of users can be provided using any integer number which is weight.

The paper organization is as follows. Section 2 explains the code construction of proposed ZCCC along with an example. In Sect. 3, the analysis of *BER* along with system design is described. Simulation results are discussed in Sect. 4. Conclusion is drawn in Sect. 5.

2 Code Construction Technique

The Sect. 2 is divided into two subsections. The code construction algorithm is given in Sect. 2.1 which is defined for weight W and number of users N. Sect. 2.2 explains code construction by using an example with $N = 3$ and $W = 2$.

2.1 Algorithm

Define the weight W and number of users N for code construction. Code Length is $L = N * W$. Code construction steps are as follows

1. Basic matrix (M) of size $2 \times W$ is defined as

$$M = \begin{bmatrix} S_1 \\ S_2 \end{bmatrix} = \begin{bmatrix} \lfloor \frac{W+1}{2} \rfloor 1s \; \lfloor \frac{W}{2} \rfloor 0s \\ \lfloor \frac{W+1}{2} \rfloor 0s \; \lfloor \frac{W}{2} \rfloor 1s \end{bmatrix}_{2 \times W} \tag{1}$$

2. Basic matrix M is repeated $N + 1$ times as

$$U = \begin{bmatrix} S_1 & .. & .. & .. & .. \\ S_2 & S_1 & .. & .. & .. \\ .. & S_2 & .. & .. & .. \\ .. & .. & .. & .. & .. \\ .. & .. & .. & S_2 & S_1 \\ .. & .. & .. & .. & S_2 \end{bmatrix}. \tag{2}$$

3. The first and last row from matrix are deleted. The first and last columns are deleted $\lfloor \frac{W+1}{2} \rfloor$ times and $\lfloor \frac{W}{2} \rfloor$ times, respectively. Fill all empty spaces with zeros.

$$U = \begin{bmatrix} S_2 & S_1 & 0 & .. & 0 \\ 0 & S_2 & .. & .. & 0 \\ .. & .. & .. & .. & .. \\ 0 & 0 & .. & S_2 & S_1 \end{bmatrix}. \tag{3}$$

Code for N users are designed with weight W.

2.2 Example

Value of users and weight are chosen as $N = 3$ and $W = 2$. Code length L is 6.

1. Find the basic matrix for given weight $W = 2$

$$M = \begin{bmatrix} S_1 \\ S_2 \end{bmatrix} = \begin{bmatrix} \lfloor \frac{2+1}{2} \rfloor 1s & \lfloor \frac{2}{2} \rfloor 0s \\ \lfloor \frac{2+1}{2} \rfloor 0s & \lfloor \frac{2}{2} \rfloor 1s \end{bmatrix}_{2\times2} \tag{4}$$

$$M_W = \begin{bmatrix} S_1 \\ S_2 \end{bmatrix} = \begin{bmatrix} 1 & 0 \\ 0 & 1 \end{bmatrix}_{2\times2} \tag{5}$$

2. Repeat basic matrix $N + 1 = 4$ times as

$$U = \begin{bmatrix} 10 & & & \\ 01 & 10 & & \\ & 01 & 10 & \\ & & 01 & 10 \\ & & & 01 \end{bmatrix}. \tag{6}$$

3. Delete the first and last rows of matrix and delete the first and last column $\lfloor \frac{2+1}{2} \rfloor = 1$ time and $\lfloor \frac{2}{2} \rfloor = 1$ time, respectively, and fill the empty spaces with zeros as

$$U = \begin{bmatrix} 01 & 10 & & \\ & 01 & 10 & \\ & & 01 & 10 \end{bmatrix} \tag{7}$$

$$= \begin{bmatrix} 1\ 10\ 00\ 0 \\ 0\ 01\ 10\ 0 \\ 0\ 00\ 01\ 1 \end{bmatrix}. \tag{8}$$

Codes are generated for 3 users using weight 2 given as User1 =110000, User2 = 001100 and User3 =000011. All generated codes have ZCC property.

3 *BER* Calculation

The direct detection (DD) technique is used to detect the unique wavelengths for each user shown in Figs. 1 and 2. By comparing other detection techniques, it has less complex structure and requires only one photodetector. It does not require any subtractor at receiver.

At the transmitter, optical signal from broadband source is filtered as per the given code to each user. These filtered signals are modulated by using Mach zehnder modulator according to the given data. Modulated data from different users are combined and send to the fiber. A photo detector is used to detect W wavelengths for each user. Figures 1 and 2, show the simulation setups for proposed ZCCC and previously reported ZCCC using optisystem 13, respectively. By comparing encoder designs of both setup, the number of filtering components are different. Proposed ZCCC requires less components. Similarly to [4, 5, 13] the Gaussian approximation is applied to system. The variance of photocurrent due to the detection of an ideally unpolarized thermal light, which is generated by spontaneous emission, can be written as

$$\langle I^2 \rangle = \langle I^2_{shot} \rangle + \langle I^2_{thermal} \rangle \tag{9}$$

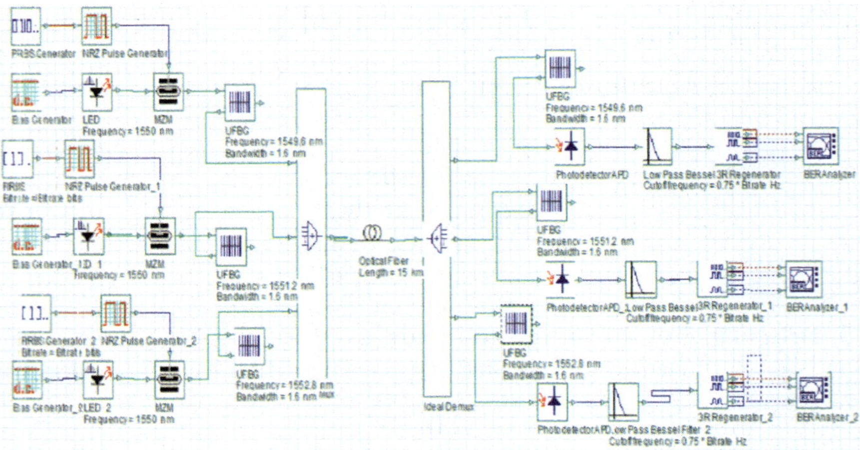

Fig. 1 Proposed ZCCC system design with weight 2 using optisystem 13

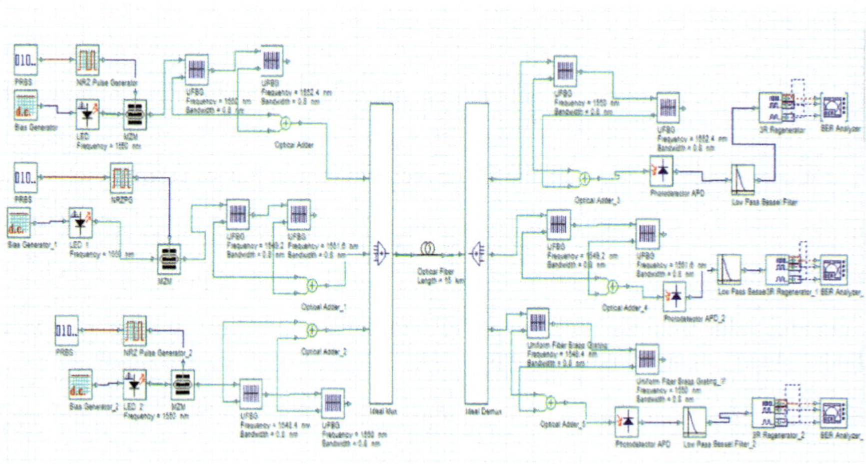

Fig. 2 ZCCC system design with weight 2 using optisystem 13

where, Shot noise $\langle I^2_{shot} \rangle$ is given as

$$\langle I^2_{shot} \rangle = 2eIB \tag{10}$$

and, Thermal noise $\langle I^2_{thermal} \rangle$ is written as

$$\langle I^2_{thermal} \rangle = \frac{4K_b T_n B}{R_L}. \tag{11}$$

At kth receiver, total power at the input of photodiode is defined for one-bit duration

$$\int_0^\infty G(\upsilon)d\upsilon = \int_0^\infty \left[\frac{P_{sr}}{\Delta\upsilon} \sum_{k=1}^{N} d_k \sum_{j=1}^{L} C_k(j) C_X(j) \left\{ u \left[\frac{\Delta\upsilon}{L} \right] \right\} \right] d\upsilon \tag{12}$$

$$= \frac{P_{sr}}{\Delta\upsilon} \frac{\Delta\upsilon}{L} \sum_{k=1}^{N} d_k \sum_{j=1}^{L} C_k(j) C_X(j). \tag{13}$$

Using correlation property as

$$= \sum_{j=1}^{L} (C_X(j) \cdot C_k(j)) \tag{14}$$

$$= \begin{cases} W & k = l \text{ same user} \\ 0 & k \neq l \text{ other users} \end{cases}. \tag{15}$$

Equation (13) can be written as

$$\int_0^\infty G(\upsilon)d\upsilon = \frac{P_{sr}W}{L}. \tag{16}$$

The photo current obtained due to the incident optical power is shown as

$$I = \mathcal{R}\int_0^\infty G(\upsilon)d\upsilon. \tag{17}$$

Put the value from Eq. (16) into Eq. (17) for total power incident at input of photo diode. The resultant photo current is

$$I = \frac{\mathcal{R}P_{sr}W}{L}. \tag{18}$$

Noise power, $\langle I^2\rangle$ is expressed as:

$$\langle I^2\rangle = 2eB\mathcal{R}\left[\frac{P_{sr}W}{L}\right] + \frac{4K_bT_nB}{R_L} \tag{19}$$

The probability of sending bit 1 at any time for each user is half, Eq. (19) is written as

$$\langle I^2\rangle = eB\mathcal{R}\left[\frac{P_{sr}W}{L}\right] + \frac{4K_bT_nB}{R_L} \tag{20}$$

The *SNR* is given as

$$SNR = \frac{\left(\frac{\mathcal{R}P_{sr}W}{L}\right)^2}{\frac{eB\mathcal{R}P_{sr}W}{L} + \frac{4K_bT_nB}{R_L}}. \tag{21}$$

The *BER* is calculated by using Gaussian approximation from *SNR* as given in [4, 5, 13] and is expressed as

$$BER = \frac{1}{2}erfc\sqrt{(SNR/8)}. \tag{22}$$

4 Simulation Results

Proposed ZCCC has same L for given W and N in comparison with ZCCC. But it has different wavelength position (1*s*) in code which effect the code performance. Simulation setup for proposed ZCCC and ZCCC are shown in Figs. 1 and 2. It has 1*s* in the form of combination of all weights (W) as shown in Sect. 2.2. Due to the combination of 1*s*, less filtering components are required at low cost. Because of that, proposed code requires less fiber Bragg gratings at encoder and decoder compare to ZCCC as shown in Figs. 1 and 2.

The parameters used for simulation are explained below. For LED, 1550 nm (wavelength), 0.05 (quantum efficiency), −10 dBm (transmitted Power), and 30 nm (bandwidth) are set. For fiber, 16.75 ps/nm-Km (dispersion), 0.2 dB/Km (attenuation), and 15 Km (length of fiber) are set. For photodiode 5 nA (dark current), 100×10^{-24} W/Hz (thermal noise), 1 A/W (responsivity), and 10 (gain) are chosen. UFBG wavelength and bandwidth are set according to code sequence for specific user and function with reflectivity of 0.99. Data rate is set to 622 Mbps. Spectral width is set to 1 nm for each chip.

Eye diagram for both are shown in Figs. 3 and 4. Q factors of 8.51 and 6.85 are achieved by proposed ZCCC and ZCCC, respectively. Figure 5 shows the *BER* plot for fiber length. *BER* is degraded as when fiber length is increased. Degradation in signal is due to dispersion and attenuation of signal when it travels through fiber channel. Proposed ZCCC design gives better performance when comparing previous designed code. Using proposed ZCCC, *BER* of $10^{-19.4}$ is obtained at fiber length of 3 Km, and it degrades to $10^{-17.5}$ at fiber length of 15 Km. *BER* of $10^{-12.8}$ and 10^{-11} is obtained at fiber length of 3 Km and 15 Km, respectively. Proposed ZCCC gives better performance for the same number of users (N), weight (W), and length (L) in comparison with ZCCC (Table 1).

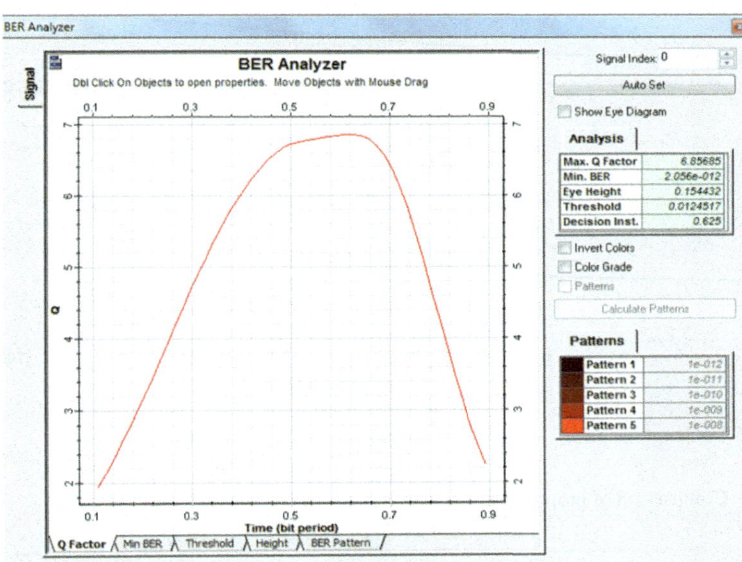

Fig. 3 Q factor for ZCCC with weight 2

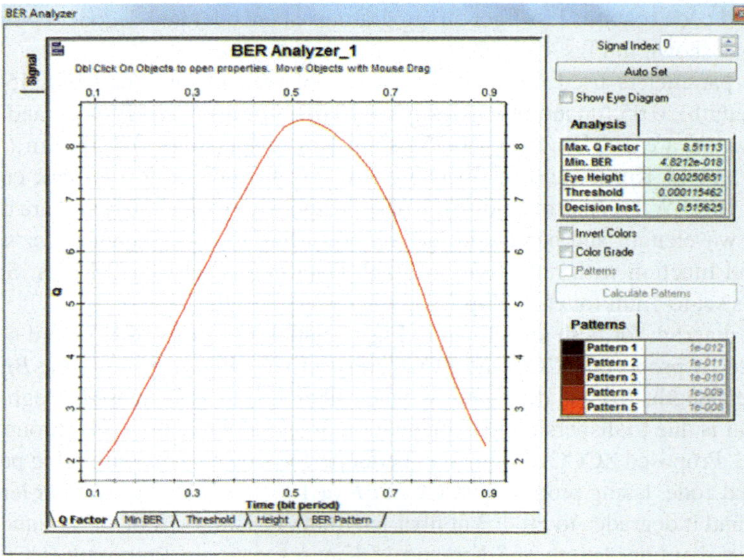

Fig. 4 Q factor using proposed ZCCC with weight 2

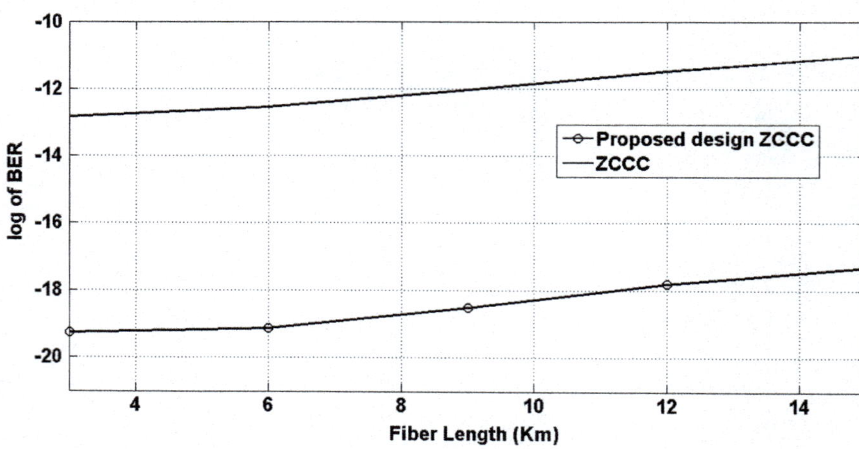

Fig. 5 Fiber length is ploted against *BER* for proposed ZCCC and ZCCC

Table 1 Comparison of proposed ZCCC and ZCCC

	Proposed ZCCC	ZCCC
Mapping	not used	used [8]
For higher weights	a single matrix Eq. 5	code transformation is used [8]
Code length	$L = N * w$	$L = L_B * 2^m$ [8]

5 Conclusion

Code construction algorithm is proposed to construct ZCCC without mapping in 3 steps. Proposed ZCCC requires less fiber Bragg gratings at encoder and decoder compare to ZCCC. Proposed ZCCC gives better performance for same number of users (*N*), weight (*W*) and length (*L*) in comparison with ZCCC. Q factors of 8.51 and 6.85 are achieved by proposed ZCCC and ZCCC, respectively. Using proposed ZCCC, *BER* of $10^{-19.4}$ and $10^{-17.5}$ are obtained at fiber length of 3 Km and 15 Km, respectively.

References

1. Kwong W, Prucnal P, Perrier P (1989) Synchronous versus asynchronous CDMA for fiber-optic LANs using optical signal processing. In: Global telecommunications conference and exhibition communications technology for the 1990s and beyond. GLOBECOM 89, vol. 2, IEEE, pp. 1012–1016
2. Zaccarin D, Kavehrad M (1993) An optical cdma system based on spectral encoding of LED. Photon Technol Lett IEEE 5(4):479–482
3. Kavehrad M, Zaccarin D (1995) Optical code-division-multiplexed systems based on spectral encoding of noncoherent sources. J Lightwave Technolo 13(3):534–545
4. Wei Z, Shalaby H, Ghafouri-Shiraz H (2001) Modified quadratic congruence codes for fiber bragggrating-based spectral-amplitude-coding optical cdma systems. J Lightwave Technolo 19(9):1274–1281
5. Wei Z, Ghafouri-Shiraz H (2002) Codes for spectral-amplitude-coding optical CDMA systems. J Lightwave Technolo 20(8):1284–1291
6. Fadhil HA, Aljunid S, Ahmad R (2009) Performance of random diagonal code for (OCDMA) systems using new spectral direct detection technique. Opt Fiber Technolo 15(3):283–289
7. Hasoon FN, Aljunid SA, Abdullah MK, Shaari S (2006) Spectralamplitude coding OCDMA systems using enhanced double weight code. J Eng Sci Technolo 1(2):192–202
8. Zahid AZG, Hasoon FN, Shaari S (2009) New code structure for enhanced double weight (EDW) code for spectral amplitude coding OCDMA system. In: International conference on future computer and communication (ICFCC), pp. 658–661
9. Anuar MS, Aljunid SA, Saad NM, Hamzah SM (2009) New design of spectral amplitude coding in OCDMA with zero cross-correlation. Opt Commun 282(14):2659–2664
10. Kumawat S, Madddila RK (2016) Generalized optical code construction for enhanced and modified double weight like codes without mapping for sacocdma systems. Opt Fiber Technolo 30:72–80
11. Abda TH, Aljunid SA, Fadhil HA, Ahmad,RA, Saad NM (2011) Modeling and simulation of multi diagonal code with zero cross correlation for SAC-OCDMA networks. In: 2nd International conference on photonics, pp. 1–5. https://doi.org/10.1109/ICP.2011.6106829
12. Rashidi CBM, Aljunid SA, Ghani F, Anuar MS (2010) Development of modified zero cross correlation code for OCDMA network. In: International conference on photonics, pp. 1–6. https://doi.org/10.1109/ICP.2010.5604406
13. Noshad M, Jamshidi K (2010) Code family for modified spectral-amplitude-coding ocdma systems and performance analysis. Opt Commun Network J IEEE/OSA 2(6):344–354

Design of Octagonal and Decagonal Lattice Photonic Crystal Fiber for Achieving Ultra Low Flattened Dispersion

Amritveer Kaur, Julie Devi, Ritu Sharma, Varshali Sharma and Santosh Chaudhary

Abstract This paper presents some key concepts of design and optimization of solid core photonic crystal fiber (PCF) based on different geometries for broadband applications. The three different geometrical structures, hexagonal, octagonal, and decagonal of photonic crystal fiber, are designed for different air filling fractions at pitch length 2 μm. The parameters affecting the light propagation through PCF structure such as effective refractive index, chromatic dispersion, mode effective area, nonlinear coefficient, and V parameter are numerically analyzed and compared using finite element method (FEM) on COMSOL Multiphysics software.

Keywords Photonic crystal fiber (PCF) · Chromatic dispersion
Effective mode area · Finite element method · V parameter

1 Introduction

Photonic crystal fibers (PCFs) are a new class of optical waveguides which offer exceptional light guiding mechanism and deliver significantly improved performance compared to conventional optical fibers. PCF is a kind of optical fiber that uses photonic crystals to form the cladding around the core of the cable [1].

A. Kaur · J. Devi · R. Sharma (✉)
Department of Electronics and Communication Engineering,
Malaviya National Institute of Technology Jaipur, Jaipur, Rajasthan, India
e-mail: rsharma.ece@mnit.ac.in

V. Sharma
Department of Electronics and Communication Engineering,
Manipal University Jaipur, Jaipur, Rajasthan, India

S. Chaudhary
Department of Mathematics, Malaviya National Institute of Technology Jaipur,
Jaipur, Rajasthan, India

V. Janyani et al. (eds.), *Optical and Wireless Technologies*, Lecture Notes
in Electrical Engineering 472, https://doi.org/10.1007/978-981-10-7395-3_5

The PCF has a microscopic array of air holes running along the length of the fiber. The light can be confined and guided through the fiber by the mechanisms of either a total internal reflection (TIR) or a photonic band gap (PBG). In contrast to conventional fibers, PCFs have additional design parameters, namely the pitch, air hole diameters, air hole arrangement, and number of rings that offers design flexibility which is not possible in conventional fibers. PCFs possess numerous properties in terms of transmission characteristics such as single mode operation, high nonlinearities, flat dispersion, high birefringence, large mode area, high negative dispersion, etc. PCFs have diverse applications in telecommunications, sensor design, design of super continuum generation(SCG), interferometry, fiber optic gyroscope, lasers, medical instruments, various polarization sensitive devices, coherent-optical detection system, and so on [2–4].

This paper discusses some of the concepts involved in the design of solid core PCF based on hexagonal, octagonal, and decagonal all three lattices of low index circular inclusions (the refractive index of these inclusions will be that of air, $n = 1$). The effect of structural parameters of PCF such as pitch (p), number of air holes, air hole diameter, air hole arrangements on key properties of PCF such as chromatic dispersion, effective refractive index, effective area, and nonlinear coefficient will also be presented. This will help in the optimization of these parameters to design PCF to achieve specific operational behavior [5–8].

2 Simulation Methodology

The propagation of electromagnetic radiations through the photonic crystal fiber is studied by the numerical technique full vector Finite Elemental Method (FEM). The studies are carried out over a radiation spectrum of wavelengths ranging from 700 to 1800 nm. The FEM technique is one of the advanced vectorial methods to study the light propagation characteristics in PCFs. In the present study, we calculate the propagation of the electric vector through the fiber over the given wavelength range with the help of the software COMSOL Multiphysics 5.0. The full vector analysis of the fiber structure is done by calculating the propagation of electric vector through the fiber by solving an eigenvalue problem drawn from Maxwell's equations using FEM in each triangular subspace in the area of the meshed fiber structure and effective refractive index (*neff*) of the PCF is obtained from the simulation result. The modified total internal reflection mechanism guides the radiation through the proposed PCFs. To lock the back reflection of the leaky modes into the core of the fiber structure, a scattering boundary condition is applied around the cladding region of the PCF. Effective mode area (*Aeff*), chromatic dispersion (*D*), nonlinear coefficient (γ), and V parameters (V) can be obtained by using values of *neff* to the following equations.

3 Design Methodology

The significant parameters for the PCF design are diameter of the holes and the distance between adjacent holes called pitch (p). The basic unit of geometry is an isosceles triangle with the vertex angle A. The length of the identical legs is taken as the pitch. The length of the side opposite to the vertex angle, calculated from the formula 2psin(A), is the spacing between the air holes on the same ring. The geometry in the cladding is satisfied by repeating the unit isosceles triangle around the core. Air holes are placed at each corner of the basic triangle. The three different geometries hexagonal, octagonal, and decagonal have been designed which are shown in Fig. 1. In hexagonal geometry (Fig. 1a), the number of air holes around central core is six and vertex angle of isosceles triangle is 60°, with all sides having length equal to pitch length (p). Similarly, in octagonal (Fig. 1b) and decagonal (Fig. 2c), the number of air holes around central core is eight and ten, respectively, and vertex angle of isosceles triangle is 45° and 36°, respectively [3, 4].

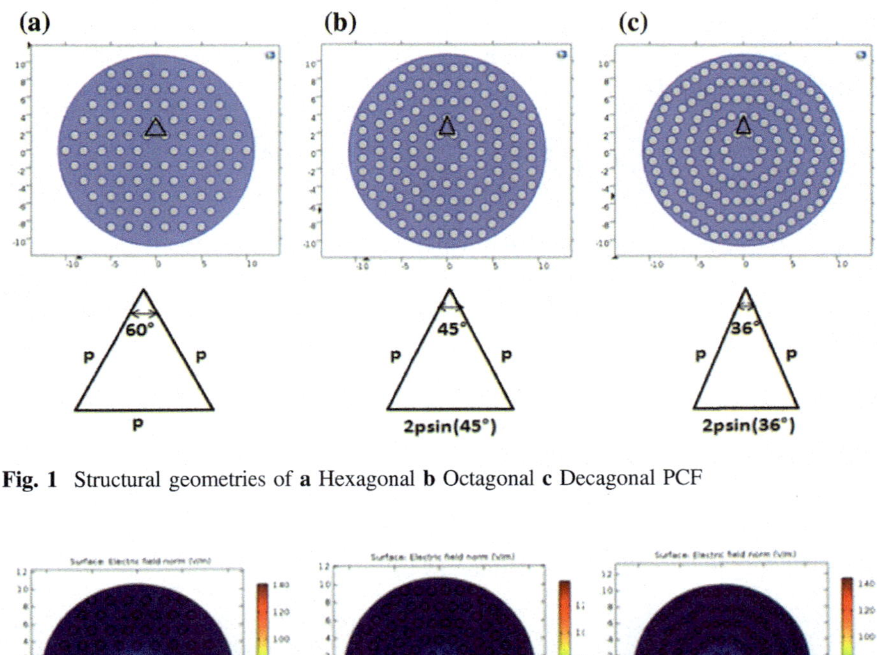

Fig. 1 Structural geometries of **a** Hexagonal **b** Octagonal **c** Decagonal PCF

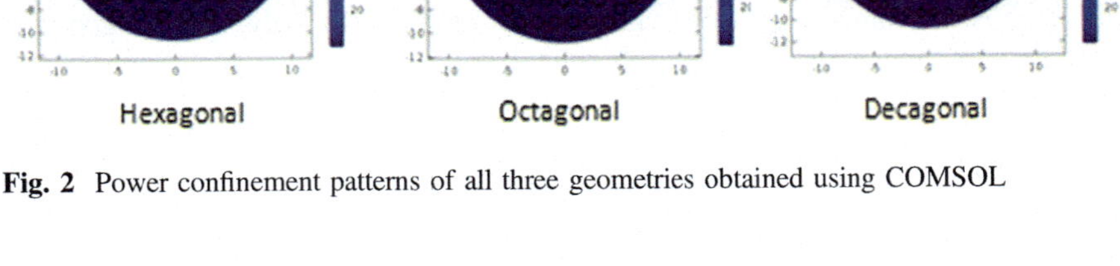

Fig. 2 Power confinement patterns of all three geometries obtained using COMSOL

3.1 Chromatic Dispersion

The dispersion (D) is proportional to the second derivative of the n_{eff}, with respect to the wavelength (λ) obtained as [9, 10].

$$D = -\frac{\lambda}{c}\frac{d^2(Re[n_{eff}])}{d\lambda^2} \tag{1}$$

where Re $[n_{eff}]$ is the real part of n_{eff}, λ is wavelength, and c is the velocity of light in vacuum. The total dispersion is calculated as the sum of the geometrical dispersion (or waveguide dispersion (D_g)) and the material dispersion (D_m) obtained as [11, 12]:

$$D(\lambda) = D_g(\lambda) + \Gamma D_m(\lambda) \tag{2}$$

where Γ is the confinement factor in silica, which is close to unity for the most practical PCFs as the modal power is almost confined in the silica with high refractive index. The effective refractive index is directly obtained from the three-term Sellmeier formula given in Eq. (3) which is further used to calculate the material dispersion D_m:

$$n^2 = 1 + \frac{B_1\lambda^2}{\lambda^2 - C_1} + \frac{B_2\lambda^2}{\lambda^2 - C_2} + \frac{B_3\lambda^2}{\lambda^2 - C_3} \tag{3}$$

where λ is operating wavelength in μm and the Sellmeier coefficients for silica are: $B_1 = 0.69616630$, $B_2 = 0.40794260$, $B_3 = 0.89747940$, $C_1 = 0.068404300$ μm, $C_2 = 0.11624140$ μm, $C_3 = 9.8961610$ μm secondly the total dispersion is obtained by Eq. (1).

3.2 Effective Mode Area (A_{eff})

The effective area of the fiber core A_{eff}, which strongly depends on core area and core-cladding index contrast, is defined as [6]

$$A_{eff} = \frac{\left(\iint |E_t|^2 dxdy\right)^2}{\iint |E_t|^4 dxdy} \tag{4}$$

where E_t is the transverse electric field and is obtained by solving an Eigenvalue problem from Maxwell equations.

3.3 *Nonlinear Coefficient* (γ)

The nonlinearity is typically only observed at very high light intensities. PCFs can confine high-intensity light. Therefore, PCFs are expected to possess high nonlinearity. The nonlinear coefficient, γ in $(\mathrm{W.km})^{-1}$, is calculated by the following equation [6]:

$$\gamma = \frac{n_2 \omega}{c A_{eff}} = \frac{n_2 2\Pi}{\lambda A_{eff}} \tag{5}$$

where n_2 is the nonlinear-index coefficient in the nonlinear part of the refractive index [6].

3.4 *V Parameter*

With the help of effective V parameter within telecom band, single mode characteristic of the fiber is verified. The V parameter for the PCF can be calculated by using the following formula [8, 13]:

$$V_{eff} = \frac{2\Pi \wedge}{\lambda} \sqrt{n_{co} - n_{eff}} \tag{6}$$

where n_{co} is the refractive index of core material, n_{eff} is the effective refractive index of the cladding, Λ is pitch size, and λ is wavelength. It is known that the condition for higher order mode cutoff is associated with $V_{eff} \leq \Pi$ at 1550 nm.

4 Simulation Results and Discussion

Design parameters used

- Pitch size = 2 μm
- Diameter of hole to pitch ratio = 0.3, 0.4, 0.5, 0.6
- Operating wavelength = 1.55 μm
- Refractive index of material used = 1.45 (Doped Silica)
- Number of rings of air holes = 5

4.1 *Effective Refractive Index*

Light propagation through the different PCFs at d/p = 0.5 and pitch length 2 μm is shown in Fig. 2. Figures 3 and 4 show the variation of the real part of effective refractive index with the operating wavelength for fibers at different d/p ratios.

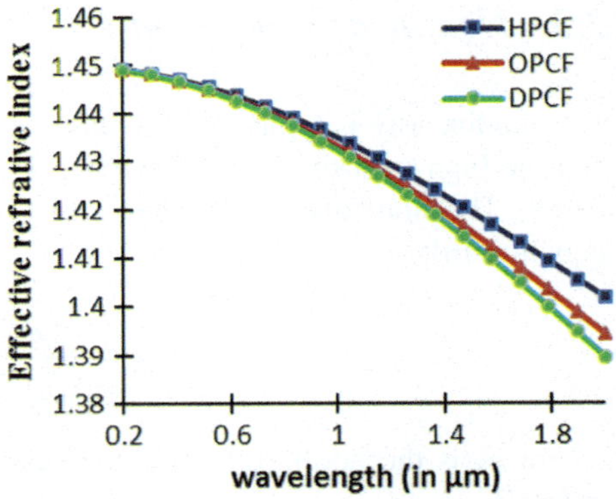

Fig. 3 Effective refractive index versus wavelength for three geometries with pitch length = 2 µm and air fraction, d/p = 0.5

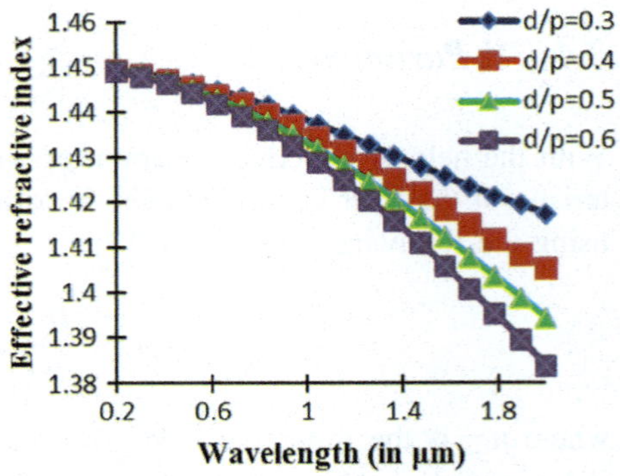

Fig. 4 Effective refractive index versus wavelength for OPCF with pitch length = 2 µm and varying air fraction (d/p)

At shorter wavelength, n_{eff} is converging toward the refractive index of silica and the modes are tend to be more confined in the core part due to its waveguide structure, as the wavelength increases n_{eff} diverge. As the geometry shifted to higher order polygon or the relative air filling fraction increases, the effective refractive index decreases relatively.

4.2 *Chromatic Dispersion*

Figures 5, 6, and 7 show the dispersion properties of three geometries hexagonal, octagonal, and decagonal, respectively. For all the structures, the value of chromatic dispersion increases with increase in air filling fraction i.e., d/p value. The zero dispersion wavelength can be shifted in visible region by selecting proper air hole sizes. The zero dispersion wavelength and dispersion value at $\lambda = 1.55$ µm are given in Tables 1 and 2.

As diameter to pitch ratio increases, relative air fraction increases, zero dispersion wavelengths get shifted to smaller wavelengths and dispersion value at

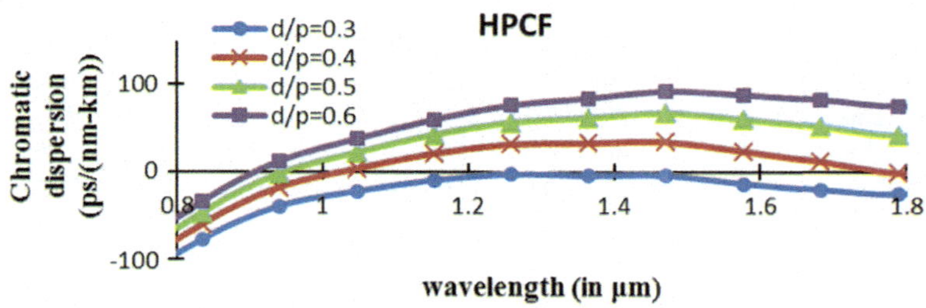

Fig. 5 Chromatic dispersion versus wavelength graph at pitch length = 2 μm with different air filling fractions (d/p) for hexagonal geometry

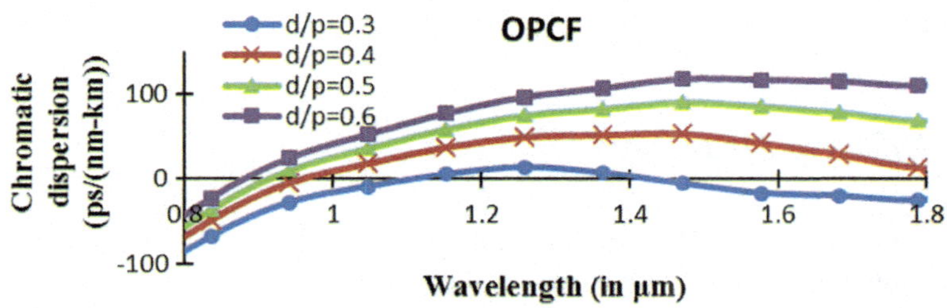

Fig. 6 Chromatic dispersion versus wavelength graph at pitch length = 2 μm with different air filling fractions (d/p) for octagonal geometry

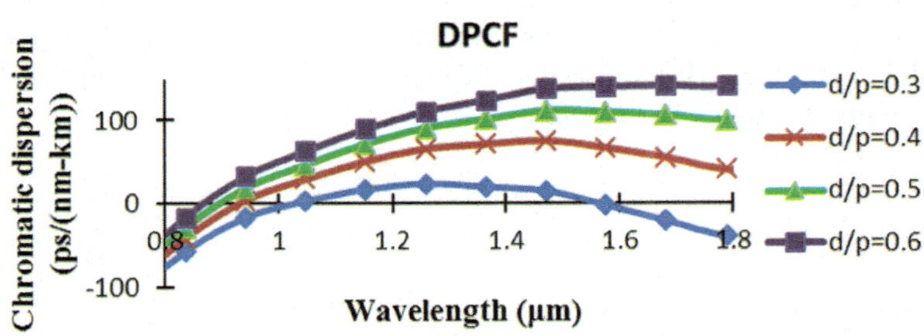

Fig. 7 Chromatic dispersion versus wavelength graph at pitch length = 2 μm with different air filling fractions(d/p) for decagonal geometry

operating wavelength shifted toward upward. As the geometry gets shifted to higher order polygon or as the air filling fraction (d/p) increases, the dispersion curve shifts upward, therefore, the values of chromatic dispersion at 1.55 μm wavelength increase. Here, the decagonal structure with d/p = 0.3 shows least dispersion of −1.74 ps/(nm-km) and gives comparative flat dispersion for the range of 1.15–1.47 μm.

Table 1 Zero dispersion wavelengths (in μm) of different geometries at different values of diameter to pitch ratio for same pitch size = 2 μm

	Hexagonal	Octagonal	Decagonal
d/p = 0.3	1.28	1.10	1.04
d/p = 0.4	1.01	0.95	0.94
d/p = 0.5	0.94	0.92	0.901
d/p = 0.6	0.91	0.88	0.87

Table 2 Chromatic dispersion (in ps/(nm-km)) at λ = 1.55 μm for different geometries at different values of diameter to pitch ratio for same pitch size = 2 μm

	Hexagonal	Octagonal	Decagonal
d/p = 0.3	−13.32	−16.30	−1.74
d/p = 0.4	24.42	42.06	67.78
d/p = 0.5	61.26	87.34	110.70
d/p = 0.6	89.07	117.81	141.50

4.3 *Effective Mode Area*

Figures 8 and 9 show the variation of effective mode area with the incident wavelength for different geometries at d/p = 0.3 and 0.5. A_{eff} increases linearly with wavelength and decreases with air filling fraction (d/p ratio) [3, 4]. If we are moving to higher order polygon geometry, the relative air filling fraction increases, the confinement of the guiding modes through the core increases, as a result effective mode area decreases. The value of A_{eff} of different geometries at operating wavelength 1.55 μm with different air filling fraction is summarized in Table 3. The variation of A_{eff} considerably affects the nonlinear property and the light gathering capacity of the PCFs. Here, decagonal PCF gives comparative less value of A_{eff} for higher filling fraction value (d/p).

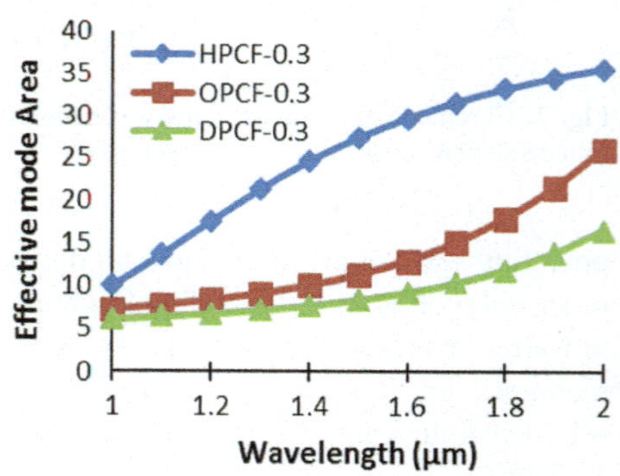

Fig. 8 Effective mode area (in μm^2) versus wavelength of all three geometries with air filling fraction(d/p) = 0.3 at pitch length of 2 μm

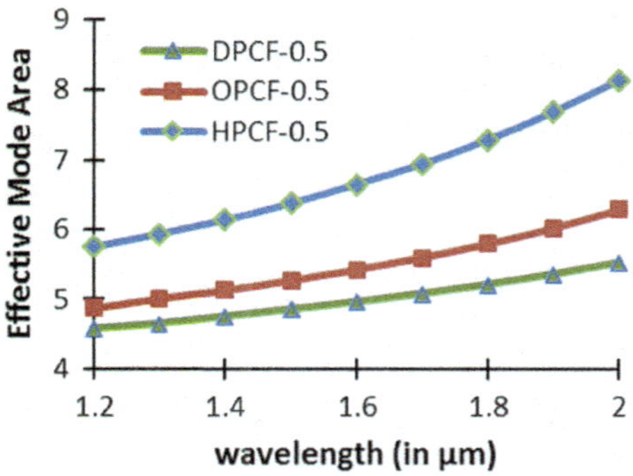

Fig. 9 Effective mode area (in μm^2) versus wavelength of all three geometries with air filling fraction (d/p) = 0.5 at pitch length of 2 μm

Table 3 Effective mode area (in μm^2) for different structures of PCF at operating wavelength of 1.55 μm corresponds to its air filling factor value for pitch size 2 μm

	Hexagonal PCF	Octagonal PCF	Decagonal PCF
d/p = 0.3	28.545	12.219	8.899
d/p = 0.5	6.518	5.348	4.911

4.4 *Nonlinear Coefficient*

Figures 10 and 11 show the nonlinear coefficient ($km^{-1}W^{-1}$) for the three geometries at d/p = 0.3. For the calculations, the refractive index coefficient for silica is taken as $n_2 = 3.09 \times 10^{-20}$ m^2/W [2]. It is observed that the nonlinear coefficient decreases with increases in wavelength. As geometry moves to higher order polygon or air filling fraction (d/p) increases, the nonlinearity increases.

The V parameters of different geometries at different d/p values are calculated using Eq. (6) (Figs. 12 and 13). As the geometry moves to higher order polygon or as d/p increases, its V parameter value increases. Here, decagonal PCF at d/p = 0.6 gives larger V parameter value, still $V_{eff} \leq \Pi$, hence all the geometries will act as single mode fibers.

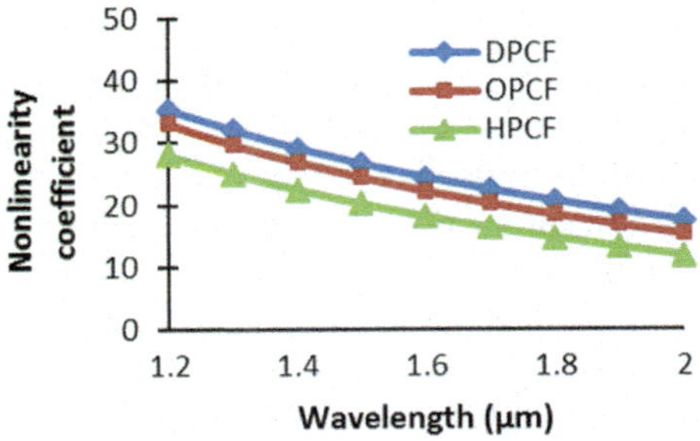

Fig. 10 Nonlinearity coefficient, γ (in $km^{-1}W^{-1}$) for three PCFs for d/p = 0.5 at pitch length = 2 μm

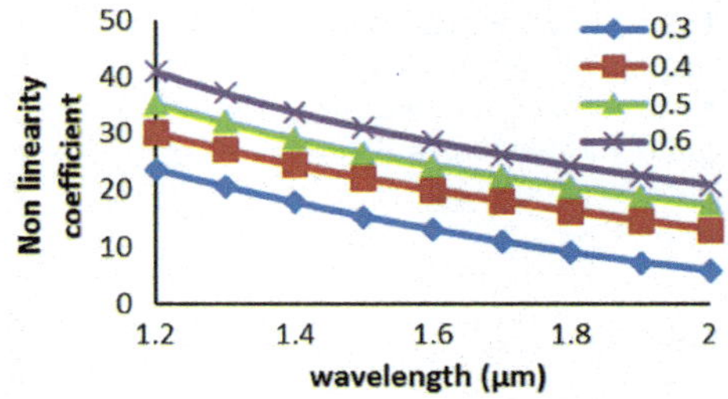

Fig. 11 Nonlinearity coefficient, γ (in $km^{-1}W^{-1}$) for three PCFs for Decagonal PCF at different d/p ratios

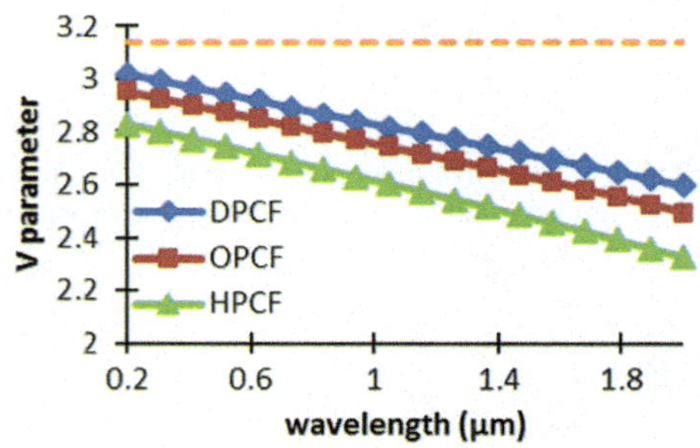

Fig. 12 V parameters for different PCFs at d/p = 0.5 and pitch size of 2 µm

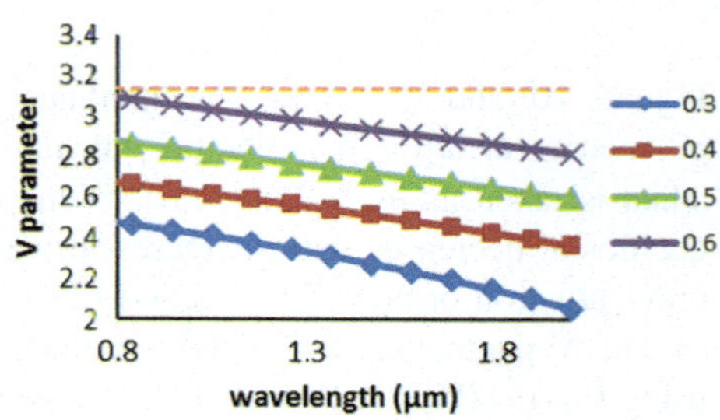

Fig. 13 V parameters of decagonal PCF for different d/p ratios at 2 µm of pitch length

5 Conclusion

It has been demonstrated that the basic properties of PCFs can be engineered by tweaking the geometries and any structural parameter such as d/p and hole arrangements, which affects the core size, indirectly effects the real part of effective index. The zero dispersion wavelengths can be obtained at desired wavelength by varying structural parameters. The decagonal PCFs have shown higher confinement, low effective mode area, higher nonlinearity as compared to other structures and its chromatic dispersion at 0.3 air filling fraction shows flattened behavior and it can be further more flattened by varying its hole size. It is very difficult for any PCF to excel in all these properties. Some properties will have to compromise in order to maximize others. The design principles presented in this paper will be of great value for the optimization of PCF parameters to achieve excellent optical properties.

Acknowledgements The author wishes to acknowledge DST Rajasthan, India for funding this Project.

References

1. Zolla F, Renversez G, Nicolet A, Zolla E (2001) Foundations of photonic crystal fibres
2. Krishna DG, Prasannan G, Sudheer SK, Pillai VPM (2015) Design of ultra-low loss highly nonlinear dispersion flattened octagonal photonic crystal fibers. Opt Photonics J 05(12):335–343. https://doi.org/10.4236/opj.2015.512032
3. Kaijage SF, Namihira Y, Begum F, Hai NH, Razzak SMA, Kinjo T, Miyagi K, Nozaki S, Zou N (2009) Highly nonlinear and polarization maintaining octagonal photonic crystal fiber in 1000 nm regions. https://doi.org/10.1109/OECC.2009.5214270
4. Habib SM, Motin MA, Hasan IM, Razzak SMA, Khan MGA (2013) Dispersion and confinement loss control with decagonal photonic crystal fibers for wide band transmission systems
5. Mortensen NA, Folkenberg JR, Nielsen MD, Hansen KP (2003) Modal cutoff and the V parameter in photonic crystal fibers. Opt Lett 28(20):1879. https://doi.org/10.1364/ol.28.001879
6. Kaur A, Gupta DP, Devra S, Singh K (2016) Photonic crystal fiber: developments and applications. Res Cell An Int J Eng Sci Print 17:2229–6913
7. Akowuah EK, Ademgil H, Haxha S (2012) Design and analysis of photonic crystal fibres (PCFs) for broadband applications. https://doi.org/10.1109/ICASTech.2012.6381077
8. Chen D, Tse M-LV, Tam H-Y (2010) Optical properties of photonic crystal fibers with a fiber core of arrays of subwavelength circular air holes: birefringence and dispersion. Prog Electromagn Res 105:193–212. https://doi.org/10.2528/pier10042706
9. Sharma R, Sharma A, Sharma V, Santosh C (2015) Hybrid square lattice elliptical air hole photonic crystal fiber with doped core. Int J Adv Electron Electr Eng (IJAEEE) 4:5
10. Sharma V, Sharma R (2016) Design of hybrid photonic crystal fiber with elliptical and circular air holes analyzed for large flattened dispersion and high birefringence. J Nanophotonics 10:8
11. Sharma R, Janyani V, Sharma A (2011) Design of elliptical air hole PCF with hybrid square lattice for high birefringence and a lower zero dispersion wavelength. Int J Comput Sci Emerg Technol 2:5
12. Sharma R, Janyani V, Bhatnagar SK (2011) Low chromatic dispersion and high birefringence investigated in elliptical air hole photonic crystal fiber. J Mod Opt 59:205–212
13. Sharma R, Janyani V, Bhatnagar SK (2011) Improved single mode property in elliptical air hole photonic crystal fiber. J Mod Opt 58:604–610

Transmission Properties of Lower Refractive Index Liquid Filled Hexagon Solid Core PCF

Shalini, Shahiruddin, Dharmendra Kumar Singh and M. A. Hassan

Abstract Lower refractive index liquid filled hexagon solid core PCF geometry is demonstrated. The optical characteristics of designed fiber can be modified by varying the geometrical parameters such as air hole diameter and keeping the pitch constant. It has been observed that low confinement loss along X and Y polarization, high birefringence, nearly zero dispersion, effective area of 42.18 μm^2 at 1.55 μm simultaneously achieved by using the commercial Rsoft software. The other optical properties such as effective refractive index, normalized frequency and nonlinear coefficient are also evaluated. With all these good features the designed model has potential applications in heat sensing and food quality control. But among the various advantages of the proposed PCF the most important is its simplified fabrication procedure.

Keywords Photonic crystal fiber (PCF) · Dispersion · Confinement loss
Total internal refraction · Photonic band gap

Shalini (✉) · Shahiruddin
Department of Electronics and Communication Engineering, Birla Institute of Technology, Patna Campus, Patna 800014, India
e-mail: singhshalini0607@gmail.com

Shahiruddin
e-mail: shahir@bitmesra.ac.in

D. K. Singh
Department of Electronics and Communication Engineering, National Institute of Technology, Patna 800005, India
e-mail: dksingh_bit@yahoo.com

M. A. Hassan
Department of Mechanical Engineering, Birla Institute of Technology, Patna Campus, Patna 800014, India
e-mail: hassan@bitmesra.ac.in

V. Janyani et al. (eds.), *Optical and Wireless Technologies*, Lecture Notes in Electrical Engineering 472, https://doi.org/10.1007/978-981-10-7395-3_6

1 Introduction

Photonic crystal fiber or micro-structured fiber is a class of optical fiber, composed of core enclosed by cladding. Claddings are formed by an array of air holes. In recent times, PCF have bring much interest among researchers due to their extremely good and unique properties over traditional optical fiber because of high design flexibility, high numerical aperture, achieve very birefringence and exhibit unique dispersion characteristics [1, 2]. Guiding mechanisms depends up the PCFs geometry and core or cladding material: index- guiding PCF which guide light in solid core through modified TIR. It works over whole transmission spectrum. PBG-guiding PCF which guide light in hollow core through PBG effect. It works over selected transmission spectrum [3].

The sensing technique depend either on selective filling liquid analyte or by simply immersing the sensing fiber into the liquid analyte [4, 5], High birefringence plays an important role in many application such as fiber-optic sensing, generate microwave pulses and polarization maintaining fiber [6], N_{eff} is the leading parameter of a PCF, whose real part holds the dispersion information whereas imaginary part contains loses, positive dispersion degrades the optical pulses. So it is necessary to achieve negative dispersion, low effect area for nonlinear application [7–9]. By increasing the number of air holes or by changing the diameter of air holes reduces the confinement losses; unsymmetrical PCF structure induced high birefringence, shows large optical nonlinearity and almost uniform low dispersion [10–12]. Circular photonic crystal fiber(C-PCF), such type of fiber can hold up high level mode transmission with larger bandwidth and lower confinement loss [13]. Different kind of polarization distribution were produced, improved quality and steady CVBs achieved, fiber filled with disulfide shows better results and such type of fiber used as optical switching [14, 15]. Abdelkader sonne et al. [16] design a double octagonal lattice photonic crystal fiber with negative dispersion of −100 to −200 (ps/nm/km) and birefringence of 10^{-2}. Jian Liang et al. [17] present a highly birefringent index guiding PCF consisting of elliptical air holes in the cladding and small holes both in the core and cladding area, which show flattened dispersion, high birefringence of 10^{-3} and low effective area. D. Paul et al. [18] present a PCFs with guiding core contains seven missing air holes. Two different composite—phosphate and crown glasses are used in the model to find the properties such as V parameter, spot size, mode area, beam divergence and dispersion. Crown PCF show more spot size and good coverage of V parameter as compare to phosphate.

In this paper, hexagon solid core PCF is proposed. The inner ring of designed fiber is refill with Paraffin oil. Geometry consists of five ring of air hole in the circular form. The designed PCF is used successfully to obtain various optical properties and it will cover all the O, E, S, C, L and U optical bands. Rsoft software

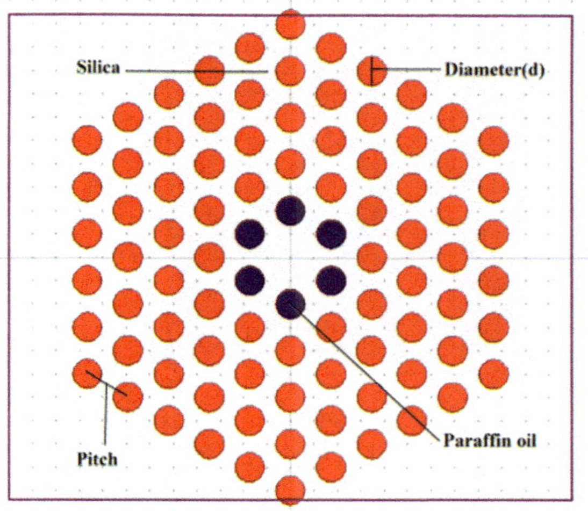

Fig. 1 Cross sectional view of designed PCF

is used for analyzing the various properties. Such type of designed is used for application such as sensor. In Sect. 1.2 PCF design will be shown. Results and various properties of PCF will be examined in Sect. 1.3.

2 Fiber Configuration

Figure 1 display the geometry of the designed hexagon solid core PCF with five rings. The solid core is created by elimination of air hole and cladding is created by five ring of circular air hole align in hexagonal lattice. The equidistant spacing (pitch) between the air hole in the cladding region has been expressed by Λ. The air hole diameter in each ring of cladding is denoted by d. The air hole diameter to pitch ratio is d/Λ and $\rho = 2\Lambda - d$ is the core length. The first inner ring of the PCF is refill with paraffin oil. Silica glass is used as a background material whose refractive index is 1.45. The refractive index of silica is greater than the refractive index of paraffin oil. Figure 2 displays the field intensity of the design fiber. From the mode profile it is observed that the air hole ring of the cladding have strong effect on the PCF properties. Mode profiles contain the value of N_{eff} which contains the information of dispersion and losses.

From the Fig. 2 it is observed that lights are mainly confined in the center of the designed solid core PCF refilled with paraffin oil in the inner ring of the PCF. On the top of the simulation result mode profile is mentioned which contain the value of N_{eff}.

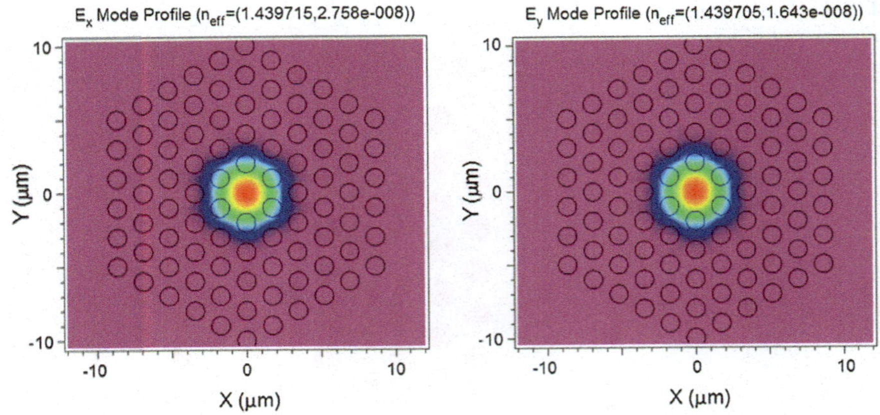

(a) E_X mode
(b) Ey mode

Fig. 2 Field intensity of **a** E_X mode and **b** Ey mode the designed PCF

3 Simulation Result

The optical characteristics of presented fiber are modified by varying the geometrical parameters, d = 1.20, 1.24, 1.28 μm and Λ = 2 μm. Low refractive index liquid is used to refill the inner ring of the proposed fiber. The simulation result covers the wavelength range of 1.2–1.7 μm. The graph of Effective refractive index is shown in Fig. 3. It is an important parameter of the PCF because it describes dispersion properties or loss of the PCF. It is observed that the curve of effective refractive index decreases linearly by increasing wavelength. Effective refractive indexes rely on the wavelength as well as on the mode in which light propagates.

The confinement loss first increases and then decreases by increasing wavelength as shown in Fig. 4. The result of the confinement loss is expressed in dB/m. The confinement loss has been calculated by using imaginary part of the effective refractive index ($\mathrm{Im}(N_{eff})$) and wavelength (λ) is shown in (1) [19]

$$L_c = 8.686 K_0 \mathrm{Im}\left(N_{eff}\right) \times 10^6 \mathrm{dB/m} \tag{1}$$

where $2\pi/\lambda = k_0$ denotes the wavenumber in free space and by solving the 20/ln (10) a constant value is obtained. Confinement loss is affected by the air hole diameter to pitch ratio. Ratio of d/Λ becomes greater as diameter of air hole increases. So it is observed that the greater d/Λ ratio results in lower confinement loss. From Fig. 4 it is found that the low confinement loss of 0.11, 0.12 dB/m along X and Y polarization is obtained for d = 1.28 μm at 1.55 μm wavelength. The curve of confinement loss first starts decreasing as wavelength increases and between 1.4 and 1.7 μm of wavelength its shows approx. Smooth horizontal line in the range of 0.1 dB/m. It is also observed that the value of confinement loss is

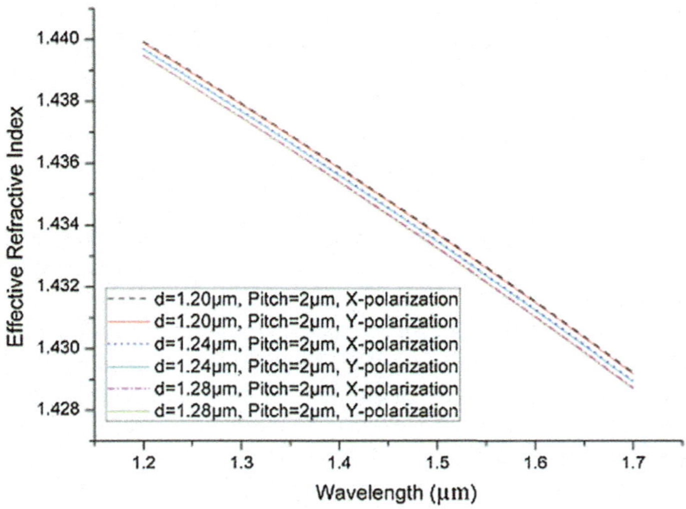

Fig. 3 Effective refractive index across X and Y mode as a function of wavelength for d = 1.20, 1.24, 1.28 μm and Λ = 2 μm

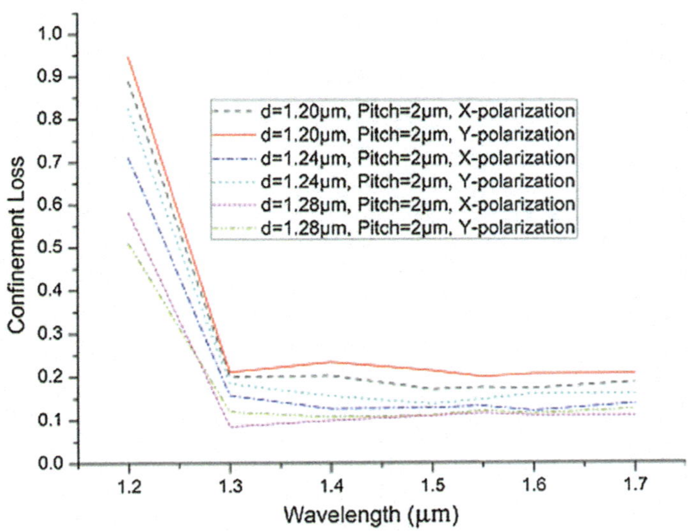

Fig. 4 Confinement losses along X and Y as a function wavelength for d = 1.20, 1.24, 1.28 μm and Λ = 2 μm

higher for smaller value of diameter of air holes. As shown in Fig. 4, the diameter of air holes increases from 1.20–1.28 μm, the value of confinement loss starts decreasing.

Dispersion is measured in ps/km-nm and in PCF we obtained negative or zero dispersion which is not possible in conventional fiber. It is calculated using the given formula (2) [17]

$$D = -\left(\frac{c}{\lambda}\right)\left(\frac{d^2 N_{eff}}{d\lambda^2}\right) \quad (2)$$

For d = 1.20, 1.24, 1.28 μm and pitch = 2 μm the dispersion have been measured and their corresponding curves are illustrated in Fig. 5. Dispersion is another important parameter of the PCF.

From the above Fig. 5 it is understandable that lowest dispersion is successfully achieved in the C-band. So, it is concluded that dispersion is more sensitive to air hole diameter. Nearly approx zero dispersion is obtained for wavelength 1.55–2 μm wavelength. From 1.2 to 1.4 μm wavelength we obtained negative dispersion. The wavelength selection is chosen from 0.7 to 2 μm for dispersion because zero dispersion is achieved between 1.4 and 1.65 μm optical bands and when it entered it up to 2 μm we will get the same zero dispersion. But from 0.7 to 1.4 μm, negative

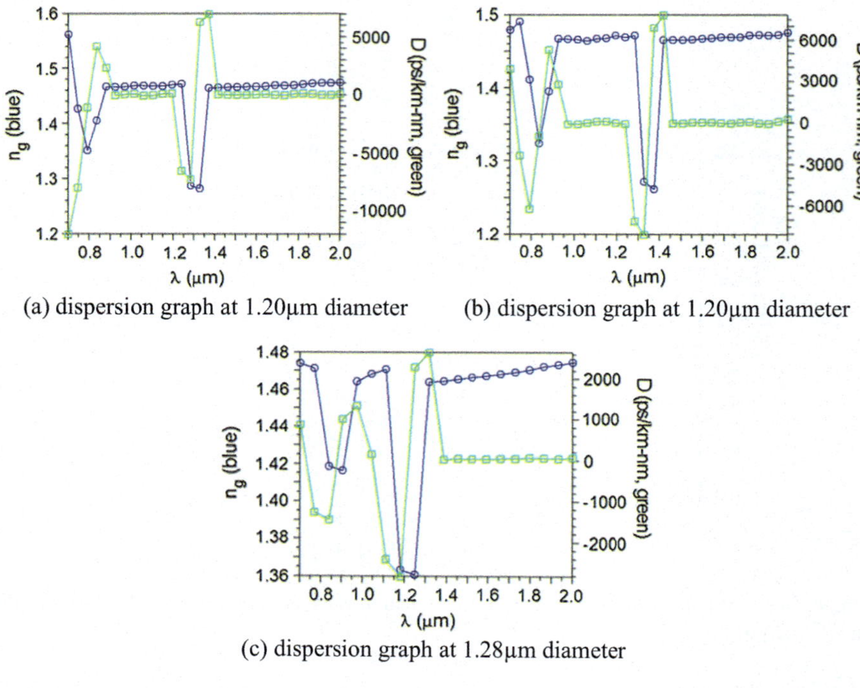

(a) dispersion graph at 1.20μm diameter

(b) dispersion graph at 1.20μm diameter

(c) dispersion graph at 1.28μm diameter

Fig. 5 Curve of dispersion as a function of wavelength at constant pitch = 2 μm **a** dispersion graph at 1.20 μm diameter **b** dispersion graph at 1.20 μm diameter **c** dispersion graph at 1.28 μm diameter

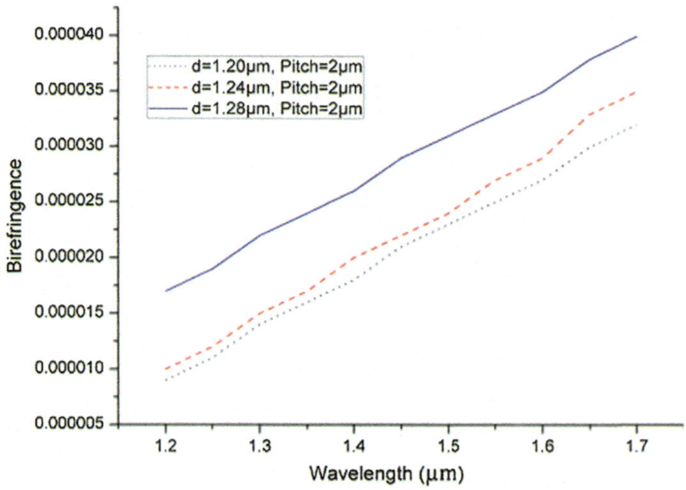

Fig. 6 Birefringence curve as a function of wavelength for d = 1.20 μm, 1.24 μm, 1.28 μm and Λ = 2 μm

and positive dispersion is achieved and it will useful for dispersion compensation fiber.

Birefringence curve increases with increases of wavelength as shown in Fig. 6. Birefringence depends upon wavelength. The birefringence of a PCF is the difference between the effective refractive index of two perpendicularly polarization modes. It is calculated using (3) [16]

$$B = \mathrm{Re}|N_{effX} - N_{effY}| \tag{3}$$

where N_{eff} along X and Y polarization.

The value of birefringence increases when we increase the diameter of the circular air holes. The highest birefringence 3.3×10^{-5} is obtained when the diameter is 1.28 μm at 1.55 μm wavelength. It means when we increase the wavelength 1.2–1.7 μm the birefringence increases and by increasing the diameter from 1.20 to 1.28 μm the birefringence also increases. The total three lines are shown in the graph of birefringence and it is observed all the three lines are moved in a similar way.

Curve of V_{eff} decreases as increases of wavelength as shown in Fig. 7. High V_{eff} may increases radiation losses at core- cladding interface. The value of normalized frequency must be less than or equal to 2.405. It is unit less parameter of PCF. It is calculated using (4) [18]

$$V_{eff} = \frac{2\pi\rho}{\lambda}\sqrt{n_{co}^2 - n_{fsm}^2} \tag{4}$$

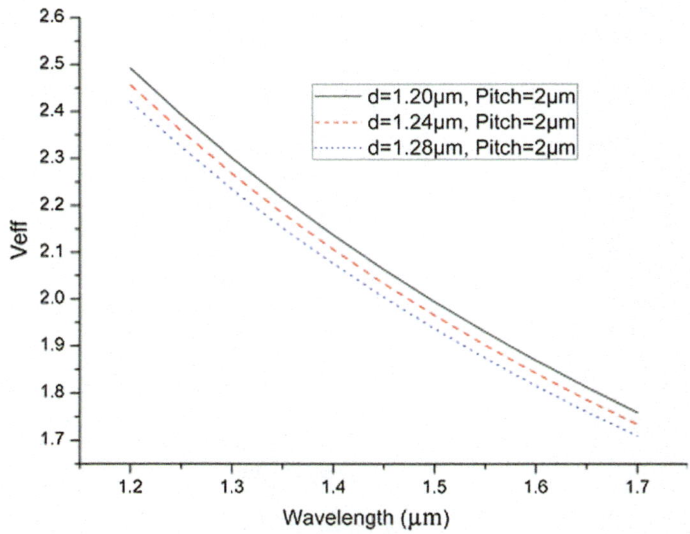

Fig. 7 Normalized frequency (V_{eff}) versus wavelength for different d and constant Λ

where ρ is the effective core radius, n^2co and n^2fsm is the effective index of core and cladding and λ denotes the wavelength in μm. The effective index of cladding is lower than the refractive index of core. V_{eff} is inversely proportional to the wavelength. It means as the wavelength increases from 1.2 to 1.7 μm wavelength the curve of V_{eff} starts decreasing. We put the value of refractive index of core and cladding in the Eq. 4. Normalized frequency is lowest when d = 1.28 μm at 1.55 μm wavelength as shown in Fig. 7. The value of the normalized frequency is higher when the diameters of the air holes are smaller than the 1.28 μm. The normalized frequency is 1.87 at 1.55 μm wavelength for 1.28 μm diameter. The curve of normalized frequency bends in a similar fashion for all the diameter of air holes.

Effectivearea of the PCF is evaluated using (5) [18]

$$A_{eff} = \pi[W_{PCF}]^2 \tag{5}$$

where W_{PCF} represent spot size. Effective area measured in μm^2. Effective area is wavelength dependent and it also depends upon the size of air holes placed in the cladding region. In Fig. 8 it is found that Effective area (A_{eff}) increases with increases of wavelength. It is also observed that for different value of d = 1.20, 1.24, 1.28 μm and pitch = 2 μm the curve of effective area smoothly increases and after a certain wavelength the curves of effective area overlapped. As the diameter of microstructure air holes increases the value of effective area slowly decreases as

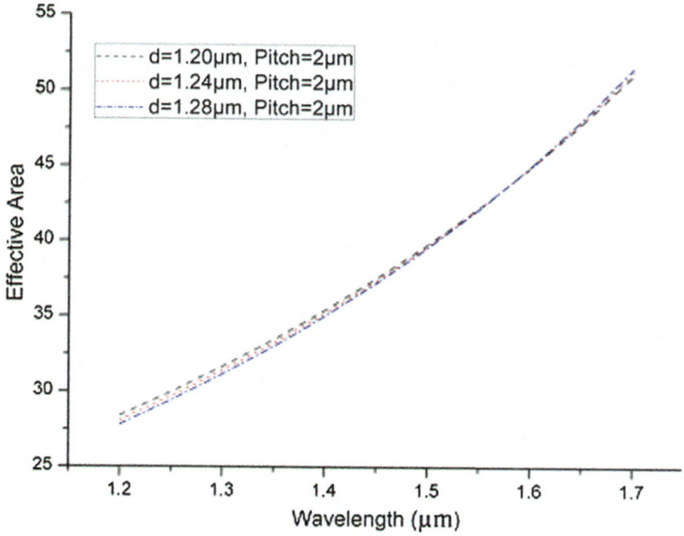

Fig. 8 Graph of effective area as a function of wavelength for different d and constant Λ

shown in Fig. 8. Means the lowest diameter of air holes shows the highest value of effective area. At 1.55 μm, the effective area is 42.25, 42.20, 42.18 μm^2 for diameter 1.20, 1.24, 1.28 μm respectively. After the decimal place the value changes and before the decimal points no changes in the value of effective area. So the plot of effective area overlapped after a certain wavelength range. Effective area and nonlinear coefficient is related with each other.

After finding the value of effective area, we easily calculated the value of nonlinear coefficient. Non-linear coefficient γ is calculated by (6) [18]

$$\gamma = \frac{2\pi}{\lambda} \frac{n_2}{A_{eff}} \tag{6}$$

where n_2 is kerr coefficient and A_{eff} denoted effective area. From the Eq. 6, it is analyzed that nonlinear coefficient inversely proportional to the λ and A_{eff}. As the effective area and wavelength increases the nonlinear coefficient decreases. The graph of nonlinear coefficient as a function of wavelength is shown in Fig. 9. Curve of nonlinear coefficient decreases in a smooth manner as wavelength increases. In the starting, the curve of nonlinear coefficient decreases and after 1.5 μm wavelength the curve of ϒ starts touches. Between 1.55 and 1.7 μm wavelengths range the graph of ϒ overlapped due to small variation in the value of nonlinear coefficient with respect to wavelength. As shown in Fig. 9, higher diameter shows higher value of nonlinear coefficient. But in case of effective area larger diameter shows

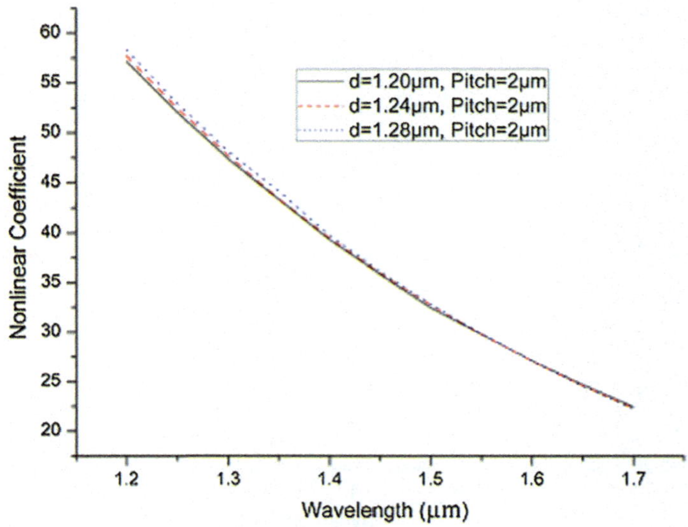

Fig. 9 Nonlinear coefficient versus wavelength

Table 1 Comparative PCF properties at different diameter

Diameter (µm)	Confinement loss (dB/m)		Birefringence	V-parameter (V_{eff})	Effective area (µm^2)	Nonlinear coefficient (W^{-1} km^{-1})
	X pol.	Y pol.				
1.20	0.17	0.19	2.5×10^{-5}	1.92	42.25	29.73
1.24	0.13	0.14	2.7×10^{-5}	1.90	42.20	29.77
1.28	0.11	0.12	3.3×10^{-5}	1.87	42.18	29.78

smaller value of effective area. It means that both are inversely correlated with each other. When the diameter is 1.20, 1.24, 1.28 µm, the nonlinear coefficient is 29.73, 29.77, 29.78 w^{-1} km^{-1} is calculated at 1.55 µm wavelength and it is analyzed that we obtained the closest value of nonlinear coefficient after 1.55 µm wavelength range. So, that between 1.55 and 1.7 µm wavelength range curves of nonlinear coefficient shows overlapping (Tables 1 and 2).

Table 2 Comparison added and shows the calculated parameter with published paper

S. no.	Author name	Dispersion	Birefringence	V-parameter	Effective area	Nonlinear coefficient	Confinement loss
1	Dunke Lu et al. [20]	21.83–21.85 ps/km-nm at 1.505–1.577 μm	–	–	–	–	0.01 dB/km at 1.505–1.577 μm
2	M. Aliramezani et al. [21]	−208.8 to −61.5 ps/km-nm at 1360–1675 nm	–	–	–	–	0.46357×10^{-2} dB/km at 1550 nm
3	Proposed work	0 ps/km-nm at 1.4–2 μm	3.3×10^{-5} at 1.55 μm	1.87 at 1.55 μm	42.18 μm^2 at 1.55 μm	29.78 w^{-1} km^{-1} at 1.55 μm	0.11 dB/m at 1.55 μm w

4 Conclusion

The best result is obtained when the diameter is 1.28 μm, for the designed hexagon solid core PCF refill with paraffin oil in the inner ring. From the simulation results it is found that the dispersion of the designed PCF is zero in the wavelength ranging from 1.2 to 1.7 μm, low confinement loss and high birefringence are successfully achieved. Other transmission properties such as effective refractive index, effective area, nonlinear coefficient and normalized frequency are also calculated using Rsoft software.

References

1. Peng T, Xu T, Wang X, Liu Z, Dai S, Liu S, Zhao Z, Pan Z (2016) Simulation and fabrication of micro-structured optical fibers with extruded tubes. Optik 127:8240–8247
2. Lagerwall JPF, Scalia G (2012) A new era for liquid crystal search: Application of liquid crystals in soft matter nano, bio and microtechnology. Curr Appl Phys 12:1387–1412
3. Li Y, Wang C-Y, Hu M, Liu B, Chai L (2007) Bandgap properties of modified square photonic crystal fibers. Phys E Low-Dimens Syst Nanostruct 39:26–29
4. Zhang Y, Zhao Y, Wang Q (2015) Measurement of methane concentration with cryptophane E infiltrated photonic crystal microcaity. Sens Actuators B Chemicals 209:431–437
5. Li J, Wang J, Cheng Y, Wang R, Zhang B, Wang H (2012) Novel large mode area photonic crystal fibers with selectively material-filled structure. Opt Laser Technol 48:375–380
6. Rui H (2016) Highly birefringent photonic crystal fiber with a squeezed core and small modal area. Optik 127:5245–5248
7. Hasan MI, Habib MS, Razzak SMA (2014) An elliptical-shaped core residual dispersion compensating octagonal photonic crystal fiber. IEEE Photonics Technol Lett 26:2047–2050
8. Chen M, Yang J, Sheng P, Tong X, Chen H (2016) Tunable microwave generation method based on birefringence photonic crystal fiber. Optik 127:5990–5999
9. Demir H, Ozsoy S (2011) Comparative study of large solid core photonic crystal fibers: dispersion and effective mode area. Optik 123:739–743
10. Kabir S, Khandokar MRH, Khan MAG (2016) Design of triangular core LMA-PCF with low-bending loss and low non-linearity for laser application. Opt Laser Technol 81:84–89
11. Abdelaziz I, Ademgil H, Abdelmalek F, Haxha S, German T, Bouchriha H (2010) Design of a large effective mode area photonic crystal fiber with modified rings. Opt Commun 283:5218–5223
12. Yang XC, Lu Y, Liu BL, Yao JQ (2016) Temperature sensor based on photonic crystal fiber filled with liquid and silver nano-wires. IEEE Photonics J 8(3)
13. Zhang H, Zhang W, Xi L, Tang X, Zang X, Zhang X (2016) A new type circular photonic crystal fiber for orbital angular momentum mode transmission. IEEE Photonic Technol Lett 28:1426–1429
14. Zhao C, Gan X, Li P, Fang L, Han L, Tu L, Zhao J (2016) Design of multicore photonic crystal fibers to generate cylindrical vector beams. J Lightwave Technol 34:1206–1211
15. Munera N, Herrera RA (2016) Modeling anon linear optical switching in a standard photonic crystal fiber infiltrated with carbon disulfide. Opt Commun 368:185–189
16. Sonne A, Ouchar A, Sonne K (2016) Improving of high birefringence with negative dispersion using double octagonal lattice photonic crystal fiber. Optik 127:8–10
17. Liang J, Yun M, Kong W, Sun X, Zhang W, Xi S (2016) Highly birefringent photonic crystal fiber with flattened dispersion and low effective mode area. Optik 122:2151–2154

18. Paul D, Biswas R, Bhattacharyya NS (2015) Investigating photonic crystal fiber within E to F communication band with different material composites. Optik 126:4640–4645
19. Xu Z, Li X, Hong Y, Liu P, Yang H, Ling W (2015) Temperature dependence of beat-length and confinement loss in an air-core photonic band-gap fiber. Opt Commun 366:38–44
20. Lu D, Liu J (2016) Broadband single-polarization single-mode operation in photonic crystal fibers with hexagonally latticed circular air holes. Lightwave Technol 34(10):2452–2458
21. Aliramezani M, Nejad MSh (2010) Numerical analysis and optimization of dual concentric core photonic crystal fiber for broad band dispersion compensation. Opt Laser Technol 42:1209–1217

Design and Analysis of Rhombus-Shaped Dual-Core Propylene Glycol Filled PCF

Neha Kumari, Shahiruddin, Dharmendra Kumar Singh and M. A. Hassan

Abstract A novel form of rhombus solid dual-core PCFs structures filled with propylene glycol for different air hole diameter is presented. The proposed fiber contains seven circular air holes. Using Rsoft FEMSIM software, the transmission properties such as confinement loss, effective mode area, chromatic dispersion, birefringence, nonlinearity, and V_{eff} are analyzed. The proposed structure shows the high birefringence of 1.07×10^{-4}, low confinement loss along X polarization is 1.8606×10^{-4} and along Y polarization is 1.5095×10^{-4} dB/m, large negative dispersion, effective area of 27.66 μm^2 at 1.55 μm. Proposed fiber has ultimate performance, and it is uncomplicated and easy to fabricate.

Keywords Photonic crystal fiber (PCF) · Dispersion · Confinement loss Air fill fraction · Normalized frequency · Effective refractive index Effective area

N. Kumari (✉) · Shahiruddin
Department of Electronics and Communication Engineering,
Birla Institute of Technology, Patna Campus, Patna 800014, India
e-mail: neha310126@gmail.com

Shahiruddin
e-mail: shahir@bitmesra.ac.in

D. K. Singh
Department of Electronics and Communication Engineering,
National Institute of Technology, Patna 800005, India
e-mail: dksingh_bit@yahoo.com

M. A. Hassan
Department of Mechanical Engineering, Birla Institute of Technology,
Patna Campus, Patna 800014, India
e-mail: hassan@bitmesra.ac.in

V. Janyani et al. (eds.), *Optical and Wireless Technologies*, Lecture Notes in Electrical Engineering 472, https://doi.org/10.1007/978-981-10-7395-3_7

1 Introduction

Photonic crystal fibers (PCFs) are silica glass fibers hold a well-ordered array of microscopic holes that stretch along the whole fiber length. PCF possesses excellent properties which are impossible to achieve in conventional fiber such as large mode area, negative or zero dispersion, low losses, and high birefringence. The PCFs can be categorized according to their core; solid core PCFs in which refractive index of core is greater than the refractive index of cladding and hollow core PCFs in which refractive index of cladding is greater than refractive index of core [1–3]. Huge-size PCF can support higher order modes, PCF filled with ethanol shows better birefringence, and C-PCF with minimized air holes is used in super continuum generation [4–6]. Hollow core liquid-filled PCF displays powerful optical properties such as nonlinearity and effective mode, and fiber with hybrid dual-core PCF supports polarization and mode conversion [7, 8]. Jianhua Li et al. [9] proposed an optical coupler by introducing two cores into PCFs. The structure contains two fiber cores that are filled with magnetic fluid and separated by an air hole. The result shows high large sensing range and high sensitivity. The single polarization single mode (SPSM) double-core PCF coupler has various benefits compared with conventional optical coupler such as more flexible design freedom, shorter coupling length [10]. Hexagonally latticed circular air hole PCF has confinement loss of 0.01 dB/km at 2.034 μm, and dispersion was calculated between (21.83; 21.85) ps/km-nm, so a bandwidth of 70 nm for 0.02 ps/km/nm amplitude of dispersion variation has been achieved [11]. A ZBLAN (ZrF_4-BaF_2-LaF_3-AlF_3-NaF) photonic crystal fiber for communication in mid-infrared is presented. The characteristics such as normalized frequency, confinement loss, flattened chromatic dispersion, effective mode area, and nonlinearity are calculated [12]. Haiming Jiang et al. presented dual-core photonic crystal fiber. DC-PCF is used in fiber filter. The effective refractive index of dual core is 1.4205 at 1.55 μm. When input port and output port are the same core, a band-stop filter can be obtained with a full-width half-maximum (FWHM) of 58 nm. The length of the filter is 1.83 mm [13]. Fiber has high negative ultra-flattened dispersion for both X and Y polarization modes. The confinement loss and low birefringence were 10^{-2} to 10 dB/m and 6.65×10^{-3} to 6.9×10^{-3} respectively, at 1.25–1.7 μm. The effective area and negative flattened dispersion for X polarization mode and for Y polarization mode at 1.25–1.7 μm are presented [14].

In this work, a rhombus solid dual-core PCF with seven rings of circular air holes is presented. Only circular air holes are used in the proposed structure. The circular holes are arranged in a rhombus lattice. Geometry consists of symmetrical air hole which makes fabrication easy, and Rsoft software is used to analyze various transmission properties of the designed geometry. Designed geometry is used successfully in various applications such as coupler, fiber filter.

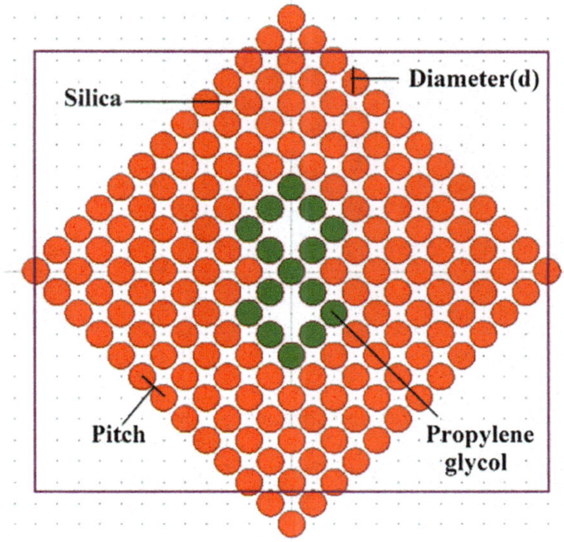

Fig. 1 Schematic diagram of rhombus dual solid core PCF

2 Configuration of Rhombus PCF

The proposed design is a dual solid core fiber. Dual solid core fiber consists of seven rings of circular air holes in the rhombus lattice. The inner ring of the designed PCF is filled with propylene glycol and silica as a background material. In this paper, three different designs are presented by varying the air filling fraction (d/Λ) and keeping the pitch (Λ) constant. By taking three values of d = 1.20 μm, 1.24 μm, 1.28 μm and constant Λ = 2 μm, we find various optical properties. The refractive index of silica is 1.45, and the refractive index of propylene glycol is 1.43. Figure 1 shows cross-sectional view of designed PCF, and in Sect. 3, the simulation results are obtained of the designed PCF.

Electric field distribution at 1.55 μm wavelength for designed fiber is shown in Fig. 2. The power is mostly intense in the center of the rhombus solid dual-core PCF filled with propylene glycol.

3 Simulation Result

Effective refractive index comprises two parts; imaginary and real. Real part contains the information of dispersion and imaginary part helps in finding the loss [15]. By increasing the d of the proposed PCF, the effective refractive index decreases. When diameter of air hole is 1.20 μm, the value of N_{eff} is high. The value of N_{eff} is low when d = 1.28 μm. Figure 3 indicates that by increasing the wavelength, N_{eff} decreases.

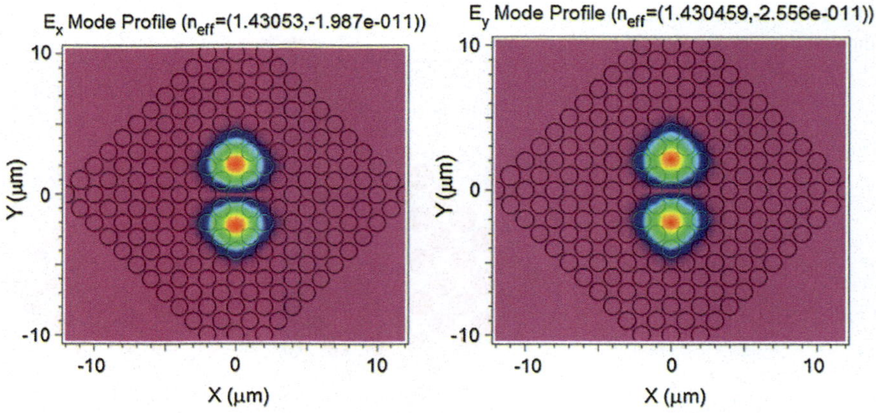

(a) x polarization (b) y polarization

Fig. 2 Field distribution along at 1.55 μm wavelength

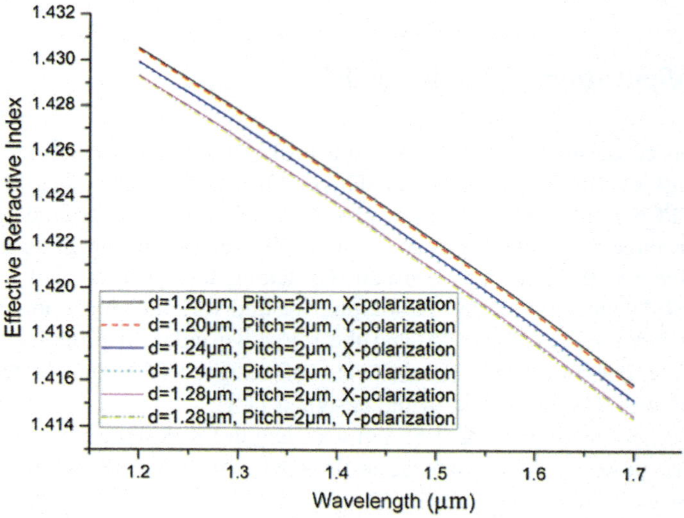

Fig. 3 Effective refractive index as a function of wavelength for x and y polarization

Confinement loss of PCF is a measure of variation of power loss from core to its cladding interface. The value of confinement loss reduces by increasing the number of air hole. It is seen that when the d is increased, the birefringence is reduced. Confinement loss of PCF is easily calculated from (1) [9]

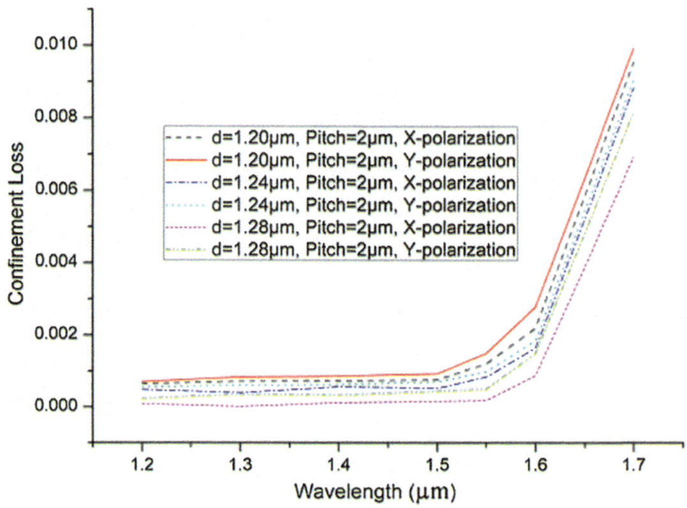

Fig. 4 Confinement loss curve as a function of wavelength for different value of diameter

$$L_c = 8.686 K_0 \mathrm{Im}\left(N_{eff}\right) \times 10^6 \, \mathrm{dB/m} \tag{1}$$

where Im (N_{eff}) is the imaginary part of the refractive index, $K_0 = 2\pi/\lambda$ is the wave number in free space. The hole of diameter is increased, and the confinement loss is reduced. When $\Lambda = 2$ μm and d = 1.28 μm, the lowest confinement loss of 10^{-4} dB/m is obtained as shown in Fig. 4.

Dispersion is a phenomenon of splitting of light which causes interference in optical communication. It can be numerically calculated as (2) [4]

$$D = -\left(\frac{c}{\lambda}\right)\left(\frac{d^2 N_{eff}}{d\lambda^2}\right) \tag{2}$$

The proposed rhombus dual-core PCF shows larger negative dispersion in the wavelength range 1.4–1.8 μm, which will cover S, C, L, and U optical bands. However, the wavelength considered from 0.7 to 1.8 μm shows the positive, negative, and zero dispersion, and it will be used as dispersion compensation fiber. It can be observed that the entire dispersion curve first goes high and then decreases with the increase of wavelength. From Fig. 5a, it is found that the dispersion is approx. −2500 ps/km-nm when d = 1.20 μm and pitch = 2 μm at 1.55 μm wavelength. Fig. 5b shows the dispersion curve as a function of wavelength when d = 1.24 μm and pitch = 2 μm. Figure 5c shows dispersion of the proposed rhombus dual solid core PCF as a function of wavelength for d = 1.28 μm. From the graph, it is observed that the dispersion becomes more negative by increasing the diameter of the air holes.

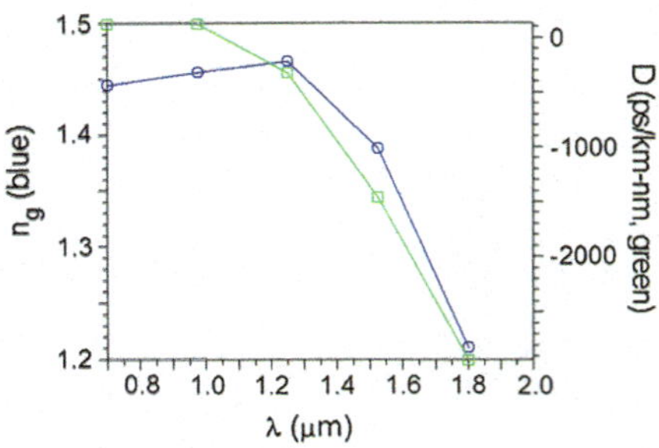

(a) d=1.20μm

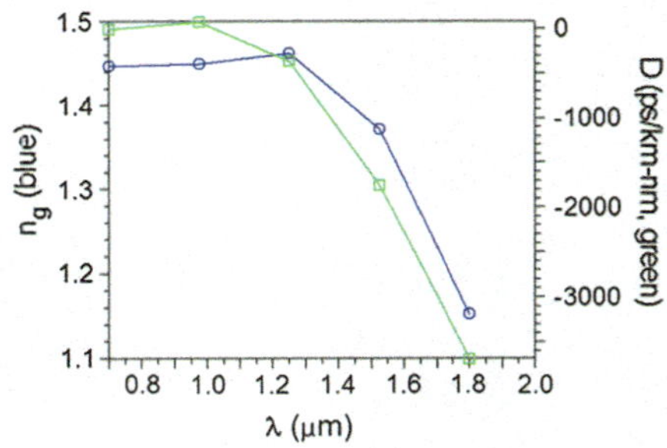

(b) d=1.24μm

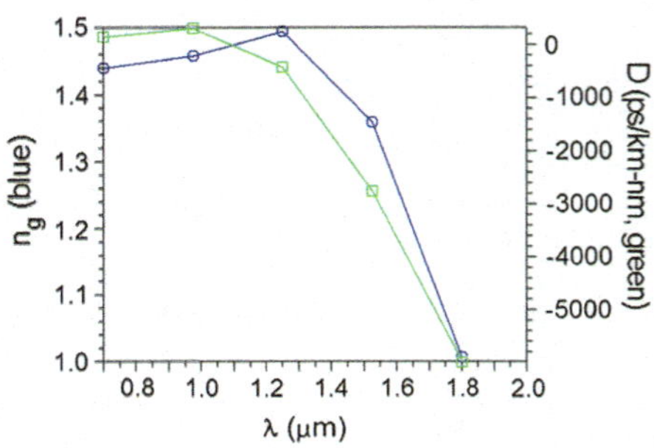

(c) d= 1.28 μm

Fig. 5 Dispersion curve at constant $\Lambda = 2$ μm

It is found that the value of dispersion is about −3500 ps/km-nm for d = 1.28 μm, and pitch = 2 μm at 1.55 μm is achieved. The impact of the variation of d to the birefringence properties is shown in Fig. 6. Birefringence is obtained for three different values of d. It can be obtained by using (3) [5]

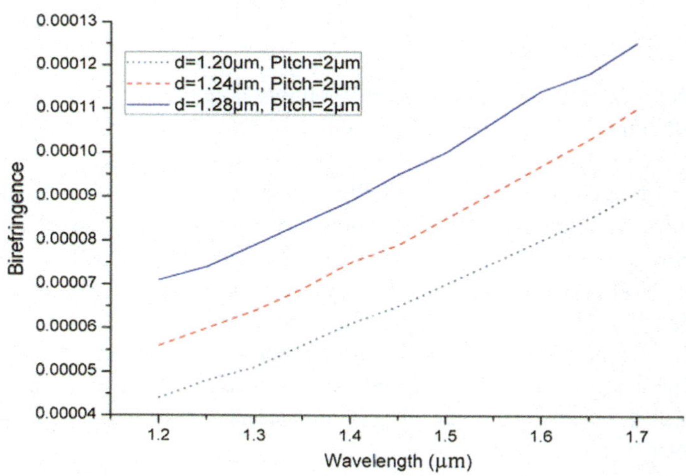

Fig. 6 Birefringence as a function of wavelength for different diameter

$$B = \mathrm{Re}|N_{effX} - N_{effY}| \tag{3}$$

It is observed that birefringence increases with the increase of wavelength. The birefringence depends on the relative structure size of the PCFs. High birefringence can be up to 10^{-4} for wavelength 1.55 μm.

Normalized frequency is calculated by using this formula (4) [12]

$$V_{eff} = \frac{2\pi\rho}{\lambda}\sqrt{n_{co}^2 - n_{fsm}^2} \tag{4}$$

where ρ is the effective core radius and n_{co}, n_{fsm} are the effective index of core and cladding. Normalized frequency (V_{eff}) of the proposed PCF as a function of wavelength is ranged from 1.2 to 1.7 μm as shown in Fig. 7. The curve of v parameter decreases with increase of wavelength. The v parameter is 1.323 at 1.55 μm wavelength.

Effective area is an important optical characteristic which can be used to calculate the nonlinear coefficient. Effective area is calculated using (5) [14]

$$A_{eff} = \pi[W_{PCF}]^2 \tag{5}$$

where W_{PCF} is spot size. The variations of effective area as a function of wavelength ranging from 1.2 to 1.7 μm for different air hole diameter are shown in Fig. 8. It is found that effective area (A_{eff}) increases with increase of wavelength. Its values continuously decreased by increasing the diameter from 1.20 to 1.28 μm.

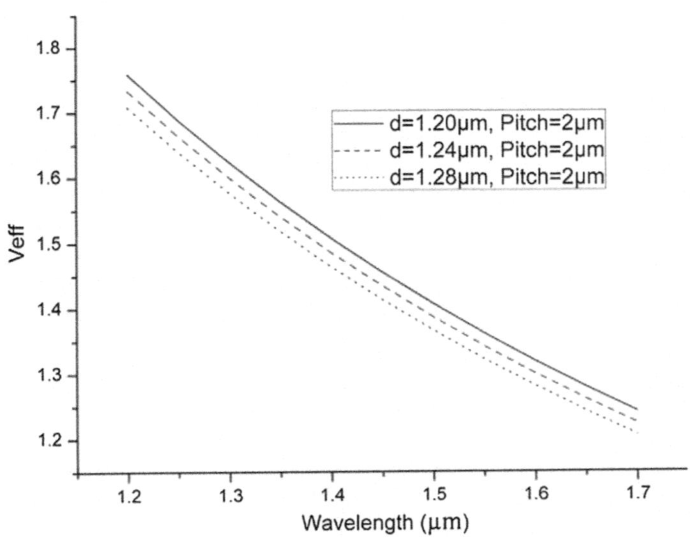

Fig. 7 V_{eff} as a function of wavelength for different d and constant Λ

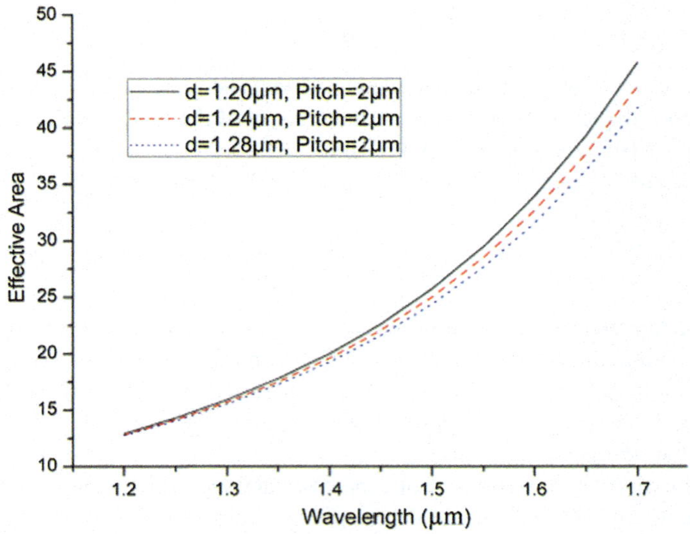

Fig. 8 Effective area as function of wavelength

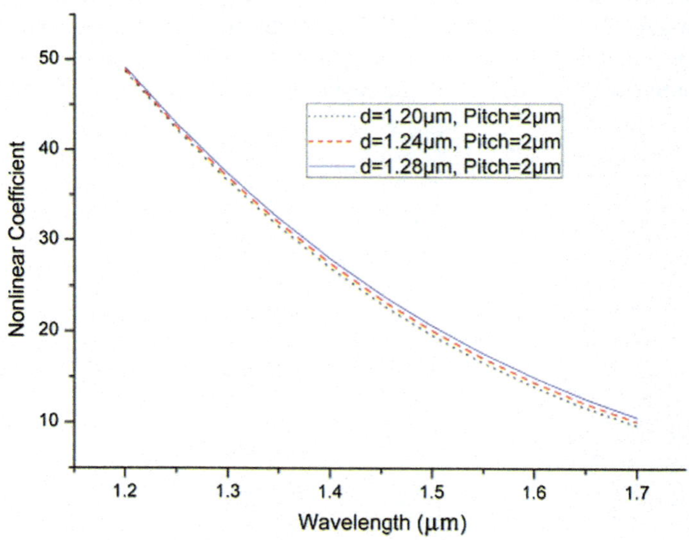

Fig. 9 Nonlinear coefficient as function of wavelength for different d

The nonlinear coefficient of the proposed fiber is shown in Fig. 9. Nonlinear coefficient can be obtained using (6) [16]

$$\gamma = \frac{2\pi}{\lambda} \frac{n_2}{A_{eff}} \quad (6)$$

where λ is the wavelength in μm, A_{eff} denotes effective area which is expressed in μm^2 and n_2 represents nonlinear refractive index coefficient of materials. It is observed that the nonlinear coefficient decreases with increase of wavelength. At 1.55 μm, the nonlinear coefficient is 17.59 w^{-1} km^{-1}.

4 Conclusion

For different values of air hole diameter at 1.55 μm wavelength, various characteristics such as highly birefringence, negative dispersion, low confinement loss of 10^{-4}, V_{eff} are obtained and also additional properties like large effective area and nonlinearity are demonstrated successfully for the design dual solid core PCF. By using the Rsoft FEMSIM software, the designed structure and all the optical characteristics are analyzed. Such PCFs are fabricated in future due to negative dispersion, low confinement loss, and other extraction of parameters for designing of coupler, fiber filter applications.

References

1. Poli F, Selleri S (2007) Photonic crystal fibers, properties and applications. Springer Series, The Netherlands
2. Razzak SMA, Namihira Y (2007) Guiding properties of a decagonal photonic crystal fiber. J Microw Optoelectron 6:44–47
3. Olyaee S, Taghipour F (2011) Design of new square-lattice photonic crystal fibers for optical communication applications. Int J Phys Sci 6:4405–4411, (2011)
4. Demir H, Ozsoy S (2011) Comparative study of large-solid-core photonic crystal fibers: dispersion and effective mode area. Optik 123:739–743
5. Hu T, Zhao Y, Zhu Y, Wang Q (2013) Experimental measurement of the temperature-birefringence characteristics of birefringent photonic crystal fiber filled with ethanol. Opt Commun 309:3–9
6. Medjouria A, Simohamed LM, Ziane O, Boudriouad A, Becer Z (2015) Design of a circular photonic crystal fiber with flattened chromatic dispersion using a defected core and selectively reduced air holes: application to supercontinuum generation at 1.55 μm. Photonics Nanostruct —Fundam Appl 16:43–50
7. Monfared YE, Ponomarenko SA (2016) Slow light generation via stimulated Brillouin scattering in liquid-filled photonic crystal fibers. Optik 127:5800–5805
8. Cai S, Yu S, Wang Y, Gao L (2016) Hybrid dual-core photonic crystal fiber for spatial mode conversion. IEEE Photonics Technol Lett 28:1041–1135
9. Li J, Wang R, Wang J, Zhang B, Xu Z, Wang H (2014) Novel magnetic field sensor based on magnetic fluids infiltrated dual-core photonic crystal fibers. Opt Fiber Technol 20:100–105
10. Li J, Duan K, Wang Y, Cao X, Guo Y, Lin X (2007) Design of a single-polarisation single-mode photonic crystal fiber double-core couple. Optik 120:490–496

11. Lu D, Liu J (2016) Broadband single-polarization single-mode operation in photonic crystal fibers with hexagonally latticed circular air holes. J Lightwave Technol 34:2452–2458
12. Tee DC, Tamchek N, Ooi CHR (2016) Numerical modeling of the fundamental characteristics of ZBLEN photonic crystal fiber for communication in 2–3 μm midinfrared region. IEEE Photonics J 8(2)
13. Jiang H, Wang E, Xie K, Hu Z (2016) Dual-core photonic crystal fiber for use in fiber filters. IEEE Photonics J 8(2)
14. Mahmud RR, Razzak SMA, Hasan MI (2016) A new photonic crystal fiber design on the high negative ultra-flattened dispersion for both X and Y polarization modes. Optik 127:8670–8677
15. Maji PS, Chaudhuri PR (2016) Studies of the modal properties of circular photonic crystal fiber (C-PCF) for high power application. Photonics Nanostruct 19:1569–4410
16. Cai W, Liu E, Feng B (2016) Dispersion properties of a photonic quasi-crystal fiber with double cladding air holes. Optik 127:4438–4442

High-sensitive Fiber Bragg Grating Sensor for Different Temperature Application

Sanjeev Kumar Raghuwanshi, Manish Kumar Thukral and Alisha Priya

Abstract The main physical quantities in fiber Bragg grating are temperature and strain. The temperature sensors for different materials have been analyzed in this work. The sensing can be possible on the range of Bragg wavelength shifts occurred by the temperature change in the medium. Thus, the temperature is measurement based on the wavelength shifting of the fiber Bragg grating. The mathematical descriptions and simulation of various parameters of FBG temperature sensor are also included in the work.

Keywords Fiber Bragg grating (FBG) · Reflectivity · Temperature sensor
Thermo-optic coefficient · Thermal expansion coefficient

1 Introduction

Fiber Bragg gratings are periodic changes of the refractive index occurred in the core of an optical fiber and formed by UV laser light under particular condition. Many functions like reflection and filtering can be performed by this device in a highly efficient manner. The most important revolution of fiber Bragg gratings is in the field of telecommunications. FBG is also used in the optical fiber sensor field. This is the simple device of a periodic modulation of the refractive index in the core of the fiber. FBG strain sensors are dependent on the property of material, and it is a complicated phenomenon as both the temperature and strain regulate the reflected wavelength of the sensor. To measure adequate temperature, we must compensate

S. K. Raghuwanshi · M. K. Thukral (✉) · A. Priya
Department of Electronics Engineering, IIT(ISM), Dhanbad, Jharkhand, India
e-mail: manish.ism14@gmail.com

S. K. Raghuwanshi
e-mail: sanjeevrus77@gmail.com

A. Priya
e-mail: priyaalisha01@gmail.com

V. Janyani et al. (eds.), *Optical and Wireless Technologies*, Lecture Notes in Electrical Engineering 472, https://doi.org/10.1007/978-981-10-7395-3_8

the strain effects on the FBG and vice versa. The Bragg grating is also used in coupling light from one propagating mode to another mode [1–4].

The central wavelength of the reflected component satisfies the Bragg relation.

$$\lambda_B = 2n_{eff}\Lambda \tag{1}$$

where λ_B is reflected wavelength of Bragg grating, n_{eff} is refractive index in the core of the fiber, Λ is period of the grating. We are considering a light wave propagation in the z-direction.

$$n(z) = n_{eff} + \Delta n \cos(2\pi z/\Lambda) \tag{2}$$

where Δn is the amplitude of the refractive index perturbation.

By using coupled mode theory, the description of reflectivity properties of a grating [5].

$$r = \frac{\sinh^2\left(\sqrt{(k^2 - \hat{\sigma}^2)}\ L\right)}{\cosh^2\sqrt{(k^2 - \hat{\sigma}^2 L^2) - (\hat{\sigma}^2/k^2)}} \tag{3}$$

where L is grating length, k is ac coupling coefficient, $\hat{\sigma}$ is general dc self-coupling coefficient, and r is reflectivity.

The FBGs have unique feature which makes its suitable to wavelength division multiplexing techniques. Thus, it can be used for multiple parameter of sensing with different Bragg wavelengths along a single fiber.

2 Spectral Reflectivity Dependence on Grating Length

We have analyzed that spectral reflectivity depends on different grating length L, 1, 2, 3, and 6 mm. At 3 mm, the reflectivity reaches maximum value. When the grating length is increased to L = 6 mm, it is as similar as of 3 mm plot with slight changes shown in Fig. 1 As the grating length is increased, bandwidth decreases.

Figure 2 shows linear behavior between the change in grating length and that of the shift in center wavelength. As noted from the graph, with the variation in the grating length, the center wavelength correspondingly shifts. When the length is increased wavelength increases.

Figure 3 also shows linearity, whenever the variation in refractive index is observed, the center wavelength also gets changed. The Bragg wavelength used for calculation of different external parameter like strain, temperature, pressure, etc. [2, 3].

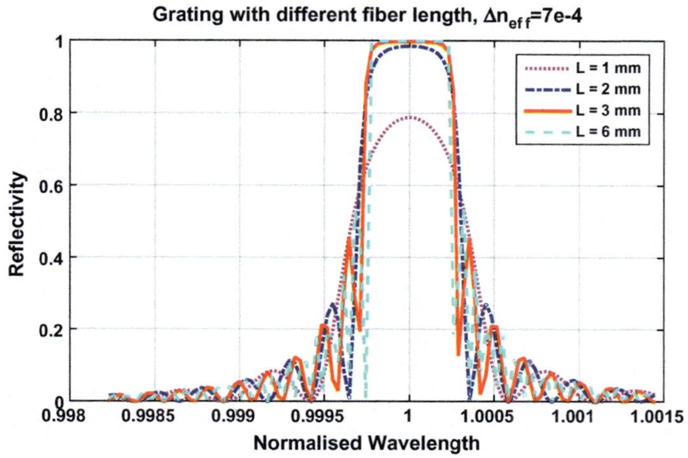

Fig. 1 Reflection spectra of gratings with different lengths

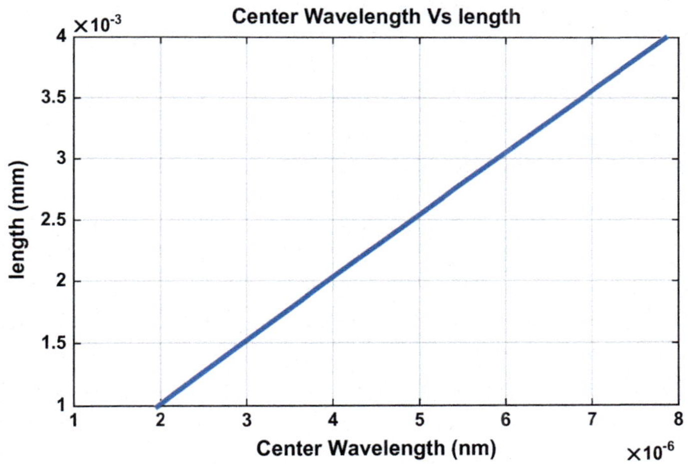

Fig. 2 Effect of grating length on center wavelength

3 Principle of Optical Fiber Bragg Grating Temperature Sensors

FBG sensors are used due to its real-time processing, high sensitivity and stability properties [6]. Due to these properties, optical sensors are widely used as compared to electrical sensors. One of the important applications appropriate for fiber sensors is in the area of temperature analysis. For temperature sensing, the Bragg

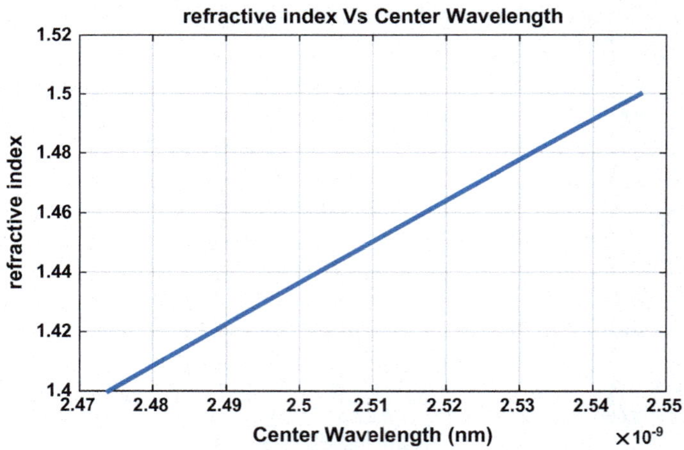

Fig. 3 Effect of refractive index on center wavelength

wavelength is a function of the temperature of material. The temperature is directly related to changes in the refractive index of the fiber along with thermal expansion coefficient of the material.

At this Bragg wavelength, there will be a peak in the reflection spectra. Due to temperature change or strain effect, there will be a shift in the Bragg wavelength. This can be written mathematically as follows:

$$\Delta\lambda_B = 2\left(n_{eff}\frac{\Lambda}{L} + \Lambda\frac{n_{eff}}{L}\right)\Delta L + 2\left(n_{eff}\frac{\Lambda}{T} + \Lambda\frac{n_{eff}}{T}\right)\Delta T \tag{4}$$

In the above equation, the first term is due to induced strain and the second term is due to temperature change.

Due to temperature only, the second term will be taken into consideration [6].

$$\Delta\lambda_B = 2\left(n_{eff}\frac{\Lambda}{T} + \Lambda\frac{n_{eff}}{T}\right)\Delta T \tag{5}$$

$$\frac{\Delta\lambda_B}{\lambda_B} = \frac{2n_{eff}\frac{\Lambda}{T}\Delta T}{\lambda_B} + \frac{2\Lambda\frac{n_{eff}}{T}\Delta T}{\lambda_B}$$

$$\frac{\Delta\lambda_B}{\lambda_B} = \left(\frac{1}{\Lambda}\frac{\Lambda}{T} + \frac{1}{n_{eff}}\frac{n_{eff}}{T}\right)\Delta T \tag{6}$$

where n_{eff} is the effective refractive index of the fiber core and Λ is the grating period. Whenever FBG is in contact with a substrate, then change in grating period is observed due to change in temperature. The variation in grating period not only occurs due to thermal expansion of the fiber but it also takes place due to strain induced by thermal expansion of the substrate.

4 The Refractive Index of the Fiber Core Changes Due to Thermo-Optic Effect

The above equation represents the temperature sensitivity of the FBG. The first term is the thermal expansion coefficient, and the second term is thermal-optic coefficient. Due to change in temperature, ΔT, the shift in Bragg wavelength, is given as [6].

$$\frac{\Delta\lambda_B}{\lambda_B} = [\alpha + \xi]\Delta T \tag{7}$$

$$\Delta\lambda_B = \lambda_B[\alpha + \xi]\Delta T \tag{8}$$

where $\xi = \left(\frac{1}{n_{eff}}\right)\left(\frac{\Delta n_{eff}}{\Delta T}\right)$ is the thermal-optic coefficient of the fiber, and $\alpha = \frac{1}{\Lambda}\frac{\Lambda}{T}$ is the thermal expansion coefficient of the substrate. For temperature sensor, it is obvious that physical interface and thermal expansion of substrate would be greater than the fiber optic length [2, 3]. The Table 1 shows the thermos-optic and thermal expansion coefficient of silica, glass and PMMA with fixed temperature range.

Here, two different material silica and Glass with their different thermal expansion coefficient and different thermo-optic coefficient are used. As shown from Fig. 4 with the change in temperature there is change in wavelength in this graph silica material shows more variation than glass material with same temperature variation.

Here, Polymethyl methacrylate (PMMA) material is used. In Fig. 5, as the temperature increases the wavelength decreases linearly. The linear relation can be used for sensing the temperature. The temperature range depends on the material used for sensing. For PMMA, from Figs. 4 and 5, we observe that silica should be preferred as it shows greater wavelength variation.

Table 1 Optical properties of the materials

Serial no	Material	Thermo-optic coefficient (ξ)	Thermal expansion coefficient (α)	Temperature (T) (°C)
1	Silica [9, 10]	$2 \times 10^{-8} K^{-1}$	$0.55 \times 10^{-60} C^{-1}$	25 – 100
2	Glass [7–9]	$-6.4 \times 10^{-6} K^{-1}$	$8.5 \times 10^{-60} C^{-1}$	25 – 100
3	PMMA [8]	$-1.1 \times 10^{-4} K^{-1}$	$73 \times 10^{-6} K^{-1}$	25 – 100

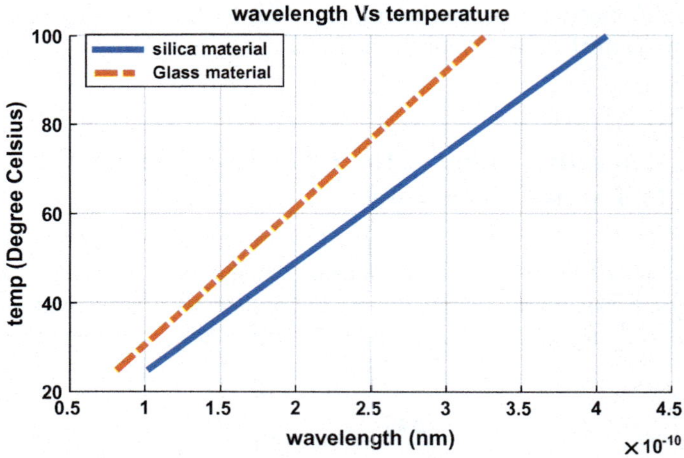

Fig. 4 Change in wavelength with change in temperature for silica and glass material

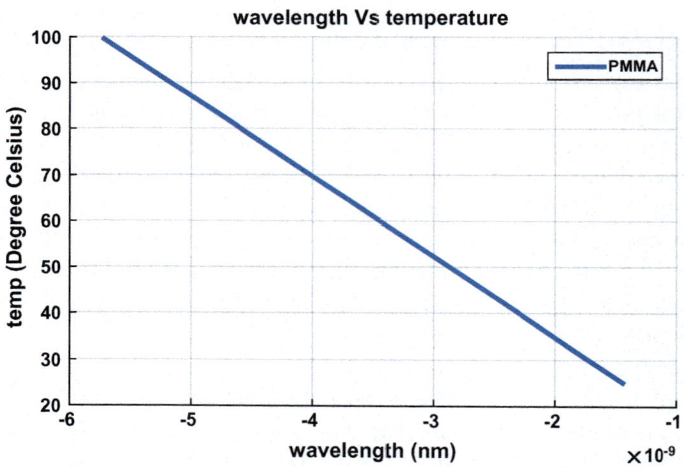

Fig. 5 Change in wavelength with change in temperature for PMMA material

5 Conclusion

The modeling and simulation with MATLAB for optical fiber Bragg grating were realized. The reflectivity and bandwidth are varied with change in grating length. The bandwidth of fiber Bragg grating was inversely proportion to the grating length. The center wavelength is analyzed with change in grating length and refractive index. When the refractive index and length are increased, the center wavelength also increases. The temperature variation of three different materials is analyzed with their different thermal-optic coefficient and thermal expansion coefficient with simulation graph to illustrate their effect. Thus, it can be used for high-sensitive temperature sensor application. WDM gives each FBG sensor to its separate wavelength range within the light spectrum. FBG result shows measurements sensor is inconsistent with light intensity losses/attenuations due to bending or transmission.

References

1. Measures RM (2001) Structural monitoring with fiber optic technology. Academic, San Diego, CA
2. Ugale SP, Mishra V (2011) Optimization of fiber Bragg grating length for maximum reflectivity. In: 2011 International conference on communications and signal processing (ICCSP), pp 28–32
3. Kim KT, Kim IS, Lee CH, Lee J (2012) A temperature-insensitive cladding-etched fiber Bragg grating using a liquid mixture with a negative thermo-optic coefficient. Sensors 12:7886–7892
4. Gupta S, Mizunami T, Yamao T, Shimomura T (1996) Fiber Bragg grating cryogenic temperature sensors. Appl Opt 25:5202–5205
5. Erdogan T (1997) Fiber grating spectra. J light tech 15(8):1277–1294
6. Kashyap R (2009) Fiber Bragg gratings. Academic, San Diego, CA
7. Lebesque HJM et al. (1965) Thermal expansion coefficient of BK 8 optical glass between 15 and 300° K. Physica 31(6):967–972
8. Suhir E, Lee YC, Wong CP (2007) Micro- and Opto-electronic materials and structures: physics, mechanics, design, reliability, packaging: Volume I Materials physics-materials mechanics. In: Bar-Cohen A, Han B, Kim KJ (eds) Chapter 2: Thermo-optic effects in polymer Bragg gratings. Springer, pp A65–A110
9. Ghosh G (1998) Handbook of thermo-optic coefficients of optical materials with applications handbook of optical constants of solids: handbook of thermo-optic. Academic Press
10. Ghosh G (1995) Model for the thermo-optic coefficients of some standard optical glasses. J non-crystal solid 189(1–2):191–196

A Classification of Emotion and Gender Using Local Biorthogonal Binary Pattern from Detailed Wavelet Coefficient Face Image

Kamaljeet Singh Kalsi and Preeti Rai

Abstract This work investigates a framework which identifies gender and emotion of the person from the face image. Gender with their expressions has a vital role in the suspect detection systems. The proposed system aids in identification of a person with their gender as male and female. Also detects gender's expression as joy and sadness. In this paper, wavelet detailed coefficient and Biorthogonal family-based system have been used simultaneously to identify gender and emotion of a face image. Detailed image local Biorthogonal binary pattern (DILBBP) has been applied for feature extraction and for classification purpose; SVM is applied. Experiments are performed on publicly available standard FERET, INDIAN FACE, and AR FACE databases. Proposed work gives acceptable classification and recognition results with less computational time.

Keywords Gender classification · Emotion detection · Detailed image local biorthogonal binary pattern (DILBBP) · Support vector machine

1 Introduction

With the evolution of human–computer interaction (HCI) technology, to meet the people's growing demand for secure, reliable, convenient, and individualized services, computer vision approaches such as face identification, gesture recognition, and perhaps more fundamental gender and expression classifier will play an increasingly important role in our daily life. Recognizing gender and emotional expression from the face image is an interesting and challenging task in computer vision. When we look at any individual's face, we concern much information about him/her, such as gender, about emotional expressions, age as well as his/her

K. S. Kalsi (✉) · P. Rai
Department of Computer Science, GGITS, Jabalpur, MP, India
e-mail: kamaljeetkalsiggits@gmail.com

P. Rai
e-mail: preetirai@ggits.org

V. Janyani et al. (eds.), *Optical and Wireless Technologies*, Lecture Notes in Electrical Engineering 472, https://doi.org/10.1007/978-981-10-7395-3_9

traditions. Emotional and gender distinctions are very easy from human point of view but quite challenging job for computer. Security system design and surveillance to find criminals and missing persons, biometric authentication systems such as intelligent systems which control the movement of people in the restricted areas and image filter-based search engines which enhanced the searches are some of the advanced applications. Gender classification along with emotional expression recognition has a great importance in order to improve the interaction between man and machine.

The aim of the developed classification system is to recognize gender and emotional expressions from the input face image. Gender classifier gives status as "male or female," whereas emotion classifier gives status of the expression as "joy or sadness" with enhanced classification accuracy involving less computation time using detailed coefficient local Biorthogonal family. The process initiates with input face images preprocessing along with feature extraction and classification. In this paper, the proposed feature extraction method is detailed image local Biorthogonal binary pattern (DILBBP), which is a combination of detailed coefficient-based wavelet transformation using Biorthogonal family and local binary pattern. For the classification of gender with expressions, support vector machine (SVM) classifier has been applied, which gives the final output as male/female and joy/sadness for the specific face image. Testing results and cross-database evaluation give appropriate results with accuracy process computational time.

The further paper is arranged as below. Section 2 gives review works. Section 3 gives the overview of proposed gender and emotion classification system. Experimental evaluation is described in Sect. 4, and finally, Sect. 5 does conclusion in the work.

2 Related Work

In the literature, number of work exist in which researcher works in the area of gender classification and emotion detection have focussed on various feature extraction and expression detection techniques on face image. Tamura et al. [1] have done some starting work to do classification of gender using face images. Baluja and Rowley [2] although obtained good classification accuracy, but scratch face images have high dimensions and are sensitive to illumination and expression variations.

Turk and Pentland [3], Rowley et al. [4], Osuna et al. [5], Schneiderman and Kanade [6], Viola and Jones [7] used appearance-based methods to detect the face. Sandeep and Rajagopalan [8], Rajagopalan et al. [9], and Saha [10] have proposed an algorithm for the detection of human face in color images. Ekman and Friesen [11] developed the facial action coding system (FACS) to code facial expressions where movements on the face are described set of action units (AU s). Some latest work on face expression recognition [12] has been used these basic expressions. In order to recognize facial expressions, Mase [13] has been used optical flow technology. Ozbudak et al. [14] use discrete wavelet transform technique to do decomposition of face images. Lyon et al. [15] do feature extraction process with the help of Gabor filter from the face image. Sun et al. [16] use local features of face to extract histograms from face using local features.

For gender classification, Shan [17] proposed a discriminative LBP-Histogram (LBPH) bins technique. Craw et al. [18] and Lanitis et al. [19] used the template matching methods to detect the face. Huchuan et al. [20] do gender reorganization with texture feature extraction using proposed pixel pattern. For the detection of the emotion and gender detection, most of the algorithm involves the feature extraction technique along with the classification process. Till now, detection of the gender and expression of face simultaneously have not gained much attention by the researchers. Only Saatci and Town [21] arranged both the classifier in cascade form. For the face detection, Pentland et al. [22] suggest PCA method based on component-based version. Yang [23] has classified single image detection method in four categories. Yang and Huang [24] used a hierarchical knowledge-based method to detect faces. Shakhnarovich et al. [25] proposed a demographic analysis and face detection system which is flexible in real time. Freund and Schapire [26] introduced a new "boosting" algorithm called AdaBoost for face gender classification.

3 Proposed System

The system proposed for emotion recognition and gender classification is shown in Fig. 1. The modules comprise of preprocessing, feature extraction, and cascaded classifiers. The proposed system has training and testing phases. During training phase, face images passes through preprocessing modules, then feature extraction method has been applied and extracted features stored in the database. These stored features are used to train the classifier. In the testing phase, features of target face image are calculated in the similar manner as in training phase and are classified using SVM classifier to find the emotion and gender of the face image.

In this work, features have been extracted by detailed image local Biorthogonal binary pattern (DILBBP), i.e., input image first passes through detailed coefficient-based wavelet decomposition method using Biorthogonal family, then passes through local binary pattern. These extracted local features of the specific test face image are used as input for classifiers for two elementary emotional states (joy, sadness) and for gender detection (male, female). These classifiers are arranged in parallel in order to optimize overall recognition performance. Gender classification and emotion recognition will be the final outcome. A computer system with this application (gender classification and expression recognition) has a wide range of applications in research areas, in the stream of communication, security, education, and telecommunication.

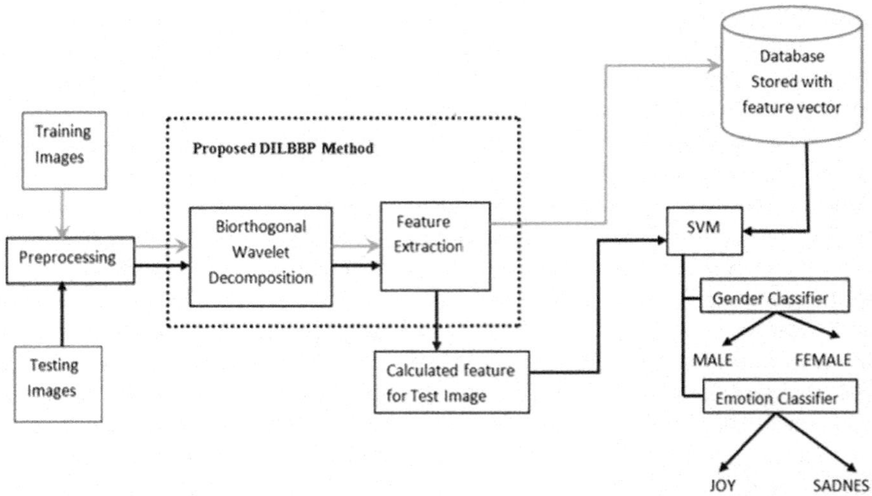

Fig. 1 Work flow of the proposed system

3.1 Preprocessing

Some of the extraneous details like hair and background of image, etc., are present in face databases. Before the feature extraction process, first the face area is to be cropped from the images available in the databases. To extract the face area, three spots, right and left corners of the eyes and center of the chin are selected to crop the face image. Centroid of the triangle (say, c) formed with these three points, and its distance from any corner (say, r) is computed. Taking "c" as center, a square with side 2r is drawn. Face image within this square is the desired cropped face image [27].

3.2 Feature Extraction Technique

Detailed image local biorthogonal binary pattern (DILBBP)

The applied feature extraction technique is DILBBP which is a combination of (detailed coefficient wavelet decomposition using biorthogonal family + local binary pattern). Wavelet decomposition is a time–frequency signal method used for analysis. Initially on the face image, wavelet decomposition method has been applied, which transforms face image into detailed coefficients face images. Then on these images, local binary pattern (LBP) has been used to extract local features of the face images.

A. *Detailed coefficient biorthogonal wavelet decomposition*

Wavelet decomposition method is a time–frequency signal analysis method. It can be used to decompose a face image into many sub-band images with different spatial resolution, frequency characteristic, and directional features [28]. In this work, by decomposing the face image up to two levels, the detailed coefficients are computed. Detailed coefficients contain the highest frequency component of an image. More than two level decomposition of face image result in information loss and hence not considered in this work.

B. *Mechanism of local binary pattern*

Then on the wavelet transformed images, formed by wavelets decomposition, a local binary pattern (LBP) method has been applied, in which face image is first divided into small regions from which local binary patterns, which are also known as histograms have been extracted. Different LBP histograms extracted from the each of the face image are finally concatenated into single feature histogram, which forms the representation of the face image. The basic LBP operator was present by Ojala et al. [29]. This operator runs with the eight neighbors of a pixel, using the value of this center pixel, which has been considered as the threshold value. On comparison with center pixel, if neighbor pixel has a higher or equal gray value than one is assigned to that pixel, else it assigned zero. For the center pixel, by concatenating all of the zeros and ones in the form of binary code, finally LBP code has been created [30].

Later extended work on LBP operator has been done, which use neighborhoods of different sizes. Here, from central pixel with radius R, a circle has been drawn. Comparison between SP sampling points and central pixel value has done, on the edge of this circle. It is essential to get bilinear interpolation, in order to extract the values of all neighborhood sampling points of variable number of pixels and radius size. For neighborhoods values, the notation (SP, R) is used. Using DILBBP method for feature extraction process, when we applied LBP technique along with biorthogonal wavelets transformation, then the computational process time of overall system activity gets reduced, which increases the efficiency of the system and also tends to improve the overall performance of the system.

3.3 Support Vector Machine

With the intention to isolates two different classes, SVM is a supervised machine learning algorithm technique for classification and regression. Working of SVM is based on hyperplanes concept that target to separate a set of objects with maximum distance [31]. SVM finds out some support vectors from the feature space which are useful to find the optimal hyperplane [32].

The standard support vector machine (SVM) takes the group of input data and predicts that only two possible categories gives an output, intending it binary linear classifier. A support vector machine forms a hyperplane in an upgraded dimensional area, which can be used for classification and regression, represents concept of hyperplane, and separation of two classes. Vapnik suggests a linear classifier based on optimal hyperplane concept and whose performance was up to mark for linearly separable databases. Later, Boser, Vapnik, and Guyon explore nonlinear classifier for non-separable database by applying the concept of kernel function [33]. This kernel function is being applied to get the hyperplane, which is optimal in non-separable database. The resultant discrimination function has some variations compare to previous one, a nonlinear kernel function has been substituted instead of dot product as follows:

$$f(x) = \sum_{i=1}^{l} \alpha_i z_i (x_i.x) + b \tag{1}$$

The proposed work employs radial basis function (RBF) given as

$$k(x_i.x) = \exp(-\gamma \|x_i - x\|^2) \quad \text{for } \gamma > 0 \tag{2}$$

4 Experimental Analysis

The task performed here has been trained and tested on Intel (R) Core (TM) i3-3120 M CPU with 2.50 GHz having 4 GB memory. MATLAB is used for all the computational process. Computational time and classification rate are the parameters on the basis of which performance is to be evaluated.

4.1 Dataset Used for Experimentation

Face database applied in this work is FERET, INDIAN FACE, and for cross-database evaluation; AR FACE database has been used. Figure 2 shows some of the images from applied databases.

Phillips and Wechsler at George Mason University put a collaborative effort for the collection of FERET [34] database, which has been collected in a semi-controlled environment. The FERET database is a collection of good quality color face images. In February 2002, INDIAN FACE [35] database which contains face images of human has been captured in Indian Institute of Technology campus Kanpur. Aleix Martinez and Robert Benavente were created an AR face database in

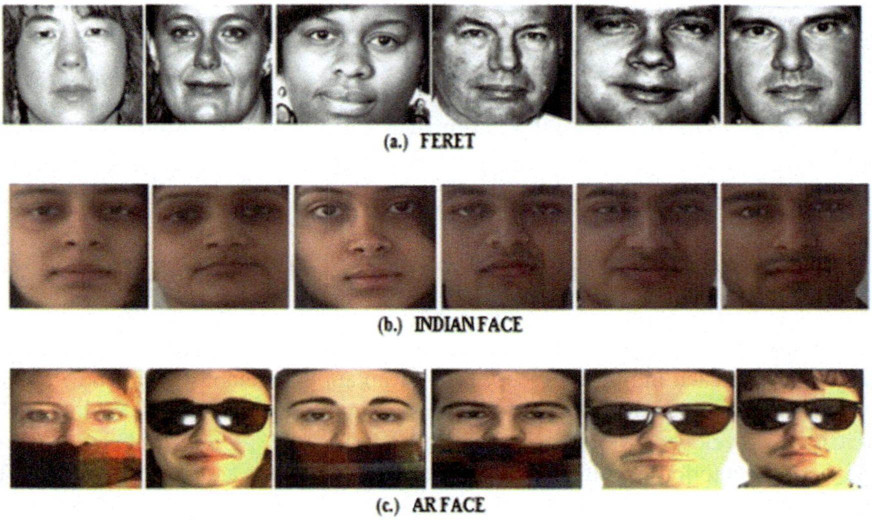

Fig. 2 Standard face image databases

the Computer Vision Center (CVC) at Universitat Autonoma de Barcelona, Spain (U.A.B). This database contains face image with varies illumination conditions and occlusion introduced through wearing sunglasses and scarf [36].

4.2 *Criteria of Performance*

Results are usually obtained with k-fold cross-validation. In k-fold cross-validation, data is divided into k-parts, out of which (k − 1) parts are used for training in rotation and the remaining one part is used for testing. The system tested for high value of k, uses more data for training and less data for testing. As the publicly available database is large enough to handle, using such 2-cross-validation process has benefit to be applied. The system is trained with 50% of images and tested on the remaining 50% images, and each sample image is used for both training and testing. This work applied twofold cross-validation which logically isolates database is divided into two sets; say B1 and B2, of equal size. The system is trained with B1 dataset, and testing is performing on B2 dataset. Classification rate (CR) has been evaluated to check the performance of gender classification systems proposed in this work.

The rate of classification is explored as the percentage of absolutely classified face images with the total number of face images in the testing dataset. Classification rate of a system is defined by

$$CR = \frac{B1}{B2} \times 100 \quad (3)$$

where B1 gives the number of test face images which has been absolutely classified, and B2 gives hints about the total count of face images in the testing set.

4.3 Evaluation of Emotion and Gender Classification

The proposed system gives two classifications output in the form of gender and emotion with each individual face images from applied database with some of the variations in illumination, pose, and expression in the captured image. With FERET database, developed system gives 92.56% and 79.33% of the accuracy rate for gender and emotion classification, respectively. With the standard INDIAN FACE database, system gives 93.33 and 80% of classification accuracy as shown in Table 1.

4.4 Cross-Database Performance on Occluded Face Database

The faces images are termed as "occluded" when person covers his/her face with hand or scarf, wear sunglasses, hold moles or suffer injury on faces. To evaluate cross-database performance, the system applied non-occluded face images for training with FERET and INDIAN FACE databases and testing is done for gender classification on occluded images using AR FACE databases. Here, the proposed system is not dependent on specific category of person, i.e., images of face used for training perspective and testing system are of different category of people. With the

Table 1 Classification rate of gender and emotion classification

Dataset	Classification rate (CR %)	
	Gender classification	Emotion classification
FERET	92.56	79.33
INDIAN FACE	93.33	80

Table 2 Cross-database classification rate with non-occluded images

Dataset	Classification rate (CR %)
Training/Testing	AR FACE dataset
FERET	70.81
INDIAN FACE	50.07

Table 3 Comparison of the proposed approach with the existing techniques for gender classification

Existing method references	Features	Classifier	Database	CR %
Annalakshmi [37]	SEWLD	SVM	FERET	91.37
	SELBP	SVM	FERET	90.14
Mashagba [38]	Geometry-based extraction	Dynamic neural network	Own dataset using sequential video	91.2
Kyu-Dae Ban [39]	MCT	AdaBoost	LFW	94.72
	MCT	AdaBoost	TIOG	87.92
Proposed method	DILBBP	SVM	FERET	**92.56**
	DILBBP	SVM		**93.33**

applied FERET and INDIAN FACE dataset, the developed work gives classification response up to 70.81 and 50.07% as shown in Table 2. Comparison of the proposed approach with the existing techniques for gender classification has been shown in Table 3.

5 Conclusion

In this proposed work, classification of gender and emotion has been done in parallel using the DILBBP method for feature extraction, which includes the biorthogonal and detailed coefficients properties of wavelet decomposition along with LBP. The work explores here gives the acceptable response for gender classification along with their face expressions. The proposed work is also tested for cross-data performances, which results efficient accuracy rate. In this work, face images have to perform overall process. The further areas which can be covered in this work are to classify gender along with their expressions from videos also.

References

1. Tamura S, Kawai H, Mitsumoto H (1996) Male/female identification from 8 × 6 very low resolution face images by neural network. Pattern Recogn 29(2):331–335
2. Baluja S, Rowley HA (2007) Boosting sex identification performance. Int J Comput Vision 71 (1):111–119
3. Turk M, Pentland A (1991) Eigenfaces for recognition. J Cogn Neurosci 3(1):71–86
4. Rowley H, Baluja S, Kanade T (1998) Neural network-based face detection. IEEE Trans Pattern Anal Mach Intel 20(1)
5. Osuna E, Freund R, Girosi F (1997) Training support vector machines: an application to face detection. In: Proceedings of the IEEE conference on computer vision and pattern recognition, pp 130–136

6. Schneiderman H, Kanade T (1998) Probabilistic modeling of local appearance and spatial relationships for object recognition. In: Proceedings of the IEEE conference on computer vision and pattern recognition, pp 45–51
7. Viola P, Jones MJ (2001) Rapid object detection using a boosted cascade of simple features. In: Conference on computer vision and pattern recognition, vol 1, pp 8–14
8. Sandeep K, Rajagopalan AN (2002) Human face detection in cluttered color and edge information. In: The Indian conference on computer vision, graphics and image processing (ICVGIP)
9. Rajagopalan A, Kumar K, Karlekar J, Manivasakan R, Patil M, Desai U, Poonacha P, Chaudhuri S (1998) Finding faces in photographs. In: Proceedings of the 6th IEEE international conference on computer vision, pp 640–645
10. Saha S (2007) A symmetry based face detection technique. In: IEEE machine intelligence unit
11. Ekman P, Friesen WV (1978) Facial action coding system: investigator's guide. Consulting Psychologists Press, Palo Alto, CA
12. Tian Y, Kanade T, Cohn J (2001) Recognizing action units for facial expression analysis. IEEE Trans Pattern Recogn Mach Intel 23(2):97–115. Carnegie-Mellon University
13. Mase K (1991) Recognition of facial expression from optical flow. IEICE Trans 3474–3483
14. Ozbudak O, Tukel M, Seker S (2010) Fast gender classification. In: IEEE international conference on computational intelligence and computing research (ICCIC), pp 1–5
15. Lyons MJ, Budynek J, Plante A, Akamatsu S (2000) Classifying facial attributes using a 2-d Gabor wavelet representation and discriminant analysis. In: Fourth IEEE international conference on automatic face and gesture recognition. IEEE, pp 202–207
16. Sun N, Zheng W, Sun C, Zou C, Zhao L (2006) Gender classification based on boosting local binary pattern. In: Advances in neural networks, Springer, pp 194–201
17. Shan C (2012) Learning local binary patterns for gender classification on real world face images. Pattern Recogn Lett 33(4):431–437
18. Craw L, Tock D, Bennett A (1996) Finding face features. In: Proceedings of the 2nd European conference on computer vision, pp 92–96
19. Lanitis A, Taylor CJ, Cootes TF (1995) An automatic face identification system using flexible appearance models. Image Vis Comput 13(5):393–401
20. Huchuan L, Huang Y, Chen Y, Yang D (2008) Automatic gender recognition based on pixel-pattern-based texture feature. J Real-Time Image Process (2008)
21. Saatci Y, Town C (2006) Cascaded classification of gender and facial expression using active appearance models. In: 7th IEEE international conference on automatic face and gesture recognition, pp 393–400
22. Pentland A, Moghaddan B, Starner T (1994), View based and modular Eigenspaces for face recognition. In: Proceedings of the IEEE conference on computer vision and pattern recognition, pp 84–91
23. Yang M (2002) Detecting faces in images: a survey. IEEE Trans. Pattern Anal Mach Intell 24 (1)
24. Yang G, Huang TS (1994) Human face detection in complex background. Pattern Recogn 27 (1):53–63
25. Shakhnarovich G, Viola P, Moghaddam B (2002) A unified learning framework for real time detection and classification. In: IEEE conference on AFG
26. Freund Y, Schapire RE (1996) Experiments with a new boosting algorithm. In: Proceedings of the 13th international conference on machine learning, pp 148–156
27. Rai P, Khanna P (2015) An illumination, expression, and noise invariant gender classifier using two-directional 2DPCA on real Gabor space. J Vis Lang Comput 26:15–28
28. Daubechies I et al (1992) Ten lectures on wavelets. SIAM 61
29. Ojala T, Pietikainen M, Harwood D (1996) A comparative study of texture measures with classification based on feature distributions. Pattern Recogn 29
30. Rahim MA, Hossain MN, Wahid T, Azam MS (2013) Face recognition using local binary patterns (LBP). Glob J Comput Sci Technol Graph Vis 13(4). version 1.0

31. Burges CJC (1998) A tutorial on support vector machines for pattern recognition. Data Min Knowl Disc 2(2):121–167
32. M. Sonka, V. Hlavac, R. Boyle, Digital image processing and computer vision, Cengage Learning, India edn (2008)
33. Vapnik VN (1998) Statistical learning theory. Wiley
34. Phillips PJ, Wechsler H, Huang J, Rauss PJ (1998) The FERET database and evaluation procedure for face-recognition algorithms. Image Vis Comput 16(5):295–306
35. Jain V, Mukherjee A (2002) The indian face database
36. Martinez AM (1998), The AR face database. CVC Technical Report
37. Annalakshmi M, Roomi SMM, Priya SS (2016) Gender recognition from face images using texture descriptors for human computer interaction. J Chem Pharm Sci 9(3)
38. Al Mashagba EF (2016) Real-time gender classification by face. Int J Adv Comput Sci Appl 7(3)
39. Ban KD, Kim J, Yoon H (2016) Gender classification of low-resolution facial image based on pixel classifier boosting. ETRI J 38(2)

Performance of Digital Communication System Over OWDP Fading Channels

Suparna Goswami and A. Dinamani Singh

Abstract The outage and Average Bit Error Rate (ABER) performance of a single-channel Digital communication system in One Wave Diffused Power (OWDP) fading are obtained. The probability density function (pdf) is used to obtain the expressions for outage probability, and ABER for coherent and non-coherent modulation. The effects of the types of coherent BPSK, Binary Frequency-Shift Keying (BFSK), M-ary PSK, Rect.Quadrature Amplitude Modulation (QAM) and non-coherent Differential Binary Phase-Shift Keying (DBPSK), BFSK and M-ary FSK, and the fading parameter K on the system performance have been studied. The results are therefore obtained and verified with the results found in literature.

Keywords ABER · Coherent modulation · Non-coherent modulation
Outage probability · OWDP

1 Introduction

The one wave diffused power (OWDP) consists of only one wave with constant amplitude. OWDP fading model is considered as a special case of two wave diffused power (TWDP) fading channel. OWDP fading channel comes across in wireless sensor network that is mounted within a cavity such as tunnel, aircraft, public buses, trains. In [1], the authors have considered the Rayleigh, Rician fading model as a special case TWDP fading model. The approximate expression of TWDP fading model is given in [1]. The bit error rate (BER) performance in TWDP fading for Binary Phase-Shift Keying (BPSK) has been obtained in [2]. The cumulative distribution function (cdf) technique is used for the Quadrature

S. Goswami (✉) · A. D. Singh
Department of ECE, NERIST, Nirjuli, Arunachal Pradesh, India
e-mail: suparnanerist@gmail.com

A. D. Singh
e-mail: ads@nerist.ac.in

V. Janyani et al. (eds.), *Optical and Wireless Technologies*, Lecture Notes in Electrical Engineering 472, https://doi.org/10.1007/978-981-10-7395-3_10

Amplitude Modulation (QAM) performance over TWDP fading channel [3]. The performance analysis of Selection Combining (SC) receiver with L-branch in κ-μ and η-μ fading channel has been done in [4]. In [5], the expression for Symbol Error Rate (SER) of the SC receiver with M-ary Phase-Shift Keying (MPSK) and MQAM has been obtained. The Symbol Error Probability (SEP) using M-ary rectangular QAM with mgf approach has been obtained over TWDP fading channel in [6]. The expression of spectral efficiency for power and rate adaptive transmission techniques with un-coded MQAM over TWDP fading channel has been presented in [7]. The evaluation of outage probability and ABER for SC receiver is obtained over TWDP fading [8]. The expression of pdf, moment generating function (mgf), and cdf has been obtained over κ-μ shadowed fading channel in [9]. In [10], the expression of capacity for different adaptive transmission with Dual-MRC over non-identical TWDP channels has been obtained. In [11], the authors have obtained the expression for outage probability and Average Symbol Error Rate (ASER) of a single-channel system over TWDP fading channel. In [12], the expression for ergodic capacity in κ-μ shadowed fading channel has been derived. The expression for average capacity of channel with four adaptive transmission techniques over log-normal distribution has been obtained in [13]. There are several works done for the performance and capacity analysis over fading channels found in the literature. From the literature study, it can be observed that no work on performance evaluation has been presented over an OWDP fading channel. This extends the idea and motivation to work on the performance measurement of digital communication system over OWDP fading channels.

The remaining part of the paper is organized as follows. The channel and system has been discussed in Sect. 2. In Sect. 3, the performance parameters, i.e., the outage probability and ABER performance over OWDP fading channel have been evaluated. In Sect. 4, the study of numerical outcomes is presented. The conclusion is given in Sect. 5.

2 Channel and System

The OWDP fading channel contains only specular components with many diffused components. The channel has been considered to be slow, frequency non-selective, with OWDP fading statistics. The envelope pdf of OWDP fading is obtained from [1]. The expression can be written as

$$f_r(r) = \sum_{i=1}^{5} a_i \left[\frac{r}{\sigma^2} e^{-\left\{ \frac{r^2}{2\sigma^2} + K \right\}} I_0\left(\frac{r}{\sigma} \sqrt{2K} \right) \right], \quad (1)$$

where $I_0()$ is termed as first kind and zeroth-order Bessel function and K is the fraction of total specular power and diffused power. The order of approximation is represented by i. By performing the square and random variable transformation, the signal-to-noise ratio (SNR) pdf of OWDP fading channel is given as

$$f_\gamma(\gamma) = \sum_{i=1}^{5} \sum_{p=0}^{\infty} a_i \frac{K^p \eta^{p+1}}{(p!)^2} e^{-K} \gamma^p e^{-\eta\gamma}, \quad (2)$$

where $\eta = \frac{K+1}{\bar{\gamma}}$.

3 Performance Analysis of OWDP Fading

In this section, the evaluation of outage probability, ABER for the coherent and non-coherent modulation of various digital modulation schemes over OWDP fading channels is done.

3.1 *Outage Probability*

The outage probability (P_{out}) is used to measure the performance of the system. The P_{out} is the probability that the instantaneous probability of error is less than the numerical value of threshold SNR, γ_{th} [14]. To obtain the outage probability expression, the cdf of the output SNR is to be evaluated. The expression of the cdf is derived from output SNR expression given in (2). The expression is obtained as

$$F(\gamma_0) = \sum_{i=1}^{5} \sum_{p=0}^{\infty} a_i \frac{K^p \eta^{p+1}}{(p!)^2} e^{-K} \int_0^{\gamma_0} \gamma^p e^{-\eta\gamma} d\gamma. \quad (3)$$

The integration is solved using [15, 3.351.1]. The expression of P_{out} is obtained from the cdf by replacing γ_0 by γ_{th}. The expression obtained can be written as

$$P_{out}(\bar{\gamma}) = \sum_{i=1}^{5} \sum_{p=0}^{\infty} a_i \frac{K^p}{(p!)^2} e^{-K} g(p+1, \eta\gamma_{th}), \quad (4)$$

where $g()$ is termed as lower incomplete gamma function.

3.2 Aber

The ABER can be evaluated by averaging the conditional bit error rate (BER) as per the modulation over the pdf of the output SNR. Mathematically, ASER is given in [14] as

$$P_e(\bar{\gamma}) = \int_0^\infty p_e(\varepsilon/\gamma) f_\gamma(\gamma) d\gamma, \tag{5}$$

where $p_e(\varepsilon/\gamma)$ is termed as conditional BER as per the incorporated modulation scheme. The expression of ABER for coherent and non-coherent modulation is obtained below.

3.2.1 Coherent Modulation

The expression for the conditional BER is given for coherent modulation in [11]. Substituting $p_e(\varepsilon/\gamma)$ and $f_r(r)$ into (5), the expression can be written as

$$P_E(\bar{\gamma}) = \sum_{i=1}^{5} \sum_{p=0}^{\infty} \frac{a_i a}{2\sqrt{\pi}} \frac{K^p \eta^{p+1}}{(p!)^2} e^{-K} \int_0^\infty \gamma^p e^{-\eta\gamma} \Gamma\left(\frac{1}{2}, \frac{b}{2}\gamma\right) d\gamma. \tag{6}$$

The integral is solved using [15, 6.455.1]. Final expression for ABER for coherent modulation can be obtained as

$$\begin{aligned} P_E(\bar{\gamma}) = \sum_{i=1}^{5} \sum_{p=0}^{\infty} a_i a\sqrt{b} \frac{K^p \eta^{p+1} e^{-K}}{2\sqrt{2\pi}(p!)^2} \frac{\Gamma(p+1.5)}{(p+1)(0.5b+\eta)^{p+1.5}} \\ \times {}_2F_1\left(1, p+1.5; p+2; \frac{2\eta}{b+2\eta}\right), \end{aligned} \tag{7}$$

where ${}_2F_1$ *()* is hypergeometric function.

3.2.2 Non-Coherent Modulation

The expression for the conditional BER is given for non-coherent modulation in [11]. Substituting $p_e(\varepsilon/\gamma)$ and $f_r(r)$ into (5), the expression can be written as

$$P_E(\bar{\gamma}) = \sum_{i=1}^{5} \sum_{p=0}^{\infty} a_i a \frac{K^p \eta^{p+1}}{(p!)^2} e^{-K} \int_0^\infty \gamma^p e^{-(b+\eta)\gamma} d\gamma. \tag{8}$$

The integral is solved using [15, 3.351.3]. Final expression for ABER for non-coherent modulation can be obtained as

$$P_E(\bar{\gamma}) = \sum_{i=1}^{5} \sum_{p=0}^{\infty} a_i a \frac{e^{-K} K^p \eta^{p+1}}{p!(b+\eta)^{p+1}}. \tag{9}$$

4 Numerical Outcomes and Analysis

Numerical calculation of the expressions of outage probability and ABER for coherent and non-coherent modulation techniques in OWDP fading channel is given in Sect. 3. Results are obtained by varying the values of parameter *K*. In Fig. 1, outage probability versus $(\bar{\gamma})$ in dB has been plotted. The plot is obtained for different values of *K*. It can be observed from the plot that the outage probability performance of the system is better when the value of *K* is higher. This is because the increased value of *K* signifies that the power of direct wave is more, and hence, increased value of *K* gives a better channel. Similarly from the plot of Figs. 2 and 3, it can be observed that the ABER performance improves with higher values of *K*. The reason behind this is stated above. In Fig. 2, the ABER versus $(\bar{\gamma})$ in dB for different coherent modulation schemes has been plotted. The plot is for different values of parameter *K*. From the plot, it can be seen that BPSK is performing better than BFSK. This is because the message is stored phase variation in BPSK, and in

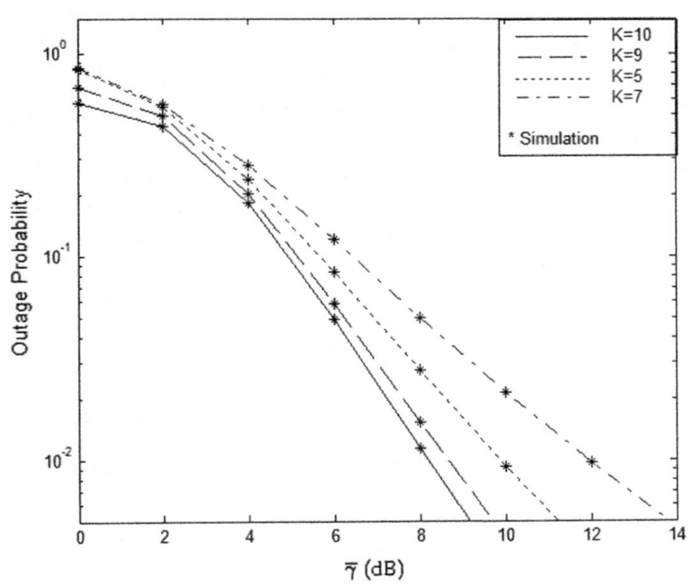

Fig. 1 Outage probability of a single-channel system in OWDP fading

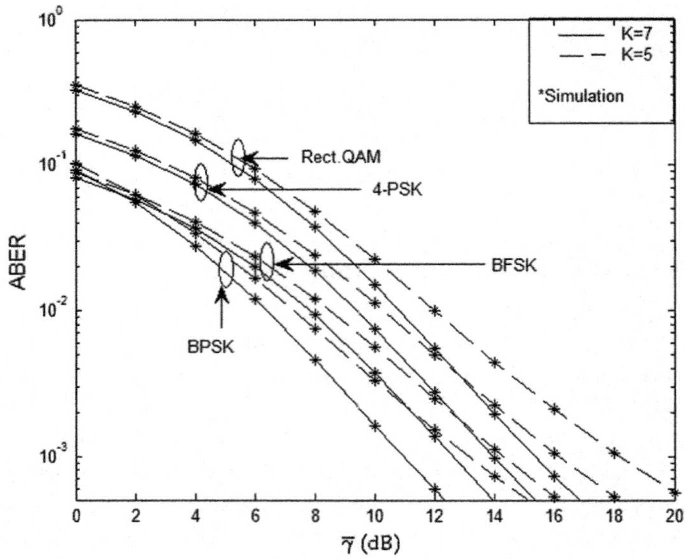

Fig. 2 ABER of single-channel coherent system in OWDP fading

BFSK, the message is stored in frequency variation. Since the noise can easily affect the frequency of the signal, the ABER performance of BPSK modulation is better than BFSK modulation. Furthermore, the performance of BFSK is better than BPSK before 2 dB. This is due to the fact that the region before 2 dB is noise dominant. Thus, the result obtained is insignificant and erroneous. At the same time, it can be observed that the ABER performance of Rect.QAM is worse than 4-PSK. This is because in QAM the message is stored in the amplitude of the carrier where the signal can be affected by the fading deeply. In Fig. 3, the ABER versus ($\bar{\gamma}$) in dB for different non-coherent modulation schemes has been plotted. The plot is for different values of parameter *K*. However, the plot can also be made for any value of *K*. It can be observed from the plot that ABER performance of DBPSK is better than BFSK, and on the other hand, the 4-FSK performance is the worst. The reason behind this is stated above.

5 Conclusion

The OWDP fading channel can be applied for such wireless sensor network that is mounted within a cavity such as tunnel, public transportation systems. In this paper, the expressions for outage probability, ABER performance for coherent and non-coherent modulation schemes of a single-channel system in OWDP fading have been derived. The expressions are obtained by using the pdf of output SNR of

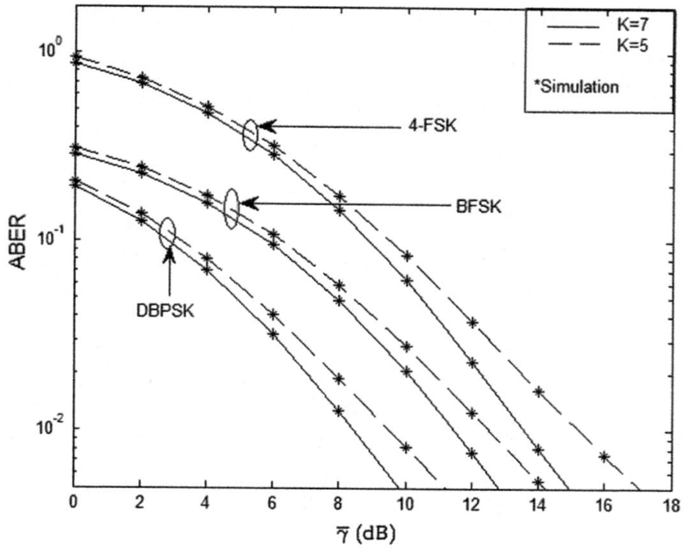

Fig. 3 ABER of single-channel non-coherent system in OWDP fading

the system. The numerical outcomes are obtained for different values of parameter K. After verification, it is ensured that the outcomes are correct.

Paper no: #1570333678

Title: Performance of Digital Communication System over OWDP Fading Channels.

Author: Suparna Goswami and Aheibam Dinamani Singh.

Response to Review 1 (Reviewer B)'s comments

We sincerely thank the esteemed reviewer for reviewing our work and for giving useful comments. Our responses (R) can be found immediately below his/her comments (C).

(1) **C**: Heading, Reference are not formatted properly a per manuscript guideline.
R: The Headings and References have been formatted properly as per manuscript guideline.
(2) **C**: Its a study paper and author have not shown any comparison with literature data.
R: The expressions of the performance parameters have been derived newly. Therefore, the comparison with any literature data is not shown in the paper.
(3) **C**: Equation are not written properly.
R: The equations are rewritten and formatted properly wherever required.

(4) **C**: The quality of simulation results is not up to level of springer conference.
R: Figs. 2 and 3 is newly plotted with different values of fading parameter for better simulation results.

Paper no: #1570333678

Title: Performance of Digital Communication System over OWDP Fading Channels.

Author: Suparna Goswami and Aheibam Dinamani Singh.

Response to Review 2 (Reviewer C)'s comments

We sincerely thank the esteemed reviewer for reviewing our work and for giving useful comments. Our responses (R) can be found immediately below his/her comments (C).

(1) **C**: Ref [1] states TWDP captures widest breadth of fading and important similarity then what is the motivation significance of using OWDP.
R: The OWDP fading model comprises one specular component and many diffused component. The OWDP fading model is a special case of TWDP fading model. We did not find any work related to this fading channel. Hence, we derived the expression newly instead of considering OWDP fading channel as a special case of TWDP fading channel.
(2) **C**: References of Eqs. 1 and 2 not stated.
R: The reference of Eq. 1 has been stated as per suggestion. The expression in Eq. 2 has been derived from Eq. 1. Therefore, the reference no. of Eq. 2 has not been stated in the paper.
(3) **C**: Reference of Eq. 5 is not [11] but [14]. Figure 3 caption is not as per the format.
R: The changes have been made in the paper as per suggestion.
(4) **C**: In Fig. 2 why before 2 dB the performance of BFSK is better than BPSK.
R: The performance of BFSK is better than BPSK before 2 dB. The reason behind this is that the region before 2 dB is noise dominant. In BFSK the information is stored in the frequency variation of the signal. The noise can easily affect the signal carrying information in frequency variation. The result obtained in this manner is insignificant and erroneous.
(5) **C**: Why K = 5 is taken for Fig. 3 Not explained. The conclusion need to be elaborated in terms of significance and application
R: The figures can be plotted for any values of K. The change has been made in the plot of Figs. 2 and 3. The figures of Figs. 2 and 3 have been made for K = 5 and K = 7. The changes have been made in the conclusion as suggested.

References

1. Durgin GD, Rappaport TS, de Wolf DA (2002) New analytical models and probability density functions for fading in wireless communication. IEEE Trans Commun 50(6):1005–1015. https://doi.org/10.1109/TCOM.2002.1010620
2. Oh SH, Li KH (2005) BER performance of BPSK receivers over two-wave with diffuse power fading channels. IEEE Trans Wirel Commun 4(4):1448–1454. https://doi.org/10.1109/TWC.2005.852129
3. Suraweera HA, Lee WS, Oh SH (Feb 2008) Performance analysis of QAM in a two-wave with diffuse power fading environment. IEEE Commun Lett 12(2):109–111. https://doi.org/10.1109/LCOMM.2008.071699
4. Subadar R, Reddy TSB, Sahu PR (2010) Performance of an L-SC receiver over κ-μ and η-μ fading channels. In: IEEE ICC proceedings. https://doi.org/10.1109/ICC.2010.5502444
5. Tan BS, Li KH, Teh KC (Dec 2011) Symbol error rate of selection combining over two wave with diffuse power fading. In: IEEE ICSPCS. Honolulu, Hawaii. https://doi.org/10.1109/ICPCS.2011.6140894
6. Lu Y, Yang N (Jan 2011) Symbol error probability of QAM with MRC diversity in two wave with diffuse power fading channels. IEEE Commun Lett 15(1):10–12. https://doi.org/10.1109/LCOMM.2010.112310.101824
7. Saikia B, Subadar R (2012) Power and rate adaptation techniques with uncoded MQAM over TWDP Fading Channel. IEEE. https://doi.org/10.1109/ICPEN.2012.6492322
8. Subadar R, Singh AD (June 2013) Performance of SC receiver over TWDP fading channels. IEEE Wire Commun Lett 2(3):267–270. https://doi.org/10.1109/WCL.2013.13.120778
9. Paris JF (2014) Statistical Characterization of κ-μ Shadowed fading. IEEE Trans Veh Technol 63(2):518–526. https://doi.org/10.1109/TVT.2013.2281213
10. Singh AD, Subadar R (2015) Capacity of dual MRC receiver with adaptive transmitters over non identical TWDP fading channels. In: International conference on advances in computer engineering and applications (ICACEA). IMS Engineering College, Ghaziabad, India. https://doi.org/10.1109/ICACEA.2015.7164673
11. Singh AD, Goswami S (2015) Performance of digital modulations over TWDP fading channel. In: International conference on electronic design, computer networks and automated verification (EDCAV), pp 135–138. https://doi.org/10.1109/EDCAV.2015.7060554
12. Moreno-Pozas L, Lopez-Martinez FJ, Paris JF, Martos-Naya E (April 2015) The κ-μ shadowed fading model: unifying the κ-μ and η-μ distributions. http://arxiv.org/abs/1504.05764[c.IT]
13. Verma PK, Soni S, Jain P, Raw RS (March 2016) Channel capacity of different adaptive transmission techniques over log-normal shadowed fading environment. In: International conference on computing for sustainable global development, IEEE conference on proceedings of the 10th INDIACom, INDIACom
14. Simon MK, Alouini M-S (2000) Digital communications over fading channels: a unified approach to performance analysis. Wiley, New York
15. Gradshteyn S, Ryzhik IM (2007) Table of integrals, series, and products, 7th edn.

Magneto-Optical Tuning of Refractive Index for Bidispersed Ferrofluid

Keyur G. Khatsuriya

Abstract Refractive index of Bidispersed ferrofluid (BDF) changes with the external magnetic field during light transport. BDF is the most appropriate "liquid magnet" for the magnetic modulation. BDF is composed of micrometer-sized magnetic sphere dispersed in a magnetic nanofluid consisting of magnetic nanoparticles in a non-magnetic liquid carrier. The variation in the relative refractive index (RRI) of BDF was studied as a function of external magnetic field. RRI increases with external magnetic field and saturates. Probability of such behavior of RRI is extinction of light. Probable reason of extinction like Mie resonance, optical limiting, and magnetic-hole theory is discussed. The study gives better insight about the role of RRI in extinction of light for BDF, which is governing factor in light.

Keywords Magnetic fluid · Refractive index · Mie resonance Optical limiting

1 Introduction

Study of optical properties of various magnetic colloids has become the most explored area of research in nowadays, as evidenced by various applications in the field ranging from medicine, biosensors, communications, and nanocircuits [1, 2]. In the presence of external magnetic field, optical properties of ferrofluid like birefringence [3], magneto-chromatics [4] can be observed. Many photonic devices engineered using optical transport like optical switches [5], tunable optical and fiber filters [6].

Refractive index of medium is an important parameter to control the optical properties of medium. Thus, magnetic modulation of refractive index is an interesting area for many researchers. A magnetic fluid or ferrofluid (FF) is a colloidal

K. G. Khatsuriya (✉)
Shantilal Shah Engineering College, Bhavnagar, Gujarat 364060, India
e-mail: keyur.physics@gmail.com

V. Janyani et al. (eds.), *Optical and Wireless Technologies*, Lecture Notes in Electrical Engineering 472, https://doi.org/10.1007/978-981-10-7395-3_11

suspension of single-domain nanoscale magnetic particles in a liquid carrier. Bidispersed ferrofluid (BDF) sometimes called *inverse* ferrofluid is composed of micrometer-sized magnetic spheres dispersed in a magnetic nanofluid consisting of magnetic nanoparticles in a non-magnetic liquid carrier like water or kerosene. BDF is one of the best engineered fluids for the study of light transport. The magnetization of FF is well described by the Langevin's theory of paramagnetism for non-interacting single-domain magnetic particles dispersed in a suitable liquid carrier [7]. In this paper, relative refractive index (RRI) of BDF is plotted against the external magnetic field for three different values of refractive indices of scatterer (n_s). The graph is explained based on extinction theory of light for BDF.

2 Theory

When external magnetic field is applied, the refractive index of the FF changes because of the magnetically induced structure formation [8]. It was reported that the refractive index of FF (n_f) increases with the external magnetic field and saturated for higher field [9]. For small particle size (typically less than 10 nm), the transmitted intensity increases with external magnetic field but decreases for the particle size greater than 10 nm. This is due to effects of van der Waals and dipole–dipole interaction [10]. In our recent work, we reported variation in refractive index of mono-dispersed FF with temperature and particle size [11]. For mono-dispersed FF, entire anisotropy in light transmission can be explained based on the structure formation in presence of external magnetic field and dominance of dipole–dipole interaction over van der Waals interaction. However, the chain formation is not possible in case of BDF and variation in refractive index due to field-induced structure formation may not be possible when particle size is more (of the order of micrometer). The magnetic modulation of refractive index for BDF can be explained, therefore, on the basis of magnetically modulated dielectric constant (ε). The refractive index is given by, $n = \sqrt{\varepsilon}$, $\mu = 1$ at optical frequency, and ε can be tuned by the external magnetic field.

We select BDF, which is suspension of nanosized magnetic particle with suitable surfactant in suitable carrier like kerosene or water in known volume fraction (approx. $10^{15}/\text{m}^3$) in which micron-size particle of known volume fraction (approx. $10^5/\text{m}^3$) is suspended. We have considered field-dependent refractive index for fluid as $n_f = \sqrt{\varepsilon_f}$ where $\varepsilon_f = \varepsilon_0 + F(\alpha)\,\varepsilon_\infty$; ε_0 is the dielectric constant at zero field, ε is dielectric constant at an infinite field (saturation value), $F(\alpha) = \alpha L^2(\alpha)/\alpha - L(\alpha)$, $L(\alpha) = \coth\alpha - \alpha^{-1}$, $\alpha = \mu H/k_B T$, μ is magnetic moment which is 5.652×10^{-19}, H is applied field (ranging from 0 to 0.12 T), k_B is Boltzmann constant, and T is absolute temperature which is 300 K. $F(\alpha)$ is derived using effective field medium, and wavelength of light is taken as 633 nm.

Light transport through BDF is affected by changing the refractive index of scatterer (n_s). The RRI (n) can be calculated by taking ratio of refractive indices of

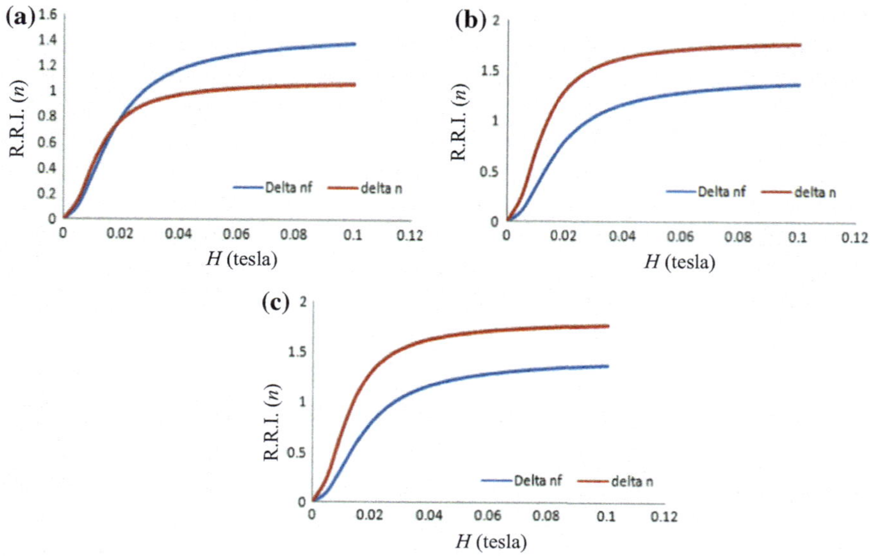

Fig. 1 **a** RRI (n) versus H for $n_s = 3$. **b** RRI (n) versus H for $n_s = 4$. **c** RRI (n) versus H for $n_s = 5$

scatterer (n_s) to the fluid (n_f). We choose three different values of refractive index for scatterer as $n_s = 3$, 4, and 5. From above-mentioned formula, RRI was calculated. Graphs of RRI versus external magnetic field are shown in Fig. 1a–c.

3 Discussion

It is noticed from the graph that saturation value of Δn increases with the increase in n_s but saturation value for Δn_f remains constant. It is also observed that Δn tends to saturate "early," when as refractive index of scatterer (n_s) increased from 3 to 5.

The possible reasons for theoretically obtained results shown in Fig. 1a–c, are explained below.

3.1 Magnetic Mie Resonance

In BDF, magnetic nanoparticles of fluid will behave as Rayleigh scatterers, whereas the micron-size magnetic particle will behave as a Mie scatterer [12]. Optical properties of such fluid change drastically when it is subjected to external magnetic field [13]. One possible reason of such behavior is a change in transmitted light intensity with the increase in refractive index of scatterer. This is due to dipolar interaction of the particles and Brownian rotation of micron-size particles in a magnetically interacting medium. With increase in the magnetic field, size of nanoparticles will increase by means of formation of very small chain-like structure which results in increase of scattering anisotropy factor $< \cos\theta >$ and extinction efficiency factor Q_{ext} [14]. The resonant behavior of scattering anisotropy factor $< \cos\theta >$ and extinction efficiency factor Q_{ext} with size parameter (ka) decreases the transmission of light as it produces morphology-dependent standing wave. This morphology-dependent structural anisotropy may be the reason for "early" saturation in Δn for large value of n_s. Thus, transmission intensity of BDF decreases due to *Mie resonance* [15, 16].

3.2 Optical Limiting due to Refractive Index Mismatch

Another possible mechanism is *optical limiting of laser light* from a colloidal suspension of spherical particles [17]. Optical limiting is observed due to nonlinear light scattering by two components of colloidal suspension. The significant mismatch between refractive indices of two suspended particles of BDF causes nonlinear light scattering in the medium which decreases the intensity of transmitted light. Photoinduced mismatch in refractive indices may be one of the reasons for the saturation of Δn at low field value for high n_s.

3.3 Magnetic Hole Theory

Patel et al. [18] have given magneto-optical extinction theory of light for BDF using the concept of *magnetic hole*. When micron-sized magnetic spheres are dispersed in the magnetized fluid, the void produced by the particle possesses an effective magnetic moment equal in size and opposite in the direction to the magnetic moment of displaced fluid. This is known as magnetic hole [19, 20]. With the increase in field magnetic moments of large particle and displaced fluid tries to align in the direction of field. This phenomenon induces two types of torque on the particle in opposite to each other. Resultant torque of these two may cause the extinction of light which, in turn saturates Δn.

4 Conclusion

We report the modulation in refractive index of BDF with external magnetic field. A new approach to understand light transport in BDF is presented. BDF is colloidal suspension of nanomagnetite and micron-size magnetic particle. Variation in change in RRI (Δn) with external magnetic field was studied theoretically for three values of refractive index of scatterer (n_s) 3, 4, and 5. A probable reason behind the modulation of RRI is extinction of light. The magnetic modulation of Δn was discussed based on magnetic Mie resonance, optical limiting, and magnetic-hole theory. It elucidates the variation in RRI with magnetic field at nanolevel. The study helps to design magnetically modulated medium for optical transport and novel photonic devices.

References

1. Haes J, Van Duyne RP (2002) A nanooptical biosensor: sensitivity of an approach based on the localised surface plasmons resonance spectroscopy of triangle silver nanoparticles. J Am Chem Soc 124:10596–10604
2. Engheta N (2007) Circuits with light at nanoscale: Optical nanocircuits inspired by metamaterials. Science 317:1698–1702
3. Scholten PC (1980) Magnetic birefringence of ferrofluids. J Phys D Appl Phys 13:L231
4. Horng HE, Hong CY, Yeung WB, Yang HC (1998) Magneto chromatic effects in magnetic fluid thin films. Appl Opt 37(13):2674–2680
5. Chen CS; Fang KL; Yang SY; Chieh JJ; Hong CY, Yang HC (2004) Tunable optical switch using magnetic fluids. Appl Phys Lett 85(23):5592–5594
6. Liao W, Chen X, Chen Y, Shengli P, Xia Y, Li Q (2005) Tunable optical fiber filters with magnetic fluids. Appl Phys Lett 87:151122
7. Resensweig RE (1985) Ferrohydrodynamics. Cambridge University Press, Cambridge
8. Yang SY, Chieh JJ, Horng HE, Hong CY, Yang HC (2004) Appl Phys Lett 84:5204
9. Pu S, Chen X, Chen Y, Liao W, Chen L, Xia Y (2005) Appl Phys Lett 86:171904
10. Rao GN, Yao YD, Chen YL, Wu KT, Chen JW (2005) Phys Rev E 72:031408
11. Khatsuriya KG, Bhatt H (2015) Temperature and size dependence of structural anisotropy in presence of magnetic field—a study to control light transport. In: International conference on NANO-15
12. Bhatt H, Patel R (2013) Optical transport in bidispersed magnetic colloids with varying refractive index. J Nanofluid 2:188–193
13. Philip J, Laskar JM (2012) Optical properties and applications of ferrofluid—a review. J Nanofluid 1:3–20
14. Philip LJ, Raj B (2008) Phy Rev E 78:031404
15. Bhatt H, Patel R, Mehta RV (2010) J Opt Soc Am A 27:873
16. Bhatt H, Patel R, Mehta RV (2012) Phys Rev E 86:011401
17. Joudrier V, Bourdon P, Hache F, Flytzanis C (1998) Appl Phys B Lasers Opt 67:627
18. Pate R, Upadhyay RV, Mehta RV (2005) Optical properties of magnetic and non-magnetic composites of ferrofluids. JMMM 300:217
19. Skeltorp AT (1983) Phys Rev Lett 51:2306
20. Parekh K, Patel R, Upadhyay RV, Mehta RV (2005) JMMM 289:311

Mach–Zehnder Interferometer-Based 3-Bit All-Optical Sequence Detector

Rakesh Ranjan, Abhishek Ranjan and Dharmendra Kumar Singh

Abstract All-optical devices can support high data rate in a communication system compared with its electronic counterparts. Several all-optical logic devices can be implemented using Mach–Zehnder interferometers (MZIs). The MZI structure uses electro-optic effect to switch light across the titanium-diffused lithium niobate optical waveguide from cross-state to bar-state and vice versa. Sequence detector is a significant sequential circuit that detects the presence of predefined bit sequence in the input bit stream. The paper presents an all-optical 3-bit sequence detector that detects the bit sequence 110 from the main bit stream. The operations of the proposed device are analysed and verified using OptiBPM software. This all-optical device could be deployed in several optical switching and networking operations.

Keywords Mach–Zehnder interferometer · Electro-optic effect Titanium-diffused $LiNbO_3$ · Beam propagation method · Optical sequence detection

1 Introduction

In today's scenario, there is a great requirement of bandwidth and contrast to this the existing electronic devices are approaching the speed limit. Optical fibre communications, accompanied by all-optical devices, could be a good substitute to the bandwidth limits of electrical devices, due to its large bandwidth capabilities.

R. Ranjan (✉) · A. Ranjan · D. K. Singh
OFC and Photonics Laboratory, Department of Electronics and Communication Engineering, National Institute of Technology, Patna, Bihar, India
e-mail: rakesh.r1804@gmail.com; rr@nitp.ac.in

A. Ranjan
e-mail: 075abhishek@gmail.com

D. K. Singh
e-mail: dksingh@nitp.ac.in

V. Janyani et al. (eds.), *Optical and Wireless Technologies*, Lecture Notes in Electrical Engineering 472, https://doi.org/10.1007/978-981-10-7395-3_12

All-optical domain is in its incubation period and several researches are still going on worldwide to explore the possibilities of several all-optical devices. These devices can be widely used as an important subsystem to perform various arithmetic and logic operations for different applications, such as optical signal/data processing and networking. Ultrahigh-speed reconfigurable optical gate architecture based on semiconductor optical amplifier (SOA) has been described [1] to perform various logic operations. Various mechanisms to implement different combinational and sequential logic circuits have been reported in literature. All-optical SR and D flip-flop employing cross-gain modulation (XGM) effect in semiconductor optical amplifiers have been designed [2]. Various schemes with and without SOA for the implementation of all-optical gates have been studied and compared in [3].

Recently, the performance analysis of Mach–Zehnder interferometer (MZI)-based all-optical switches has been done [4]. MZI structures, basically titanium-diffused lithium niobate waveguide, can efficiently acquire intensity modulation from phase modulation. The phase modulation can be realized by electro-optic effect from the electrodes associated with MZI structure. The MZI structures can act as extremely useful building block for several optical devices and circuits, such as optical modulators and logic devices. Several all-optical switches and logic circuits using MZIs such as logic gates, flip-flops, shift registers, counters and parity generator have been proposed in [5–8]. Feedback is a necessary obligation for the implementation of sequential circuits, as they require previous outputs. This can be achieved by using lithium niobate ($LiNbO_3$) switches-based guided-wave optical time delay network [9]. This device can provide a maximum delay of 2.772 ns in the step of 44 ps. A new D flip-flop configuration based on combination of SOA and MZI has been suggested with very low complexity in [10].

In several optical computing and signal processing, detection of some specific bit sequence is an important aspect to execute the next stages of operation. Sequence detector can perform this function efficiently by recognizing the predefined bit sequence in the main bit stream. It can be useful in designing various electrical counterpart operations. In this paper, the operation of an all-optical 3-bit sequence detector is presented using the MZIs. The design has been made up by cascading MZIs. The layout and the analysis of the proposed design have been done using the help of beam propagation method (BPM). The schematic and working principle of the all-optical 3-bit sequence detector has been presented in Sect. 2. The design is presented for a specific bit sequence '110', but it can be extended for any length of bit sequence. The implementation and simulation results obtained through BPM are presented in Sect. 3. Section 4 comprises conclusion.

2 Three-Bit Sequence Detector Using Mach–Zehnder Interferometers

Figure 1 shows the schematic diagram of the 3-bit sequence detector using all-optical flip-flops and gates. Thirteen MZIs are being used to implement this sequential circuit. Different logic gate and D flip-flop operations are achieved by connecting different MZIs. MZIs 1, 2 and 3, 4 form two AND gate logics. MZIs 5, 6 form OR gate logic. MZIs 7, 8 form first D flip-flop (D_1) and MZIs 9, 10, 11 form second D flip-flop (D_2). D_1 and D_2 are the electrical equivalent signals connected to the second electrode of MZIs 7 and 9, respectively. Therefore, $D_1 = Q_2 \cdot (A + Q_1)$ and $D_2 = A$. Six combinations of photo detector and amplifier are used to generate the electrical equivalent of the optical signal to drive the second electrode of MZI. An optical signal is continuously being fed in first input port of the MZIs 1, 3, 5, 6, 7, 9 and 12. 'A' is the input signal in the electrical form. This input electrical signal (A) is applied to second electrode of MZIs 1 and 9. It is the input D_2 for the MZI 9. A clock signal (CLK), which is in the electrical form, is being applied to second electrodes of MZIs 8, 10 and 11.

2.1 Working Principle of the Proposed Sequence Detector

Sequence detector detects the presence of the exact bit sequence in the main bit stream. The proposed all-optical sequence detector detects the bit sequence 110 in the main input stream (A). Block arrows in the Fig. 1 show the possible optical paths within the waveguide. The optical signal from first output port of MZI 7 is applied to first input port of MZI 8. The signal from first output port of MZI 8 (i.e. Q_1) is feedback after 1-bit time delay to its second input port. Also, electrical equivalent of Q_1 is applied to second electrode of MZIs 3 and 13. In MZI 9, output port 1 is connected to first input port of MZI 10, and output port 2 is connected to first input port of MZI 11. In both MZIs 10 and 11, the first output port is feedback to second input port after 1-bit time delay, respectively. Electrical equivalent of output port 1 of MZI 10 (i.e. Q_2) drives the MZIs 2 and 4, while the electrical equivalent of output port 1 of MZI 11 (i.e. $\bar{Q}_2$) drives MZI 12. First output port of MZI 3 is connected to first input port of MZI 4, and the electrical equivalent of first output port of MZI 4 is connected to second electrode of MZI 5. Similarly, first output port of MZI 1 is connected to first input port of MZI 2, and electrical equivalent of first output port of MZI 2 is connected to second electrode of MZI 6. The first output ports of MZIs 5 and 6 are combined to implement OR gate logic, and the electrical equivalent of this (i.e. D_1) is connected to second electrode of MZI 7. The output port 1 of MZI 12 is connected to input port 1 of MZI 13 which logically performs the AND gate logic between Q_1 and $\bar{Q}_2$. The optical output is acquired through first output port of MZI 13. Presence of optical signal at the output port shows the presence of the desired bit sequence (110) in the input bit stream (A). Figure 2 shows the timing diagram for bit sequence 1110, which shows all the intermediate logics for the proposed device. In this bit sequence, the desired bit sequence (110) is appearing at 2nd, 3rd and 4th clock pulses and the corresponding logical high output is present in the 5th clock pulse. The timing diagram is generated in MATLAB Simulink software to perform the transient analysis for each logic state configuration and the output is verified in simulation (10 states for each configuration).

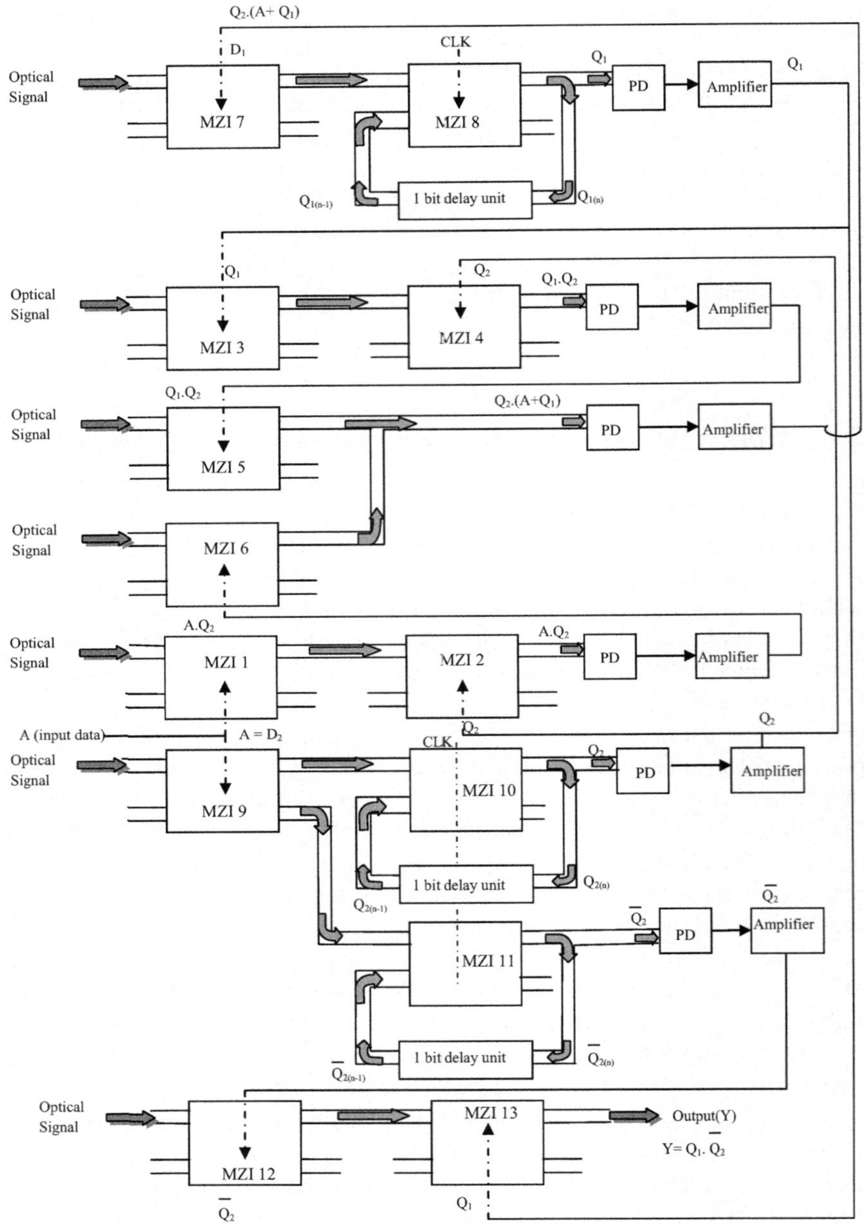

Fig. 1 Schematic diagram of 3-bit (110) sequence detector

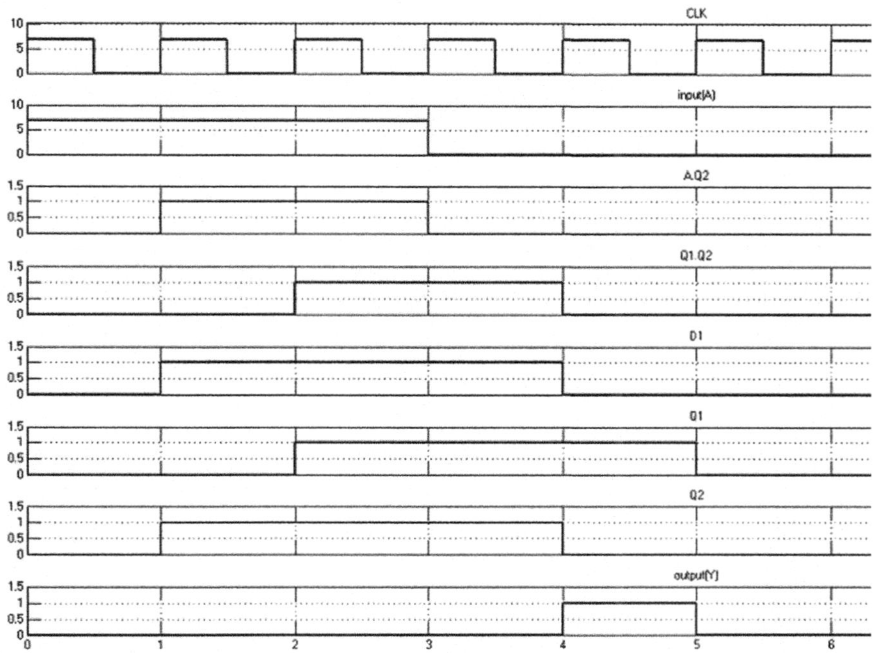

Fig. 2 Timing diagram for 3-bit (110) sequence detector

3 Implementation and Simulation Results Using BPM

Beam propagation method-based OptiBPM software has been used to analyse the proposed structure of 3-bit sequence detector. BPM is a powerful numerical technique to study the propagation of light signal through the slow varying optical waveguides. BPM is mainly applied to describe field profile throughout the waveguide structure provided the distribution of refractive index over the complete waveguide is known [11]. Figure 3 shows the layout for the 3-bit sequence detector using the OptiBPM. Here, input electrical signal is applied to second electrodes of MZI 1 and MZI 9. A common clock is applied to the MZIs 8, 10 and 11. For the sake of simplicity, an input bit sequence of length four (1110) is considered and applied to the input port A. The simulative analysis has illustrated that when the desired bit sequence 110 arrives at the port A, the optical output is obtained at first output port of MZI 13, i.e. the output (Y). In OptiBPM, only one active source can be applied at a time. All the incoming optical signals in the proposed sequence detector are from the same light source. Hence, all the light signals are coherent, and it is logical and consistent for all the optical data sequence. For the present design of 3-bit sequence detector, the Ti-strip thickness is varied from 0.046 to 0.053 μm for the operating wavelength of 1.3 μm. The MZI structure has been optimized for the extinction ratio ($\geq$ 25 dB) with very minimal loss. The optimal

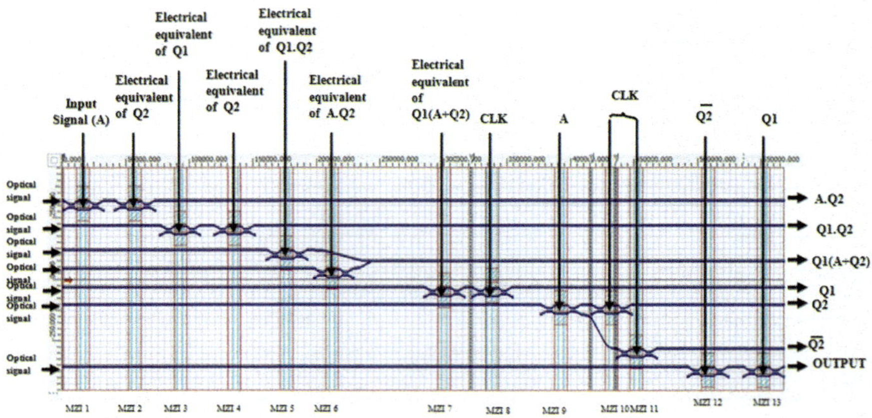

Fig. 3 BPM layout for the proposed 3-bit (110) sequence detector

performance of the MZI with lower BER in all-optical domain is anticipated with the higher extinction ratio. The simulated results for all the five clock pulses are discussed below.

Case 1: Input data (A) = 1, 1st clock signal = 1

Prior to arrival of the first clock pulse, it is assumed that all the flip-flops are reset ($Q_1 = Q_2 = 0$), so 0 V is applied to the second electrode of MZIs 2, 3, 4 and 13, while, 6.75 V to MZI 12. Input sequence A = 1, so 6.75 V is applied to the second electrode of MZIs 1 and 9. CLK = 1, so 6.75 V is applied to second electrode of MZIs 8, 10 and 11. Since D_1 is OR logic of Q_1 AND Q_2 and A AND Q_2, $D_1 = 0$. Hence, 0 V is applied to second electrode of MZI 7. D_2 = A, so 6.75 V is applied to MZI 9. The output from first output port of MZI 8 is Q_1 and that of MZIs 10 and 11 are Q_2 and $\bar{Q}_2$. The simulation result is shown in Fig. 4a.

Case 2: Input data (A) = 1, 1st clock signal = 0

$Q_1 = Q_2 = 0$, so 0 V is applied to the second electrode of MZIs 2, 3, 4, 13 and 6.75 V to MZI 12. Input sequence A = 1, so 6.75 V is applied to the second electrode of MZIs 1 and 9. CLK = 0, so 0 V is applied to second electrode of MZIs 8, 10 and 11. D_1 is OR logic of Q_1 AND Q_2 and A AND Q_2, $D_1 = 0$. Hence, 0 V is applied to second electrode of MZI 7. D_2 = A, so 6.75 V is applied to MZI 9. The output from first output port of MZI 8 is Q_1 and that of MZIs 10 and 11 are Q_2 and $\bar{Q}_2$. The simulation result is illustrated in Fig. 4b.

Case 3: Input data (A) = 1, 2nd clock signal = 1

$Q_1 = 0$, so 0 V is applied to the second electrode of MZI 3 and 13. $Q_2 = 1$, so 6.75 V is applied to second electrode of MZIs 2 and 4, while 0 V to MZI 12. Input sequence A = 1, so 6.75 V is applied to the second electrode of MZIs 1 and 9. CLK = 1, so 6.75 V is applied to second electrode of MZIs 8, 10 and 11. Since

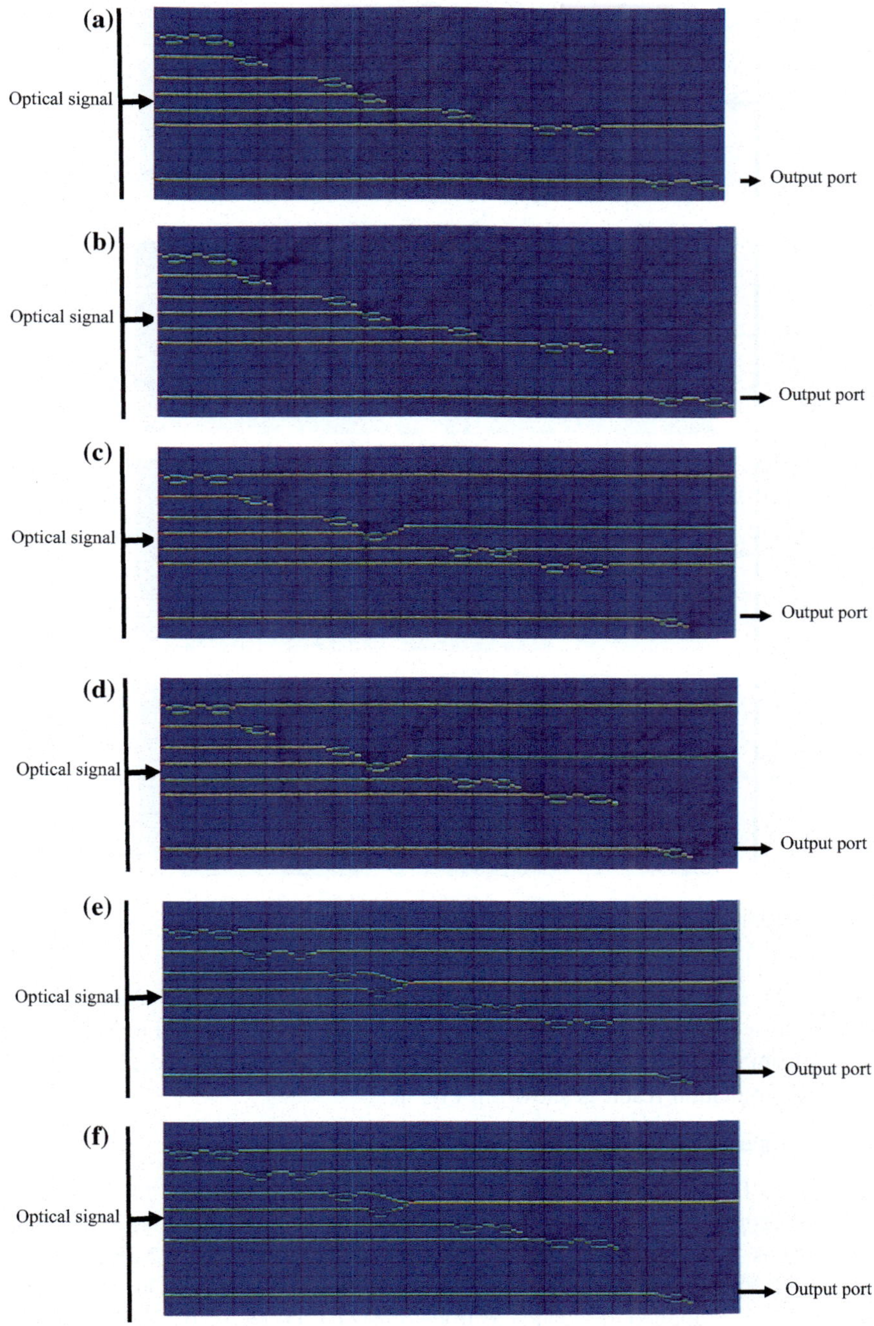

Fig. 4 **a** Simulation result of the proposed sequence detector for 1st clock up signal. **b** Simulation result of the proposed sequence detector for 1st clock down signal. **c** Simulation result for 2nd clock up signal. **d** Simulation result for 2nd clock down signal. **e** Simulation outcome of the proposed device for 3rd clock up signal. **f** Simulation outcome of the proposed device for 3rd clock down signal. **g** Result for the proposed 3-bit sequence detector for 4th clock up signal. **h** Result for the proposed 3-bit sequence detector for 4th clock down signal. **i** Simulation result of the proposed 3-bit sequence detector at 5th clock up signal. **j** Simulation result of the proposed 3-bit sequence detector at 5th clock down signal

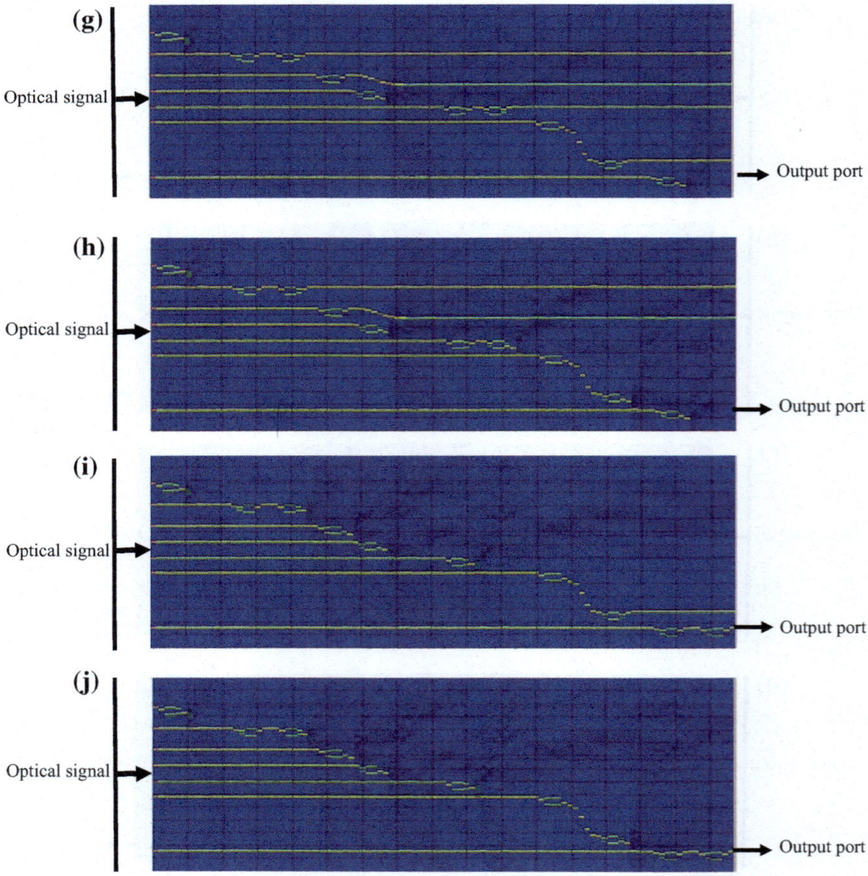

Fig. 4 (continued)

flip-flop D_1 is OR logic of Q_1 AND Q_2 and A AND Q_2, $D_1 = 1$. Hence, 6.75 V is applied to second electrode of MZI 7. $D_2 = A$, so 6.75 V is applied to MZI 9. The output from first output port of MZI 8 is Q_1 and that of MZIs 10 and 11 are Q_2 and $\bar{Q}_2$. The simulation result is shown in Fig. 4c.

Case 4: Input data (A) = 1, 2nd clock signal = 0

$Q_1 = 0$, so 0 V is applied to the second electrode of MZIs 3 and 13. $Q_2 = 1$, so 6.75 V is applied to second electrode of MZIs 2 and 4, while 0 V to MZI 12. Input sequence A = 1, so 6.75 V is applied to the second electrode of MZIs 1 and 9. CLK = 0, so 0 V is applied to second electrode of MZIs 8, 10 and 11. As $D_1 = 1$, hence 6.75 V is applied to second electrode of MZI 7. $D_2 = A$, so 6.75 V is applied to MZI 9. The output from first output port of MZI 8 is Q_1 and that of MZIs 10 and 11 are Q_2 and $\bar{Q}_2$. The simulation result is shown in Fig. 4d.

Case 5: Input data (A) = 1, 3rd clock signal = 1

$Q_1 = 1$, so 6.75 V is applied to the second electrode of MZIs 3 and 13. $Q_2 = 1$, so 6.75 V is applied to second electrode of MZIs 2 and 4, while 0 V to MZI 12. Input sequence A = 1, so 6.75 V is applied to the second electrode of MZIs 1 and 9. CLK = 1, so 6.75 V is applied to second electrode of MZIs 8, 10 and 11. $D_1 = 1$, hence 6.75 V is applied to second electrode of MZI 7. $D_2 = A$, so 6.75 V is applied to MZI 9. The output from first output port of MZI 8 is Q_1 and that of MZIs 10 and 11 are Q_2 and $\bar{Q}_2$. The simulation result is shown in Fig. 4e.

Case 6: Input data (A) = 1, 3rd clock signal = 0

$Q_1 = 1$, so 6.75 V is applied to the second electrode of MZIs 3 and 13. $Q_2 = 1$, so 6.75 V is applied to second electrode of MZIs 2 and 4, while 0 V to MZI 12. Input sequence A = 1, so 6.75 V is applied to the second electrode of MZIs 1 and 9. CLK = 0, so 0 V is applied to second electrode of MZIs 8, 10 and 11. $D_1 = 1$, hence 6.75 V is applied to second electrode of MZI 7. $D_2 = A$, so 6.75 V is applied to MZI 9. The output from first output port of MZI 8 is Q_1 and that of MZIs 10 and 11 are Q_2 and $\bar{Q}_2$. The simulation result is illustrated in Fig. 4f.

Case 7: Input data (A) = 0, 4th clock signal = 1

$Q_1 = 1$, so 6.75 V is applied to the second electrode of MZIs 3 and 13. $Q_2 = 1$, so 6.75 V is applied to second electrode of MZIs 2 and 4, while 0 V to MZI 12. Input sequence A = 0, so 0 V is applied to the second electrode of MZIs 1 and 9. CLK = 1, so 6.75 V is applied to second electrode of MZIs 8, 10 and 11. D_1 is OR logic of Q_1 AND Q_2 and A AND Q_2, $D_1 = 1$. Hence, 6.75 V is applied to second electrode of MZI 7. $D_2 = A$, so 0 V is applied to MZI 9. The output from first output port of MZI 8 is Q_1 and that of MZIs 10 and 11 are Q_2 and $\bar{Q}_2$. The simulation result is shown in Fig. 4g.

Case 8: Input data (A) = 0, 4th clock signal = 0

$Q_1 = 1$, so 6.75 V is applied to the second electrode of MZIs 3 and 13. $Q_2 = 1$, so 6.75 V is applied to second electrode of MZIs 2 and 4, while 0 V to MZI 12. Input sequence A = 0, so 0 V is applied to the second electrode of MZIs 1 and 9. CLK = 0, so 0 V is applied to second electrode of MZIs 8, 10 and 11. $D_1 = 1$, hence 6.75 V is applied to second electrode of MZI 7. $D_2 = A$, so 0 V is applied to MZI 9. The output from first output port of MZI 8 is Q_1 and that of MZIs 10 and 11 are Q_2 and $\bar{Q}_2$. The simulation result is illustrated in Fig. 4h.

Case 9: Input data (A) = 0, 5th clock signal = 1

$Q_1 = 1$, so 6.75 V is applied to the second electrode of MZIs 3 and 13. $Q_2 = 0$, so 0 V is applied to second electrode of MZIs 2 and 4, while 0 V to MZI 12. Input sequence A = 0, so 0 V is applied to the second electrode of MZIs 1 and 9. CLK = 1, so 6.75 V is applied to second electrode of MZIs 8, 10 and 11. $D_1 = 0$, hence 0 V is applied to second electrode of MZI 7. $D_2 = A$, so 0 V is applied to MZI 9. The output from first output port of MZI 8 is Q_1 and that of MZIs 10 and 11 are Q_2 and $\bar{Q}_2$. The simulation result is shown in Fig. 4i. Optical signal through the output port can be analysed from the simulation results. Since optical signal is present, it signifies the detection of the desired bit sequence (110) in the main input sequence.

Case 10: Input data (A) = 0, 5th clock signal = 0

$Q_1 = 1$, so 6.75 V is applied to the second electrode of MZIs 3 and 13. $Q_2 = 0$, so 0 V is applied to second electrode of MZIs 2 and 4, while 0 V to MZI 12. Input sequence A = 0, so 0 V is applied to the second electrode of MZIs 1 and 9. CLK = 0, so 0 V is applied to second electrode of MZIs 8, 10 and 11. $D_1 = 0$, hence 0 V is applied to second electrode of MZI 7. $D_2 = A$, so 0 V is applied to MZI 9. The output from first output port of MZI 8 is Q_1 and that of MZI 10 and 11 are Q_2 and $\bar{Q}_2$. The simulation result is shown in Fig. 4j. Since for clock down pulse, there is no change in the output, optical signal is still present at output port signifying the detection of bit stream 110 in the input bit stream. Simulation results for the different clock pulses are shown below.

The major drawback of the proposed sequence detector is the use of a larger number of MZIs, hence, the fabricated design may be bulky in size. The future researches on the development of an effective mathematical model for the current sequence detector can incorporate the BER and bit rate investigations.

4 Conclusion

Sequence detectors are used to detect the presence of specific bit sequence in main bit stream. In this paper, a 3-bit (110) all-optical sequence detector using MZIs has been reported. The design of the proposed device is analysed and verified using beam propagation method. Simulative results have demonstrated that whenever the desired bit sequence (110) arrives in the main bit sequence, the optical output is detected. The proposed device can be extended for any length of bit sequence. This device could be a module in advanced optical networks and computer systems to support high data rate operations and WDM applications.

References

1. Kaur S, Kaler RS (2012) Ultrahigh speed reconfigurable logic operations based on single semiconductor optical amplifier. J Opt Soc Korea 16:13–16
2. Kaur S, Kaler RS (2014) All optical SR and D flip-flop employing XGM effect in semiconductor optical amplifiers. Optik—Int J Light Electron Opt 125:865–869
3. Pallavi S, Tripathi DK, Jaiswal S, Dixit HK (2014) All-optical logic gates: designs, classification, and comparison. Adv Opt Technol 2014:1–13
4. Singh G, Janyani V, Yadav RP (2012) Modeling of a high performance Mach-Zehnder interferometer all optical switch. Optica Applicata 42:613–625

5. Raghuwanshi SK, Kumar A, Chen NK (2014) Implementation of sequential logic circuits using the Mach-Zehnder interferometer structure based on electro-optic effect. Opt Commun 333:193–208
6. Kumar S, Raghuwanshi SK, Kumar A (2013) Implementation of optical switches using Mach-Zehnder interferometer. Opt Eng 52:1–9
7. Kumar A, Kumar S, Raghuwanshi SK (2014) Implementation of XOR/XNOR and AND logic gates by using Mach-Zehnder interferometers. Optik—Int J Light Electron Opt 125:5764–5767
8. Kumar S, Singh G, Bisht A, Amphawan A (2015) A design of D flip-flop and T flip-flop using Mach-Zehnder interferometers for high-speed communication. Appl Opt 54:6397–6405
9. Murphy EJ, Adda TF, Minford WJ, Irvin RW, Ackerman EI, Adams SB (1996) Guided-wave optical time delay network. IEEE Photon Technol Lett 8:545–547
10. Chattopadhyay T, Reis C, Andre P, Teixeira A (2012) Theoretical analysis of all-optical clocked D flip-flop using a single SOA assisted symmetric MZI. Opt Commun 285:2266–2275
11. Wartek MS (2013) Computational photonics—an introduction with MATLAB, pp 288–315. Cambridge University Press

Improving Dynamic Range of RoF System Using Dual-Drive Mach-Zehnder Modulator

Joseph Zacharias, V. Civin and Vijayakumar Narayanan

Abstract A novel radio over fiber (RoF) system incorporating a dual-electrode dual-parallel Mach-Zehnder modulator (DE-DPMZM) is used to transmit vector signal to minimize the distortions that are caused by the Mach-Zehnder modulator (MZM) nonlinearity. The MZMs are biased such that the offset voltage in both the MZMs is opposite. A 30 MSym/s 16-QAM signal is transmitted after single-sideband modulation, and the detected signal is analyzed for its error vector magnitude (EVM) performance. The measured EVM is 8.363% which shows there is much improvement in the EVM performance. The system is cost-effective since it rectifies the nonlinearities of the Mach-Zehnder modulator using the simplest optical methods and the EVM% shows that proposed biasing results in higher dynamic range.

Keywords RoF · Dynamic range · Nonlinearity

1 Introduction

The radio over fiber (RoF) is widely used to cope up with the increased bandwidth demand in recent times. It is an emerging field in communication owing to its advantages such as interference cancellation and low losses. Optical modulation using external modulator is widely used in RoF system design. The most common external modulator is Mach-Zehnder modulator which modulates the carrier by controlling the amplitude of the data signal. But the Mach-Zehnder modulator

J. Zacharias (✉) · V. Civin
Department of Electronics and Communication, Rajiv Gandhi Institute of Technology, Kottayam, Kerala, India
e-mail: joseph.zacharias@rit.ac.in

V. Narayanan
Department of Electronics and Communication, Government Engineering College, Barton Hill, Thiruvananthapuram, Kerala, India

V. Janyani et al. (eds.), *Optical and Wireless Technologies*, Lecture Notes in Electrical Engineering 472, https://doi.org/10.1007/978-981-10-7395-3_13

suffers from inherent nonlinearity property which results in the formation of harmonics and subsequent power loss during the modulation.

This could also result in nonlinear distortion during the modulation processes [1]. Latest systems are using complex modulation methods, and they have increasing demand for linearity in the system [2]. The research in this area mainly focuses on improving the linearity of the MZM and to improve the quality of the received signal.

Various schemes have been developed to analyze the intermodulation distortion compensation and to increase the spurious-free dynamic range of the systems.

The schemes include generation of optical single-sideband modulation that eliminates the second- and third-order harmonics [3, 4] using polarization modulator [5, 6] using dual-parallel MZM [7]. But many of the schemes are aimed at analyzing the double-sideband modulation (DSB). Earlier schemes were aimed at measuring the intermodulation distortion 3 (IMD3) [8] and spurious free dynamic range (SFDR) [9] based on two-tone analysis. The application of real-time signals or vector signals in the systems was not considered in these papers.

In this paper, we propose a novel technique in which a vector signal can be transmitted via an optical fiber with less distortion. This novel method consists of a DE-DPMZM which is biased in such a way so that the offset voltages in both the MZMs are opposite, resulting in improved performance measurements.

Moreover, the SSB modulation is found to be superior to the DSB modulation which encourages its use in the RoF systems. A 30 MSym/s 16-QAM signal is applied as the data signal in the analysis and the recovered signal is tested for its EVM% value. The measured error vector magnitude (EVM) is 8.363% which is much more improved than the conventional scheme. This shows that the usage of DE-DPMZM in the proposed biasing results in higher dynamic range for the vector signals.

2 Principle

The conventional method for single-sideband (SSB) modulation and transmission of optical signal is shown in Fig. 1.

This scheme consists predominantly of the MZM which is biased to modulate the signal in SSB modulation. The modulated signal is transmitted via an optical fiber and is beaten at the photodetector. This is followed by demodulation to obtain the signal at the receiver. The nonlinearity of the MZM results in formation of harmonics during the modulation which in turn results in a large EVM value. This proves that the dynamic range for the vector signal is very low. The schematic diagram of the proposed system is shown in Fig. 2. The output of the laser diode (LD) is split and is used as the carrier for both the MZM.

The 30 MSym/s 16-QAM signal is applied to DE-MZM1 while the DE-MZM2 is been applied with the phase shifted version of this signal. The MZMs are biased to produce the SSB modulation. The bias voltage of both the MZMs is given such

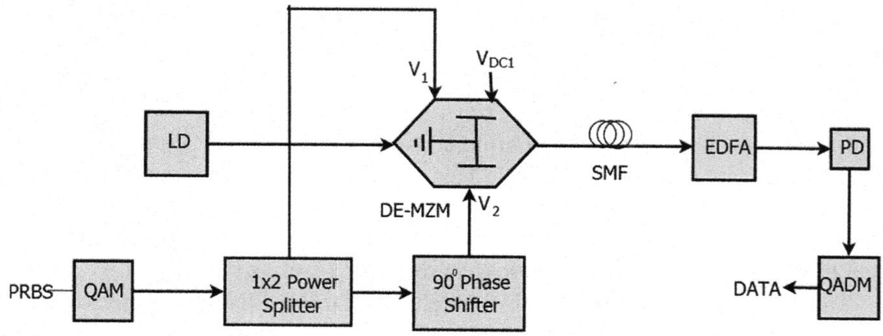

Fig. 1 Conventional system [9] using DEMZM. LD laser diode, SMF single mode fiber, EDFA erbium-doped fiber amplifier, PD photodetector, QAM quadrature amplitude modulator, QADM quadrature amplitude demodulator

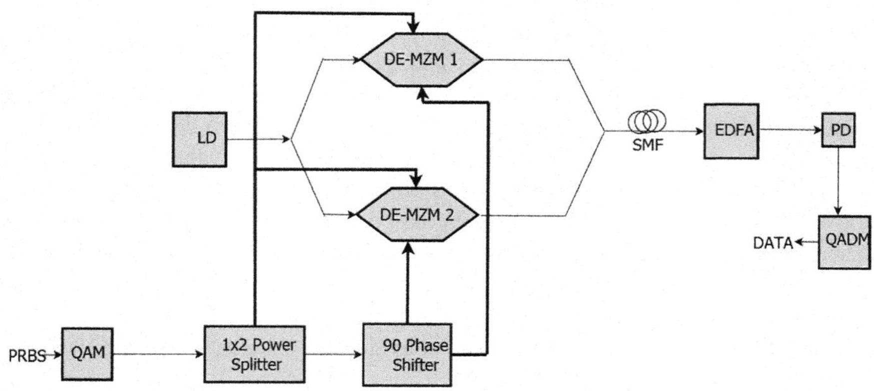

Fig. 2 Experimental setup of the proposed scheme

that opposite offset voltage is formed and the harmonics will be deleted on adding the signals from both the DE-MZM. The resultant signal is amplified by an EDFA and beaten by a PIN diode. This signal is demodulated using a quadrature amplitude demodulator (QADM) to obtain the recovered signal which is analyzed for its performance measurements.

3 Simulation and Results

Figure 2 shows the simulation setup for the proposed scheme. Optisystem 14 is used to simulate the project. The laser diode generates lightwave at a wavelength of 1553.6 nm with a power of 13.2 dB which acts as the optical carrier for both the

MZMs. Optical 1 × 2 splitter is used to divide the output of the laser diode to these external modulators. A $2^7 - 1$ length pseudorandom binary sequence (PRBS) is given as the baseband data to the QAM modulator. The DE-MZM 1 is applied with a bias voltage 1 and a ground in the other arm (arm 2) while DE-MZM2 is applied with the bias voltage at arm 2 and ground at arm1. There is a 90° phase shift for the QAM signal applied to DE-MZM2. This generates the resultant output that gets added to remove the harmonics in the system. The resultant signal is transmitted through an optical fiber having dispersion coefficient of 16.75 ps/nm/km and fiber length of 20 km. The attenuation loss of the fiber is set as 0.2 dB/km.

The signal is beaten by a PIN photodetector. The PIN photodetector has a responsivity of 1 A/W. The QAM data is recovered by coherent detection applying the same frequency of sine and cosine waves used for modulation. The distinction between the conventional scheme and the use of DP-DEMZM is mainly the use of more than one MZM to modulate the data signal. This scheme preserves the linearity that is required in the process of external modulation.

EVM% is measured in both the cases by varying the input power. This data is used to analyze the performance of both the systems. Figure 3 shows the EVM performance comparison of both the schemes.

EVM% is measured in both the cases by varying the input power. This data is used to analyze the performance of both the systems. Figure 3 shows the EVM performance comparison of both the schemes. The EVM value gets decreased below 10 dBm for both the schemes and when the applied input power increases above 10 dBm, the EVM value gets increased. The EVM% for the conventional scheme is found out to be 8.508% and that for the proposed scheme is found out to be 8.363%. This shows that the proposed system performs better than that of the conventional system.

The BER curves of the proposed and conventional scheme are shown in Fig. 4.

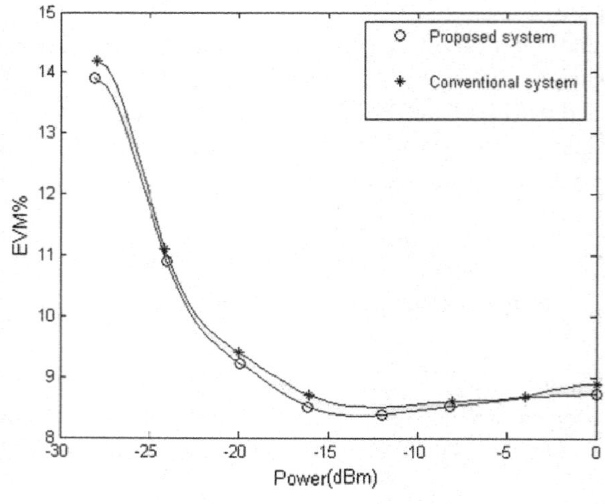

Fig. 3 Error vector magnitude (EVM)% of conventional and proposed schemes

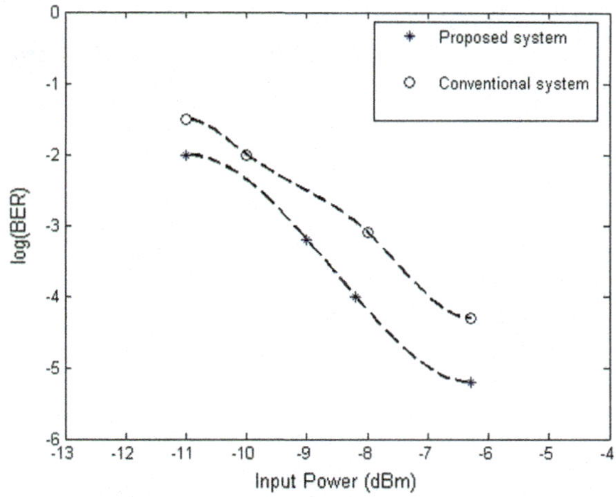

Fig. 4 BER versus received optical power for proposed and conventional systems

The BER values increase as the input power gets decreased. It can be inferred that the proposed system is better in terms of BER analysis when compared with the conventional system.

The constellation diagram of the input QAM is given in Fig. 5 and that of the output constellation is given in Fig. 6. The eye diagram of the I component and Q component is, respectively, given in Fig. 7 and Fig. 8. The eye diagrams show clear openings that result in a dependable system based on the performance measures.

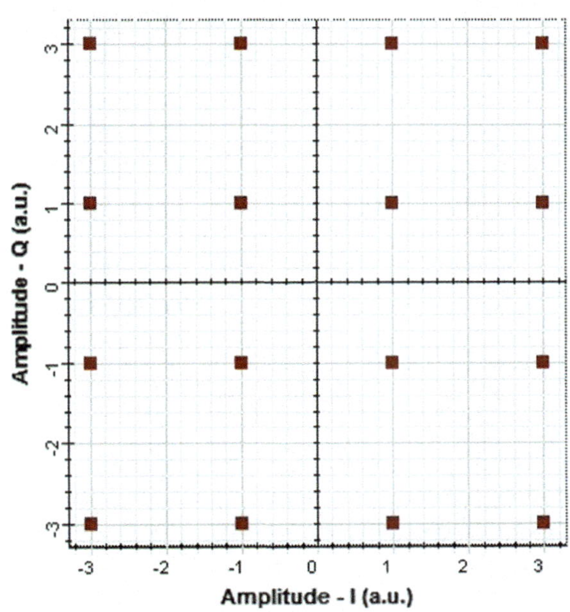

Fig. 5 Constellation of input QAM

Fig. 6 Constellation of output QAM

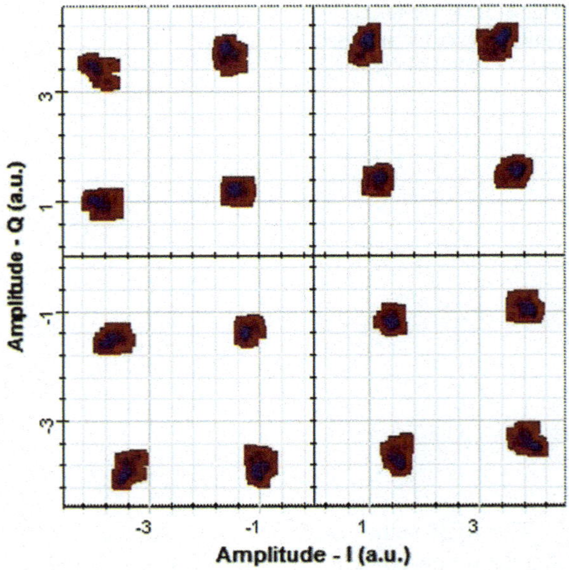

Fig. 7 Eye diagram of I component

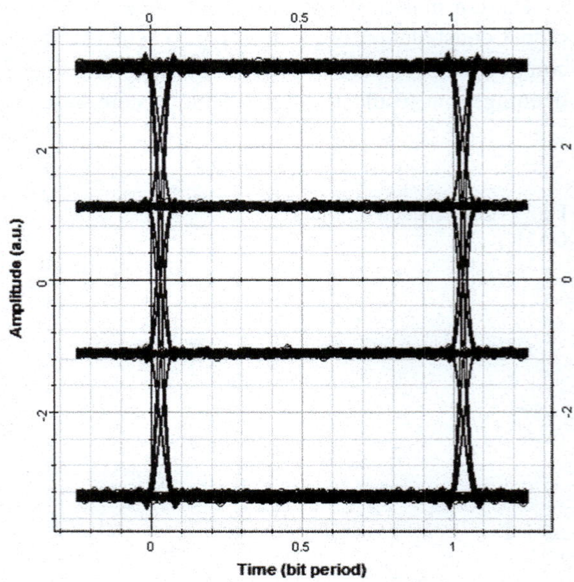

The proposed system needs the proper bias setup for the DEMZMs to avoid bias drifting problem. Bias control techniques [10] need to be incorporated so that the optimum bias point will not change for the system.

Fig. 8 Eye diagram of Q component

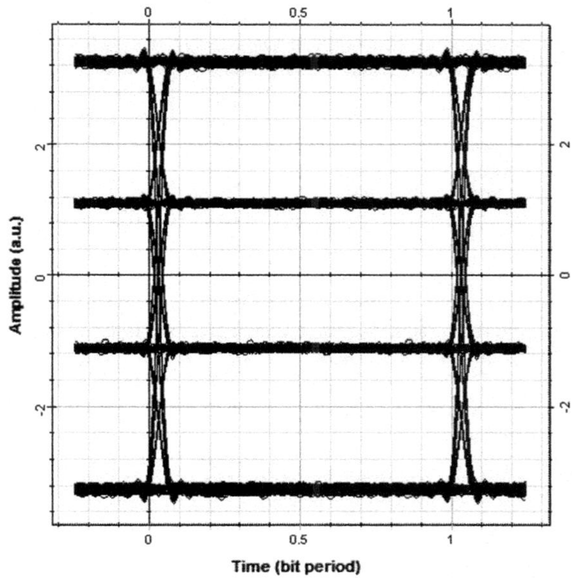

4 Conclusion

In short, we propose a system which preserves the linearity of a RoF system during the modulation and transmission of vector signal. The proposed scheme is analyzed for vector signal such as QAM signal. The scheme helps to get rid of the nonlinearity property of the MZM that generates harmonics during modulation. The EVM% is plotted for varying input power, and the results show that the scheme is viable for vector signal transmission. The EVM% is found out to be 8.363% for the 16-QAM RoF system. The proposed system has superior performance when compared with the conventional systems based on the BER curve also. This method proves to be a reliable method to transmit other complex vector signals when using MZM as the external modulator.

References

1. Korotky SK, de Ridder RM (1990) Dual parallel modulation schemes for low distortion analog optical transmission. IEEE J Sel Areas Commun 8(7):1377–1381
2. Kim SK, Liu W, Pei Q, Dalton LR, Fetterman HR (2011) Nonlinear intermodulation distortion suppression in coherent analog fiber optic link using electro-optic polymeric dual parallel Mach-Zehnder modulator. Opt Express 19(8):7865–7871
3. Shimotsu S, Oikawa S, Saitou T, Mitsugi N, Kubodera K, Kawanishi T, Izutsu M (2001) Single side-band modulation performance of a $LiNbO_3$ integrated modulator consisting of four-phase modulator waveguides. IEEE Photon Technol Lett 13(4):364–366

4. Liu X, Liu Z, Li J, Shang T (2010) Performance improvement of optical single sideband signal using an integrated Mach-Zehnder modulator. Fiber Integr Opt 29(6):453–465
5. Chen X, Li W, Yao J (2013) Microwave photonic link with improved dynamic range using a polarization modulator. IEEE Photon Technol Lett 25(14):1373–1376
6. Li W, Yao J (2013) Dynamic range improvement of a microwave photonic link based on bi-directional use of a polarization modulator in a Sagnac loop. Opt Express 21(13):15692–15697
7. Li S, Zheng X, Zhang H, Zhou B (2010) Highly linear radio-over-fiber sys-tem incorporating a single-drive dual-parallel Mach Zehnder modulator. IEEE Photon Technol Lett 22 (24):1775–1777
8. Yao J (2009) Microwave photonics. J Lightwave Technol 27(3):314–335
9. Sun J, Yu L, Zhong Y (2015) A single sideband radio-over-fiber system with improved dynamic range incorporating a dual-electrode dual-parallel Machzehnder modulator. Opt Commun 336:315–318
10. Fu Y, Zhang X, Hraimel B, Liu T, Shen D (2013) Mach-Zehnder: a review of bias control techniques for Mach-Zehnder modulators in photonic analog links. IEEE Microw Mag 14 (7):102–107

Positioning LED Panel for Uniform Illuminance in Indoor VLC System Using Whale Optimization

Ishwar Ram Kumawat, Satyasai Jagannath Nanda and Ravi Kumar Maddila

Abstract In 5G, indoor visible light communication is going to use panel of white light emitting diodes (LED) to provide high-speed communication as well as illumination. In this system, it is important to maintain uniform illuminance as non-uniformity creates bright and dark spots, which may cause discomfort for the occupants. The uniform illuminance depends upon the position and semi-angle of radiation associated with the LED panel. The objective is to optimize the position of the LED panels so that the received power and SNR at all locations in the room are uniform. The optimization task is carried out using a new metaheuristics whale optimization algorithm. Simulation study is performed in a room with dimension 5 × 5 × 3 m, having 4 to 6 LED panels (each with 60 × 60 LEDs). Results reveal that the proposed approach obtained superior uniformity (received power and SNR achieved in dB) compared to that achieved by particle swarm optimization.

Keywords Visible light communication · Uniform illumination
Whale optimization algorithm · Particle swarm optimization

1 Introduction

In upcoming fifth-generation wireless technology, visible light communication is going to become an essential component for indoor communication as well as illumination [1]. This system uses white light emitting diode (LED) panel which

I. R. Kumawat (✉) · S. J. Nanda · R. K. Maddila
Department of Electronics and Communication Engineering, Malviya National Institute of Technology, Jaipur 302017, Rajasthan, India
e-mail: kumawatishwar54@gmail.com; 2015pwc5359@mnit.ac.in

S. J. Nanda
e-mail: nanda.satyasai@gmail.com

R. K. Maddila
e-mail: rkmaddila.ece@mnit.ac.in

V. Janyani et al. (eds.), *Optical and Wireless Technologies*, Lecture Notes in Electrical Engineering 472, https://doi.org/10.1007/978-981-10-7395-3_14

acts as both transmitter and illuminator [2–4]. In a given indoor environment, the users' need is for high-speed data download and to obtain uniform illuminance. To achieve it, there is need for uniform SNR and power distribution [5, 6] across the indoor environment.

Komine and Nakagawa [7] proposed a model with white light emitting diode panel for indoor VLC transmitter and illuminator. The authors considered a pre-defined fixed position of LED panel in a 5 × 5 × 3 m room and analyzed the effect of interference and reflection on illuminance. Nguyen et al. [8] developed a simulink model considering the same position of LED panel for the same room size as in [7], to analyze the effect of RMS delay spread and illuminance distribution in the room. Recently, Wang et al. [9] used evolutionary algorithm (EA) to maximize the minimum SNR to achieve effective communication in indoor VLC communication.

In this manuscript, the effective position of LED panels (4 to 6 in numbers) in a 5 × 5 × 3 m room is determined to achieve uniform illuminance (received power) and high-speed download (improving SNR). The optimization task is accomplished by a recently developed metaheuristics in 2016 by Mirjalili and Lewis [10] known as whale optimization algorithm (WOA). It is inspired by the bubble-net hunting method of the Humpback whale. The WOA is applied for developing an early warning system for air quality measurement in various cities of China [11], optimizing the size of skeletal structures [12] and minimizing the fuel consumption of a vehicle considering various traffic scenarios [13].

The rest part of the paper is organized as follows. Section 2 describes the problem of placing LED panel for uniform illuminance and improving SNR. The steps associated with solving the problem using WOA are narrated in Sect. 3. The simulation studies carried out using a number of LED panels, parameters used for optimization, results achieved using proposed WOA and another comparative algorithm particle swarm optimization(PSO) are highlighted in Sect. 4. The conclusions drawn are given in Sect. 5.

2 Positioning LED Panel for Uniform Illuminance and Improving SNR

A rectangular room of 5 × 5 × 3 m is considered for analysis as the same dimension is standardized by [7, 8]. A basic structure of using indoor VLC system in this room using four LED panels (each panel comprises of 60 × 60 rectangular array) is shown in Fig. 1. Only line of sight communication (LOS) is considered for analysis. The white LED panels are installed at the ceiling, at height of 3 m from

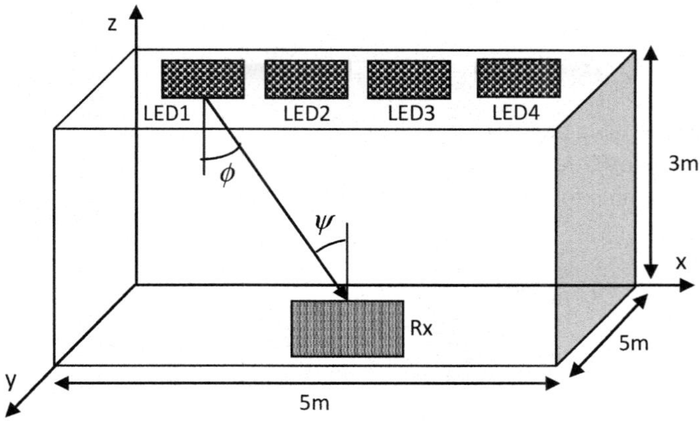

Fig. 1 Model of an indoor visible light communication environment

the floor. In this manuscript, the (x, y) coordinate of the panel is determined by the WOA and PSO algorithms.

The radiation intensity of each LED panel is described in [7], which is given by

$$R_0(\phi) = (m_l + 1)\cos^{m_l}(\phi)/2\pi, \phi \in [-\pi/2, \pi/2] \tag{1}$$

The ϕ is irradiance angle (angle between axis normal to LED panel and receiver) and m_l is mode of emission (Lambertian mode). The relation between mode of emission and semi-angle at half power is discussed in [14] given by

$$m_l = -\ln 2/\ln(\cos\phi_l/2) \tag{2}$$

Assuming the power emitted by a single LED panel is P_{LED}, the overall transmitted power by the panel is

$$p_t = P_{LED} \times R_0(\phi) \tag{3}$$

The channel transfer function of a line of sight signal in VLC is described in [14], which is given by

$$h_{LOS} = \frac{A_{rx}}{d^2} R_0(\phi) \times \cos(\psi); 0 \le \psi \le \psi_c \tag{4}$$

The ψ is incidence angle (angle between LED panel and axis normal to the receiver). The ψ_c is the field of view of the receiver. The total transmitted power by the LED panels in line of sight communication is

$$P_{LOS} = \sum_{k=1}^{K} P_t \times h_{LOS} \tag{5}$$

The receiver has a photodiode with an optical filter and a concentrator. An optical concentrator is used to increase the area of the photodiode (larger area of the photodiode is desirable to compensate attenuation). The gain of optical concentrator is

$$g(\psi) = \begin{cases} \left(\frac{n}{\sin(\psi_c)}\right)^2 & if \quad 0 \le \psi \le \psi_c \\ 0 & if \quad \psi > \psi_c \end{cases} \tag{6}$$

The optical power obtained at the receiver is given by

$$P_{Rx} = P_{Los} \times T(\psi) \times g(\psi) \tag{7}$$

where $T(\psi)$ is the optical filter's transmission coefficient. The photodiode converts the received optical power into electric current. The current at the output is $i = P_{Rx} \times r$ where r is the responsivity of photodiode.

The SNR at the receiver is defined as

$$SNR = \frac{i^2}{\sigma^2_{Total}} \tag{8}$$

where σ^2_{Total} is the variance of shot noise, described in [14] is given by

$$\sigma^2_{Total} = \sigma^2_{shot} + \sigma^2_{amp} \tag{9}$$

The degree of uniformity (U_D) represents amount of variations in received optical power on the floor. It is defined by the ratio of maximum optical power to the minimum optical power at the receiver plane [5, 15].

$$U_D = \frac{p_{\max}}{p_{\min}} \tag{10}$$

The illuminance expresses the brightness of an illuminated surface. The luminous intensity in angle ϕ is given by

$$I(\phi) = I(0)\cos^m(\phi) \tag{11}$$

The $I(0)$ is center luminous intensity of an LED panel. A horizontal illuminance E_{hor} at a point (x, y) is given by

$$E_{hor} = \frac{I(\phi)}{D_d^2} \cos(\psi) \quad (12)$$

where D_d is the distance between a LED panel and the detector surface.

3 Solution Using Whale Optimization

3.1 Fitness Function of the Problem

The aim is to determine the (x, y) coordinate of the white LED panels which minimize the degree of uniformity defined in (10). Thus, the fitness function in this problem expressed in dB is given by

$$U(dB) = 10 \log_{10} \frac{P_{\max}}{P_{\min}} \quad (13)$$

where $P_{\max}$ and $P_{\min}$ are the maximum and minimum values of received optical power. The $U(dB)$ is minimized using WOA and PSO algorithm.

3.2 Whale Optimization Algorithm

The WOA is a metaheuristic algorithm developed by Mirjalili and Lewis [10] by observing the natural behavior of Humpback whale. The algorithm has both exploitation and exploration operations to effectively determine the optimal solution in a search space. The exploitation job is carried out by whale's bubble-net attacking method. Two types of methods are modeled for the attacking one: shrinking encircling method and spiral updating position method. The exploration is done using the whale search for pray mechanism. In this mechanism, the Humpback whales analyze location of other whales and search for pray randomly. The position of search agent is updated by randomly chosen best result. The exploration and exploitation processes are balanced based on a coefficient vector given by

$$\vec{A} = 2\vec{a}.\vec{r} - \vec{a}$$

where $\vec{a}$ decreases linearly in range [2, 0]. The $\vec{r}$ is a random vector in range [0, 1]. The best solutions are chosen if $\left|\vec{A}\right| < 1$, while random search is selected if $\left|\vec{A}\right| > 1$.

The pseudocode of the WOA algorithm for obtaining the minimal value of $U(dB)$ is described as follows:

Pseudocode of WOA Algorithm:

```
Initialize the whale population X_i ( i=1,2,......n)
Fitness of each searching whale
For each i: 1≤i≤n
Calculate the corresponding fitness function U_i
End for
/* obtain the current best searching whale X* */
While (t < Max_itr)
    For each searching whale
            /* Update a,A,c,l and p */
        If (p< 0.5) then
            If (|A|<1) then
                X(t + 1) = X*(t) − A.D
            Else if (|A|>1) then
                Select a whale (X_rand) randomly
                X(t + 1) = X_rand − A.D
            End if
        Else if (p>0.5) then
            X(t + 1) = D'.e^bl.cos(2πl) + X*(t)
        End if
    End for
        Check the range of each searching whale and amend it
        Calculate the fitness of each searching whale U_i
    /* Update the best searching whale X* */
    Set t= t+1
    Return X*
End while
```

4 Simulations and Discussion of Results

The simulation of the proposed indoor VLC model is carried out in MATLAB 2013a environment. The program is allowed to run in an Intel i3 laptop with 4 GB ram and 1.90 GHz clock frequency with windows 10, 64-bit operating system. The indoor room size is 5 × 5 × 3 m. Numbers of transmitter white LED panels are varied from 4 to 6. Each white LED panel comprises of a 60 × 60 array. The irradiance angle ϕ at transmitter and incidence angle ψ at receiver both are taken as 70°. The center luminous intensity $I(0)$ is taken as 0.73 cd. The transmission coefficient of the LED panel $T(\psi)$ is taken as 1.

In the optimization task using WOA algorithm the initial population of whales is taken as 20, number of generation for run is taken as 100. The algorithm is compared with particle swarm optimization (PSO) [16] with initial swarm size 20 and number of generation 100. In PSO, the c1 and c2 values are taken as 2.05, w value is 0.4.

The degree of uniformity achieved in room with optimal locations of LED panels determined by WOA and PSO is presented in Table 1, and the SNR values

Table 1 Degree of uniformity achieved at room with optimal locations of LED panels determined by WOA and PSO

Number of LED panels	Optimal LED Panel locations with WOA	Degree of uniformity (dB)	
		WOA	PSO
4	(1.1184,0.3032,2.15) (4.7447,0.8283,2.15) (0.4211,3.9302,2.15) (3.9927,4.5277,2.15)	**1.9602**	2.1122
5	(4.0226,0.6150,2.15) (3.0465,2.7494,2.15) (4.6939,4.6397,2.15) (0.2834,4.4316,2.15) (0.0574,0.4762,2.15)	**2.3504**	2.4089
6	(1.2183,2.4798,2.15) (0.4265,4.7003,2.15) (0.1910,0.2871,2.15) (4.2413,0.3531,2.15) (4.0169,4.47693,2.15) (4.1693,2.6143,2.15)	**2.1020**	2.4778

Table 2 SNR values achieved at room before and with optimal locations of LED panels determined by WOA and PSO

Number of LED panels	SNR (in dB) before optimization LED panel location as in [7]	SNR (in dB) after LED panel location at optimal positions using metaheuristics	
		WOA	PSO
4	75.3251	**111.50**	78.9438
5	78.4195	**111.89**	82.0843
6	80.7360	**114.53**	84.3833

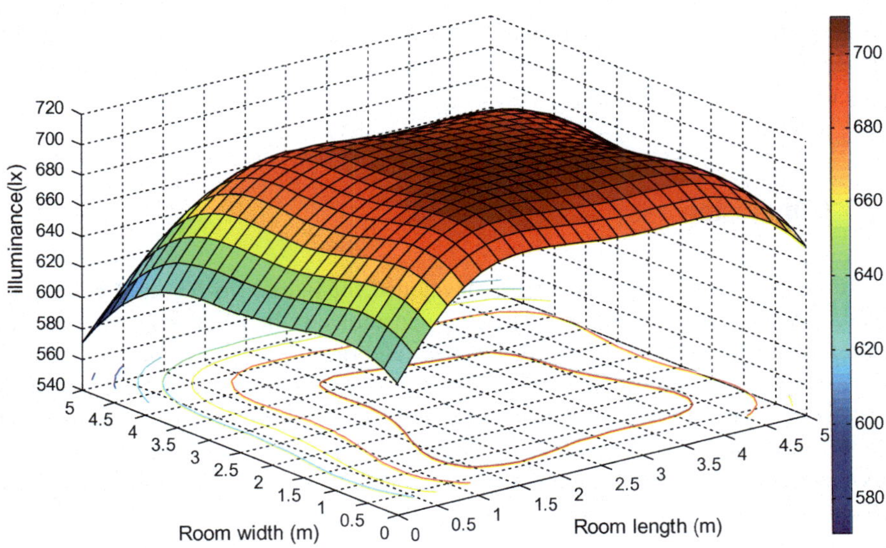

Fig. 2 Distribution of illuminance in the room obtained with four LED panels placed at optimal locations using WOA

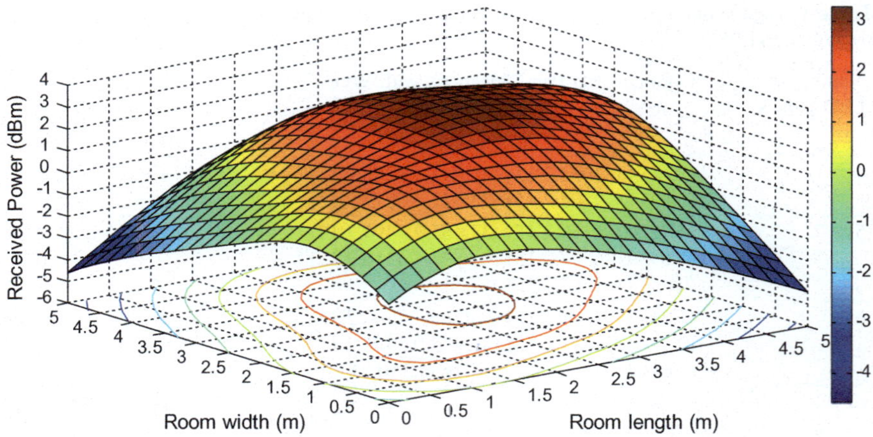

Fig. 3 Distribution of received optical power in the room obtained with four LED panels placed at optimal locations using WOA

achieved are presented in Table 2. The best results achieved are highlighted in bold letters. The ideal uniformity value for this system should be 1 dB. From Table 1, it is observed that by increasing the number of LED panels does not contribute to achieve uniform illuminance in the room; however, from Table 2, it is helpful in improving the SNR.

The distribution of illuminance and received optical power in the room obtained for four LED panels placed at optimal locations using WOA is shown in Figs. 2 and 3, respectively. As per ISO standardization, illuminance of this system [7] should vary between 300 and 1500 (lx) for official work. From Fig. 2, the average illuminance of the proposed simulated system is 650 (Lx) which is within the range of the benchmark.

5 Conclusion

In this paper, a simulation model is developed for indoor visible light communication using white LED panels. The (x, y) positions of the LED panels were set in the roof of the room using WOA and PSO algorithms to achieve uniform illuminance and enhancing the data download speed (by improving the SNR). Results reveal that the performance achieved by the WOA algorithm is superior than that achieved with PSO. The obtained results with WOA are also better than those achieved without using optimization, i.e., with a predefined position used by Komine and Nakagawa in 2004.

Acknowledgements This research work is supported by Institute Fellowship to I.R. Kumawat by Ministry of HRD, Govt. of India to pursue his Master in Technology in Electronics and Communication Engineering from MNIT Jaipur.

References

1. Grobe L, Paraskevopoulos A, Hilt J, Schulz D, Lassak F, Hartlieb F, Kottke C, Jungnickel V, Langer KD (2013) High-speed visible light communication systems. IEEE Commun Mag 51 (12):60–66
2. Komine T (2001) Basic study on visible-light communication using light emitting diode illumination. In: Proceedings of 8th International Symposium on Microwave and Optical Technology, Montreal, Canada, pp 45–48
3. Komine T, Nakagawa M (2003) Integrated system of white LED visible-light communication and power-line communication. IEEE Trans Consum Electron 49(1):71–79
4. Tanaka Y, Komine T, Haruyama S, Nakagawa M (2003) Indoor visible light data transmission system utilizing white LED lights. IEICE Trans Commun 86(8):2440–2454
5. Bakalidis GN, Glavas E, Tsalides P (1996) Optical power distribution in wireless infrared LANs. IEEE Proc Commun 143(2):93
6. Yang H, Lu C (2000) Infrared wireless LAN using multiple optical sources. IEEE Proc Optoelectron, 147(4), 301–307
7. Komine T, Nakagawa M (2004) Fundamental analysis for visible-light communication system using LED lights. IEEE Trans Consum Electron 50(1):100–107
8. Nguyen HQ, Choi JH, Kang M, Ghassemlooy Z, Kim DH, Lim SK, Kang TG, Lee CG (2010) A MATLAB-based simulation program for indoor visible light communication system. In: Proceedings of IEEE 7th International Symposium on Communication Systems Networks and Digital Signal Processing (CSNDSP), pp 537–541
9. Wang L, Wang C, Chi X, Zhao L, Dong X (2017) Optimizing SNR for indoor visible light communication via selecting communicating LEDs. Optics Commun 387:174–181
10. Mirjalili S, Lewis A (2016) The whale optimization algorithm. Adv Eng Softw 95:51–67
11. Xu Y, Yang W, Wang J (2017) Air quality early-warning system for cities in China. Atmos Environ 148:239–257
12. Kaveh A, Ghazaan MI (2016) Enhanced whale optimization algorithm for sizing optimization of skeletal structures. Mech Based Des Struct Mach, 1–18
13. Horng MF, Dao TK, Shieh CS, Nguyen TT (2017) A Multi-objective optimal vehicle fuel consumption based on whale optimization algorithm. In: Advances in intelligent information hiding and multimedia signal processing: proceedings of the twelfth international conference on intelligent information hiding and multimedia signal processing, Kaohsiung, Taiwan, 371–380
14. Kahn JM, Barry JR (1997) Wireless infrared communications. Proc IEEE 85(2):265–298
15. Alexander SB (1997) Optical communication receiver design. Bellingham, Washington, USA: SPIE Optical Engineering Press, 66
16. Clerc M, Kennedy J (2002) The particle swarm-explosion, stability, and convergence in a multidimensional complex space. IEEE Trans Evol Comput 6(1):58–73

A Novel Method for Distortion Suppression in Radio over Fiber Communication Systems Using FBG

Anu Wilson, R. Pradeep and N. Vijayakumar

Abstract In this paper, a linearization method that reduces nonlinear distortion and increases receiver sensitivity of radio over fiber (RoF) links is discussed. The design also addresses the RoF system without significantly adding great expense and complexity to the system. The proposed method is capable of suppressing both second- and third-order distortions despite the RF carrier that is used. Additionally, it is shown to surpass the traditional RoF systems in terms of BER and receiver sensitivity.

Keywords RoF · Nonlinear distortion · Linearization · EAM
FBG

1 Introduction

There has been a never-ending need for high-capacity transmission systems, and radio over fiber (RoF) is a technology that disseminates RF analog or digital baseband signals through the wide bandwidth of an optical fiber so as to procure high data rates [1]. The general RoF architecture is shown in Fig. 1 in which each base station is connected back to a corresponding control station through fiber optic cables. The wireless signals are received by using a transceiver, and the RF signal will be modulated onto an optical carrier allotted to that particular base station.

A. Wilson (✉) · R. Pradeep
Fiber Optics Laboratory, Department of ECE, College of Engineering Trivandrum, Thiruvananthapuram, India
e-mail: anuwilson117@gmail.com

R. Pradeep
e-mail: pradeep@cet.ac.in

N. Vijayakumar
GEC Barton Hill, Thiruvananthapuram, Kerala, India
e-mail: dr.nvkr@gmail.com

V. Janyani et al. (eds.), *Optical and Wireless Technologies*, Lecture Notes in Electrical Engineering 472, https://doi.org/10.1007/978-981-10-7395-3_15

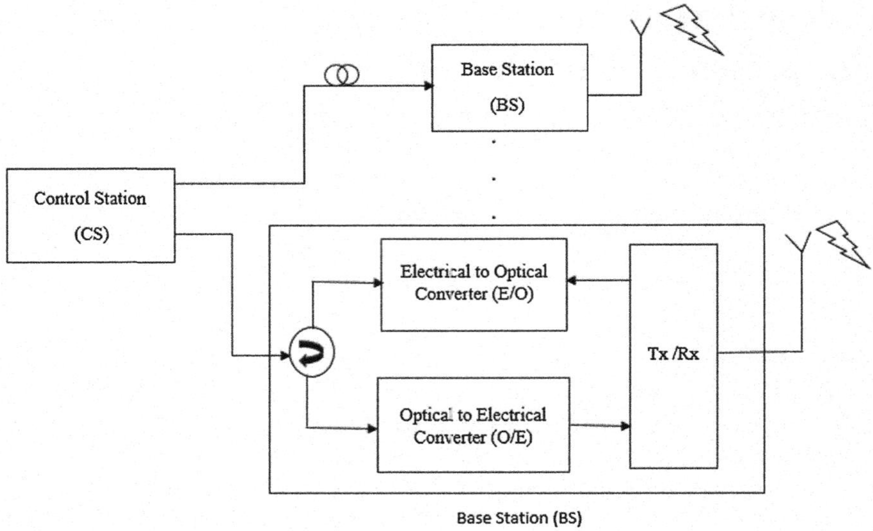

Fig. 1 Basic block diagram of RoF

Many a number of optical carriers can be modulated in this way, and then they are optically coupled together before transmitting back to the central station.

RF signals are modulated onto light in a process known as optical subcarrier modulation or optical analog modulation. The RF signal, which is made up of m-RF carriers, is fed into an electro-optical modulator (EOM), which in turn modulates a CW laser set to the frequency of F. It is then picked up by a photodetector, and the obtained electrical signal is simultaneously fed to different band-pass filters (BPF). The initial RF carrier frequency is made the center frequency of the band-pass filters present so that the authentic information can be recovered. For this purpose, homodyne or heterodyne detection can also be employed, which uses phase-locked loop (PLL) and low-pass filter (LPF) to recover the original data transmitted. DWDM can be done using this concept where many sets of complex RF signals may be modulated with separate wavelengths and EOMs [2]. The major advantages of RoF communication systems include depressed attenuation loss, generous bandwidth, and freedom from radio-frequency interference.

Since RoF involves analog modulation and detection of light, it can be considered as an analog transmission system. Therefore, signal impairments, such as distortion which are major drawbacks in analog communication systems, are important in RoF systems as well. These impairments limit the dynamic range (DR) of the RoF links, which will affect the system performance. DR is a very important parameter for cellular communication systems since it determines the system capacity. The power received from a mobile unit, which is close to the base station can be much higher than the power received from a mobile unit, which is several kilometers away, within the same cell and therefore, the power received varies widely.

2 Nonlinearity Issues

Nonlinear distortion (system nonlinearity) is engendered by the nonlinear characteristics of the optical modulator as well as the frequency response of the optical transmitter, receiver, and the other optical components present, and it deteriorates the sensitivity of the receiver leading to a poor bit error rate, which in turn leads to an increase in the RoF system cost. The nonlinear transfer function of the optical modulator is found to be a major limiting factor for analog optical communications.

Even though the nonlinear transfer function of the modulator was not a great issue when used in digital optical communications, it has become a limiting factor for analog optical communications since in analog communication the objective is the successful recovery of continuous waveforms, which can now be severely degraded by nonlinear distortion [3]. Here, the harmonics generated by the electro-absorption modulator alone is considered. The electro-absorption modulator exhibits second- and third-order nonlinearities (based on the device's DC biasing). Its output power equation is as follows:

$$P_{out} = P_{in} e^{-f(v(t))} \tag{1}$$

where P_{in} is the average laser input power, $f()$ is the transfer characteristics of the modulator, and v(t) is the DC bias. Using Taylor series expansion for the exponentials, we get the output power as

$$P_{out}(v(t)) = P_{in}\left[1 - f(v(t)) + \frac{f^2(v(t))}{2!} - \frac{f^3(v(t))}{3!} + \ldots\right] \tag{2}$$

It can be seen from the above equation that both second- and third-order nonlinearities are introduced by using EAM (Fig. 2).

The connection between the output power of an EAM and its applied reverse voltage is depicted in Fig. 3. The voltage required to switch the modulator between the on and off states is between 1.5 and 4 V. And also, the dynamic reduction in output power comes in the range of 11 to 13 dB. Due to the change in refractive index on the application of an electric field, nonlinearities are introduced.

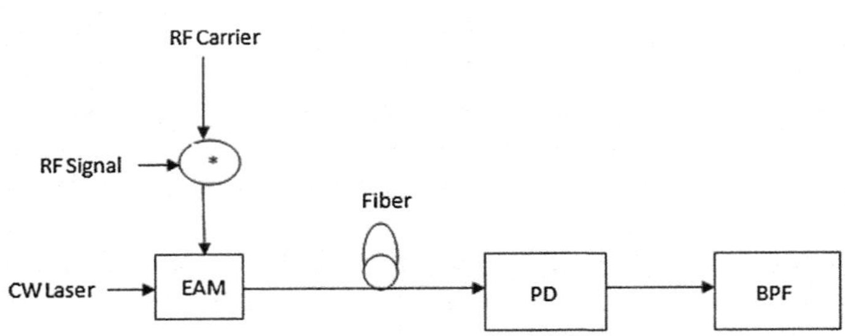

Fig. 2 EAM in RoF communication systems

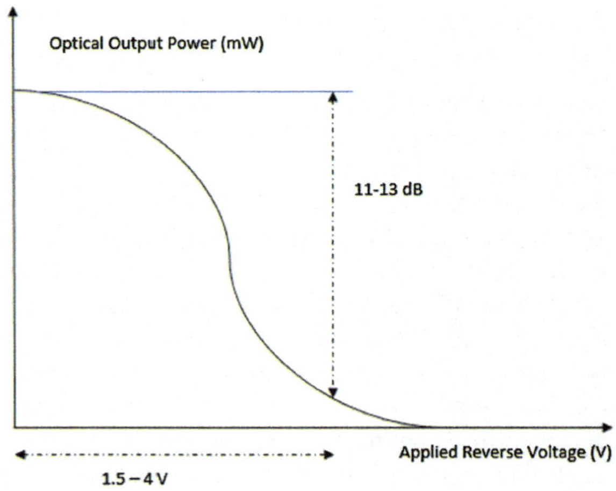

Fig. 3 Switching characteristics for a nonlinear EAM

3 Proposed System

Previously a single wavelength balanced system had been developed to suppress the nonlinearities existing in the RoF communication systems, but it suppresses only the second-order distortions [4]. Another method employing asymmetric Mach-Zehnder modulator (MZM) has been shown to increase third-order intermodulation distortion.

Here, an FBG-based balanced RoF system, which makes use of an EAM and two cascaded nonlinearly chirped fiber Bragg gratings (FBGs), is proposed. FBGs (fiber Bragg gratings) have the privilege that it has dispersion in reflection and transmission as well as being linear. Distinct optical path lengths are introduced by reflecting each individual wavelength in a dispersed pulse at different positions in the grating. Therefore, it provides the possibility for neutralizing dispersion in long-haul fiber links, but it demands the use of a 3 dB coupler or an optical circulator. It will be shown later that the proposed system will perform as well in reducing the nonlinear distortions as the single wavelength balanced system and will be cost effective due to the presence of a single EAM. It uses only one wavelength, optical modulator, and fiber. This balanced system separates the sidebands (both USB and LSB) generated by modulation together with the optical carrier before forwarding to the photodetector. The time delay is introduced to one of the sidebands by means of two cascaded nonlinearly chirped FBGs.

The initial FBG is given a sizeable bandwidth as well as group velocity dispersion so as to insert a comparable dead time for the optical subcarrier and the second-order distortion components, while the second FBG will have a bandwidth and GVD to insert a disparate dead time for third-order distortion components.

The net effect of the time delays is to provide a phase shift of x for the subcarrier and second-order distortion components and a phase shift of y for the third-order distortion components [5]. Figure 4 schematically shows the proposed tunable FBG-based balanced system. A laser source at 1553 nm (1931.17 THz) is used, and

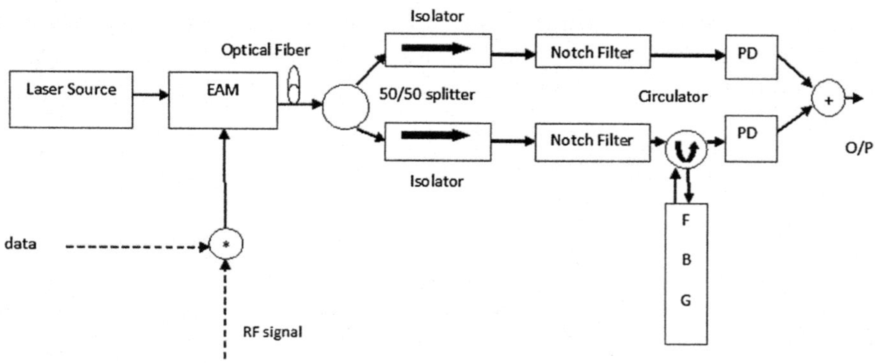

Fig. 4 Schematic of the proposed system

rays are injected into an EAM having nonlinear response characteristics. Using subcarrier modulation (SCM), the main optical carrier and DSB subcarrier signals are produced and are relayed through a single mode fiber of 10 to 50 km length, and having a chromatic dispersion parameter of 15 ps (nm-krn).

After transmission, the two subcarriers are filtered out together with half the optical carrier power by means of a 50:50 optical splitter. The two notch filters used are implemented by using FBGs and is used to remove the unwanted optical subcarrier components, i.e., USB (U) along the upper branch and LSB (L) along the lower branch. The carrier and U subcarrier are delivered to a photodiode of the receiver along the upper fork.

In the other branch, a time delay line realized with the help of a circulator and a nonlinearly chirped FBG is inserted, before being fed to the other photodiode of the optical receiver. Three polarization controllers are required to minimize any polarization mismatch at the output of the photodetector [10]. The current obtained as the photodetector output is given by the expression,

$$I_d = \left[I_1^U + I_2^U + I_3^U\right] - \left[I_1^L \cos\theta_1 + I_2^L \cos^2\theta_2 + I_3^L \cos^3\theta_3\right] \tag{3}$$

In which the angles $\theta_{1,}$ $\theta_{2,}$ and θ_3 are the phase difference of the currents between the forks. I_1^U and I_1^L are the carrier currents in the upper and lower branches, I_2^U, I_2^L are the currents due to second-order distortions and I_3^U, I_3^L are the currents due to third-order distortion, respectively. If θ_1 and θ_2 are equal to π and θ_3 equal to 2π, then the detected current becomes:

$$I_d = \left[I_1^U + I_2^U + I_3^U\right] - \left[I_1^L \cos\pi + I_2^L \cos^2\pi + I_3^L \cos^3 2\pi\right] \tag{4}$$

$$I_d = \left[I_1^U + I_2^U + I_3^U\right] - \left[-I_1^L + I_2^L + I_3^L\right] \tag{5}$$

$$I_d = I_1^U + I_1^L \tag{6}$$

That is the signal carrier currents add up, and the currents from the distortion components get canceled. The detected current at the output will be the sum of the carrier currents in the two branches.

4 Simulated Results

The circuit was simulated using OptiSystem (version 14) software. Its key features include component library, mixed-signal representation, advance visualization tools, hierarchical simulation with subsystem and multiple layouts to name a few. Here, the system performance before and after distortion suppression is analyzed, and better performance is obtained, which can be seen from the eye diagram and BER plots after applying the above-mentioned technique.

The optical spectrum at the fiber end after transmission is taken and shown in Fig. 5a. It shows distortions on either side of the spectrum. Figure 5b shows the compensated spectrum showing dips in power levels at the distortion regions. Hence, the distortions are suppressed effectively. The bottom graph represents noise (shown in green) (Fig. 6).

Betterment in eye diagram is observed, and the bit error rate obtained by using BER analyzer of OptiSystem shows a BER of the order of 10^{-9} for the uncompensated case and 10^{-12} after compensation. Hence, the BER also improves after employing the distortion compensation technique. Overall, there is an improvement in the system performance (Fig. 7).

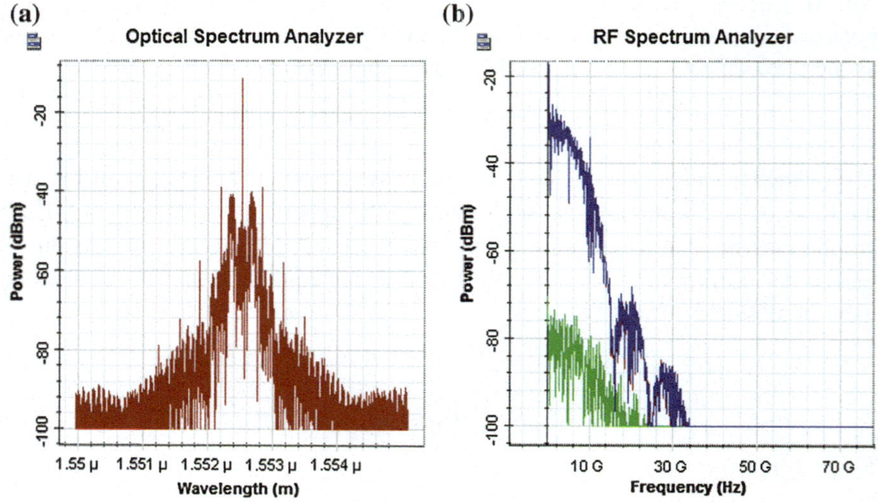

Fig. 5 **a** Optical spectrum at fiber link end after transmission showing second- and third-order nonlinearities and **b** RF spectrum at the photodetector end after distortion suppression showing dips (shown in blue) in power at the points were nonlinearities previously observed

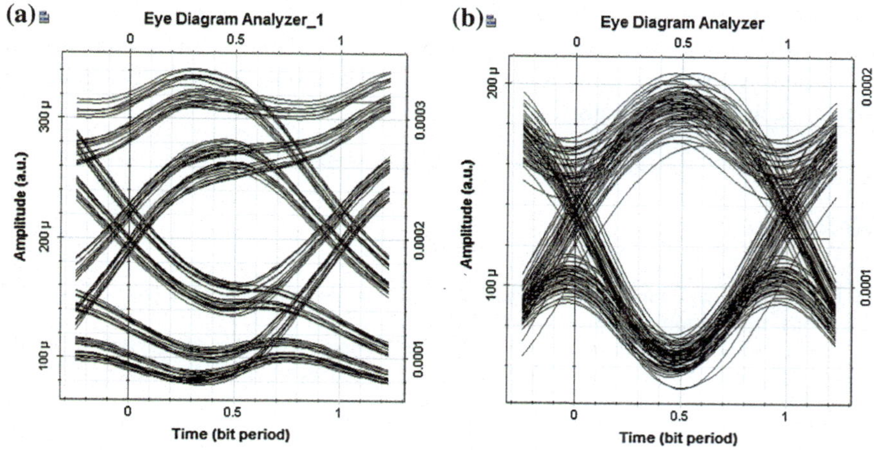

Fig. 6 **a** Distorted eye diagram before compensation and **b** Eye diagram after distortion compensation

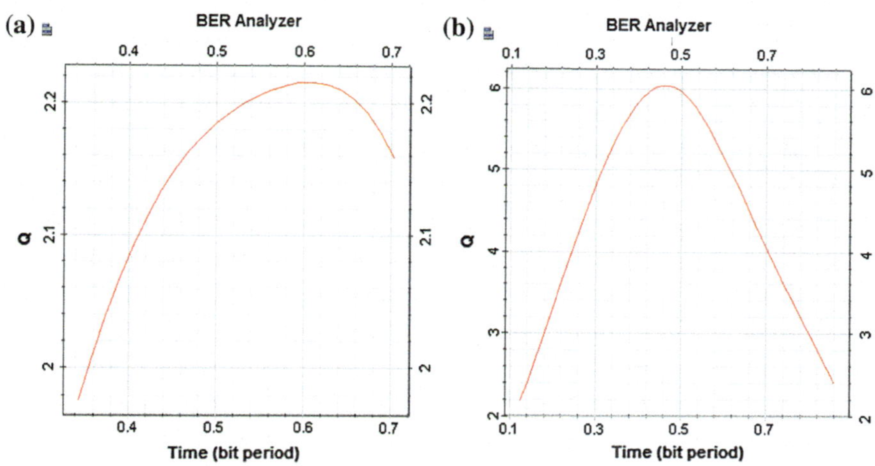

Fig. 7 **a** Bit error rate before distortion compensation and **b** Bit error rate after compensation

5 Conclusions

In this paper, an FBG-based balanced system has been presented, which has the ability to subdue the effect of second- and third-order nonlinear distortions. There are minimum restrictions on the modulator bias that can be used. This allows the selection of a bias, which will optimize the optical RF signal subcarrier power. Altogether, the system is economical because of the use of only a single optical modulator. As demonstrated using simulation, the suggested system provides a better eye diagram and BER. The suppressed nonlinear distortions allow improved receiver sensitivity and reduced multi-subcarrier interference. But regrettably, the suggested system uses DSB transmission, which is not as spectrally efficient as single sideband (SSB) transmission.

References

1. Pradeep R, Vijayakumar N (2015) A novel single wavelength balanced technique to suppress second order distortion in RoF system. ICCC, Trivandrum, pp 429–432
2. Pradeep R, Seena R, Vijayakumar N (2014) A full duplex radio over fiber system using fiber bragg grating filter. In: ICCSC first international conference, IEEE conference publication, pp 139–143
3. Mochii T, Shiva A, Hai-Han L, Chang-Jen W, Ting-Chieh L (2015) A bidirectional wireless-over-fiber transport system. J Light Technol 7(6)
4. Jung HD, Lee GW, Han SK (2001) Nonlinearity suppression of electro absorption modulator through dual-parallel modulation. Microwave Technol Lett 29:2–5
5. Agarwal GP (2002) Fiber-optic communication systems. Wiley, New Jersey
6. Johnson LM, Roussell HV (1988) Reduction of intermodulation distortion in interferometric optical modulators. Optical Lett 13:928–930
7. Hui R, Zhu B, Huang R, Allen Demarest CA, Richards D (March 2002) Subcarrier multiplexing for high speed optical transmission. J Light Technol 20(3)
8. Zin AM, Bongsu MS, Idrus SM, Zulkifli N (2010) An overview of radio over fiber technology. IEEE
9. Fernando XN (2014) Radio over fiber for wireless communications from fundamentals to advanced topics. Wiley

Digitized RoF Technique for Multiservice Fiber-Radio Systems

I. V. Reshma, R. Pradeep and N. Vijayakumar

Abstract An efficient and systematic hybrid fiber-radio scheme is demonstrated. The system is mainly based on digitized RF over fiber technique. The systems use a single analog-to-digital converter (ADC) and digital-to-analog converter (DAC) to frequency translation instead of mixers and local oscillators. To improve the transmission of analog radio over fiber (RoF) links, a digitized radio over fiber (DRoF) scheme is used. For improving the performance of the overall system, electronic ADC is replaced by photonic ADC. The system then shows the improvement in terms of BER and bandwidth performance.

Keywords Radio over fiber · Hybrid fiber-radio · Photonic ADC

1 Introduction

In recent days, there is greater bandwidth requirement from subscribers who are continuously using large data for Facebook, WhatsApp, etc. in their day-to-day life. This has led to the integration of optical network and wireless broadband access. The bandwidth capacity of optical fiber is combined with the mobility and ubiquity of wireless networks, thus creates a powerful system. Radio over fiber (RoF) is an emerging technology for access network. Two main classification of RoF is analog RoF and digital RoF. Analog RoF suffers from intermodulation distortions (IMD),

---

I. V. Reshma (✉) · R. Pradeep
Department of Electronics and Communication, College of Engineering Trivandrum, Thiruvananthapuram, Kerala, India
e-mail: reshmavasudevan10@gmail.com

R. Pradeep
e-mail: pradeep@cet.ac.in

N. Vijayakumar
GEC Barton Hill, Thiruvananthapuram, Kerala, India

V. Janyani et al. (eds.), *Optical and Wireless Technologies*, Lecture Notes in Electrical Engineering 472, https://doi.org/10.1007/978-981-10-7395-3_16

and the dynamic range of analog links decreases linearly with the increasing length of the optical fiber while a digitized RoF transmission can maintain its dynamic range independent from the fibers transmission length [1]. As a result, the DRoF can be based on low-cost digital hardware with a high dynamic range capable of supporting long distances.

The digitized RF interfaces could be easily integrated with existing and future broadband optical and high-speed networks. Therefore, the DRoF technique is potentially a promising solution for the future low-cost high-performance hybrid fiber-radio implementation. The performance of the system can be further improved by using photonic ADC instead of electronic converters.

A downlink scheme using digitized RF over fiber technique without the use of physical mixers and local oscillators is demonstrated [2]. Here, they design a simple base station by using single ADC and DAC. The DRoF system can be designed by using the technique envelope delta sigma modulation in which the tolerance to jitter introduced by optical components increases [3]. An all-digital transmitter can be used to implement a flexible centralized digital radio and also able to perform the baseband processing and transmission of radio signals at radio frequencies (RF) over an optical fiber. The digital signal transmission is based on the simple on–off digital modulation, reducing the limitations of conventional analog radio over fiber systems [1].

In the proposed system, digital radio over fiber system is designed by replacing the electronic ADC by photonic ADC. Photonic conversion is done by optically sampling and electronically quantizing technique.

The remaining of this paper is organized as follows: Sect. 2 discusses radio over fiber system architecture. Section 3 deals with the basic single service DRoF system and its working. Section 4 discusses the photonic analog-to-digital converter and the DRoF system including PADC is introduced. Section 5 introduces the DRoF system with photonic ADC. Section 6 provides the experimental simulations and network performance evaluation and discussions leading to a precise conclusion presented in Sect. 7.

2 Radio Over Fiber System

Radio over fiber (RoF) is an emerging technology where light from a laser source is modulated by a high-frequency RF signal. The modulated signal is then transmitted over an optical fiber link to facilitate wireless access. In other words, radio signals are carried over fiber-optic cable. In traditional way, each protocol needs different equipment at every location of the antenna. Where in the case of RoF, a single antenna can process all the signals and transmitted through a single fiber to the location. In that central station, the equipment converts the signals. The main application of RoF includes television networks and satellite base stations.

RoF is used for multiple purposes, such as in cable television (CATV) networks and in satellite base stations. The term RoF is usually applied when this is done for wireless access.

Radio over fiber is actually an analog optical link which transmits modulated RF signals. It transmits RF signal to and from central station (CS) and to base station (BS). The principal requirements of ROF link architecture are duplex operation (downlink–uplink), reasonable length, and high-performance optical components. Enhanced cellular coverage, lower attenuation losses, higher capacity, larger bandwidth, and immunity to radio frequency interference are the major advantages of RoF system. The basic configuration of a ROF link is shown in Fig. 1, where the system consists of a central station and remote access unit (RAU) connected by a single mode fiber.

In RoF systems, wireless signals are converted to optical signals before being radiated through air. The Wi-Fi, 5G equipment are centralized in one place, with remote antennas attached via fiber optic serving all protocols: This is the main advantage of RoF systems. It will decrease the equipment and maintenance cost of the network. RoF technology enables convergence of fixed and mobile networks. In RF over fiber architecture, a data-carrying RF (radio frequency) signal with a high frequency is imposed on a light wave signal before being transported over the optical link. Therefore, wireless signals are transmitted to base station in the optical domain itself, and then, it is converted to electrical domain before being radiated by antenna. As a result, no frequency up–down conversion is required at the various base stations. Thus, resulting in simple and rather cost-effective implementation is enabled at the base stations.

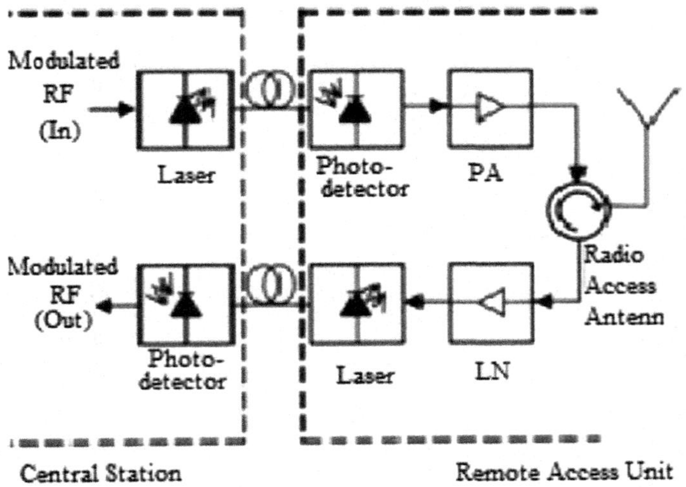

Fig. 1 Basic RoF system

3 Basic DRoF Single Service Systems

While analog RoF link suffers from some impairment, the digitized RoF appears as an alternative to the next generation of RoF systems. The digitization of an RF signal consists in the production of a sampled digital signal in a serial form that can be modulated on an optical carrier. Modulation of the digital signal onto an optical carrier minimizes the nonlinear effects originated from the optical-to-electrical conversion function presented in analog RoF [3].

In order to not use considerably high-sample rates at the ADC/DAC components, such as the bandpass sampling technique, are generally applied to the RF signal. All the functions and protocols can still be centralized at the central office, and the digital signal can be recovered by direct detection at the RAUs. The DRoF inherits benefits from analog RoF and is capable of maintaining its dynamic range independent from the fibers transmission length until the received signal goes below the sensitivity of the link. Other improvement presented in DRoF systems is that it utilizes less transmission power to achieve the same performance of the analog RoF systems. Figure 2 shows the macroview of the DRoF system. The major differences to the RoF system are the ADC and DAC components. All processes are performed through the following steps: sampling, normalization, quantization, and codification. The sampling rate to digitalize the analog signal is selected applying Nyquist criterion wherein the sampling frequency must be at least twice the signal bandwidth of the highest bandwidth. After that, the signal is normalized to ADC dynamic range to prevent that signal samples are recovered out of limit in Fig 2. In the quantization, the continuous signal is discretized according to levels defined by ADC resolution. At the end of this process, the signal has a new representation in discrete-time and amplitude domain. The signal is converted to a binary sequence and then encoded using a non-return-to-zero (NRZ) encoder. After that, the bit

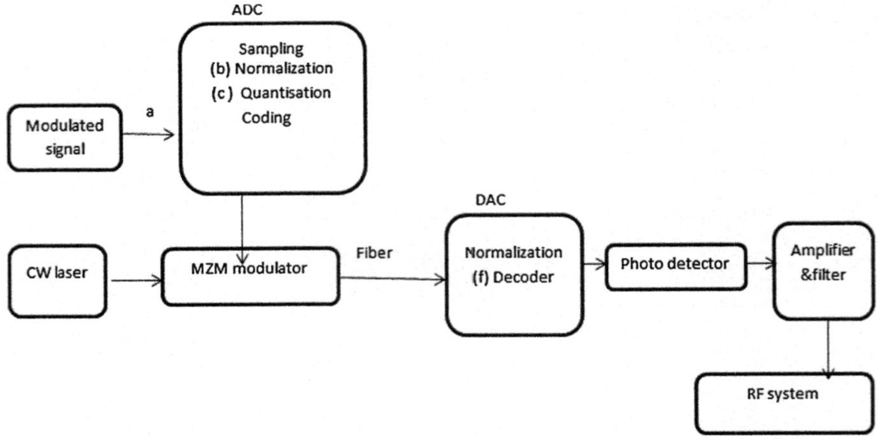

Fig. 2 Basic single service DRoF system

stream is modulated using MZM. Each sample is converted into a bit stream sequence according to ADC resolution. Therefore, the overall data rate on an optical link is the product of bit rate and resolution (data rate * resolution). Higher data rates impose limits to the performance of the link and increase system costs. The trade-off between optical link versus ADC resolution has a significant challenge related to practical DRoF implementations and should be carefully considered since the cost-effectiveness may be expensive compared to RoF systems. In the base station, the digital bit stream is detected using a PIN and processed to get the parallel digital data. The DAC is employed to convert this parallel bit stream to an analog signal. This process is shown in the steps f and g of the Fig. 2, and it is performed through normalization and decoder. Similar to the process in ADC, the signal is first normalized for dynamic range to avoid the samples to be recovered out of the limit. Finally, in the last stage, the samples are grouped in parallel bit sequences. In the frequency domain, this analog signal has spectral replicas of the original carrier frequency (Fig 3).

The performance of DRoF mainly depends on the analog-to-digital converter and digital-to-analog converter. Analog-to-digital conversion can be done in electrical domain or optical domain. Electronic ADC has less bandwidth. This will reduce the speed and thus reduces the overall system performance. This is the main drawback of the existing systems. Since the optical signal has the bandwidth in Terahertz range the ADC used for signal conversion should be a high-speed device. The proposed system includes the high-speed ADC such as photonic ADC, thereby improving the performance of the overall system. Analog-to-digital conversion architecture employing photonic technique offers a great deal of versatility with including very broad bandwidth, precision timing, and low loss signal transport over optical fiber.

Various disadvantages of analog radio over fiber link are nonlinearity and chromatic dispersion. Disadvantages of analog optical links are inherently suffer from intermodulation distortions arising from the nonlinearity of both microwave

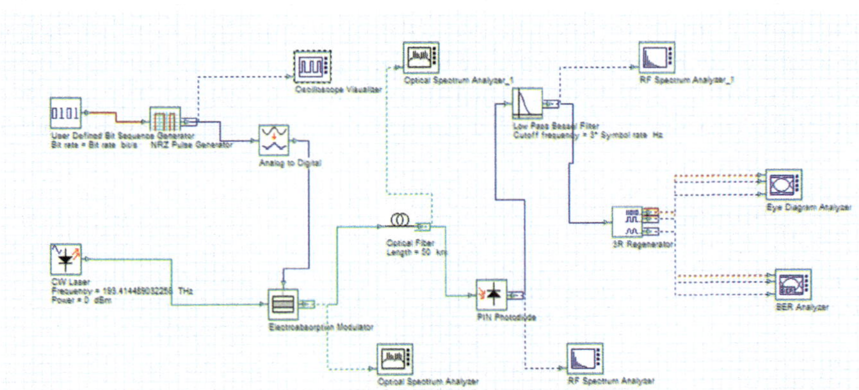

Fig. 3 Simulation setup

and optical components that make up the optical link. The dynamic range of an analog optical link decreases linearity with the increasing length of the optical fiber link due to the attenuation in the optical fiber. Compare with the analog radio over fiber, the digital RoF has the advantage that the architecture of RoF network is achieved, and it can maintain its dynamic range independent of the fiber distance. Digital RF over fiber can be based on low-cost digital transmitters and receivers. It has dynamic range which can be sustained over long distances in comparison with that of analog optical links.

4 Photonic Microwave-to-Digital Converter

Many avionic systems are being driven to operate over multiple frequency bands and deliver accurate processed information with low latency in an increasingly cluttered electromagnetic environment. Digital signal processing (DSP) can provide the versatility necessary to achieve these results, however only after the accurate conversion of the analog signal to a useful digital form. Analog-to-digital conversion architectures employing photonic techniques over a great deal of versatility with numerous benefits including very broad bandwidths, precision timing, and low loss signal transport over optical fiber [1]. Combined with the physical advantages related to fiber transport including electromagnetic interference immunity, signal multiplexing, and low size and weight; interests in photonic analog-to-digital systems have been pursued with great promise.

Analog-to-digital conversion systems, utilizing photonic techniques, capable of the direct digitization of microwave frequency input signals, which we refer to as microwave-to-digital converters (MDCs). State-of-the-art electronic analog-to-digital converters with more than a few bits of effective resolution are generally limited to input bandwidths less than a few GHz. Oliveira et al. [4] the current method for handling input signals of higher frequency is shown in Fig. 4 where a microwave local oscillator and down conversion hardware convert a narrowband portion of the frequency spectrum around the microwave signal to an intermediate frequency (IF in Figure lea) within the bandwidth of an electronic analog-to-digital converter (ADC) [5]. Note we also include a data management and DSP subsystem, performing data interleaving, equalization, and other processing functions, typically

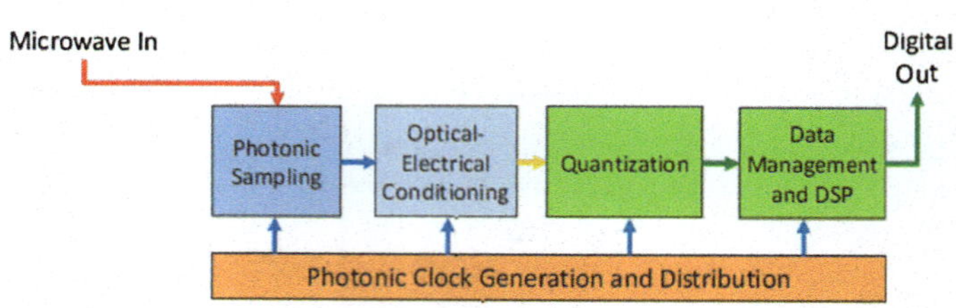

Fig. 4 Photonic microwave-to-digital converter

required for achieving useful digital output for application-specific processing. Application-specific digital output can take on many forms and, for our purposes, will be grouped using the required input and output bandwidths, which is not necessarily symmetric.

The conventional microwave down conversion architecture of Fig. 4 supports symmetric narrowband input–output waveform content for a single set of hardware. The vision for a photonic MDC is the direct sampling of broadband microwave input and conversion to useful digital output. Considerations for photonic implementations will be reviewed covering numerous techniques that take advantage of the precision timing and wide bandwidths of photonic sampling, as shown in Fig. 4. For symmetric input–output MDC bandwidth applications, sampling at twice the highest frequency input is required, and numerous architectures have been demonstrated utilizing time interleaving of many electronic ADCs coupled with photonic sampling. More recently, rate-scalable parallel architectures have been demonstrated using parametric synchronous sampling. For asymmetric input–output MDC bandwidth applications, the focus has been on using high-frequency photonic sampling to handle the direct sampling of microwave signals at rates much lower than twice the highest frequency [6]. Approaches in this category include uniform down sampling and compressive sampling. Recent demonstrations of photonics-based compressive sensing systems fall into two categories: one where the random sampling pattern is introduced in time, and the other in frequency using frequency-to-time mapping. This paper will review the architectures and recent demonstrations relating to photonic microwave-to-digital conversion and explore their promise and design issues.

5 DRoF System Using Photonic ADC

A DRoF system using photonic ADC is proposed and presented in Fig. 5. The architecture includes mach zehnder modulator as it works as an optical sampling and modulation device at the CS. A photonic ADC (PADC) by using mode-locked laser and MZM is used in the system. This system can scale the timing jitter of the laser sources to the femtosecond level [7, 8]. So the resolution bandwidth of the PADC system is far better than that of electronic ADC system. In Fig. 5, the light from the laser source is modulated by using RF signal from the photonic analog-to-digital converter [9]. The PADC is driven by the output of the mode-locked laser. Mode-locked laser transmits the narrow pulses which correspond to the input RF signal. The RF signal is first sampled by the MLLD and fed to photonic ADC; as done in the case of electronic quantization. The modulated optical signal is transmitted through an optical fiber and at the receiver side the optical digital signal is converted by using photonic DAC. The signal is then converted to the analog domain by using photodetector and passed through bandpass filter, and the signal is analyzed in BER analyzer and eye diagram analyzer.

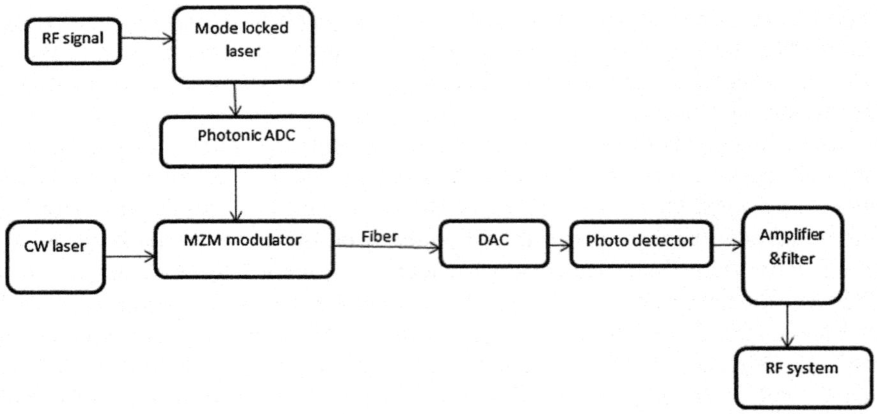

Fig. 5 DRoF system using photonic ADC

6 Simulation Results

The circuit was simulated using OptiSystem software. OptiSystem (Version 14) is the tool used here, and it is innovative, rapidly evolving, and powerful software design software that enables its users to plan and simulate almost any type of optical link in the transmission layer of a broad spectrum of optical networks. Here, the eye diagram of both analog RoF and digital RoF is calculated. Digital RoF has the better performance compare with analog RoF. Eye pattern of both the system is shown in Fig. 6. DRoF has the main advantage of nonlinearity and dynamic range performance.

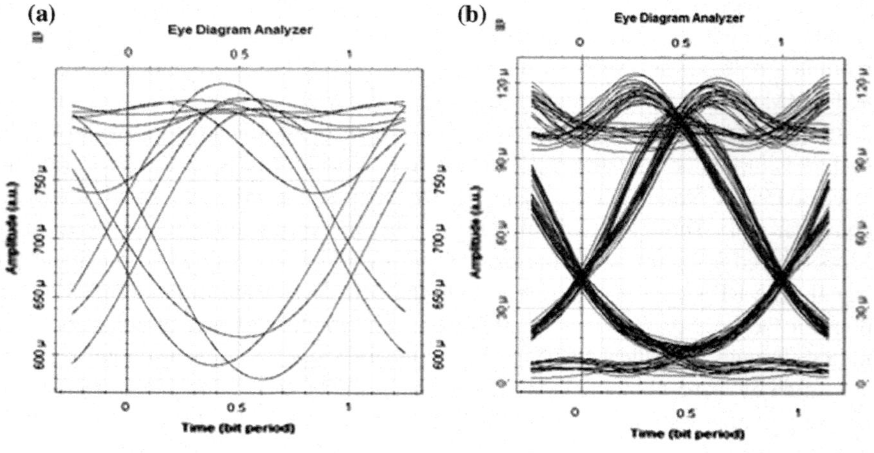

Fig. 6 Eye diagram of **a** Analog RoF system, **b** DRoF system

Simulation result of electrical and photonic ADC is also shown in Fig. 7. For the same input data, the power levels of both ADCs are calculated and compared. Electrical spectrum shows that the power level of photonic ADC is higher than that of electrical ADC. Performance of photonic ADC can be analyzed by increasing the bit rate and calculating the corresponding power. For 1, 3, and 5 Gb/s data rates, signal power is observed as −30, −15, and 10 dBm, respectively. Also observed that at each data rate photonic ADC has better power level than electronic ADC. Figure 8 shows the BER plot of DRoF system with electronic ADC and photonic ADC. The DRoF system with electronic ADC presents the lower BER equal to 106 keeping this level until 20 km of fiber length.

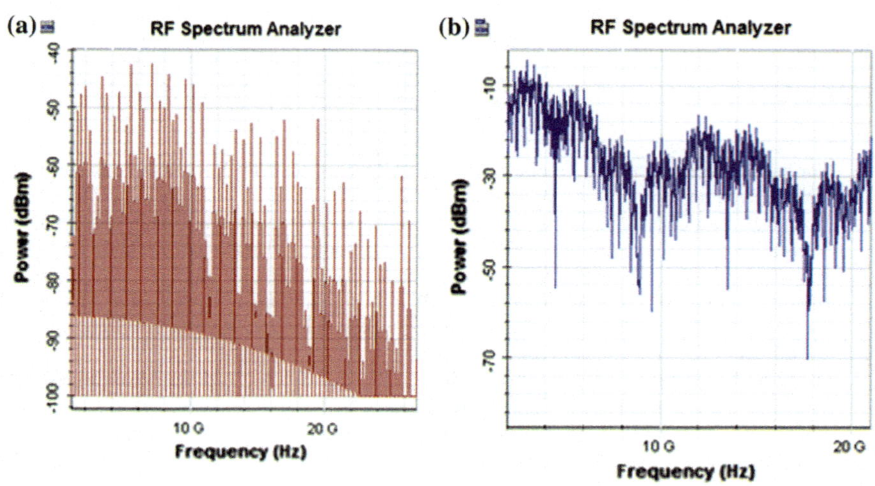

Fig. 7 Electrical spectrum of **a** Electronic ADC **b** Photonic ADC

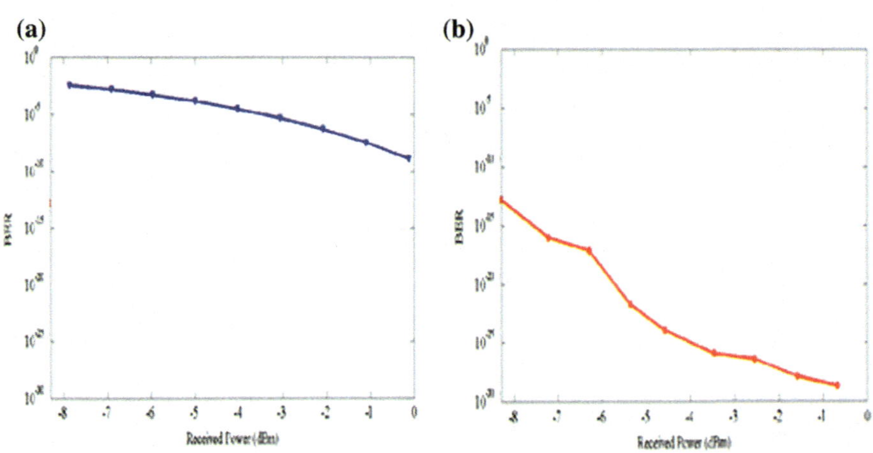

Fig. 8 BER plot of **a** DRoF system with electronic ADC **b** DRoF system with photonic ADC

7 Conclusion

Digital radio over fiber link for single radio service is designed and analyzed the system performance with eye diagram. Simulated the existing system using OptiSystem and analyzed its performance. Existing DRoF technique uses low-speed electronic ADC. Photonic analog-to-digital conversion architectures have the potential to provide significant advantages in bandwidth, timing precision, and timing stability. In our work, we have focused on designing a system architecture that will maximally realize the advantages of photonics while retaining the ability to utilize the high performance and maturity of electronic digitization and processing.

References

1. Cordeiro RF, Oliveira ASR, Vieira J All-Digital Transmitter with RoF Remote Radio Head. Conference Paper
2. Yizhuo B, Lim C, Nirmalathas A (2011) Experimental demonstration of multi-service hybrid fiber-radio system using digitized RF-over-fiber technique. J Lightwave Technol 29(14), 15 July 2011
3. Hori S, Yamase T, Tanio M, Kaneko T (2016) A digital radio-over-fiber downlink system based on envelope delta-sigma modulation for multi-band/mode operation.Conference Paper 2016 IEEE
4. Oliveira RS, Moritsuka NS, Santos RC, Almeida RP, Teixeira A (2016) Low cost digital radio over fiber system. Conference Paper 2016 IEEE
5. Valley GC, Hurrell JP Photonic analog-todigital converters: fundamental and practical limits. Electronics and Photonics Laboratory the Aerospace Corporation, El Segundo, CA, pp 90009–2957
6. Clark Jr TR, Kalkavage JH, McKenna TP (2015) Recent progress in photonic Analog-to-Digital converters. Johns Hopkins University Applied Physics Laboratory
7. Kim J, Chen J, Cox J, Kartnr FX Attosecond-resolution timing jitter characterization of free-running mode-locked lasers. Opt Lett 32, 3519–3521
8. Bartels A, Diddams SA, Ramond TM, Holberg L (2003) Modelocked laser pulse trains with sub femtosecond timing jitter synchronized to an optical reference oscillator. Opt Lett 28, 663–665, 67 5
9. Juodawlkis PW, Twichell JC (2001) Optically sampled analog-to-digital converters. IEEE Trans Microw Theory Tech 49(10), Oct 2001

Study of Microstructure and Microhole Fabrication on PDMS by Femtosecond Laser Direct Writing Technique for Development of Miniaturized Photonic Devices

Sanyogita, Amar Ghar, Utpal Das and P. K. Panigrahi

Abstract Microhole drilling and microstructures fabrication on polymer material have been carried out using femtosecond laser. The characterization of the device has been performed using optical microscope and profilometer. Parametric study on effect of laser pulse energy, scan speed, and depth of focusing has been presented. The findings from this study can be utilized to improve the performance and quality of many photonic devices.

Keywords Photonic devices · Femtosecond laser fabrication · Microhole Microstructure and PDMS

1 Introduction

Polymers are good substitute to fused silica glass for photonic devices due to their flexibility, low cost, and ease of processing [1]. Femtosecond laser microfabrication is a very encouraging technology for fabrication of waveguides, channels, microholes, grating, optofluidic devices, and sensors in polymers and glasses due to its nonlinear light-material interaction [2, 3]. Demand for fabrication of microholes with high aspect ratio has been recently raised due to potential applications in microfluidic photonic devices, combustion chambers, and surgical needles [4]. Fabrication with picosecond, nanosecond or longer pulsed laser is still limited due to the redeposition of melt, cracks, and heat effected zone in the surrounding area [3, 4]. Recently, the development of femtosecond (fs) laser sources has given an alternative method for microscale and nanoscale fabrication. Femtosecond laser-based microfabrication technique has unique nonlinear absorption property that is quite different from conventional laser-based fabrication methods.

Sanyogita (✉) · A. Ghar · U. Das · P. K. Panigrahi
Center for Lasers and Photonics, IIT, Kanpur 208016, UP, India
e-mail: sanyogi@iitk.ac.in

V. Janyani et al. (eds.), *Optical and Wireless Technologies*, Lecture Notes in Electrical Engineering 472, https://doi.org/10.1007/978-981-10-7395-3_17

This paper experimentally investigates the influence of incident femtosecond laser writing parameters, i.e., average laser fluence, depth of fabrication, and writing speed on the fabrication of microstructures and microholes inside and on the surface of PDMS, and identifies the ideal fabrication conditions for fabricating microstructures and microholes with better quality.

2 Experimental Setup

PDMS is an optical transparent polymer which has many advantages to fabricate microstructures, such as flexibility, biocompatibility, thermal stability, and non-toxicity. The preparation process of the PDMS sheets is similar as mentioned by Jiale Yong [5, 6]. Dimension of the PDMS sheet used for fabrication is 5 cm × 3 cm × 4 mm. Figure 1 shows the experimental setup used in the present study for fabrication process. SeRF^{TM} fiber oscillator-based and regeneratively amplified Ti: Sapphire laser (CLRK-MXR) have been used for the fabrication. It produces 120 fs laser pulses with 1 kHz rep rate and 775 nm central wavelength. Beam size at the focal point of objective lens (10X) is 4 micron. Laser pulse energy varies between 2.5 and 6.5 μJ in this experiment which is sufficiently lower than ablation threshold of substrate PDMS material. Laser beam has been focused inside the sample using 10X objective lens. Sample placed on XY stage which is moved with 1-micron resolution. Laser focus is adjusted below the front surface of PDMS up to 600-micron depth, and sample is moved along XY direction using translation stage with speed varying from 50 to 140 μm/s.

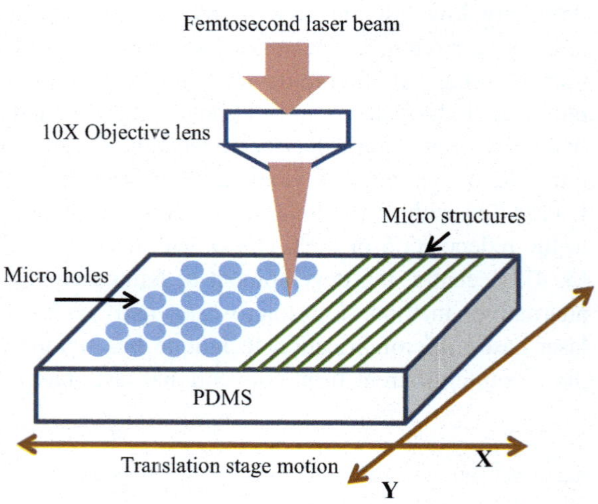

Fig. 1 Schematic diagram of the experimental setup for fabrication of microstructures and microholes

3 Results and Discussions

3.1 Microstructure Fabrication

The fabrication process in PDMS with low rep rate (1 kHz) femtosecond laser pulses with varying laser pulse energy, writing speed, and depth of laser focus is described in this section.

3.1.1 Effect of Laser Pulse Energy

Figure 2a shows microstructure images (from A to E) fabricated at pulse energies of 2.5, 3.5, 4.5, 5.5, and 6.5 μJ. Translation speed has been fixed at 50 μm/s. All microstructures have been fabricated 500 μm below the front surface of PDMS. Variation in width of microstructure with change in pulse energy is shown in Fig. 2b. When microstructures are fabricated inside PDMS material, width of microstructure increases with increase in pulse energy but at higher pulse energy, it is observed that the rate of width change decreases.

3.1.2 Effect of Translation Speed

To study the effect of writing speed on quality of microstructure, PDMS sample is translated with different speed of 20, 50, 80, 110, and 140 μm/s at fixed laser power of 4.5 μJ. It is observed that width of microstructure decreases from 11 to 8 μm with increase in speed of translation stage from 20 to 140 μm/s as shown in Fig. 3.

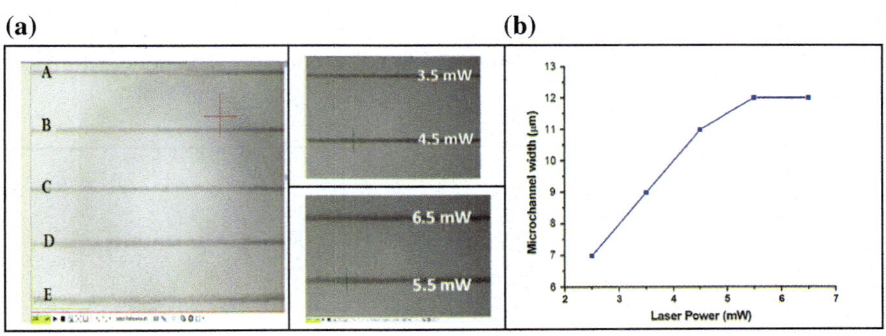

Fig. 2 **a** Transmission image of microstructure as a function of laser pulse energy (A. 2.5 mW, B. 3.5 mW, C. 4.5 mW, D. 5.5 mW, E. 6.5 mW), **b** Variation of microstructure width as a function of laser pulse energy

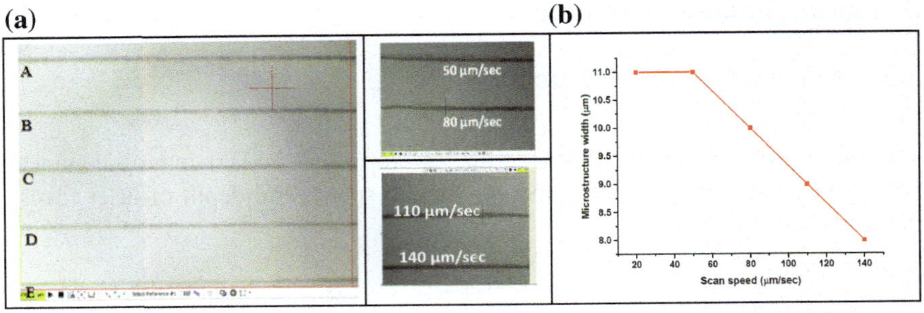

Fig. 3 **a** Transmission image of microstructure as a function of writing speed (A. 20 μm/s, B. 50 μm/s, C. 80 μm/s, D. 110 μm/s, E. 140 μm/s), **b** Variation of microstructure width as a function of writing speed

3.1.3 Effect of Depth of Fabrication

To study the effect of depth of focusing below front surface of PDMS sample on microstructure fabrication, microstructure inscribed inside sample at different depth as 200, 300, 400, 500, and 600 μm below the substrate surface as shown in Fig. 4a. Microstructures are fabricated at speed of 50 μm/s with pulse energy of 50 μJ in focus at a wavelength of 775 nm. Figure 4b shows that change in structure width decreases as depth below the surface of PDMS increases. Here, depth refers to focus point of femtosecond laser beam inside the sample. In this way, waveguide like microstructures has been fabricated.

3.2 Microholes Array Fabrication on PDMS

Microholes investigated in this paper have been drilled using a fs laser of 1 kHz rep rate at 775 nm central wavelength and 120 fs pulse duration. The hole drilling was

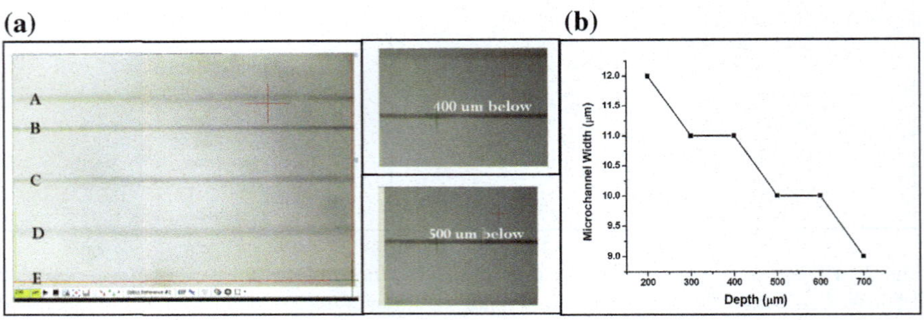

Fig. 4 **a** Transmission image of microstructure width as a function of depth location below the substrate surface. (A. 200 μm, B. 300 μm, C. 400 μm, D. 500 μm, E. 600 μm), **b** Variation in microstructure width as a function of depth location below substrate

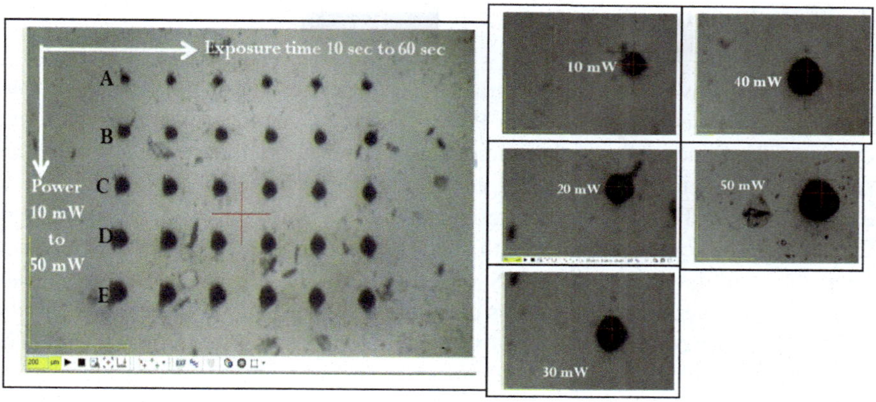

Fig. 5 Microholes array fabrication on PDMS with varying power and exposure time

performed with different laser power and exposure time. Diameter of microholes was measured using microscopy, and depth of microholes was confirmed by Stylus Profiler (KLA Tencor).

Microholes have been fabricated on the surface of PDMS by varying laser pulse energy from 10 to 50 μJ with exposure time of 10 s, 20 s, 30 s, 40 s, 50 s, and 60 s, respectively, as shown in Fig. 5. Circular shaped microholes with high aspect ratio have been obtained. It was observed that diameter and depth of microholes increase with increase in laser pulse energy and exposure time as shown in Fig. 6a and b. The effect of exposure time on depth of microhole is lower in low power regime compared to that in high power regime.

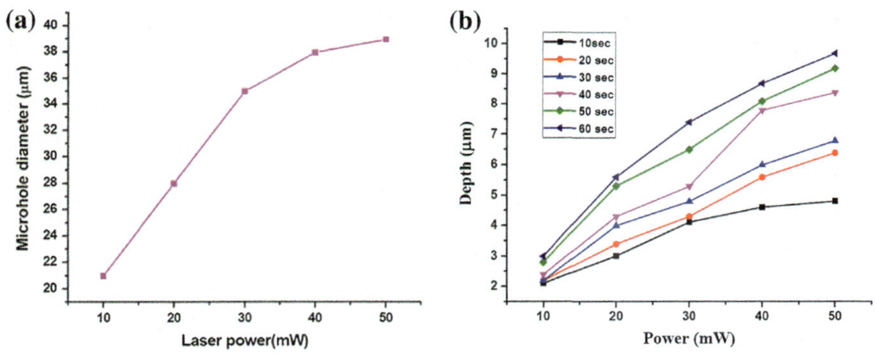

Fig. 6 **a** Variation in microhole diameter as a function of laser power, **b** Variation of microhole depth as a function of laser power

4 Conclusion

In conclusion, we have demonstrated the dependence of microstructure and microhole parameters fabricated in PDMS using the femtosecond laser micromachining. The dimensions of the microstructure and microhole for this process are strongly dependent on the applied energy, scan speed, depth of fabrication, and exposure time. The microstructure dimension varying stepwise with depth of focusing indicates range of depth below the focusing surface for required microstructure dimension in PDMS which provides an ideal window for 3D fabrication. The holes produced by less exposure time with high laser pulses exhibit a smaller diameter compared to long exposure time at same laser writing power. The microstructure width increases with increase in laser power and decreases with increase in scan speed.

Acknowledgements The experiential work leading to concluding the result presented in this paper has been performed at the Femtosecond facility of CELP IIT Kanpur.

References

1. Sugioka K, Cheng Y, Midorikawa K (2005) Three-dimensional micromachining of glass using femtosecond laser for lab-on-a-chip device manufacture. Appl Phys A 81:1–10
2. Zhao X, Shin YC (2011) Femtosecond laser drilling of high-aspect ratio micro channels in glass. Appl Phys A Mater Sci Process 104(2):713–719
3. Darvishi S, Cubaud T, Longtin JP (2011) Ultrafast laser machining of tapered micro channels in glass and PDMS. Opt Lasers Eng 50(2):210–214
4. Huang H (2003) Micro-hole drilling with femtosecond fiber laser. In: Proceedings of SPIE Photonics, vol 8607–19
5. Yong J et al (2013) Femtosecond laser weaving superhydrophobic patterned PDMS surfaces with tunable adhesion. J Phys Chem C 117:24907–24912
6. Yong J et al (2013) Controllable adhesive superhydrophobic surfaces based on PDMS microwell arrays. Langmuir 29:3274–3279

First-Principle Calculations of Optical Properties of $LiInTe_2$ at Different Pressures

S. Chandra and V. Kumar

Abstract First-principle calculations are performed to calculate the values of dielectric constant, refractive index, and energy-loss spectra for $LiInTe_2$ semiconductor at different pressures. The calculated values are compared with available experimental and reported values. A fairly good agreement has been obtained between them.

Keywords First-principle calculations · Optical properties

1 Introduction

Lithium-containing ternary chalcopyrite semiconductors have become more important due to their diverse structural, electronic, and optical properties [1, 2]. These chalcopyrites have wide applications in light-emitting diodes (LEDs) [3, 4], solar cells [5, 6], nonlinear optical (NLO) devices [7], optical frequency conversion [8], and many more applications in mid- and far-IR regions [9, 10]. Different experimental and theoretical approaches have been made to explain different properties of $LiInTe_2$ semiconductors [11–15]. $LiInTe_2$ crystal structure has been discussed by Hönle et al. [11] using experimental programs SHELXTL [16] and ORFFEE [17]. Other workers [12–15] have performed the first-principle calculations for structural, electronics, second harmonic generation and optical properties of $LiInTe_2$. Here, first-principle calculations have been performed for the optical properties such as dielectric constant, refractive index, extinction coefficient, and energy-loss spectra of $LiInTe_2$ at 0, 3, and 5 GPa pressures. Our calculated values are well agreed with the available experimental and reported values at 0 GPa pressure.

S. Chandra (✉) · V. Kumar
Department of Electronics Engineering, Indian Institute of Technology
(Indian School of Mines), Dhanbad 826004, India
e-mail: chandra.satish1989@gmail.com; satish.2015dr0118@ece.ism.ac.in

V. Kumar
e-mail: vkumar52@hotmail.com

V. Janyani et al. (eds.), *Optical and Wireless Technologies*, Lecture Notes
in Electrical Engineering 472, https://doi.org/10.1007/978-981-10-7395-3_18

2 Computational Details

First-principle calculations have been performed to calculate various optical parameters of $LiInTe_2$ using Cambridge Sequential Total Energy Package (CASTEP) code and Material Studio 7.0 software [18]. The calculations are based on generalized gradient approximation (GGA) with adding the Perdew–Burke–Ernzerhof (PBE) [19]. Ultra-soft pseudopotential (USP) [20] has been applied with plane-wave cutoff energy of 350 eV over the Brillouin zone of the 2 × 2 × 2 Monkhorst–Pack mesh. Structure has been optimized by applying Broyden, Fletcher, Goldfarb, and Shanno (BFGS) scheme [21] at ultra-fine quality for convergence threshold. Lattice constants (a = b = 6.398 and c = 12.460) [15] and atomic position have been taken for optimizing $LiInTe_2$ semiconductor having I-42d space group.

3 Result and Discussion

The dielectric function of $LiInTe_2$ has been calculated under different pressures and shown in Figs. 1 and 2, respectively, for the real and imaginary parts. Figure 1 shows that $\varepsilon_1(0) = 7.10644$ which is slightly less than the reported values mentioned in Table 1. The calculated values using GGA methods always overestimate the value of energy gap (E_g) and hence smaller $\varepsilon_1(0)\{= (1 + (\hbar\omega_p/E_g)^2)\}$ [22]. The calculated values of $\varepsilon_1(0)$ increase with increase in pressure and found to be 7.10644, 6.84310, and 7.20202, respectively, at 0 GPa, 3 GPa, and 5 GPa pressures. An increase in $\varepsilon_1(\omega)$ has been observed and reaches to maximum values 1.73212 eV, 2.07055 eV, and 2.11935 eV, respectively, at 0, 3, and 5 GPa pressures. The peak values of the curves are marked by cross (×) sign, and expanded part is shown in small box on right top of figures. This shows that the peak shifts toward high energy side as pressure increases. Further, sharp decrease in $\varepsilon_1(\omega)$ is observed at a minimum value of 5.94744 eV in the range of 0–15 eV. The calculated values of imaginary part of dielectric function $\varepsilon_2(\omega)$ have been shown in Fig. 2 at 0 GPa, 3 GPa, and 5 GPa pressures. Figure 2 shows that the absorption starts from 1.26245 eV, 1.57497 eV, and 1.62457 eV, respectively, at 0 GPa, 3 GPa, and 5 GPa pressures, and corresponding peaks have been observed at 4.57623 eV, 4.65275 eV, and 4.61933 eV which have been marked with cross (×) sign. This also depicts that the absorption edge moves toward higher energy region with increase in pressures.

The refractive index $n(\omega) = \left\{\sqrt{\varepsilon_1(\omega)}\right\}$ [23] has been plotted in Fig. 3 in energy span 0–15 eV and the value of refractive index n(0) = 2.66579 at 0 GPa using GGA PBE which is slightly lesser than the reported value mentioned in Table 1. Table 1 shows that refractive index n(0) of $LiInTe_2$ increases with increase in pressure which further shows that all refractive indices occur in infrared (IR) region. This shows high applications of $LiInTe_2$ in IR region. The extinction coefficient

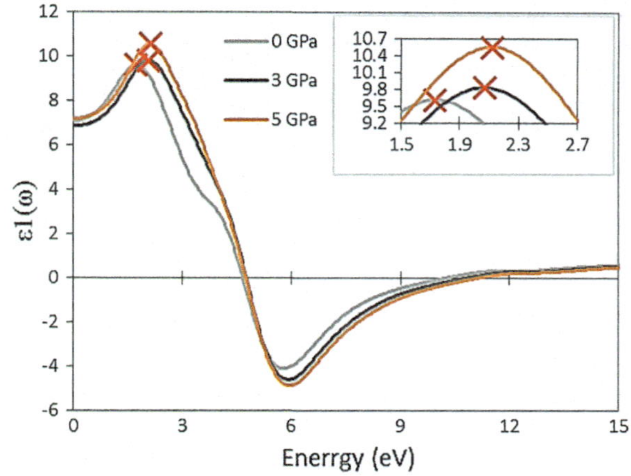

Fig. 1 Real part of dielectric function of $LiInTe_2$ and expanded part is shown in small box

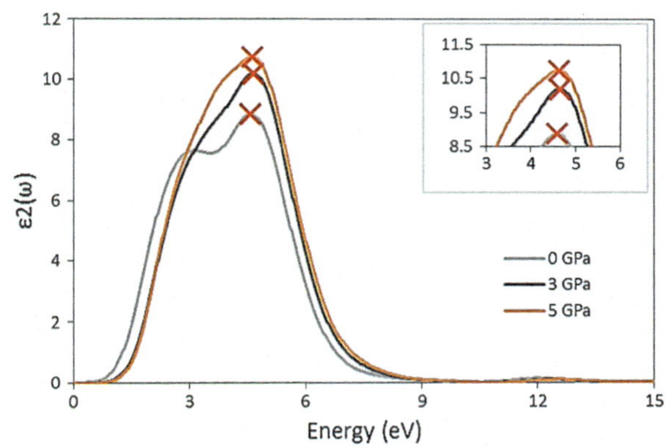

Fig. 2 Imaginary part of dielectric function of $LiInTe_2$ and expanded part is shown in small box

Table 1 Dielectric constant $\varepsilon_1(0)$, refractive index n(0), and plasmon energy ($\hbar\omega_p$) of $LiInTe_2$ under different pressures

Pressures (GPa)	$\varepsilon_1(0)$		n(0)		$\hbar\omega_p$ in eV
	This work	Reported	This work	Reported	This work
0	7.10644	8.12[a], 7.78[a], 7.90[b]	2.66579	2.85[a], 2.79[a], 2.81[b]	10.21227
3	6.84310		2.61593		10.78222
5	7.20202		2.68366		11.10364

[a]Reference [14], [b]Ref. [26]

k(ω) curves have been plotted in Fig. 4 using GGA PBE method. The peaks of k(ω) have been obtained at 5.33292 eV, 5.30482 eV, and 5.42661 eV and shown in Fig. 4, respectively, at 0 GPa, 3 GPa, and 5 GPa pressures. This shows that the

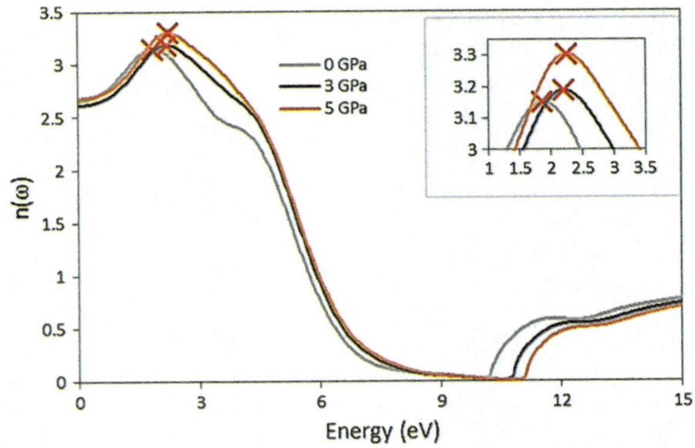

Fig. 3 Refractive index of $LiInTe_2$ and expanded part is shown in small box

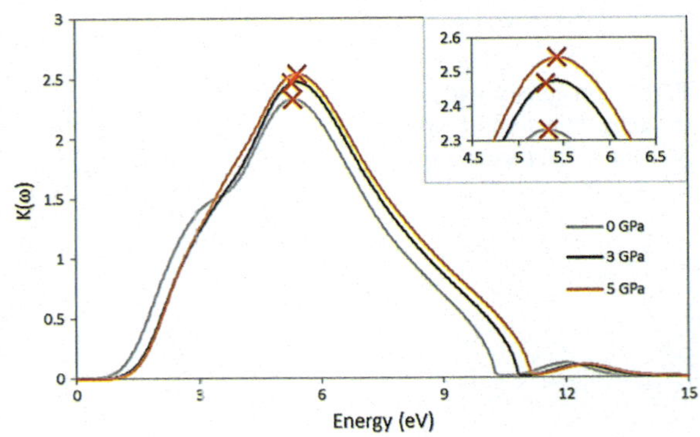

Fig. 4 Extinction coefficient of $LiInTe_2$ and expanded part is shown in small box

extinction coefficient k(ω) peak moves toward the high energy side with increase in pressures.

The energy-loss spectrum tells about the loss of energy due to fast traversing electron in the semiconductor [24], and highest loss-function point depicts the plasmon energy ($\hbar\omega_p$) of the semiconductor. At this point, the dielectric function tends to zero and corresponds to the trailing edge of R(ω) [25]. Figure 5 shows that the highest loss-function points occur at 10.21227 eV, 10.78222 eV, and 11.10364 eV for $LiInTe_2$, respectively, at 0 GPa, 3 GPa, and 5 GPa, which show that the plasmon energy also moves toward the higher energy side as pressure increases. Here, plasmon energy has been calculated first time for $LiInTe_2$ materials for different pressures.

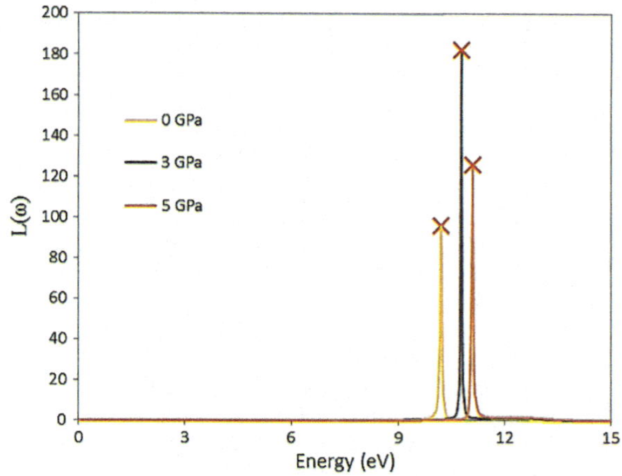

Fig. 5 Energy-loss spectra of $LiInTe_2$

4 Conclusion

In this paper, first-principle calculations have been performed successfully to calculate $\varepsilon_1(\omega)$, $\varepsilon_2(\omega)$, $n(\omega)$, $k(\omega)$, and $L(\omega)$ for $LiInTe_2$ using GGA PBE method within the density functional theory (DFT) at 0 GPa, 3 GPa, and 5 GPa pressures. At 0 GPa pressure, the calculated values of $\varepsilon_1(0)$ and $n(0)$ have good compatibility with the available experimental and reported values. The values of these parameters at 3 GPa and 5 GPa are not available in literature for the comparison and reported for the first time in this paper. The values of $\varepsilon_1(0)$ and $n(0)$ increase with the increase in pressure. The values of other optical parameters such as dielectric constants, extinction coefficient, and energy-loss spectra have been also calculated. In almost all the cases, our calculated values are found to have good compatibility with the experimental and reported values.

References

1. Parthe E (1964) Crystal chemistry of tetrahedral structures. Gordon and Breach, New York
2. Goryunova NA (1965) The chemistry of diamond-like semiconductors. Chapman and Hall, New York
3. Shay JL, Schiavone LM, Buehier E, Wernick JH (1972) Spontaneous and stimulated emission spectra of $CdSnP_2$. J Appl Phys 43:2805–2810. https://doi.org/10.1063/1.1661599
4. Wagner S, Shay JL, Tell B, Kasper HM (1973) Green electroluminescence from CdS–$CuGaS_2$ heterodiodes. Appl Phys Lett 22:351. https://doi.org/10.1063/1.1654669
5. Magesh M, Arunkumar A, Vijayakumar P, Anandha Babu G, Ramasamy P (2014) Investigation of optical property in LiInSe2 single crystal grown by Bridgman Stockbarger method using stepper translations for mid IR laser application. Opt Laser Technol. 56:177–181. http://dx.doi.org/10.1016/j.optlastec.2013.08.003
6. Kazmerski LL(1983) Ternary-compound thin-film solar cells. II Nuovo Cimento D 2:2013–2028. https://doi.org/10.1007/BF02457903

7. Levine BF (1973) Bond-charge calculation of nonlinear optical susceptibilities for various crystal structures. Phys Rev B 7:2600–2626
8. Hopkius FK (1995) Laser Focus World 31:87
9. Shay JL, Wernick TH (1975) Ternary chalcopyrite semiconductors: growth, electrical properties and applications. Pergamon, Oxford
10. Boeknke UC, Kuhu G, Frolova FI, Paukov IE, Neumann H (1988) Heat capacity of LiInTe2. J Thermal Anal 33:205–209. https://doi.org/10.1007/BF01914601
11. Hönle W, Kühn G, Neumann H (1986) The crystal structure of $LiInTe_2$. Zeit Annorg Allg Chem 532:150–156. https://doi.org/10.1002/zaac.19865320121
12. Kosobutsky AV, Basalaev YuM (2010) First principles study of electronic structure and optical properties of $LiMTe_2$ (M = Al, Ga, In) crystals. J Phys Chem Solids 71:854–861. https://doi.org/10.1016/j.jpcs.2010.03.033
13. Kosobutsky AV, Basalaev YuM, Poplavnoi AS (2009) Lattice dynamics of chalcopyrite semiconductors $LiAlTe_2$, $LiGaTe_2$ and $LiInTe_2$. Phys Status Solidi B 246:364–371. https://doi.org/10.1002/pssb.200844283
14. Ma C-G, Brik MG (2015) First principles studies of the structural, electronic and optical properties of LiInSe2 and LiInTe2 chalcopyrite crystals. Solid State Commun. 203:69–74. http://dx.doi.org/10.1016/j.ssc.2014.11.021
15. Reshak AH, Brik MG (2016) Strong second harmonic generation in LiInX2 (X = Se, Te) chalcopyrite crystals as explored by first-principles methods. J Alloys Compds 675:355–363. http://dx.doi.org/10.1016/j.jallcom.2016.03.104
16. Sheldrick GM: SHELXTL. Cambridge 1975, unverfoffentlicht
17. Busing WR, Martin' KD, Levy JA (1974) ORFEE-3 ORNL-TM-306, Oak Ridge, Tenn. Oak Ridge National. Laboratory, USA
18. Segall MD, Philip JD, Lindan MJ, Probert CJ, Pickard PJ, Hasnip S, Clark J, Payne MC (2002) First-principles simulation: ideas, illustrations and the CASTEP code. J Phys Condens Matter 14:2717–2743. http://iopscience.iop.org/article/10.1088/0953-8984/14/11/301
19. Perdew JP, Burke K, Ernzerhof M (1996) Generalized gradient approximation made simple. Phys Rev Lett 77:3865–3869
20. Vanderbilt D (1990) Phys Rev B 41:7892–7895
21. Fischer TH, Almlof J (1992) General methods for geometry and wave function optimization. J Phys Chem 96:9768–9774
22. Penn DR (1960) Phys Rev 128:2093–2097
23. Moss TS (1950) A relationship between the refractive index and the infra-red threshold of sensitivity for photoconductors. Proc Phys Soc B 63:167–176. doi:http://iopscience.iop.org/article/10.1088/0370-1301/63/3/302
24. Bouhemadou A, Khenata R (2007) Ab initio study of the structural, elastic, electronic and optical properties of the antiperovskite $SbNMg_3$. Comput Mater Sci 39:803–807
25. Saniz R, Ye LH, Shishidou T, Freeman AJ (2006) Structural, electronic, and optical properties of NiAl3: First-principles calculations. Phys Rev B 74:014209–1: 014209–7. https://doi.org/10.1103/PhysRevB.74.014209
26. Ma T-H, Zhuang Z-P, Ren Y-L (2012) First-principles calculations of optical and mechanical properties of $LiBX_2$ (B = Ga, In; X = S, Se, Te). Acta Phys Sin 61:197101. https://doi.org/10.7498/aps.61.197101

A Top-Down Approach for Fabrication of Nanorods on GaN-Based LEDs Using Self-Assembled Ni

Alka Jakhar, Manish Mathew, Ashok Chauhan, Kuldip Singh, Vijay Janyani and Nikhil Deep Gupta

Abstract In optoelectronic devices such as LEDs, LASERs, strain-induced piezoelectric polarization occurs in the structure because of large lattice mismatch occurs between sapphire substrate and GaN layers. This degrades the performance of these devices due to strain-induced piezoelectric polarization. Nanostructuring can reduce this stress due to reduction in quantum-confined stark effect (QCSE) and strain-releasing effect. We demonstrate a top-down approach for fabricating GaN nanorods on LED using self-assembled nickel as an etching mask during reactive ion etching (RIE). The self-assembled Ni islands were formed by rapid thermal annealing at 850 °C for 1 min followed by RIE. The shape and morphology of GaN nanorods were studied using scanning electron microscopy (SEM). We realized vertical array of nanorods with dimensions varying from 90–200 nm.

Keywords Gallium nitride (GaN) · Nanorods · Top-down approach
Nanomasking · Quantum-confined stark effect (QCSE)

1 Introduction

Direct wide bandgap group III-nitride-based semiconductors (GaN, AlN, InN) have attracted much attention for optoelectronic devices and high power/temperature electronics [1]. The bandgap of these materials ranges from 0.7 eV (InN), 3.4 (GaN) to 6.2 eV (AlN) which almost covers the entire visible and deep UV wavelengths. The group III-nitrides also possess several characteristics which make

A. Jakhar (✉) · V. Janyani · N. D. Gupta
Electronics and Communication Department, Malaviya National Institute of Technology, Jaipur, Rajasthan, India
e-mail: 2015pwc5135@mnit.ac.in

M. Mathew · A. Chauhan · K. Singh
Optoelectronic Devices Group, Council of Scientific and Industrial Research-Central Electronics and Engineering Research Institute (CSIR-CEERI), Pilani, Rajasthan, India

V. Janyani et al. (eds.), *Optical and Wireless Technologies*, Lecture Notes in Electrical Engineering 472, https://doi.org/10.1007/978-981-10-7395-3_19

them attractive for solid-state lightning [2] such as high-thermal conductivity, low-dielectric constant, fairly high-bond strength, and high liability as operated under harsh environment. Sapphire is the most commonly used substrate for nitride-based devices because of its wide availability, ease of handling, low cost, and stability at higher temperature. Lattice mismatch between sapphire and GaN layers leading to compressive stress which results in strain-induced piezoelectric polarization and quantum-confined stark effect, thus limits the efficiency of devices based on these material systems [3]. The inherent built-in piezoelectric and spontaneous polarization in GaN-based alloys because of lattice mismatch and thermal strain leads to bending of energy bands in quantum wells. This is known as quantum-confined stark effect (QCSE). These titled bands cause large deformation of normal quantum wells which results in the decrease of stimulated radiative recombination due to the decrease of overlapping between electron and hole overlapping which, in turn, deteriorated the efficiency of optoelectronic device [4]. One-dimensional (1D) semiconductor nanostructures such as rods, wires, tubes have recently attracted much interest for electronic and optoelectronic devices because of strain relief and quantum confinement due to reduction in quantum-confined stark effect, which results in increase in radiative recombination efficiency in multiple quantum wells of LEDs and LASERs [5–9]. Nanostructuring on the surface leads to high surface to volume ratio which results in increase in active area for optoelectronic devices.

1.1 Methods of Fabrication of Nanostructures

Two methods have been used for fabrication of nanostructures, i.e., bottom-up and top-down approaches. The bottom-up nanofabrication approaches related to the construction of multifunctional nanostructural materials and devices by the self-assembly of atoms or molecules [10]. This methodology seeks to have molecular or atomic components built up into more complex nanoscale assemblies or directed self-assemblies based on complex mechanisms and technologies [9, 10]. The mostly used bottom-up fabrication methods include molecular self-assembly, vapor–liquid–solid (VLS) growth, sol–gel nanofabrication, vapor phase deposition of nanomaterials [10]. However, all these methods are relatively complicated. These techniques have some associated challenges such as lack of control over the doping profile, site uniformity, quality of the reactants, surface preparation for the controlled deposition of atoms; this can lead to non-optimal material quality and less flexibility in material design such as heterostructures and doping [10, 11].

The top-down approach corresponds to using nanofabrication tools that are controlled by external experimental parameters to create nanoscaled structures/ functional devices with the desired shapes and characteristics starting from larger dimensions and reducing them to the required values [12–15]. The various techniques include conventional lithography, e-beam lithography, optical lithography, nanoimprint lithography, self-assembled nanomasking, and nanosphere lithography. High-aspect ratio of nanorods can be achieved by using lithography, but these are expensive, difficult for production of large-scale dense nanostructures, low throughput, and slow process [12]. Nowadays, the most widely used techniques are self-organized nanomasking and nanosphere lithography because these are cost-effective, simpler, and applicable to nanofabrication over large area substrates.

Herein, we report the successful fabrication of GaN nanorods using self-assembled Nickel (Ni) islands and reactive ion etching with good structural uniformity. The mean size and density of the Ni nanoislands depend on the Ni layer thickness, annealing time, and annealing temperature [16].

2 Experimental Details

The GaN-based LED samples used in this study were grown by metal organic chemical vapor deposition (MOCVD) on the sapphire substrate. In MOCVD, Trimethylgallium (TMGa), Trimethylindium (TMIn), Ammonia (NH_3), Trimethylaluminium were used as sources of Ga, In, N, Al, respectively. Biscyclopentadienyl-magnesium (Cp_2Mg) and Disilane (Si_2H_6) were used as p and n dopant, respectively. The structure consists of 30 nm thick GaN nucleation buffer layer, a 4 μm thick highly conductive Si-doped GaN layer, then a stack of 5 InGaN/GaN multi-quantum well with quantum well and barrier thickness of 2 nm and 13 nm, respectively. Finally, 120-nm-thick Mg-doped GaN layer followed by high-doping p-GaN with 5-nm thickness. After this, a 100 nm thin SiO_2 layer was deposited by using plasma-enhanced chemical vapor deposition (PECVD) which behaves as an insulating layer. The samples were initially cleaned by using standard three-stage solvents (Trichloroethylene, Acetone, and Methanol) and finally dry by blowing nitrogen. Before the samples were loaded into the furnace, pre-metallization etch process was carried out in HCL and deionized water

(DI) in 1:1 ratio and thoroughly rinsed in DI water in order to remove native oxides from the sample surface. In order to fabricate nanorods, first a 10-nm-thick nickel (Ni) layer was deposited by using e-beam evaporation system. After this, the samples were undergone to rapid thermal annealing (RTA) at 850 °C for 1 min under nitrogen ambient in order to form Ni nanoislands on the top surface of LED sample which acts as masking for RIE. In the next step, RIE of SiO_2 was performed under the gas mixture of $CF_4/O_2 = 35/2$ sccm for 2 min. Finally, the etching of GaN was carried out with reactant gases BCl_3, Cl_2, Ar of 10 sccm, 3 sccm, 2 sccm, respectively. The etching rate was about 80 nm/min and bias power was 75 W with RF frequency of 13.56 MHz. Finally, SiO_2 and Ni were etched using HF to fabricate GaN nanorods. The process for fabrication of GaN nanorods is shown in Fig. 1. After that the dimensions and density of nanorods were examined by scanning electron microscopy (SEM). It is possible to obtain nanometer-sized nickel islands by proper selection of Ni metal thickness, annealing temperature, and annealing time.

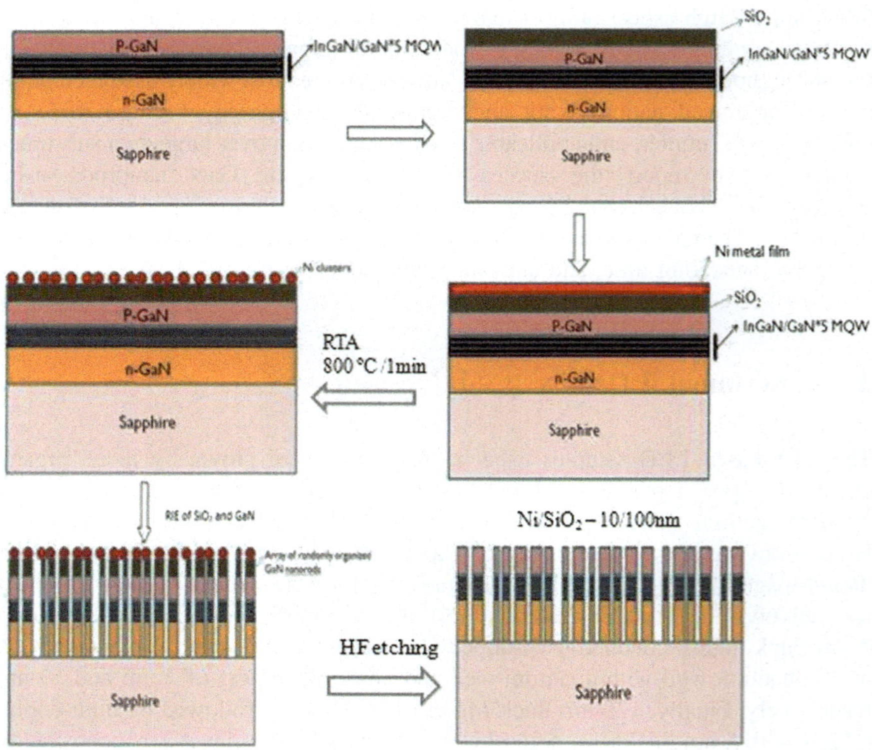

Fig. 1 Schematic diagram showing processing steps involved in top-down fabrication of GaN nanorods

3 Result

The shape and morphology of the GaN-based nanorods were observed using scanning electron microscopy (SEM). The top view image of the GaN nanorods for different magnification (×50,000 and ×20,000) at 30 kV high voltage was shown in Fig. 2. From the image, it was found that not only isolated nanoislands but cluster of nanorods was also formed. We obtained nanorods with diameter varying from 90–200 nm.

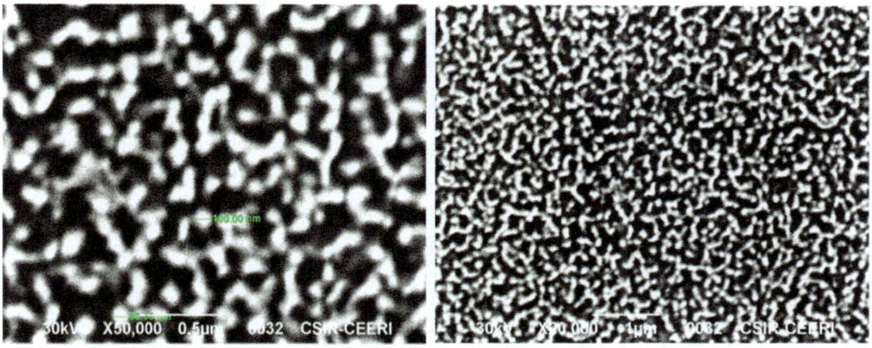

Fig. 2 SEM images of GaN nanorods at two different magnifications (×50,000 and ×20,000)

4 Conclusion

We report the fabrication of GaN nanorods-based LED by top-down approach. Self-assembled Ni nano-clusters were used as an etching mask during RIE. The nanostructures on GaN-based LED can enhance the efficiency because of strain relaxation effect and quantum confinement which reduces quantum-confined stark effect. Thus, GaN-based nanorod LEDs with higher radiation efficiency are expected to be a good candidate for future solid-state lightening devices development.

Acknowledgements This work was carried out in Council of Scientific and Industrial Research-Central Electronics And Engineering Research Institute (CSIR-CEERI), Pilani. The author would like to thank Dr. Manish Mathew for providing the experimental support. The author is thankful to Dr. Vijay Janyani for his valuable suggestions during this work.

References

1. Reddy N Parvathala et al (2016) Enhanced luminescence from GaN nanopillar arrays fabricated using a top-down process. Nanotechnology 27(6):065304
2. Nakamura S, Mukai T, Senoh M (1994) Candela-class high-brightness InGaN/AlGaN double-heterostructure blue-light-emitting diodes. Appl Phys Lett 64(13):1687–1689
3. Chiu CH, Lin DW, Li ZY, Chiu CH, Chao CL, Tu CC, Kuo HC, Lu TC, Wang SC (2010) Improvement in crystalline quality of InGaN-based epilayer on sapphire via nanoscaled epitaxial lateral overgrowth. Jpn J Appl Phys 49(10R):105501
4. Abdullah RA, Ibrahim K (2013) Effects of quantum confined stark effect and well thickness on optical properties of double quantum wells violet InGaN laser diodes. Opt-Int J Light Electron Opt 124(4):292–296
5. Ryou JH, Yoder PD, Liu J, Lochner Z, Kim H, Choi S, Kim HJ, Dupuis RD (2009) Control of quantum-confined stark effect in InGaN-based quantum wells. IEEE J Sel Top Quant Electron 15(4):1080–1091

6. Ramesh V, Kikuchi A et al (2010) Strain relaxation effect by nanotexturing InGaN/GaN multiple quantum well. J Appl Phys 107(11):114303
7. Fujii T, Gao Y, Sharma R, Hu EL, DenBaars SP, Nakamura S (2004) Increase in the extraction efficiency of GaN-based light-emitting diodes via surface roughening. Appl Phys Lett 84(6):855–857
8. Huh C, Lee KS, Kang EJ, Park SJ (2003) Improved light-output and electrical performance of InGaN-based light-emitting diode by microroughening of the p-GaN surface. J Appl Phys 93 (11):9383–9385
9. Huang HW, Chu JT, Kao CC, Hseuh TH, Lu TC, Kuo HC, Wang SC, Yu CC (2005) Enhanced light output of an InGaN/GaN light emitting diode with a nano-roughened p-GaN surface. Nanotechnology 16(9):1844
10. Ji Q, Honma I, Paek SM, Akada M, Hill JP, Vinu A, Ariga K (2010) Layer-by-layer films of graphene and ionic liquids for highly selective gas sensing. Angew Chem 122(50):9931–9933
11. Ariga K, Ji Q, McShane MJ, Lvov YM, Vinu A, Hill JP (2011) Inorganic nanoarchitectonics for biological applications. Chem Mater 24(5):728–737
12. Biswas A, Bayer IS, Biris AS, Wang T, Dervishi E, Faupel F (2012) Advances in top–down and bottom–up surface nanofabrication: Techniques, applications & future prospects. Adv Coll Interface Sci 170(1):2–27
13. Xu Q, Rioux RM, Dickey MD, Whitesides GM (2008) Nanoskiving: a new method to produce arrays of nanostructures. Acc Chem Res 41(12):1566–1577
14. Fourkas JT (2010) Nanoscale photolithography with visible light. J Phys Chem Lett 1 (8):1221–1227
15. Schmid GM, Miller et al (2009) Step and flash imprint lithography for manufacturing patterned media. J Vac Sci Technol B: Microelectron Nanometer Struct Proces Meas Phenom 27(2):573–580
16. Huang HW, Kao CC, Hsueh TH, Yu CC, Lin CF, Chu JT, Kuo HC, Wang SC (2004) Fabrication of GaN-based nanorod light emitting diodes using self-assemble nickel nano-mask and inductively coupled plasma reactive ion etching. Mater Sci Eng B 113 (2):125–129
17. Kumar A, Latzel M, Tessarek C, Christiansen S, Singh R (2013) Characterization of GaN nanorods fabricated using Ni nanomasking and reactive ion etching: a top-down approach. J Nano-Electron Phys 5(2):2001–1

MgO Doped Lithium Niobate Waveguides Based All Optical Modulator

Sanjay Kumar, Ghanshyam Singh, Vijay Janyani, Oleh Buryy, Ubizskii Serhij, Sugak Dmytro and Manish Tiwari

Abstract This paper reviews performance of all-optical modulator based on Magnesium oxide-doped Lithium Niobate ridge waveguides to achieve low $V_\pi L$ at 1550 and 1330 nm wavelengths. The modulation is carried by Mach-Zehnder interferometer structure having 0.4 cm long push-pull arm configuration. The proposed structure is simulated by using beam propagation method. The results of optimization may help in designing electrically tunable optical switch to operate with low switching voltage or an external intensity modulator for optical communication applications.

Keywords Ridge waveguides · Mach-Zehnder interferometer (MZI) MgO-doped lithium niobate · External modulator

---

S. Kumar (✉) · G. Singh · V. Janyani
Department of Electronics and Communication Engineering, Malaviya National Institute of Technology, Jaipur, India
e-mail: 2015pwc5344@mnit.ac.in

G. Singh
e-mail: gsingh.ece@mnit.ac.in

V. Janyani
e-mail: vjanyani.ece@mnit.ac.in

O. Buryy · U. Serhij · S. Dmytro
Institute of Telecommunications, Radioelectronics and Electronic Engineering, Lviv Polytechnic National University,
Lviv, Ukraine
e-mail: oburyi@polynet.lviv.ua

U. Serhij
e-mail: gubask@polynet.lviv.ua

S. Dmytro
e-mail: dm_sugak@yahoo.com

M. Tiwari
Department of Electronics and Communication Engineering,
Manipal University Jaipur, Jaipur, India
e-mail: manish.tiwari@jaipur.manipal.edu

V. Janyani et al. (eds.), *Optical and Wireless Technologies*, Lecture Notes in Electrical Engineering 472, https://doi.org/10.1007/978-981-10-7395-3_20

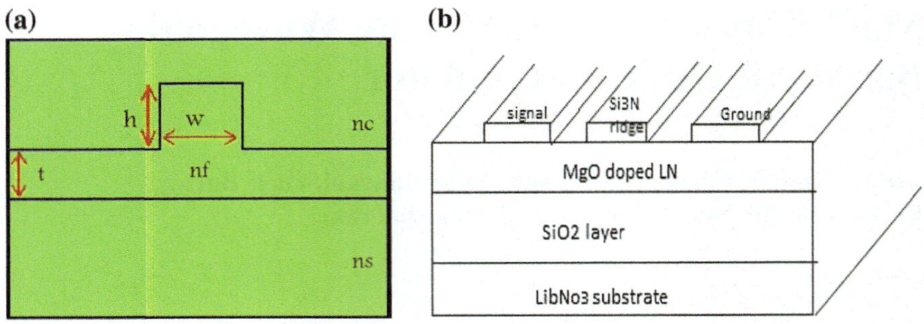

Fig. 1 **a** Ridge waveguide structure **b** ridge waveguide-based optical modulator; Ref. [6]

1 Introduction

In last few decades, Lithium Niobate becomes an extremely useful and a high interest of material for photonic device fabrication and research because of its excellent electro-optic coefficients, low losses, and linear response to the applied electrical field. Optical signal modulation is the basic requirement for optical communication to carry electrical signal in the optical domain. In general, Lithium Niobate modulator is fabricated by Ti in diffusion or proton exchange process which is not very much suitable because of low refractive index contrast, larger mode size, and have very large push-pull MZI arm length which results in very high $V_{\pi}L$ [1]. In order to overcome these limitations, ridge waveguide modulator structure, which employs Lithium Niobate on insulator [2] showing in Fig. 1a, has been widely used. Recently, Jin et al. [3] proposed ridge waveguide modulator structure having Lithium Niobate thin film and ridge of silicon nitride which shows $V_{\pi}L \sim 3$ Vcm. Undoped Lithium Niobate has a non-stoichiometric structure with a high concentration of intrinsic defect and shows "optical damage" of the refractive index while carrying high power which is obvious in WDM system [4]. This problem can be severed if the device is used for optical switching; therefore, suppression of optical damage is an important task. The intrinsic defect of undoped LN can be compensated by the addition of external dopants. It is shown that doping of MgO in Lithium Niobate (MgO-LN) can reduce "optical damage" of refractive index besides that doping of MgO also increases the electro-optic coefficients of Lithium Niobate [4, 5].

2 Theory and Design Analysis

Figure 1b shows the device structure in which first an insulating layer of the SiO_2 height of 2 μm has been grown on LN substrate and then a thin layer of 0.73 μm of LN is deposited on the insulator. Thin film of LN is doped with MgO by using

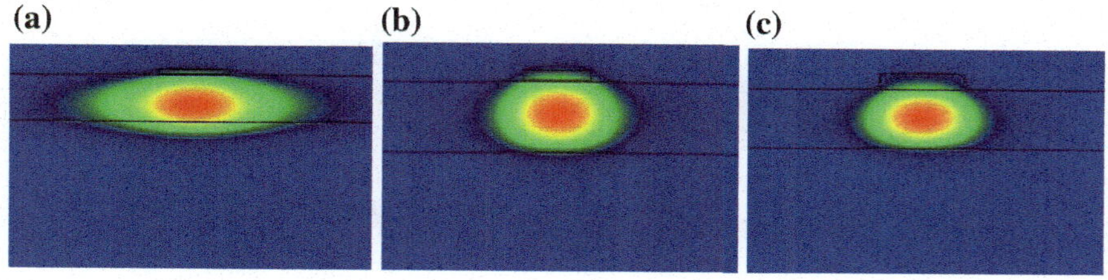

Fig. 2 Field propagation in waveguide for different ridge height **a** h = 100 nm, **b** h = 200 nm, **c** h = 300 nm, while the width has been kept fixed as w = 3 μm

diffusion process that enhances the electro-optic coefficient of LN film and suppresses the "optical damage" problem of refractive index; then, a layer of silicon nitride is developed inform of ridge to confine the field propagation inside the core, having width "*w*" μm and height "*h*" μm. The field confinement inside the core depends on the dimension of insulator ridge as shown in Fig. 2.

So we choose here ridge of width 3 μm and height 0.3 μm for the maximum confinement of field inside the core. The refractive index used for the MgO-doped LN film governed by Sellmeier's equation [7].

$$n_e^2 - 1 = A\lambda^2/(\lambda^2 - B) + C\lambda^2/(\lambda^2 - D) + E\lambda^2/(\lambda^2 - F) \tag{1}$$

where A = 2.2454, B = 0.01242, C = 1.3005, D = 0.05313, E = 6.8972, F = 331.33. Value calculated for MgO-doped LN for extra ordinary ray at 1330 nm is 2.136 and at 1550 nm is 2.129.

3 Discussion

External intensity modulation of the optical signal is done by using MZI structure on BPM simulator platform [8]. Beam propagation method (BPM) solves Maxwell's equation by finite difference method rather than partial derivative method [9]. The 3 dB MMI coupler is used rather than Y-branch splitter to split power in both push-pull arms of MZI. The whole structure is created by using the ridge waveguide structure, as soon as the optical signal is launched in one of the port; it gets divided equally in both arm of MZI due to 3 dB coupler. The electrode is placed at both sides of one arm to create the relative phase shift between the optical signals traveling in MZI arms. When the relative phase shift is 0, the power completely transfers to port 1 (BAR mode, Fig. 3a), and when the relative phase shift is π, the power completely transfers to port 2 as shown in Fig. 3b (CROSS mode). Theoretically, V_π can be calculated as [10].

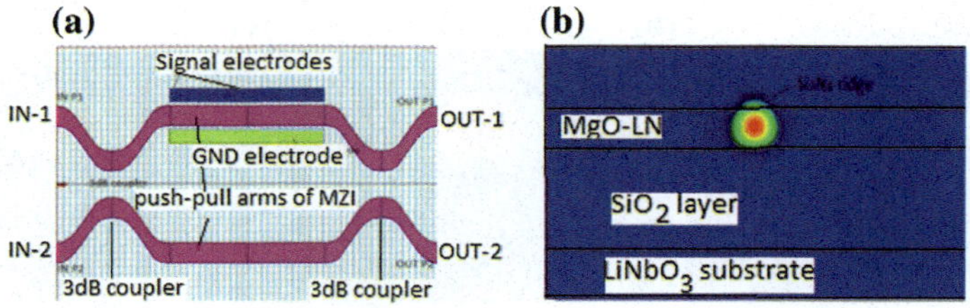

Fig. 3 **a** MZI modulator. **b** Signal propagation in modulator

Table 1 Comparison of results of modulator based on different structures/waveguides

Waveguides	$V_\pi L$ (Vcm)	Extinction ratio (dB)
MgO-LN (λ = 1330 nm)	2.2	18
MgO-LN (λ = 1550 nm)	2.4	15
Si3N4/LN [3]	3	13
Ta2O5/LN film [11]	4	20
A-Si/LN [12]	8.8	20
ChG/Si/LN [13]	3.8	13

$$V_\pi = d\ \lambda / 2 r_{33} n_e^3 L \Gamma \tag{2}$$

where d is the separation distance between signal and ground electrode, λ is operating wavelength, r_{33} is electro-optic coefficient of MgO-LN, n_e is the refractive index of extraordinary ray travel in thin film, L is the length of the electrode (4000 μm), and r is the optical confinement factor inside MgO-LN film. Du et al. [5] reported 11% increment in electro-optic coefficient (r_{33}), so we use here r_{33} 34.5 pm/v. Table 1 depicts the comparison of the calculated result with different structures proposed previously [3, 11–13] (Fig. 4).

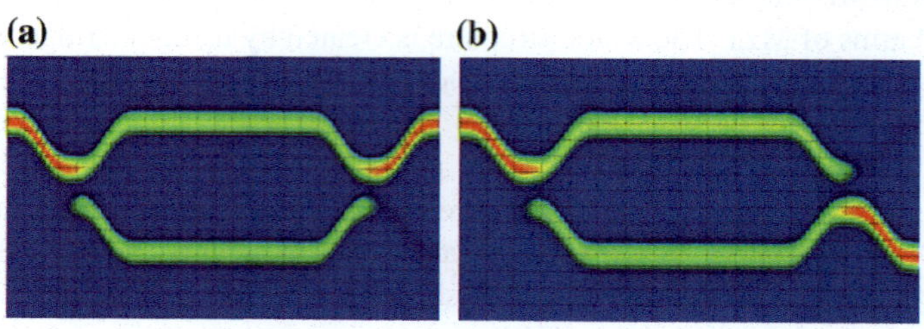

Fig. 4 Optical power transmission **a** in BAR mode **b** in CROSS mode

4 Conclusion

$V_{\pi}L$ for an all-optical modulator based on MgO-doped LN ridge waveguides for its operation with wavelengths 1550 nm and 1330 nm is calculated as 2.4 Vcm and 2.2 Vcm, respectively. The values reported here are promising as compared to another type of structures; however, these results are obtained from Opti BPM simulator software; therefore, further improvements can be achieved by redefining the MgO-doped LN channel waveguide structuring. Doping of MgO also suppresses the "optical damage" effect that means the same structure can also be used for optical switching.

Acknowledgements The authors are grateful to India–Ukraine inter-governmental science and technology cooperation program between the MNIT Jaipur (India) and the Lviv National Polytechnique Institute, Lviv (Ukraine): Project sanction no: INT/RUS/UKR/P-15/2015.

References

1. Tanaka K, Suhara T (2015) August. fabrication of 0.7 μm^2 ridge waveguide in ion-sliced $LiNbO_3$ by proton-exchange accelerated etching. In: Conference on lasers and electro-optics (OSA), p 26I1_5
2. Saha S, Yohanes SS, Jun D, Danner A, Tsang M (2016) Fabrication and characterization of optical devices on lithium niobate on insulator chips. Proc Eng 140:183–186
3. Jin S, Xu L, Zhang H, Li Y (2016) $LiNbO_3$ thin-film modulators using silicon nitride surface ridge waveguides. IEEE Photon Technol Lett 28(7):736–739
4. Choubey RK, Sen P, Sen PK et al (2006) Optical properties of MgO doped $LiNbO_3$ single crystals. Opt Mater 28(5):467–472
5. Du WY, Zhang ZB, Ren S, Wong WH et al (2016) Electro-optic coefficients of Li-deficient MgO-doped $LiNbO_3$ crystal. Rev Sci Instrum 87(9):096105
6. Ismail MM, Meor MA et al (2012) Buried vs. ridge optical waveguide modeling for light trapping into optical fiber. Int J Eng Innovative Technol (IJEIT) 2(1):273–278
7. Zelmon DE, Small DL, Jundt D (1997) Infrared corrected Sellmeier coefficients for congruently grown lithium niobate and 5 mol.% magnesium oxide–doped lithium niobate. J Opt Soc Am B 3319–3322
8. Singh G, Janyani V, Yadav RP (2011) Modeling of a 2 $\times$ 2 electro-optic Mach-Zehnder Interferometer optical switch with s-bend arms. Photon Lett Pol 3(3):119–121
9. Beam propagation method: technical background and tutorials (2007) Waveguide optics modeling software system, Version 9.0, Optiwave Inc
10. Yariv A (1989) Quantum electronics, 3rd edn. Willey Publication. ISBN: 978-0-471-60997-1
11. Rabiei P, Ma J, Khan S et al (2013) Heterogeneous lithium niobate photonics on silicon substrates. Opt Exp 21:25573–25581
12. Chen L, Wood MG, Reano RM (2013) 12.5 pm/V hybrid silicon and lithium niobate optical microring resonator with integrated electrodes. Opt Exp 21:27003–27010
13. Rao A, Patil A, Chiles J et al (2015) Heterogeneous microring and Mach-Zehnder modulators based on lithium niobate and chalcogenide glasses on silicon. Opt Express 23(17): 22746–22752

Optimized 2 × 1 Multiplexer Based on Reversible Logic Using Titanium-indiffused Lithium Niobate Channel Waveguides

Harsh Kumar, Sanjeev Jain, Manish Tiwari, Oleh Buryy, Ubizskii Serhij, Vijay Janyani and Ghanshyam Singh

Abstract In this paper, we propose an optimized 2 × 1 all-optical reversible multiplexer using Ti-indiffused Lithium Niobate channel waveguiding structure. The proposed device offers high-speed conversion and acceptable losses. Information is not lost as the inputs are also available at the output in reversible logic. The power consumption in MZI is optimal with the interferometric arm length L = 14000 μm and the width w = 8 μm. The proposed MUX can be used in all-optical logic networks.

Keywords Lithium niobate · Mach–Zehnder interferometer (MZI) Reversible logic · 2 × 1 multiplexer · PIC

1 Introduction

All-optical photonics integrated circuits are an evolving topic of research in the field of optical communication, and a large number of devices have been proposed. Multiplexer (MUX) is a universal combinational logic circuit, which acts as the key element of various encoders, logic selectors for secure optical digital

H. Kumar (✉) · V. Janyani · G. Singh
Department of Electronics and Communication Engineering, Malaviya National Institute of Technology, Jaipur, India
e-mail: 2015pwc5305@mnit.ac.in

S. Jain
Department of Electronics and Communication Engineering, Government Engineering College, Bikaner, India

M. Tiwari
Department of Electronics and Communication Engineering, Manipal University, Jaipur, India

O. Buryy · U. Serhij
Institute of Telecommunications, Radioelectronics and Electronic Engineering, Lviv Polytechnic National University, Lviv, Ukraine

V. Janyani et al. (eds.), *Optical and Wireless Technologies*, Lecture Notes in Electrical Engineering 472, https://doi.org/10.1007/978-981-10-7395-3_21

communication. All-optical multiplexer has already been simulated and fabricated using different technologies [1]. The most effective design of multiplexer is using reversible logic. In the reversible logic gates, the number of input and the number of outputs are equal, and the input vector is reconstructed as the output [2]. Higher order MUX is designed using the lower order MUX using the reversible logic [3, 4]. The SOA-based MZI switch is the key element in designing of all-optical multiplexer [5, 6]. In this paper, 2 × 1 MUX based on SOA-MZI is designed and optimized using Ti-indiffused Lithium Niobate (Ti-LN) channel waveguide structure. LN is used as substrate because its electro-optic (EO), acousto-optic (AO) and piezoelectric coefficients are very high [7]. Passive devices such as beam splitter, beam combiner and a low-powered SOA-based Mach–Zehnder interferometer switch are the basic elements of the photonics integrated circuit (PIC) of multiplexer. The MZI switch is designed and optimized using beam propagation method, and it is the only active element of the PIC of multiplexer. The four beam splitter and two beam combiner are also used.

2 Working Principle of SOA-MZI Switch

MZI structure can be used to design an EO-switch, consisting two interferometric arms, i.e. SOAs, two 3 dB couplers, and the switching action takes place based on the relative phase change between them [7, 8]. The relative phase shift and the output power at port 1 and 2 can be calculated as:

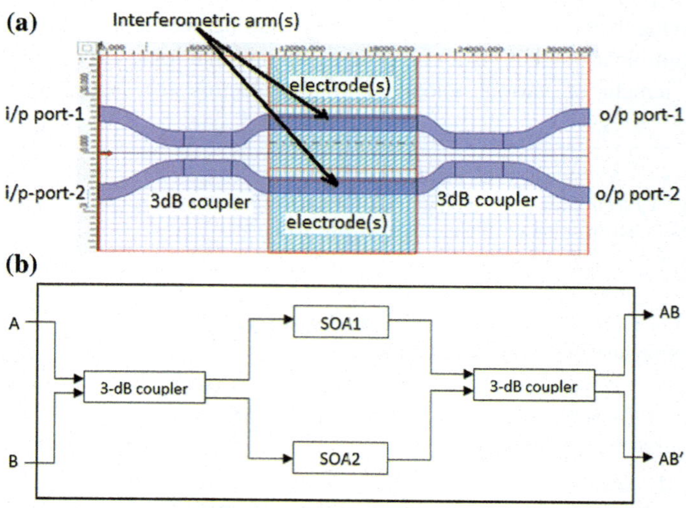

Fig. 1 a Schematic view of MZI: Ref. [9, 10], b logical block diagram of SOA-MZI

$$P_{out1} = P_{in} sin^2 \frac{\Delta\varnothing}{2} \quad (1)$$

$$P_{out2} = P_{in} cos^2 \frac{\Delta\varnothing}{2} \quad (2)$$

$$\Delta\varnothing = \frac{\pi}{V_\pi} V \quad (3)$$

where V_π is the electrode voltage corresponding to the phase difference, $\Delta\varnothing = \pi$, and P_{in} is input power. The schematic diagram of MZI is shown in Fig. 1a. The logic block diagram of SOA-MZI is shown in Fig. 1b with the input signal as A and control signal as B. The output we are getting at bar port (o/p port 1) is AB and at cross port (o/p port 2) is AB'.

3 2 × 1 MUX Using Reversible Logic

The basic 2 × 1 MUX has two inputs, one select line and one output, and can be designed using the truth table given in Table 1. The logical block diagram is given in Fig. 2 as it is designed on reversible logic so the number of inputs and outputs is equal. X and Y are the MUX inputs, S is a select line, Z is the MUX output, and A is an additional input to make the number of inputs and outputs equal and called as ancilla input, which is at logic 0.

3.1 Implementation of Multiplexer

The 2 × 1 reversible MUX is implemented and simulated by using design tool: Optisystem [11]. The SOA-MZI is designed using Ti-indiffused LN waveguide in OptiBPM [11], and by using its scattering parameters, it is exported into Optisystem

Table 1 Truth table for 2 × 1 MUX

Inputs			Outputs
X	Y	S	Z
0	0	0	0
0	0	1	0
0	1	0	0
0	1	1	1
1	0	0	1
1	0	1	0
1	1	0	1
1	1	1	1

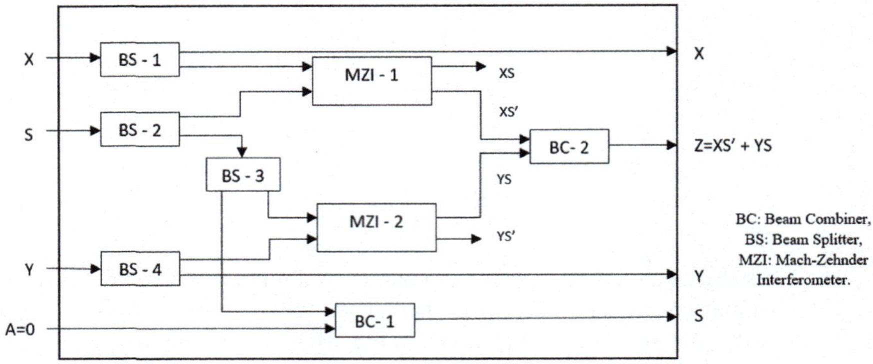

Fig. 2 Logical block diagram of 2 × 1 reversible MUX

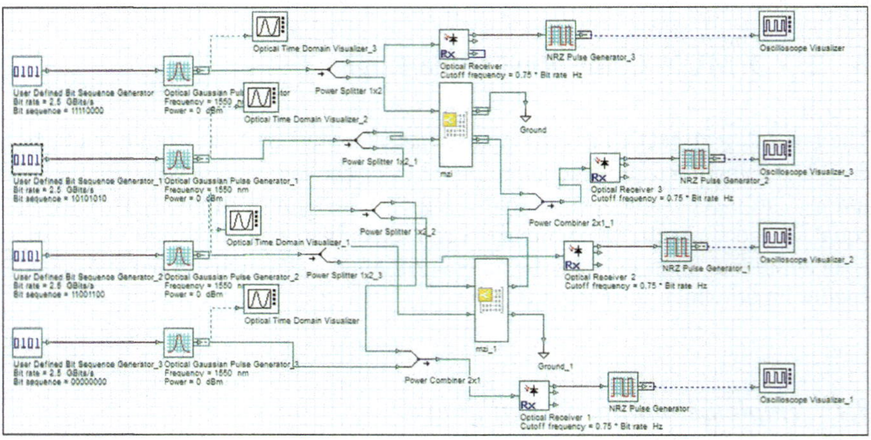

Fig. 3 Proposed MUX design layout

to simulate the complete simulation circuit of MUX. The layout of design is shown in Fig. 3, which consists of two MZI switches, four beam splitters and two beam combiners with four input/output ports, two unused outputs of MZI and one ancilla input (A = 0). The simulation SOI-MZI is done with the central wavelength of 1.55 μm and the TE polarized light wave.

3.2 Result and Discussion

The inputs are given to the MUX as X = 11110000, S = 10101010, Y = 11001100 and A = 0. The output we are getting as X = 11110000, Z = 11011000, Y = 11001100 and S = 10101010. The waveforms of input signal are in optical domain and shown in Fig. 4a, and the outputs are in digital domain and shown in Fig. 4b. The output power of the MZI switch can be optimized by varying interferometer arm length L and width of interferometer W. The variation of output

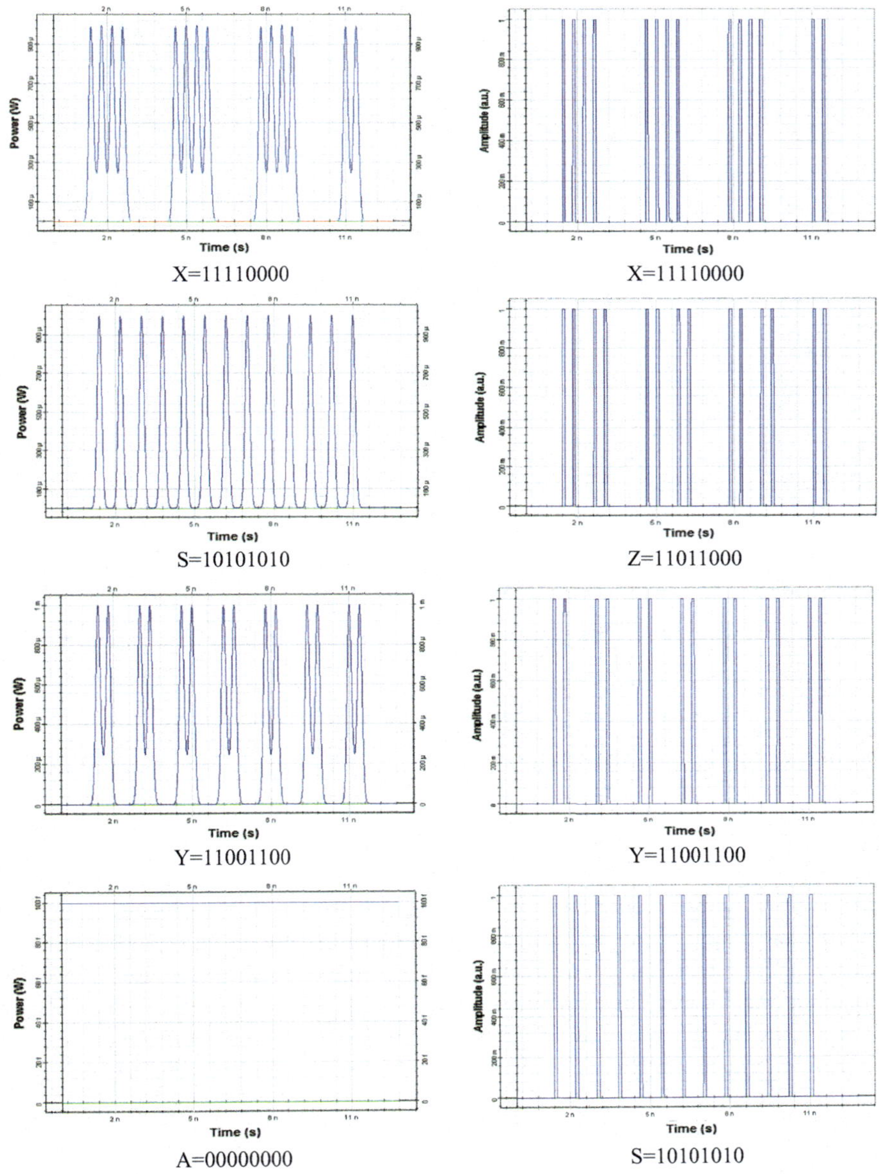

Fig. 4 Input and output waveforms of MUX

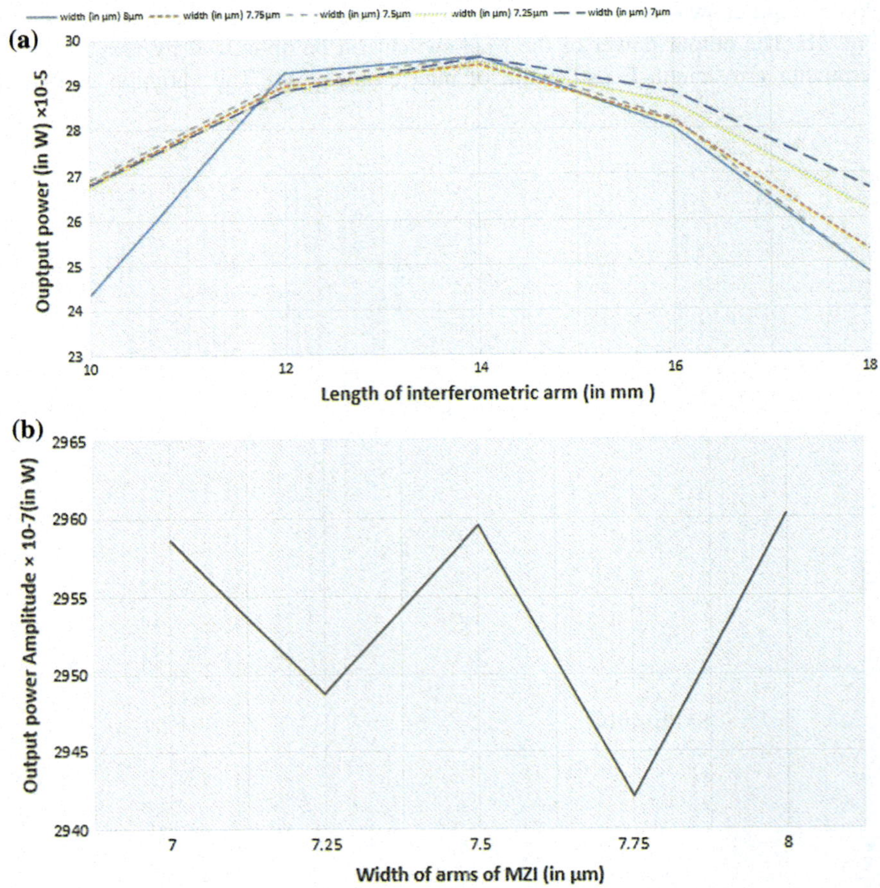

Fig. 5 **a** Output power versus L with W constant, **b** output power versus W with L constant

power by varying L at different values of W is shown in Fig. 5a, and the variation of power by varying W with keeping L constant is shown in Fig. 5b. As seen from Fig. 5a, the output power is maximum for interferometer length L = 14000 µm, and from Fig. 5b, the output is maximum for interferometer width W = 8 µm. So we can conclude that power consumption in MZI is optimized for interferometer length L = 14000 µm and width W = 8 µm. In future, many other optimizations can be done to fabricate the PIC of reversible 2 × 1 MUX with least die size and least power consumption.

4 Conclusion

In this paper, all-optical 2 × 1 MUX is designed and simulated based on reversible logic using Ti-indiffused Lithium Niobate-based Mach–Zehnder interferometer (MZI) switches. It has one ancilla line input to make the MUX reversible with four input and four outputs. It has two unused outputs that can be used in future for some other functioning. The power consumption in MZI is optimal with L = 14000 μm and W = 8 μm, as it is the only active device in the PIC of MUX. Hence, the overall power of MUX is optimized. As reversible logic is becoming a popular emerging paradigm because of its application in various emerging technologies, it can be used to design many logical circuits like adders, subtractors and shifters.

Acknowledgements The authors are grateful to India–Ukraine intergovernmental science and technology cooperation programme between the MNIT Jaipur (India) and the Lviv National Polytechnique Institute, Lviv (Ukraine): Project sanction no: INT/RUS/UKR/P-15/2015.

References

1. Shahar A (2007) All optical multiplexer. US patent no 7,174,069
2. Garipelly R, Kiran PM, Kumar AS (2013) A review on reversible logic gates and their implementation. Int J Emerg Technol Adv Eng 3(3):417–423
3. Kamalika D, Sengupta I (2014) All optical reversible multiplexer design using Mach-Zehnder interferometer. In: 27th International Conference on VLSI design and 2014 13th international conference on embedded systems, 2014. IEEE xplorer, pp 539–544. ISSN: 1063-9667. https://doi.org/10.1109/VLSID.2014.100
4. Kumar S, Chauhan C (2016) Design of reversible multiplexer using electro-optic effect inside lithium niobate-based Mach-Zehnder interferometers. Opt Eng 55(11):115101
5. Katti R, Prince S (2015) Implementation of a reversible all optical multiplexer using Mach-Zehnder interferometer. In: 2015 IEEE international conference on signal processing, informatics, communication and energy systems (SPICES), IEEE xplorer, pp 1–4. ISSN: 978-1-4799-1823-2
6. Ahmed MZ, Imran Ali SM et al (2016) Robust reversible multiplexer design using Mach-Zehnder interferometer. IRJET 3(3). ISSN: 2395-0072
7. Birnie DP (1993) Analysis of diffusion in lithium niobate. J Mater Sci 28(2):302–315. https://doi.org/10.1007/BF00357800
8. Singh G, Janyani V, Yadav RP (2011) Modeling of a 2 × 2 electro-optic Mach-Zehnder Interferometer optical switch with s-bend arms. Photon Lett Pol 3(3):119
9. Singh G, Janyani V, Yadav RP (2012) Modeling of a high performance Mach-Zehnder interferometer all optical switch. Opt Appl 42(3):613–625
10. Kumar H, Kumar L, Janyani V, Oleh B, Serhij U, Singh G (2016) Gray to binary code converter using Ti-indiffused lithium niobate based Mach-Zehnder interferometer. In: 3rd international conference on opto-electronics and applied optics, August 18–20, 2016
11. Waveguide optics modeling software system "Technical background and tutorials" version 8.0, 2nd edn (2006) Optiwave Inc

Quantum Well Width Effect on Intraband Optical Absorption in Type-II InAs/AlSb Nano-Scale Heterostructure

Nisha Yadav, Garima Bhardwaj, S. G. Anjum, K. Sandhya, M. J. Siddiqui and P. A. Alvi

Abstract In this paper, we have studied theoretically the effect of width variation of quantum well on energy dispersion curves and transverse electric (TE) and transverse magnetic (TM) Intraband optical absorption coefficients in type-II InAs/AlSb nanoscale heterostructure by utilizing eight bands Kohn–Luttinger Hamiltonian. The outcomes of the calculations made in this work suggest that the optical absorption and transition energies can be enhanced by reducing the width of quantum well (active) region of the designed heterostructure. One more observation is that the polarization modes have no effect on the behaviour of change in transition energy with change in well width.

Keywords InAs · AlSb · Heterostructure · Optical absorption Intraband transition

1 Introduction

The InAs/AlSb material system has been reported a very potential candidate among other materials for the application of pure electronics, for example, in fabrication of InAs/AlSb-based high electron mobility transistor (HEMT) and for optoelectronics [1–3]. Since the heterojunction between InAs and AlSb materials has a lattice-matched semiconductor junction of the type-II band alignment at the interfaces; therefore, the InAs/AlSb-based heterostructure has been attractive to ultrafast switching electronic devices due to its large conduction band offset and high

N. Yadav · K. Sandhya · P. A. Alvi (✉)
Department of Physics, Banasthali University, Rajasthan 304022, India
e-mail: drpaalvi@gmail.com

G. Bhardwaj
Department of Electronics, Banasthali University, Rajasthan 304022, India

S. G. Anjum · M. J. Siddiqui
Department of Electronics, F/o Engineering and Technology,
Aligarh Muslim University, Aligarh 202002, UP, India

V. Janyani et al. (eds.), *Optical and Wireless Technologies*, Lecture Notes in Electrical Engineering 472, https://doi.org/10.1007/978-981-10-7395-3_22

mobility [4]. Recently, it has been shown that type-II InAs/AlSb nanoscale heterostructure can be transformed into type-I structure, thereby increasing the interband optical gain [5]. The type-I heterostructures based on InN/InAlN material system have been studied in detail by using eight band k.p method [6]. Most of the type-II heterostructures based on ternary semiconductor compounds have been studied and found suitable for lasing action leading to a very high optical gain. For example, type-II InGaAs/GaAsSb nanoscale heterostructure have shown a very optical gain of the order of ~9000/cm [7].

In the following section of the paper, we have discussed device structure, theory applied and the outcomes of the calculations done in the work.

2 Device Structure Information and Theory

In the proposed type-II InAs/AlSb structure, there is single QW of InAs layer having width ~15 nm sandwiched between the barriers of the AlSb material having width ~5 nm. The barrier regions are doped with n-type dopants having concentration ~$1.5 \times 10^{18}/\text{cm}^3$. This structure is lattice matched because of its fabrication on InAs substrate. In such a heterostructure, we have negligible interband dipole matrix elements due to very small spatial separation between conduction band electrons of and the valence band holes of barriers.

Next, in order to know the effect of width variation of quantum well on optical absorption, we have simulated the four heterostructures of different well width but with unchanged the width of barrier region. Now, to study the optical absorption of all the heterostructure, we have solved the eight bands Kohn–Luttinger Hamiltonian for all structures separately for getting the carrier's wavefunctions and their probability density within the conduction band followed by momentum matrix elements and finally the optical absorption of the heterostructure of different well width.

The required expression of the absorption coefficient of the heterostructure can be given as [8];

$$\alpha(\omega) = \frac{4\pi^2 e^2}{n_r c m_o^2 V \omega} \sum_K \left|\hat{e} \cdot p_{ba}\right|^2 \delta(E_b - E_a + \hbar\omega)(f_a - f_b)$$

where c is speed of light in vacuum and n_r is the refractive coefficient. In the above equation, the functions f_a and f_b are the Fermi–Dirac distribution functions for energy states "a" and "b" and $\hat{e}$ is polarization vector of electric field, and it is taken as $\hat{e} = \hat{x}$ for TE mode; while for TM mode, it is taken as $\hat{e} = \hat{z}$. The quantity p_{ba} represents the momentum matrix element.

3 Results and Discussion

The eight bands Kohn–Luttinger Hamiltonian have been solved to obtain the required electronic wavefunctions associated with energy states in conduction band of quantum well (InAs) region of InAs/AlSb nanoscale heterostructure. These calculations have been done for four InAs/AlSb heterostructures of well width of 4, 6, 10 and 15 nm. By knowing the wavefunctions, the probability density of the electrons and discrete energy states within the conduction band of the quantum well region have also been calculated. Figure 1 shows the dispersion of energy states in conduction band of InAs quantum well of width (a) 15 nm, (b) 10 nm, (c) 6 nm, (d) 4 nm. We have shown the dispersion of conduction band energy states only, because of the intraband transitions between conduction band energy states. Only two types of intraband transitions are found responsible for achieving the optical absorption in the InAs/AlSb heterostructure. These transitions are as e_1 to e_2 and e_2 to e_3 including spins of the electrons (according to the selection rule). From Fig. 1a, b, c and d, it is clear that on reducing the well width from 15 to 4 nm, the dispersed energy states lose their parabolocity nature. Moreover, the energy states are also found to shift towards higher energy values with reducing the well width. One more important observation is that the value of *k* vector, at which the discrete energy states meet on moving from $k_{||} = 0$, also increases with reducing the well width.

After calculating the discrete energy states, we calculate the momentum matrix elements in order to evaluate the optical absorption coefficients. In Figs. 2 and 3, the calculated optical absorption coefficients within the TM and TE polarization modes for type-II InAs/AlSb nanoscale heterostructure have been plotted. The difference of TE and TM polarization modes can be seen on the values of optical absorption coefficients. From these figures, it can be predicted that the peak absorption falls down on increasing the well width from 4 to 15 nm. So, for achieving the higher optical absorption, the smaller value of quantum well width is required. Further, the behaviour of reducing the peak value of optical absorption with increasing the well width within both the polarization modes is summarized in Fig. 4a and b. Figure 4a and b shows that the behaviour of reduction of peak absorption with increasing the well width is exponential in nature. The well width also affects the transition energies and lasing wavelength of the heterostructure. This effect is shown in Fig. 5. The transition energy is found to decrease exponentially with increasing the well width. From Fig. 5, it is clear that the change in transition energy due to change in well width within both TE and TM modes for type-II InAs/AlSb nanoscale heterostructure is same, and thus there is no effect of polarization modes on the behaviour of change in transition energy with change in well width.

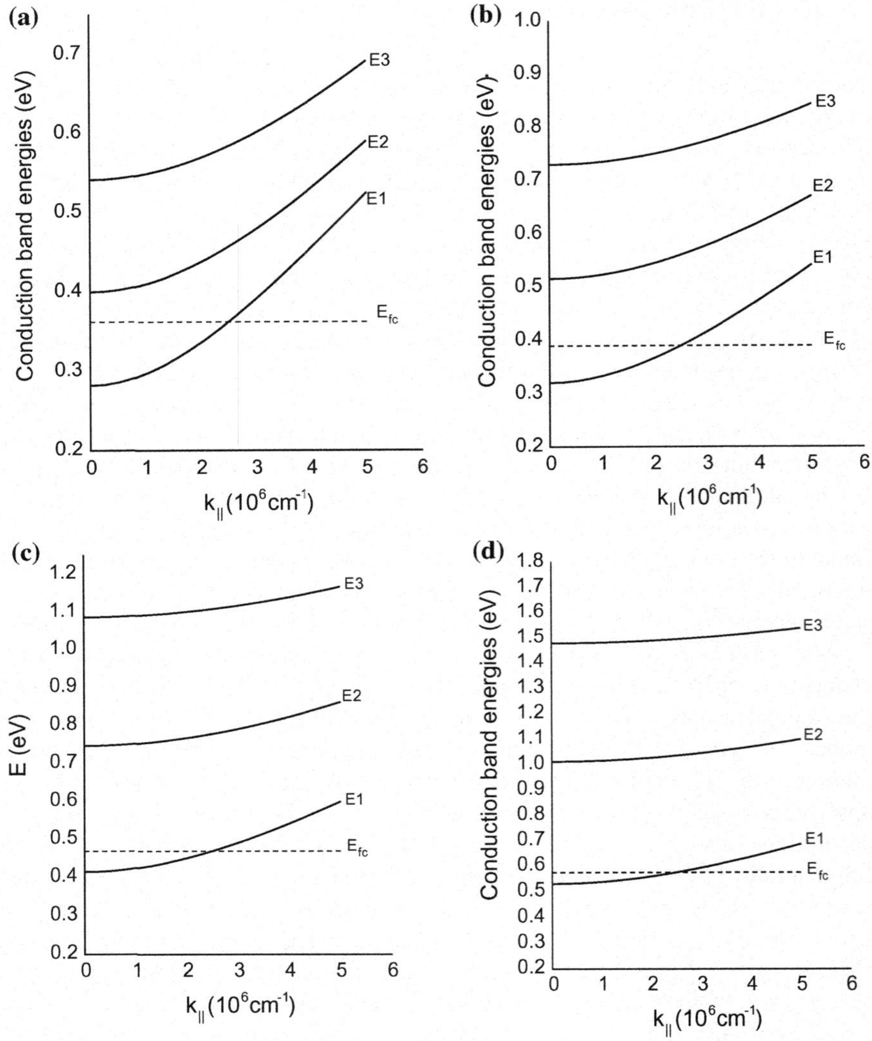

Fig. 1 Dispersion of energy states in conduction band of InAs quantum well of width **a** 15 nm, **b** 10 nm, **c** 6 nm, **d** 4 nm

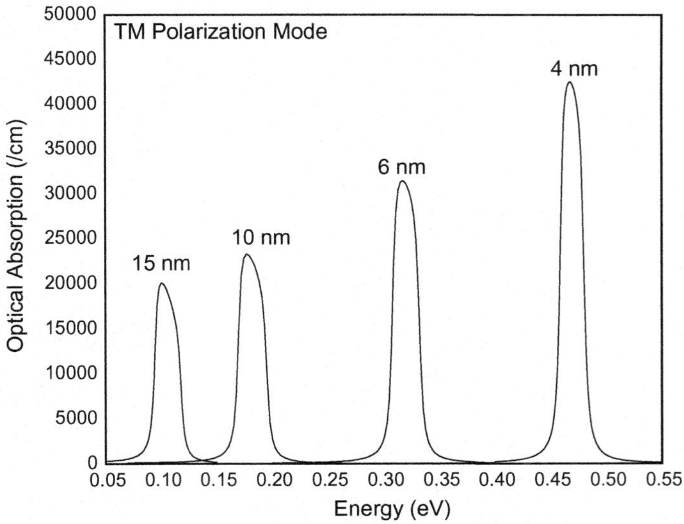

Fig. 2 Optical absorption (with in TM mode) of type-II InAs/AlSb nanoscale heterostructure

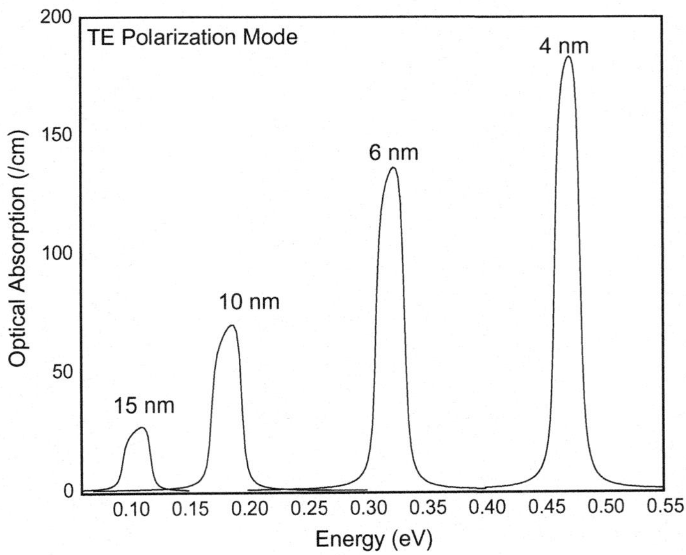

Fig. 3 Optical absorption (with in TE mode) of type-II InAs/AlSb nanoscale heterostructure

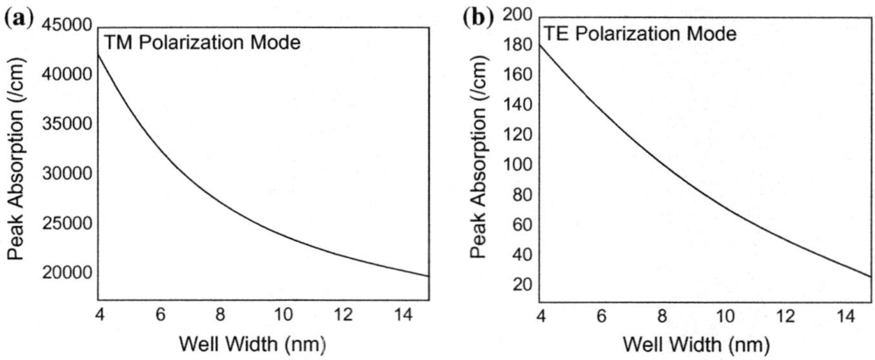

Fig. 4 Behaviour of peak absorption coefficients **a** with in TM mode, **b** with in TM mode for type-II InAs/AlSb nanoscale heterostructure

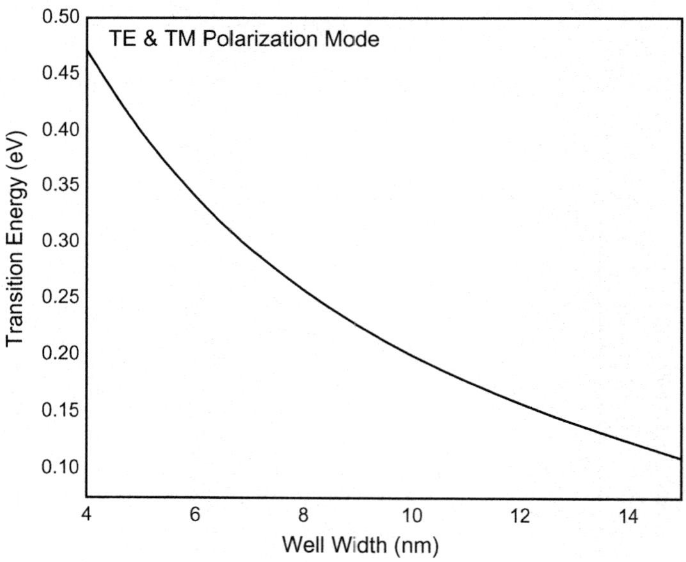

Fig. 5 Change in transition energy for both TE and TM modes for type-II InAs/AlSb nanoscale heterostructure

4 Conclusion

The eight bands Kohn–Luttinger Hamiltonian were solved to find the electronic wavefunctions and their probability density within the conduction band of the quantum well of the InAs/AlSb nanoscale heterostructures of different well width. According to the calculations, it was found that the intraband optical absorption and transition energies are enhanced by reducing the width of quantum well (active) region of the designed heterostructure. It has also been observed that the polarization modes have no effect on the behaviour of change in transition energy with change in well width.

Acknowledgements P. A. Alvi and Nisha Yadav are grateful to "Banasthali Center for Research and Education in Basic Sciences" under CURIE programme supported by the DST, Government of India, New Delhi. Authors are also thankful to Dr. Konstantin I. Kolokolov (Physics Department, Moscow State University, Moscow, Russia) for supporting the research work.

References

1. Moschetti G, Zhao H, Nilsson P-Å, Wang S, Kalabukhov A, Dambrine G, Bollaert S, Desplanque L, Wallart X, Grahn J (2010) Anisotropic transport properties in InAs/AlSb heterostructures. Appl Phys Lett 97:243510
2. Bennett BR, Magno R, Boos JB, Kruppa W, Ancona MG (2005) Antimonide-based compound semiconductors for electronic devices: a review. Solid State Electron 49:1875
3. Klokolov KI, Ning CZ (2003) Doping induced type-II to type-I transition and interband optical gain in InAs/AlSb quantum wells. Appl Phys Lett 83(8):1581
4. Ren S-F, Shen J (1997) Ab initio pseudopotential calculations of InAs/AlSb heterostructures. J Appl Phys 81:1169
5. Alvi PA (2017) Transformation of type-II InAs/AlSb nano-scale heterostructure into type-I structure and improving interband optical gain. Phys Status Solidi B. https://doi.org/10.1002/pssb.201600572
6. Alvi PA, Gupta S, Sharma M, Jha S, Rahman F (2011) Computational modeling of novel InN/ $Al_{0.30}$ $In_{0.70}$ N multilayer nano-heterostructure. Physics E 44:49
7. Nirmal HK, Yadav N, Rahman F, Alvi PA (2015) Optimization of high optical gain in type-II $In_{0.70}$ $Ga_{0.30}$ As/GaAs$_{0.40}$ Sb$_{0.60}$ lasing nano-heterostructure for SWIR applications. Superlattices and Microstruct 88:154–160
8. Chuang SL (1995) Physics of optoelectronic devices. Wiley, NY

Multi-material Photonic Crystal Fiber in MIR Region for Broadband Supercontinuum Generation

Shruti Kalra, Sandeep Vyas, Manish Tiwari and Ghanshyam Singh

Abstract In this paper, we have numerically investigated a multi-material photonic crystal fiber and simulated 1.6–4.2 μm mid-infrared supercontinuum generation. This *mid-infrared* broadband supercontinuum is observed for photonic crystal fiber of 100 mm length, when pumped with 85 fs laser pulses operating at 2.5 μm and peak power pulse is 350 W. The design of photonic crystal fiber with borosilicate and As_2S_3 glass has broad and flat dispersion profile with two zero dispersion wavelengths and high nonlinearity which helps in the generation of broadband supercontinuum.

Keywords PCF (Photonic crystal Fiber) · Chromatic dispersion A_{eff} (Effective mode area) · Supercontinuum generation (SCG)

1 Introduction

In the past few years, on the basis of the huge amount of development in fiber optic systems, the research has been conducted into the development of fiber and fiber-based devices suitable for near and mid-infrared wavelengths for different applications [1, 2]. As a resource, the demand of fiber-based supercontinuum

S. Kalra · S. Vyas
Department of ECE, Jaipur Engineering College & Research Centre, Jaipur, India
e-mail: shrutikalra.ece@jecrc.ac.in

S. Vyas
e-mail: vyas.sandeep@vitej.ac.in

M. Tiwari
Department of ECE, Manipal University, Jaipur, India
e-mail: manishtiwari@ieee.org

G. Singh (✉)
Department of ECE, Malaviya National Institute of Technology, Jaipur, India
e-mail: gsingh.ece@mnit.ac.in

V. Janyani et al. (eds.), *Optical and Wireless Technologies*, Lecture Notes in Electrical Engineering 472, https://doi.org/10.1007/978-981-10-7395-3_23

(SC) source [3, 4] has increased for both the visible and IR wavelengths due to the distinct applications in many areas. It has wide application in the field of telecommunication [5], to name few more are spectroscopy [6], microscopy [7], optical coherence tomography (OCT) [8], low coherence interferometry [9], wavelength division multiplexing (WDM) for high powers [10], the frequency metrology [11], and many more. Photonic crystal fiber (PCF) also known as micro-structured optical fibers (MOF) [12], in general built up by a central solid defect region that is called the core, encircled by multiple air-holes in a regular triangular/square lattice with the same air-hole diameter placement, has become an important requirement for supercontinuum source due to the availability of suitable pump sources [13]. The chalcogenide glass-MOF depicts some special properties in the mid-IR region, which makes them the best choice to be considered to realize devices for application in this wavelength region [14–16]. The dispersion profile of these photonic crystal fibers can be controlled easily by varying the diameter of air-hole (d) and the distance between the holes to hole (Λ) [17]. Controllability of chromatic dispersion in photonic crystal fiber is a major challenge for practical applications such as optical communication systems, nonlinear optics, dispersion compensation, and many more. Afterward different micro-structured optical fibers with large dispersion properties, for example, the wavelength of zero dispersion shifted to the visible and near-IR wavelength [18], and an ultra-flat chromatic dispersion [19, 20] has been examined. A PCF in which the zero dispersion wavelengths (ZDW) is blue shifted to the shorter wavelength and the dispersion profile with flat anomalous property had been well-realized for soliton construction in the visible wavelength window and accordingly expanded for supercontinuum generation (SCG) [21]. The second interesting method to achieve SCG broadband is based on normal flat dispersion profile. Fibers having such profiles dispersion are expected for the production of unique spectral profiles such as high spectral consistency and stability [22].

In this paper, we report that all-normal, ultra-flat, near-zero dispersion profiles can be obtained by the photonic crystal fiber structure with background As_2S_3 and replacing first internal ring and center air-holes by borosilicate glass. With the introduction of borosilicate, internal air-holes in the ring resulted in change in the effective index helping us to obtain desired dispersion profile keeping diameter of both air-hole and borosilicate rods identical. In 2015, Maji et al. [23] proposed a similar design of PCF for SCG in mid-IR range. With the proposed design, Maji investigated all-normal flattened dispersion of −0.087 ps/nm/km near 2.8 μm wavelength and supercontinuum generation in region of mid-IR for 740 nm bandwidth. Here, in this paper, we improved the PCF design and achieved dispersion curve in the anomalous dispersion region. Our optimized design is suitable for 1.6–4.2 μm SCG bandwidth with pump wavelength at 2.5 μm.

In this paper, the sequence is under mentioned. Sect. 2 of the paper presents the study of modal properties of multi-material chalcogenide PCF designed with As_2S_3 and borosilicate. In order to determine the values of refractive index and few other parameters of the fiber, we have used full modal analysis method along with and solutions of second-order sellmeier equation. The supercontinuum generation results in few millimeters (mm) long multi-material photonic crystal fiber are discussed in Sect. 3. The supercontinuum generation is analyzed using generalized nonlinear Schrödinger equation.

2 Method of Analysis

To analyze the optical properties of photonic crystal fiber, the finite-difference time-domain (FDTD) method is used. The total dispersion or chromatic dispersion (D_c) is the addition of waveguide dispersion (D_W) and material dispersion (D_M) i.e. $D_c(\lambda) = D_M(\lambda) + D_W(\lambda)$. Chromatic dispersion $D_c(\lambda)$ can be calculated using the following relation.

$$D_c(\lambda) = -\frac{\lambda}{c}\frac{d^2 n_{eff}}{d\lambda^2} \tag{1}$$

In the given relation, c is speed of the light in vacuum, λ is the wavelength of operation, and n_{eff} is the effective index (real part) which can be defined as $n_{eff} = \beta/2\pi$ where β is propagation constant [24]. We can calculate the wavelength-related refractive index of As_2S_3 glass and borosilicate material by following sellmeier Eqs. (2) and (3), respectively, as [23].

$$n^2(\lambda) = 1 + \frac{1.898367\lambda^2}{\lambda^2 - 0.15^2} + \frac{1.922297\lambda^2}{\lambda^2 - 0.25^2} + \frac{0.87651\lambda^2}{\lambda^2 - 0.35^2} + \frac{0.11887\lambda^2}{\lambda^2 - 0.45^2} + \frac{0.95699\lambda^2}{\lambda^2 - 27.3861^2} \tag{2}$$

and

$$n^2(\lambda) = 1 + \frac{0.967\lambda^2}{\lambda^2 - 0.020452} + \frac{0.00511\lambda^2}{\lambda^2 - 107.9261} + \frac{0.000233\lambda^2}{\lambda^2 - 0.34938} \tag{3}$$

The nonlinear coefficient (γ) can be evaluated by the equation given below [24]

$$\gamma = \frac{2\pi n_2}{\lambda A_{eff}} \tag{4}$$

where n_2 is the refractive index of nonlinear material, λ is pumped wavelength and A_{eff} is effective area of fundamental mode. A high value of nonlinear coefficient (γ) is desirable for broadband supercontinuum generation, which can be obtained by carefully selecting a material which offers a high nonlinear refractive index and designing the photonic crystal fiber with small effective mode area (A_{eff}). The effective area of the propagating mode (A_{eff}) of the photonic crystal fiber can be determined with the given relation [24].

$$A_{eff} = \frac{\left(\int_{-\infty}^{\infty} |E|^2 dxdy\right)^2}{\int_{-\infty}^{\infty} |E|^4 dxdy} \quad (5)$$

where, the electric field distribution is E. The confinement loss is calculated by relation [24] CL = 8.686 k_O $Im(n_{eff})$, where the imaginary part of effective refractive index is $Im(n_{eff})$ and $k_O = 2\pi/\lambda$ is wave number in the free space [25].

The following generalized nonlinear Schrödinger equation (GNLSE) numerically relates the pulse propagation in the photonic crystal fiber. The output pulse envelop ($A(z, t)$) equation has been calculated using the split-step Fourier method.

$$\frac{\partial A}{\partial z} + \frac{\alpha}{2}A - \left(\sum_{n\geq 2} \beta_n \frac{i^{n+1}}{n!}\frac{\partial^n A}{\partial t^n}\right) = i\gamma\left(1 + \frac{i}{\omega_0}\frac{\partial}{\partial t}\right)\left[A\ (z,t) \int_{-\infty}^{\infty} R(t')|A\ (z,t\ - t')|^2 dt'\right] \quad (6)$$

In the expression above, γ is nonlinear coefficient, β_n is the nth derivative of the propagation constant β, z, and t are distance and time, respectively, τ is the envelope group velocity retarded time traveling and $A(z, t)$ is optical field envelope. In this equation, left-hand side part is used to study linear propagation and right-hand side part is used to study the nonlinear propagation effects [24].

The Raman response function of material is expressed as the contribution of both instantaneous electronic and vibrational Raman contributions [24].

$$R(t) = (1 - f_r)\delta(t) + f_r h_R(t) \quad (7)$$

This is to our knowledge that photo-induced effects play a vital role contributing to regulate the optical properties of the chalcogenide materials, namely photo-darkening and photo-induced anisotropy [26]. It can be observed that on increase in the input power with increase in function of the irradiation time, then Raman gain spectrum varies and portrays oscillatory changes and gets vanished after reaching to final saturation state [26]. The ($h_R(t)$) is the temporal Raman response and in relation to Raman gain spectrum can be expressed as [27]

$$g_R(f) = \frac{2\omega_p}{c} n_2 f_r \mathrm{Im}[H_R(f)] \quad (8)$$

where, pump frequency is ω_p, imaginary part of the Fourier transform of temporal raman response $h_R(t)$ is $\mathrm{Im}[H_R(f)]$ [28], and $h_R(t)$ is expressed as

$$h_R(t) = \frac{\tau_1^2 + \tau_2^2}{\tau_1 \tau_2^2} \exp\left(-\frac{t}{\tau_2}\right) \sin\left(\frac{t}{\tau_1}\right) \quad (9)$$

where, $\tau_1 = 15.5$ fs and $\tau_2 = 230.5$ fs for As_2S_3-based glasses. Here, fractional contribution has been got to be equal to 0.2 [23]. *Kramers-Kroning* relation can be expressed as

$$f_r = \frac{\lambda}{2\pi^2 n_2}\int_0^\infty \frac{g_R(f)}{f}\,df \tag{10}$$

In this paper, we have considered hyperbolic secant pulse as input pulse, which can be verbalized as

$$A(z=0,t) = \sqrt{P_0}\,\mathrm{sec\,h}\left(\frac{t}{t_0}\right)\exp\left(-i\frac{C}{2}\frac{t^2}{t_0^2}\right) \tag{11}$$

In the expression, peak power of input pulse is P_0 and $t_0 = t_{FWHM}/1.7627$.

3 Proposed PCFs Design and Analysis

Here, a new idea is proposed to obtain high refractive index by the replacement of the air-holes with material from a series of high refractive index. To obtain this tailored application, particular fiber designs for mid-IR applications, we have considered an arsenic sulfide (As_2S_3)-based photonic crystal fiber structure which consists of four rings with hexagonal geometry. The proposed design has four rings forming the cladding, where the air-hole of inner ring and center air-hole is replaced with borosilicate rods as highlighted in red color in the schematic of the PCF design in Fig. 1 with a background of As_2S_3. The diameter of each air-hole is kept constant in the proposed design.

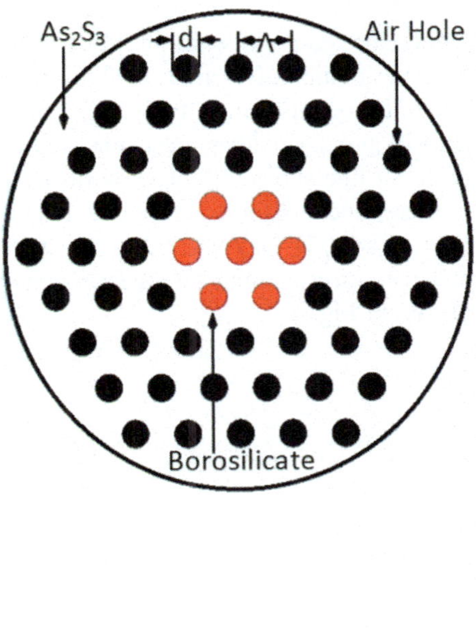

Fig. 1 Schematic design of proposed photonic crystal fiber

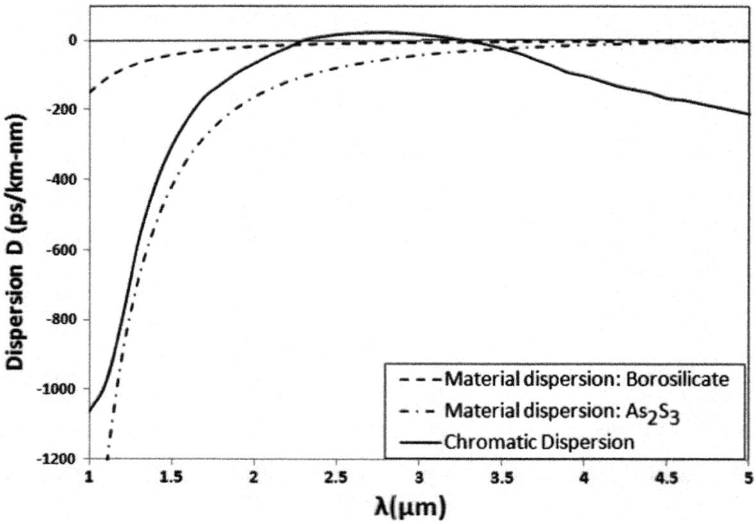

Fig. 2 Dispersion plot of proposed photonic crystal fiber

The chalcogenide As_2S_3-based photonic crystal fiber offers high nonlinearity coefficient (γ) making it most popular and suitable for SCG. The dispersion of the fiber is calculated by investigating the effective refractive index (n_{eff}) of the fundamental mode of the photonic crystal fiber. The proposed photonic crystal fiber design consists of four rings with uniform fixed $d/\Lambda = 0.5310734$ ratio.

Fig. 2 displays the variation in chromatic dispersion and material dispersion change with wavelength of the proposed fiber for photonic crystal fiber. The values of pitch and air-hole diameter were selected as $\Lambda = 1.77$ μm and $d = 0.94$ μm respectively, to obtain a flat dispersion profile for the range 1.5–4.3 μm. The two zero dispersion wavelengths (ZDW) obtained from dispersion plot were 2.27 μm and 3.3 μm. A flat and low dispersion profile helps enhancing the SCG. It can reduce the temporal walk-off effect at the time of spectral boarding process [29]; here the pump wavelength of 2.5 μm was used in the anomalous dispersion for photonic crystal fiber. In Fig. 2, the black solid line shows the chromatic dispersion (D_c) of proposed fiber, dash line shows material dispersion of borosilicate, and dash-dot line shows the material dispersion of As_2S_3 glass.

Here, the nonlinear refractive indices of As_2S_3-based chalcogenide glass is $n_2 = 4 \times 10^{-18}$ m^2/W at 2.5 μm [30]. The variation in effective mode area (A_{eff}) and the corresponding nonlinear coefficient of the proposed fiber have been displayed in Fig. 3.

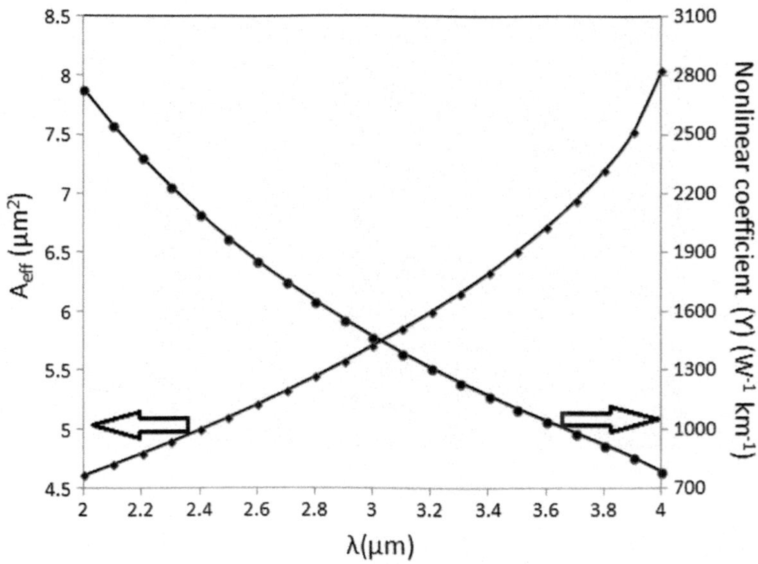

Fig. 3 Effective mode area and nonlinear coefficient of PCF

4 Supercontinuum Generation in Proposed PCF

The proposed fiber design was simulated for SCG. The spilt-step Fourier method was used to solve GNLSE [24]. A PCF of 100 mm length was selected and was pumped with optical pulses of 2.5 μm for the numerical simulation. At this pump wavelength, the value of A_{eff} obtained was 5.10019 μm^2 and γ was 1972 W^{-1} km^{-1}. The pumped wavelength 2.5 μm is in anomalous dispersion regime, and flat dispersion is obtained for the range from 1.5 to 4.3 μm. The increase in anomalous regime pumping results in spectral broadening in long wavelength side. The solitons fission and Raman soliton self-frequency shift are the main reasons for this. In the initial stage, the broadening of pulse is mainly due to SPM, and as the pulse propagates a few millimeters, the spectrum breaks into two major peaks; after moving outward from the original pump wavelength, the number of sub-peaks in between two major peaks keeps on increasing continuously. The pulse duration is 85 fs (FWHM) with the peak power as 350 W. Figures 4 and 5 portray spectral and temporal spans of SCG in the As_2S_3-based PCF. The spectral escalation of input pulse for different input power has been displayed in Fig. 4. The SC spectral broadening is depicted in Fig. 6.

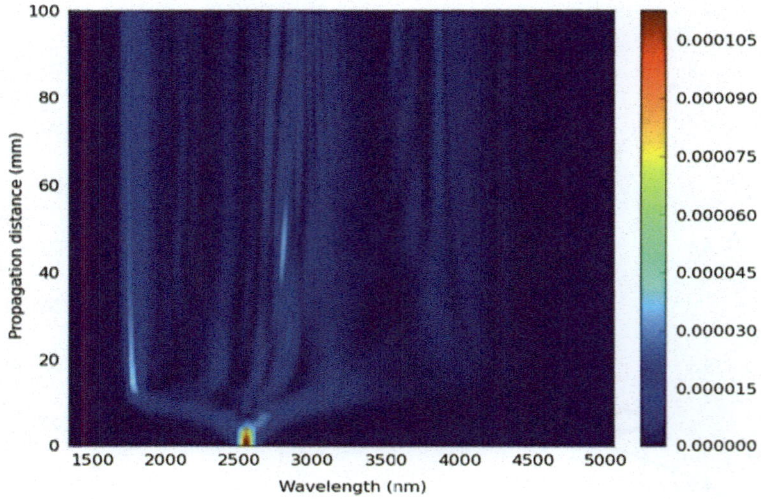

Fig. 4 Spectral escalation of SC in PCF designed

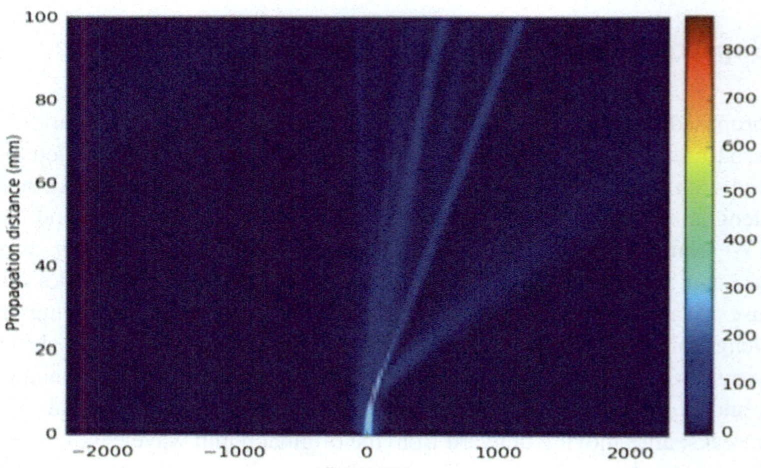

Fig. 5 Temporal escalation of SC in PCF designed

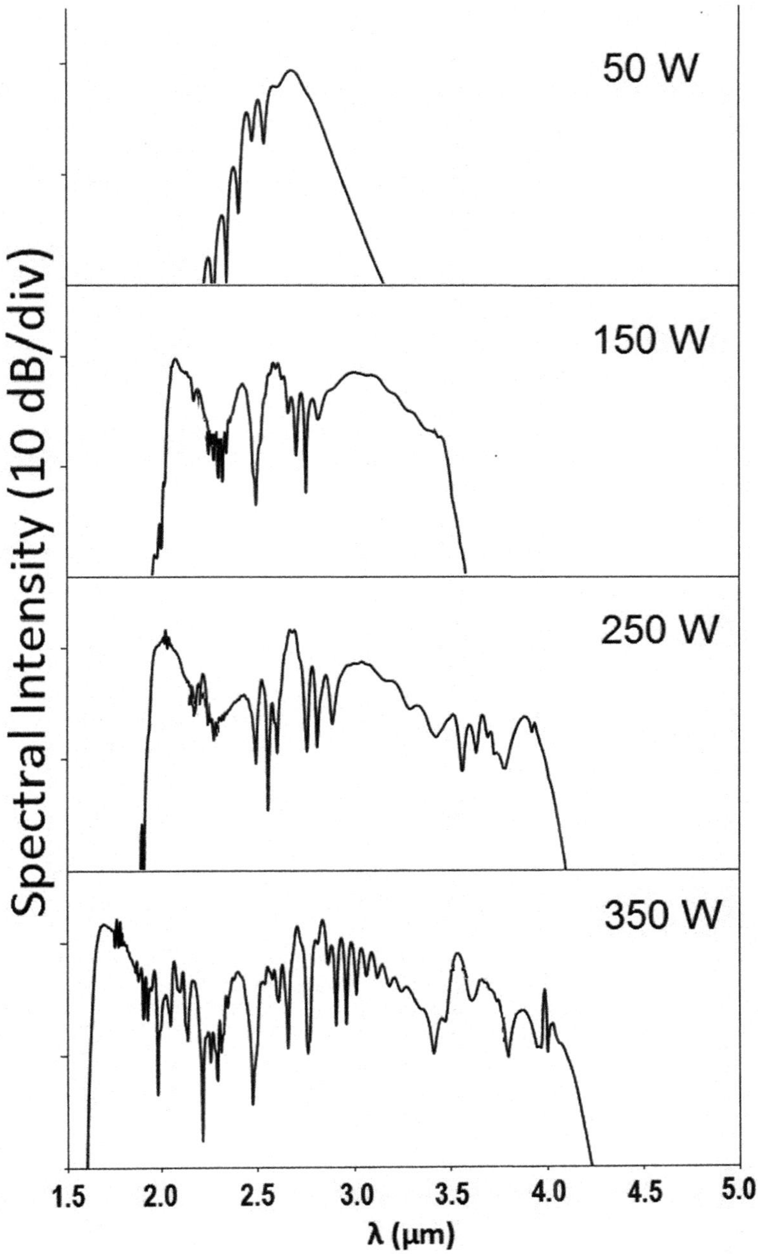

Fig. 6 Spectra of SC for different input power

5 Conclusion

In the paper presented here, a multi-material photonic crystal fiber with hexagonal lattice arrangement of air-hole extending to 100 mm length is investigated for the generation of supercontinuum. The range for flat dispersion of the proposed photonic crystal fiber design is from 1.5 to 4.3 μm. The nonlinear coefficient (γ), effective area (A_{eff}), and dispersion at the wavelength (λ) 2.5 μm are 1972 W^{-1} km^{-1}, 5.10019 μm^2 and 18.5545 ps/km/nm, respectively, for the design of the fiber. The photonic crystal fiber was numerically simulated for SCG using 85 fs pump pulses of 350 W peak powers at a wavelength 2.5 μm. The results of simulation generate supercontinuum for the bandwidth values from 1.6 to 4.2 μm with 350 W peak powers.

Acknowledgments The research work proposed in the paper is the carried out under the collaborative activity of the faculty members of Malaviya National Institute of Technology, Jaipur, and Manipal University, Jaipur.

References

1. Barh A, Ghosh S, Agrawal GP, Varshney RK, Aggarwal ID, Pal BP (2013) Design of an efficient mid-IR light source using chalcogenide holey fibers: a numerical study. J Opt 15:035205
2. Barh A, Ghosh S, Varshney RK, Pal BP (2013) An efficient broadband mid-wave IR fiber optic light source: design and performance simulation. Opt Express 21:9547–9555
3. Dudley JM, Gentry G, Coen S (2006) Supercontinuum generation in photonic crystal fiber. Rev Modern Phys 78(4):1135–1184
4. Vyas S, Tanabe T, Tiwari M, Singh G (2016) Ultraflat broadband supercontinuum in highly nonlinear $Ge_{11.5}As_{24}Se_{64.5}$ photonic crystal fibres. Ukr J Phys Opt 17:132–139
5. Tamura KR, Kubota H, Nakazawa M (2000) Fundamentals of stable continuum generation at high repetition rates. IEEE J Quantum Electron 36:773–779
6. Hult J, Watt RS, Kaminski CF (2007) High bandwidth absorption spectroscopy with a dispersed supercontinuum source. Opt Express 15:11385–11395
7. Kaminski CF, Watt RS, Elder AD, Frank JH, Hult J (2008) Supercontinuum radiation for applications in chemical sensing and microscopy. Appl Phys B 92:367–378
8. Hartl I, Li XD, Chudoba C, Ghanta RK, Ko TH, Fujimoto JG, Ranka JK, Windeler RS (2001) Ultrahigh-resolution optical coherence tomography using continuum generation in an air-silica microstructure optical fiber. Opt Lett 26:608–610
9. Shi K, Li P, Liu Z (2007) Broadband coherent anti-Stokes Raman scattering spectroscopy in supercontinuum optical trap. Appl Phys Lett 90:141116
10. Morioka T, Kawanishi S, Mori K, Saruwatari M (1994) Transformlimited, femtosecond WDM pulse generation by spectral filtering of gigahertz supercontinuum. Electron Lett 30:1166–1168
11. Reichert J, Udem T, Hänsch TW, Knight JC, Wadsworth WJ, Russell PStJ (2000) Optical frequency synthesizer for precision spectroscopy. Phys Rev Lett 85:2264–2267
12. Russell PStJ (2006) Photonic-crystal fibers. J Lightwave Technol 24:4729–4749
13. Saitoh K, Koshiba M, Mortensen NA (2006) Nonlinear photonic crystal fibers: pushing the zero-dispersion towards the visible. New J Phys 8:207–215

14. Cardinal T, Richardson KA, Shim H, Schulte A, Beatty R, Le Foulgoc K, Meneghini C, Viens JF, Villeneuve A (1999) Non-linear optical properties of chalcogenide glasses in the system As-S-Se. J Non-Cryst Solids 256:353–360
15. Zakery A, Elliott SR (2003) Optical properties and applications of chalcogenide glasses: a review. J Non-Cryst Solids 330:1–12
16. Boudebs G, Cherukulappurath S, Guignard M, Troles J, Smektala F, Sanchez F (2004) Linear optical characterization of chalcogenide glasses. Opt Commun 230:331–336
17. Vyas S, Tanabe T, Tiwari M, Singh G (2016) Chalcogenide photonic crystal fiber for ultraflat mid-infrared supercontinuum generation. Chin Opt Lett 14:123201
18. Gander MJ, McBride R, Jones JDC, Mogilevtsev D, Birks TA, Knight JC, Russell PStJ (1999) Experimental measurement of group velocity dispersion in photonic crystal fibre. Electron Lett 35:63–64
19. Ferrando A, Silvestre E, Andrés P, Miret JJ, Andrés MV (2001) Designing the properties of dispersion flattened photonic crystal fibers. Opt Express 9:687–697
20. Reeves WH, Knight JC, Russell PStJ, Roberts PJ (2002) Demonstration of ultra-flattened dispersion in photonic crystal fibers. Opt Express 10:609–613
21. Dudley JM, Taylor JR (2010) Supercontinuum generation in optical fibers. Cambridge
22. Heidt AM (2010) Pulse preserving flat-top supercontinuum generation in all-normal dispersion photonic crystal fibers. J Opt Soc Am B 27:550–559
23. Maji PS, Chaudhuri PR (2015) Design of all-normal dispersion based on multi-material photonic crystal fiber in IR region for broadband supercontinuum generation. Opt Soc Am 54:4042–4048
24. Agarwal GP (2007) Nonlinear fiber optics, 4th edn. Academic
25. Vyas S, Tanabe T, Singh G, Tiwari M (2016) Broadband supercontinuum generation and Raman response in $Ge_{11.5}As_{24}Se_{64.5}$ based chalcogenide photonic crystal fiber. In: IEEE international conference on computational techniques in information and communication technologies (ICCTICT), pp 607–611. https://doi.org/10.1109/ICCTICT.2016.7514651
26. Salem AB, Cherif R, Zghal M (2011) Raman response of a highly nonlinear As_2Se_3-based chalcogenide photonic crystal fiber. Progress in electromagnetics research symposium proceedings, pp 20–23. Marrakesh, Morocco
27. Kärtner FX, Dougherty DJ, Haus HA, Ippen EP (1994) Raman noise and soliton squeezing. J Opt Soc Am B 11:1267–1276
28. Vyas S, Tanabe T, Tiwari M, Singh G (2016) Mid-infrared supercontinuum generation in $Ge_{11.5}As_{24}Se_{64.5}$ based chalcogenide photonic crystal fiber. In: IEEE international conference advances in computing, communications and informatics (ICACCI), pp 2547–2552. https://doi.org/10.1109/ICACCI.2016.7732436
29. Yuan W (2013) 2–10 μm mid-infrared supercontinuum generation in As_2Se_3 photonic crystal fiber. Laser Phys Lett 10:095107
30. Wei D-P, Galstian TV, Smolnikov IV, Plotnichenko VG, Zohrabyan A (2005) Spectral broadening of femtosecond pulses in a single-mode As-S glass fiber. Opt Express 13:2439–2443

WDM-Based Visible Light Communication System

Anil Kumar Jangir, Baria Dipikaben Manharbhai and Ravi Kumar Maddila

Abstract Visible light communication (VLC) is a wireless technology that uses visible light as a carrier to transmit the signal. Visible white light is the combination of three basic colors that are red, green and blue. These different colors are generated by using different color LEDs which are operated at different wavelengths according to their respective colors. These LEDs are operated by different data signals, and then by using the wavelength division multiplexing (WDM), these optical modulated data signals are converted into a single optical multiplexed data stream. At the receiver side, this optical data stream is divided into different colors or wavelengths by using color filters and then by using photo detectors and hence, their respective data are decoded. Due to WDM data is transmitted in parallel form so the higher data rate can be achieved. In this paper, a WDM-based VLC system is demonstrated on Opti-System software. The achievable data rate is up to 100 Mbps with the maximum bit error rate of 5.3×10^{-8} which is less than the pre-forward error correction limit i.e., 3.8×10^{-3}.

Keywords Visible light communication · Wavelength division multiplexing LED · LASER · Orthogonal frequency division multiplexing

1 Introduction

As the technology is improving in the wireless field, the traffic in the radio spectrum increases rapidly, and this results in a huge congestion. Most of the communication-based applications and systems which are implemented without using any wire, use radio spectrum for communication purpose. The radio spectrum is in the range of 3 kHz–3 THz. There is another wide-band spectrum which can also be used for communication purpose that is visible light spectrum which is in

A. K. Jangir (✉) · B. D. Manharbhai · R. K. Maddila
Department of Electronics and Communication Engineering,
Malaviya National Institute of Technology, Jaipur, India
e-mail: 2015pwc5445@mnit.ac.in; 2015pwc5334@mnit.ac.in

V. Janyani et al. (eds.), *Optical and Wireless Technologies*, Lecture Notes in Electrical Engineering 472, https://doi.org/10.1007/978-981-10-7395-3_24

the range of 428–750 THz or in terms of wavelength from 400 to 750 nm. There are numerous advantages of visible light spectrum over the radio spectrum i.e., it is a license-free broad spectrum, free from electromagnetic interference, less hazardous to human eyes, transmission speed is higher than radio waves, etc. Also the light can be enclosed in a small geographical area, so the privacy and security can also be improved [1]. Firstly, Alexander Graham Bell had transmitted the optical signal to his newly invented photo phone in 1880 [2]. The research on visible light spectrum (VLC) is initiated by VLC consortium, Japan [3] and the first VLC standard is given by Institute of Electrical and Electronics Engineering (IEEE) as IEEE802.15.7 [4].

In visible light communication system, LEDs or LASERs are used as sources and photo diodes and photo transistors are used as detectors. The signal which is to be transmitted is modulated in the light form which is visible to human eyes and transmitted through free space. As the technology of LED is improving, fluorescent bulbs are replaced by LEDs in building infrastructure and vehicle lights that is much suitable infrastructure for VLC [5]. The data rate of the VLC system depends on some factors like modulation techniques used to modulate the information signal, switching speed of light source and responsibility of photo detector. Most commonly used light source in VLC is white LED which fulfills two requirements that are illumination and communication. There are two techniques to get the white light. The first one is by passing the output of blue LED through yellow phosphor i.e., YAG (Yellow Atrium Grating) but it has limited bandwidth up to 400 MHz [1] because of the presence of the remaining carriers when it turns ON from OFF. The second one is by combining the output of red, green and blue LEDs, this white light has high bandwidth and also provides the parallel transmission. Nowadays some special purpose LEDs are also available like resonant cavity LED has higher bandwidth as compared to general purpose LED. LED and laser have switching speed in the range of MHz and GHz that cannot be perceived by human eyes. The LED has broad spectral width but the laser has narrow spectral width so it cannot be used for indoor illumination purpose and only used for communication. There are some another techniques to improve the data rate of the VLC system. By using red and green laser diodes at the transmitter, side and pre-amplifier and adaptive filtering at receiver side, data rate up to 500 Mbps can be achieved [6]. The combination of RGB luminary LED and discrete modulation technique (DMT) provides the data rate of 813 Mbps [7]. A high-speed 680 nm laser diode with 600 micro-watt transmitted power and illumination of 250 lx can be used to generate the data rate of 2.5 Gbps [8]. Data rate of 3 Gbps can be achieved by using double optical diversity (DOD) technique in time and frequency, i.e., the transmission of a signal with its delayed version by multiplying with orthogonal frequencies simultaneously is DOD [4]. A quadrature amplitude modulation (QAM) with orthogonal frequency division multiplexing (OFDM) is more popular to increase the data rate in VLC system, e.g., when a laser diode is driven by a 4-QAM OFDM signal then the data rate of 10 Gbps can be achieved [9], a 12.5 Gbps data rate can be achieved by using vertical cavity surface emitting laser (VCSEL) diode with 16-QAM OFDM signal and if eight channels similar to this are combined and form a MIMO system then it is possible to get 100 Gbps data rate [10].

Q-factor which is a measure of goodness of a system and can be given by the ratio of power stored to the power dissipated in the circuit reactance and resistance [11].

$$\text{i.e. } Q = \frac{P_{stored}}{P_{dissipated}} \tag{1}$$

Higher value of Q-factor corresponds to narrower bandwidth. This relation is given by,

$$Q = \frac{f_c}{BW} \tag{2}$$

And we also know that

$$f_c = \frac{c}{\lambda} \tag{3}$$

Hence, from the Eqs. (2) and (3), we can write

$$Q = \frac{c}{BW * \lambda} \tag{4}$$

Q-factor characterizes the signal based on signal to noise ratio. In practical system, it is determined by the difference between signal level and noise level. Distance of the signal level from the noise level can be determined from,

$$DSNR = 10 \log\left(\frac{P_i}{N_i}\right) + 10 \log\left(\frac{B_m}{B_r}\right) \tag{5}$$

P_i average power of the optical signal [W]
N_i average power of the noise signal [W]
B_m spectral width [nm]
B_r reference BW

2 WDM-Based VLC System

Wavelength division multiplexing is a technique to transmit more than one data stream simultaneously through a single channel. In this, the complete channel is divided into different slots according to different wavelengths and each source of signal is allocated a separate wavelength slot. All signals travel simultaneously on the same channel, and at the receiver side, the data streams are distinguished and forwarded to the respective destinations. In VLC system, the channel has the spectrum of 400 to 750 nm, i.e., an optical domain. The light of different wavelengths can be generated by using different color sources. Every color has different wavelengths. If we consider white light, then the purpose of indoor lightening as well as communication can be fulfilled and it contains the complete visible spectrum. If the output of red, green and blue LED is combined then it generates low-quality white light but the combination of red, orange, yellow, green, violet, blue and indigo results in good quality white light and the channel utilization is also increased. As the spectral width of LED is broad, the separation between the channels needs to be maintained in order to prevent them from intersymbol interference (ISI). But if pointed lasers are used which have very narrow spectral width and the channel spacing is around 1–2 nm then the channel can be divide into more slots and higher number of data streams can be transmitted simultaneously. This is referred as dense wavelength division multiplexing (DWDM). Due to this, the data rate is also enhanced because of simultaneous transmission of different data signals and higher switching speed of laser.

Here, a system shown in Fig. 1 is demonstrated on Opti-System software. In this, different signals are generated in the form of square wave. These signals modulate the light by switching the states of eight different color LEDs. In this particular work, the intensity modulation technique is used i.e., the intensity of output light is varying according to the amplitude of applied input signal. The output light from these different color LEDs is combined by using passive optical devices like lenses, resulting in white light. This is referred as optical wavelength division multiplexing or color division multiplexing. This multiplexed signal is transmitted through free space. In this case, the transmission distance is limited and it requires LOS path but the achievable data rate is very high. Here, the system is simulated for the distance of 1500 m, suitable for indoor environment. At receiver

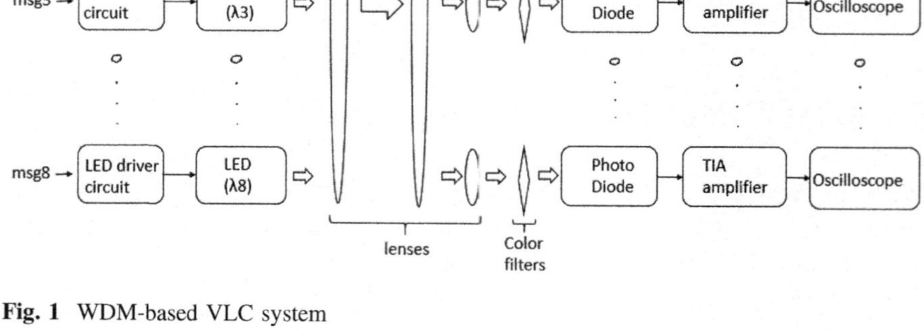

Fig. 1 WDM-based VLC system

side, by using different color filters, each color light is separated from the white light. These color filters are passive optical components and act as the band pass filter for their respective color wavelengths. Then by using different photo detectors for each color signal, the transmitted light is detected and converted into electrical domain and compared with the pre-FEC limit. As the achievable BER is less than the FEC limit, the system is acceptable.

This system can be easily developed on hardware. Different color LEDs are driven individually by using different data signals. According to their wavelengths, different color filters and photo diodes are used for each LED. To increase the distance between transmitter and receiver optical power level has to be increased which can be achieved by using high-power LEDs, special LEDs (Resonant Cavity coupled LED) or LED arrays.

3 Simulation Results

The achievable data rate for this system is 100 Mbps with BER less than the FEC limits i.e., 3.8×10^{-3} for the distance which is varied from 100 to 1700 m. The spectrum of white light containing the all color wavelengths is as shown below (Fig. 2):

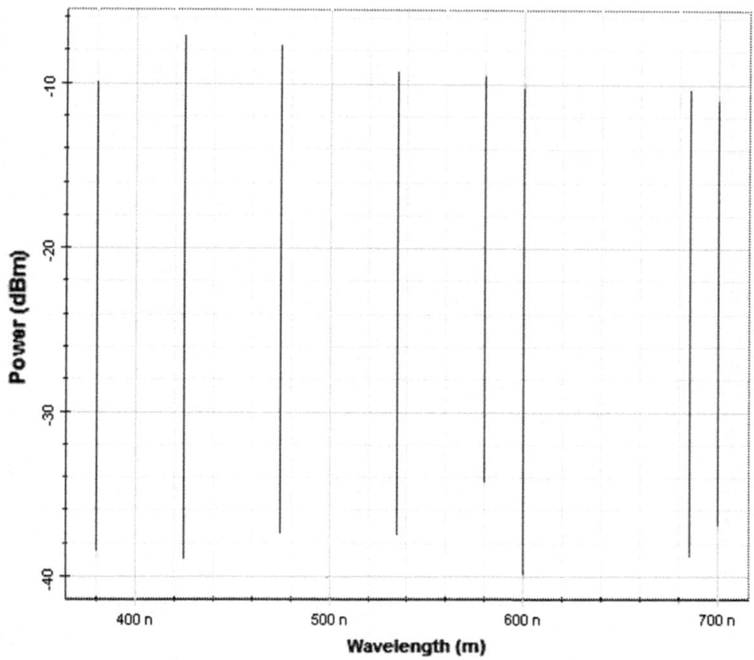

Fig. 2 Spectrum of white light

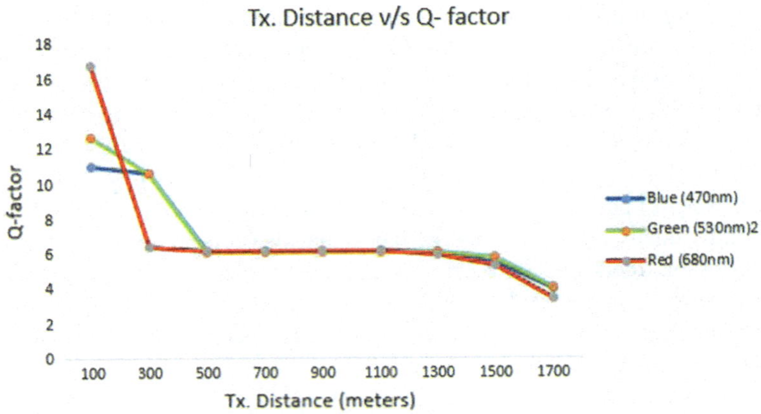

Fig. 3 Plot of transmission distance versus Q-factor

Resulting Q-factor plot for different colors with respect to transmission distances is shown below. It shows that as the transmission distance increases the BER increases. The best BER plot is of 530 nm wavelength of green color (Fig. 3).

4 Conclusion

This system is demonstrated for transmission distance of 1500 m, and we are getting the data rate of 100 Mbps with the maximum bit error rate of 5.3×10^{-8}, which is very low as compare to pre-forward error correction limit. The optical power level can be increased by using the LED arrays so the transmission distance can also be increased. It also can be increased by using pointed laser diodes which provide the high-intensity monochromatic light, which also increases the data rate. The data rate can also be increased by improving modulation techniques. One of the most suitable techniques for VLC is OFDM with QAM modulation. As the 4-QAM OFDM modulation technique provides the data rate of 10 Gbps, by using 16, 32, 64, 128 or 256—QAM OFDM signal can be further increased. By applying the WDM, multiplexing with higher QAM OFDM signal data rate of more than 100 Gbps can be achieved.

References

1. O'Brien D, Collins S, Chun H, Faulkner G, Rajbhandari S, Watt A, Henderson R (2014, October). Visible light communications: improving data rate, link margin and field of view. In: 2014 IEEE photonics conference (IPC). IEEE, pp 128–129

2. Komine T (2005) Visible light wireless communications and its fundamental study. Doctoral dissertation, PhD dissertation, Keio-University
3. Sindhubala K, Vijayalakshmi B (2015) Design and implementation of visible light communication system in indoor environment. ARPN J Eng Appl Sci 10(7):2882–2886
4. Sewaiwar A, Han PP, Chung YH (2015) 3-Gbit/s indoor visible light communications using optical diversity-schemes. IEEE Photonics J 7(6):1–9
5. Haruyama S (April 2013) Visible light communication using sustainable LED lights. In: 2013 Proceedings of ITU kaleidoscope: building sustainable communities (K-2013). IEEE, pp 1–6
6. Lin WY, Chen CY, Lu HH, Chang CH, Lin YP, Lin HC, Wu HW (2012). 10 m/500 Mbps WDM-visible-light-communicationsystems. Opt Express 20(9):9919–9924
7. Vučić J, Kottke C, Habel K, Langer KD (March 2011) 803 Mbit/s visible light WDM link based on DMT modulation of a single RGB LED luminary. In: Optical fiber communication conference and exposition (OFC/NFOEC), 2011 and the national fiber optic engineers conference. IEEE, pp 1–3
8. Yuanquan W, Nan C (2014) A high speed bi-directional visible light communication system based on RGB-LED. Chin Commun 11(3):40–44
9. Singh S, Bharti R (2014) 163 m/10 Gbps 4QAM-OFDM visible light communication. IJETR 2:225–228
10. Chang CH, Li CY, Lu HH, Lin CY, Chen JH, Wan ZW, Cheng CJ (2014) A 100-Gb/s multiple-input multiple-output visible laser light communication system. J Lightwave Technol 32(24):4121–4127
11. Ivaniga T, Ivaniga P (2014) Evaluation of the bit error rate and Q factor in optical networks. IOSR J Electron Commun Eng 9(6):01–03
12. Manivannan K, Raja AS, Selvendran S (2016) Study of the impact of receiver aperture diameter, LED electron carrier life time and RC time constant on visible light communication using opti-system simulation. Int J Adv Engg Tech VII(I):375, 378
13. Pei Y, Zhu S, Yang H, Zhao L, Yi X, Wang JJ, Li J (2013) LED modulation characteristics in a visible_light_communication_system. Opt Photonics J 3(02):139
14. Tanaka H, Umeda Y, Takyu O (Jan 2011) High-speed LED driver for visible light communications with drawing-out of remaining carrier. In: 2011 IEEE radio and wireless symposium (RWS). IEEE, pp 295–298

BER Comparison of OFDM System with the Mapping of Different Modulation Techniques Over Conventional Modulation Technique

Mandvi and Ashish Kumar Ghunawat

Abstract This paper presents the comparative BER performance of OFDM using BPSK, QPSK, and QAM. There is no advantage of OFDM-BPSK over conventional BPSK, but OFDM-QPSK performs well over a QPSK after an SNR value of 7 dB. OFDM with M-QAM performs well over a conventional QAM at high SNR (greater than 15 dB) values. So, nowadays in 4G mobile communication, we are using OFDM with the mapping of higher order QAM system to achieve high data rate at slightly high SNR value. OFDM with higher order QAM is used where proper SNR value is available (to take the advantage of high SNR and high data rate).

Keywords BER · E_b/n_o · SNR · OFDM · BPSK · QPSK
QAM · MATLAB software

1 Introduction

OFDM is the most popular technique for parallel data transmission in wireless communication system. OFDM is widely used in all 4G wireless communication systems. Multicarrier modulation technique forms the basis of OFDM system [1]. In OFDM system, a single high stream of data is divided into lower data rate streams with closely spaced subcarriers. Subcarriers are used to transmit data on several parallel channels. OFDM has been used for various broadcast standards such as Digital Video Broadcasting (DVB) and Digital Audio Broadcasting (DAB) standards [2]. OFDM is also used in wireless local area standards such as

Mandvi (✉) · A. K. Ghunawat
Department of Electronics and Communication Engineering, Malviya National Institute of Technology, Jaipur, India
e-mail: 2015pwc5308@mnit.ac.in

A. K. Ghunawat
e-mail: ashishghunawat@gmail.com

V. Janyani et al. (eds.), *Optical and Wireless Technologies*, Lecture Notes in Electrical Engineering 472, https://doi.org/10.1007/978-981-10-7395-3_25

IEEE802.11g and 802.11n. The orthogonality between subcarriers minimizes interference. The orthogonality between subcarriers is achieved if the OFDM symbol duration is equal to the reciprocal of the frequency separation between subcarriers. OFDM is computationally an efficient technique by utilizing the IFFT and FFT techniques to implement modulation and demodulation. To eliminate the effect of intersymbol interference from the previously transmitted symbol, cyclic prefix has to be used [3]. The receiver is designed in such a way that it discards the cyclic prefix sample.

In recent years, a lot of research has been done in this area. BER performance was analyzed using OFDM, and it showed that if we are not considering the effect of spectral efficiency, then lower order modulation technique is good for use in communication system [4]. But if spectral efficiency is considered, then we must use higher order modulation techniques. It is also shown that the lower order modulation techniques result in less BER at receiver, and hence, it results in the improved system performance [5]. With the help of error correcting codes using AWGN channel, the BER gets improved [6]. To reduce the BER with the help of error correcting codes, the complexity gets increased (because it is difficult to choose such a code word) and spectral efficiency gets reduced and results in reduced data rate. In OFDM technique, to remove the ISI, we already used the cyclic prefix to maintain the orthogonality between subcarriers. We can improve the BER by reducing the peak to average power ratio (that is already available in literature, high peak to average power ratio problem occurs due to a large number of subcarriers that are added up at the output of IFFT) with the help of proper hybrids technique.

In this paper, we use MATLAB, which is interactive simulation software, to find out which technique is preferred for a certain range of SNR value and to evaluate the BER versus SNR performance of OFDM-BPSK, OFDM-QPSK, and OFDM-M-QAM using Rayleigh fading channel.

2 OFDM System Model

In OFDM technique, a high input data rate stream is divided into a number of parallel lower data streams. These parallel lower data streams are transmitted using various orthogonal subcarriers [6]. The division of a serial data stream into a number of parallel lower data streams results in increased symbol duration, and thus, the problem of intersymbol interference (ISI) is going to be reduced. OFDM makes the task for managing the intersymbol interference much easier because of the insertion of cyclic prefix. Insertion of cyclic prefix makes the OFDM symbol duration much longer than the single-carrier frequency division multiplexing, so the coherence bandwidth is greater than the signal bandwidth. OFDM makes the frequency selective fading channel into flat fading channel. So in OFDM system, frequency selective fading channel gets converted into flat fading channel.

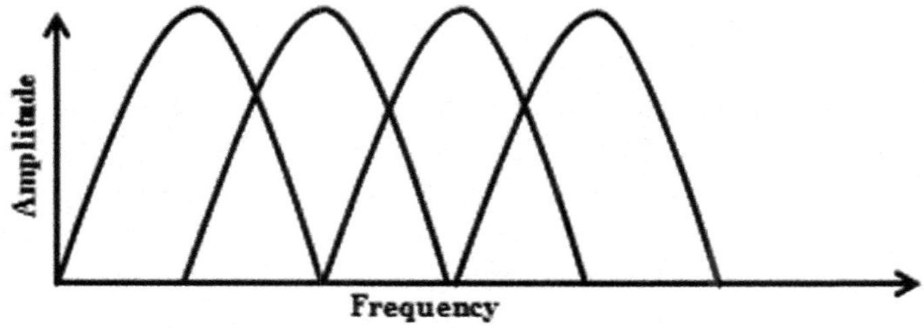

Fig. 1 Orthogonal multicarrier technique (OFDM)

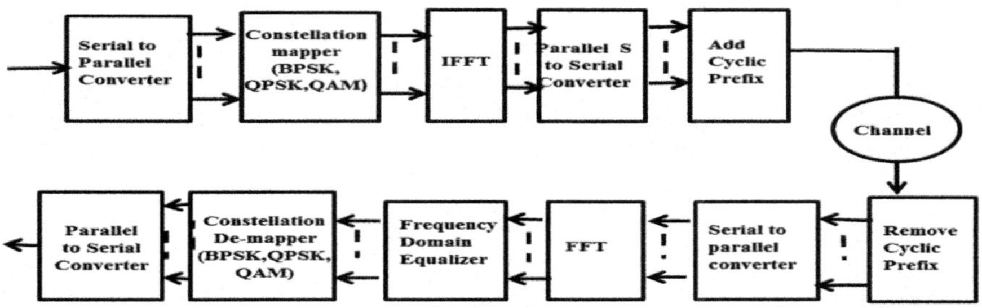

Fig. 2 OFDM system model

OFDM system makes the signal orthogonal in the frequency domain which eliminates the effect of intersymbol interferences as indicated in Fig. 1. These subcarriers experience ‘flat fading’ because the subcarrier bandwidth is less than the coherence bandwidth of the mobile channel.

Figure 2 shows an OFDM system model using the inverse fast Fourier transform and fast Fourier transform for modulation-demodulation. Fast Fourier transform is a fast algorithm to perform Fourier transform. In OFDM model, the input serial stream of data is mapped into complex data symbols (with the help of different modulation techniques) with symbol rate $1/T_S$ (where T_S is the OFDM symbol period). The serial to parallel converter used in this model gives an output X_0 to X_{n-1} that consists of N complex symbols. The output then passes through an IFFT having N points. IFFT has a rectangular window whose length is $N * T_s$. IFFT gives different complex samples X_0 to X_{n-1}.

The output of IFFT is in the form of samples. These samples are added to the front of the serial data stream and then converted into a serial data stream to produce the OFDM transmit symbol of length T = N * Ts. Copies of the last part of frequency selective fading, which introduce intersymbol interference, prefix it with cyclic prefix to eliminate the effect of intersymbol interference. After that, OFDM symbols are transmitted through the Rayleigh fading channel.

In 4G mobile communication system, the signals are received maximum time through various reflection, diffraction, and scattering. If there is no dominant line of sight path between transmitter and receiver, we used Rayleigh fading channel, and if there exists at least one line of sight path, Ricians fading channel is used. In Rayleigh fading channel, we assume that the signals that pass through the channel will vary in a random manner and follow the Rayleigh distribution. At the receiver side, the inverse transmitter phenomenon is done for obtaining the transmitted data and the cyclic prefix (CP) is removed prior to the FFT processing.

3 BER Performance of OFDM System

For an OFDM signal with bit duration Td, the total time period of an OFDM signal is given as $T_d + T_{CP}$, where T_{CP} is the time period of the cyclic prefix adder. So the relation between the energy of OFDM signal and bit energy E_b is given as

$$E_b \times T_d = E_S(T_d + T_{Cp}) \tag{1}$$

Now, consider additive white Gaussian noise (AWGN) n, with a two-sided noise spectral density No, the variance is given as

$$Variance = \frac{N_O}{2} \tag{2}$$

And the OFDM symbol energy can be expressed as,

$$E_S = \log_2(M) \times R_C \times R_b \tag{3}$$

where for QPSK M = 4 and for 16-QAM M = 16, for BPSK M = 2. R_C is the code rate of the system and Eb is the bit energy.

Simplifying the Eq. (3) and dividing it by N_o, the expression is given as

$$\frac{E_b}{N_0} = \frac{E_S}{\log_2(M) \times R_C \times No} \tag{4}$$

The relation between the symbol energy and bit energy in terms of the number of used subcarriers (n DSC) and FFT size (n FFT) is as follows [9]

$$\frac{E_S}{N_0} = \frac{E_b}{N_O} \times (n\,DSC \div N\,FFT) \times \left(\frac{T_d}{(T_{CP} + T_{CP})}\right) \tag{5}$$

The above equations are used to obtain the relation between symbol energy and bit energy during the MATLAB coding.

4 Simulation Result

This paper presents the performance of BER using BPSK, QPSK, and M-QAM (here we have taken M = 16, 64, 128) digital modulation with OFDM technique using Rayleigh fading channel and comparison of it with the BER performance of the conventional modulation techniques (BPSK, QPSK, M-QAM) over the same fading channel. The performance of BER has been investigated by using MATLAB coding. We have taken the following OFDM system parameters (based on IEEE 802.11a specifications)—Number of subcarriers used (N DSC) is 52, FFT sampling frequency is 20 MHz, Spacing between subcarriers is 312.5 kHz, Duration of cyclic prefix, T_{CP} is 0.8 μs, Duration of data symbol, T_d is 3.2 μs, Total symbol duration, T_S is 4 μs, and Methods used for modulation are BPSK, QPSK, and M-QAM.

The BER performance of an OFDM-BPSK over conventional BPSK using Rayleigh fading channel is shown in Fig. 3. From the MATLAB results, we can observe that the OFDM-BPSK modulation has no advantage over a conventional BPSK modulation scheme. Because the BER performance of OFDM-BPSK and conventional BPSK are same. So there is no advantage of using OFDM-BPSK over conventional BPSK scheme.

The BER performance of an OFDM-QPSK over conventional QPSK using Rayleigh fading channel is shown in Fig. 4. From the results, it is clear that OFDM-QPSK performs well over QPSK after the SNR equals to 7 dB. From the result, it is concluded that OFDM-QPSK is used where SNR value is greater than 7 dB.

The BER performance of an OFDM-16QAM over 16-QAM using Rayleigh fading channel is shown in Fig. 5. From the results, it is clear that OFDM-16QAM performs well over conventional 16-QAM as after SNR equals to 15.5 dB, the BER is less in comparison with that of 16-QAM.

The BER performance of an OFDM-64QAM over 64-QAM using Rayleigh fading channel is shown in Fig. 6. From the result, it is clear that OFDM-64QAM performs well over conventional 64-QAM as after SNR equals 17 dB, the BER is less in comparison with that of 64-QAM.

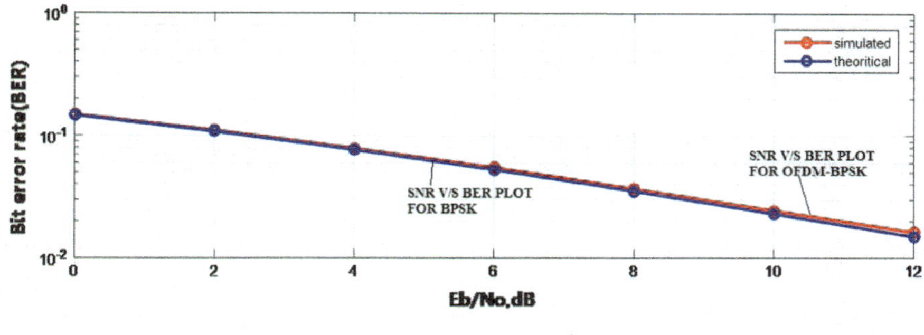

Fig. 3 Performance of OFDM-BPSK over BPSK modulation using Rayleigh fading channel

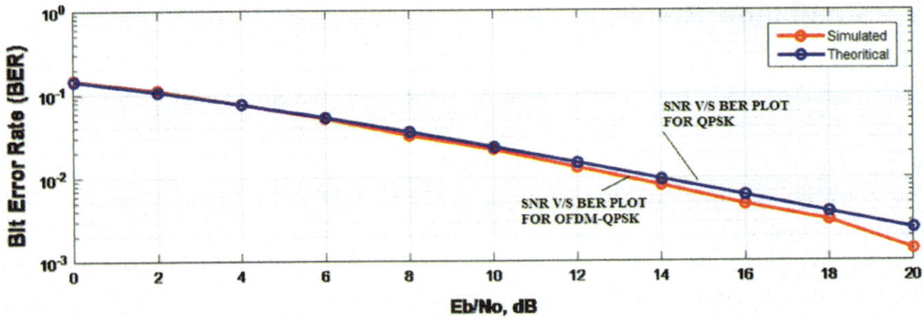

Fig. 4 Performance of OFDM-QPSK over QPSK modulation using Rayleigh fading channel

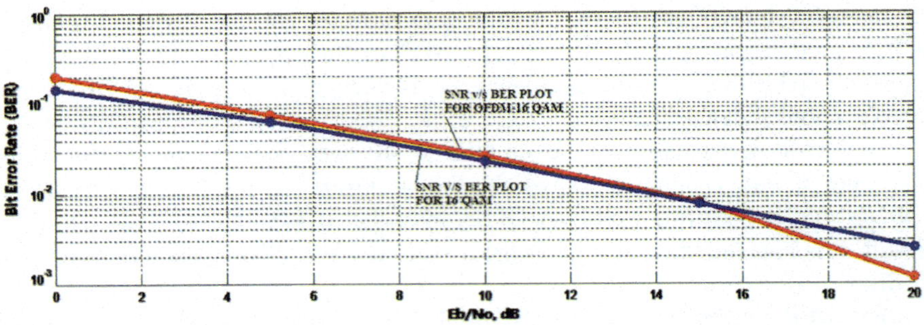

Fig. 5 Performance of OFDM-16 QAM OVER 16 QAM using Rayleigh fading channel

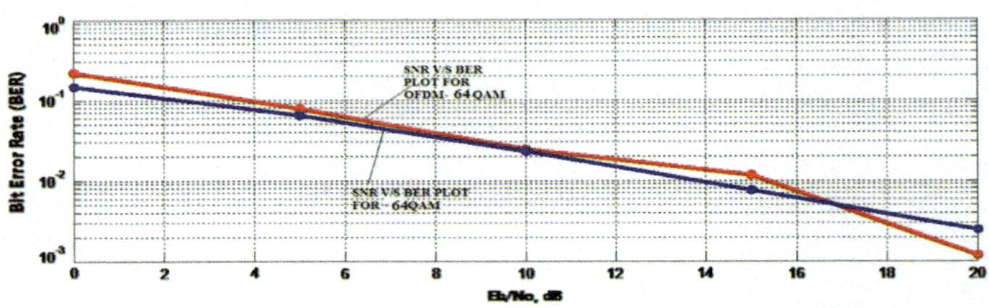

Fig. 6 Performance of OFDM-64 QAM Over 64-QAM using Rayleigh fading channel

The BER performance of an OFDM-28QAM over 128-QAM using Rayleigh fading channel is shown in Fig. 7. From the result, it is clear that OFDM-128QAM performs well over 128-QAM as after SNR equals 17.5 dB, the BER is less in comparison with that of 128-QAM.

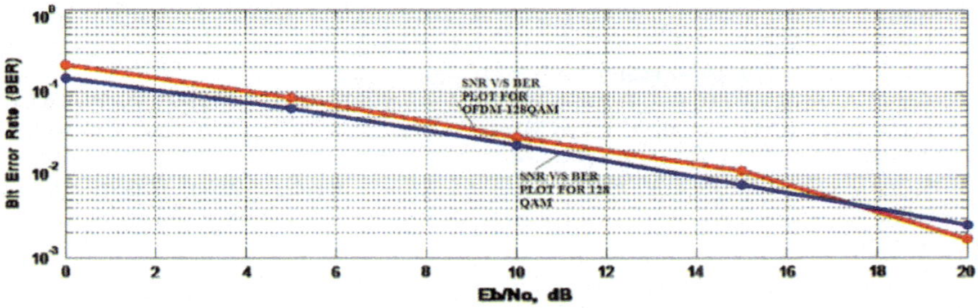

Fig. 7 Performance of OFDM-128 QAM Over 128-QAM using Rayleigh fading channel

5 Conclusion

From the MATLAB coding results, it is concluded that OFDM-QAM modulation performs well over QAM modulation technique with a high value of SNR using Rayleigh fading channel. As we are using higher order QAM with OFDM, the number of bits per symbol is increased, so the BER is increased. But in mobile communication, we need more data rate which is only possible with higher order modulation. OFDM performs well over conventional modulation techniques because of the insertion of cyclic prefix. As the number of cyclic prefix increases the BER gets reduced (the effect of intersymbol interference gets reduced) results data rate going to be reduced. Instead of cyclic prefix, if another alternative would have been found, then it might have been resulted in high spectral efficiency. As clear from the above results, OFDM with higher order QAM performs well over QAM at high value of SNR (SNR value greater than 15 dB), and that is why we are using OFDM with higher order of QAM modulation technique with high value of SNR in wireless communication system.

References

1. Joshi M, Dubey AN, Panda DK (2015) Analysing various fading channels using different modulation techniques under IEEE 802.16 Standard. In: IEEE sponsored 2nd international conference on electronics and communication system (ICECS)
2. Usha SM, Nataraj KR (2013) BER performance of digital modulation schemes with and without OFDM model for AWGN, Rayleigh and Rician channels. Int J Sci Res (IJSR). ISSN (Online): 2319-7064, Index Copernicus Value
3. Kalaivani P, Puviyarasi T, Raju SS, Suresh S (2016) Performance analysis of OFDM system with QPSK for wireless communication. IOSR J Electron Commun Eng (IOSR-JECE) 11 (3):33–37, Ver. I. e-ISSN: 2278-2834, p-ISSN: 2278-8735
4. Khan MA, Pal S, Jose A (2015) BER performance of BPSK, QPSK & 16 QAM with and without using OFDM over AWGN, Rayleigh and Rician fading channel 4(7)
5. Kumar P, Kumari A (2016) BER analysis Of BPSK, QPSK, 16-QAM & 64-QAM based OFDM system over Rayleigh fading channel. IOSR J Electron Commun Eng (IOSR-JECE) 11(4):66–74, Ver. III. e-ISSN: 2278-2834, ISSN: 2278-8735

6. Sharma V, Shrivastav A, Jain A, Panday A (2012) BER performance of OFDM-BPSK, -QPSK, - QAM over AWGN channel using forward error correcting code. Int J Eng Res Appl (IJERA) 2(3):1619–1624. ISSN: 2248-9622
7. Ghosh A, Majumder B, Paul P, Mullick P, Thakurta IG, Ghosh SK (2012) Comparative BER performance of M-ary QAM-OFDM system in AWGN & multipath fading channel. Int J Comput Sci Eng (IJCSE) 4(06). ISSN:0975-3397
8. Namdeo S, Rani R (2013) Designing and performance evaluation of 64 QAM OFDM system. IOSR J Electron Commun Eng (IOSR-JECE) 5(6):97–105. e-ISSN: 2278–2834, p-ISSN: 2278-8735
9. Kaur G, Garg S (2015) Performance evaluation of OFDM system for different modulation techniques on the basis of bit error rate and peak to average power ratio. Commun Appl Electron (CAE) 1(2). Foundation of Computer Science FCS, New-York, USA. ISSN: 2394-4714
10. Shirisha N, Balamunaiah K, Munirathnam S (2012) A simulation and analysis of ofdm system for 4G communications. Int J Adv Res Comput Eng Technol (3)

Optical Investigation of Cu Diffusion Depth in $LiNbO_3$ Crystals Under High-Temperature Treatment

Dmytro Sugak, Ihor Syvorotka, Oleh Buryy, Uliana Yakhnevych, Natalia Martynyuk, Serhij Ubizskii, Ghanshyam Singh, Vijay Janyani and Harsh Kumar

Abstract The paper presents the results of investigations of optical absorption spatial changes in congruent bulk $LiNbO_3$ crystals in 300–1500 nm spectral region along *X*, *Y*, and *Z* directions caused by high-temperature annealing in presence of copper ions. Additional optical absorption was measured at different depths that allows to estimate the spatial distribution of dopant ions. As it is shown, the spatial distributions of absorption caused by copper ions (Cu^+ and Cu^{2+}) have got maxima at the distances about 400–600 μm from the crystal edge. The formation of copper oxide ultrafine nanoparticles in the surface layer of $LiNbO_3$ is confirmed by XRD analysis.

Keywords Lithium niobate · Redox treatment · Diffusion · Quasi-chemical reactions · Optical absorption

1 Introduction

Lithium niobate crystal ($LiNbO_3$) is one of the most frequently used optoelectronic materials. The photorefractive properties of lithium niobate have given the possibility to its using for optical data storage [1]. Increase of $LiNbO_3$ photorefractive sensitivity can be achieved by its doping by metal ions with valency changed under the influence of light or thermo-chemical treatment (Fe, Mn, Cu and their combinations). Usually doping is carried out during crystal growth [2]. The diffusion method is also used at the dopants incorporated into the crystal during high-temperature treatment in an appropriate environment containing metal ions.

D. Sugak (✉) · O. Buryy · U. Yakhnevych · N. Martynyuk · S. Ubizskii
Semiconductor Electronics Department, Lviv Polytechnic National University, Lviv, Ukraine
e-mail: dm_sugak@yahoo.com

D. Sugak · I. Syvorotka
Scientific Research Company 'Carat', Lviv, Ukraine

G. Singh · V. Janyani · H. Kumar
Malaviya National Institute of Technology, Jaipur, India

V. Janyani et al. (eds.), *Optical and Wireless Technologies*, Lecture Notes in Electrical Engineering 472, https://doi.org/10.1007/978-981-10-7395-3_26

Particularly, the diffusion of Cu ions is used for photorefractive elements formation as well as for the modification of refractive index in $LiNbO_3$ near-surface layers for obtaining of integral optics structures [3–6].

Doping of $LiNbO_3$ by copper ions is realized in various ways. Particularly, the crystals are immerged into the copper salt melts [3, 4], the copper film is deposited on the crystal surface; before the thermal treatment [7, 8], the crystals are covered by CuO-containing powders before annealing [5]. Investigation of the dopants diffusion in lithium niobate by optical methods is important scientific and practical task, because it gives the additional information about the composition and homogeneity of diffusion layers, allows to determine the diffusion coefficients and to define the optimal technological conditions of high-temperature treatment (temperature, duration, and environment) for obtaining of the elements with reproducible properties. Moreover, the absorption spectra allow to distinguish the charge state of the dopant in $LiNbO_3$ crystals. However, usually the optical absorption of Cu-doped lithium niobate was registered in the directions which coincide with the direction of diffusion, i.e., the obtained spectra represent only the average changes of crystal optical properties. Only in paper [9], the absorption changes are measured in the direction normal to the diffusion one for the 1-mm-thick $LiNbO_3$:Cu wafers. Also the optical microscopy was used for studies of diffusion layers in the direction normal to diffusion one [7]; particularly, the thicknesses of these layers were determined.

The peculiarities of anisotropy of copper ions diffusion in lithium niobate were studied insufficient, not only for different crystallographic directions, but also for the orientation of spontaneous polarization vector.

The aim of our work is the determination of the Cu ions space distributions in different crystal-physics directions of bulk $LiNbO_3$ samples by optical spectrometry method. The different sources of copper ions were used during high-temperature treatment.

2 Experimental Details

The optical absorption changes, caused by diffusion of ions, were registered in longitudinal (the directions of diffusion and light propagation are the same) and transversal (directions of diffusion and light propagation are perpendicular) geometries of the experiment. All samples were in forms of parallelepiped with dimensions $7(X) \times 15(Y) \times 32(Z)$ mm^3 cut from congruent $LiNbO_3$ crystals grown at SRC 'Carat' [10]. The faces of the samples were polished. Reduction of the samples was carried out in furnace, Nabertherm (Germany), at that the samples were covered by CuO powder. The rate of temperature increasing was 5 K/min, and its stability ensured by the temperature controller is about 0.1 K. Annealing was carried out in air at 1073 K in three stages (3 + 6 + 12 h). After each stage, the samples were cooled and their transmission spectra were registered in Z-direction. After the last stage of annealing, the samples were cut into wafers (1 mm thick)

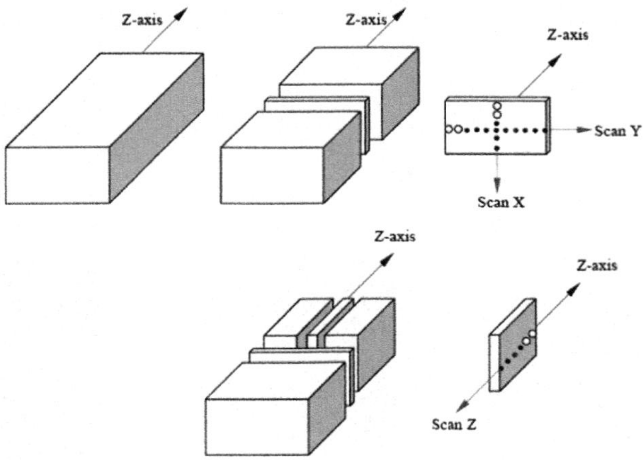

Fig. 1 Scheme of annealed samples cutting for investigation of coloration depth after treatment

used for registration of the optical absorption in (*X*, *Y*, and *Z*) directions perpendicular to the direction of diffusion, as it is proposed in [11]. The scheme of samples cutting is shown in Fig. 1. The cut plates were then polished to the thickness of 1 mm.

The spectra were measured by UV3600 Shimadzu spectrophotometer in the 300–1500 nm region with the specially designed device allows to scan by light beam through the aperture of 50 μm in radius with the step of 20 μm.

The photographs of wafer surfaces were obtained by polarizing microscope ECLIPSE LV100 POL (Nikon, Japan) in the interference phase contrast mode.

The wafer surface layers are also investigated by XRD technique (rays diffractometer DRON-3 (CuK_α irradiation, $\lambda = 1.54056$ Å) in the reflectance mode.

3 Results and Discussion

After annealing the shift of the absorption edge of $LiNbO_3$ to longer wavelengths and formation of the weak broad band with a maximum near 1000 nm are observed in the longitudinal geometry of the experiment. Both the values of the edge shift and the absorption at 1000 nm increase with increasing of the duration of annealing that in general coincides with the results of [4]. In the transversal geometry of the experiment, the band with a maximum of nearly 400 nm was also observed in additional absorption spectra. According to [5], this band can be associated with the intervalence transition $Cu^+ \rightarrow Nb^{5+}$, whereas the band in the vicinity of 1000 nm —with the intracenter transition in Cu^{2+} ion.

The spatial distributions of the additional absorption at 400 and 1000 nm are shown in Fig. 2. As it is followed from our investigations, both Cu^+ and Cu^{2+} ions incorporate into the crystal, however, their distributions are more complex than the

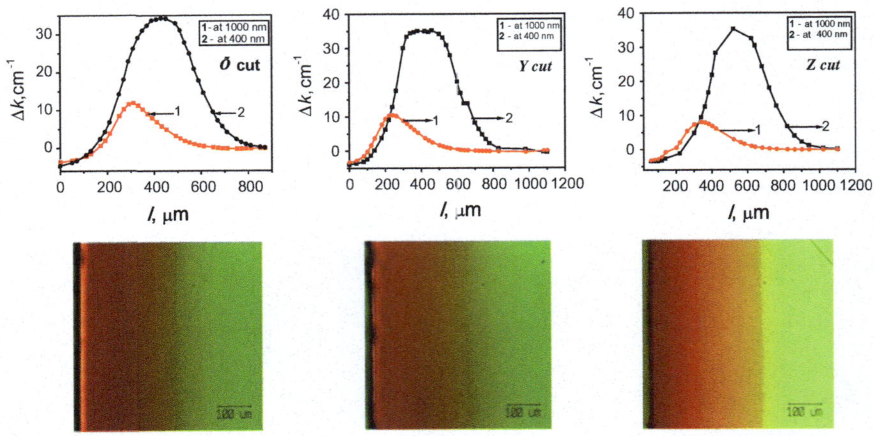

Fig. 2 Upper row—the dependencies of the absorption coefficient at 400 and 1000 nm on depth; lower raw—the photographs of the diffusion regions for different crystal-physics directions (50-fold increase)

ones predicted by conventional diffusion models [12]. As shown in the figure, the maxima for 1000 nm (Cu^{2+} ions) are observed in the absorption spatial distributions at the distances of 250 (*X*), 300 (*Y*) and 350 (*Z*) μm from the wafer edge. For the wavelength of 400 nm (Cu^{+} ions), the position of the maximum also depends on the crystallographic direction: it is located near 400 μm for *X*-direction, near 500 μm for *Y* and near 600 μm for *Z*-direction. At the photographs of the annealed wafer (Fig. 2), the higher depth of coloration in *Z* direction can also be observed, whereas for *X* and *Y* directions, the differences in coloration depths are hardly distinguished. It must be mentioned that observed decreasing of optical absorption near the crystal edge cannot be explained by crossing of the diaphragm through the edge because the depths of the maxima are about an order higher than the diameter of diaphragm (50 μm).

The spatial distributions of copper ions were determined in accordance with the approach of paper [13]. If the absorption band is nearly Gaussian in shape, the concentration of absorption centers can be calculated in accordance with the formula

$$N\left(\mathrm{cm}^{-3}\right) = 8.7 \times 10^{16} \frac{n}{\left(n^2+1\right)^2} \frac{1}{f_o} K_{\max} \Delta E, \tag{1}$$

where n is the refractive index calculated in accordance with Sellmeier formula for $LiNbO_3$ [14], $f_o = 4 \times 10^{-2}$ for Cu^{+} (400 nm) and 2×10^{-4} for Cu^{2+} (1000 nm) are the oscillator strengths, K_{max} is the measured peak absorption coefficient in cm^{-1} and ΔE is the width of absorption band at a half-maximum in eV. The calculated distributions of copper ions concentrations are shown in Fig. 3 for different crystal-physics directions in $LiNbO_3$.

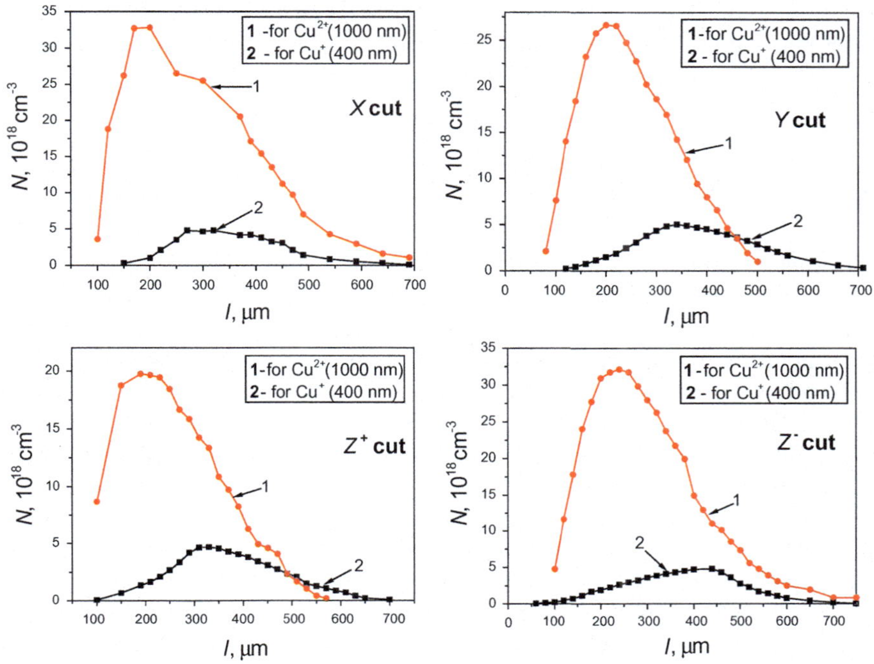

Fig. 3 The spatial distributions of Cu^+ and Cu^{2+} ions concentrations in lithium niobate crystal for different crystal-physics directions

As shown in Fig. 3, the concentration of Cu^{2+} ions is higher than the one of Cu^+. The maxima on the spatial distributions of concentrations probably caused due to the recharging processes takes place during annealing. Probably, copper incorporates into the near-surface regions of the crystal due to $Cu^+ \leftrightarrow Li^+$ and $Cu^{2+} \leftrightarrow 2Li^+$ exchange processes [3, 4, 15]. The incorporation of copper in neutral charge state and their further recharging, as well as the recharging process $Cu^+ \rightarrow Cu^{2+}$ cannot also be excluded.

The XRD investigations showed that the additional reflections are observed in the X-ray patterns of the surface regions in comparison with the ones for the inner parts of the samples. The additional reflections are also observed in [4] where the Cu ions diffusion was intensified by heating of $LiNbO_3$ crystal in Na_2SO_4–$CuSO_4$ melt. The authors of [4] assumed that these reflections belong to $CuNb_2O_6$, $CuNbO_3$ compounds or to $Cu_xLi_{1-x}NbO_3$ solution. The halo is also observed in our XRD experiments that is obviously caused by formation of scattering centers. In accordance with Scherer formula, the dimensions (diameters) of these centers are equal to 1.3, 0.6 and 1.1 nm (for *X*, *Y* and *Z* directions correspondingly). This observation also indicates the anisotropy of copper incorporation into lithium niobate crystal. Based on the aforementioned results, we suppose that during annealing of $LiNbO_3$ in presence of copper oxide, the ultrafine nanoparticles of

CuO are formed in the surface layers of the crystal. However, the definitive determination of the nature of these nanoparticles as well as their influence on Cu ions diffusion requires additional investigations.

4 Conclusions

The optical absorption spatial changes in congruent bulk $LiNbO_3$ crystals in 300–1500 nm spectral region along *X*, *Y*, *Z* directions caused by high-temperature annealing in presence of copper ions are investigated. As it is followed from our measurements, the spatial distributions of absorption as well as of copper ions concentrations have got a maximum for all crystal-physics directions of the $LiNbO_3$ crystal. The maxima are located at the distances about 250–600 μm from the crystal edge. The presence of these maxima indicates that the characters of the processes taking place during annealing are more complex than the one predicted by conventional diffusion models. Such a behavior suggests that the quasi-chemical reactions take place during annealing, and the compounds with participation of niobium and incorporated ions are probably formed. The determination of mechanisms of these reactions required further investigations at different temperatures and durations of annealing, different copper ions concentrations, etc. The formation of copper oxide ultrafine nanoparticles in the surface layer of $LiNbO_3$ is confirmed by XRD analysis.

Acknowledgements The work was supported by Ukrainian Ministry of Education and Science (scientific research project DB/EMSh) and joint Indian-Ukrainian project M/201-2017. The authors are grateful to Dr. Iwan Solskii (SRC 'Carat') for growing the crystal for investigations and to Mr. Vasyl Baluk for XRD measurements.

References

1. Volk T, Wöhlecke M (2009) Lithium niobate: defects, photorefraction and ferroelectric switching. Springer
2. Arizmendi L, Cabrera JM, Agullo-Lopez F (1984) Defects induced in pure and doped LiNbO3 by irradiation and thermal reduction. J Phys C Solid State Phys 17:515–529
3. Ivanov VSh, Ganshin VA, Korkishko YuN (1992) Analysis of ion exchanged Me:$LiNbO_3$ and Cu:$LiTaO_3$ waveguides by AES, SAM, EPR and optical methods. Vacuum 43:317–324
4. Caccavale F, Sada C, Segato F, Bogomolova LD, Krasil'nikova NA, Korkishko YN, Fedorov VA, Morozova TV (2000) Copper-lithium ion exchange in $LiNbO_3$. J Mater Res 15:1120–1124
5. Kar S, Bartwal KS (2008) Cu^{2+} ion in-diffusion in congruent $LiNbO_3$ single crystals. Mater Lett 62:3934–3936
6. Kobayashi T, Muto K, Kai J, Kawamori A (1979) EPR and optical studies of $LiNbO_3$ doped with Cu^{2+} ions. J Mag Res 34:459–466
7. Mikami O (1974) Cu-diffused layers in $LiNbO_3$ for reversible holographic storage. Opt Comm 11:30–32

8. Noda J, Saku T, Uchida N (1974) Fabrication of optical waveguiding layer in $LiTaO_3$ by Cu diffusion. Appl Phys Lett 25:308–310
9. Buse K, Jermann F, Krätzig E (1994) Infrared holographic recording in $LiNbO_3$:Cu. Appl Phys A Solids Surf 58:191–195
10. Solskii IM, Sugak DYu, Vakiv MM (2013) Growing large size complex oxide single crystals by Czochralski technique for electronic devices. Acta Phys Pol A 124:314–320
11. Yatsenko A, Pritulenko A, Yevdokimov S, Sugak D, Syvorotka I, Sugak Yu, Solskii I, Vakiv M (2015) The influence of annealing in saturated water vapor on $LiNbO_3$ crystals optical and electrical properties. Solid State Phenom 230:233–237
12. Crank, J.: The mathematics of diffusion. Clarendon Press, 1975
13. Dexter DL (1956) Absorption of light by atoms in solids. Phys Rev 101:48–55
14. Kuzminov YuS (2012) Lithium niobate crystal. Viva Books
15. Bobrov YuA, Ganshin VA, Ivanov VSh, Korkishko YuN, Morozova TV (1991) Formation of Cu:$LiTaO_3$ exchanged waveguides and their investigations. Phys Stat Sol A 123:317–324

Invisible Image Watermarking by DWT-HUFFMAN-EZW Methodology

Vipra Bohara and Sandeep Toshniwal

Abstract In information security world, invisible image watermarking is the important part. Many invisible image watermarking algorithms fail to achieve the good quality-based watermarking with high security. In this paper, we are working to improve the performance of the invisible watermarking. We are proposing a DWT-EZW-based watermarking. In the results session, we are showing the comparison between DWT-FWHT-SVD and DWT-Huffman-EZW-based watermarking. Comparison is shown for peak signal to noise ratio (PSNR) and mean square error (MSE).

Keywords DWT · Decomposition · Affine · DWT-FWHT-SVD Fibonacci-Lucas · Administration

1 Introduction

For the strong protection against the unwanted copyright, several methods are now using with best possible security for digital products like image, video, data or audio significantly. The digital-based image plays a significant role in the age of information transmission technology. Whereas it is now very susceptible to being a copy by duplicate owner, and after copying it become difficult to separate copied content from the real image. Thus, in the growing Internet use, the traditional digital image watermarking technology does not have the capability to provide security from intentional or may unintentional attacks. The digital-based content can be safe

V. Bohara (✉) · S. Toshniwal
Electronics Communication, Kautilya Institute of Technology and Engineering, Jaipur, India
e-mail: viprabohara@gmail.com

S. Toshniwal
e-mail: toshniwal.sandeep@gmail.com

V. Janyani et al. (eds.), *Optical and Wireless Technologies*, Lecture Notes in Electrical Engineering 472, https://doi.org/10.1007/978-981-10-7395-3_27

by watermarking. Many of the methods were already put forward to increase the required quality factor of the watermark data. Now researchers start to utilize evolutionary techniques for this digital-based watermarking. So for hiding information of multimedia content using digital image, watermarking is very useful way associated with trouble of copyright protection. The planting activity of supplementary statistics into an image to build assertion about the image. This extra statics is known as a watermark. Its identification code may be the invisible or visible type; from both types, the visible type of the watermark can be remote without difficulty from the main cover of the digital image. But remote the invisible type of the watermark is very difficult from the cover of the digital image. This feature of the invisible watermark due to the integral multiple components of the host image after inserting the watermark.

2 Related Work

In the embedded design, digital image watermarking method is defined as the spatial type domain and transform type of domain. In the special type of domains, LSB (Least Significant Bit) type of methods are implementing. In this review, some DWT-based and composite transform-based methods are presented. Discrete wavelet transform (DWT) provides the multi-type of resolution and provides good reconstruction image. In paper [3–7], Chemtoo Wong et al. presented a DWT-based image watermarking. In many papers, authors are designing three transforms so that they can take benefit for the individual transform. In paper [8], Keta Roval et al. presented a method for combination of DWT-DCT-based methodology. The combined DWT-DCT-based methodology is also presented in a paper [9]. In paper [10], Haiiong Shi et al. presented the combination of DWT and SVD. In paper [11], Priyanka Singh et al. used hybrid DWT-SVD-based color image watermarking. In paper [11], image watermarking was performed between YUV and YIQ color spaces. This methodology is checked at 512 $\times$ 512 image section so that we can get the PSNR and MSE values, respectively. This method gives better results in Y color spaces. In paper [12], Lai and Tsai presented DWT- and SVD-based methodology and proved that security is high of watermarking against the different attacks. In paper [13], Somira Lagzion et al. presented a paper for redundant discrete wavelet transform (RDWT)-SVD methods. In this, data get embedded in low-, middle-, and high-frequency DWT-based sub-bands. In paper [14], Xiangui Kong et al. presented DWT-DFT-based image watermarking so that security can achieve against image compression attacks. In paper [15], Sivavenkateswara Rao et al. presented a paper for image watermarking in which DWT, DCT, and SVD algorithms are used for optimizing the results.

3 Proposed Methodology

For improve the performance of the watermarking, we are using DWT-Huffman-EZW based encryption and decryption for improve the security of the invisible watermarking. In this method, cover image and secret image are converted into HH, HL, LH, and LL by using DWT. Embedded zero tree wavelet (EZW) is used for designing zero tree as shown in Fig. 2. Huffman encoding method is used for encoding the watermarking. It is called DWT-Huffman-EZW (DHE)-based image watermarking.

3.1 Discrete Wavelet Transform

Discrete wavelet transform (DWT) is presented as a multi-resolution of an image. Multi-resolution-based image representation provides a simple framework for interpreting the image information. DWT divides the image into low-frequency bands and high-frequency bands. Now low frequency based quadrant again divide into low-frequency sub band and high frequency based sub-bands. The process will be continuing until entire image will not get decomposed.

DWT divides the image into four parts. One part is finding the low frequency of input image. Top right part of Fig. 1 contains the horizontal details of the image. Bottom left box contains the vertical details of the original image. Bottom right box contains the high-frequency information of the image. Low-frequency-based coefficients have to be more robust because this part contains more information of the original image. IDWT is used to target the reconstructed image from the decomposed image.

The digital wavelet transform is scalable in nature. DWT is mostly used in image watermarking due to its special localization and multi-resolution technique.

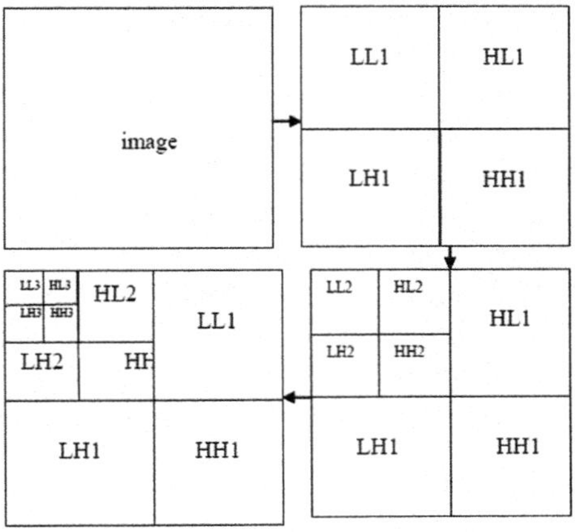

Fig. 1 Discrete wavelet decomposition

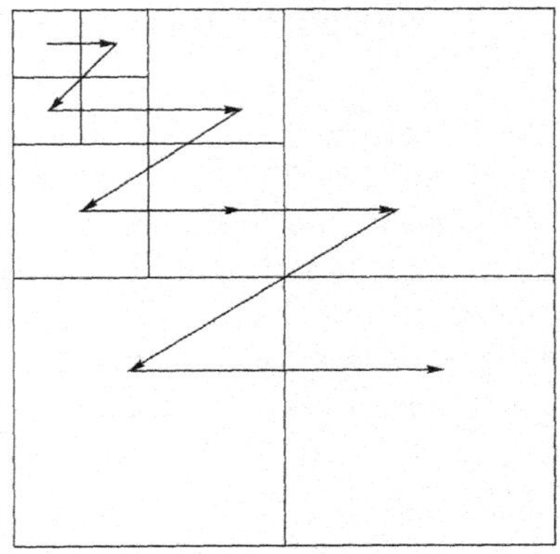

Fig. 2 Scanning pattern in EZW

Excellent spatial localization property is very important to match the area for the cover image in which watermark gets embedded efficiently (Fig. 2).

3.2 EZW (Embedded Zero Tree Wavelet)

Shapiro designed embedded zero tree wavelet (EZW) algorithm. EZW algorithm is used to find the best order of transmission wavelet coefficient. Progressive embedding is used for the transmission of significant bits. Discrete wavelet transform (DWT) is applied to the image, and wavelet coefficients are received T_0—T_{N-1}.

$$T_0 = \frac{|C_{max}|}{2} \tag{1}$$

where C_{max} is the biggest coefficient.

Embedded zero tree wavelet is coded by the wavelet transform sub-band coding. The wavelet transformed coefficients contain low energy, so the decomposition level gets increased. As the level of decomposition get to increase, the complexity and minute details get increase.

In EZW algorithm, all coefficients of the matrix are scanned to find the maximum value of the coefficient. The value of C_{max} is 57 form the Fig. 3. We can say that maximum value of coefficient mostly finds from the lowest frequency Sub-band (LL Band). So we have to check the entire coefficients of the matrix to get the maximum value coefficients. It will be able to reduce computational complexity. Threshold T is used to show the position of the bit. Current quantization width bit is T. In every iteration, the width is divided by 2, so that we can find another less significant bit. In EZW algorithm every iteration keep dominates pass (Principal Stage) which is followed by subordinate pass (Secondary Stage).

56	-35	52	8	5	12	-10	5
-29	25	14	-14	3	1	5	-2
15	15	2	-9	8	-10	6	12
-10	-6	-11	7	7	-2	4	5
-2	12	-1	47	6	6	-2	3
0	3	-2	1	1	-4	3	1
0	-4	8	-4	4	5	3	3
5	14	4	3	-2	5	-4	1

Fig. 3 DWT coefficients

The coefficients of the dominant list will check the value from the current threshold value. For generating the results, a digital magnitude comparator is used. If the value of the coefficient find significant then

- The sign will be determined.
- It will change into P(00) or N(01)
- The magnitude of coefficient appended to the subordinated.

Assign Z (10) and T (11) as an insignificant wavelet coefficient. According to this, the coefficients wavelet is coded as follow:

- Coefficient will be defined as **P**, if the coefficient is positive and coefficient absolute value is greater than the threshold.
- Wavelet coefficient will be defined as **N**, if the coefficient is negative and the absolute value is greater than the threshold.
- Wavelet coefficient will be defined as **Z**, if the value of the coefficient is less than threshold but its children contain any coefficient whose absolute value is greater than the threshold.
- **T** if the coefficient is less than Threshold and it all children also has all coefficient absolute value less than the threshold.

Table 1, is showing to the symbols which we will receive after apply EZW (Embedded Zero tree Wavelet) algorithm from the wavelet coefficients of figure and

Table 1 Symbols of the wavelet coefficients for the dominant pass

Threshold	Dominant pass results
32	P N Z T P T T T T Z T T T T T T T P T T
16	Z T N P T T T T T T T T
8	Z Z Z Z Z P P N P P N T T N N P T P T T N T T T T T T T P T T T P T T T T T T T T T P T T T T T T T T T T T T
4	Z Z Z Z Z Z Z T Z T Z N Z Z Z Z P T T P T P P T P N P T N T T T T T P T P N P P P P T T T T T P T P T T T P N P
2	Z Z Z Z Z T Z Z Z Z Z T P Z Z Z T T P T T T T N P T P P T T P T T T N P P N T T T T P N N P T T P T T P P T T T

Table 2 Bits of the wavelet coefficients for the subordinate pass

Threshold	Dominant pass results
32	1010
16	100111
8	0011100111100010001110
4	01010001111101010100101100011000000 1100000
2	11011100011011000000000111110010100000011001000101011011

in Table 2, obtained bits are showing which are receiving after applying EZW (Embedded Zero-tree Wavelet) algorithm at wavelet coefficients like Fig. 3.

As per hardware point of view, we have four symbols. Each symbol will take two bits for representation. So now

- P will be coded as 00,
- N will be coded as 01,
- Z will be coded as 10,
- T will be coded as 11,

Now, these symbols, P, N, Z, and T, are also encoded using Huffman coding technique for further compression. At the time of decoding, the EZW process will be reversed as mentioned above.

(1) Initialization: the entire coefficient will initialize by zero. The coefficients list is proceeded by the empty list. T_n gives the threshold of this matrix.
(2) Principal Stage: Principal list code will read one by one. I that every code will be anyone from the [P, N, Z, T].

    - Give the value of coefficient as T_n, if the code is P and in the list of processed coefficients (its Position) add the coefficients.
    - Give the value of the coefficient as T_n, if the code is N. Now in the list of processed coefficients (its Position) add the coefficients.
    - But if the code is Z and T then we have to do nothing.

(3) Secondary Stage: Now get the codes of the secondary list. It will match to the list which already processed the coefficients.
    For every coefficient, check the value of 0 and 1. If the secondary list corresponding term is 0, then do nothing and if the value is 1 then modify its value. Add $1/2(T_n\text{-}1)$ in it, if the coefficient value is positive. If it is negative then subtract the same quantity from it.
(4) Threshold Update: new threshold can be calculated with $T_n + 1 = (1/2)\ T_n$.

3.3 Huffman Coder Architecture

Huffman coding is the process for compressing the lossless data. It represents the data in fewer bits and prepares a table which contains the information of the mapping data between code word and real data for encoding. Huffman coding is of two types

- Adaptive Huffman Coding
- Static Huffman Coding

After applying DWT and EZW, we consider that symbol T comes more easily as compared to another three symbols. Now, we have to assign one bit to the symbol. It can be 0 or 1. According to the presence of the symbol Z, N and P the binary bits value will be assigning as

- P will encode as 1110
- N will encode as 110
- Z will encode as 10
- T will encode as 0

We can also use these bits

- P will encode as 0001
- N will encode as 001
- Z will encode as 01
- T will encode as 1

Symbol P, N, Z and T bits string will store in the RAM. The decoder will use to read these bits strings and create a code and a length for every symbol. Huffman encoded codes will represent like 1110, 110, 10 and 0 or 0001, 001, 01, 1 for the P, N, Z and T symbols. In the Huffman coding length represents the number of bits. In the system, the length of the symbol is 4 while the length of the symbol N is 3. The coding bits string will store in 11 bits register. Whenever Register is filled with 8 Bits, then Huffman encoded strings will able to generate the output. Eight-bit strings are selected because it can show two symbols at least if both are the P symbols.

Algorithm

Step 1: Select the cover image and secret image for invisible image watermarking.
Step 2: Now apply DWT at both the cover and secret image.
Step 3: Apply EZW encoding at the secret image.
Step 4: Apply Huffman encoding at EZW output to get the bit stream at the output.
Step 5: Now generate a new HH sub-band from the Huffman encoding output.
Step 6: Apply inverse DWT at target image with new HH sub-band of step 5 and show watermark image.

Step 7: Apply Huffman decoding and EZW decoding on the watermarked bit stream.
Step 8: Apply DWT on watermarked image.
Step 9: Append the output of EZW decoding to form extracted watermarked for HH sub-band.
Step 10: Apply the inverse DWT to form the extracted image.
Step 11: Apply the mean adjustment to increase the intensity of the image.

4 Results

In the results session, we show the results of the DWT-Huffman-EZW-based image watermarking.

Figure 4 shows the target image for the image watermarking. It will be our cover image, and Fig. 5 shows the secret image which we have to insert in the image watermarking.

Figure 6 shows the watermarked image by using DWT-Huffman-EZT methodology, and Fig. 7 shows the extracted image by DWT-Huffman-EZT methodology from the image watermark.

Table 3 shows the comparison between DWT-FWHT-SVD and DWT-Huffman-EZW methodology. PSNR is getting increase and MSE is getting decrease.

Fig. 4 Watermarking cover image

Fig. 5 Secret image

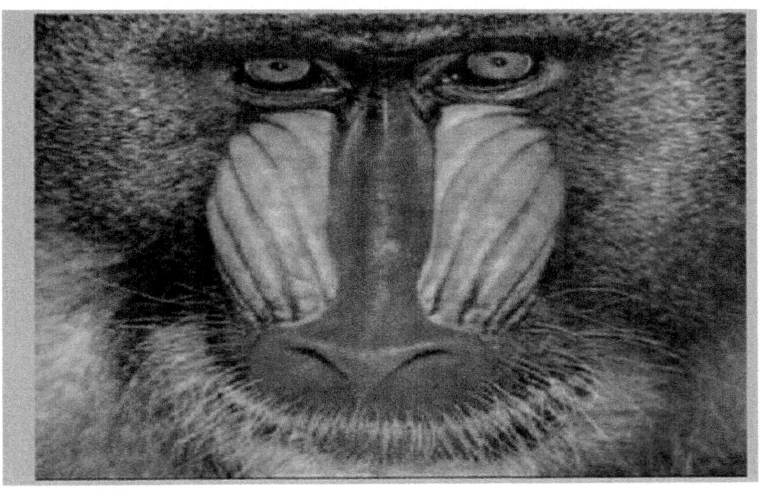

Fig. 6 Watermark image

Fig. 7 Extracted images

Table 3 Comparison Table

Methods	MSE	PSNR
DWT-FWHT-SVD	0.1190	57.3764
DWT-Huffman-EZW	0.0017	75.9425

5 Conclusion

In this paper, we are working for the Invisible Image Watermarking. We are improving the performance of the Image Watermarking. For improve the performance, we are using DHE (DWT-Huffman-EZW) based Image watermarking. As we can see from the results the PSNR value is getting increase for DWT-Huffman-EZW based Image Watermarking.

References

1. Urvoy M, Goudia D, Autrusseau F (2014) Perceptual DFT watermarking with improved detection and robustness to geometrical distortions. IEEE Trans Inf Forensics Secur 9 (7):1108–1119
2. Kamran M, Farooq M (2012) An information-preserving watermarking scheme for right protection of EMR systems. IEEE Trans Knowl Data Eng 24(11):1950–1962
3. Wang C, Ni J, Huang J (2012) An informed watermarking scheme using hidden Markov model in the wavelet domain. IEEE Trans Inf Forensics Secur 7(3):853–867
4. Bi N, Sun Q, Huang D, Yang Z, Huang J (2007) Robust image watermarking based on multi band wavelets and empirical mode decomposition. IEEE Trans Image Process 16(8):1956–1966
5. Kundur D, Hatzinakos D (1999) Digital watermarking for telltale tamper proofing and authentication. In: Proceeding of IEEE, vol 87, no 7, pp 1167–1180
6. Zareian M, Tohidypour HR (2013) Robust quantization index modulation-based approach for image watermarking. IET Image Process 7(5):432–441
7. Senthil V, Bhaskaran R (2008) Digital image watermarking using edge detection and wavelets with robustness analysis against jpeg compression attacks. In: International conference on innovations in information technology, pp 583–587, 16–18
8. Raval K, Zafar S (2013) Digital watermarking with copyright authentication for image communication. In: IEEE international conference on intelligent systems and signal processing (ISSP), pp 111–116
9. Kasmani SA, Naghsh-Nilchi A (2008) A new robust digital image watermarking technique based on joint DWT-DCT transformation. In: IEEE third international conference on convergence and hybrid information technology, pp 539–544
10. Shi H, Wang N, Wen Z, Wang V, Zhao H, Yang Y (2012) An RST invariant image watermarking scheme using DWTSVD. In: IEEE international symposium on instrumentation, measurement, sensor network and automation (IMSNA), pp 214–217
11. Singh P, Agarwal S, Pandey A (2012) A hybrid DWTSVD based robust watermarking scheme for color images and its comparative performance in YIQ and YIV color spaces. In: 3rd IEEE international advance computing conference (IACC), pp 1213–1218
12. Lai CC, Tsai CC (2010) Digital image watermarking using discrete wavelet transform and singular value decomposition. IEEE Trans Instrum Meas 59(11):3060–3063
13. Lagzian S, Soryani M, Fathy M (2011) Robust watermarking scheme based on RDWT-SVD: embedding data in all sub bands. IEEE, pp 48–52
14. Kang X, Huang J, Shi Q, Van Lin Y (2003) A DWT-DFT composite watermarking scheme robust to both affine transform and JPEG compression. IEEE Trans Circ Syst Video Technol 13(8):776–786
15. Rao VS, Shekhawat RS, Srivastava VK (2012) A DWT-DCT-SVD based digital image watermarking scheme using particle swarm optimization. IEEE students' conference on electrical, electronics and computer science (SCEECS), pp 1–4

Nanocavity-Coupled Waveguide Photonic Crystal Biosensor for Detection of Hemoglobin Concentration in Blood

Shivani Ameta, Arvind Sharma and Pawan Kumar Inaniya

Abstract Biosensor is an analytical device which is used to detect the biological analyte which can be antibodies, enzymes, cells, or biomolecules. The biosensor has many applications like in biomedical research, environmental monitoring, and healthcare. In this paper, we have designed a 2D photonic crystal biosensor with nanocavity coupled with a photonic crystal waveguide. This structure possessing hexagonal symmetry with an area of $19 \times 13\ \mu m^2$ is considered. Over the wafer of air, dielectric rods of Si which have refractive index of 3.47772, radius of 210 nm, and lattice constant of 700 nm are introduced. This design is based on hemoglobin-filled nanocavity coupled with a waveguide. The biosensor can detect hemoglobin concentration in blood. Hemoglobin concentration in blood with 0, 10, and 20 g/dL is detected through the simulation and analysis. For the analysis, FDTD method is used. Change in refractive index causes a spectral shift in transmission spectra is observed. This property is useful in sensing application by considering different concentration with different refractive indices. In this work, design is allowed to operate in the wavelength range of 1500–1600 nm.

Keywords FDTD · Hemoglobin biomolecules · Nanocavity-coupled waveguide Nanocavity · Photonic crystal · Plane wave expansion (PWE)

S. Ameta (✉) · A. Sharma · P. K. Inaniya
Department of ECE, Government Women Engineering College, Ajmer, Rajasthan, India
e-mail: shivani.ameta@gmail.com

A. Sharma
e-mail: gweca.arvind@gmail.com

P. K. Inaniya
e-mail: pawan.inaniya@gmail.com

V. Janyani et al. (eds.), *Optical and Wireless Technologies*, Lecture Notes in Electrical Engineering 472, https://doi.org/10.1007/978-981-10-7395-3_28

1 Introduction

Hematological disorders are responsible for the many mortal diseases such as leukemia, Hemophilia, Hepatitis B and C, anemia, and diabetes. From the report of the Harvard School of Public Health and the World Economic Forum (September-2011), 63% of all deaths worldwide are responsible for killing over 36 million people each year due to some diseases like cancers, cardiovascular diseases, respiratory diseases, diabetes [1]. Prevention and protection from these diseases can be done through blood analysis. Normally, blood analysis is based on laboratory techniques where techniques are time-consuming, complicated, and tedious since samples of blood are collected, conserved, and transported to the laboratories for analysis which require long-term investments along with human involvement [2]. Human blood contains 78% water and 22% solids and is composed of more than 4000 different kinds of components. Blood plasma contains 55% of human blood fluid and proteins like albumin, fibrinogen, and other contents like glucose, minerals, vitamins, hormones, red blood cells (RBCs which contain hemoglobin), enzymes, and white blood cells (including leukocytes and platelets). It is difficult to acquire optical behavior of blood because of the highly concentrated hemoglobin content of the cell and interfering waves from neighboring cells [2] due to some conditions like oxygen saturation [3], osmolality [3, 4], or shear rate. Generally, the time taken by the conventional techniques is in the range of 12 to 72 h. Therefore, cost of analysis as well as analysis time both increases. So, there is a need for biosensor which can detect and analyze the hemoglobin concentration in blood accurately and in a short period of time [2]. The normal range of blood hemoglobin level in men is from 14.0 to 17.5 (mean 15.7) g/dL and in women is from 12.3 to 15.3 (mean 13.8) g/dL. A low hemoglobin level is known as anemia or low red blood count, and its range is less than 12–13 g/dL. A high hemoglobin concentration is known as polycythemia or high red blood cell count, and its range is greater than 15–18 g/dL. According to the report of the World Health Organization (WHO), 1.62 billion people are suffering from anemia [5]. In present days, hemoglobin is measured by collecting blood by pricking the finger and transferring it to a cuvette, and after that, the cuvette is put in an analyzer and the hemoglobin is measured spectrophotometrically [6]. Photonic crystal bandgap structures can be utilized for hemoglobin detection. Since permittivity of biological molecules is greater than that of air and water, the propagation velocity of electromagnetic waves will reduce when they pass through these biological molecules and the signals resulting due to interaction of biological elements, and analytes can be easily detected, measured, and quantified by biosensor which can be further used for medical applications [7, 8]. The purpose of this paper is to develop a lab-on-chip photonic crystal-based sensor which is easy to fabricate and can be scalable and reconfigurable. This modeled and designed chip can be able to detect hemoglobin concentrations in blood sample. This sensor can give accurate spectral signatures for different concentrations of hemoglobin present in blood sample. The designed biosensor has the advantages of easy fabrication, label-free detection method, acceptable resolution, wide detection range, and rapid response time.

2 Theory

In this paper, a nanocavity coupled with a waveguide photonic crystal-based biosensor is proposed. Our waveguide-based biosensor coupled with a nanocavity provides a good quality factor by inducing line defect, a linear waveguide is created, and broad PBG is obtained. The photonic band structure can be referred as 'optical insulator' where transmission of the electromagnetic waves in certain range of frequencies is prohibited, i.e., for some wavelengths, no light propagates through photonic crystal, and is obtained by plotting the resonant frequency against the 'k'-wave vector [9] which is of no use for sensing. A point defect can be introduced either by missing a hole, by changing its dimensions, or by changing its material. Periodicity of the structure breaks through the defects. This property is important for sensing application since a strong light–matter interaction is achieved through provided unique light-guiding mechanism by localizing light [10]. Our main concern of optimization is to give a noticeable shift in resonant wavelength with a minute increase in RI. Proposed biosensor makes use of this resonant wavelength as it shows variation with change in the refractive index of cavity or by changing its surrounding structure. Our ultimate goals are to achieve high Q factor and high sensitivity simultaneously, where how much output of sensor is varying with minute changes in input is measured by sensitivity [11, 12].

3 Designing of a Biosensor

This design involves a photonic crystal built with dielectric rods of Si which have a radius of 210 nm, a lattice constant of 700 nm, and a refractive index of 3.47772 over the wafer of air (refractive index of air 1) Si rods are placed. With these parameters, a photonic crystal possessing a hexagonal symmetry with an area of $19 \times 13\ \mu m^2$ is designed as shown in Fig. 1. TE photonic bandgap in the range of ω_1€ [319, 444 nm] with $\Delta\omega_1 = 125$ nm and ω_2€ [587, 753 nm] with $\Delta\omega_2 = 166$ nm is obtained after plane wave expansion (PWE) simulation as shown in Fig. 2. The propagation of TE mode is strictly forbidden within these spectral regions. The periodicity of the structure is breaks through introducing some defects or by variation of refractive index causes some modes are allowed to propagate through the structure within the photonic bandgap. For sensing applications, this property is used for observing quality factor and resonance value.

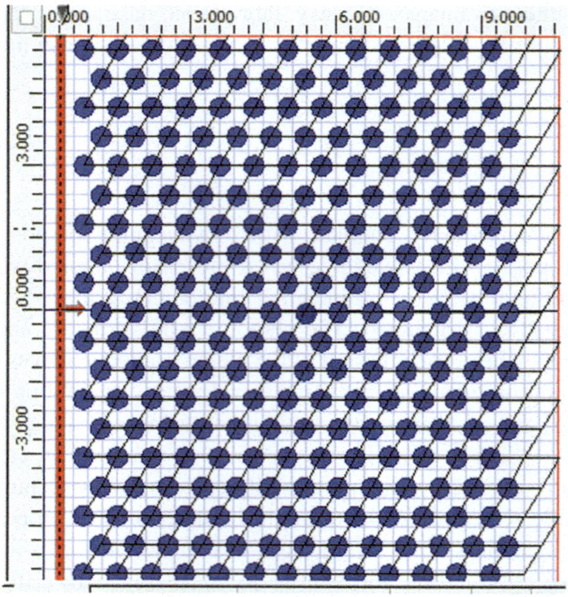

Fig. 1 Design of a two-dimensional photonic crystal (hexagonal symmetry) with dielectric rods of Si (η = 3.47772) over the wafer of air (η = 1)

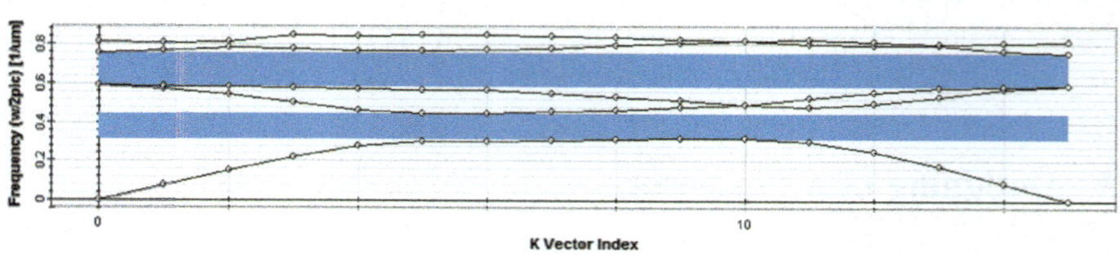

Fig. 2 TE band diagram for the photonic crystal structure considered in Fig. 1

4 Nanocavity-Coupled Waveguide PC Biosensor

The propagating light wave in the structure is trapped through nanocavity. In our design, defect is created by changing the radius of the center hole. The rods are introduced in the wafer of air, and the background sample is replaced by the blood when sensor device is dipped in blood sample. The light that passes through the photonic crystal from one end will interact with the hemoglobin in blood and gets detected from another end. The propagation of the light will vary in the photonic crystal according to the refractive indices of the blood analyte. The structure possessing hexagonal symmetry by a nanocavity coupled with a waveguide (same

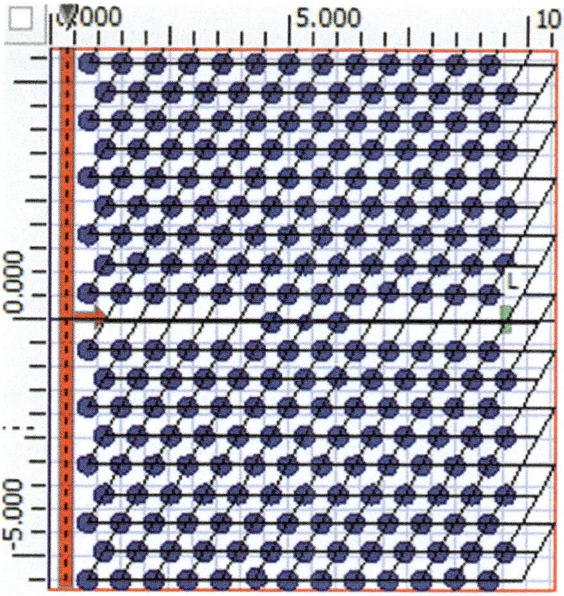

Fig. 3 Nanocavity-based biosensor with a radius of 160 nm introduced within photonic crystal (dielectric rods of Si over the wafer of air)

parameters as that of structure in Fig. 1) is shown in Fig. 3. The radius of the center hole is changed from 210 to 160 nm and creates nanocavity.

5 Result

This nanocavity is filled with hemoglobin (η = 1.351, η = 1.4, η = 1.425). Power spectrum for all these cases is observed and compared at an input wavelength of 1550 nm. Results obtained show that as the wave propagates, it gets absorbed by hemoglobin molecule, and therefore, the intensity of light wave gets reduced as shown in (Figs. 3, 4, 5, 6) (Table 1).

Table 1 Input refractive index and output transmitted power for different concentrations of hemoglobin present in blood

Blood sample with hemoglobin concentration (g/dL)	Refractive index	Resonant wavelength (um)	Monitor value (a.u.)	FWHM (nm)	Quality factor (Q)
0	1.351	1.553	0.654	0.1	1553
10	1.4	1.556	0.656	0.1	1556
20	1.425	1.573	0.659	0.11	1430

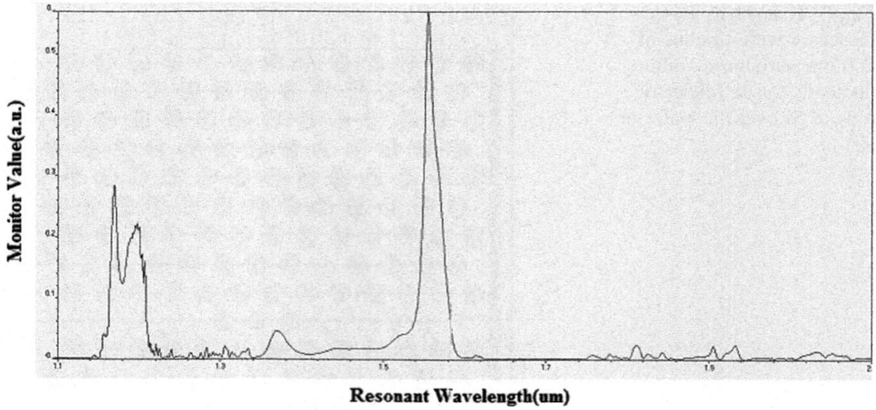

Fig. 4 Power spectrum for 0 g/dL concentration of hemoglobin in blood sample

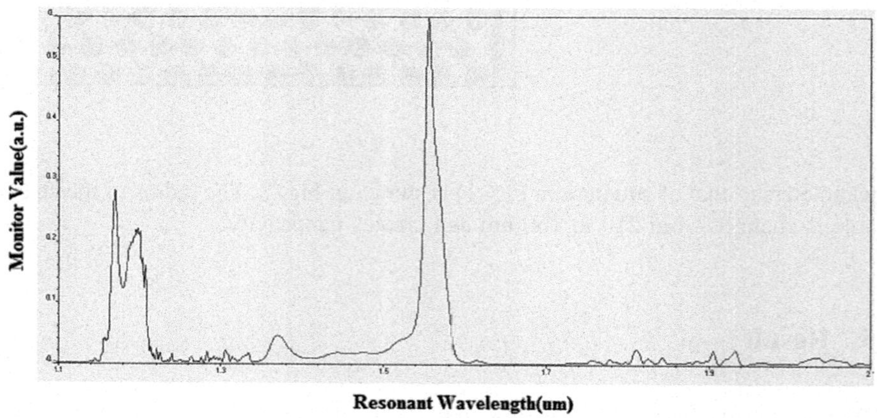

Fig. 5 Power spectrum for 10 g/dL concentration of hemoglobin in blood sample

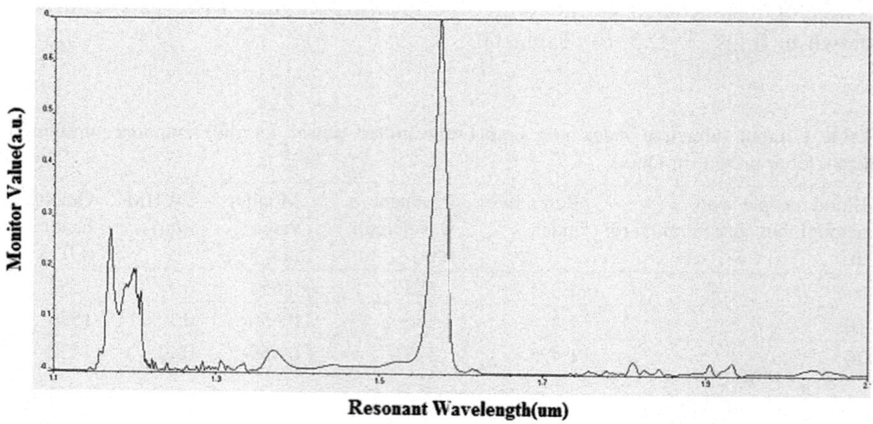

Fig. 6 Power spectrum for 20 g/dL concentration of hemoglobin in blood sample

6 Conclusion

This design is proposed based on nanocavity-coupled photonic crystal waveguide. The sensor designed is cost-efficient, small and can be fabricated as lab-on-chip. The proposed sensor is highly sensitive since it is able to sense even minute variations in the input refractive index of hemoglobin concentration present in blood. It provides label-free accurate detection and very effective noninvasive method for monitoring the hemoglobin level in blood since no reagents are added in the blood so there are no chances of any error in the output. The proposed sensor which is detecting high and low hemoglobin level in blood is very helpful in detection of anemia or polycythemia. Proposed nanostructured stand shows large values of Q factor. The insertion of nanocavity in waveguide causes sharp resonance in the transmission spectra which makes it possible to realize hemoglobin level detection in blood. The analysis is performed using FDTD method. The presence of different hemoglobin concentration in blood samples is detected by measuring the resonance shift in transmission spectra of proposed nanocavity-coupled photonic crystal waveguide.

References

1. A report by the World Economic Forum and the Harvard School of Public Health, September 2011. www.weforum.org/EconomicsOfNCDappendix
2. Sharmal P, Gudaguntj FD, Sharan P (2014) An analysis of quality factor for different bioanalytes by using photonic crystal based sensor. IEEE. ISBN No. 978-1-4799-3914-5/14
3. Roggan A, Friebel M, Dorschel K, Hahn A, Muller G (1999) Optical properties of circulating human blood in the wavelength range 400–2500 nm. Biomed Opt 4(1):36–46
4. Bayer R, Çaglayan S, Günther B (1994) Discrimination between orientation and elongation of RBC in laminar flow by means of laser diffraction. In: Proceedings of SPIE 2136, pp 105–113
5. http://www.who.int/vmnis/anaemia/prevalence/summary/anaemia_data_status_t2/en
6. Noninvasive Anemia Screening Technology Opportunity Assessment Prepared for the Merck for Mothers Program, February 2012
7. Fan X, White IM, Shopova SI, Zhu H, Suter JD, Sun Y (2008) Sensitive optical biosensors for unlabeled targets: a review. Anal Chim Acta 620(1–2):8–26
8. Sharma P, Roy SK, Sharan P (2014) Design and simulation of photonic crystal based biosensor for detection of different blood components. In: Proceedings of IEEE, pp 171–176
9. Meade RD, Rappe AM, Brommer KD, Joannopoulos JD (1992) Existence of a photonic band gap in two dimensions. Appl Phys Lett 61(4):495–497
10. Chhabra M, Selwal C (2014) Design of a photonic crystal biosensor using DNA filled microcavity and ring cavity coupled with waveguide. IEEE. ISBN NO. 978-1-4799-3140-8/14/
11. Yang D, Tian H, Wu N, Yang Y, Yuefeng J (2013) Nanoscale torsion free photonic crystal pressure sensor with ultra-high sensitivity based on side-coupled piston-type micro-cavity. Sens Actuators A: Phys 199:30–36
12. Chao CY, Fung W, Guo LJ (2006) Polymer microring resonators for biochemical sensing applications. IEEE J Sel Top Quantum Electron 12(1):134–142

Survey on FSO Communication System—Limitations and Enhancement Techniques

Essar Farooq, Anupam Sahu and Sachin Kumar Gupta

Abstract Free-space optical (FSO) communication is a line-of-sight communication in which the data modulates the infrared or visible laser light through the air from transmitter to receiver. FSO ensures fast, secure, and reliable data transmission for which it is widely considered as a next-generation high-speed wireless communication technology. FSO systems are mostly the "last mile solution" for commercial systems of metro networks between fixed sites. Because of the inherent benefit of mobility, fast deployment, security, and high data rates, FSO systems are being considered for military systems. The properties of FSO system such as license-free band, cost-effectiveness, and high-bandwidth access allows it to gain rapid popularity as a means of transferring data at high rates over small distances. However, FSO does suffer from certain limitations as well; mostly caused due to atmospheric turbulence that highly degrades its performance. In this paper, various limitations of FSO and the possible means to mitigate their effect are discussed.

Keywords FSO · Atmospheric turbulence · Modulation technique
Photodetector · OOK · DPSK

E. Farooq · S. K. Gupta (✉)
Department of Electronics and Communication Engineering, Shri Mata Vaishno Devi University, Kakryal, Katra 182320, Jammu & Kashmir, India
e-mail: sachin.rs.eee@iitbhu.ac.in

E. Farooq
e-mail: 15mmc006@smvdu.ac.in

A. Sahu
Department of Electronics and Communication Engineering, Madan Mohan Malaviya University of Technology, Gorakhpur 273010, Uttar Pradesh, India
e-mail: anupam.or.sahu@gmail.com

V. Janyani et al. (eds.), *Optical and Wireless Technologies*, Lecture Notes in Electrical Engineering 472, https://doi.org/10.1007/978-981-10-7395-3_29

1 Introduction

The FSO has been researched upon since Graham Bell's photophone in 1880. Bell made voice signal to modulate sun's radiation and transmitted it over the distance of about 200 m. He used a parabolic mirror with selenium cell at its focal point as a receiver. However, this experiment did not go well due to the crudity of the devices used and the intermittent nature of the sun's rays. However, the fortune of FSO changed after the discovery of laser in 1960. Nippon Electric Company (NEC) built the very first FSO laser link in Japan in 1970 in order to handle commercial traffic. This link was a full duplex He–Ne laser FSO and was of 14 km length. National Aeronautics and Space Administration (NASA) and European Space Agency (ESA) have also researched FSO in deep-space applications with programs such as the Mars Laser Communication Demonstration (MLCD) and the Semiconductor Laser Inter-satellite Link Experiment (SILEX), respectively. FSO technology finds various applications in communications networks as well, where a connectivity gap exists between two or more points. This technique also provides a cost-effective optical wireless connectivity and a faster Return on Investment (ROI) for enterprises and mobile carriers. As the demand for greater bandwidth for enterprise and mobile carrier subscribers is ever-increasing, there is a critical need for FSO-based products so that a balance of throughput, distance, and availability is obtained [1]. However, optical wave propagation through the air experiences several hindrances like the fluctuation in amplitude and phase due to atmospheric turbulence and the irradiance fluctuation called scintillation. Scintillation is caused because of the thermal gradients and turbulence within the optical path incurred due to the variation in air temperature and density. Zones differing in density act as lenses which scatter light away from its intended path. These effects can cause scattered laser beam to travel along different paths and then recombine. Atmospheric turbulence degrades the performance of FSO to great extent. Researchers have worked in many ways in order to mitigate the effect of atmospheric turbulence. Different modulation formats have been tested in order to find the one which is least affected by this phenomenon.

2 General Block Diagram of FSO System

The general block diagram of FSO is shown in Fig. 1. It simply comprises transmitter, channel, and receiver. The transmitter sends modulated laser beam through free space, which is received, demodulated, and detected at the receiver.

In an FSO system, the message signal is modulated in the modulator block that essentially is an intensity modulation (IM). Several modulation techniques have been implemented on FSO system such as OOK, BPSK, QPSK, PPM. In the intensity modulation, the modulated signal is used to vary the intensity of the light coming out of the laser source. The transmitter telescope directs the modulated laser

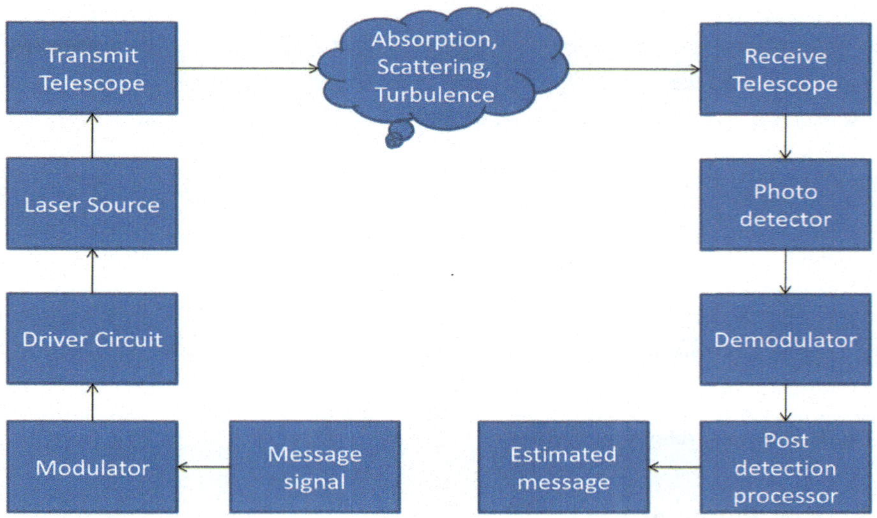

Fig 1 Block diagram of FSO system

beam toward the receiver telescope. The receiver comprises a photodetector that can be a PIN or an APD. The photodetector converts the light signal into corresponding electrical signal so that it can further be demodulated and processed.

As FSO is a LOS communication, there should be no obstruction between transmitter and receiver. The architecture of FSO can be summarized as:

Transmitter: In transmitter, the intensity modulation is done in which the intensity of light is varied depending upon the modulating signal. The intensity of laser is varied by controlling the laser current in the driver circuit. The transmit telescope is used to transmit the intensity-modulated light signal with high directivity. The light source is chosen depending upon the wavelength and properties.

Channel: The modulated light signal gets degraded in the channel. The channel used in FSO is free space, that is, air. Atmospheric turbulences such as fog, rain, snow cause the scattering, absorption, and fading of the transmitted signal which eventually degrades the performance of the FSO system.

Receiver: At the receiver, the receive telescope collects the modulated light signal and the photodetector converts the light signal into corresponding electrical signal. The detected electrical signal is then demodulated and processed in order to retrieve the transmitted data (Fig. 2).

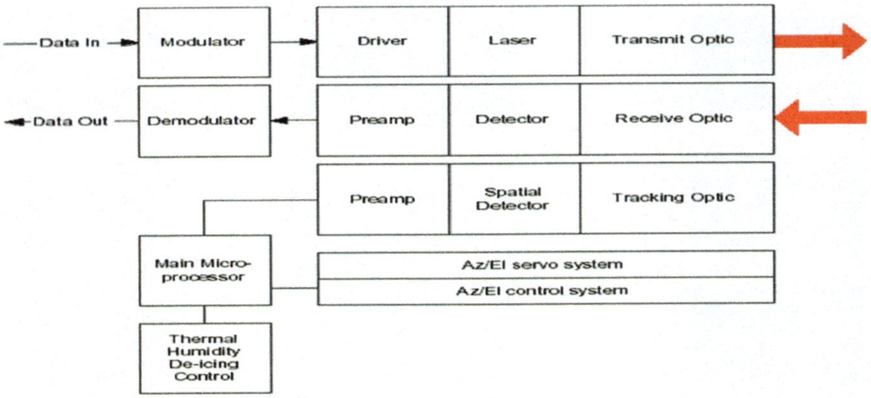

Fig. 2 FSO major subsystem

3 Key Points of FSO

Number of key points in terms of advantages and applications are there that has significantly attracted the researchers' attention and become vibrant area of research among academia and industrial people.

3.1 Advantages: FSO Communication Has Got a Lot of Advantages like

- High-Bandwidth Availability: FSO has got a 2000THz data bandwidth which offers a very high channel capacity [23].
- Narrow Beam Width: As FSO uses laser at the transmitter end, the beam width is very narrow which provides high-level secrecy that cannot be achieved by RF communication [23].
- Unlicensed Spectrum: FSO communication takes place in ISM band, which is an unlicensed band. Hence, FSO is free from spectrum distribution which makes it convenient and cost-effective [23].
- Low Installation Cost: The installation cost of FSO is very low as compared to RF for the same data rate [23].
- Easy Deployment and Redeployment: The installation time in FSO is very low. It takes as less as four hours to install an FSO. In addition, it can easily be redeployed from one place to another [23].

3.2 *Applications: The Number of Potential Applications in Various Fields of FOS Communication System Has Been Observed as Mentioned Below*

- Inter-Campus Connectivity: There is a heterogeneous network traffic (i.e., voice, data, fax, multimedia traffic, etc.) that overwhelm the typical connections, be it in the school/university campuses or the corporations. FSO systems have the capability of providing ultrahigh speeds with low cost unlike dedicated fiber optic connections [2].
- Monitoring and Video Surveillances: Nowadays, video cameras are deployed almost everywhere; it may be in the supermarkets, airports, malls, and so on for public safety and in military applications. Deployment of video cameras is easy. However, the conventional wireless technology is incapable of providing high throughput required for video streams. FSO technology supports high-quality video transmission, thereby proving to be an efficient alternative.
- Back-haul for Cellular Systems: In cellular systems, FSO can prove to be of high use in carrying the traffic of cellular telephone with high speed as well as fast data rates from antenna towers to the PSTN, thereby increasing the speed of transmission [2].
- Redundant Link and Disaster Recovery: As FSO links can be deployed easily and in less time, this advantage of FSO can prove to be extremely helpful in natural disasters, terrorist attacks, and emergency situations in which the communication infrastructure may have been damaged or became unreliable. The practical example of the FSO deployment as a redundant link was witnessed after 9/11 terrorist attacks in New York City, where FSO links were rapidly deployed in the area for financial corporations which were left out with no landlines.
- Military Access: FSO has got the advantage of being secure and undetectable as it uses narrow beam width for data transmission. Hence, it can provide secure connection in large areas with minimal planning and less deployment time, which makes it most suitable for military access [3].
- Point-to-Point and Point-to-Multipoint Communication: FSO communication is possible in point-to-point links, for example, between two shops, two buildings, and also in point-to-multipoint links, for example, from satellite to ground for short as well as long-distance communication [4].
- Last Mile Access: Last mile has always been a bottleneck for most of the telecommunication, cable television, and Internet industries. FSO can provide a solution to this problem by deploying it in the last mile along with other networks [2].
- Storage Area Network (SAN): SAN is a network that provides access to consolidated block-level data storage. FSO links can be used to form a SAN [5].

Typically, the best use scenarios for this technology are:

- In campuses, FSO is providing fast ethernet or gigabit ethernet speeds for LAN-to-LAN connections.
- In metropolitan area network, FSO provides LAN-to-LAN connections.
- It converges voice data connection.
- For certain specific events or purposes, FSO provides temporary network installation.
- In disaster recovery, FSO provides re-establishment of high-speed connection quickly.
- FSO may be used to upgrade existing wireless technologies.
- FSO provides an effective backup for important fiber connections.

3.3 *Limitations: In Spite of Number of Advantages and Applications of FSO, It Has Some Limitations that May Lead to Challenges*

- Atmospheric Disturbances: A major impairment on FSO links is caused due to the atmospheric turbulence. It is the result of random variations in the refractive index caused due to inhomogeneities in temperature and pressure. In clear weather conditions, the atmospheric turbulence causes the intensity fluctuation of the received signal called fading which is also known as scintillation in optical communication terminology [6, 7]. Turbulence is caused by the inhomogeneities of temperature as well as pressure in the atmosphere and can, thereby, severely degrade the link performance, particularly for link distances of 1 km or longer. The performance of this technology depends strongly on certain parameters like atmospheric conditions between transmitter and receiver, parameters of the link such as length, operating wavelength. Moreover, other factors such as fog, rain, atmospheric gases, and aerosols also result in beam attenuation due to photon absorption and scattering [8].
- Pointing Error: Apart from scintillation effects, the performance of the FSO link degrades because of the building sway, which introduces a pointing error (PE) between transmitter (Tx) and receiver (Rx), representing a particular problem in urban areas, where FSO equipment is placed on high-rise structures [9, 10].
- Physical Obstructions: Since FSO is a line-of-sight communication, any obstruction in the path can interrupt or affect the communication. These obstructions may be in the form of birds, tall trees and buildings, etc. [11]. However, interruption of this kind is temporary, and transmissions are easily and automatically resumed.

- Geometric Losses: These losses are the form of optical beam attenuation that is caused due to the spreading of beam and results in the reduction of power level of signal as it travels from transmitter to receiver [12].
- Background Radiation: Last but not least, background radiation, also called background noise or ambient noise, can degrade the performance of FSO links. In fact, in addition to the useful signal, the receiver lens also collect some undesirable background radiations that may consist of direct sunlight, reflected sunlight, or scattered sunlight from hydrometeor or other objects.

3.4 Possible Solutions

The above limitations pose a great challenge in the working of FSO; however, many ways have been found to counter these limitations. The primary way to counter fog in the deployment of FSO is through a network design that shortens FSO link distance and adds network redundancies. While to combat absorption, use of spatial diversity with appropriate power levels may help to maintain the required level of network availability. Also, Light Pointe's multibeam system (spatial diversity) is found useful to address physical obstructions, atmospheric turbulence as well as scintillation [11]. Also, further performance enhancement may be obtained by using appropriate channel model(gamma-gamma distributed channel model), multiplexing technique(OFDM), and by the use of optical amplifier(EDFA) in place of electrical amplifier [15, 22].

3.5 Reported Works

There has been lot of research going on in order to mitigate the limitations of FSO and enhance its performance. In this context, various fields have been pointed out where researchers are working in order to overcome the effect of atmospheric turbulence. These fields can be broadly classified as:

Modulation Techniques: Several researchers have chosen different modulation techniques to see the effect of atmospheric turbulence on their performance. In [13], different modulation techniques like OOK, Subcarrier BPSK, and Q-ary PPM in turbulence regime are compared. It was found that the performance of BPSK was better than all others except that when the order of Q in Q-ary PPM was increased then its performance also got improved. Wang et al. [14] proposed that Space Diversity Reception Technique (SDRT) and advanced modulation formats can successfully mitigate the transmission impairments of the atmospheric turbulence. In this paper, modulation formats, which were compared, are OOK, DPSK, and DQPSK. They observed that in strong turbulent scenario, the OOK and DPSK formats can have high SDRT gains in the order of 19.5 and 20.3, respectively, at the

BER of 10^{-3}. In addition, it was also found that in strong turbulent scenarios, DPSK and DQPSK formats have same BER performance under the same symbol rate. Simplicity is the advantage of OOK while DPSK format, which encodes information in its phase, can mitigate severe effect of scintillation to some extent. Moreover, in the comparison of the binary formats like OOK and DPSK, the DQPSK format doubles the spectral efficiency.

However, BPSK SIM offers improved performance across all range of turbulence variances [15, 16]. The performance of FSO systems with APD detectors is severely impacted by turbulence, and an optimum average APD gain must be used to avoid excessive APD noise in the receiver. In these systems, it was found that differential amplitude PPM has better performance than that of other schemes for same peak power [17, 18].

Diversity Techniques: An increase in the number of apertures increases the gain and mitigates the effect of atmospheric turbulence. The performance of SISO FSO links is severely degraded from turbulence. The results in [14, 15, 19, 20] demonstrated that significant performance gains can be obtained when multiple apertures are used at transmitter and receiver, thereby enhancing the quality of FSO.

Performance Enhancement: The performance of FSO system gets significantly improved by using OFDM technique in place of traditional TDM-FSO link. In [21], an increase in gain and improvement in receiver sensitivity is obtained in OFDM-FSO link as compared to TDM-FSO link. The performance of FSO system is affected by the type of amplifier used. The optical pre-amplifier provides better performance as compared to the electronic amplification. In electronic amplification, the amplification is done after the photodetector while as in pre-amplification, the amplification takes place before the photodetector by means of erbium-doped fiber amplifier (EDFA) [22].

Channel Model: Various statistical models have been used in order to describe the optical channel characteristics with respect to the atmospheric turbulence strength like Log-normal, Gamma-Gamma (G-G), I-K, K, Negative Exponential, Rician Log-normal Distribution. There are three main statistical models which are used to describe the atmospheric turbulence channel that are Log-normal, K, and G-G distributed channel model. The G-G distribution out of the three is used to model atmospheric turbulence irradiance fluctuation in both weak as well as strong turbulence regimes. Where for the limit of strong turbulence, Negative Exponential distribution is adopted [15]. Generally, the Log-normal distribution is used in case of the weak atmospheric turbulence regime [15, 16, 23].

Pointing Error (PE): The gain of FSO system increases when multiple apertures are used at transmitter or receiver. However, the PE which is generated by building sway degrades the performance of FSO, eradicating the benefits obtained by using multiple apertures. In [24], the effect of PE in Gamma-Gamma fading atmospheric fluctuation of FSO MIMO system is discussed. Two schemes are studied, namely Equal Gain Combining (EGC) and Maximal Ratio Combining (MRC). MRC was found out to be more impregnable for large PE than EGC scheme. However, for low to moderate PE, EGC is preferred over MRC due to its low implementation complexity.

4 Conclusion

With the ever-increasing demand for bandwidth and high data rates from the heterogeneous networks, the FSO can be the efficient way out. Terrestrial FSO links with transmission rates of 10 Gbps (assuming a range of few hundred meters) are already in the market, and the speeds of recent experimental FSO systems are promising even more. To further enhance the performance of FSO, tremendous amount of research work has been done and is still going on. This paper provides the overview of FSO communication and its numerous advantages, applications, and limitations. The insight is provided into the research that has been done in order to mitigate its limitation and enhance its performance.

References

1. Ghassemlooy Z, Popoola WO (2010) Terrestrial free space optical communication. In: Mobile and wireless communications network layer and circuit level design, Optical Communication Research Group, NCR Lab, Northumbria University, Newcastle Upon Tyne, UK, 355–391
2. Willebrand HA, Ghuman BS (2001) Fiber optics without fiber. IEEE Spectr 38(8):40–45
3. Shaulov G, Patel J, Whitlock B, Mena P, Scarmozzino R (Oct. 2005) Simulation assisted design of free space optical transmission systems. In: Military communications conference (MILCOM), Atlantic city, NJ, USA, 1–5
4. Sahbudin RKZ et al (2013) Performance of SAC OCDMA-FSO communication systems. Optik-Int J Light Electron Opt 124(17), 2868–2870. http://psasir.upm.edu.my/28583/
5. Kaufmann J (2011) Free space optical communication: an overview of applications and technologies, Boston IEEE Communication Society Meetings
6. Andrews L (2004) Atmospheric optics, SPIE Optical Engineering Press
7. Andrews L, Phillips R, Hopen C (2001) Laser beam scintillation with applications, SPIE Optical Engineering Press
8. Gagliardi R, Karp S (1995) Optical communications. Wiley, New York
9. Gappmair W (2011) Further results on the capacity of free-space optical channel sin turbulent atmosphere, IET Commun, 5(9), 1262–1267
10. Lee IE, Ghassemlooy Z, Ng WP, Khalighi MA (2013) Joint optimization of a partially coherent Gaussian beam for free-space optical communication over turbulent channels with pointing errors. Opt Lett 38(3):350–352
11. FSO History and Technology. http://www.laseroptronics.com/index.cfm/id/57–66.htm
12. Vigneshwaran S, Muthumani I, Raja AS (Feb. 2013) Investigations on free space optics communication system. In: IEEE International Conference on Information and Embedded System (ICICES), Chennai, India, 819–824
13. Barua B, Barua D (Dec. 2011) Evaluate the performance of FSO communication link with different modulation techniques under turbulent condition. In: 14th International conference of computer and information technology, 1–5
14. Wang Z, Zhong WD, Lin FC (Dec. 2009) Performance comparison of different modulation formats over FSO turbulence links with space diversity reception techniques. IEEE Photon J 1 (6), 277–285
15. Popoola WO, Ghassemlooy Z (2009) BPSK subcarrier intensity modulated free-space optical communications in atmospheric turbulence, J Light Technol 27(8), 967–973
16. Tang X, Rajbhandari S, Ghassemlooy Z, Kandus G (July 2010) Performance of BPSK subcarrier intensity modulation free- space optical communications using a log-normal

atmospheric turbulence model. In: IEEE Symposium on Photonics and Optoelectronic (SOPO), 1–4
17. Kiasaleh, K (Sep. 2005) Performance of APD-Based, PPM Free-Space Optical Communication Systems in Atmospheric Turbulence, IEEE Transaction on Communications, 53(9), 1455–1461,
18. . Gopal, P., Jain, V. K., Kar, S.: Performance of OOK and variants of PPM in APD based free space optical communication systems, 1–6. http://www.photonicsindia.org/Final/20report/PoojaGopal.pdf
19. Peppas K, Nistazakis HE, Assimakopoulos VD, Tombras GS (2012) Performance analysis of SISO and MIMO FSO communication systems over turbulent channels, Chapter-17, Opt Commun Intech, 415–438
20. Tsiftsis TA, Sandalidis HG, Karagiannidis GK, Uysal M (2009) Optical wireless links with spatial diversity over strong atmospheric turbulence channels, IEEE Trans Wireless Commun 8(2), 951–957
21. Kumar P, Srivastava A (Apr 2015) Enhanced performance of FSO link using OFDM and comparison with traditional TDM- FSO link. In: IEEE International Broadband and Photonics Conference, 65–70
22. Cao Q, Brandt-Pearce M, Wilson SG, Brown CL (Dec. 2006) Free space optical MIMO System using an optical pre- amplifier. In: IEEE Global Telecommunications Conference (GLOBECOM), 1–5
23. Zhu X, Kahn JM (Aug. 2002) Free-space optical communication through atmospheric turbulence channels. IEEE Trans Commun, 50(8), 1293–1300
24. Bhatnagar MR, Ghassemlooy Z (2016) Performance analysis of gamma–gamma fading FSO MIMO links with pointing errors. J Light Technol, 34(9), 2158–2169

Optical Properties of Photonic Crystal Fibers

Ashish Kumar Ghunawat, Anjali Jain, Kumari Nikita, Manish Tiwari and Ghanshyam Singh

Abstract We present a brief study of the photonic crystal fibers and recent developments and modifications presented to them for betterment of the optical parameters. The properties birefringence, nonlinearity, confinement loss, and chromatic dispersion are discussed, and the fiber structures of the fiber designed from the past few years have been studied. The PCFs can be seen as the future of a number of optical applications owing to its low chromatic dispersion, low confinement loss, high birefringence, and nonlinearity, and, the best of all, its flexibility to modify its structural aspects to manipulate these properties as desired.

Keywords Nonlinearity · Birefringence · Confinement loss
Chromatic dispersion · Optical parameters

1 Introduction

PCFs have been emerged as a new research activity in the scientific community due to its unique properties of light guidance. Its structural flexibility allows easy manipulation of optical properties like birefringence, nonlinearity, confinement loss, and dispersion [1]. Fiber with hollow cores exhibits high potential in communication applications owing to their higher damage threshold [2]. Conventional step-index fibers guide light only by TIR mechanism, and thus, the refractive index (RI) of core must be greater than cladding. Standard SMF causes bend loss of about 0.5 dB at 1550 nm due to small RI difference which limits its applications. So the curiosity of trapping light inside a hollow core gives birth to PCFs. PCFs not only carry more power but also act as a superior host for multiple cores and rare earth

A. K. Ghunawat (✉) · A. Jain · K. Nikita · M. Tiwari · G. Singh
Department of Electronics and Communication, MNIT Jaipur,
JLN Marg, Jaipur 302017, Rajasthan, India
e-mail: ashishghunawat@gmail.com

V. Janyani et al. (eds.), *Optical and Wireless Technologies*, Lecture Notes
in Electrical Engineering 472, https://doi.org/10.1007/978-981-10-7395-3_30

ions [3, 4]. Along with TIR mechanism, PCF also uses photonic bandgap effect; that is, light is propagated through defects of its crystals which is formed by eliminating a few central capillaries [5].

2 Brief History

The evolution of PCFs started in early 1970s with a central core connected to a tubular cladding shell structure. However, it was difficult to fabricate such fibers with MCVD which was an overgrowing fabrication technique at the time. In 1995, the formulation of PCF via fiber drawing tower was introduced. The solid core photonic bandgap fiber eventually progressed leading to the first working PCF in 1998 [6].

3 PCF Structures and Developments

When PhC waveguides were used for communication purposes, high losses were there. Hence, the idea of the development of fibers encompassing holes in fiber structure was proposed [7]. Initially, it was proposed to use silica instead of GeO_2 in the fiber core and this idea was widely accepted as it decreases the attenuation by 0.04 dB/km [8]. Later, multicomponent materials like silicate, chalcogenide [3], ytterbium [9], tellurite, Si-nc rod [10], benzene [11], chalcogenide glass (As2S3) [12], and AsSe2 glass [13] were used as fiber materials owing to their large leverage in producing more output power, birefringence, and nonlinearity than any conventional silica glass fibers.

Various fabrication methods like multiple thinning, extraction process, and using silicon nanowire were adopted; however, hole symmetry is disturbed while employing thinning process especially for high d/Λ (> 0.6) ratio [3, 14].

Eventually, fibers having different geometries were proposed. Starting with a fiber structure with polymer core and elliptical air hole, cladding with squeezed triangular lattice was designed [15]. Later, PCFs with different number of air hole rings (n_r) in cladding were introduced like a PCF made of pure silica having nr = 5 (radii of air holes = 1.4 μm) [16, 17], asymmetric fiber with nr = 5 [18], nr = 4 when pitch = 2.2 μm [19], nr = 4 when Λ = 1.96 μm [20], and nr = 3 [21, 22]. Results have established the fact that as nr increases, PCF undergoes improved chromatic dispersion, confinement loss, and birefringence [23, 24].

With the advancements in PCFs, a structure made of pure silica core with a missing air hole and rotational symmetry was introduced having beat length 6 times shorter as compared to conventional fibers. The large core size, pitch (Λ = 1.96 μm), and hole diameter (d = 0.40 μm) facilitate high birefringence of about 3.7×10^{-3} at 1540 nm wavelength [25]. Birefringence when calculated through FEM with perfect boundary conditions further increased to 4×10^{-3} for degenerate fundamental modes when the beat length was taken to be 0.385 mm [4, 26]. The properties of PCFs were also studied by varying the numerical aperture and filling fraction. For this, a design with pitch = 7.2 μm and filling fraction = 0.15 was fabricated and various relations were obtained using finite element method at 632 nm wavelength which shows that PCFs have enormous potential in single-mode regime [27].

Fiber attenuation per unit length was also a key factor for introducing structural changes in PCF [2]. Bragg fibers offer attenuation of 10 dB/km at 0.4–2.2 mm wavelength range. Later, a 20 μm diameter hollow core hexagonal lattice fiber was designed by removing 19 central capillaries which marked an immense decrease in attenuation by 11.3 dB/km [5].

For the enhancement of effective area, an index-guided large mode area (LMA) PCF was proposed comprising of hexagonal lattice of air holes [28, 29]. Two designs (A and B) were proposed both having hole diameter from inner to outermost hole ring as 1.5, 1 and 6 μm, respectively, but having refractive index of air holes = 1 and 1.4507 for designs A and B, respectively. Results have shown that the proposed design B can exhibit $A_{eff} = 3000\ \mu m^2$ along with low confinement loss; hence, this result marks the fact that both confinement losses and effective mode area can be improved simultaneously [30].

To improve the nonlinear characteristics, SBS slow light structures are realized. A highly nonlinear PCF (length = 25 m) of core radii as 0.9 μm and air hole diameter as 0.89 μm was designed. Results verified that delay efficiency of proposed fiber (2.8 ns/W/m) was 10 times more as compared to SMF (0.25 ns/W/m) due to small effective area, hence paving a long path for PCF to achieve lower pump power with shorter fiber length along with large delay efficiency [31]. For the enhancement of slow light nonlinear effects, a silicon PhC waveguide (waveguide length = 0.9 mm) with triangular lattice of air holes (hole radii 118.9 nm) by emitting one row of air holes was designed [32].

For ultra-flattened chromatic dispersion, an eight-ringed HNL-PCF structure was designed (nonlinear coefficient > 33 $W^{-1}\ km^{-1}$) and to form concentric cores dimensions of 1st and 4th rings were decreased. This structure exhibits a dispersion of −1.65 to −0.335 ps/nm/km for $\lambda = 1.45$ to 1.65 μm [33]. Similarly for the same purpose a modified hexagonal PCF design with pitch $\Lambda = 3.0$ μm was proposed which along with dispersion, decreases confinement loss too (< 10^{-4} dB/km) [34].

Holes are assumed perfectly circular by this time [26]. Initially with silicon PCF with spiral lattice, nonlinear coefficient obtained was 162 $W^{-1}m^{-1}$ [35]. Later, an elliptical–spiral PCF was designed which helps in modifying the birefringence, nonlinearity, and chromatic dispersion to an enormous extent [36, 37].

Also, a honeycomb PCF structure was proposed in which some air holes were eliminated from cladding portion (Fig. 1).

This structure was widely effective in reducing the confinement loss, hence finding its widespread use in communication applications [38].

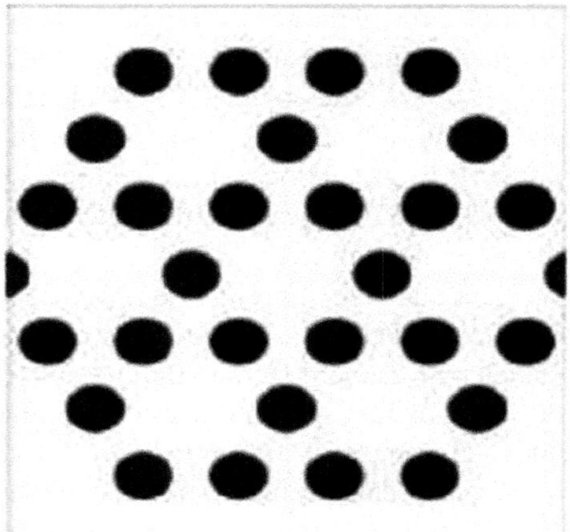

Fig. 1 Honeycomb PCF structure

Also, PCF having triangular and square lattice air holes in the cladding was designed using multipole method. Two air holes were removed from the transverse section through the center. In this fiber, a birefringence of 10^{-2} order was observed which leads to the conclusion that as numbers of asymmetries are increased in the air holes of inner rings of fiber, modal birefringence also increases [13, 39]. Another square lattice structure having six rings of cladding cells was proposed. This structure has high nonlinear coefficient (14 $W^{-1}m^{-1}$) at 1.55 μm, and near zero dispersion was achieved at filling fraction of 0.15 [40]. Similarly, structures with tetragonal lattice [1, 41], five layers of air holes [42], four layers with circular–elliptical air holes confining in square lattice [23] were proposed, and their variations of confinement losses and birefringence were recorded. Results have established the fact that PCFs having tetragonal lattice are more birefringent irrespective of the geometry of the cladding [43, 44].

Further a PCF with hexagonal lattice of circular air holes in cladding and rectangular line up of 4 circular air holes in the core was designed. This structure was analyzed through FEM, and numerical results indicated that all circular PCFs can also achieve ultrahigh birefringence (greater than 0.01) and ultralow confinement loss (less than 0.002) [45]. Similarly, when DC-Hy PCF was used with the same symmetry, the structure exhibits birefringence = 3.79×10^{-2} and confinement loss = 0.007 dB/m [46].

An octagonal PCF has more air holes in cladding region, thus lowering the cladding RI_{eff} which further leading to low confinement loss [47]. Hence, a PCF structure consisting of octagonal outer ring and hexagonal inner ring was proposed. 4 holes were omitting on the 1st ring and 2 on the 2nd ring and $\Lambda 0 = 3.6$ μm, distance between each rings $\Lambda = 2$ μm, and diameter of air holes were d1 = 1.8 μm, d2 = 1.4 μm, d3 = 0.6 μm in different rings, respectively. This structure gives high birefringence of 0.001 within 1.5–1.65 μm wavelength [48]. Results have established the fact that as d/Λ increases, nonlinearity, NA increases and confinement loss decreases [49, 50].

To control the dispersion and dispersion slope properties, fractal PCF structure was proposed, and in this, refractive index of the core was higher as compared to its surrounding area. Results have shown improved dispersion values [51].

Not only gases (air) but liquids also possess a lot of nonlinear properties; hence, solitons were also introduced in PCFs for applications like supercontinuum generation, pulse delivery, and wavelength conversion [52, 53].

Tapering of PCF is also introduced to modify birefringence and dispersion properties. It is found that when taper is maximum, i.e., around 50%, birefringence is improved to 5×10^{-3} owing to the uneven flaring of the mode field [9]. Another analysis of PCF is performed by making the surrounding walls of PCF as perfect magnetic conductor and by incrementing the count of air hole rings in the center which confines electric field better in the center. Results established the relation that effective mode index decreases as λ wavelength increases, and increases as air hole diameter decreases and pitch increases [54].

4 Properties of PCF

4.1 *Birefringence*

The birefringence is an optical property of any material characterized by polarization and direction of light propagation. Birefringence usually occurs in PCFs due to rotational asymmetry of the fiber structure [21]. The birefringence is defined as the difference value of the effective refractive index of fundamental x- and y-polarization modes, which can be determined by the following equation:

$$B = |n^{x}\mathrm{eff} - n^{y}\mathrm{eff}|,$$

where n^{x}eff and n^{y}eff are the effective refractive indices of x- and y-polarization modes. High birefringence is preferred in PCFs as they retain polarization state of a traveling light wave [55]. The birefringence values for different PCF structures are shown in Table 1.

4.2 *Nonlinearity*

Nonlinearities paved the way for many advantageous optical effects. Nonlinearities originate when different optical fields, waves, and vibrations simultaneously interact with one another. Nonlinearities also vary with birefringence and chromatic dispersion. Nonlinearities can have disadvantageous effects on the optical communication but they open up numerous applications [56]. The development of modified PCFs has led to the improvement in the value of nonlinearity [50]. The nonlinearity values for different PCF structures are shown in Table 2.

Table 1 Birefringence of modified PCF structures

Filling fraction	Fiber material	Fiber structure	Birefringence
0.2	Silica glass	A silicon microfiber structure with nanoscale low index elliptical slot rod inside	0.2006
0.86	Silica glass	All circular air holes in the cladding and elliptical air holes in core area	0.022
0.9	Silicon	silicon PCF and spiral lattice	0.24886
0.7	Chalcogenide glass	Spiral PCF with elliptical air hole	10^{-2}
0.80	Silica	Three-ring solid core hexagonal structure	0.00625
0.95	Silica	Five-ring dispersion compensating hybrid cladding PCF	0.0379
0.59	Silica	Elliptical-hole microstructure optical fiber (MOF)	0.0035
–	Led-silicate glass	Variable spiral soft glass PCF	0.04233
–	Soft glass (SF57)	Elliptical–spiral soft glass (SF57) photonic crystal fiber	0.05554
0.764	Silica	Triangular array of air holes around the core	0.004
0.7	Silica	PCF with hexagonal inner ring and octagonal outer ring	0.001
–	Silica	Adiabatic tapering of PCF with 50% of the taper	0.005
0.76	Fused silica	A rectangular array of four elliptical air holes and a square array of air holes in core	> 0.01
0.80	Fused silica	Circular air holes in the cladding and elliptical air holes in the fiber core	0.014

4.3 *Chromatic Dispersion*

Waveguide dispersion and material dispersion are collectively called chromatic dispersion. Since the PCFs work in single mode, only chromatic dispersion is caused.

This could be directly calculated by the following relation:

$$D = (-\lambda/c) * \left(d^2 \mathrm{Re}[n_{eff}]\right)/d\lambda^2,$$

where $\mathrm{Re}[n_{eff}]$ is the real part of effective refractive index and c is velocity of light in vacuum [55]. The chromatic dispersion values for different PCF structures are shown in Table 3.

Table 2 Nonlinearity and corresponding fiber structures

Filling fraction	Fiber material	Fiber structure	Non linearity
0.735	Silica	An octagonal PCF (OPCF) with five air hole rings in the cladding	54.74 W^{-1} km^{-1}
0.86	Silica glass	PCF with large circular air holes in the cladding and (three) elliptical air holes in the core	50 W^{-1} km^{-1}
0.90	Silica	Air holes arranged in a triangular lattice with pure silica core	25.32 W^{-1} km^{-1}
0.7	Chalcogenide glass	Spiral PCF with elliptical air hole	26739.42 $W^{-1}m^{-1}$
0.50	Silica	A circular PCF with five rings of air holes	38.38 W^{-1} km^{-1}
0.50	Silica	A circular PCF with four rings of air holes	14.47 W^{-1} km^{-1}
0.60	Silica	A silicon nanowire-embedded microstructure	3648 $W^{-1}m^{-1}$
–	Soft glass (SF57)	An elliptical–spiral photonic crystal fiber	3089 $W^{-1}km^{-1}$

Table 3 Chromatic dispersion for different fiber structures

Filling fraction	Fiber material	Fiber structure	Chromatic dispersion
0.2	Silica glass	A silicon microfiber structure with nanoscale low index elliptical slot rod inside	1.2358×10^3 ps/(nm · km) to 0 ps/(nm km)
0.752	Fused silicon	A spiral lattice structure with six spiral arms of three air hole rings around an elliptical slot in core	0.91 ps/(nm · km)
0.50	Silica	A circular PCF with five rings of air holes	−150 ps/(km · nm)
0.50	Silica	A circular PCF with four rings of air holes	−13.26 ps/(km · nm)
0.6	Silica	Octagonal structure is proposed having five air holes rings in the fiber	0 ± 0.1 ps/nm · km
0.74	Silica	PCF having air hole shape as fractal shape	1.1×10^{-10}ps/(km · nm)

4.4 Confinement Loss

The cladding of PCFs results in reduced optical confinement which is termed as the confinement loss. The PCF, being very flexible, is optimized to get a low confinement loss [21]. The confinement loss for a PCF is calculated by the given equation:

$$CL = 8.686 * (2\pi\lambda) * Im[n_{eff}] \times 10^6,$$

where $Im[n_{eff}]$ is the imaginary part of the effective refractive index. The confinement loss values for different PCF structures are shown in Table 4.

Table 4 Confinement loss for different fiber structures

Filling fraction	Fiber material	Fiber structure	Confinement loss
0.735	Silica	An octagonal PCF (OPCF) with five air hole rings in the cladding	0.06521 dB/km
–	Silica	Slow light photonic crystal waveguide with fluid infiltration	150 dB/cm
0.86	Silica glass	PCF structure which has elliptical air holes in core and large circular air holes in the cladding	$<10^{-3}$ dB/km
0.90	Silica	PCF design in which air holes arranged in a triangular lattice, with pure silica core	0.044 dB/m
0.80	Silica	PCF with three-ring solid core hexagonal structure	4.84×10^{-15} dB/km
0.95	Silica	Five-ring dispersion compensating hybrid cladding PCF	0.007 dB/m
0.50	Silica	A circular PCF structure with five rings of air holes	2.357×10^{-10} dB/m
0.50	Silica	A circular PCF design with four rings of air holes	1.4×10^{-5} dB/m
0.57	Silica	PCF structure with two four-ring square solid cores, one with circular air holes and other with elliptical air holes	1.8 dB/m (2-ring), ~ 0(5, 6 rings)
0.60	Silica	Silicon nanowire-embedded microstructure	7×10^{-7} dB/km
–	Soft glass	An elliptical–spiral photonic crystal fiber	0.1 dB/km
0.6	Silica	A honeycomb PCF	0.1×10^{-4} dB/km
0.49	Silica	A three-ring hexagonal PCF	0.1 dB/km
0.764	Silica	PCF with core surrounded by a triangular array of air holes	0.051 dB/km
0.82	Silica	PCF with a rectangular array of four elliptical air holes in the fiber core	10^{-4} dB/km
–	Silica	Adiabatic tapering of PCF with 50% of the taper	<0.002 dB/km
0.76	Fused silica	A rectangular array of four elliptical air holes and a square array of air holes in core	<0.002 dB/km
0.80	Fused silica	PCF with circular air holes in the fiber cladding and elliptical air holes in the fiber core	10^{-3} dB/km

5 Conclusion

In this paper, various types of photonic crystal fibers are discussed which have been modified for a number of desired parameters. We see that the flexibility of PCF can be used to modify the fiber and manipulate the desired optical properties, namely birefringence, nonlinearity, confinement loss, and chromatic dispersion. This ability opens up the portal to numerous applications such as telecommunication, super-continuum generation, imaging, and designing a fiber with desired confinement loss and dispersion.

References

1. Chau Y-F, Liu C-Y, Yeh H-H (2010) A comparative study of high birefringence and low confinement loss photonic crystal fiber employing elliptical air holes. In: Fiber cladding with tetragonal lattice Progress In Electromagnetics Research B, vol 22, 39, 52
2. Jonathan CK (2003) Photonic crystal fibres. Nature 424(14)
3. Philip St. JR (2006) Photonic-crystal fibers. J Light Technol 24(12)
4. Ju J, Jin W, Demokan MS (2003) Properties of a highly birefringent photonic crystal fiber. IEEE Photon Technol Lett, 15(10)
5. Buczynski R (2004) Photonic crystal fibers, Acta Physica Polonica A 106(2)
6. Russell et al (2003) Photonic crystal fibers, Philip Sci 299, 358
7. AliakbarEbnali H, Christoph P, MajidEbnali H, Christian K (2015) A proposal for loss engineering in slow-light photonic crystal waveguides. J Light Technol 33(9)
8. Masaaki H (2011) Future of transmission fiber Volume 3, Number 2, April 2011
9. Sourabh R, Kajal M, Partha RC (2009) Effect of tapering realistic photonic crystal fiber in tailoring birefringence and dispersion properties. ICOP 2009-International Conference on Optics and Photonics
10. Jianfei L, Junqiang S, Mingdi D, Yi Q (Feb. 15 2014) Highly nonlinear dispersion-flattened slotted spiral photonic crystal fibers. IEEE Photon Technol Lett 26(4)
11. Pranaw K, Madhusmita S, Sambuddha B Propagation characteristics of benzene filled photonic crystal fibers with high birefringence. In: International conference on electrical, electronics, and optimization techniques (ICEEOT)
12. Revathi S, Srinivasa R, Rizwan AS (2014) Highly nonlinear and birefringent spiral photonic crystal fiber, Advanc Opto Electron Hindawi Publishing Corporation, Article ID 464391
13. Shuguang L, Yanfeng L, Yuanyuan Z, Guiyao Z, Ying H, Lantian H (2008) Correlation between the birefringence and the structural parameter in photonic crystal fiber. Opt Laser Technol 40, 663–667
14. Gunasundari E, Senthilnathan K, Sivabalan S, Abdosllam MA Large dispersion and high nonlinearity in silicon nanowire embedded photonic crystal fiber, SR/S2/LOP-0014/201
15. Peng S (2014) A polymer photonic crystal fiber with high and flattened birefringence Optik 125 1330–1332
16. Roy S, kayser SF, Azmaeen T (2016) Design and optimization of a single mode octagonal photonic crystal fiber for high negative dispersion and high nonlinearity. In: 5th International Conference on Informatics, Electronics and Vision (ICIEV)
17. Xiaoqi L, Weimin S, Feiru W (2011) Design and optimization of ring cladding photonic crystal fiber with high nonlinearity and low zero dispersion wavelength. IEEE 978-1-4577-0796-4/11/$26.00 ©2011

18. Huseyin A, Shyqyri H, FathiAbdel M (2010) Highly nonlinear bending-insensitive birefringent photonic crystal fibres. Sci Res Eng 608–616
19. Daru C, Linfang S (Feb 15, 2007) Ultrahigh birefringent photonic crystal fiber with ultralow confinement loss. IEEE Photon Technol Lett 19(4)
20. Wenbin L, Ningliang L, Zhihua L, Peixiang L (Nov 1, 2012) Highly birefringent elliptical-hole microstructure fibers with low confinement loss. J Light Technol 30(21)
21. Yuan-Fong C (2014) Design of high birefringence photonic crystal fiber with three ring circular air holes in fiber cladding and two ring elliptical air holes in fiber core. Int J Comput Commun Instrument Eng (IJCCIE) 1(1) ISSN 2349–1469 EISSN 2349-1477
22. Mishra SS, Vinod KS (2011) Highly birefringent photonic crystal fiber with low confinement loss at wavelength 1.55 μm, Optik 122 1975–1977
23. Amir A, Revathi S, Srinivasa R, Abhijith C (2013) Modelling of circular photonic crystal fiber structure for high non-linearity. vol. 2(3) ISSN (Print): 2278–8948
24. Lin A, Zheng Z, Zheng L, Yang L, Tao Z, Jiangtao C (2009) Ultra-wideband single-polarization single-mode photonic crystal fiber with high nonlinearity and low dispersion, OSA/ACP
25. Ortigosa-Blanch A, Knight JC, Wadsworth WJ, Arriaga J, Mangan BJ, Birks TA, P. St. J. Russell (Sep 15, 2000) Highly birefringent photonic crystal fibers. Opt Lett 25(18)
26. Toulouse J (Nov 2005) Optical nonlinearities in fibers: review, recent examples, and systems applications. J Light Technol 23(11)
27. Jacob RF, Peter MW, Skovgaard, JesBroeng Numerical aperture of single-mode photonic crystal fibers. IEEE Photon Technol Lett
28. Jan S, Miloslav F, Vladimír T, Pavel R (June 2011) Principle of photonic crystal fibers. 2(2)
29. Timothy JA, Ivan B, Koen C (2016) Nonlinear optics near the fundamental limit: Introduction. J Opt Soc Am B 33(12)
30. Abdelaziz F, AbdelMalek HA, Haxha S, Gorman T, Bouchriha H (Oct 1, 2010) Senior Member, IEEE, Enhanced effective area photonic crystal fiber with novel air hole design, J Light Technol 28(19)
31. Yin W, Wei Z, Yidong H Jiangde P (2006) SBS Slow light in high nonlinearity photonic crystal fiber. Opt Soc Am
32. Christelle M, Bill C, Majid E-H, Christian G, Benjamin JE, Thomas PW, Liam O'Faolain, Thomas FK (2009) Slow light enhancement of nonlinear effects in silicon engineered photonic crystal waveguides. Opt Soc Am
33. Yamiao W, Xia Z, Xiaomin R, Long Z, Xiaolong L, Yongqing H (2009) Ultra-flattened chromatic dispersion photonic crystal fiber with high nonlinearity for supercontinuum generation, SPIE-OSA-IEEE vol. 7630 76301F-1
34. Abdur Razzak M, Yoshinori N, Feroza B, Kazuya M, Shubi K, Nguyen HH, Tatsuya K, Nianyu Z Dispersion-flattened modified hexagonal photonic crystal fibers with low confinement loss. Opt Rev 14(4), 165–168
35. Tianye H, Jianfei L, Songnian F, Tang M, Shum P, Deming L (2014) Slot spiral silicon photonic crystal fiber with property of both high birefringence and high nonlinearity. 6(3)
36. Chengcheng G, Jian W (2012) Elliptical–spiral photonic crystal fibers with wideband high birefringence. Large Nonlin Low Dispers, 4(6)
37. Chengcheng G, Jian W, Junqiang S, Huifeng W (2012) Achieving wide-band high birefringence, large nonlinearity and low dispersion using variable-spiral soft glass photonic crystal fibers. In: 17th Opto-Electronics and Communications Conference (OECC 2012)
38. Mishra SS, Vinod KS (2011) Comparative study of fundamental properties of honey comb photonic crystal fiber at 1.55 μm wavelength. J Micro Optoelectron Electromagn Appl, 10(2)
39. Soumya B, Sudheer SK, MahadevanPillai VP (2013) Design and simulation of highly nonlinear and low loss square photonic crystal fiber. In: 4th International Workshop on Fiber Optics in Access Network (FOAN)
40. Wenjuan L, Yumin L, Zhongyuan Y, Lihong Han, Ding N (2010) Design of high nonlinearity fiber with square lattice novel microstructure photonic crystal fiber, 978–1-4244-4964-4/10/ $25.00

41. Jianfei L, Fan Y, Yingmao X, Xinghua W, Tianye H, Zuzhou X, Fangguang K (Sep 1, 2015) Ultrahigh birefringent nonlinear slot silicon microfiber with low dispersion. IEEE Photon Technol Lett 27(17)
42. Jiyoung P, Jongki K, Songrae L, Soan K, Kyunghwan O (2010) Ultra-flattened dispersion and high nonlinearity in a square lattice photonic crystal fiber. In: 15th OptoElectronics and Communications Conference (OECC2010) Technical Digest
43. Tianye H, Jianfei L, Songnian F, Tang M, Shum P Deming L (2014) Slot spiral silicon photonic crystal fiber with property of both high birefringence and high nonlinearity. 6(3)
44. Jouri AM, Simohamed LM, Boudrioua A, Ziane O, Hassani B, Dadi A (2012) Analysis of indexed-guided highly birefringent photonic crystal fiber employing different cladding geometries. Centre Universitaired 'EL OUED, J Fundam Appl Sci 4(1), 32–38
45. Lin A, Zheng Z, Zheng L, Tao Z, Jiangtao C (Aug 1, 2009) Ultrahigh birefringent photonic crystal fiber with ultralow confinement loss using four airholes in the core. J Light Technol 27 (15)
46. Hasan MI, SelimHabib M, SamiulHabib M, AbdurRazzak SM (2014) Highly nonlinear and highly birefringent dispersion compensating photonic crystal fiber. Opt Fiber Technol 20 32–38
47. Saeed O, Fahimeh T (2012) Doped-core octagonal photonic crystal fiber with ultra-flattened nearly zero dispersion and low confinement loss in a wide wavelength range. Fiber Integr Opt 31, 178–185
48. Feifei S, Meicheng L, Yu Z, Liancheng Z (2009) High birefringence photonic crystal fiber design. SPIE, 7493 749351–1
49. Dhanu KG, Prasannan G, Sudheer SK. MahadevanPillai VP (2015) Design of ultra-low loss highly nonlinear dispersion flattened octagonal photonic crystal fibers. Opt Photon J 5, 335–343
50. Zhen-KF, Shu-GL, Hai-LC, Qiang L, Guo-WA, Shuo L (2015) A highly nonlinear and birefringent photonic crystal fiber with zero dispersion at 2 μm eye-safe spectral window. J Mod Opt 62(8), 676–681
51. Olyaeea S, Sadeghib M, Taghipoura F (2012) Design of low-dispersion fractal photonic crystal fiber. Int J Opt Photon (IJOP) 6(1), Winter-Spring
52. Knight JC, Skryabin DV (Nov 12, 2007) Nonlinear waveguide optics and photonic crystal fibers. Opt Exp 15(23), 15365
53. Matthew GW, Kevin C, Rodrigo AC, Frédéric G, William JW, Andrey VG, Dmitry VS, Jonathan CK (June 1, 2009) Solitons in hollow core photonic crystal fiber: engineering nonlinearity and compressing pulses. J Light Technol 27(11)
54. Jalal Uddin M, Shah Alam M (2008) Dispersion and confinement loss of photonic crystal fiber. Asian J Informat Technol 7(8): 344–349

Measurement of Parameters of Frequency-Locked Two-Mirror Laser Resonator

E. R. Aarathy, Arpit Rawankar and N. S. Kumar

Abstract A Fabry Perot cavity, resonating cavity or optical resonator is an arrangement of mirrors that forms a standing wave cavity resonator for light waves. During resonance, a slight misalignment in position can result in high losses associated with Q-factor. In this experiment, position feedback and temperature feedback systems are used for the stable operation of resonator. The peizo actuator-controlled mirror stages are used to compensate for misalignment while RTD-based temperature feedback system is used to control experiment temperature. Measurement of FSR, finesse and enhancement factor of laser resonator is done during stable resonance condition. An optical cavity is formed with a gold-coated coverslip. The parameters of gold slip cavity are obtained with high-resolution spectrograph with the broadband super continuum incident light.

Keywords FSR · FWHM · Finesse · APD · RTD · VI LabVIEW · COMSOL

1 Introduction

The most common types of optical cavities consist of two facing plane (flat) or spherical mirrors. The simplest of these is the plane-parallel or Fabry Perot cavity, consisting of two opposing flat mirrors as shown in Fig. 1. Plane-parallel resonators are commonly used in microchip, micro-cavity lasers and semiconductor laser. The plane-parallel resonator is also the basis of Fabry-Perot interferometers. When a laser beam incident within the cavity, there is both the constructive and destructive interference taking place. If the integer number of half wavelength of incident laser beam equals to the mirror spacing, constructive interference takes place and light is

E. R. Aarathy (✉)
Department of Physics, Amrita University, Kollam, Kerala, India
e-mail: aarathysby.93@gmail.com

E. R. Aarathy · A. Rawankar · N. S. Kumar
Tata Institute of Fundamental Research, Mumbai, India

V. Janyani et al. (eds.), *Optical and Wireless Technologies*, Lecture Notes in Electrical Engineering 472, https://doi.org/10.1007/978-981-10-7395-3_31

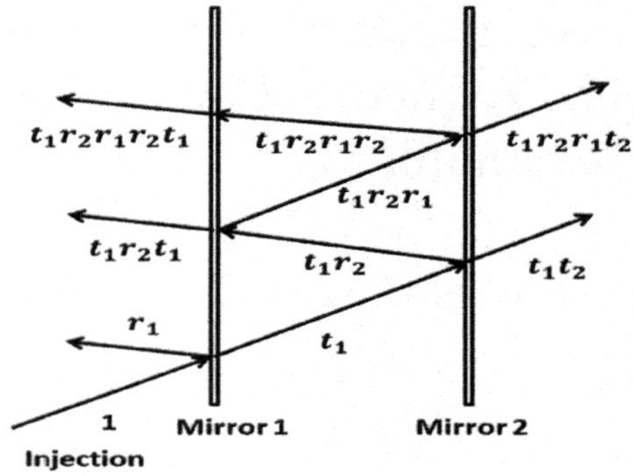

Fig. 1 Basic Fabry-Perot cavity model

transmitted out of the cavity. During destructive interference, no light is transmitted. Transmittance of a Fabry-Perot cavity reduced to an Airy function [1] is given by the equation:

$$T = \frac{1}{1 + F\sin^2\left(\frac{\delta}{2}\right)} \tag{1}$$

where

$$\delta = 2\pi l/\lambda \tag{2}$$

Here L is length of the cavity, λ is wavelength of the incident light, F is finesse of the optical cavity. From Eq. (1), if $\delta = N\pi$ then $T = 1$ and the cavity transmits the incident light.

2 Parameters of Compact Resonators

2.1 *Finesse*

When a laser beam is injected into the two mirror cavity, inside the cavity the laser wave reflects back and forth and results in the effective power. The power enhancement takes place only when our cavity is in the resonance condition of a standing wave. Finesse represents the sharpness of our resonance width, and it is defined by the reflectivity of the two cavity mirrors [2]

$$F = \frac{\pi R_{eff}}{1 - R_{eff}^2} \tag{3}$$

$$R_{eff} = \sqrt{R_1 R_2} \tag{4}$$

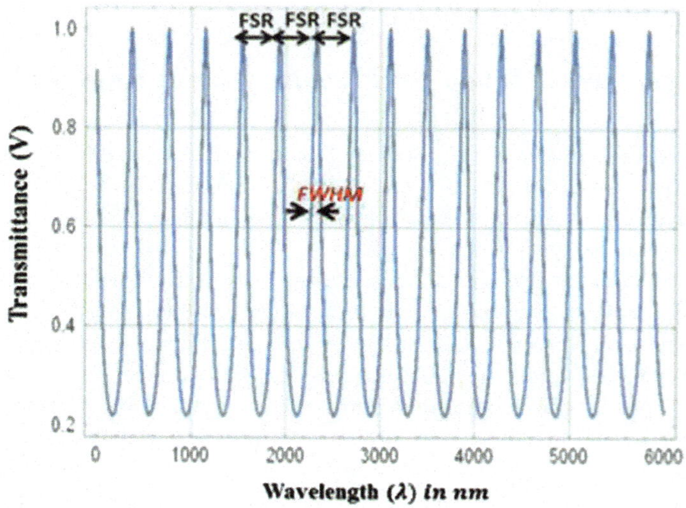

Fig. 2 Transmittance of laser resonator

where R_{eff} is the effective reflectivity of the resonator. Therefore, total finesse is given by

$$F = \frac{\pi\sqrt{R_1R_2}}{1 - R_1R_2} \tag{5}$$

Experimentally, the finesse is measured by finding the ratio of the free spectral range (FSR) to full width half maximum (FWHM) of the Airy function (Fig. 2).

$$F = FSR/FWHM \tag{6}$$

2.2 *Enhancement Factor*

The laser wave inside the cavity reflects back and forth and builds up the effective power. The power enhancement is realized only when the cavity satisfies the resonance condition of a standing wave. The enhancement factor is given by Eq. [3]:

$$\text{Enhancement Factor} = \frac{T_1(1 + R_2)}{(1 - \sqrt{R_1R_2})^2} \tag{7}$$

where T_1 and T_2 are transitivity of two mirrors. Both mirrors of cavity are 50% reflectance and 50% transmittance. The theoretical enhancement factor from Eq. (7) is calculated as 3.

3 Experimental Set-ups and Measurements

3.1 *Experimental Set-up for Frequency-Locked Resonator*

The IR LASER (780 nm), from Topia Photonics Series DL100 was used, with average power up to 30 m. Laser beam was first collimated using 50 and 40 mm plano-convex lenses which increased the spot size. In this experiment, a broadband dielectric beam splitter which is made of N-BK7 glass is used, 25.4 mm in diameter, 6.1 mm thick, with a surface flatness of λ/10. Both beam splitters are set on a stable mounts. Piezo actuator is used to control the cavity length D, and it supports the second beam splitter. Its dynamic range is about 3 μm with sensitivity of 3 nm/V. We can adjust the transverse position of first mirror by micrometres. So the cavity's optical axis can be aligned to that of the laser beam. Both mirrors used in this experiment have reflectivity of 50% for 780 nm wavelength (Fig. 3).

The cavity length is changed by the piezo actuator to observe several FSRs, and output is observed at APD [4]. These waveforms are represented by Airy function. From these, finesse is calculated by finding ratio of the FSR and FWHM of the standing wave pattern. The input power, coupled power and transmitted power for two mirror cavity are measured. The enhancement factor is measured as ratio of stored power to coupled power.

3.2 *Experimental Set-up for Gold Slip Resonator*

The super continuum laser, from Fianium SC400-4, with average power up to 4 W was used. The power spectrum changes with the input power of pump laser. The ocean optics-HR4000 spectrometer was used having a range from 200–1100 nm (Fig. 4).

A coverslip of 160 μm thickness, with refractive index of 1.5 and 0.04 reflectivity, was used for the observation of the Fabry-Perot resonances. The coverslips were cleaned using the methanol and nitrogen blow, since these resonances are sensitive to the dust particle. The laser beam was collimated using 25.4 and 500 mm plano-convex lens, to increase the spot size. This collimated beam was used for illuminating the coverslip. Then the transmitted intensity is measured using

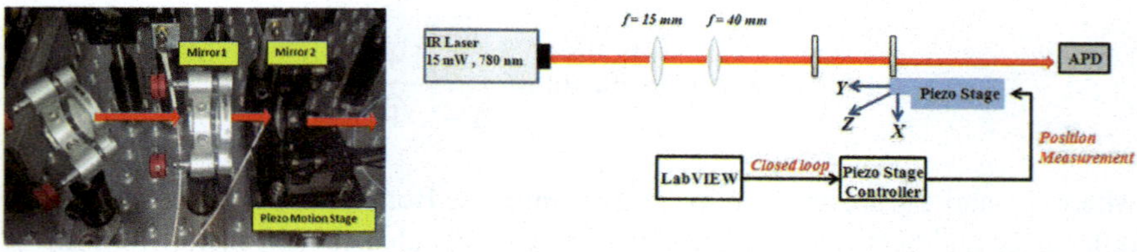

Fig. 3 Experimental set-up block diagram of the experimental set-up

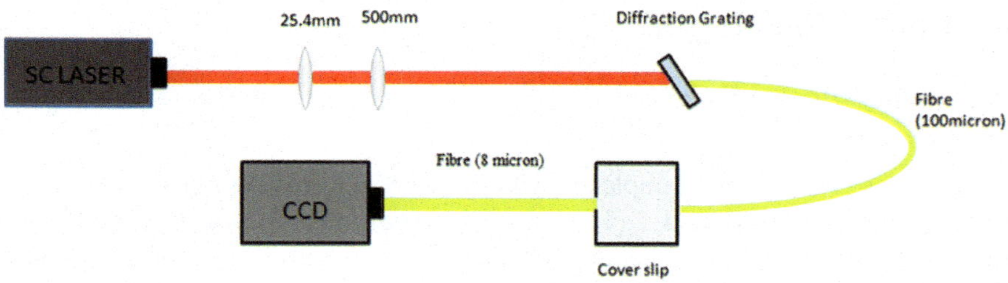

Fig. 4 Block diagram of experimental set-up

the CCD of the spectrometer having diffraction grating of 1200 lines/mm and Blaze wavelength at 500 nm [5]. Again this process was repeated for the gold-coated coverslip. The measured transmittance for the non-coated and one side gold-coated coverslip in a wavelength range of 540–560 nm using Acton spectrograph is shown below: Fig. 5, 6.

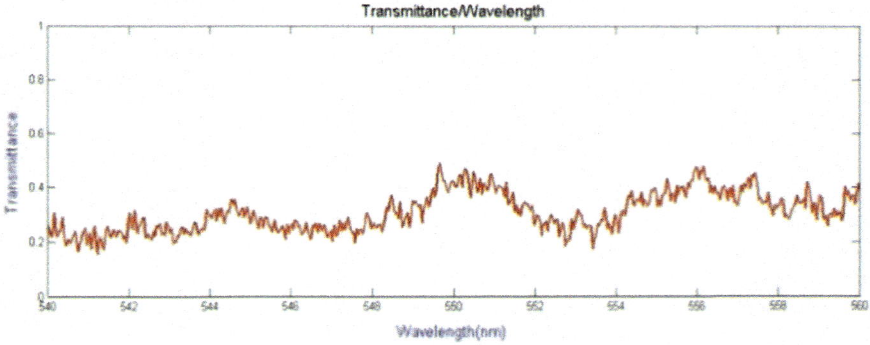

Fig. 5 Transmittance of a non-coated coverslip

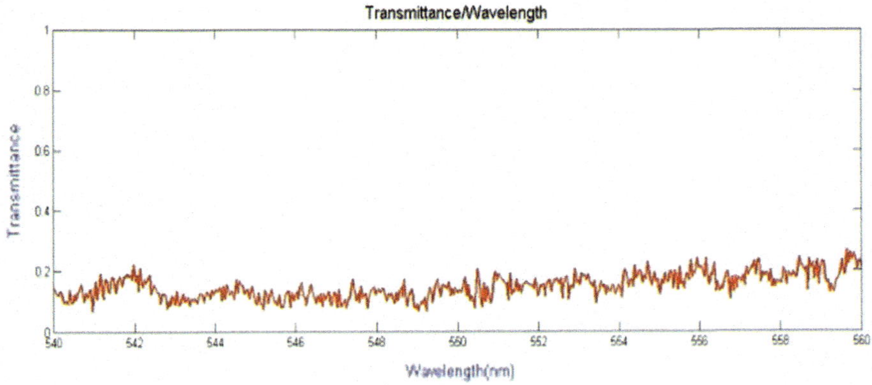

Fig. 6 Transmittance of a gold-coated coverslip

4 Comsol Simulations

4.1 Modelling of Cavity

Here, we use FEM method implemented in COMSOL Multiphysics to simulate the experimental set-ups 1 and 2. Here, we are describing two configurations with air-glass-air interface and glass-air-glass interface. Refractive index of air and glass is taken as 1 and 1.5, respectively. Periodic boundary conditions are used to simulate the infinite interface of the cavity. From the input and output field, S-parameters are determined and hence we get the transmittance and reflectance parameters.

Figure 7 shows the normalized field distribution (|E|). First region shows the interference pattern because of the incoming and reflected waves from the cavity, whereas interference pattern in the second region is due to the reflected waves from the two glass-air interfaces [6], and the last region is constant intensity as there are no reflected waves in this region. Glass air glass cavity also shows a similar field distribution, apart from the fact that the intensity in the cavity region is higher in glass air glass cavity.

Figure 8 shows the similar oscillations as that of experiment. Here the minima are far apart when compared to experimental observations. This is because the cavity size used in the FEM is smaller than the actual cavity size.

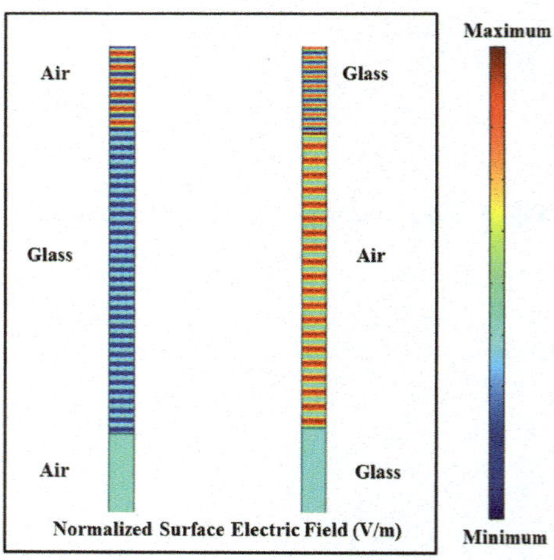

Fig. 7 Normalized surface field distribution for air-glass-air and glass-air-glass cavities

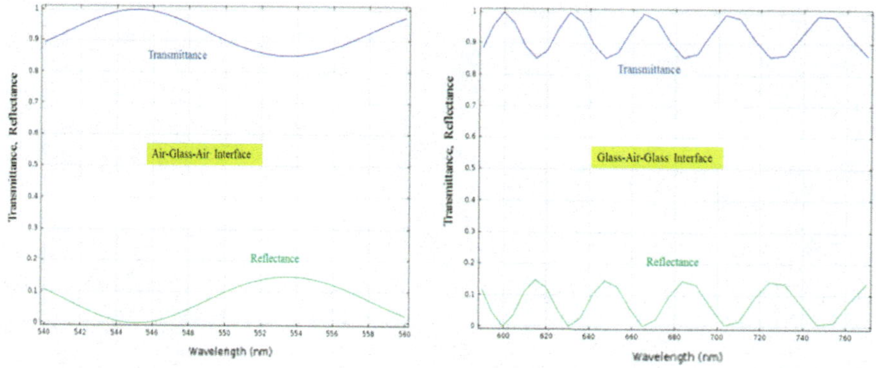

Fig. 8 Transmittance and reflectance spectra of air-glass-air interface and glass-air-glass interface

5 Feedback Control System

5.1 *Position Feedback Control System*

A piezo actuator-controlled motion stage "Piezo system Jena" [7] is used for the high accuracy positioning of the optical cavity. In this set-up motion stage model NV 40/3 is used, which has three axes piezo amplifier in a tabletop unit. The piezo actuator stage is moved in the x-, y- and z-direction. The resolution of piezo stage is in nm or sub nm range. The IR LASER ($\lambda = 780$ nm), Topia Photonics Series DL100 is used in this experiment with average power up to 30 mW. Laser beam was first collimated with series of convex lens to reduce the divergence of laser beam. The Fabry-Perot cavity is constructed with two dielectric mirrors which are made up of N-BK7 silica material. The diameter of the mirrors is 25.4 mm. Both mirrors are 6.1 mm thick, with a surface flatness of $\lambda/10$. The mirrors of optical cavity are mounted on specially designed mounts. These mounts maintain the overall rigidity of system, thus provide the stability from external vibrations [8].

The dynamic range of piezo actuator-controlled motion stage is 100 μm with sensitivity of 3 nm/V. We can adjust the transverse position of first mirror by micrometre stage; thus we can align the cavity's optical axis with that of the laser beam. Both mirrors used in this experiment have reflectivity of 70% for 550 nm wavelength of incident laser. The LabVIEW program triggers the piezo actuator motion stage with step size of 1 nm for the closed loop operation, and optical cavity is locked at highest intensity point of peak as shown in Fig. 9. It is clear from Fig. 9 that after feedback is turned on there are no fluctuations of transmitted light intensity from Fabry-Perot cavity [9].

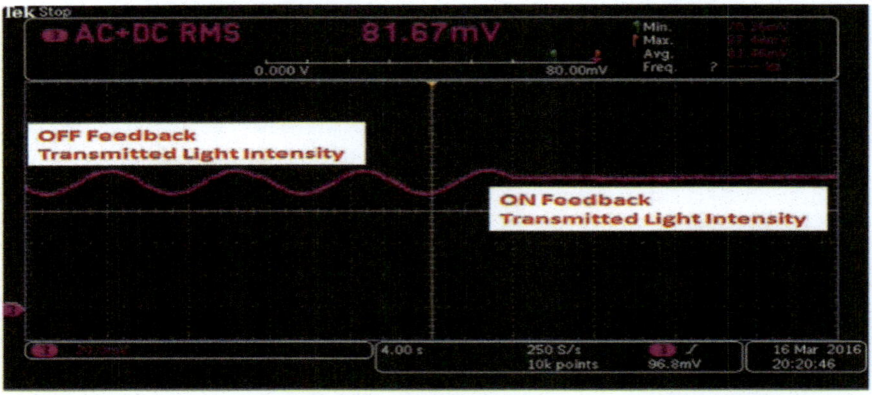

Fig. 9 Transmitted light intensity with change in cavity length and stabilization of transmitting light intensity

5.2 Temperature Feedback Control System

A resistance temperature (RTD) is a temperature sensing device whose resistance increases with temperature. An RTD consists of a wire coil or deposited film of pure metal. RTD is a passive device; a current must be passed through the device to produce a measurable voltage. Compared to other temperature devices, the output of an RTD is relatively linear with respect to temperature [7].

Signal conditioning is generally required to interface an RTD to a measurement device such as a plug-in data acquisition (DAQ) board. Signal conditioning produces current excitation for the RTD, amplifies the output signal and filters the signal to remove unwanted noise. Signal Conditioning eXtensions for Instrumentation (SCXI) is a signal conditioning front end used to plug-in DAQ boards. The SCXI-1121 is used for temperature feedback which is an isolated amplifier and multiplexer module with four input channels. RTDs are connected to the SCXI-1121. A LabVIEW program is used to monitor the voltages that the RTDs generate. Scaling of voltage measurement to a temperature reading is done in LabVIEW code. LabVIEW has a virtual instrument (VI) that uses the Callendar-Van Dusen equation to scale your voltage reading to a temperature reading. Callender-Van Dusen equation used to calibrate sensors is given as:

$$R_t = R_0\left[1 + At + Bt^2 + C(t - 100)^3\right] \tag{8}$$

where R_t resistance of the RTD at temperature "t", R_0 is resistance of RTD at zero temperature and A, B, C defines the Callender-Van Dusen coefficients.

As shown in Fig. 10, the main RTD sensor is kept in between the two mirrors of cavity. RTD 1 is placed outside the set-up box; RTD-2 and RTD-3 are placed inside the set-up box. RTD sensor 4 is placed near a halogen light source which is acting

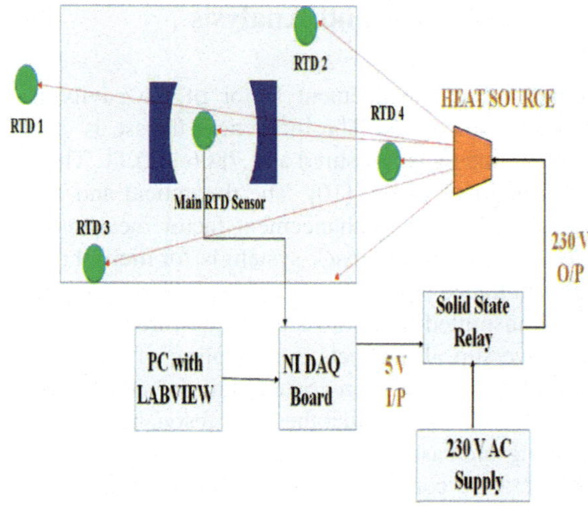

Fig. 10 Feedback control system for temperature stabilization

as a heat source. The temperature detected at main sensor RTD is read by DAQ board. A set point temperature is fixed in LabVIEW code which is 23 °C for this experiment. Whenever the overall temperature of set-up is below the set point temperature, the DAQ triggers the solid state relay (SSR) which turns ON the halogen lamp. Whenever the temperature is above the set point temperature, the SSR turns OFF the lamp. The stability obtained with temperature feedback is 0.5 °C as shown in Fig. 11.

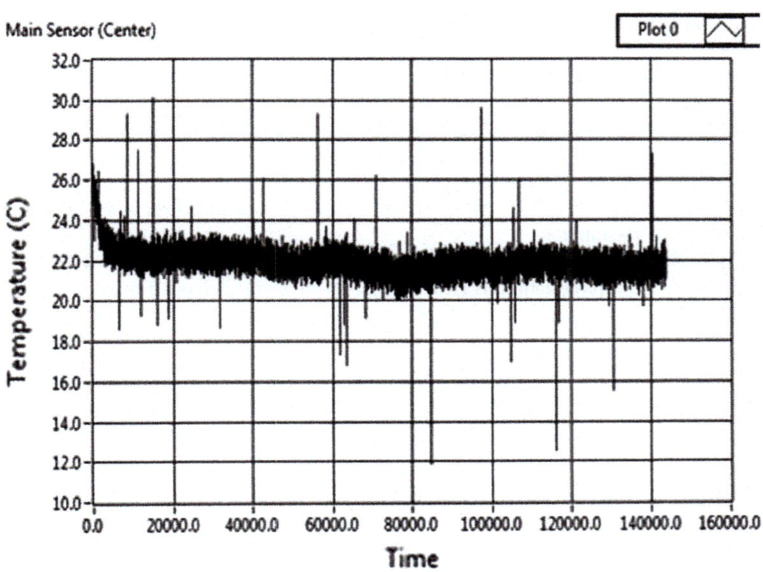

Fig. 11 Temperature stability of whole system with feedback

6 Conclusion and Analysis

Finesse and enhancement factor of the cavity is measured with both feedback system turned ON. The theoretical finesse is calculated as 2.1, while the experimental finesse is measured as 1.7806 ± 0.11. The experimental finesse is calculated with an error of 6% [10]. The theoretical and experimental finesse calculated are considerable. The enhancement factor measured experimentally is 1.66. Stability achieved in this feedback system is for more than one hour. In the gold slip cavity, transmittance observed is very small, and it can be due to the improper coupling of the transmitted light into the optical fibre (8 μm diameter core) and also a small tilt of coverslip also affects the coupling of the fibre. From the transmittance versus wavelength of the non-coated and one side gold-coated coverslip (thickness about 22 nm), we observe that the inner resonances are contributed by coverslip. The gold coating increases the contrast of resonance. The both position control system and temperature control system are implemented successfully. For the stabilization of the cavity, both position and temperature are controlled. The stability obtained with temperature feedback is within error of 0.5 °C.

References

1. Siegman AE (1986) Lasers. University Science books, Mill Valley, California
2. Rawankar Arpit et al (2013) Design studies on compact four mirror laser resonator with mode-locked pulsed laser for 5 mm laser wire. Nucl Instrum Methods Phys Res Sect A 700:145–152
3. Yosuke H Experimental studies of a low emittance electron beam in the KEK-ATF damping ring with a laser wire beam profile monitor
4. Amnon Y, Pochi Y (2007) Photonics-optical electronics in modern communication, 6th ed., New York
5. Baaha EAS, Malvin CT Resonator optics-fundamentals of photonics
6. Thygarajan, Ajoy G, Lasers-fundamentals & application
7. Piezojena system. www.piezosystemjena
8. Anya MD (September 17, 2009) Laser frequency stabilization with optical cavities. University of Washington
9. Hodgson Norman (1997) Horest Weber. Springer, Optical Resonators, pp 41–42
10. Taylor JR (1997) Introduction to error analysis. University Science Books

Photonic Crystal Fiber: A Review

Priyanka Sidhar, Poonam Singal and Shefali Singla

Abstract With the tremendously growing demand for internet, developments in optical fiber technology are rapid. These fibers offer low propagation loss for longer lengths, high transmission rate, and high channel capacity and bandwidth. However, traditional fibers have very rigid design rules such as limited core diameter for single-mode operation and cut-off wavelengths. Photonic crystal fibers are a new class of fiber, which along with the benefits of conventional fibers provides unique properties such as endlessly single mode operation, no cut-off wavelength, and single mode operation with larger core diameters. PCFs work in two different modes: index-guiding mode and photonic bandgap mode. These fibers can be used in ophthalmology, dental imaging, military applications, and many more areas. This paper provides an overview of photonic crystal fiber and its different modes of operation and applications.

Keywords Modified TIR · Photonic bandgap · Crystal fibers
Holey fibers

1 Introduction

Optical fiber is a thin hair like structure made up of silica or glass consisting of a core surrounded by the cladding, which is followed by the jacket. The light is guided through this optical fiber by the method of total internal reflection, which requires the refractive index of the core to be greater than cladding [1, 2]. The cross-sectional view of the fiber is shown in Fig. 1.

P. Sidhar (✉) · P. Singal · S. Singla
Deenbandhu Chhotu Ram University of Science & Technology, Murthal, Haryana, India
e-mail: priyanka.sidhar@gmail.com

P. Singal
e-mail: singal.poonam@rediffmail.com

S. Singla
e-mail: shefalisingla1@gmail.com

V. Janyani et al. (eds.), *Optical and Wireless Technologies*, Lecture Notes
in Electrical Engineering 472, https://doi.org/10.1007/978-981-10-7395-3_32

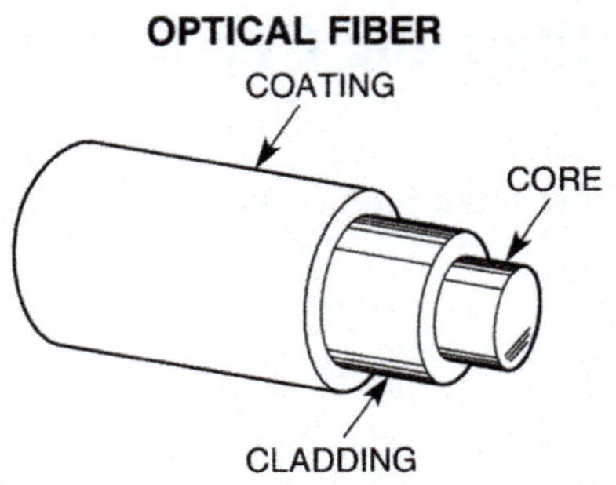

Fig. 1 Conventional optical fiber (*Source* www.newport.com) [11]

These fibers have low loss near the current optical window (1550 nm) of approximately 0.2 dB/km and offer a large bandwidth of around 50 THz. However, these fibers have certain limitations such as cut-off wavelength, limitations on core diameter for single mode operation, intermodal losses, etc. Photonic crystal fiber is a new class of fibers, which overcomes all these limitations and also provides the advantages of conventional fibers.

The term photonic crystal fiber was first coined by Phillips St. J. Russel in early 1990s. These fibers combine the properties of photonic crystals and conventional fibers and thus possess unique properties. PCF contains a network of capillaries along its length, which is laid down to form a fixed lattice structure [3, 4]. Depending on the geometry, these fibers are categorized into holey fibers or index-guiding fibers and photonic band gap fibers. Holey fibers have a solid core, and the photonic bandgap fibers have a hollow core.

2 Holey Fibers

This fiber contains a solid core surrounded by a network of capillaries arranged along the length of fiber to form a lattice structure (Fig. 2).

The airholes surrounding the core lower the effective refractive index of the surrounding media thus making these fibers suitable for conventional operation, i.e., total internal reflection. These fibers thus do not require a different medium for

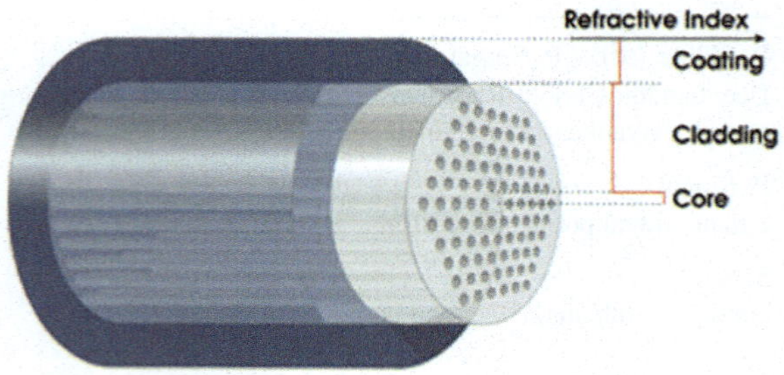

Fig. 2 Holey fiber (*Source* www.photonics.com) [13]

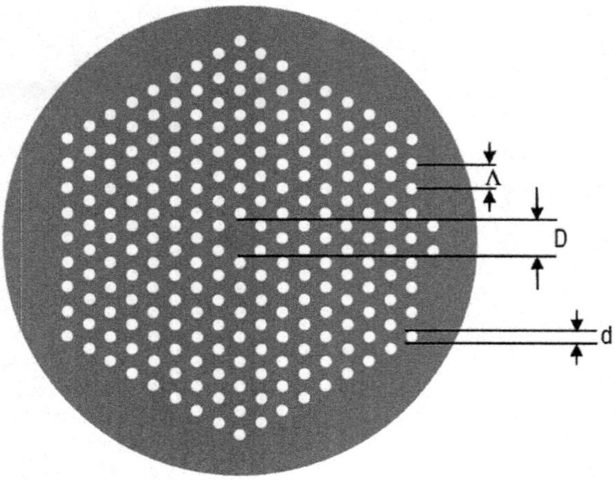

Fig. 3 Holey fiber parameters definition (*Source* www.newport.com) [11]

cladding. The higher order modes leak through the holes, and thus the fiber acts as a single-mode fiber for a long range of operating optical frequencies. Figure 3 defines the parameters of holey fibers.

2.1 Lattice Pitch

It is the distance between the holes in the lattice structure and is denoted by ʌ. The pitch of the fiber must be less than the wavelength of operation.

2.2 Core Diameter

It is the diameter of the central solid part of the fiber and is denoted by **D**.

2.3 Hole Size

It is the diameter of the holes in the capillary network and is denoted as **d**.

By changing these parameters, the fiber can be used as a single mode for the complete optical frequency range. Also, the large core diameter does not lead to multi-mode operation.

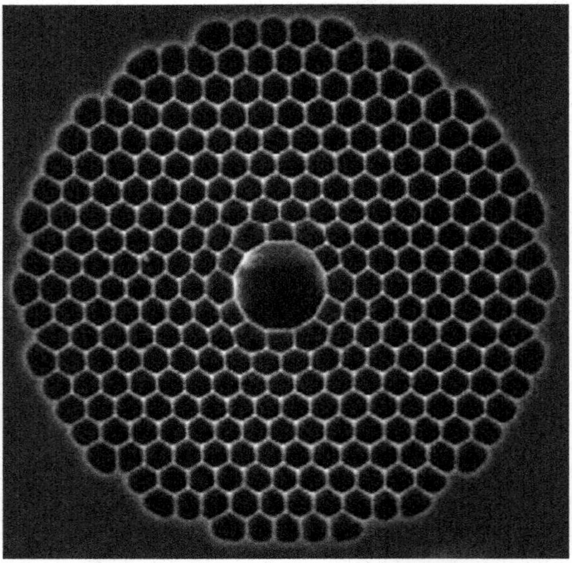

Fig. 4 Photonic bandgap fibers (*Source* www.newport.com) [11]

3 Photonic BandGap Fibers

These are the fibers with a hollow core surrounded by the capillary network of holes. Thus, the core refractive index becomes lower than the surrounding media, and the conventional principle of operation for fibers do not apply to this fiber. Due to the presence of hollow core, there exists a photonic bandgap for the fiber, which blocks a band of frequencies when transmitted along its length. The optical bandgap for a fiber can be changed by changing the geometrical parameters such as core diameter, lattice pitch, airhole refractive index, and lattice structure. Figure 4 shows the photonic band gap fiber or the hollow fiber.

4 Generation of Photonic BandGap

Photonic crystals are optical periodic nanostructures containing medium of high dielectric contrast stacked together in a periodic manner. These nanostructures change the motion of photons incident on them. The photons are either reflected or refracted or transmitted depending on the wavelength and the periodic structure depth. The wavelengths, which are reflected in phase with the incident light, get added to it, and thus, light travels through the medium with minimum absorption. The wavelengths which are reflected out of phase with the incident light cancel them out and thus do not pass through the crystal. Figures 5 and 6 depict the in phase and out of phase reflection.

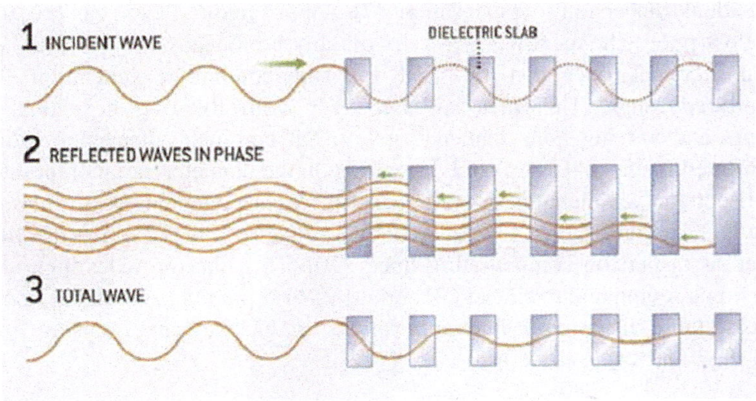

Fig. 5 In phase reflected wave in photonic crystals (*Source* www.quantum.phys.uni-sofia.bg) [12]

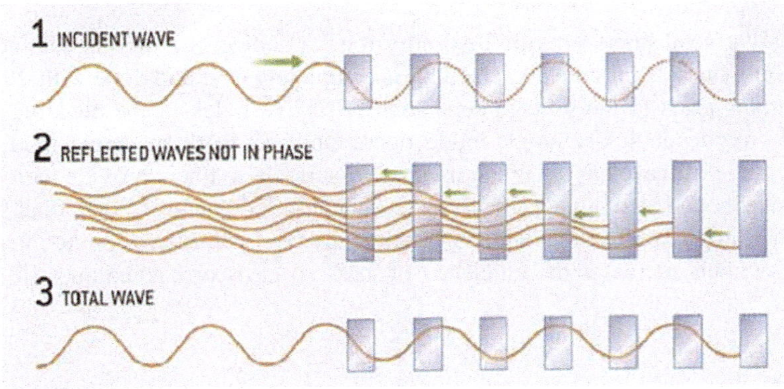

Fig. 6 Out of phase reflected wave in photonic crystals (*Source* www.quantum.phys.uni-sofia.bg) [12]

5 Applications of PCF

A highly nonlinearly designed PCF having four layers of airholes with first layer diameter 0.46 μm and the remaining layers having a diameter of 0.80 μm with lattice pitch defined to be 0.87 μm can be used for broadband supercontinuum generation, which is used in ophthalmology, dermatology, and dental imaging [5]. PCFs with butterfly lattice structure and central core part doped with GeO_2 and fiber Bragg grating inscribed in the core can be used as pressure sensors [6]. These kinds of sensors are temperature insensitive and have higher sensitivity for the measured quantity in comparison with the conventional sensor technology. A dual-core PCF can be used to obtain ultrahigh negative dispersion value. A PCF-in-PCF structure exhibits ultrahigh negative dispersion of around −474.5 ps/nm/km for a broad spectrum with a wavelength ranging from 1360 to 1690 nm and thus can be used for residual dispersion compensation [7]. Opto-Fluidic PCFs can be used in microflow photo-chemical reactors [8]. A highly nonlinear $Ge_{11.5}As_{24}Se_{64.5}$ based photonic crystal fiber can be used for supercontinuum generation for the mid-infrared range. This material-based PCF with five-ring hexagonal lattice structure and varying hole diameters give a flat chromatic dispersion profile for mid-infrared range (1–10 μm) and depending on the diameter variation possess two zero dispersion wavelengths in that range [9]. A chalcogenide As_2Se_3 glass-based PCF with hexagonal lattice and square lattice structure with 0.2 μm as the pitch can be used as dispersion compensating fibers [10]. This fiber provides high negative dispersion as compared to silica PCF in the 1.25–1.6 μm range. With the research in progress, these fibers will find applications in bio-medicine, sensing, material processing, and many more areas.

6 Advantages of PCF

Photonic crystal fibers provide flexibility for the choice of material. The need for different materials for core and cladding is completely over and done with. Presence of airholes leads to the voltage-dependent refractive index of the cladding. These fibers provide endlessly single-mode operation, i.e., work in single mode for a broad range of wavelength. When used as sensors, these fibers provide temperature insensitivity and also higher sensitivity to the measured quantity. Also, fiber sensors are small in size, light in weight, and chemically inert [6]. Micro-fluidic integration of these fibers increases the efficiency of mass spectroscopy techniques [8].

7 Conclusion

In this paper, we discussed photonic crystal fiber. We also discussed various properties of these fibers, which are in stark contrast with the conventional fibers. These fibers have great flexibility in their design and thus lend themselves immense application fields. These fibers have the ability to replace the existing fiber network. The ongoing research aims at bringing the losses offered by these fibers to a practically implementable value. Also, the focus of current research is on different materials used for fabrication of PCF. With new properties being discovered every other day and growing network of the fiber technology, these fibers have a bright future ahead.

References

1. Senior JM (2009) Optical fiber communications-principles and practice, 3rd edn. Pearson Publications
2. Keiser G (1991) Optical fiber communications, 2nd edn. Tata McGraw Hill Publications
3. Photonic crystal fiber. www.wikipedia.org
4. Russel PSt (2003) Science 299:358
5. Begum F, Namihira Y (2012) Photonic crystal fiber for medical applications. In: Yasin M (ed) Recent progress in optical fiber research. InTech. ISBN: 978-953-307-823-6
6. Popp J, Matthews DL, Tian J, Yang C-C (eds) (2011) Optical sensors and biophotonics III. In: Proceedings of SPIE-OSA-IEEE Asia communications and photonics, vol 8311, 831102
7. Tee DC, Abu Bakar MH, Tamchek N, Mahamd Adikan FR (2013) Photonic crystal fiber in photonic crystal fiber for residual dispersion compensation over E + S + C + L + U wavelength bands. IEEE Photon J. https://doi.org/10.1109/JPHOT.2013.2265980. 1943-0655/$31.00 2013
8. DerNaturwissenschaftlichenFakultät der Friedrich-Alexander-Universität Erlangen-NürnbergzurErlangung des Doktorgrades Dr. rer. nat. Optofluidic photonic crystal fibres for biomedical research in fibra
9. Vyas S et al (2016) Ultraflat broadband supercontinuum in highly nonlinear Ge11.5As24Se64.5 photonic crystal fibres. Ukr J Phys Opt 17(3):132–139. https://doi.org/10.3116/16091833/17/3/132/2016
10. Dabas B, Sinha RK (2010) Dispersion characteristic of hexagonal and square lattice chalcogenide As_2Se_3 glass photonic crystal fiber. Opt Commun 283(7):1331–1337
11. Newport Corporation. https://www.newport.com
12. Department of Quantum Electronics, Sofia University, Bulgaria. http://quantum.phys.uni-sofia.bg
13. Photonic Media. http://www.photonics.com

Slow-Light Enhanced Second Harmonic Generation in Lithium Niobate Photonic Crystal Waveguides

Zaineb Gharsallah, Makni Sana, Monia Najjar and Vijay Janyani

Abstract In this research study, we propose a slow-light-based photonic crystal waveguide consisting of a combination of circular and elliptic airholes with a background material of lithium niobate $LiNbO_3$ (refractive index n = 2.211). By modifying the radii of the closest row to the waveguide in each side, we demonstrate a high value of normalized delay-bandwidth product, equal to 0.60. Using finite-difference time-domain method (FDTD), a significant increase of second harmonic generation efficiency of about 0.14 at a moderate power is observed when analyzing the nonlinear performance of the designed slow-light structure.

Keywords Slow light · NDBP · Photonic crystal waveguide
Nonlinear effects · Second harmonic generation

Z. Gharsallah (✉) · M. Sana · M. Najjar
National Engineering School of Tunis Communications Systems Laboratory (SysCom), University of Tunis El Manar, Ariana 1002, Tunisia
e-mail: gharsallah.zaineb@hotmail.fr

M. Sana
e-mail: maknisana87@gmail.com

M. Najjar
e-mail: monianajjar@yahoo.fr

M. Najjar
Higher Institute of Computer Science of El Manar (ISI), University of Tunis El Manar, Ariana, Tunisia

Z. Gharsallah · V. Janyani
Department of Electronics & Communication, Malaviya National Institute of Technology, Jaipur, India
e-mail: v.janyani.ece@mnit.ac.in

V. Janyani et al. (eds.), *Optical and Wireless Technologies*, Lecture Notes in Electrical Engineering 472, https://doi.org/10.1007/978-981-10-7395-3_33

1 Introduction

There has been a large interest to slow light using photonic crystal because of its application for switches, buffers, sensors, and delay lines, also nonlinear devices and processes [1]. It is known that in slow-light devices, increased optical energy density by means of the spatial compression of an optical pulse enhances nonlinear phenomena like Raman scattering, self-phase modulation, and second harmonic generation. Slow light can be used to increase the efficiency of nonlinear effects [2] over a wide bandwidth range [3]. The explanation of this effect is that the presence of slow light can engineer the dispersion and realize a 'flat band.' One way to achieve this dispersion relation, for example, is by altering the size, position, or the form of first or few rows near the line defect in the photonic crystal waveguide. One of the most widely used materials in electro-optic devices is lithium niobate ($LiNbO_3$) due to its high electro-optic coefficient and large nonlinear effects. The fabrication technology of $LiNbO_3$-based devices is adequately mature and well established [4]. In this paper, we present a nonlinear slow-light waveguide, and we calculate its second harmonic generation (SHG) to compare with that of the fast waveguide with the same nonlinear material. Our paper is structured as follows: In part 2 of our work, we define the background and the structural analysis. Thereafter, the attention is focused on the slow-light conception and results in Sect. 3. The SHG and nonlinear performance are added in Sect. 4 using finite-difference time-domain method. Finally, the conclusion and some future works are presented in Sect. 5.

2 Background and Structural Analysis

Normalized delay-bandwidth product (NDPB) is an important factor specifying the efficiency of slow light. It is the multiplication's result of the average of group index per frequency range where n_g is restricted by ±10% variation.

$$\mathrm{NDBP} = n_g * (\Delta\omega/\omega_0) \tag{1}$$

Here, n_g is the group index that qualifies how much the light is delayed [5] and $(\Delta\omega/\omega_0)$ is normalized central frequency of light dividing normalized bandwidth. More significant specification for slow-light structures is the group velocity. We can calculate it from the dispersion curve by using the following formula:

$$V_g = \partial\omega/\partial k = c/n_g \tag{2}$$

where ω indicates the angular frequency, k is the wave vector, and c expresses the velocity of light in free space.

For the determination of NDBP, we carry out with the procedure reported below:

We start first by calculating the normalized frequency variation limits:

$$\Delta\omega = \omega_h - \omega_l \tag{3}$$

where $\omega_h = \max(\omega_c)$, $\omega_l = \min(\omega_c)$, and ω_c indicates the range of normalized frequency within the group index n_g is regarded as steady with n_g variation ±10% range.

Second, from the band structure diagram, we pulled out the central frequency ω_0 considered in the middle frequency of the bandgap. In the final step, we determined the NDBP value as a ratio between variables obtained previously.

$$NDBP = n_g * (\Delta\omega/\omega_0) \tag{4}$$

Our basic structure consists of airholes with a triangular array in lithium niobate ($LiNbO_3$) slab with refractive index 2.211. By using Scanner Rsoft Method [6], we get the bandgap width against ratio shown in Fig. 1a, which allows choosing an optimal value of holes radius.

To justify our radius choice, we plotted the gap width versus the ratio, which is defined as radius over the period. Generally, we select the ratio corresponding to the maximum gap width, but in our case, we chose smaller ratio values to avoid the holes intersection while changing their forms. Therefore, the suitable ratio is 0.35 for gap width 0.09.

To calculate the bandgap map, we used plan wave expansion method with a radius equal to 0.35 * period. For the given structure, the bandgap is obtained for only transverse magnetic (TM) with a band limited between 0.33 and 0.42 for normalized frequency (a/λ) as shown in Fig. 1b. Thus, the period 'a' is determined for middle value of the bandgap.

Period (a) = (0.33 + 0.42)/2 * 1.55 = 0.58125.

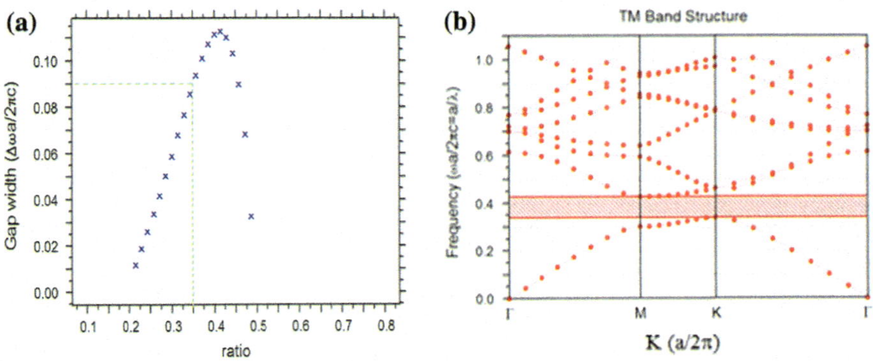

Fig. 1 **a** Bandgap width versus ratio. **b** Bandgap of our basic structure

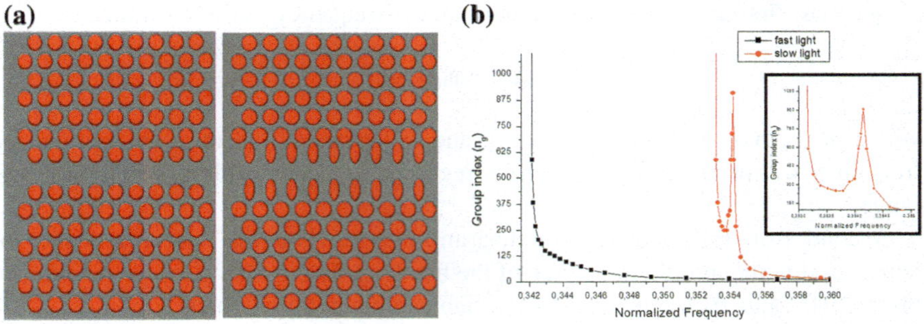

Fig. 2 **a** Waveguide structure for fast and slow light. **b** Group index versus normalized frequency

3 Fast and Slow Light

In our proposed structure, we remove the central row of holes to create a photonic crystal waveguide with one line defect. As we show above, the guided wave is TM mode with a frequency lying in the bandgap.

The photonic crystal waveguide without any form modification represents the fast (normal) light structure (Fig. 2a). Otherwise, the slow device is created by changing the radius of the two first rows bordering the waveguide. The light speed in the two studied structures is analyzed by plotting group index versus normalized frequency (Fig. 2b).

From the above curve, we observe that in the case of slow-light structure, there is a band of frequencies where n_g is high and relatively constant, which mean the light is delayed for a band of normalized frequency. For further assessment of structures' performance, we conducted calculations of the product NDBP. The results are summarized in Table 1.

The obtained normalized delay-bandwidth product for slow-light structure is greater than that of the fast light structure. Furthermore, the NDBP of our proposed is more efficient than the results published in [9–12] and shown in Table 2.

Table 1 Comparison among the proposed structures

Structure	n_g	$\Delta\omega$	$\Delta\omega/\omega_0$	NDBP
Fast light	12	0.00525	0.014	0.17
Slow light	250	0.0009	0.0024	0.60

Table 2 Comparison of slow-light performance between proposed structure and published results

Structure	n_g	$\Delta\omega/\omega_0$	NDBP
[7] Mirjalili 2012	39.44	0.0148	0.59
[8] Nagesh 2012	20.87	0.0179	0.37
[9] Kumar 2015	36.60	0.0120	0.43
[10] Our previous work	50	0.0084	0.42
Present work	250	0.0024	0.60

4 Second Harmonic Generation

Second harmonic generation (SHG) has been reported in photonic crystal (PhC) waveguides [11] made in III–V semiconductors. Due to their large second order, these crystals present a strong nonlinear enhancement. Nevertheless, the wavelengths of electronic bands limit the nonlinear conversion efficiencies because of absorption phenomena. The nonlinear quadratic materiel $LiNbO_3$ has a wide transparency window of about 1220 nm centered at 1550 nm. Therefore, it presents an alternative to generate second-order nonlinear PhC devices [12]. In this section, we will demonstrate an improvement of the second harmonic generation in nonlinear slow-light structure. The second harmonic is generated in the above structures (Fig. 2b) by introducing the nonlinear parameters of $LiNbO_3$. We opted for the following parameters: Nonlinear material $LiNbO_3$ with second-order susceptibility equals to 6.88×10^{-6} m/V [13] and a Gaussian modulated continuous input signal with wavelength 1.55 μm.

Figure 3a displays the evolution of the SHG intensity observed for different input power values of 10, 20, 30, 40, and 50 W/m. It is seen that SHG intensity for the slow-light structure is higher than obtained for fast light structure. In fact, by using the fast light structure, we can obtain a high SHG intensity with $P_{in} = 50$ (W/m), but for the slow-light structure, the input power can be much lower (20 W/m). Figure 3b illustrates SHG intensity in case of slow-light structure and fast-light structure using a value of 50 W/m as input power. It is revealed that SHG intensity in nonlinear structure is 650 units, and the SHG in slow-light nonlinear waveguide is 1050 units at a wavelength of 775 nm. We consider the value of SHG at wavelength 775 nm because the conversion of light from its original wavelength to its second harmonic at a wavelength twice the frequency and half the wavelength [14].

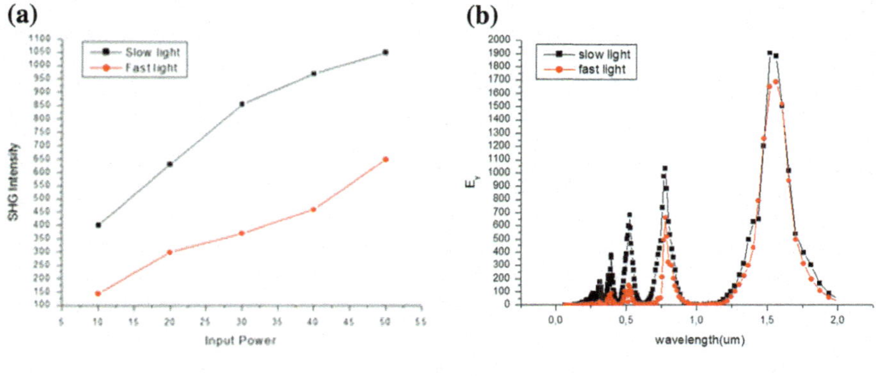

Fig. 3 **a** SHG intensity over input power. **b** The intensity of SHG in the two waveguides slow and fast light

So, we remark that slow-light nonlinear structure gives intensive SHG than that given by the fast light nonlinear structure. So the SHG performance is enhanced by using a suitably designed nonlinear slow-light structure.

5 Conclusion

In conclusion, we analyzed and assessed the slow-light performance of two studied photonic crystal structures based on $LiNbO_3$. Furthermore, we investigated the nonlinear effect in the slow-light structure to enhance the generation of second harmonic. For the results, the minimization of applied power to generate the second harmonic is achieved by using nonlinear slow light. Further structure dimension optimization will be considered as a future work.

References

1. Krauss TF (2008) Why do we need slow light? Nat Photon 2:448
2. Li J, White TP, Faolain LO, Gomez-Iglesias A, Krauss TF (2008) Systematic design ofat band slow light in photonic crystal waveguides. Opt Express 16:622–6232
3. Baba T (2008) Slow light in photonic crystals. Nat Photon 2:465–473
4. Alibart O, Auria V, Micheli M et al (2016) Quantum photonics at telecom wavelengths based on lithium niobate waveguides. J Opt Inst Phys (IOP) 18(10):104001
5. Krauss TF (2007) Slow light in photonic crystal waveguides. J Phys D: Appl Phys 2666–2670
6. Rsoft software. https://optics.synopsys.com/rsoft/. Version 2014
7. Mirjalili SM, Kambiz A (2012) Light property and optical buffer performance enhancement using particle swarm optimization in oblique ring-shape-hole photonic crystal waveguide. In: Photonics global conference
8. Janrao N, Zafar R, Vijay J (2012) Improved design of photonic crystal waveguides with elliptical holes for enhanced slow light performance. Opt Eng 51(6):064001–064007
9. Sharma A, Mukesh K (2015) Flat band slow light in silicon photonic crystal waveguide with large delay bandwidth product and low group velocity dispersion. IET Opt 9(1):24–28
10. Zaineb G, Monia N, Vijay J (2016) Slow light optimization in symmetric photonic crystal waveguide with elliptical holes. In: 22nd Asia-Pacific conference on communications, pp 144–147
11. Coquillat D, Vecchi G, Comaschi C, Malvezzi AM, Torres J, dYerville ML (2005) Enhanced second- and third-harmonic generation and induced photoluminescence in a two-dimensional GaN photonic crystal. Appl Phys Lett 873
12. Diziain S, Geiss R, Zilk M, Schrempel F, Kley E, Tünnermann A, Pertsch T (2013) Second harmonic generation in free-standing lithium niobate photonic crystal L3 cavity. Appl Phys Lett 103
13. Sana M, Monia N, Taoufik A (2016) Optimisation of second harmonic generation using photonic crystal cavity. In: 22nd Asia-Pacific conference on communications
14. de Dood MJA (2006) Second-harmonic generation. Huygens Laboratorium 909a

Design Optimization of a Highly Birefringent and Highly Nonlinear Silicon Photonic Crystal Fiber

Ashish Kumar Ghunawat, Ritesh Chandra, Manish Tiwari and Ghanshyam Singh

Abstract This paper presents an optimum design for highly birefringent and highly nonlinear photonic crystal fiber based on modified spiral photonic crystal fiber structure. The PCF structure includes an elliptical slot in the core of the fiber. The finite element method is used to investigate the guiding property. The simulation results show that nonlinear coefficient at 1.55 μm wavelength for X and Y polarization is 2951 and 1379, respectively. This design also offers high birefringence up to 0.292219 at 1550 nm wavelength. In addition to this, confinement factor is also reported and discussed.

Keywords Spiral photonic crystal fibers · Birefringence · Nonlinear coefficient Confinement factor · Finite element method

1 Introduction

Recently, photonic crystal fibers have attracted significant attention due to their flexibility in the design, outstanding optical properties, and various applications. PCF properties include endlessly single mode operation, high birefringence, high nonlinearity, ultra-flattened dispersion characteristics, low confinement loss, and so on [1–5]. Because of these properties, PCF has applications in fiber lasers, communications, nonlinear devices [6], generation of slow light [7] etc. High birefringent PCFs can be used as single-polarization mode fibers or polarization maintaining fibers. PCFs maintain the state of polarization which proves usefulness of PCFs in polarization mode dispersion (PMD) compensation. Furthermore, birefringent PCFs have very long list of potential applications in optical sensor

A. K. Ghunawat (✉) · R. Chandra · G. Singh
Department of ECE, Malaviya National Institute of Technology, Jaipur, Rajasthan, India
e-mail: akghunawat.ece@mnit.ac.in; ashishghunawat@gmail.com

M. Tiwari
Department of ECE, Manipal University, Jaipur, India

V. Janyani et al. (eds.), *Optical and Wireless Technologies*, Lecture Notes in Electrical Engineering 472, https://doi.org/10.1007/978-981-10-7395-3_34

design, electro-optical modulation, and signal processing systems [3]. One of the main reasons to enhance nonlinearity is that power consumption of optical device can be reduced. Nonlinear coefficient in PCF can be realized by reducing the effective area of core of the PCF or using material with high nonlinearity [8]. The problem lies in the field confinement ability of the core for wavelength of operation. Nonlinear PCFs can be extensively used for many practical applications like supercontinuum generation, optical parametric amplification, and all-optical wavelength conversion [9]. Optical nonlinearities can be very useful for a number of applications, such as distributed in-fiber amplification [10] and pulse regeneration, wavelength conversion, optical monitoring, multiplexing and demultiplexing, and switching [11].

Many reports have been published about PCFs to achieve high nonlinearity till now. Using soft glass with liquid crystal [12], employing background material of high refractive index and high nonlinearity in the [2, 9], employing high refractive index slot in the core of PCF [9] and to increase the birefringence elliptical slot or elliptical air holes are introduced in the fiber core [2, 3, 9]. In recent, due to fabrication technique advancement and material compatibility with traditional fiber structures, silicon optical fibers have attracted attention. In this paper, spiral PCF structure is designed to achieve high birefringence and large nonlinearity. Due to these properties, fiber will have applications in realization of all-optical signal processing devices. For accurate analysis of birefringence and nonlinearity, finite element method is applied.

2 Structure and Design of Slot Spiral PCF

The cross-sectional view of elliptical-slot spiral PCF used in simulation is shown in the Fig. 1. The circular air holes in the cladding are arranged in the spiral lattice with air holes having diameter of d_1, d_2, d_3, and d_4. Spiral lattice has six arms, each having four air holes; the first ring of circular holes has a radius of r. The air hole arranged in an arm is at an angular increment of θ than the previous one. Diameter of air holes is enlarged step by step for good mode confinement. The background material in PCF is chosen as silicon; in addition to this, an elliptical slot of silicon nano-crystal is embedded in the core of the PCF with semi-minor axis of a and semi-major axis of b. This slot introduces asymmetry between the two axes and hence induces birefringence. The presence of elliptical slot in core region offers better mode confinement hence we are getting ultrahigh nonlinearity.

In comparison with conventional fiber, the elliptical-slot spiral PCF offers increased freedom in tailoring birefringence and nonlinearity. The modal distribution for fundamental modes of fiber at 1550 nm is shown in Fig. 2a and b with fiber parameters as Pitch(Λ) = 0.4 μm, a = 0.03 μm and b = 0.2 μm, d_1 = 0.17 μm, d_2 = 0.27 μm, d_3 = 0.34 μm, and d_4 = 0.43 μm. The effective index of X and Y polarized modes is 2.3505 and 2.6427, respectively. Mode is well confined in

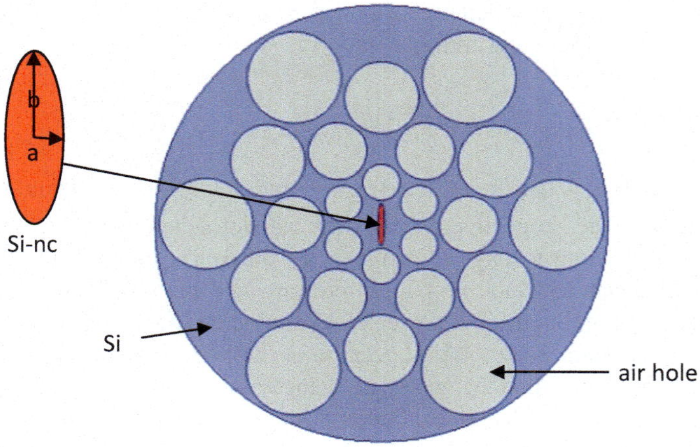

Fig. 1 Cross-sectional view of slotted spiral PCF

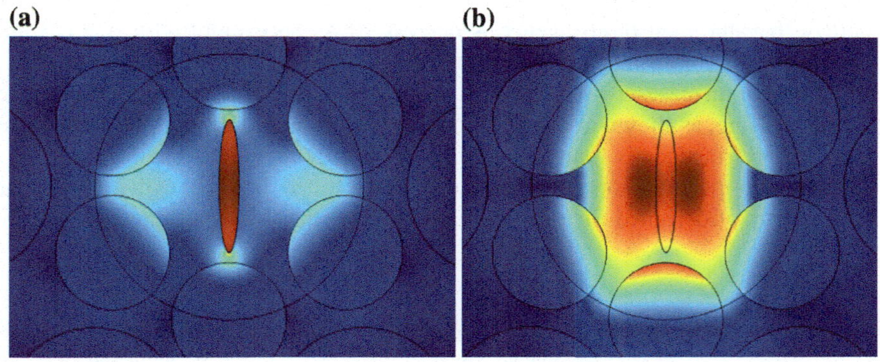

Fig. 2 Field distribution of fundamental modes for **a** X polarization and **b** Y polarization at 1550 nm for vertically aligned Si-NC slot

elliptical-slot region. In comparison with previous structures, the elliptical-slot spiral PCF structure provides high birefringence with high nonlinearity.

3 Simulation Results and Discussion

In this work, to analyze the spiral PCF, a full-vector finite element method (FEM) is used. The fiber design is simulated using COMSOL Multiphysics software. The FEM directly solves the Maxwell equations to obtain optimum value of the effective refractive index, n_{eff}. Once we get the value of n_{eff}, birefringence B can be calculated by taking the difference of effective indices of two fundamental modes and it is given by following formula

$$B = \left| n_{eff}^{x} - n_{eff}^{y} \right| \quad (1)$$

where n_{eff}^{x} and n_{eff}^{y} indicate the refractive index of x-polarization mode and y-polarization mode, respectively. The elliptical slot significantly breaks the symmetry of PCF leading to large effective index difference between the two fundamental modes. Due to which very high birefringence can be achieved.

As shown in Fig. 3a, birefringence values are increasing as the fiber parameter b increases for constant a = 0.03 μm, because of the large asymmetricity in fiber. In Fig. 3b, birefringence first increases as a increases. But after 0.036 μm, as we increase a, birefringence decreases. Because for small values of a maximum field is confined in slot region but for higher values of a field enhancement weakens. For optimum values of design parameters, the value of birefringence is found to be 0.292219 at 1550 nm wavelength (Fig. 4).

As shown in Fig. 5a, birefringence increases as we increase b, but if we further increase b value then birefringence is decreased. As an example, for b = 0.3 μm after 1550 nm, birefringence values are lower than for b = 0.24 μm. Similarly, birefringence increases as we increase a but after 0.036 μm birefringence values are start decreasing, as shown in Fig. 5b. For optimum values of design parameters, the value of birefringence is found to be 0.26239 at 1550 nm wavelength.

Another important property of photonic crystal fibers is nonlinearity. Nonlinear coefficient γ of designed fiber can be calculated by the following equation

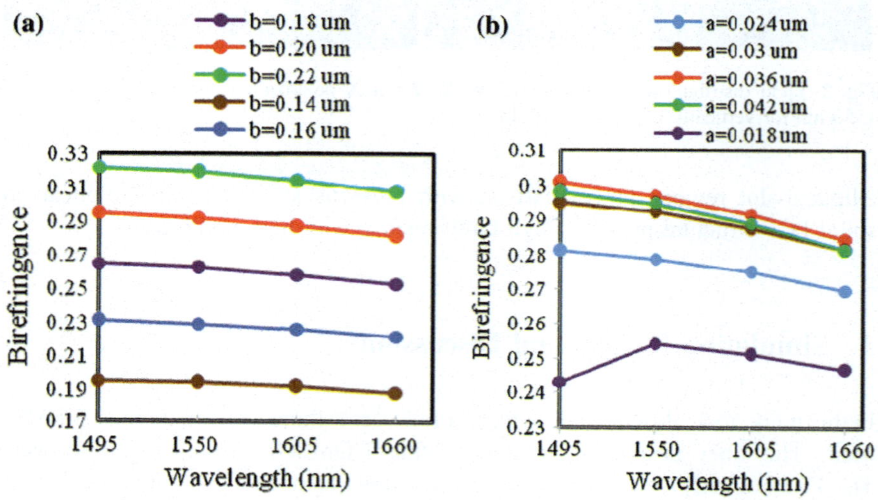

Fig. 3 Birefringence versus wavelength for vertical aligned slot in core with variation in **a** semi-major axis b for a = 0.03 μm and **b** semi-minor axis a for b = 0.2 μm

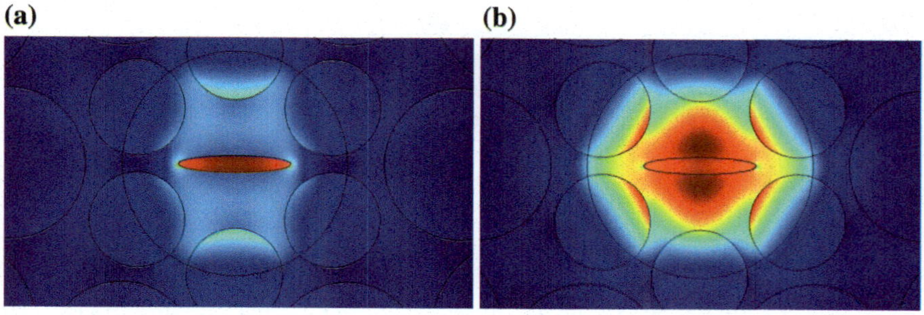

Fig. 4 Field distribution of fundamental modes for **a** Y polarization and **b** X polarization at 1550 nm for horizontally aligned Si-NC slot

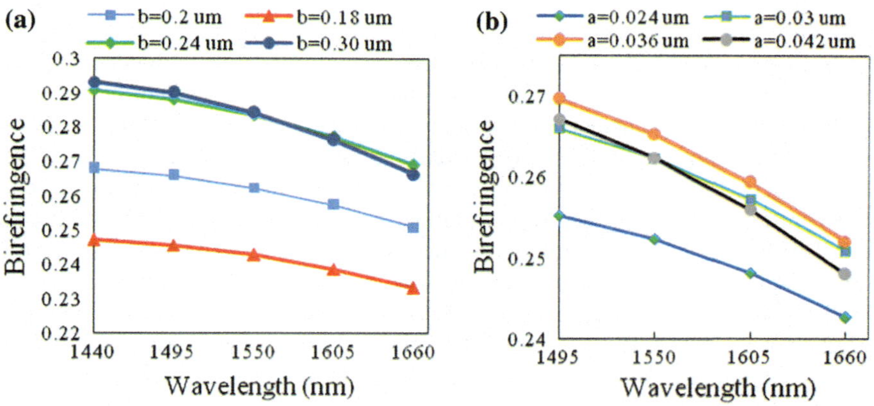

Fig. 5 Birefringence versus wavelength for horizontally aligned slot in core with variation in **a** semi-major axis b for a = 0.03 μm and **b** semi-minor axis a for b = 0.2 μm

$$\gamma = \frac{2\pi n_2}{\lambda A_{eff}} \tag{2}$$

where n_2 is nonlinear refractive index of material used, λ is wavelength of operation. Here, the nonlinear refractive indices of Si and Si-nc are 5×10^{-18} m²/W and 1×10^{-16} m²/W, respectively [18]. A_{eff} is effective area of PCF which is given by

$$A_{eff} = \frac{\left(\iint_{-\infty}^{\infty} E^2 dxdy\right)^2}{\iint_{-\infty}^{\infty} E^4 dxdy} \tag{3}$$

Effective area of PCF depends on the confinement of electric field in the core of the fiber. Small effective mode area leads to large nonlinearity that would have many applications such as supercontinuum generation, soliton pulse transmission, four-wave mixing (FWM), and many nonlinear applications.

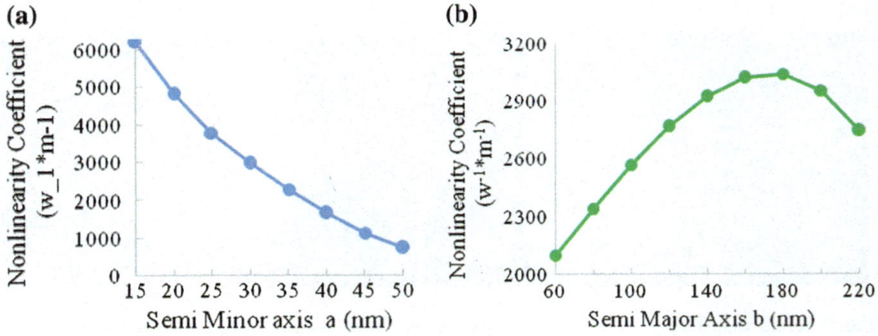

Fig. 6 Nonlinearity variation of fundamental mode at 1550 nm wavelength with variation in **a** semi-minor axis a and **b** semi-major axis b

In Fig. 6, nonlinear coefficient γ of PCF for both fundamental modes is shown with varying semi-minor axis a and semi-major axis b. It is apparent from Fig. 6a that the nonlinear coefficient value decreases as we increase a. The reason is that as we are increasing a, the effective mode area is increasing and hence the light confinement ability or confinement factor decreased. Figure 7a variation of confinement factor with a is also supporting the last statement. In Fig. 6b and 7b, the effect of b on nonlinearity and confinement factor is presented. As we increase b, nonlinearity and confinement factor increase but when b is greater than 200 nm both the values start decreasing. At 1550 nm, we have found that the nonlinear coefficient γ is 2951 $W^{-1}m^{-1}$ and 1379 $W^{-1}m^{-1}$ for x-polarized and y-polarized mode, respectively, and nonlinearity decreases with increase in wavelength.

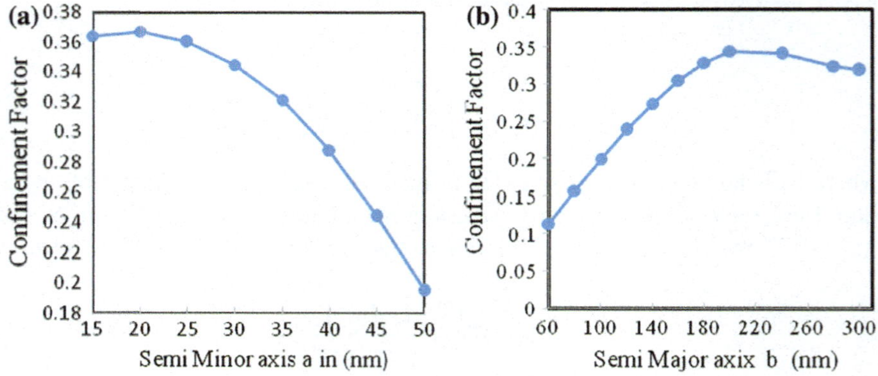

Fig. 7 Confinement factor variation at 1550 nm wavelength with variation in **a** semi-minor axis a and **b** semi-major axis b

4 Conclusion

We can conclude that ultrahigh nonlinearity and high birefringence can be attained by introducing a highly nonlinear elliptical slot in core region. It has been shown through numerical results that by optimizing the fiber parameters, nonlinear coefficient of fundamental mode and birefringence at the wavelength of 1550 nm can be as high as 2951 $W^{-1}m^{-1}$ and 0.292219, respectively. This high birefringence can be used to reduce the polarization mode dispersion effect and many other applications such as sensing applications.

References

1. Birks T, Knight J, Russell P (1997) Endlessly single mode photonic crystal fiber. Opt Lett 22 (13):961–963
2. Liao J, Sun J, Du M, Qin Y (2014) Highly nonlinear dispersion-flattened slotted spiral photonic crystal fibers. IEEE Photon Technol Lett 26(4)
3. Rabiul Hasan Md, Shamim Anower Md, Imran Hasan Md (2016) A polarization maintaining single mode photonic crystal fiber for residual dispersion compensation. IEEE Photon Technol Lett. https://doi.org/10.1109/LPT.2016.2572141
4. Samiul Habib M, Selim Habib M, Hasan MI, Razzak SMA (2014) A single mode ultra-flat high negative residual dispersion compensating photonic crystal fiber. Opt Fiber Technol. https://doi.org/10.1016/J.YOFTE.2014.03.005
5. Franco MAR, Serrão VA, Sircilli F (2008) Microstructured optical fiber for residual dispersion compensation over s + c + l + u wavelength bands. IEEE Photon Technol Lett 20 (9)
6. Russell PStJ (2006) Photonic crystal fibers. J Lightwave Technol 24(12):4729–4749
7. Sinha RK, Kumar A, Saini TS (2016) Analysis and design of single-mode As_2Se_3-chalcogenide photonic crystal fiber for generation of slow light with tunable features. IEEE J Sel Top Quantum Electron 22(2)
8. Zhang QM, Li M, Hao Q, Deng DH, Zhou H, Zeng HP, Zhan L, Wu X, Liu LY, Xu L (2010) Fabrication and characterization of on-chip optical nonlinear chalcogenide nanofiber devices. Opt Lett 35:3829–3831
9. Liao J, Yang F, Xie Y, Wang X, Huang T, Xiong Z, Kuang F. Ultrahigh birefringent nonlinear slot silicon microfiber with low dispersion. IEEE Photon Technol Lett. https://doi.org/10.1109/LPT.2015.2443986
10. Bottacini M, Poli F, Cucinotta A, Selleri S (2004) Modeling of photonic crystal fiber raman amplifiers. J Lightwave Technol 22(7)
11. Toulouse J (2005) Optical nonlinearities in fibers: review, recent examples, and systems applications. J Lightwave Technol 23(11)
12. Yang T, Wang E, Jiang H, Zhijia H, Xie K (2015) High birefringence photonic crystal fiber with high nonlinearity and low confinement loss. OSA Opt Express 23(7):6
13. Hameed MFO, Obayya SSA, El-Mikati HA (2011) Highly nonlinear birefringent soft glass photonic crystal fiber with liquid crystal core. IEEE Photon Technol Lett 23(20)
14. Huang T, Liao J, Fu S, Tang M, Shum P, Liu D (2014) Slot spiral silicon photonic crystal fiber with property of both high birefringence and high nonlinearity. IEEE Photon J 6(3)
15. Imran Hasan Md, Abdur Razzak SM, Samiul Habib Md (2014) Design and characterization of highly birefringent residual dispersion compensating photonic crystal fiber. J Lightwave Technol 32(23)

16. Kalashnikov VL, Sorokin E, Sorokina IT (2007) Raman effects in the infrared supercontinuum generation in soft-glass PCFs. Appl Phys B 87:37–44. https://doi.org/10.1007/s00340-006-2545-y
17. Hasan MI, Selim Habib M, Samiul Habib M, Abdur Razzak SM (2013) Highly nonlinear and highly birefringent dispersion compensating photonic crystal fiber. Opt Fiber Technol. https://doi.org/10.1016/J.YOFTE.2013.11.005
18. Koos C, Jacome L, Poulton C, Leuthold J, Freude W (2007) Nonlinear silicon-on-insulator waveguides for all-optical signal processing. Opt Express 15:5976–5990

Comparison of Logic Built-in-Self Test Techniques Based on FPGA in Verilog

Ravi Gupta and Kriti Suneja

Abstract Logic built-in self test (BIST) is the part of the circuit, dedicated to test the given circuit for correctness of its operation. In this paper, different techniques of test pattern generation, the first element of BIST, have been compared based on the critical parameters of delay, power consumption and hardware utilization. The random patterns are generated for International Symposium on Circuits and Systems' (ISCAS) benchmark circuit using linear feedback shift register (LFSR). The device used for hardware implementation is Spartan 2E—xc2s50e-7ft256 in Xilinx ISE. Simulation results are obtained in ModelSim 5.4, and power consumption analysis is done using XPower analyzer in Xilinx. The fault coverage results are obtained in MATLAB to grade the test patterns for their testing efficiency. The detailed analysis of the results and the corresponding plots are provided in support of our argument.

Keywords Fault modelling · Output response analyzers · Cellular automata
Gray counter · Fault coverage

1 Introduction

During the manufacturing process, defects may be introduced into electronic systems which are not easy to detect. These defects can be short circuits, missing component or damaged components.

Logic BIST is the part of the circuit, dedicated to test the given circuit for correctness of its operation. It consists of test pattern generator (TPG), output response analyzer (ORA) and circuit under test (CUT). Its prime importance lies in

R. Gupta (✉)
Government Engineering College Bharatpur, Bharatpur, Rajasthan, India
e-mail: raviguptabtp@gmail.com

K. Suneja
MNIT, Jaipur, India
e-mail: suneja212@gmail.com

V. Janyani et al. (eds.), *Optical and Wireless Technologies*, Lecture Notes in Electrical Engineering 472, https://doi.org/10.1007/978-981-10-7395-3_35

applications which are life-critical or defence-critical; for example, in aerospace machines, medical instruments, etc. A fault is an illustration of a defect and is used in computer programs for analyzing defects in electronic components.

Fault simulation is the process of simulating a circuit with defects and comparing it with the correct circuit without defects. Thus, it is used to grade the test patterns in fault detection. The quality of test patterns is determined by fault coverage [1, 2], which is the ratio of detected faults to the total number of simulated faults.

Test generation is the process of generating an efficient set of vectors that will accomplish high fault coverage for a given fault model. Since it is difficult to produce vectors targeting all possible faults that could occur during the manufacturing process, automatic test pattern generators (ATPGs) operate on an abstract representation of defects. Random test generation (RTG) is one of the simplest methods for generating vectors. Vectors are randomly generated and fault simulated on the circuit under test (CUT). Because no specific fault is targeted, the complexity of RTG is low [3]. Conventionally, linear feedback shift registers (LFSRs) are used to generate test patterns for BIST. However, the switching activities of test patterns generated by LFSRs are very high which in turn lead to more power dissipation [4]. In this work, the trade-off among hardware, power and delay has been shown with the help of four different types of TPGs-counter, LFSR, circular automata and Gray counter.

2 Test Pattern Generators

The linear feedback shift register (LFSR) is used as a test pattern generator in BISTs [4, 5]. The only external signal necessary to generate the test patterns in LFSR is the clock. If 2^n patterns are used to test an n input combinational gate, it is called *Exhaustive Testing* which generates 100% single and multiple stuck-at faults coverage. However, this process is time-consuming; thus, a subset of 2^n is used where n is huge. Data produced by an LFSR depend on its characteristic equation, which is defined in the form of its feedback. The random numbers produced are used to identify the physical faults in an IC.

Cellular automata is another test pattern generation technique for BIST, which is useful in the sense that it produces more random sequences than LFSRs and does not require global feedback. It is a collection of cells with forward and backward connections. Each cell can connect only to its local neighbours (adjacent left and right cells). The connections are expressed as rules. Assuming cell i can only talk with its neighbours $i - 1$ and $i + 1$ (Fig. 1)

$$\text{Rule 90: } x_i(t+1) = x_{i-1}(t) + x_{i+1}(t) \tag{1}$$

$$\text{Rule 150: } x_i(t+1) = x_{i-1}(t) + x_i(t) + x_{i+1}(t) \tag{2}$$

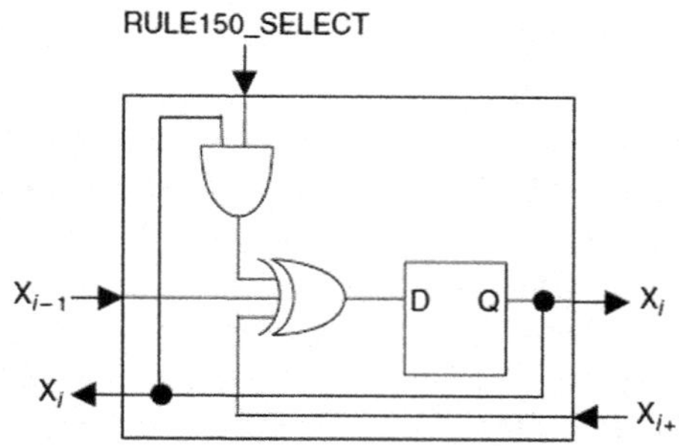

Fig. 1 Universal cellular automata cell

Counters can also be used as test pattern generators as they produce sequences of length 2^n. However, the number of test patterns required for the given fault coverage will always be high. Also, the switching activity is considerably large, which is not desirable from power consumption point of view. An alternative to counters is *gray counter*, where there is a single bit transition between two consecutive patterns. Thus, the transitions are minimal, and power consumption is low.

Let us assume that the n-bit output of counter is C_{n-1} C_{n-2}…C_1 C_0. Gray encoder is used to encode the counters output C[n − 1:0] so that two successive values of its output G[n − 1:0] will differ in only one bit. Gray encoder can be implemented by following equations:

$$
\begin{aligned}
&G[0] = C[0] \text{ xor } C[1]\\
&G[1] = C[1] \text{ xor } C[2]\\
&\ldots\\
&G[n-2] = C[n-2] \text{ xor } C[n-1]\\
&G[n-1] = C[n-1].
\end{aligned}
\tag{3}
$$

In this paper, we have shown all the hardware and software results. The device used for hardware implementation is Spartan 2E—xc2s50e-7ft256 in Xilinx ISE. Simulation results are obtained in ModelSim 5.4, and power consumption analysis is done using XPower analyzer in Xilinx. The fault coverage results are obtained in MATLAB to grade the test patterns for their testing efficiency.

Spartan 2E has the following specifications:

No. of slices-768;

No. of flip-flops-1536;

No. of look up tables (LUTs)-1536;

No. of input/output blocks (IOBs)-182;

and global clocks (GCLKs)-4

Circuit under test (CUT) used is ISCAS benchmark 74 × series' 4-bit magnitude comparator.

3 Results

Following four kinds of test pattern generators were discussed: linear feedback shift registers (LFSRs), cellular automata, counters and gray counter. For N = 3 bits, the number of slices used in above four TPGs is 6, 2, 3, and 6, respectively, as shown in Fig. 2. However, as the data size increases, the area overhead of counters increases more rapidly as compared to LFSRs for the given fault coverage (Fig. 3).

Power consumption is an important factor in BIST. Thus, if the supply voltage is constant, then by reducing the number of transitions at the input of CUT, dissipation of power can be minimised considerably. In Table 1, we can see that the power consumption in Gray counter is lowest because here, only one bit changes between two consecutive patterns while for circular automata, it is highest because the randomness in it is much more than any of the other TPGs.

Delay is again an important factor in BIST as it is necessary for TPGs to generate patterns to match the speed of the circuit under test. In Table 1, we can see that there is not much difference in the delays of the above mentioned TPGs because the number of flip-flops used in all the four TPGs is same. However, the routing delay in cellular automata is less than that of LFSR (13% less for N = 3 and 57% less for N = 10).

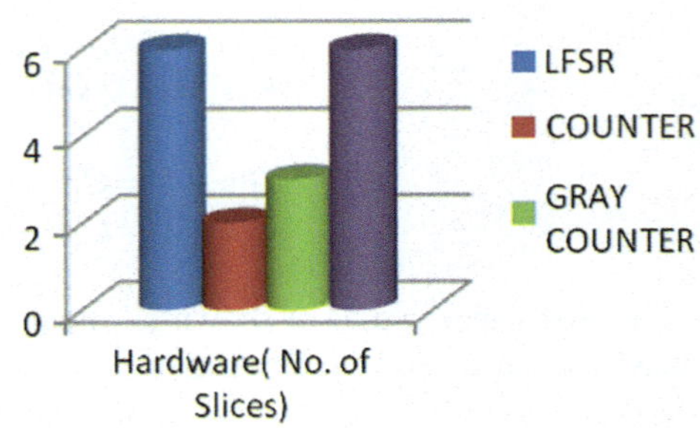

Fig. 2 Hardware utilization in terms of number of slices

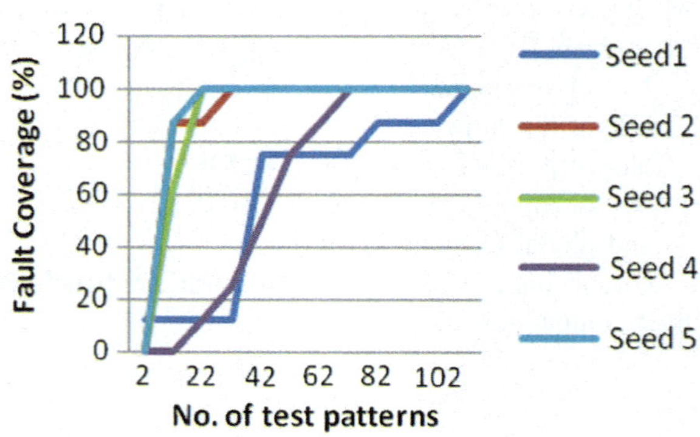

Fig. 3 Fault coverage versus number of test patterns

Table 1 Comparison table of four TPG

N	Delay (n s)				Hardware (no. of slices)				Power consumption (mWatt)			
	LFSR	Counter	GC*	CA*	LFSR	Counter	GC	CA	LFSR*	Counter	GC	CA
5	4.135	5.762	4.284	3.838	10	9	6	9	843.413	813.543	112.925	843.413
6	4.225	5.814	4.336	3.838	12	11	7	11	858.595	814.732	114.966	858.595
7	4.315	5.866	4.388	3.838	15	12	8	13	873.778	828.682	118.283	873.778
8	4.405	5.918	4.440	3.838	17	15	10	15	888.961	829.871	120.325	888.961
9	4.495	5.970	4.492	3.838	19	15	10	17	904.143	843.822	123.642	904.143
10	4.585	6.022	4.544	3.838	21	18	12	19	919.326	845.010	125.683	919.326
11	4.675	6.074	4.596	3.838	23	18	12	21	934.509	858.961	129.001	934.509
12	4.765	6.126	4.648	3.838	25	21	14	23	949.691	860.149	131.042	949.604
13	4.855	6.178	4.700	3.838	28	21	14	25	964.874	874.100	134.359	964.874

GC*—gray counter CA*—cellular automata LFSR*—linear feedback shift register

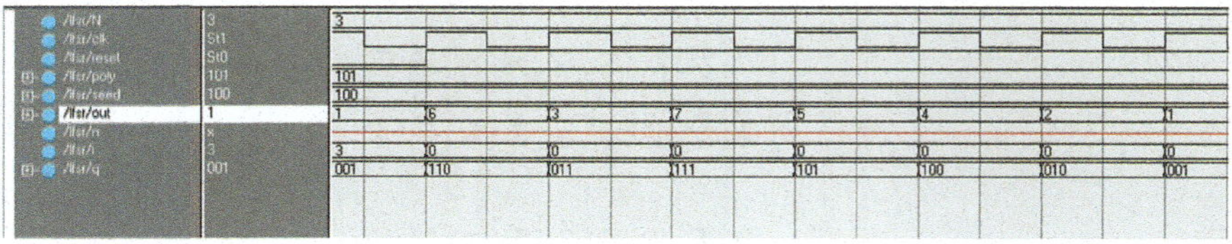

Fig. 4 ModelSim simulations of LFSR

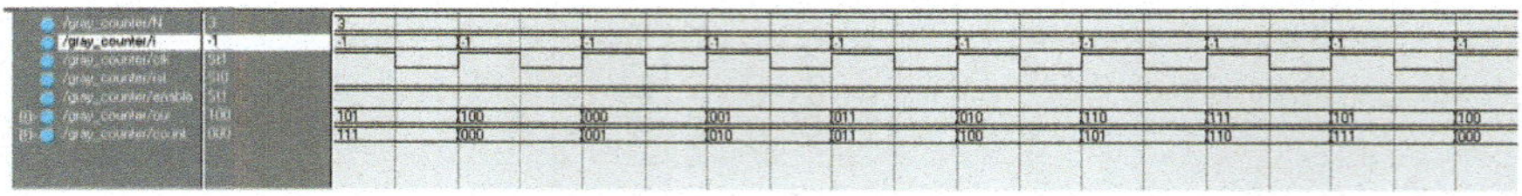

Fig. 5 ModelSim simulations of gray counter

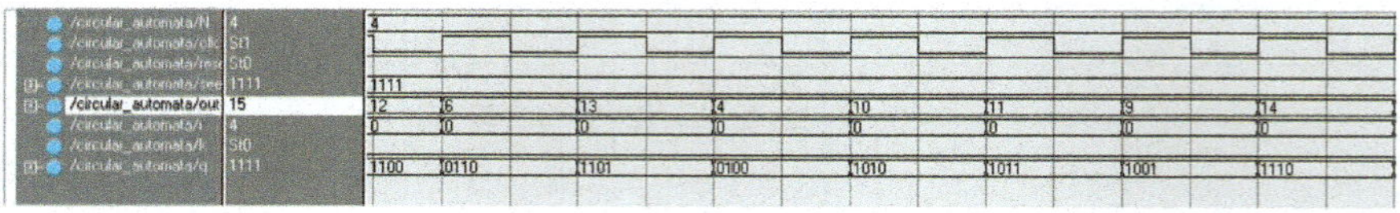

Fig. 6 ModelSim simulations of circular automata

ModelSim simulations are performed to verify the functionality of LFSR. Figure 4 shows the simulation results for N = 3, seed = 100 and poly = 101. ModelSim simulations are performed to verify the functionality of gray counter as single bit transition test pattern generator. Figure 5 shows the simulation results.

A four-stage CA generated by alternating rules 150 (on even cells) and 90 (on odd cells) gives a maximal-length sequence of 15 distinct states as listed in Fig. 6.

The number of test patterns required to attain a target fault coverage varies with the CUT and the seed applied in LFSR. As shown in Fig. 3, when the seed applied is "00000000101", 100% fault coverage is obtained in 19 test patterns, while when the seed applied is "11111111111", 100% fault coverage is obtained in 102 test patterns. The more the number of patterns, the more would be the delay and power consumption. Thus, right choice of seed is critical in BIST.

4 Conclusion and Future Scope

In this paper, various approaches of test pattern generation for logic BIST are studied and compared in terms of area complexity, delay, power consumption and fault coverage. The results obtained from Xilinx implementation with the device xc2s50e-7ft256 in which, we have generated. VCD file post-simulation and XPower for power consumption calculation show that gray counters are 71.2% more efficient than circular automata for test pattern generation. However, if delay is critical parameter in an application, then circular automata performs 57% better than LFSRs. So, the trade-off between power consumption and delay has to be considered. If there is a requirement of large test pattern, i.e., number of bits/pattern is high, then circular automata is a preferred choice over LFSR because of the absence of global feedback. In this work, a 32-bit programmable BIST architecture was implemented with LFSR and MISR and tested for ISCAS benchmark circuit. The fault modelling was done for stuck-at faults, and fault coverage for different number of test patterns was found in MATLAB for different seeds. The number of test patterns required to attain target fault coverage varies with the CUT and the seed applied in LFSR. As shown in, when the seed applied is "0000000101", 100% fault coverage is obtained in 19 test patterns, while when the seed applied is "1111111111", 100% fault coverage is obtained in 102 test patterns. The more the number of patterns, the more would be the delay and power consumption. Thus, right choice of seed is critical in BIST.

The ModelSim simulations shown in experimental results depict the correctness of the functionality of various modules used in BIST architecture. Suggestions for future work in this area would be to use genetic algorithm to select the best seed for maximum fault coverage in a minimum number of test patterns, to improve upon the adaptations made to the logic and routing BIST methodologies, to use it for digital signal processors, analogue to digital converters, etc.

References

1. Williams TW, Brown NC (1981) Defect level as a function of fault coverage. IEEE Trans Comput C-30(12):987–988
2. Thaker PA (2000) Register transfer level fault modelling and test evaluation technique for VLSI circuits, PhD thesis. The George Washington University, May 2000
3. Wang LT, Wu CW, Wen X (2006) VLSI test principles and architectures: design for testability. Morgan Kaufmann, 777 p
4. Navabi Z (2011) Digital system test and testable design: using HDL models and architectures. Springer Science + Business Media, LLC 2011, 433 p
5. Bushnell ML, Agrawal VD (2000) Essentials of electronic testing for digital, memory, and mixed-signal VLSI circuits. Kluwer, Dordecht

Network Selection by Vertical Handoff in Heterogeneous Vehicular Network Using Fuzzy MADM-TOPSIS

Sangama Bhadouria and Rakesh Roshan

Abstract The prime requirements of vertical handoff in the heterogeneous vehicular network are to ensure the quality of service requirements and to provide seamless connectivity to all mobile stations. In this research paper, Multiple Attribute Decision Making (MADM) classical methods are used like the Technique for Order Preference by Similarity to Ideal Solution (TOPSIS) to make the handoff decisions process quickly and efficiently. Our method called 'MADM-TOPSIS Scheme' is the fusing of TOPSIS method with Fuzzy Logic Sets and Rules, in order to lessen the Blocking Probability. Performance results are thus compared with the classical fuzzy TOPSIS and with an approach of the base paper. Our work is able to determine properly that by reducing the blocking probability, the handoff probability will increase that will improve the network performance in the heterogeneous vehicular system. After differentiating to classical Fuzzy TOPSIS, our work scheme shows an upgrade of 0.10 and 20% with base paper.

Keywords Vertical handover decision · MADM-TOPSIS · Fuzzy logic
Handover probability · Blocking probability

1 Introduction

Considering a present scenario, the services of wireless communication network and technology is upgrading tremendously, since the desires and demands of cell phone users while traveling from one place to another is increasing both in real, non-real-time services, and also in different Traffic Classes [1] at minimum cost. In heterogeneous network, all the Radio Access Technology (RATs) will cohabit as one to upgrade the Quality of Service (QoS) and network performance.

S. Bhadouria (✉) · R. Roshan
Pranveer Singh Institute of Technology, Kanpur, India
e-mail: Heysangama@gmail.com

R. Roshan
e-mail: rakeshroshan11dec@gmail.com

V. Janyani et al. (eds.), *Optical and Wireless Technologies*, Lecture Notes in Electrical Engineering 472, https://doi.org/10.1007/978-981-10-7395-3_36

All the mobile stations should be deployed in such a way that they are aware of all the networks around and can switch its connection to the best network. To allow the best connectivity to applications anywhere at any time, many vertical handover decisions [2, 3] strategies have been proposed. Vertical Handover Decision (VHD) algorithms are required to be developed, in order to give seamless connectivity between various wireless networks overlapped [4]. Our main aim is to reduce the handover blocking probability as given in [5]. Handoff blocking probability should be measured in terms of traffic intensity as it measures the number of channels available and the load given in [6] which are the prime requirements to lessen the blocking probability.

The remainder of this paper is organized as follows: Below section will give the short introduction of vertical handover, including state of the art about existing MADM-TOPSIS method and also deals with fuzzy logic, second last Sect. 2 will show the proposed methodology with simulation parameters, and last Sect. 3 will conclude the whole paper.

1.1 Vertical Handover Decision

Vertical handover decision [3] is a bit complicated process as it depends on an increasing number of parameters; thus, Multiple Attribute Decision Making (MADM) algorithm has been proposed to sort out this problem, further MADM has been classified in several groups [7]. Moreover, there are many Decision-Making Algorithms (DMA), which use the merits of both MADM and Fuzzy methods [2, 4, 8]. In this paper, we are using MADM-Fuzzy method [9] and introduced our 'MADM-TOPSIS Scheme' which is based on MADM-Fuzzy TOPSIS [4] approach as it has the capability of finding the optimal radio access technology in real-time condition. The goal is to have lower handover blocking probability so that seamless connectivity will be achieved while dealing with different wireless technologies for real-time application. The developed Scheme will exploit the merits of Fuzzy rules and Sets, [2, 9] for intelligent Fuzzy decision when limited amount of information is given as input, and MADM-Fuzzy TOPSIS methods will deal with the number of parameters and available network.

1.2 MADM-TOPSIS Method

Monil et al. [9] MADM decision algorithm is based on several parameters like traffic classes, seamless handover, and bandwidth. Moreover, some criteria are based on user preferences handling fuzzy logic and imprecise, fuzzy method is used for comparing several different parameters or attributes. The MADM system is basically divided into two steps: (1) To changing fuzzy numbers in real numbers and (2) Full use of classical MADM methods given in [1, 7, 10] for ranking of available networks. TOPSIS method that is used in our work deals with the handover decision and has the capability of selecting the available network of the closest one with the ideal solution (taking farthest from the worst-case solution).

1.3 Fuzzy Logic Method

It is a tool to model complex and indefinite systems. The fuzzy Logic main function is to establish a computational method and to solve those problems which require human intelligence such as verbal and non-verbal reasoning and secondly to maintain effective trade-off [4, 11, 12].

2 Proposed Methodology

There are mainly two types of fuzzy inference system (FIS) [11, 13] (A) MAMDANI FIS and (B) SUGENO FIS. In our work, MAMDANI FIS is used which incorporates simple rule based on Fuzzy Logic, for example, 'if A and B then C' this approach is basically used in solving the control problem rather modeling the mathematical system. Here we are using four input parameters (A) Data Rate [DR] (B) Received Signal Strength Indicator [RSSI] (C) Interference Rate [IR] (D) Traffic Rate [TR] and one single output.

A method for an intelligent handover decision mechanism among different radio access networks where the available network set is obtained dynamically at the mobile client comprises of four steps as explained by the flowchart in [4]. Fuzzy logic is associated with linguistic variables with membership functions that take values in the interval [0, 1]. A **fuzzy set** is a set with such a set membership function. The process of taking an observation and creating a fuzzy set from it is called **fuzzification**. A collection of rules referring to a particular system is known as a **fuzzy rule base**. Moreover, it is easier to take a crisp decision if the output is represented as a single scalar quantity. The conversion of a fuzzy set to a single crisp value is called **defuzzification**. The Fuzzy rule-based mechanism that gives a decision regarding the network to be selected is depicted in Fig. 1. The Fuzzy rule-based system takes the input parameter values of a network and evaluates its handover probability. The input crisp values are first given to the fuzzification module where these input values are converted into membership values using membership functions. The membership functions are considered as triangular functions with three different regions: low, medium, and high for each input parameter which gives the fuzzy set rules incorporating 81 laws (FIS is created with each input parameter using value that can be low, medium, high thus creating its table it gives 81 fuzzy sets known as 81 laws) shown in Fig. 2. The membership values are used for decision making to give an output membership value.

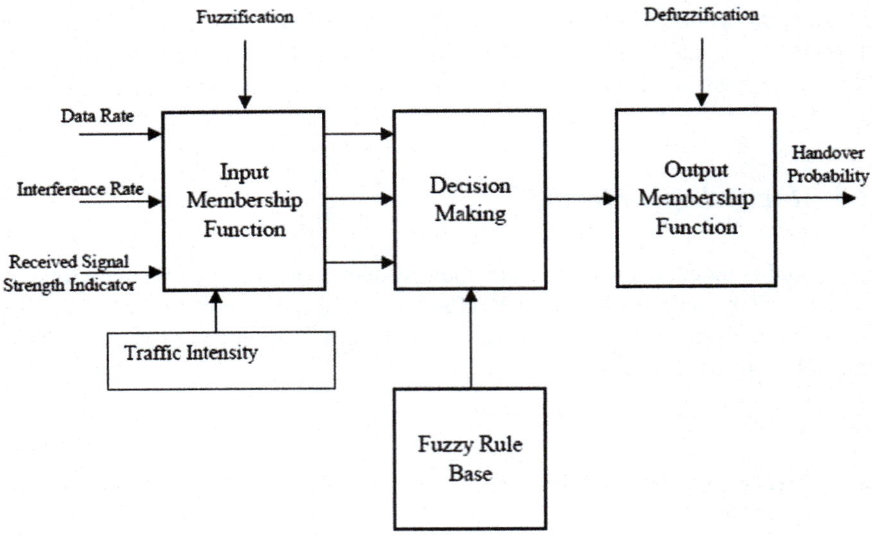

Fig. 1 Block diagram of MADM-TOPSIS scheme

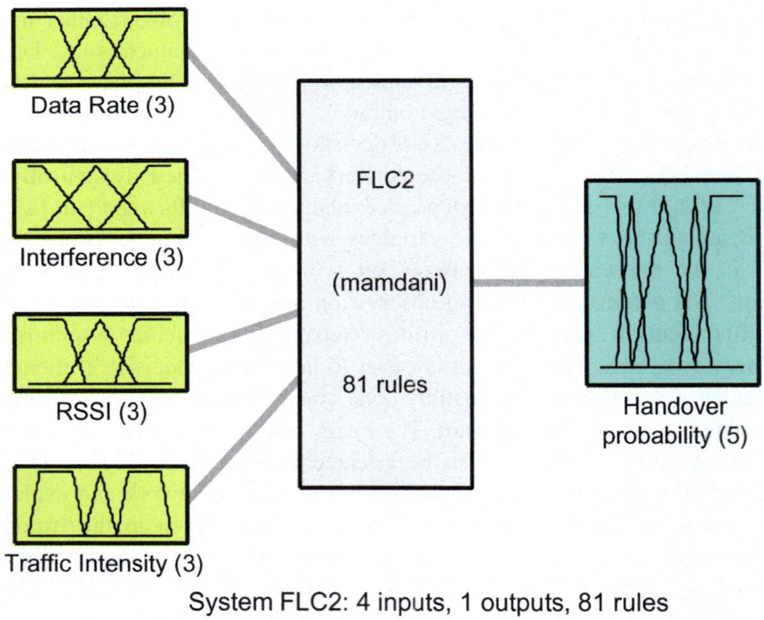

Fig. 2 Proposed fuzzy-based scheme for vertical handoff

2.1 Simulation Parameters

See Table 1.

2.2 Comparison Parameters

As the number of users is increasing on an available network as given in Fig. 3, the load on base stations will increase which will slow down the handoff process. Vertical handoff blocking probability is the main matrix of the handover process; thus, if handover algorithm triggers the handover quickly and efficiently, then surely the handoff blocking probability will increase (Table 2).

Table 1 Parameters used in MADM-TOPSIS scheme

Simulation platform	MATLAB
Data rate	Up to 20 MBPS
Interference rate	−20 to 20 DB
RSSI	−90 to −70 DB
Traffic intensity	Erlang
Fuzzy logic	MAMDANI
Number of inputs	4
Number of outputs	1

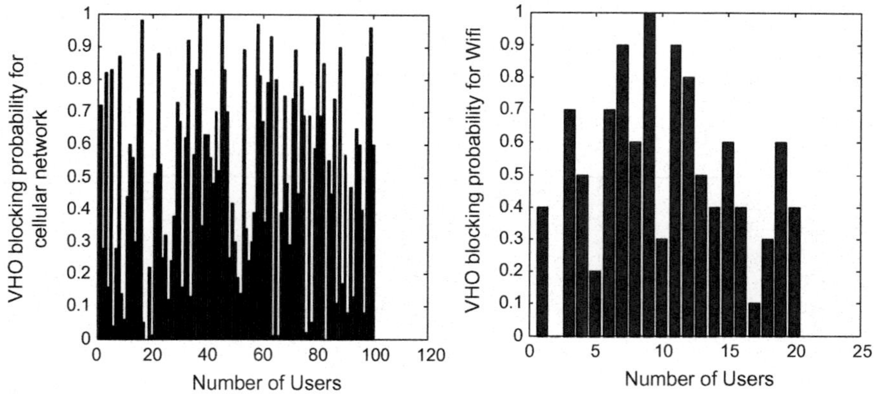

Fig. 3 VHO blocking probability of cellular system and Wi-Fi system versus number of users

Table 2 Comparison parameters with the base paper

	Ayen ben Zineb (Fuzzy MADM-based vertical handover algorithm for enhancing network performances) [4]	MADM-TOPSIS scheme
Algorithm	Vertical handoff decision	Vertical handoff decision
Approach	Classical MADM using TOPSIS method	Classical MADM using TOPSIS method
Parameters	Received signal strength, data rate, speed, and mobile battery level	Received signal strength indicator, data rate, interference rate, and traffic intensity
Results	Network Selection, handover probability percentage	handover probability percentage

2.3 *Justification of MADM-TOPSIS*

Traffic intensity is defined as the ratio of time during which the network is available for the user to the time network is occupied by the user or it basically tells us the number of channels available and the load, Fig. 4 shows the blocking probability for two and three networks as blocking probability is mentioning the load on network with respect to the traffic intensity. On reducing the handoff blocking probability, the total carried traffic of the system will also increase.

In calculating the handoff blocking probability, one can not only consider the number of users but also they have to consider the available number of channels and the load on the system which is nothing but the traffic intensity so by this statement, we are able to justify our results that MADM-TOPSIS scheme reduced the blocking probability up to 20% in comparison with the Fuzzy TOPSIS proposed approach, Ayen Ben Zinab [4], and 0.10% in comparing with the classical Fuzzy method, Fig. 5 shows the comparing results of our scheme.

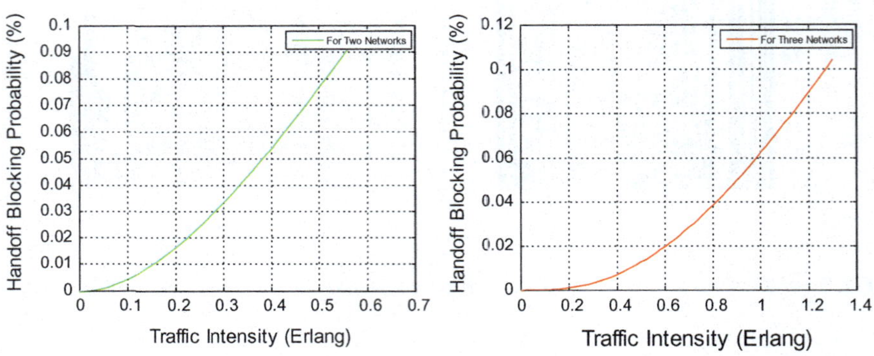

Fig. 4 Blocking probability of two and three networks with respect to traffic intensity

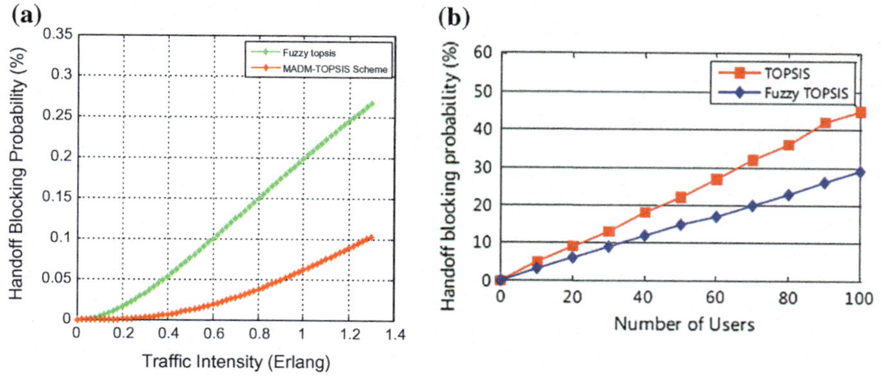

Fig. 5 **a** Comparison results of MADM-TOPSIS scheme with Classical approach and **b** comparison with Ayen Ben Zinab [4]

3 Conclusion

In this work, we proposed a Fuzzy TOPSIS-based MADM-TOPSIS algorithm for vertical handover in the heterogeneous vehicular network, which has the capability to choose the best network among the networks while accessing different radio access technology. The simulation results show that MADM-TOPSIS scheme has the ability to reduce the vertical handover blocking probability up to 0.10% as compared to the classical Fuzzy TOPSIS approach and also reduced it up to 20% in comparing with the paper dealing with Fuzzy TOPSIS approach.

References

1. Rikli, N-E (2012) Design of a fuzzy based handover function for mobile terminal with real time traffic over heterogeneous wireless networks. In: 2012 IEEE international conference on wireless information technology and systems (ICWITS), Nov 2012, pp 1–4
2. Thumthawatworn T, Tillapart P (2015) Enhanced fuzzy-based handover decision system design for wireless mobile networks. In: 2015 seventh international conference on ubiquitous and future networks in IEEE explorer, Sapporo, pp 491–496
3. Boussen S, Tabbane N, Tabbane S, Krief F (2014) A context aware vertical handover decision approach based on fuzzy logic. In: fourth international conference on communications and networking, ComNet-2014 in IEEE explorer, Hammamet, pp 1–5
4. Ben Zineb A, Ayadi M, Tabbane S (2015) Fuzzy MADM based vertical handover algorithm for enhancing network performances. In: 2015 23rd international conference on software, telecommunications and computer networks (SoftCOM), Split in IEEE explorer, pp 153–159
5. Diaba SY, Emmanuel A, Oyibo AM (2015) Performance analysis of queuing priority schemes in cellular communication. Int J Adv Res Comput Commun Eng 4(1):232–236. https://doi.org/10.17148/IJARCCE.2015.4151

6. Siddiqui AF, Kumar P, Tiwari RG (2016) Reducing handoff blocking probability in wireless cellular networks using auxiliary stations and TDMA, 1249–1253 Impact Factor value: 4.45, ISO 9001:2008 Certified Journal
7. Ning L, Wang Z, Guo Q, Jiang K (2013) Fuzzy clustering based group vertical handover decision for heterogeneous wireless networks. In: 2013 IEEE wireless communications and networking conference (WCNC), Shanghai, pp 1231–1236
8. Tie L, Liao H, Du Z (2006) A vertical handover decision algorithm based on fuzzy control theory. In: International multi-symposiums on IEEE—computer and computational sciences, vol 2, pp 309–313
9. Monil MAH, Qasim R, Rahman RM (2013) Speed and direction based fuzzy handover system. In: 2013 IEEE international conference on fuzzy systems (FUZZ-IEEE), Hyderabad, pp 1–8
10. Aziz A, Rizvi S, Saad NM (2010) Fuzzy logic based vertical handover algorithm between LTE and WLAN. In: IEEE 2010 international conference on intelligent and advanced systems, Kuala Lumpur, Malaysia, pp 1–4
11. Parsian S (2006) Decision making in next generation networks using fuzzy systems. Iran Telecommunication Research Center
12. P. Munoz, Barco R, de la Bandera I, Toril M, Luna-Ramirez S (2011) Optimization of a fuzzy logic controller for handover-based load balancing. In: 2011 IEEE 73rd vehicular technology conference (VTC Spring), Budapest, pp 1–5
13. Deswal S, Gupta D (2015) Effective network handover using fuzzy inference for heterogeneous mobility management. Int J Sci Eng Technol Res (IJSETR) 4(7):2379–2382

Investigations into Polarisations in Dielectric Resonator Antenna

Ashok Kumar and Rajveer S. Yaduvanshi

Abstract This paper presented investigations into polarization for designing antennas. The left-hand and right-hand polarization switchable cylindrical DRA has been investigated using two crossed slots of unequal lengths. They couple energy through an aperture-coupled microstrip feed line. Circular polarization with left-hand and right-hand circular polarizations can be achieved with the proposed antenna. Operating frequency of the antenna is 5.8 GHz. The antenna is designed to introduce polarization diversity, i.e., the signal can be transmitted at the same frequency while switching from LHCP to RHCP and vice versa. Hence, the designed antenna has frequency reuse features embedded in it. Measured and simulated results of VSWR are below 2, and axial ratio is below 3 dB. The polarization diversity of this designed antenna plays a vital role in wireless networks to minimize fading and multipath. The simulated and measured results of cylindrical DRA using TMM-13 substrate have been found to be in close proximity. This antenna can be used for WLAN, WiMAX, and Wi-Fi applications.

1 Introduction

There is a need of antenna with polarization diversity in today's wireless communication system. A literature survey has been shown for polarization mostly on patch antennas. Work has been done on wideband dual-polarization patch antenna with directional coupler [1]. Wong worked on capacitive-coupled feed and slot-coupled feed [2, 3]. With a pair of L-shaped slots, the broadband dual-frequency operation of circular

A. Kumar (✉)
University School of Information Communication & Technology, Guru Gobind Singh Indraprastha University, Dwarka 110078, Delhi, India
e-mail: ashok.k@ipu.ac.in

R. S. Yaduvanshi
Electronics and Communication Engineering, Ambedkar Institute of Advanced Communication Technologies & Research, Geeta Colony, 110092 New Delhi, Delhi, India
e-mail: yaduvanshirs007@gmail.com

V. Janyani et al. (eds.), *Optical and Wireless Technologies*, Lecture Notes in Electrical Engineering 472, https://doi.org/10.1007/978-981-10-7395-3_37

patch antennas is also designed [4]. Work has also been done in designing broadband matching of dual-linear probe-fed microstrip patch antenna [5]. Most of the work on DRA has been done using a circular polarizing antenna. Complicated feeding network is required to excite two orthogonal modes with a 90-degree phase difference [6]. In the earlier work, by making use of complex geometries of the DRA or through the attachment of parasitic patches, truncating patch antenna or loading the antenna/ground with slots, stubs, slits, and shorting walls, circular polarization (CP) operation was achieved [7]. The proposed work has the novelty of polarization control of DRA antenna with frequency reuse. The designed antenna can achieve right-hand circular polarization or left-hand circular polarization with a single feeding structure. LHCP to RHCP can be used to avoid any polarization miss match [8–11]. The reported work on polarization is given in Table 1. The antenna reported is simulated using a soft high-frequency simulator (HFSS). The complete paper is organized as follows. Section 2 provides antenna design structure. Section 3 describes mathematical formulations. Simulation and experiment results have been presented in Sect. 4. The result is concluded in Sect. 5.

1.1 Previous Work in Tabular Format

Table 1 Survey work on polarization

Paper name	Both LHCP/ RHCP possible in single antenna?	Complexity in design	Method of achieving circular polarization
A novel patch antenna with switchable slot (PASS) dual-frequency operation with reversed circular polarizations	Yes	Medium	Switching PIN diodes are used (when diode is open circuit, either of the polarization is achieved, when short circuit, the later is achieved)
A reconfigurable microstrip antenna for switchable polarization	Yes	Medium	Same as above
A reconfigurable patch antenna using switchable slots for circular polarization diversity	Yes	Medium	Same as above
A reconfigurable slot antenna with switchable polarization	Yes	Hard	Same as above
An RHCP/LHCP switchable slot line-fed slot ring antenna	Yes	Hard	Same as above
Circularly polarized antenna with switchable polarization sense	Yes	Hard	Same as above
Novel compact circularly polarized square microstrip antenna	NO	Easy	Feed position is changed by 90° to achieve both types of polarizations

2 Design Specifications of CDRA

See Fig. 1 and Table 2.

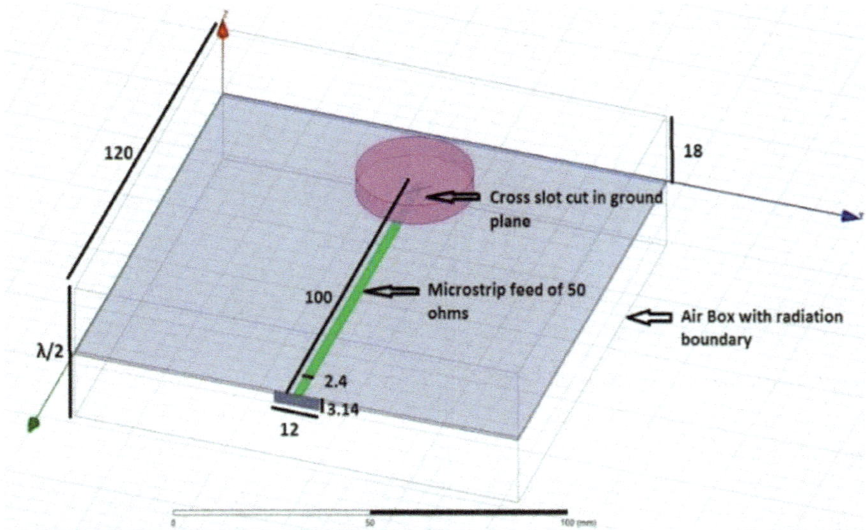

Fig. 1 CDRA design structure using HFSS

Table 2 Dimensions of antenna

S. No	Element	Dimensions (mm)
1	Substrate	120 * 120 * 0.7874 (εr Roger RT/duroid 5880 = 2.2)
2	Ground plane	120 * 120
3	Microstrip line 50 Ohms	2.4 * 100
4	Slot 1	11.4 * 0.3
5	Slot 2	8.9 * 0.3
6	Stub 1	4.7
7	Stub 2	4.2
8	Cylindrical DRA radius	15.2 (εr Roger TMM 13i = 13)
9	Cylindrical DRA height	5.1
10	Radiation box	120 * 120 * 36

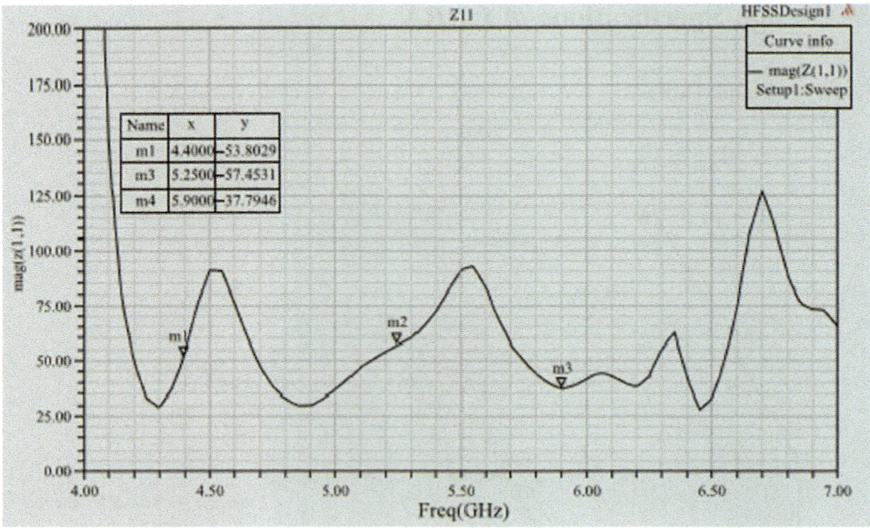

Fig. 2 Impedance of antenna is 50 Ohm

(a) (b)

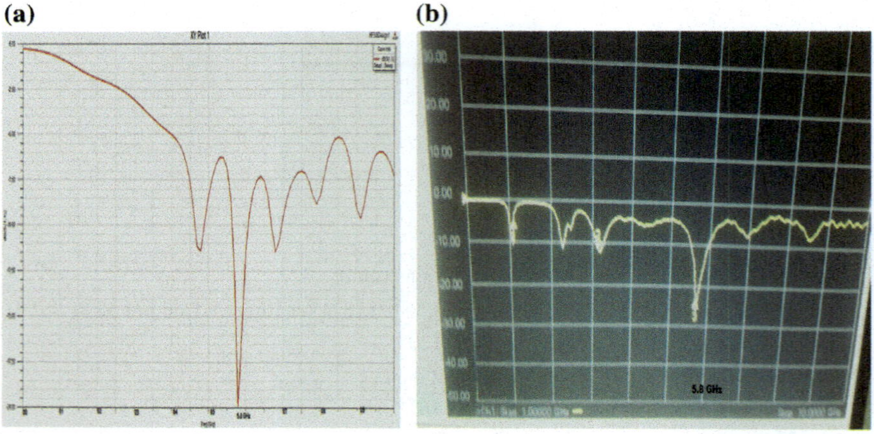

Fig. 3 **a** simulated results and **b** measured results of S_{11} (reflection coefficient) of the antenna are below −10 dB at operating frequency

3 Mathematical Formulations

Eqs. (1) and (2) sacrifice z-oriented fields:

$$\varepsilon(z;t) = \hat{a}x\,\varepsilon x\,(z;t) + \hat{a}y\,\varepsilon y\,(z;t) \tag{1}$$

(a)

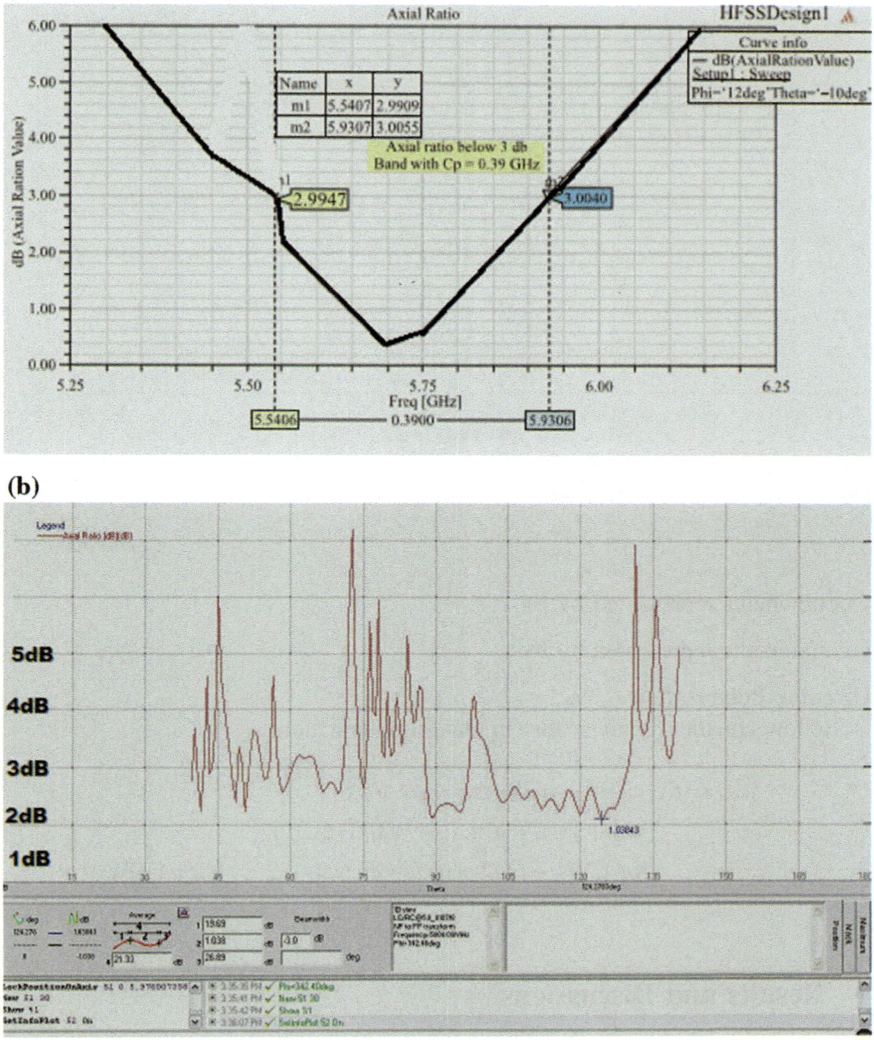

(b)

Fig. 4 **a** axial ratio below 3 dB at operating frequency, **b** axial ratio versus theta at operating frequency, 5.8 GHz

The instantaneous components are related to their complex counterparts by

$$\begin{aligned}\mathcal{E}x(z;t) &= Re[Ex - ej(\omega t + kz)] = Re[Exoej(\omega t + kz + \phi x)] = Exo\cos(\omega t + kz + \phi x)\mathcal{E}y(z;t)\\ &= Re[Ey - ej(\omega t + kz)] = Re[Eyoej(\omega t + kz + \phi y)] = Eyo\cos(\omega t + kz + \phi y);\end{aligned} \quad (2)$$

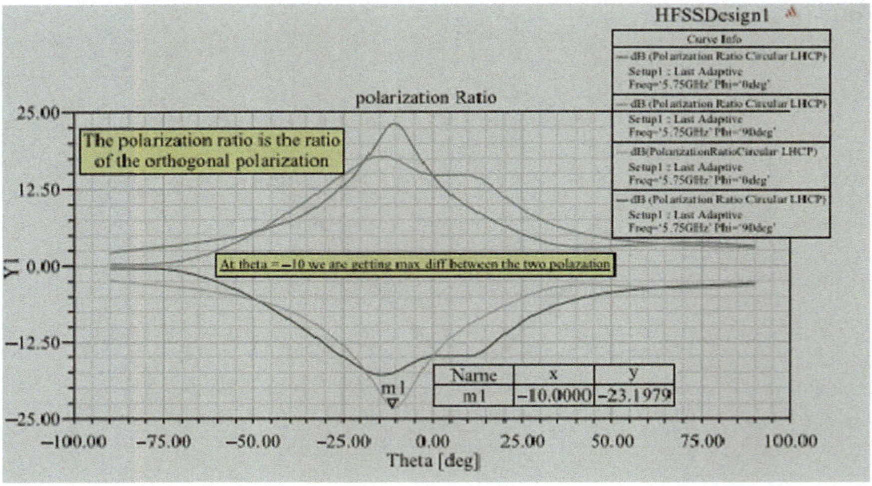

Fig. 5 Polarization ratio LHCP/RHCP is showing good cross- and co-polarization level

x components represented by Ex

y components represented by Ey

Circular Polarization

Bellow equation is given for circulation polarizations.

$$\begin{aligned} |\mathcal{E}x| = |\mathcal{E}y| \Rightarrow \text{Exo} &= \text{Eyo}\,\Delta\phi = \phi y - \phi x \\ &= \{+(1, 2+\text{n})\pi;\quad \text{n} = 0, 1, 2\ldots\} \\ &\text{for CW}\{-(1, 2+\text{n})\pi;\quad \text{n} = 0, 1, 2\ldots\}\text{ for CCW} \end{aligned} \tag{3}$$

4 Results and Discussions

Axial ratio is defined as the ratio of two orthogonal components of E-field. It can also be defined as the ratio of major axis to minor axis of polarization. In the circularly polarized antenna, the orthogonal components of E-field are of same magnitude. Ideally, axial ratio should be one, i.e., 0 dB, but practically, if we are getting till 3 dB, it is acceptable. Power is reflected when proper impedance matching is not present between an antenna and receiver. Standing wave along the transmission line is caused due to reflected voltage wave.

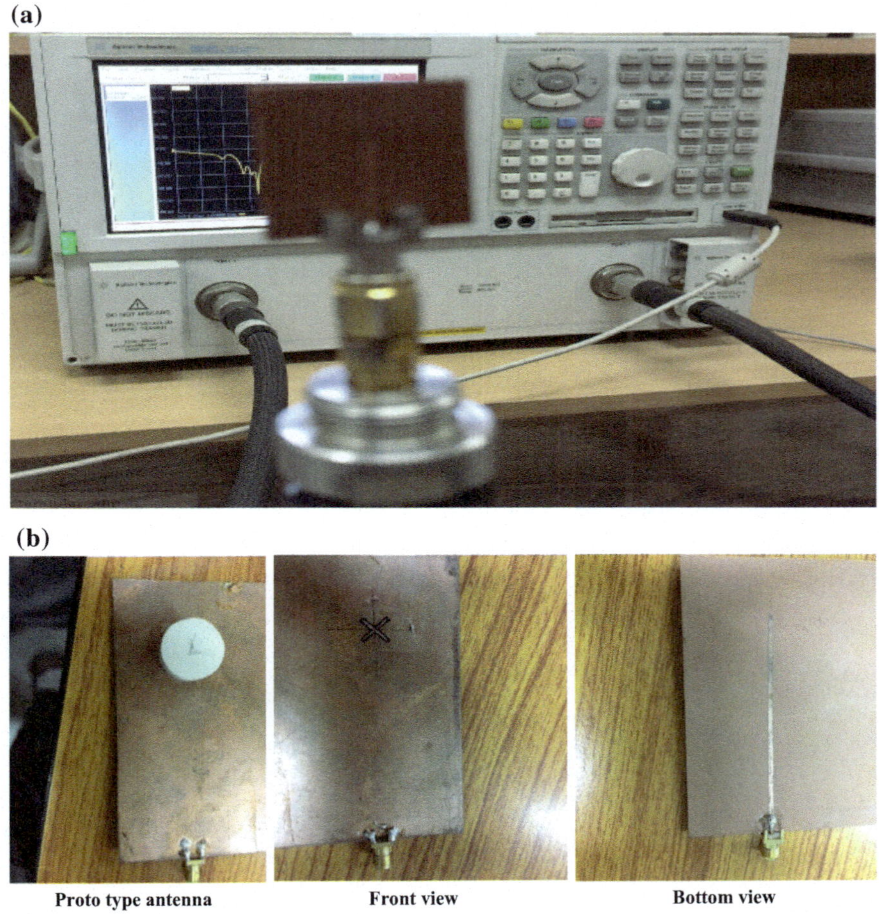

Fig. 6 **a** VNA is used for measurements on prototype antenna, **b** photograph of fabricated CDRA antenna

HFSS-simulated results have been obtained. Simulated return loss, input impedance, VSWR, axial ratio patterns, co- and cross-polarization LHCP gain, radiation pattern plot, and polarization ratio are given in Figs. 2, 3, 4, 5, 6, 7, and 8. Figure 2 shows that impedance is matched. The VSWR is below 2, which is in the acceptable range. The axial ratio is below 3 dB; it is acceptable. The cross-polarization level is −15 dB down the co-polarization level; the field is rotating in a counter-clockwise direction toward −90°, hence the antenna is left-hand circular-polarized. Similar results are obtained for RHCP configuration Table 3.

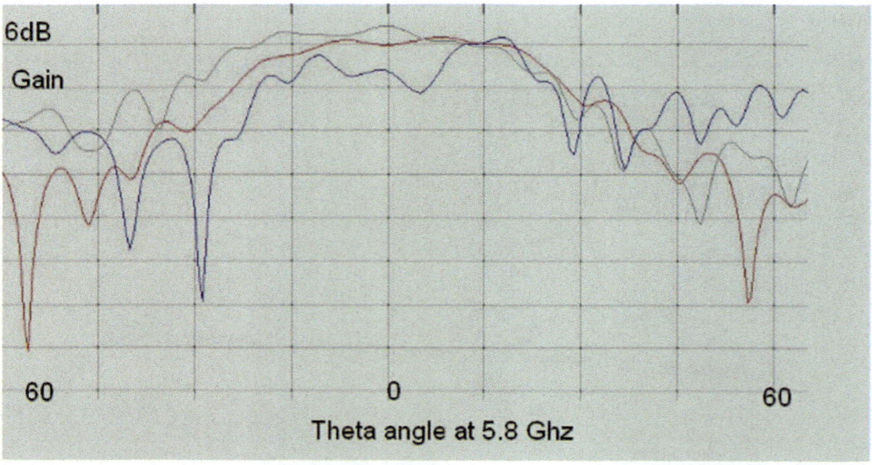

Fig. 7 Anechoic chamber measured results of radiation pattern with peak gain 6 dB at operating frequency

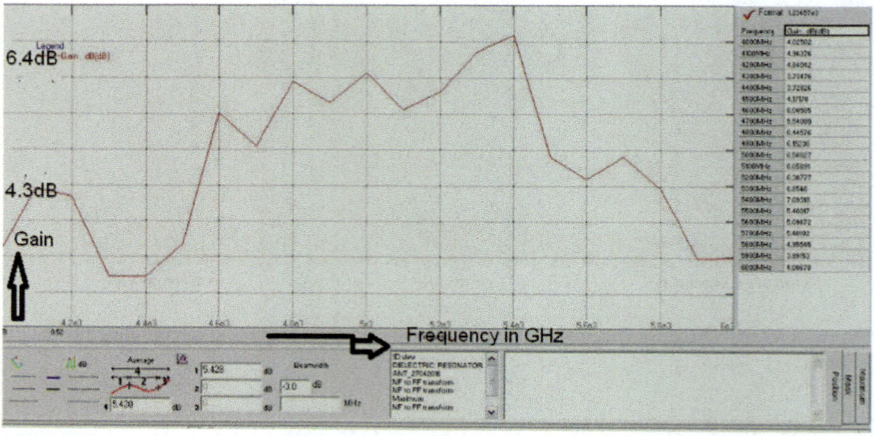

Fig. 8 Measured gain with experimentation

Table 3 Return loss (simulated and measured)

S. No	Results	Frequency (in GHz)	S_{11} (in dB)
1.	Simulated results	5.8	−20.0
2.	Measured results	5.8	−24.5

5 Conclusion

Anechoic chamber measured results on CDRA have been compared with HFSS-simulated results. These are showing good agreement. TMM-13 has been used to develop a prototype. A circularly polarized DRA antenna for polarization diversity has been designed. To achieve the circular polarization, two crossed slots of unequal length to couple energy from an aperture-coupled microstrip line have been used. The designed antenna can achieve right-hand circular polarization or left-hand circular polarization with a single feeding structure. The quality of polarization has been estimated by measuring the axial ratio, input impedance, VSWR, co- and cross-polarization gain. In the designed antenna, switching between RHCP and LHCP is achieved at the same frequency. Circularly polarized wave is robust as compared to linearly polarized wave; hence, antenna misalignment can be compensated with the use of circular polarization. The designed antenna can be used to impart secrecy and polarization diversity by specific switching between RHCP and LHCP. As a future work, investigations can be done on other resonator geometries like elliptical, octagon, and hexagon for wideband performance analysis. This polarized antenna can be used in defense for imparting security and frequency diversity with signal robustness in Ku-band.

References

1. Lau KL, Luk KM, Lin D (2002) A wideband dual polarization patch antenna with directional coupler. IEEE Antennas Wirel Propag Lett 1:186–189
2. Wang KX, Wong H (2015) A circularly polarized antenna by using rotated-stair dielectric resonator. IEEE Antennas Wirel Propag Lett 14:787–790
3. Wong Kin-Lu, Lee Gwo-Yun, Chiou Tzung-Wern (2003) A low-profile planar monopole antenna for multiband operation of mobile handsets. IEEE Trans Antennas Propag 51(1): 121–125
4. Jui-Han Lu (2003) Broadband dual-frequency operation of circular patch antennas and arrays with a pair of L-shaped slots. IEEE Trans Antennas Propag 51(5):1018–1024
5. Jaworski G, Krozer V (2004) Broadband matching of dual-linear polarisation stacked probe-fed microstrip patch antenna. IET Electron Lett 40(4)
6. Chung K, Nam Y, Yun T, Choi J (2006) Reconfigurable microstrip patch antenna with switchable polarization. ETRI J 28(3):379–382
7. Nishiyama E, Egashira S, Sakitani A (1992) Stacked circular polarized microstrip antenna with wide band and high gain. In: Antennas and propagation society international symposium
8. Leung KW, Lim EH, Fang XS (2012) Dielectric resonator antennas: from the basic to the aesthetic. Proc IEEE 100(7):2181–2193
9. Zou M, Pan J (2015) Investigation of resonant modes in wideband hybrid omnidirectional rectangular dielectric resonator antenna. IEEE Trans Antennas Propag 63(7):3272–3275
10. Sulaiman MI, Khamas SK (2011) A singly fed wideband circularly polarized dielectric resonator antenna using concentric open half-loops. IEEE Antennas Wirel Propag Lett 10:1305–1308
11. Kishk AA, Huang W (2011) Size-reduction method for dielectric-resonator antennas. IEEE Antennas Propag Mag 53(2):26–38

A Dielectric Resonator Inspired Displacement Sensor

Anuj Kumar Ojha and A. V. Praveen Kumar

Abstract Preliminary experimental studies on a cylindrical dielectric resonator (DR) excited by a microstrip line are presented, with the purpose of proposing the structure for sensing small displacements. A 50 Ω microstrip line excites a DR of dielectric constant 24, diameter 19.43 mm, and height 7.3 mm. The arrangement is placed in the open environment so that the $HEM_{11\delta}$ mode is excited as the fundamental mode. Optimum impedance matching occurs at a resonant frequency of 3.4 GHz with a reflection coefficient of −40 dB. Studies on the sensitivity of resonant frequency to offsets in the feed position are performed, and 28.5 MHz/mm is measured in the input range of −1 to 6 mm. This feature which is undesired in conventional applications may be used for displacement sensing applications.

Keywords Dielectric resonator · Displacement sensing · Frequency sensitivity
Impedance matching · Microstrip line

1 Introduction

In the last three decades, dielectric resonators (DRs) have gained substantial attention in microwave applications due to their low loss, small size, ease of excitation, and multi-modal features [1]. In addition to the classic oscillator and filter applications, DRs are also proposed as radiators when operated at certain low-Q modes [2]. Quite recently, DRs have found application in sensing gas, strain, temperature, and humidity [3–7]. The strain sensor proposed in [3] is based on displacement measurement using the $TE_{01\delta}$ mode of a cylindrical DR. The above sensor has the DR placed between the two parallel, highly conducting metal plates.

A. K. Ojha (✉) · A. V. Praveen Kumar
Department of Electrical and Electronics Engineering, BITS Pilani,
Pilani Campus, Pilani, Rajasthan, India
e-mail: ojha.anuj11@gmail.com

A. V. Praveen Kumar
e-mail: praveen.kumar@pilani.bits-pilani.ac.in

V. Janyani et al. (eds.), *Optical and Wireless Technologies*, Lecture Notes
in Electrical Engineering 472, https://doi.org/10.1007/978-981-10-7395-3_38

The change in position of the upper movable plate shifts the resonant frequency of the DR. In the present paper, the $HEM_{11\delta}$ mode of a cylindrical DR is excited with a microstrip line. For this, the DR is placed in open environment with a symmetrically placed microstrip feed. Frequency sensitivity of this mode to changes in the DR position on the feed gives motivation for proposing this structure or its other iterations for displacement measurement. Such measurements are quite useful for implementing pressure or strain sensors.

2 Principle of Operation

DRs with high dielectric constant ($\varepsilon_r \gg 1$) are generally narrowband devices (true for certain DR modes only), hence are highly sensitive to impedance mismatch errors caused by external disturbances. In most of the planar excitation schemes, impedance matching is achieved by adjusting the position of the DR relative to the transmission line. Circuits that use DRs usually require the DR to be glued precisely either to a substrate or to a ground plane. This is generally done manually which is likely to create errors in the optimum feed position. Consequently, the power reflected from the DR increases due to (1) a lowering in the coupling and/or (2) a shift in the resonant frequency. In open-ended transmission lines such as a microstrip line, the major effect is the shift in resonant frequency that can be tracked by using a vector network analyzer (VNA). This allows measurement of small displacements coupled to the DR externally that causes a change in the optimum DR position with respect to the line.

3 The Geometry

The DR used for measurements has the following parameters—dielectric constant, $\varepsilon_r = 24$, loss tangent, $\tan\delta = 0.002$, diameter, $2a = 19.43$ mm, and height, $h = 7.3$ mm. The microstrip line (50 Ω) is designed on a FR-4 substrate of dielectric constant 4 and loss factor 0.02. Resulting line has a length of 55 mm and width of 3.2 mm etched on the substrate of size 115 mm versus 115 mm. Choice of a microstrip line offers simplicity in both the design and the optimization of impedance matching. The symmetric positioning of the DR about the line is an important aspect for setting the polarization of the excited $HEM_{11\delta}$ mode. Commercial simulation software HFSS version 13 is used for optimizing the feed position, and a VNA is used for measurement. Radiation patterns are measured in a standard anechoic chamber. Photograph of the DR with the microstrip line is shown in Fig. 1.

Optimum impedance matching is indicated by a minimum of the reflection coefficient response ($|S_{11}|$ vs. frequency) as shown in Fig. 2. Corresponding distance between the DR center and the open end of the line is 7.72 mm. Measured

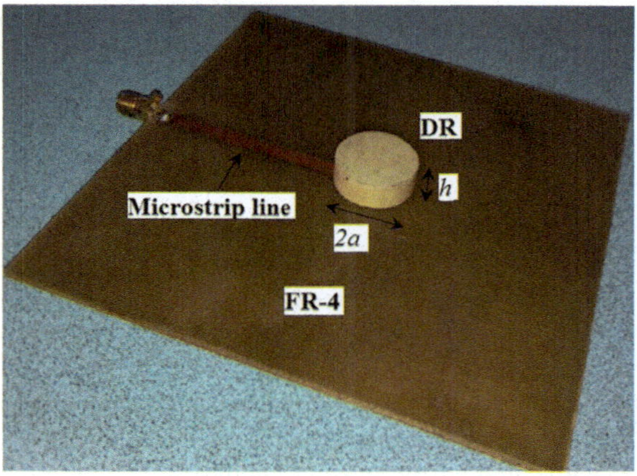

Fig. 1 Photograph of the cylindrical DR excited by a microstrip line

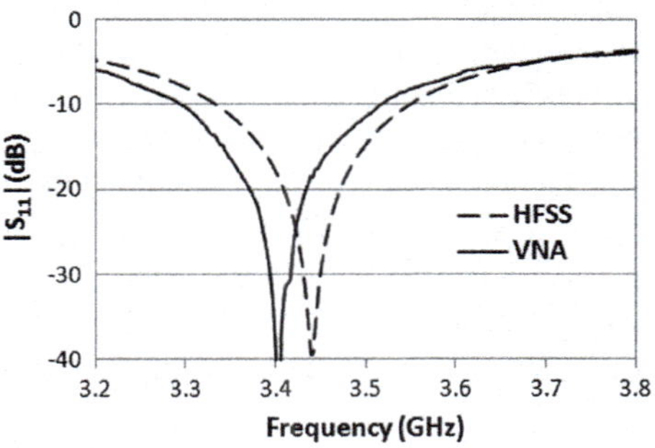

Fig. 2 Simulated and measured reflection coefficient of the DR excited by the microstrip line

resonant frequency is 3.4 GHz with $|S_{11}| \sim -40$ dB. The frequency is very close to the simulated value of 3.43 GHz.

Measured radiation patterns at 3.4 GHz are shown in Fig. 3. The patterns are broadside in nature with cross-polarization levels better than −12 dB in either plane. This confirms the $HEM_{11\delta}$ mode operation. For the measurement, following settings are adopted—the source power = 0 dBm, gain of the transmitter horn =

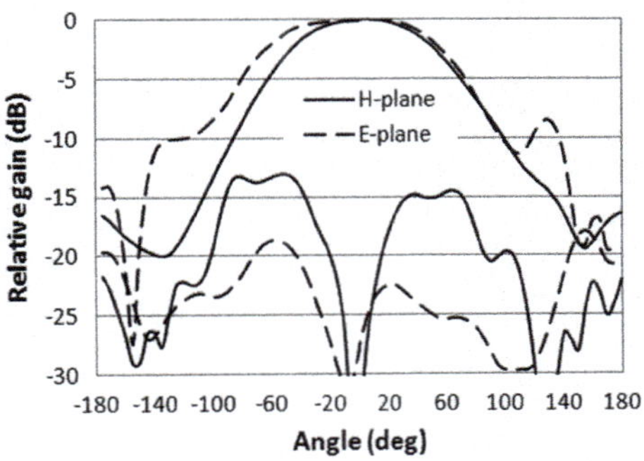

Fig. 3 Measured radiation pattern of the DR excited by the microstrip line at 3.4 GHz

9.2 dBi, total cable loss = 12 dB, and the line of sight distance between the antennas = 2 m. Using these data, peak gains are computed using Friis formula as 5.15 dB in the H-plane and 5.45 dB in the E-plane.

4 Sensitivity Study

Parametric study of the feed point optimization, i.e., $|S_{11}|$ versus *dx*, where *dx* is the relative position of the DR on the feed with respect to that for optimum matching, is very important from the point of view of this paper. Usually, in antenna applications, the position for minimum $|S_{11}|$ is found either numerically or experimentally, and the DR is glued to that position. Remaining data on frequency sensitivity are usually omitted by antenna engineers. For the present case, measured frequency sensitivity with DR offset is shown in Fig. 4.

For positive *dx*, as the DR moves away from the tip of the microstrip toward the source end, additional reactance is added by the extra overlap length (of the line) to the DR impedance which causes an increase in the frequency. For negative *dx*, the frequency obviously decreases but the coupling also decreases as there is negligible overlap between the line and the DR. So there is only one offset that can be measured which is for $dx = -1$ mm. The resonant frequencies are extracted and plotted as a function of *dx* in Fig. 5. It shows a monotonically increasing function with a sensitivity calculated as 28.5 MHz/mm from the linear fit. Thus, the DR offset *dx* can be sensed by using the above calibration curve or in other words the

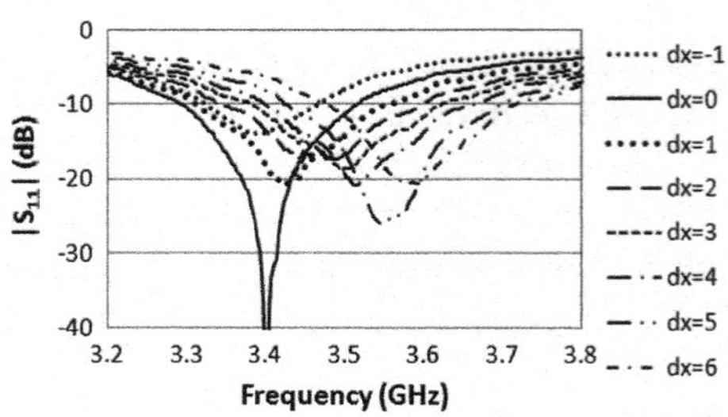

Fig. 4 Variation of the reflection coefficient with the DR offset, *dx* in mm ($dx = 0$ mm is the optimum position)

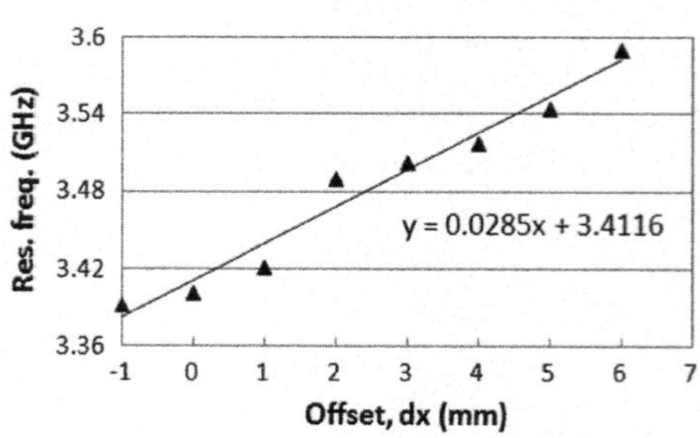

Fig. 5 Variation of resonant frequency with the DR offset, *dx* ($dx = 0$ mm is the optimum position), along with the linear fit

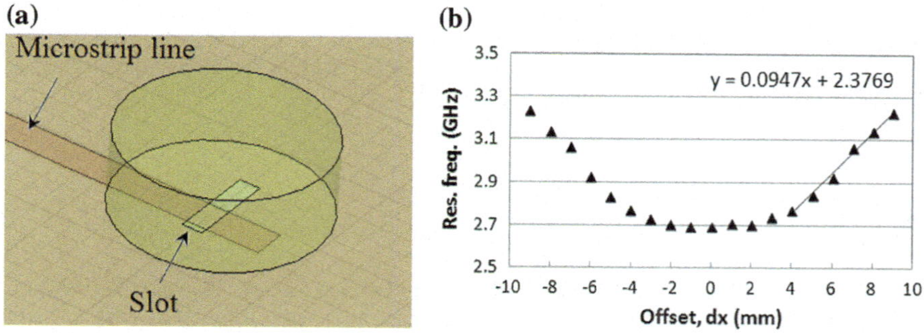

Fig. 6 **a** Geometry of the microstrip slot fed DR. **b** Variation of resonant frequency with the DR offset, dx ($dx = 0$ mm is the optimum position), along with the linear fit from $dx = 4$–9 mm

DR can be used as a displacement sensor. In a practical sensor arrangement, the DR movement must be precisely controlled only in the direction of the feed, and some coupling mechanism must be designed between the external stimulus and the DR.

The excitation scheme can also affect the sensor characteristics. For example, if the same DR is fed by a microstrip slot (length = 10 mm, width = 2 mm) as shown in Fig. 6a, then the simulated sensitivity curve will be defined for both positive and negative offsets. This is shown in Fig. 6b where a near perfect symmetry about the optimum matching position is observed. In addition, the linearity is quite good in the range $dx = 4$–9 mm with a sensitivity of 94.7 MHz/mm which is about three times that of the simple microstrip line feed. This high sensitivity is expected from a resonant feed such as the slot.

5 Conclusion

Sensitivity of a cylindrical DR excited by a microstrip line is studied for the possible use of the structure as a displacement sensor. High dielectric constant DRs are quite sensitive to the position of the feed, failing which the impedance matching is disrupted. For certain excitation schemes such as the open-ended microstrip line, the position errors also cause a shift in resonant frequency of the DR. This resonance can be used to calculate the amount of displacement of the DR on the line. Experimental studies reveal that a linearized sensitivity of 28.5 MHz/mm can be achieved using a DR of dielectric constant 24 and geometrical ratio 1.33. Simulations show that the sensitivity and linearity can be dramatically improved by using alternative excitation schemes such as a microstrip slot. Future work will focus on the experimental confirmation of the above aspect as well as the role of the geometrical ratio (a/h) and the dielectric constant of the DR in deciding the sensor performance.

Acknowledgements This work was supported by the Department of Science and Technology (DST), India through the FIST program under Grant Ref: SR/FST/ETI-346/2013. Authors thank Dr. Ananjan Basu and Mr. Rajesh Kumar Singh of CARE, IIT Delhi for offering the facilities and assistance for characterizing the antenna.

References

1. Kajefez D, Guillon P (1998) Dielectric Resonator, 2nd edn. Nobel, Atlanta, GA, USA
2. Petosa A, Ittipiboon A (2010) Dielectric resonator antennas: a historical review and the current state of the art. IEEE Antennas Propag Mag 52(5):91–116. https://doi.org/10.1109/MAP.2010.5687510
3. Aftab T, Yousaf A, Hoppe J, Stoecklin S, Ostertag T, Reindl L (2015) A parallel plate dielectric resonator as a wireless passive strain sensor. In: Proceedings of IEEE sensors application symposium (SAS), Zadar, Croatia, pp 1–6
4. Kubina B, Schusler M, Mandel C, Mehmood A, Jakoby R (2013) Wireless high-temperature sensing with a chipless tag based on a dielectric resonator antenna. In: Proceedings of IEEE sensors, Baltimore, MD, USA, pp 1–4
5. Friedt J-M, Boudot R, Martin G, Ballandras S (2014) Probing a dielectric resonator acting as passive sensor through a wireless microwave link. Rev Sci Instrum 85(9):094704
6. Boccard JM, Aftab T, Hoppe J, Yousaf A, Hütter R, Reindl LM (2016) High-resolution, far-field, and passive temperature sensing up to 700 °C using an isolated ZST microwave dielectric resonator. IEEE Sensors J 16(3):715–722
7. Kubina B, Mandel C, Schussler M, Jacoby R (2015) Compact quasichipless harmonic radar sensor with a dielectric resonator antenna. In: IEEE MTT international microwave symposium (IMS), pp 1–3. https://doi.org/10.1109/MWSYM.2015.7166820

ENL-AODV: Energy and Load-Based Routing Protocol in Ad Hoc Networks

Nandanwar Chetan Damodar, Manu Elappila, Anurag Patro and Suchismita Chinara

Abstract A mobile ad hoc network (MANET) is a network with no infrastructure, operating on wireless mobile nodes. MANET consists of quickly deployable, independent as well as self-configuring nodes with no centralized administration. There is no fixed topology in the network and the nodes have limited energy and computing resources. Jitter, a small random variation in timing, is widely used in between the periodic transmission of the control message in wireless communication protocols. It is an especially important technique during route discovery process when a process may cause a situation where adjacent nodes have to broadcast concurrently, then the use of jitter makes a protocol able to avoid concurrent packet transmissions over the same channel by neighboring nodes in the network. In AODV, a small delay during the flooding of a control message is used during route discovery process to avoid simultaneous packet transmission by neighboring nodes, which might result in the collision between these packets. The proposed energy and load-based protocol (ENL-AODV) introduces energy and load factor in the calculation of jitter while forwarding of route requests (RREQs), making it select the path with enough energy to transfer the data packet. As simulation results describe, ENL-AODV improves the efficiency of ad hoc networks, increases packet delivery ratio, throughput, and network lifetime, and also decreases average end-to-end delay.

Keywords MANET · AODV · ENL-AODV · Residual energy · Load · Jitter

N. C. Damodar · M. Elappila (✉) · A. Patro · S. Chinara
Department of Computer Science and Engineering, National Institute of Technology, Rourkela 769008, India
e-mail: manuelappila@gmail.com

N. C. Damodar
e-mail: n.chetan.damodar@gmail.com

A. Patro
e-mail: anuragpatro92@gmail.com

S. Chinara
e-mail: suchi.nitrkl@gmail.com

V. Janyani et al. (eds.), *Optical and Wireless Technologies*, Lecture Notes in Electrical Engineering 472, https://doi.org/10.1007/978-981-10-7395-3_39

1 Introduction

A mobile ad hoc network is a collection of freely moving wireless mobile nodes. The mobile nodes participate in forwarding packets in the multi-hop fashion without any existing centralized administration. The aim of architecture is to provide communication services to end user. MANETs have applications in military operations, battlefields, wireless sensor networks, wireless mesh networks, etc. In MANETs [1], a mobile node can act as a router and also a relay node. Thus, the failure of any node can mostly affect the whole performance of the network.

The most critical issue for MANET is the battery power of mobile nodes, as it is in a limited amount [2]. With less battery power, the node will be an overhead to the network as it will lead to extra delays and reduced network lifetime. So in routing, utilizing the battery power of nodes in a proper way and increasing the lifetime of a network is an important issue.

Routing protocols are mainly of two types, proactive and reactive. In proactive routing protocol [3], nodes require to exchange routing information periodically, and by that they compare routes between any nodes in the network. This routing mechanism consumes a lot of energy resources, and bandwidth may be wasted which seems undesirable in MANETs where the resources are limited. On the other hand, reactive routing protocol [4] mostly uses on-demand routing discovery, i.e., nodes do not exchange routing information periodically but find routes only when it needs for the exchange of information between two nodes. Due to continuous topology changes in MANET, links between nodes are not permanent. Also, if a node does not have sufficient energy to forward the packet, then it will drop that packet. Packet drops affect the overall performance of the network. Among all these, rapid energy depletion of a node leads to exhaustion and decreases the network lifetime. AODV is a widely used routing protocol which is primarily based on DSDV [5] and the improvement over on-demand routing techniques in DSR. If network load is light, AODV works efficiently while its performance decreases in high load conditions. That is because, it only selects the path that is shortest without considering the energy and traffic load of the nodes. So while selecting the routes, it is very necessary to examine residual energy and load of nodes.

In this paper, we used energy as well as load situation of nodes as metrics and proposed an energy and load-based AODV routing protocol (ENL-AODV). This modified AODV is primarily based on energy level and load balancing situation of the protocol.

The remaining paper is organized as follows. Section 2 explains ad hoc on-demand routing protocol (AODV) in MANETs. Section 3 explains a proposed modification to AODV, which applies energy and load level mathematical model. Section 4 contains simulation results and performance evaluation of modified AODV compared with existing AODV using NS-2 (Network Simulator 2). Section 5 concludes the paper.

2 Overview of AODV Routing Protocol

In this section, ad hoc on-demand routing protocol is examined. Routing discovery is initiated only when it is needed and routes are maintained in the routing table only when they are active and are in use.

2.1 AODV

AODV is derived from dynamic source routing (DSR) and destination sequenced distance vector (DSDV) routing protocols. The on-demand discovery of route, hop-by-hop routing, and route maintenance are inherited from DSR and node sequence numbering method from DSDV. Routing discovery stages of AODV are as follows:

Whenever the source node has data packets to send, it searches for the route in its routing cache table [6]. In the routing cache table, if there is a section that contains a route to the destination node, then data packets are sent by the source node. However, the source node initiates the route discovery by broadcasting an RREQ packet. When an intermediate node gets an RREQ packet, it does the following operations.

1. If the node is not the destination node, then move to Step 3. Otherwise, execute Step 2.
2. If the destination node is getting this packet for the first time, it inserts the requesting source address field of the RREQ into the routing table and the value of requesting broadcast ID into the broadcast ID cache table. Then, it sets up the reverse route with its last hop node and go to Step 7. Otherwise, the RREQ packet will be discarded by the destination node.
3. According to the values in the source address field and broadcast ID field of the RREQ, it looks for its broadcast ID cache table. In case, there is a broadcast ID cache entry which has the same value of request broadcast ID field and request source address field as those of the RREQ, and move to Step 6. Otherwise, move to Step 4.
4. Since the intermediate node is getting the RREQ for the first time, it will insert the value of RREQs requested source address field and the value of RREQs requested broadcast ID field into the broadcast ID cache table, and then set up the reverse route with its last hop node.
5. If the node has a route to the destination node, then move to Step 7, else the node will randomly make a delay [7]. As soon as the delay has arrived, the node will broadcast an RREQ. Then, move to Step 2.
6. Since the node already got same RREQ before, it will discard the packet.
7. The node sends an RREP to the source node. The establishment of the route is done.

2.2 Jitter Technique for Request Broadcasting in AODV

In route discovery process of AODV, RREQ packets are flooded into the network to find a route to the destination. While due to concurrent transitions by adjacent nodes, there is a probability of collisions at a wireless channel which may lead to packet loss. To prevent collisions at the wireless channel, the RFC5148 [8] recommends the use of jitter in flooding of RREQ packets. It recommends that each node which has a packet to be broadcasted holds the packet for a small random delay before flooding into the network. Holding a packet for short delay helps in avoiding the collision in the case of simultaneous transmissions. The RFC5148 suggests that the delay should be chosen such that it should follow uniform random distribution lying between 0 and a maximum jitter value, J_{max}. Usage of uniform random distribution for jitter leads to the delay inversion [7], i.e., selection of longer paths instead of short one with a minimum hop count.

2.3 Limitation of AODV

For route establishment, AODV considers paths with minimum hop count. When the number of communications increases, this consideration may lead to some drawbacks. So, it is better to incorporate different parameters that have a significant impact on network availability [5] and lifetime. Moreover, energy conservation is an imperative requirement in a wireless network. If a node that involved in a route establishment has less amount of residual energy, it may lead to a route breakage. Also, this will impose an unfavorable impact on the network lifetime. To manage these issues, the remaining energy level of the nodes has to be considered in the route establishment process. Thus, we propose a technique that considers remaining energy and traffic load of nodes, while making the routing decision.

2.4 ENL-AODV Routing Protocol

As discussed in the previous section, AODV does not consider the remaining energy and load situation of nodes in route discovery stage, which leads to unstable routes. In comparison with AODV, ENL-AODV routing protocol considers remaining energy and load situation of nodes in route discovery stage. In AODV whenever source wants to deliver some data packet to the destination, it starts route discovery process by flooding RREQ to its neighboring nodes. When flooding is initiated, nodes generate a reverse path depending on neighboring nodes that send an RREQ message. Finally, when RREQ message arrives at the destination node, a forward path is created. At intermediate nodes, while flooding the RREQ packet, nodes hold it for a random delay. And when the delay timer has expired, it again broadcasts to its neighbors.

This delay is introduced for avoiding collisions due to simultaneous packet transmissions by neighboring nodes. Collisions may lead to packet drops which might affect the whole performance of the network.

In this paper, we have modified the delay (jitter) induced at an intermediate node while broadcasting RREQ to its neighbors. In AODV, the introduced delay is totally random. We have changed the selection of induced delay at each intermediate node in the following fashion. While discovering the routes, the generation of delay for broadcasting the RREQ at the node is depending on the energy and traffic load level of the node. Taking load and energy as a parameter RREQ broadcasting delay is calculated using precise mathematical model. Also, we reduced the flooding of redundant RREQ while nodes are in lower energy state, by dropping route requests at the node whenever its energy is below the threshold energy. If a node has the route to the destination, it will consider its residual energy level and will check if it is greater than the threshold value. If the remaining energy is below the threshold, it will drop the RREQ and go back to route discovery again. Otherwise, it will calculate the delay considering the traffic load level and energy parameters. On the expiry of the delay timer, the node will broadcast the RREQ packet.

3 Mathematical Model of Proposed Protocol

In this section, we will introduce the precise mathematical model to calculate the delay in routing discovery process of the ENL-AODV routing protocol.

3.1 Residual Energy Consideration of Node

We suppose the selected route should have enough energy for data transmission, so that when the source node decides to send the data packets, the residual energy will enable the packet not to drop during transmission and the network not to be disconnected rapidly. Hence, when the source node wants to transmit a data packet to the destination it broadcasts a route request, and then an intermediate node which has received the RREQ will make a delay. The delay will depend on the amount of the current remaining energy of the node such that on the next broadcast the node with enough residual energy will be preferentially selected. The mathematical model to calculate the delay based on residual energy is as follows.

Whenever node sends or receives a data packet, the energy needed for transmission is measured by power required for data transmission by the node and handling time.

$$Energy = Power \times Time$$

Handling time is defined as follows

$$Time = 8 \times PacketSize/BandWidth$$

Let,

- Transmission Energy: E_t
- Receiving Energy : E_r
- Transmission Power : P_t
- Receiving Power : P_r

We have

$$E_t = P_t \times Time$$

$$E_r = P_r \times Time$$

Therefore, the total energy consumed by node (E_{total}) is

$$E_{total} = E_t + E_r$$

We suppose current residual (remaining) energy of a node is E_{res} varying between $[0, E_{total}]$ and defined threshold energy E_{thr}, which is 30% of total energy (E_{total}). The energy weight for calculation of delay, W_{en}, is

$$W_{en} = 1 - E_{res}/E_{total}$$

where $0 \leq W_{en} \leq 1$.
Whenever a node has to forward RREQ to its neighboring nodes, it will consider its residual energy and check if it is below or above the threshold. So, RREQ will be dropped if $E_{res} \leq E_{thr}$, else will calculate the delay based on residual energy and load level.

3.2 Load Level Consideration of Node

The current number of packets in the buffer queue of a node which is usually referred as load level of the node has influenced on broadcasting the control packets during route discovery.
Assume that the maximum capacity of buffer queue is Q_{max} packets and current packet number in the buffer queue is Q_{curr}. So, the load level weight W_l for calculation of delay is

$$W_l = Q_{curr}/Q_{max}$$

where $0 \leq W_l \leq 1$, which describes that larger the W_l is heavier the load of the node.

3.3 The Delay Model

As in the proposed routing protocol ENL-AODV, the delay depends on two metrics, i.e., residual energy and load level of the node. So by combining both the approached discussed above, we have mathematical model for delay (T),

$$T = \left(e \times W_{en} + l \times W_l\right) \times T_c$$

where e and l are constants and $0 \leq e \leq 1, 0 \leq l \leq 1$ and T_c is a delay constant, so while selecting delay constant it should be kept in the mind that it should not be too large. In the proposed algorithm, it is chosen as 0.01.

4 Performance Evaluation and Results

In this section, we compared the ENL-AODV routing protocol with existing AODV to analyze the performance of proposed protocol. We prototyped ENL-AODV on NS-2, which is a discrete event-driven network simulator. It supports various mobile ad hoc network routing algorithms. In our simulation model, we prototyped MANET networks over a wireless channel with two-ray ground reflection propagation model. Drop-tail/priority queue is used as queuing model with 802.11 MAC protocol.

4.1 Simulation Environment

In simulations, we used virtual network field of area 500 m × 500 m with a dynamic topology where nodes are free to move in any direction. Random waypoint model is used as mobility model with two-ray ground radio propagation model. Communication range for each node is set as 250 m, and at application layer CBR traffic generator is applied to produce data packets at the rate of 1 Mbps and bandwidth is 2 Mbps. The value of e, l and Tc is taken as 0.5, 0.5 and 0.01, respectively, while the simulation runs for 150 s. The maximum capacity of each nodes buffer queue is taken as 50. With a particular goal to imitate the ruleless development in this present reality, the setdest command is used to get ruleless random movement of node creating a random topology. Analysis is made on velocity of mobile nodes as 5 m/s.

4.2 Results

We simulated both AODV and ENL-AODV protocols using NS-2 with simulation environment described above and analyzed them on most commonly used

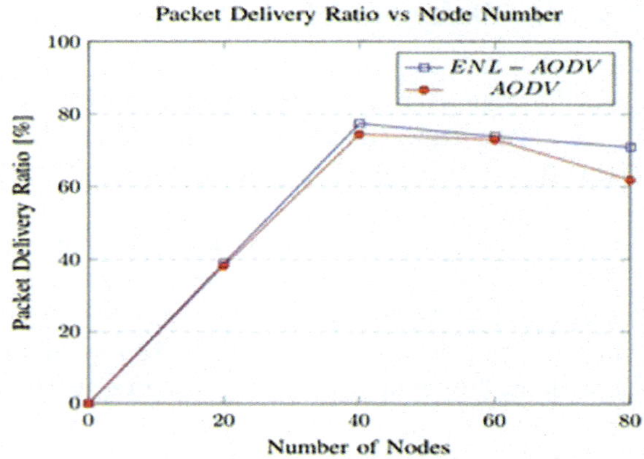

Fig. 1 Packet delivery ratio

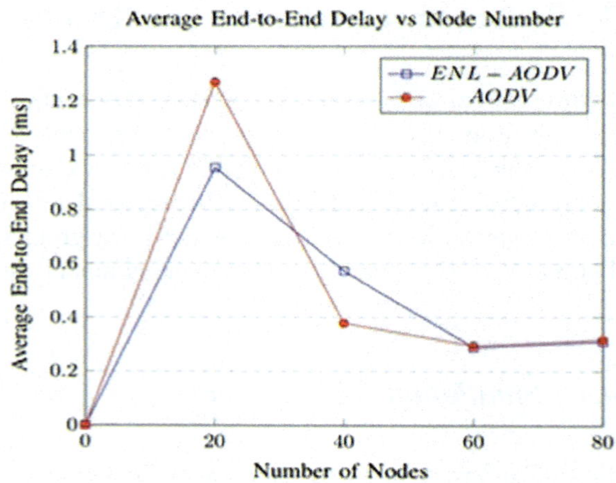

Fig. 2 Average end-to-end delay

performance metrics: packet delivery ratio, average end-to-end delay, throughput, and network lifetime. The simulation results are shown in Figs. 1, 2, 3, and 4.

Packet Delivery Ratio: It is defined as the ratio of data packets delivered to destinations to those originated at sources. Higher packet delivery ratio reflects the better performance of routing protocol. Figure 1 shows simulation results of packet delivery ratio of ENL-AODV and AODV with a different number of nodes. If we compare the average growth rate of both the protocols, the delivery ratio of ENL-AODV is an average of 5.5% higher than AODVs, which clearly implies that ENL-AODV optimizes the network performance.

Average End-to-End Delay: It is the average time taken by the data packet to reach the destination. It also includes delay caused by route discovery process and a queue in data packet transmission. Lower the end-to-end delay reflects the better performance of routing protocol. Figure 2 shows simulation results of average end-to-end delay of ENL-AODV and AODV with a different number of nodes. According to

Fig. 3 Network throughput

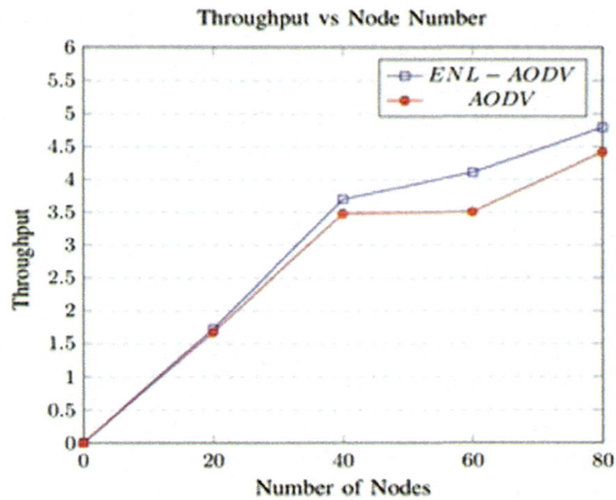

Fig. 4 Network lifetime

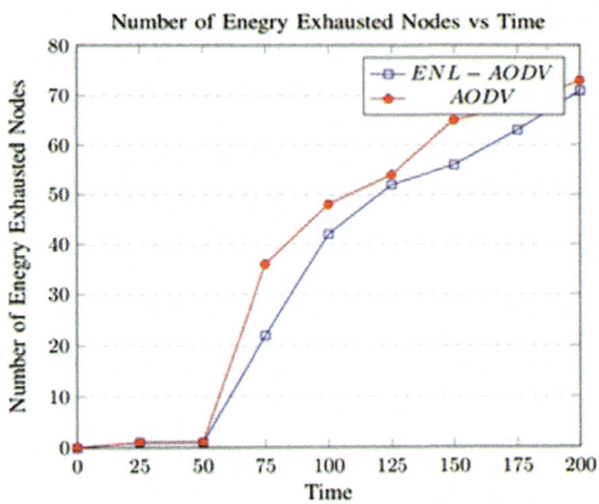

the calculation average reduction in packet delay in the case of ENL-AODV is about 5.8% lower than that of AODV which clearly implies an increase in the performance of the network.

Throughput: The network throughput is the average rate of successful message delivery over a communication channel, i.e., channel utilization, while packets are transmitted between the source and the destinations. Figure 3 shows the simulation results of throughput vs number of nodes. If we compare the average growth rate of both the protocols, the throughput of ENL-AODV is an average of 9.48% higher than AODVs.

Network Lifetime: Network lifetimes are a measure of running time of network, i.e., the number of energy exhausted nodes during simulation. In NS-2, we have carried out simulation over 200 s and number of exhausted nodes at different time are measured. Figure 4 shows the simulation results of the number of energy exhausted nodes under the maximum speed of 5 m/s. It clearly indicates that ENL-AODV performs better and has lesser number of energy exhausted node in comparison with existing AODV, which means ENL-AODV has higher network lifetime.

5 Conclusion

We proposed ENL-AODV routing protocol, which is energy- and load-based routing protocols. ENL-AODV is a modification of AODV protocol with the consideration of residual energy and load level of a node in route discovery process. In mobile ad hoc networks, the energy of participating nodes and nodes capacity to process the number of packets during data transfer plays a critical role in performance of a network. So, we have prototyped ENL-AODV on NS-2 and compared it with existing AODV. To compare both the protocols, we used performance metrics such as packet delivery ratio, throughput, average end-to-end delay, and network lifetime and found our proposed protocol, ENL-AODV performing better than existing AODV routing protocol. As a result, we improved packet delivery ratio, throughput, and average end-to-end delay. Since it does not induce any increase in the control overhead to the nodes and also extends the network lifetime, the proposed protocol works better in the networks with high traffic load.

References

1. Tseng Y-C, Hsieh T-Y (2002) Fully power-aware and location-aware protocols for wireless multi-hop ad hoc networks. In: Eleventh international conference on, computer communications and networks, 2002. Proceedings, pp 608–613. IEEE
2. Kim J-M, Jang J-W (2006) AODV based energy efficient routing protocol for maximum lifetime in manet. In: Advanced int'l conference on telecommunications and int'l conference on internet and web applications and services (AICT-ICIW'06), pp 77–77. IEEE
3. Clausen T, Jacquet P (2003) Optimized link state routing protocol (OLSR). Technical report
4. Perkins C, Belding-Royer E, Das S (2003) Ad hoc on-demand distance vector (AODV) routing. Technical report
5. Khelifa S, Maaza ZM (2010) An energy multi-path AODV routing protocol in ad hoc mobile networks. In: 2010 5th international symposium on, I/V communications and mobile network (ISVC), pp 1–4. IEEE
6. Feng Z, Wang L, Gao X (2013) An improved routing protocol ad-AODV based on AODV. In: International conference on information science and computer applications (ISCA'13). Atlantis Press
7. Yi J, Fuertes JAC, Clausen T (2013) Jitter considerations in on-demand route discovery for mobile ad hoc networks. In: 2013 16th international conference on network-based information systems, pp 170–177. IEEE
8. Clausen T, Dearlove C, Adamson B (2008) Jitter considerations in mobile ad hoc networks (manets). Technical report

Layered Software Defined Networking

Sudarshan Maurya, Naveen Kumar Tiwari and S. C. Gupta

Abstract Over last few decades, computer network has grown exceedingly complex due to advent of newer network protocols and exponential expansion of its size. A new network architecture named software defined networking (SDN) simplifies network by decoupling data plane and control plane. As per SDN, data plane is solely concerned with data forwarding while control plane takes care of logic, processing and management-related functions. But this concept is still evolving; on the other hand, we have witnessed the success of OSI reference architecture and TCP/IP suite. One of the prominent reasons of their success is the layered architecture, where each layer serves its upper layer and uses services of the layer below. On top of that, entire architecture can be viewed as a layered framework allowing proper abstraction and division of responsibilities. Thus, we are proposing seven-layer architectures for software defined networking, viz. physical layer, device link layer, service abstraction layer, worker layer, management layer, presentation layer and application layer. SDN controller consists of three layers, namely service abstraction layer, worker layer and management layer. Device link layer and presentation layer serves as south bound and north bound interfaces, respectively, of the controller. Application layer comprises of applications which are running on the controller, and physical layer consists of physical and/or virtual switches. The key idea of this architecture is to increase cohesion and reduce coupling between the immediate layers.

Keyword Software defined networking · Network architecture
SDN · Controller design

S. Maurya (✉)
Government Mahila Engineering College, Ajmer, Rajasthan, India
e-mail: maurya.sudarshan@gmail.com

N. K. Tiwari
Cognizant Technology Solutions, Noida, Uttar Pradesh, India
e-mail: naveen.tiwari.iitd@gmail.com

S. C. Gupta
IIT Delhi, New Delhi, India
e-mail: scgupta@cse.iitd.ac

V. Janyani et al. (eds.), *Optical and Wireless Technologies*, Lecture Notes
in Electrical Engineering 472, https://doi.org/10.1007/978-981-10-7395-3_40

1 Introduction

As the computer network becomes complex over due course of time, network developers redefine it as software defined network; they separate control and data plane which formerly reside in the forwarding switch. In the data plane, bare metal switches are there which do the basic forwarding of data as per the forwarding rules stored in the forwarding table. If entry is not defined for certain data packet, then that packet is forwarded to controller via south bound APIs.

Control plane is the basic logic processing element also termed as controller, which generates forwarding rules for the commodity switches over the network according to network status and push these forwarding rules to the switches. Control plane can be thought of as the heart of entire SDN architecture. It provides an abstract view of the entire network to application layer, through which they can develop applications controlling data flow. Control plane is capable of tracing and managing network architecture. Communication between data plane and control plane is accompanied according to OpenFlow specification. Controller is directly attached to every switch in the network thus enhancing programmability; any new SDN protocol can be deployed on the switches by simply placing at the controller following series of forwarding rules in terms of packets passed to the switches.

1.1 Need of Generic SDN Architecture

Current family of SDN controllers is going through evolution of multiple controllers using various architectural designs. Some of the popular controllers like beacon [1], floodlight [2] and opendaylight [3] do have some architectural similarities. But there is no detailed standardization of specifications unlike OpenFlow specifications, which is well defined and properly documented [4].

1. **Lacking Standards**: current popular controllers are not designed according to some generic architecture. Each controller has its own architectural design which may differ from others. Therefore, for developing an application or extending a controller module, then one has to design according to specific controller. If there is a common architecture for controllers, then one can achieve controller-independent applications and modules development.
2. **Multiple Vendors**: there are multiple players in the market, supporting and backing their own products. They all followed different design strategies results in different architectures. On the other hand, they do not usually share crucial information among themselves.
3. **More Clarity Required**: SDN is evolving every day; therefore, each vendor might have a different perception about it. As we are considering controller as the heart of the SDN, if all vendors design their controller upon common platform, then they have more clarity and unanimity over SDN architecture.

1.2 Why Layered Design?

There are few points observed from existing OSI and TCP/IP suites which motivate the layered approach:

1. **Abstraction**: layers have some specific responsibility to fulfil, and it has to serve its upper layer; lower layers hide the details of implementation and provide cleaner interface to its upper layer so that they focus on the job rather than the implementation. Similar to application programs running on the operating system doesn't worry about the system call details.
2. **Simplicity**: layered designs are simple to understand and implement; as we seen in the past history of networking, TCP/IP is layered one and still very popular because of its modular simplicity and robustness.
3. **Scalability**: contrary to monolithic model, layered design can be scale out on a different group of servers [5]. As in case of n-tier architecture, a tier can be moved to new servers when load increases [6].

2 Motivation

Layered architectures promise a great deal of flexibility while assigning the responsibilities to individual layers. One of the most powerful concepts of the software engineering works here, i.e. abstraction which simplifies the roles and responsibilities of layers. On top of that, there are huge differences among the SDN vendors who provide controller for software defined networks. Some of them are good for academic research while some of them are suitable for industrial purposes. Layered framework cannot be thought without redesigning the controller in the same way, that is why we have integrated the controller along with our layered design. Two of the layered network architectures that motivated us to design the SDN architecture in a layered manner are as follows:

2.1 OSI

One of the very successful seven-layer networking architectures is ISO-OSI. All the seven layers are well defined and properly abstracted which made it very famous and successful. Each layer refines the data packet transportation from application data to segment and datagrams following frames and signals.

2.2 *TCP/IP*

Similar to OSI model, TCP/IP is also a layered architecture with reduced number of layers. Various layers of the OSI have been combined at various points to form this architecture. Both of the above models exhibit common layered approach to network architecture which is more reliable, modular and robust.

3 Layer Description and SDN Controller Modules

For software defined network, our proposed reference architecture consists of layered design in which each layer serves its upper layer and upper layer directs its lower layer. And layers consist of modules which have definite role to play.

3.1 *Physical Layer*

It is outside of the core controller; it consists of physical switches and/or virtual switches. Whenever a host initiates a flow, then packet first arrives at the physical layer, i.e. switches. If switches have entry in forwarding table for this packet, it is forwarded accordingly otherwise sent to controller.

3.1.1 Physical/Virtual Switch

Bare metal switches may be physical or virtual uniformly supported by OpenFlow standard, just used for forwarding the packets according to their table. If entry is not defined, then default action is to send it to the controller (Figs. 1 and 2).

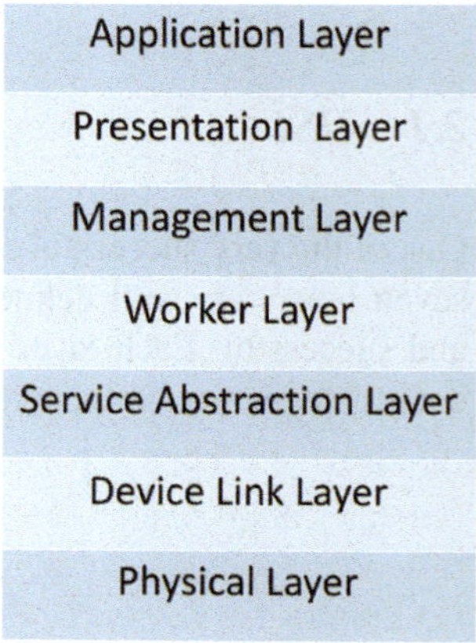

Fig. 1 SDN layered architecture (seven-layer architecture)

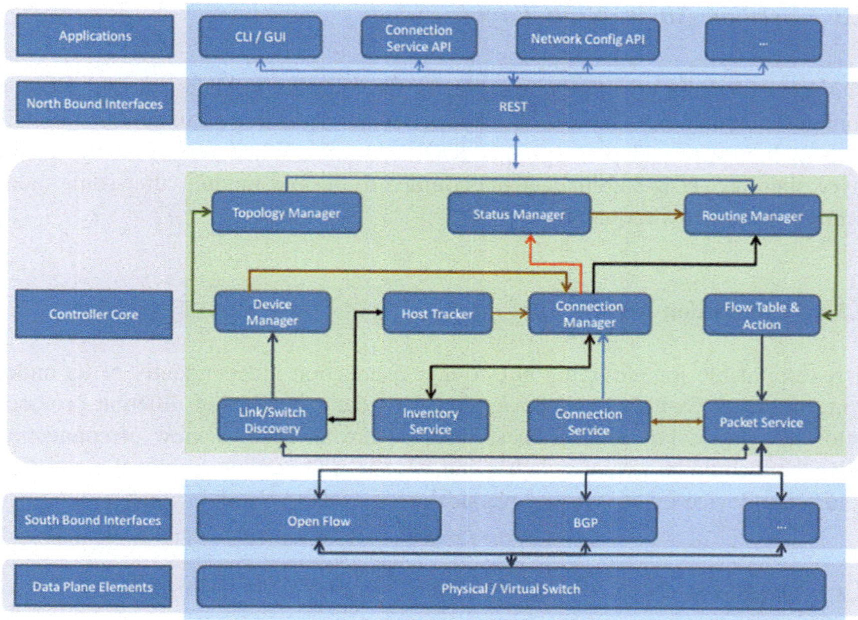

Fig. 2 SDN architecture modules

3.2 *Device Link Layer*

Device link layer links networking devices to the controller, and this is south bound API for connecting switches to controller which is responsible for transmitting actual packets from the controller to network hardware and vice versa. It consists of communication protocols that link switch and controller, e.g. Open Flow. If switches do not have rules installed to forward a packet, then that packet is forwarded to the controller using appropriate protocols and then forwarding rules, generated by the controller, are sent back to switches using this layer. Furthermore, switch configuration and initialization information are passed by this layer.

3.2.1 OpenFlow

It is standard protocol for SDN to communicate between switch and controller.

3.2.2 Other Standards (BGP)

For inter-domain communication, network connection is responsible for sharing this information, but the protocol actually carrying these packets is described here. One that is very common is border gateway protocol.

3.3 Service Abstraction Layer

This layer provides an abstract interface to all controller modules for initialization, configuration and management of network devices by hiding all unnecessary details such as hardware and protocol specification details. When packet arrives at this layer, then packet is automatically forwarded to desired module after some modification in packet if required.

3.3.1 Connection Service

It is responsible for providing hop-to-hop connection independently of its underlying protocol. Switches and devices may be connected using different protocols below this layer, but connection services ensure an abstract view of connection, since there could be virtual switches and VLANs and provide graphical view of the connected device by abstracting physical or virtual LANs and switches.

3.3.2 Packet Service

Provides simple data packet service between switches and controller; all packet sent to controller are handled by this service, and all packet sent to switches are also handled by this service. In other words, we may call it as control packet service which includes flow request messages sent to controller, grant messages from the controller.

3.3.3 Switch/Link Discovery

While connecting to a network, a switch is configured to a master and a set of slave controller IPs. Recent OpenFlow specification allows a switch to connect with multiple controllers. When switch establishes connection to the controller, it sends its features in response to the controller's feature request packet. Links are discovered after switch discovery; controller asks the switches to flood link layer discovery protocol (LLDP) and broadcast domain discovery protocol (BDDP) through all the ports. DPID and port of sender are used by the discovery protocol. LLDP determines the direct links between the switches, whereas BDDP is used to describe the broadcast domain of the switch.

3.3.4 Accounting

Accounting is done with a database tracking all the relevant information of everything happening, viz. traffic pattern, latency, congestion, etc., of links.

3.4 Worker Layer

Worker layer is responsible for end-to-end reliable device connection, configuration, tracking and action. This layer serves as factory worker to serve its managers at higher layers. This layer provides basic functionalities using the services of service layer, which are then used by management layer in complex decision-making. When packet arrives at worker layer, it generates forwarding rules with the help of network status information, topology and routing algorithm, and sends these forwarding rules to its lower layer to be inserted at the switch in forwarding table. It also takes cares of the end-to-end flow control for the initialization, completion and termination of the flow.

3.4.1 Connection Manager

It can pass abstract network graph information to controllers, residing up in the hierarchy, to form global view of network graph from the collage of domain view of each controller. It allows protocol-specific connectivity between hosts. Connections exhibited here are logical and abstract, e.g. TCP may establish a connection for some reliable transfer; UPD may perform quick reply request.

3.4.2 Device Management

Device management is responsible for initialization, configuration and management of devices like host and switch; all network hardware is managed here to provide cleaner interface to its upper layers. All the features provided by the registered devices are stored here like switches information provided by the switch/link discovery module.

3.4.3 Host Tracker

It keeps track of registered devices. Whenever a device is on move, its logical address should not change in order to support continuous communication. That is why it maintains a database of hosts registered to network. Mobile device information is kept in the database along with its varying logical address.

3.4.4 Flow Table and Action

It has a set of rules, which describe action to be taken whenever a new packet arrives. Packets are forwarded or discarded according to rules dictated by the routing. Default action is to accept by the controller when no entry is present. Whenever a flow path is established by the controller, an entry is made in this table and kept until flow timeout.

3.5 Management Layer

Management layer is responsible for managing network connection, detecting and configuring topology, network status and routing management. This is the most abstract layer in the stack; it handles high-order decision-making while using lower layers as subroutines written to fulfil larger goals. This layer receives network information from its lower layers and generates a network map and overall network status. This network map and status can be shared with its peer and/or parent controller. It also provides direction to its lower layer to generate forwarding rules according to network map and status.

3.5.1 Topology Manager

It is responsible for creating abstract network graph based on the information sent by the switch and link manager services. It also maintains various properties associated with the links and switches such as bandwidth, queue length to detect average waiting.

3.5.2 Status Manager

It is responsible for keeping track of the network such as congestion over certain network path which can be avoided by redirecting some traffic from other path using status information. Status information exposed here can be used in network learning to path selection of important data and transactions.

3.5.3 Routing Manager

It is responsible for deciding route for packets as per the network configuration using topological architecture and status of the network. Rules are formulated here to forward, store and discard packets. It maintains an abstract table here which is formulated globally in the network domain based on information collected by forwarding table of individual switches.

3.6 Presentation Layer

Presentation layer is responsible to provide interface for application to connect and use controller services and configure controller to manage network. It also acts as a medium to transfer the information to/from controller like CLI/GUI application that provides interface to access controller via command line or graphical user interface to view network status and push commands to the controller.

3.6.1 North Bound Interface

This provides APIs to applications for communicating with the controller. These APIs can be used to innovate and facilitate efficient orchestration and automation of network to align with needs of different applications via SDN programmability.

3.7 Application Layer

Application layer provide GUI/CLI interface to user and custom application for user to configure and setup network and its devices using controller as per the users requirement.

3.7.1 Connection Service API

This is responsible for categorizing the network path, viz. highways, sideways and Dedicatedways. Highways refers to high-bandwidth links which can be used to transmit huge data over a long range; sideways are low-bandwidth links used to transmit data locally in order to avoid high traffic density at the highways, whereas DedicatedWays are the dedicated links allotted to two or more parties to provide high *QOS* with minimum possible latencies. DedicatedWays can serve crucial importance in banking transaction where each millisecond is critical.

3.7.2 Network Configuration API

This is responsible for overall network configuration and provides support to any SDN-based network service for its installation, configuration and execution and is also responsible for overall virtual network configuration and provides support to any SDN-based virtual network service for its installation, configuration and execution. It is also responsible for service mapping between the physical and virtual networks.

3.7.3 CLI/GUI

This layer is also responsible for providing command or graphical user interface to controller to manage, e.g. start or stop controller, installation of application, etc.

4 Inter-Module Communication Algorithms

There are two prominent use cases encountered frequently, i.e. link/switch discovery and route discovery. Excerpts of the actual algorithms are summarized as follows:

Algorithm 1 Link-Switch Discovery

```
function SwitchHello () At Switch i
    ∀i, sendToController(pkt(i, OFPF_HELLO))
    waitForRegistration() // for the switch to be registered
    ∀i, ∀j, i ∈ Switch(), j ∈ Port(i), sendToSwitch(pkt(i, j))
function PacketService (pkt(i,j))
    pktType ← getPacketType(pkt)
    switch pktType do
        case OFPF_HELLO
            sendToSwitch(pkt(i, ALL, LLDP)) //Discover neighbours via LLDP packets
            DeviceManager(ADD_SWITCH, pkt)
        case LLDP
            DeviceManager(ADD_LINK, pkt)
function DeviceManager (ActionType, source, destination)
    pktType ← getPacketType(pkt)
    switch pktType do
        case ADD_SWITCH
            deviceID ← getDeviceID(pkt)
            registerDeivece(deviceID)
            topologyManager(ADD_SWITCH, deviceID)
        case ADD_LINK
            source ← getsourceMAC(pkt)
            destination ← getDestinationMAC(pkt)
            DeviceManager(ADD_LINK, sourceID,
                                  destinationID)
```

4.1 Link/Switch Discovery

Every switch, after induction to the network, sends a hello message to the controller which in turn received by the PacketService. After identifying the packet type, PacketService asks DeviceManager to add this switch and directs the switches to find neighbouring switches. In this way, link information is also passed to the PacketService and eventually to the DeviceManager.

4.2 Route Discovery

Algorithm 2 Route Discovery

```
function SwitchForward () At Switch k
    ∃i, ∃jFlow(pkt(i, j)) // a flow from i → j
    if Entry[Flow(i, j)] ∈ RoutingTable(k) then
        sendToSwitch(pkt(k, i, NextHop[Flow(i, j)], DATA))
    else
        sendToController(pkt(k, Flow(i, j), ROUTE_REQ))
function PacketService (pkt(i,j))
    pktType ← getPacketType(pkt)
    switch pktType do
        case ROUTE_REQ
            pendingRequests ← pendingRequests ∪ Flow(i, j)
//put the request in a queue
            ConnectionService(pkt)
        case FORWARD_RULES //arrived from FlowTabel
            if ForwardingRules_{i,j,k} ∈ pendingRequest_{i,j,k} AND QueuedResponse_{i,j,k} AND RoutePossible_{i,j,k} then
                sendToSwitch(pkt(k, ForwardingRules_{i,j,k}, DATA))
        case ROUTE_POSSIBLE
            if RoutePossible_{i,j,k} then
                RoutePossible_{i,j,k} ← true
function ConnectionService(pkt) (
    ∃i, ∃j, ∃k, i ∈ source[Flow(i, j)]
                j ∈ destination[Flow(i, j)]
                k ∈ current[Flow(i, j)]
    Route_Possible ← ConnetionManager(i, j, k, RoutingParameters)
    PacketService(Route_Possible)
function ConnetionManager (source, dest, current, RoutingParameters)
    path ← RoutingManager(source, dest, current, RoutingParameters)
    if path = ∅ then
        ConnectionService(path)
    else
        InventoryService(path)
        StatusManager(path)
        ConnectionService(path)
function RoutingManager (source, dest, current, RoutingParameters)
    topology ← getNetworkTopolgy()
    nexthop ← RoutingAlgorith(source, dest, current, getStatistics(), RoutingParameters)
    ForwardingRules_{source,dest,current} ← createNewRule(nexthop, getStatistics())
    {FlowTable, Action} ← {FlowTable, Action} ∪ ForwardingRules_{source,dest,current}
    FlowTableAction(ForwardingRules_{source,dest,current})
function FlowTableAction(ForwardingRules)
    store(ForwardingRules)
    PacketService(pkt(source, dest, current))
```

Similar to link discovery, route discovery involves sending a route request packet to the controller. PacketService, after identifying the packet type, create a pending request queue and inform the ConnectionService. Then, ConnectionService asks the ConnectionManager whether a path exists with given parameters and promulgate the same to PacketService. ConnectionManager then asks the

RoutingManager about the path with given parameters and creates respective entries to InventoryService and StatusManager if found one. On the other end, RoutingManager finds the next hop after consulting with other management layer modules and creates appropriate entry in the forwarding table and asks the PacketService to packetize and sends it back to the switch.

5 Conclusions

This design evolved from the study of some existing popular controllers and proposed design gives us a basic fundamental framework for software defined networking architecture. The proposed architecture consists of layers with basic essential modules to deliver proper distribution of roles and responsibilities. Each layer has specific set of primitives to be implemented, evolve from proper abstraction and modularity which is very similar to OSI and TCP/IP evolution methology. Although, proposed architecture looks clean at the moment, but implementation may differ.

References

1. Erickson D (2013) The beacon OpenFlow controller, August 16
2. http://docs.projectfloodlight.org/
3. http://www.opendaylight.org/
4. Khondoker R, Zaalouk A, Marx R, Bayarou K Feature-based comparison and selection of software defined networking (SDN) controllers
5. Yasunobu C, Kazushi S OpenFlow controller architecture for largescale SDN networks
6. Dixit A, Hao F, Mukherjee S, Lakshman TV, Kompella R (2013) Towards an elastic distributed SDN controller, sigcomm
7. https://www.opennetworking.org/
8. http://h17007.www1.hp.com/docs/networking/solutions/sdn/devcenter/06z_-_HP_SDN_Controller_Architecture_TSG_v1_3013-10-01.pdf
9. Das S, Parulkar G, McKeown N SDN based unified control architecture
10. de Tejada Muntaner GR (2012) Evaluation of OpenFlow controllers, October 15
11. Kepes B (2014) SDN meets the real-world: implementation benefits and challenges
12. Lin G (2012) Industry perspectives of SDN: technical challenges and business use cases
13. Monaco M, Michel O, Keller E (2013) Applying operating system principles to SDN controller design
14. Mogul JC, AuYoung A, Banerjee S, Popa L, Lee J, Mudigonda J, Sharma P, Turner Y (2013) Corybantic: towards the modular composition of SDN control programs, September

Resource Allocation at MAC to Provide QoS for Cognitive Radio Networks

Kalpana Naidu, Meenakshi Tripathi and Ramesh Babu Battula

Abstract Cognitive radio (CR) improves spectrum efficiency by using the sub-carriers that are unused by the primary users. Because of doppler shift, secondary user (SU) sub-carriers create interference to the primary user (PU) sub-carriers. The aim is to choose the sub-carriers for SU and also allot powers to the chosen sub-carriers in order to minimize the transmission delay. At the same time, SU transmitter has to conserve its battery power, and the interference induced to the PU should be within the endurable limits. Sub-carrier allocation and power allocation are done using the sub-gradient iterative algorithm. Simulation results indicate that as the number of sub-carriers assigned to the SU increases, transmission delay is reduced.

Keywords Cognitive radio · Delay · Latency · MAC · Resource allocation OFDM

1 Introduction

Cognitive radio (CR) senses and then discerns from the surrounding environment adaptively that licensed primary users (PU) are not using some of the frequency bands. CR uses these unutilized frequency bands to transmit its own data so that spectrum utilization is improved [1–9].

Orthogonal frequency division multiplexing (OFDM) is the appropriate method to be used by CR. This is in view of the fact that (i) SU (Secondary User or CR user) can easily use sub-carriers that are not used by PU, (ii) SU can have the provision to change its sub-carriers adaptively as per the usage of the sub-carriers by the PU,

K. Naidu (✉)
Department of ECE, Indian Institute of Information Technology-Kota, Kota, India
e-mail: kalpana@iiitkota.ac.in

M. Tripathi · R. Battula
Department of CSE, Malaviya National Institute of Technology, Jaipur, India
e-mail: mtripathi.cse@mnit.ac.in

R. Battula
e-mail: rbbattula.cse@mnit.ac.in

V. Janyani et al. (eds.), *Optical and Wireless Technologies*, Lecture Notes in Electrical Engineering 472, https://doi.org/10.1007/978-981-10-7395-3_41

(iii) The current technologies LTE and Wi-fi use OFDM. This reckons the CR to sense easily the sub-carriers used by the PU.

Even though SU detects that there are many sub-carriers available for its use, SU cannot use all those unused sub-carriers as many of the sub-carriers may not have enough channel-gain to transit data from SU transmitter to SU receiver. For this reason, SU transmitter has to use only the sub-carriers that can maximize the capacity (or amount) of transmitted data.

Besides, when SU transmitter transmits its data, side-lobes of SU sub-carriers may create interference to the PU sub-carriers because of the frequency discrepancy or timing discrepancy occurring as the result of doppler shifts [3, 4, 6]. But, if SU transmitter picks out the sub-carriers that are having the best channel-gains in order to maximize the capacity of its data transfer, then those sub-carriers induce more interference to the PUs. If the interference created to PU exceeds the PU's bearable levels, then PU's data will be distorted which is not at all suggestible since the PU is licensed user.

Hence, the trade-off comes in between the SU selecting the sub-carriers and the interference induced to the PU. Consequently, SU has to select the sub-carriers which will optimize the amount of SU's data transfer and at the same time produces only the tolerable interference to PU.

Additionally, SU cannot allocate as much power as it wants to the chosen sub-carriers since battery power of SU is limited. Thus, SU follows power budget while doing power allocation to the selected sub-carriers.

Moreover, SU transmitter sends either speech or video signal or data packets to the receiver SU. According to the type of data it transmits, SU transmitter has to take care of the latencies. For instance, speech signal should have very small transmission delay whereas video signal can have more delay when compared to the latency of speech signal. Data packets can experience higher delay but should reach the receiver SU with more accuracy [10].

This paper addresses all the above issues while doing resource allocation (i.e., channel allocation and power allocation). This paper further gives the amount of transmission delay produced for the certain values of PU tolerable interference levels. Besides, this paper provides the relation between the capacities of the sub-carriers and the transmission delays. In addition, this paper indicates the number of sub-carriers to be chosen in order to provide a particular transmission delay.

2 Existing Work

In [11], quality of service (QOS) related to the application layer (i.e., multimedia intra-refreshing rate) is optimized by using markovian modeling. But, medium access control (MAC) layer-related QOS parameters like latency are not considered in [11].

The authors of [12] have proposed a method for carrying out the resource allocation to do the video streaming in LTE network. Here, resource allocation is done depending on the signal to noise ratio (SNR), packet delay threshold, and the number of packets that are there in the buffer waiting for transmission. But, it is presumed here that all the LTE users proclaim the downlink SNR estimates to the base station after every 0.5 msec. Feeding back SNR values to the base station may consume time. Also, speech packets are transmitted as voice over internet protocol (VOIP) packets in LTE. If the network is busy, then sufficient resources may not be allocated to the speech packets which increases the transmission delay of speech packets.

In [13], SU transmitter communicates with the SU receiver through a relay node. The packets should reach the CR receiver within the specified time delay while maximizing the capacity of transmitted packets. Along with the delay QoS constraint, another constraint on the amount of interference that can go to the PU from the CR transmissions is taken. But, [13] does not consider preserving the battery power of CR.

All the drawbacks of the above erstwhile algorithms are subdued in this paper.

3 System Model

Both Primary User (PU) and Secondary user (SU) are presumed to be employing OFDM. When PU is not using the sub-carriers of OFDM, source SU makes use of those vacant sub-carriers and transmits its own packets.

B is the bandwidth of each sub-carrier. Further, the bandwidth of sub-carrier (or sub-carrier spacing = Δ f) is set to 15,000 Hz (as per LTE systems). If P_i is the power allocated to the ith sub-carrier, then number of bits transmitted on the ith sub-carrier is [3]:

$$C_i = B\log_2(1 + \frac{P_i h_i}{\sigma^2}) \text{ bits/sec} \tag{1}$$

where $N_i = \frac{\sigma^2}{h_i}$ is the noise level of ith sub-carrier with σ^2 representing the additive white Gaussian noise (AWGN) variance and h_i stands for the channel gain of ith sub-carrier from SU transmitter to the SU receiver.

Transmitted signal from SU transmitter can have the maximum delay of T before the signal reaches SU receiver. In this paper, resource allocation is done to keep the transmission delay with in the maximum allowable delay (T) limit.

Out of the total N sub-carriers that are there in OFDM, PU uses P sub-carriers. In the remaining sub-carriers, SU chooses the best S sub-carriers to get the transmission delay less than T.

Number of bits transmitted in a second for the transmitted signal over all the S sub-carriers is $\sum_{i=1}^{S} C_i$ bits/sec. If the transmitted signal has a total of B_S bits to be transmitted, then time taken to transmit all these B_S bits is $\frac{B_S}{\sum_{i=1}^{S} C_i}$ seconds. From (1), capacity (C_i) increases as the power allocated to the sub-carrier increases. Once capacity (total number of transmitted bits) becomes more, total delay to transmit the signal is reduced.

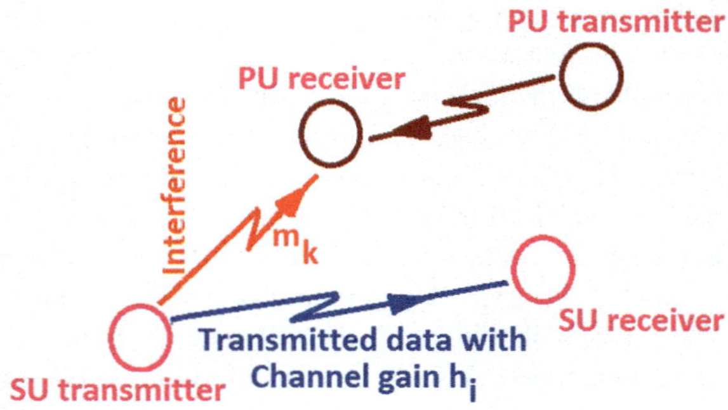

Fig. 1 Interference created from SU transmitter to PU receiver

But if the power assigned to the sub-carrier increases, then interference to the PU sub-carriers will be increased (because of the side-lobes of sub-carriers used by secondary users, interference comes to the PU sub-carriers in OFDM as shown in Fig. 1) [14, 15].

So, we cannot increase the powers of the sub-carriers as we wish. But, as the power assigned to the sub-carrier reduces, capacity is reduced. Total delay to transmit a signal is increased if capacity is reduced.

Similarly, capacity also increases if the number of sub-carriers increases. But, we may not be able to allocate more number of sub-carriers since PU may claim the sub-carriers at any time.

Hence, there is conflict between reducing the transmission delay and reducing the interference to the PU by doing the channel (or sub-carriers) allocation and power allocation.

m_k is the channel gain of kth sub-carrier from SU transmitter to PU receiver. Then, the interference created from the ith sub-carrier of the SU transmitter to the kth sub-carrier of the PU receiver is [15],

$$R_{ki} = \int_{k\text{th}\, sub-carrier} T_s sinc^2\{(f - f_i)T_s\} \cdot df \tag{2}$$

where f_i represents the ith sub-carrier's center frequency and $T_s = \frac{1}{\Delta f}$.

Consequently, the interference induced from SU transmitter to the PU receiver is [4, 15]

$$V = \sum_{k=1}^{P} m_k \sum_{i=1}^{S} P_i R_{ki} \tag{3}$$

Equation (3) can be rewritten as:

$$V = \sum_{i=1}^{S} v_i P_i \tag{4}$$

with v_i taken as:

$$v_i = \sum_{k=1}^{P} m_k R_{ki} \tag{5}$$

The problem is allocating the sub-carriers and then optimal powers to the SU chosen sub-carriers in such a way that transmission delay of the signal is within the limits, i.e., $\left(\frac{B_S}{\sum_{i=1}^{S} C_i} \leq T\right)$. Further, SU transmitter's battery power is conserved by limiting the total power assigned to all the SU sub-carriers to P_B. Besides, the interference induced to the PU receiver should not exceed its endurable interference-level (V_L).

Mathematically, the optimization problem is written as:

$$\underset{\{P_i, h_i\}}{Min} \left(T - \frac{B_S}{\sum_{i=1}^{S} C_i}\right) \tag{6}$$

subject to

$$\sum_{i=1}^{S} P_i \leq P_B \tag{7}$$

$$and \qquad V \triangleq \sum_{i=1}^{S} v_i P_i \leq V_L \tag{8}$$

Here, C_i is taken from (1).

After re-arranging the Eq. (6), the optimization problem in (6–8) can be rewritten as:

$$\underset{\{P_i, h_i\}}{Max} \left(\frac{T}{B_S} \sum_{i=1}^{S} C_i - 1\right) \tag{9}$$

subject to

$$\sum_{i=1}^{S} P_i \leq P_B \tag{10}$$

$$and \qquad V \leq V_L \tag{11}$$

Channel allocation and power allocation are done to (9–11) as given below.

4 Proposed Solution

The Lagrangian equation for (9–11) is taken as:

$$L(P_i, \lambda, \mu) = \frac{T}{B_S} \sum_{i=1}^{S} B \log_2 \left(1 + \frac{P_i h_i}{\sigma^2}\right) - 1$$
$$-\lambda\left(\sum_{i=1}^{S} P_i - P_B\right) - \mu\left(\sum_{i=1}^{S} v_i P_i - V_L\right) \tag{12}$$

Initially, Karush–Kuhn–Tucker (KKT) conditions are applied to the above lagrangian equation to get the equation for powers. Then, sub-gradient iterative water-filling algorithm [15] can be used to do the sub-carriers allocation and also to do the optimum power allocation for the chosen sub-carriers.

Differentiating the lagrangian equation w.r.t. P_i and equating the equation to zero, we get

$$P_i = \frac{T}{B_S}\frac{1}{\lambda + \mu v_i} - \frac{\sigma^2}{h_i}, i = 1\ to\ S. \tag{13}$$

To determine the number of sub-carriers to be allocated to the SU and also to do the power allocation to the selected sub-carrier, following Algorithm 1 is used.

Algorithm 1 does the sub-carrier[1] allocation and power allocation as follows: Sub-carriers with highest channel gains (h_i) are considered in order to choose the better sub-carriers out of all the available sub-carriers. For the chosen sub-carriers, powers are calculated with the initialized lagrange multipliers (λ and μ are initialized in such a way that $\sum_{i=1}^{S} P_i \le P_B$ and $V \le V_L$ restraints are not satisfied). For the computed powers, it is checked if the power budget and interference constraints are fulfilled. If not, then λ and μ values are upgraded with the step sizes $\frac{1}{P}$ and $\frac{1}{S}$, respectively. This updation continues until all the constraints are satisfied. With the obtained sub-carrier powers, capacity is calculated. If the capacity is not enough to get the transmission delay less than the time limit (T), then more number of sub-carriers are taken, and the procedure is repeated until transmission delay becomes less than T.

5 Simulation results

Simulations are done in MATLAB software. Using rayleigh fading, channel gains h_i, i $\le$ 500, and m_k, k $\le P$ are obtained. Values given in Table 1 are used while doing simulations.

---

[1]Both λ and μ are initialized to 10^6.

Algorithm 1 Sub-carrier allocation and power allocation

Require: Inputs required are: Total number of available sub-carriers N; PU sub-carriers P; channel gains h_i, i $\leq N$; σ^2, Required Transmission delay T, sub-carrier spacing B, number of bits to be transmitted B_S

Ensure: Outputs are number of SU sub-carriers S, and allocated powers P_i, i $\leq S$.

1: Take number of sub-carriers $S = 1$ initially. i $= S$ and $g = [h_j]$, j $\leq N$.
2: Take $u_i = \max\{g\}$ as the highest of all the channel gains in order to select the best sub-carrier.
3: Remaining sub-carriers $g = g - u_i$.
4: Initialize powers $P_j = 0$, j $\leq S$ and Capacities $C_j = 0$, j $\leq S$.
5: **while** ($\frac{T}{B_S} \sum_{j=1}^{S} C_j$ - 1) ≤ 0 **do**
6:     $S = S + 1$ and i = i + 1. Do the steps 2–4.
7:     Initialize λ & μ values.
8:     Calculate R_{kj}; k $\leq P$, j $\leq S$ values for the selected S sub-carriers as per (2).
9:     Compute v_j, j $\leq S$ using (5).
10:     Calculate powers of sub-carriers using $P_j = \frac{T}{B_S} \frac{1}{\lambda + \mu v_j} - \frac{\sigma^2}{u_j}$, j $\leq S$.
11:     Calculate the interference to the PU using: $V = \sum_{j=1}^{S} v_j \cdot P_j$.
12:     Take difference between the PU interference and the PU tolerable interference level V_L as $D_I = V - V_L$.
13:     Also, take the difference between sum of allocated powers and the power budget as $D_p = \sum_{j=1}^{S} P_j - P_B$.
14:     **while** ($D_p > 0$) or ($D_I > 0$) **do**
15:         **if** ($D_p > 0$) **then**
16:             $\lambda = \lambda + \frac{D_p}{S}$.
17:         **end if**
18:         **if** ($D_I > 0$) **then**
19:             $\mu = \mu + \frac{D_I}{P}$.
20:         **end if**
21:         Calculate P_i values for the updated λ and μ values.
            Do the steps 12–13.
22:     **end while**
23:     calculate capacities $C_j = B \log_2(1 + \frac{P_j u_j}{\sigma^2})$, j $\leq S$.
24: **end while**

Table 1 Parameters used for the simulations

Parameters	Values
Power budget (P_B)	1 W
PU's enumerable interference level (V_L)	0.1 W
σ^2	0.01
Sub-carrier spacing (Δf)	15,000 Hz
Carrier frequency (f_C)	2412 MHz
SU sub-carrier frequencies (f_i) are	$f_i = f_C \pm n\Delta f$, n = 1, 2, …

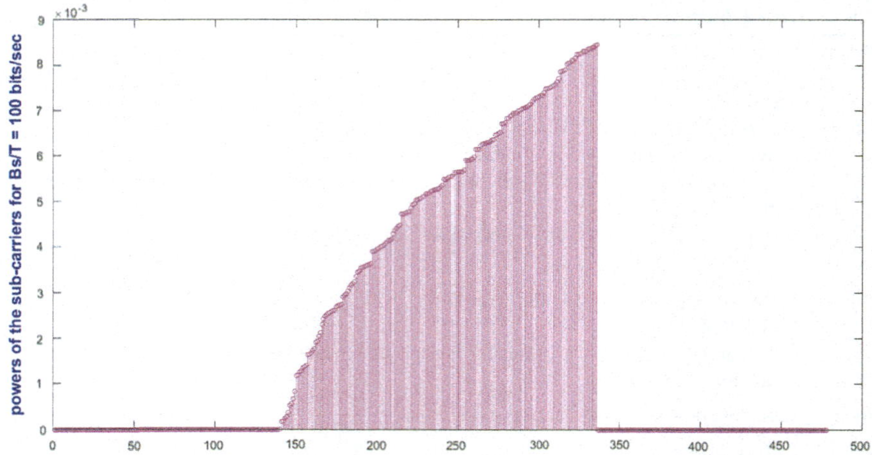

Fig. 2 SU sub-carrier powers obtained for $\frac{B_S}{T} = 100$ bps with $S = 197$ & $P = 2$

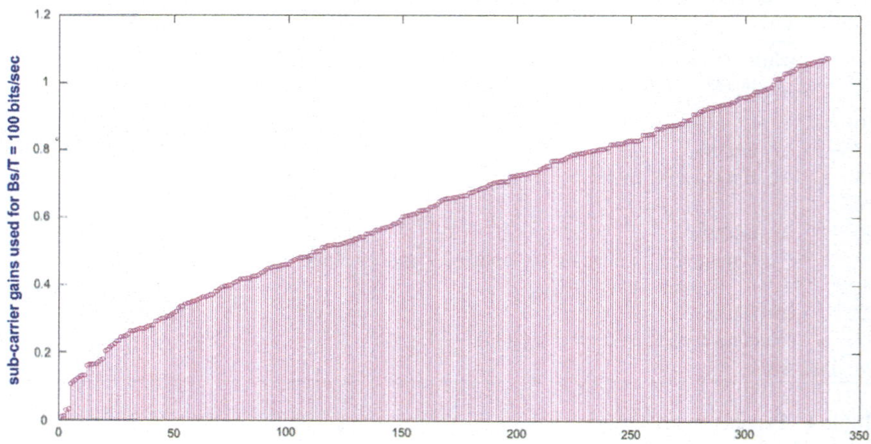

Fig. 3 Channel gains for $\frac{B_S}{T} = 100$ bps with $S = 197$ & $P = 2$

When SU transmitter takes up 197 sub-carriers and PU transmitter utilizes two sub-carriers ($S = 197$ & $P = 2$), power allocation done for the SU sub-carriers is shown in Fig. 2 for $\frac{B_S}{T} = 100$ bps. Channel gains (h_i) for $S = 197$ SU sub-carriers and $\frac{B_S}{T} = 100$ bps are shown in Fig. 3.

Similarly, Fig. 4 presents the power allocation done for SU sub-carriers for $S = 264$ and $P = 2$ with $\frac{B_S}{T} = 140$ bps. Figure 5 shows the SU sub-carrier channel gains (h_i) for $S = 264$ and $P = 2$ with $\frac{B_S}{T} = 140$ bps.

From Figs. 2, 3, 4, and 5, we can tell that powers of SU sub-carriers are proportional to the SU sub-carrier channel gains. Also, from Figs. 3 and 5, we can tell that

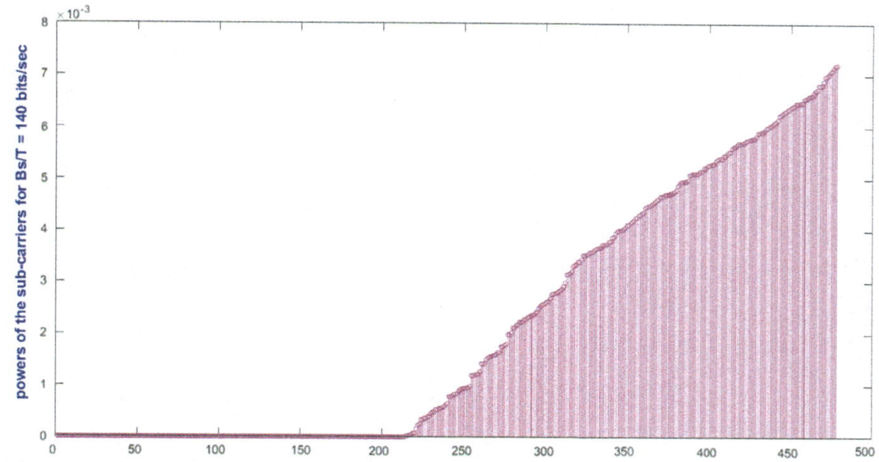

Fig. 4 SU sub-carrier powers obtained for $\frac{B_S}{T} = 140$ bps with $S = 264$ & $P = 2$

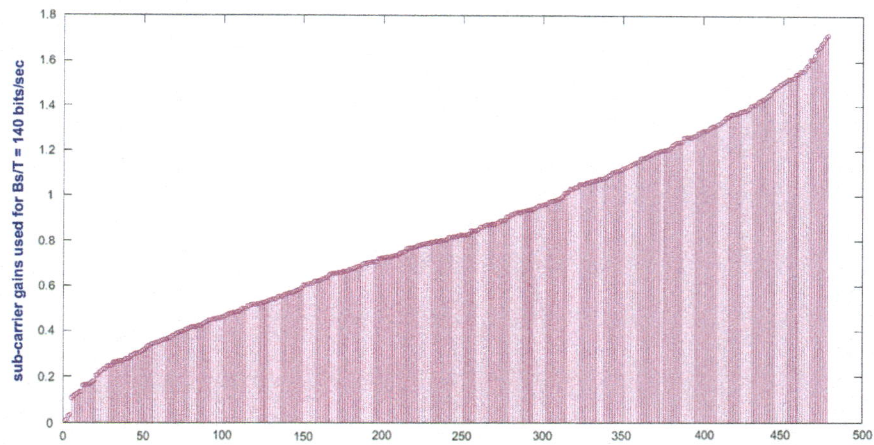

Fig. 5 Channel gains for $\frac{B_S}{T} = 140$ bps with $S = 264$ & $P = 2$

as $\frac{B_S}{T}$ increases, number of SU sub-carriers S increases. This is implied from the fact that as $\frac{B_S}{T}$ increases, time delay limit (T) reduces. That is, transmission should be done at a faster rate to meet the lesser time delay limit. Here, capacity should be increased. To increase the capacity, number of SU sub-carriers should be more, i.e., number of sub-carriers to be assigned to the SU is inversely proportional to the time delay limit (T).

Capacity versus $\frac{B_S}{T}$ is shown in Fig. 6. This figure also implies that as $\frac{B_S}{T}$ increases or as time delay limit (T) reduces, more number of sub-carriers are to be allotted in order to transmit more amount of data or to increase the capacity of SU transmitter.

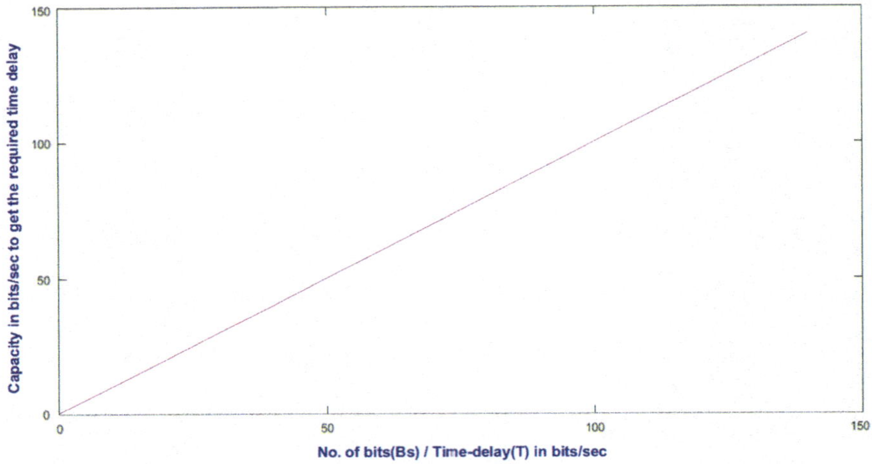

Fig. 6 Capacity for different $\frac{B_S}{T}$ values

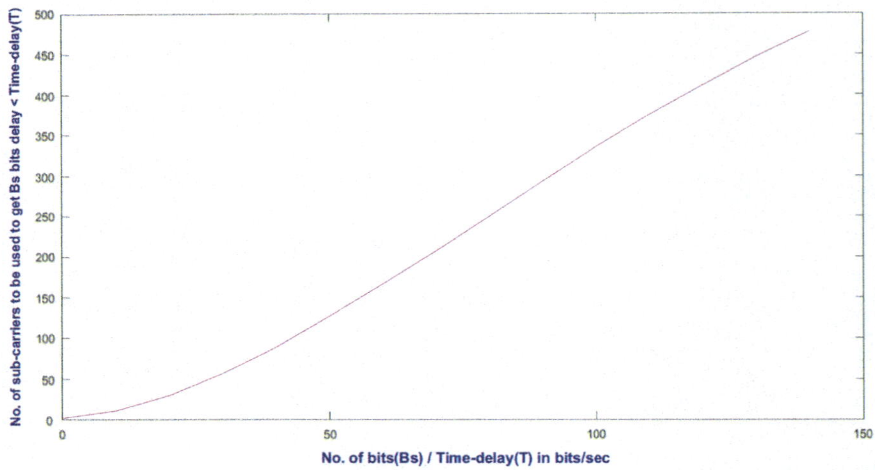

Fig. 7 Number of sub-carriers required for different $\frac{B_S}{T}$ values

Figure 7 shows the number of sub-carriers required to meet the different time delay limits (T). From this figure, we can observe that if T is less, then within lesser time, entire data is to be transmitted. Number of sub-carriers required becomes more if the entire data is to be transmitted with in lesser amount of time.

Figure 8 shows the amount of interference to the PU for different time delay limits (T). From this figure, we can observe that as the transmission time is coming close to T, required time to transmit B_S bits becomes more. As the required time limit increases, the number of sub-carriers to be allotted to transmit B_S bits is reduced.

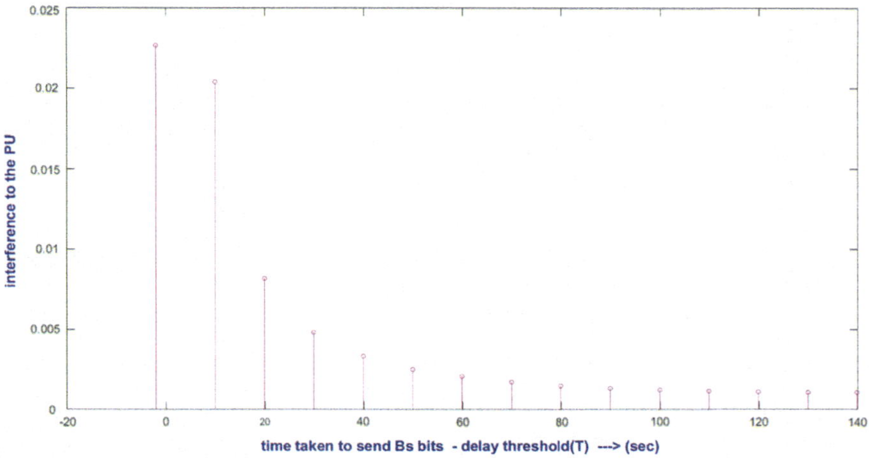

Fig. 8 Amount of interference to the PU for different $\frac{B_S}{T}$ values

If SU transmitter uses lesser number of sub-carriers, then interference to the PU becomes less.

6 Conclusion

Out of the available sub-carriers with different channel-gains, which sub-carriers to be allocated to SU is found out. Also, power allocation to the chosen sub-carriers is calculated. We have observed that sub-carriers which are having more channel-gains are allocated with more powers to reduce the transmission delay of the signal. As the required delay limit reduces, more number of sub-carriers are allocated and interference to the PU increases.

References

1. Naidu K, Khan MZA, Desai UB et al (2014) A study on white and gray spaces in India. In: Mishra AK, Johnson DF (eds) White space communication: advances, developments and engineering challenges, pp 49–73
2. Naidu K, Khan MZA (2016) A fast algorithm for solving cave-filling problems. In: Proceedings of IEEE 84th vehicular technology conference: VTC 2016-Fall, Montreal, Canada, 18–21 Sep 2016
3. Naidu K, Khan MZA, Hanzo L (2016) An efficient direct solution of cave-filling problems. IEEE Trans Commun 64(7):3064–3077
4. Kalpana N, Khan MZA, Desai UB (2011) Optimal Power allocation for secondary users in CR networks. In: Proceedings of IEEE ANTS 2011, Bangalore, India, Dec 2011

5. Naidu K (2016) Fast computation of waterfilling algorithms. https://www.raiith.iith.ac.in/2604/
6. Naidu K (2015) Fast computation of generalized water-filling problems. IEEE Signal Process Lett 22(11):1884–1887
7. Naidu K, Kumar R, Vikas V (2017) The fastest possible solution to the weighted water-filling problems. In: Proceedings of 7th IEEE IACC-2017, Hyderabad, India, 5–7 Jan 2017
8. Naidu K, Battula RB (2016) Quick resource allocation in heterogeneous networks. In: Communicated to 'Wireless Networks', Springer Journal, submission id: WINE-D-16-01202, Dec 2016
9. Kalpana N, Khan MZA (2015) Weighted water-filling algorithm with reduced computational complexity. In: Proceedings of ICCIT Conference 2015, May 2015
10. Wikimedia (2011) Latency (engineering). https://www.en.wikipedia.org/wiki/Latency_(engineering)
11. Yu FR, Sun B, Krishnamurthy V, Ali S (2011) Application layer qos optimization for multimedia transmission over cognitive radio networks. Wireless networks. Springer, vol 17, p 371383, Feb 2011
12. Ramli HAM, Sandrasegaran K, Basukala R, Patachaianand R, Xue M, Lin CC (2010) Resource allocation technique for video streaming applications in the lte system. In: Proceedings of wireless and optical communications conference (WOCC), 2010 19th Annual, May 2010, pp 1–5
13. Musavian L, Assa S, Lambotharan S (2010) Effective capacity for interference and delay constrained cognitive radio relay channels. IEEE Trans Wirel Commun 9(5):1698–1707
14. Mahmoud HA, Yucek T, Arslan H (2009) OFDM for Cognitive radio: merits and challenges. IEEE Trans Wirel Commun 16(2):6–15
15. Thumar V, Tej GPS, Nadkar T, Merchant SN, Desai UB (2010) Adaptive Power allocation for secondary users in Cognitive Radio Networks. In: Proceedings of 2010 second UK-India-IDRC international workshop on cognitive wireless systems (UKIWCWS 2010)

Performance Improvement in Inter-satellite Optical Wireless Communication to Mitigate Losses Using Diversity Technique

Guddan Kumari and Chetan Selwal

Abstract The superior network quality with highest speed and lowest cost is the requirement of the next-generation wireless communication system. Many technologies came into existence, which we have experienced in various applications. Establishment of network between satellites using optical wireless communication is a promising solution to achieve successful communication goal. In this paper, we have designed and simulated inter-satellite optical link using diversity technique to achieve the bit-rate of 30 Gbps for the pointing error of 1.1 μrad and additional losses of 3 dB. The proposed system is analyzed on the basis of performance parameters such as quality factor, BER, and eye diagram obtained. The proposed system will be advantageous to fulfill the needs of next-generation system.

Keywords Is-OWC · Diversity · BER · Line-of-sight · Pointing, acquisition, and tracking (PAT) · Q factor

1 Introduction

The contemporary spreading of cellular system has raised the complexity in processing algorithm and forced the current technologies to its curb. Due to spectrum congestion and complexity faced in terms of bandwidth, modern era has adapted an advantageous technology that is optical wireless communication [1]. This technology has gained significant interest due to its unique features such as unlicensed spectrum, large bandwidth, high security, low transmission power consumption, easily deployable, and high data rate [2]. Inter-satellite linking plays an important role in global coverage network which is the captivating target of the

G. Kumari (✉) · C. Selwal
Department of Electronics and Communication Engineering,
Government Women Engineering College, Ajmer, India
e-mail: guddankumari2008@gmail.com

C. Selwal
e-mail: selwal.chetan@gmail.com

V. Janyani et al. (eds.), *Optical and Wireless Technologies*, Lecture Notes in Electrical Engineering 472, https://doi.org/10.1007/978-981-10-7395-3_42

next-generation wireless communication. The performance of OWC depends on the link configuration. The line-of-sight link is the best option for inter-satellite communication in terms of high speed and low bit-error-rate performance [3].

Over several advantages, line-of-sight link suffers from some disadvantages such as pointing, acquisition, and tracking loss. These losses arise due to small misalignment or beam divergence which is caused due to vibration or noise produced by the base motion of the satellite. The typical value of pointing error is 1 μrad which is to be maintained for permissible error-free communication [4]. The performance of PAT associated with the system depends on the seamless contribution among optical/electrical, closed-loop control and geometrical mapping. The ordinates and line-of-sight linking require accurate tracking by the uses of beacon signal and quadrant detector [1]. The closed-loop tracking using servomotors is very efficient method to achieve accuracy because it locks the beacon signal from receiver [5–7].

Several solutions have been investigated to mitigate the problems and its effect associated with the inter-satellite link. The link performance is analyzed with BER and Q factor obtained from the system under certain parametric conditions. Hashim et al. proposed an Is-OWC system where he achieved a bit-rate of 1 Mbps for the distance of 5000 km using NRZ modulation technique, which turned into advantageous step for the wavelength of 1550 m [8]. Sodnik et al. optimized the performance of Is-OWC system using BPSK modulation [9]. These investigations have been taken into consideration to design a system with high bit-rate and large coverage distance with maximum achievable quality factor with low BER under certain parametric conditions.

We have designed and simulated an inter-satellite OWC system with diversity technique which will improve the reliability of the system. The analysis of the designed system can be done by performance parameters such as Q factor and eye diagram. The proposed system is targeted to achieve Q factor greater than 6.8 and BER as low as 10^{-9}.

The systematic arrangement of this paper is as follows: Sect. 2 consists of design consideration; Sect. 3 discusses the proposed system; Sect. 4 discusses the results obtained from the simulation of the proposed model; and Sect. 5 summarizes and concludes this paper.

2 Design Consideration of the Proposed System

The important procedure to design a system is to choose particular parameters based on the application of the designed system. Our system design is considered for inter-satellite communication where the medium of transmission is free space. The functional parameters of the system are analyzed by Friis transmission equation which can be expressed as follows [10]:

$$P_R = P_T G_T \eta_T \eta_R G_R L_T L_R (\lambda/4\pi R)^2 \quad (1)$$

Similarly, the transmitter and receiver pointing loss factors are given as follows: [12]

$$L_T = \exp\left(-G_T \Theta_T^2\right) \quad (4)$$

$$\text{And } L_R = \exp\left(-G_R \Theta_R^2\right) \quad (5)$$

Equations (1)–(5) show the dependency of the system parameters on each other which is to be maintained for the proper functioning of system.

2.1 *Pointing, Acquisition, and Tracking (PAT)*

The pointing, acquisition, and tracking (PAT) systems must have accuracy for satellite link because small misalignment can cause large pointing error. The geometric misalignment due to PAT causes divergence in the signal at some angle called beam divergence angle. This is mitigated by increasing the antenna aperture size (Fig. 1).

The maintenance of line-of-sight link is the subject of PAT system whose performance depends on the seamless collaboration among optical/electrical, closed-loop tracking and geometrical mapping subsystem. These methods are used to block the beacon signal from the remote receiver. For long distance, line width of laser is chosen to get narrower beam which should be used to avoid sufferings from high geometric loss.

2.2 *Diversity*

The selection of diversity technique for the application of optical wireless communication is used to combat error bursts, co-channel interference and fading.

Fig. 1 Diagrammatic representation of beam divergence caused by PAT

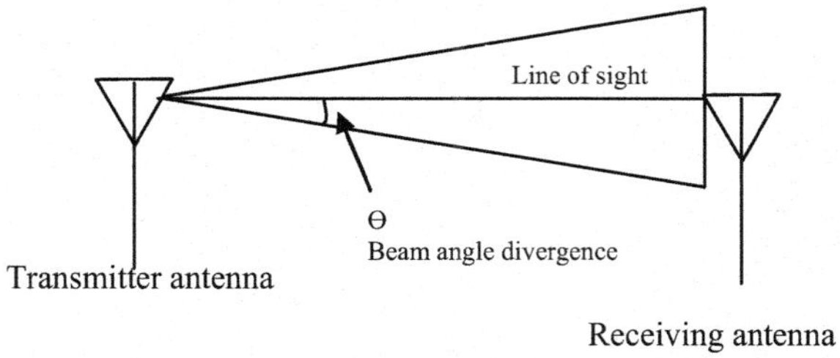

Diversity is the technique employed to improve the reliability of the system by sending multiple copies of data sent across fading channel. The diversity gain increases as the replicas of the signal received at the receiver are de-correlated [1].

3 The Proposed System

The block representation of the Is-OWC system with diversity is shown in Fig. 2. The system consists of four independent channels through which multiple copies of the signal are transmitted. The system is simulated using OptiSystem 12.

The system consists of satellite as a transmitter and receiver with the free space transmission medium. The information to be transmitted which is in the form of random sequence is converted into NRZ pulse. For long-haul communication, NRZ is the best option because it provides minimum optical bandwidth and minimum optical peak per bit interval for applied average power. NRZ pulse is modulated with continuous wave laser beam using Mach–Zehnder modulator. The optical signal generated from MZ modulator is provided to 1:4 fork to create multiple copies of the signal to be transmitted through independent channels. The signal

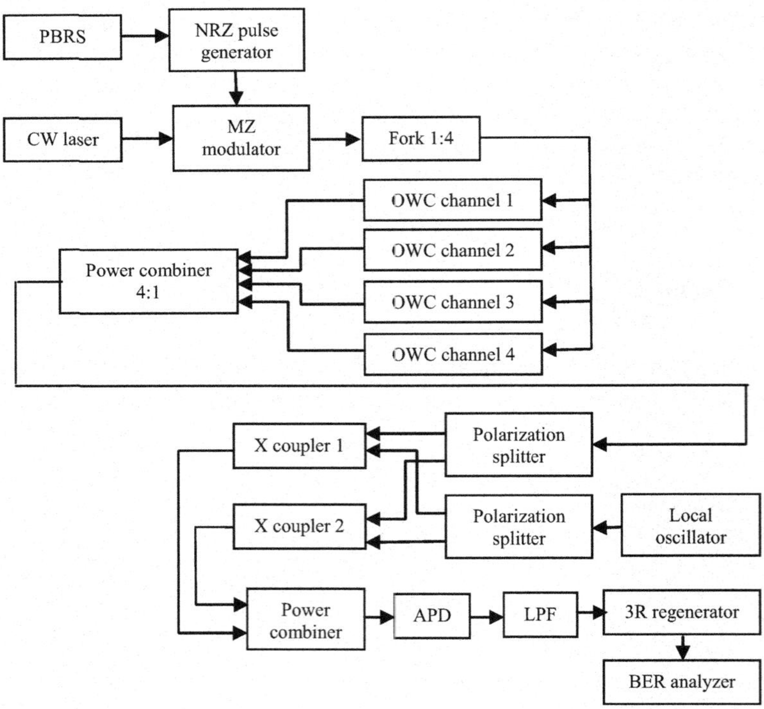

Fig. 2 Block diagram of the proposed system

Table 1 Parameters of the proposed system

Parameter	Value	Parameter	Value
Power	30 dBm	Local oscillator power	11 dBm
Data rate	30 Gbps	Tx/Rx pointing error	1.1 μrad
Modulation type	NRZ, RZ	Additional loss	3 dB
Line width	0.1 MHz	Dark current	10 nA
Sequence length	128 bits	Responsivity	1 A/W
Sample per bit	64	Frequency	193.1 THz
No. of samples	8192	Tx/Rx aperture	30 cm
Tx/Rx efficiency	0.8	Coverage distance	6000 km

transmitted through channel with different power levels is combined by the 4:1 power combiner. The operating wavelength chosen for this system is 1550 nm. The other parametric considerations are shown in Table 1.

Receiver consists of polarization beam splitter which performs the selection of appropriate component of the signal at the input port and sends each polarization component on each of the output ports [1]. The signal from polarization beam splitter is fed to cross-coupler where the optical signal is combined and then split. The detection section of the receiver consists of avalanche photodiode, low-pass filter along with 3R generation to get the eye diagram in BER analyzer.

4 Result

The proposed system is designed for the achievable bit-rate of 30 Gbps with 30 dBm input power at the distance of 6000 km. The results obtained from the system are analyzed on the basis of performance parameters such as Q factor and BER by varying several transmission conditions and input parameters. The dependencies of the several parameters are shown by the graphical representations which are as follows.

4.1 Consequence of Input Power

From the study presented above in Eq. (1), input power is directly proportional to transmission distance and gain of the system. Hence, we can decisively say that for larger power, highest gain can be achieved for the large coverage distance. The graph is plotted by varying distance and input power, which is shown in Fig. 3.

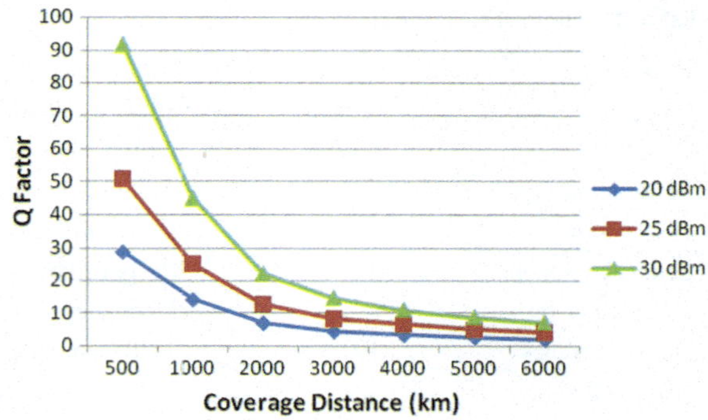

Fig. 3 Graph of Q factor versus distance versus input power for the bit-rate of 30 Gbps

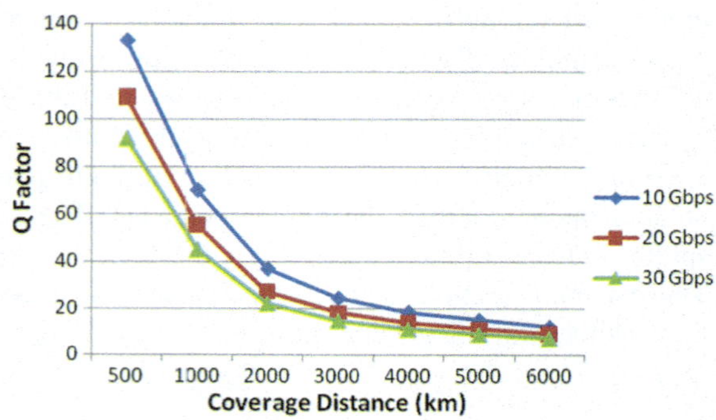

Fig. 4 Graph of Q factor versus bit-rate versus coverage distance with input power of 30 dBm

4.2 Impact of Bit-Rate

Bit-rate is the important parameter which is taken into consideration by the network users. As bit-rate increases, processing speed of the system becomes fast and hence decreases the coverage distance and the gain or Q factor of the system. The result obtained decisively states that higher bit-rate can be achieved at the cost of lower gain and decreased range (Fig. 4).

4.3 Consequence of Tx/Rx Antenna Aperture

The pointing and additional losses in the channel cause divergence in the signal. If we consider small aperture, there is loss in the signal. Hence for large aperture, loss in the signal is minimized which leads to the enhancement in the gain and coverage distance of the signal. Elaborated description of the statement is shown below through graph (Fig. 5).

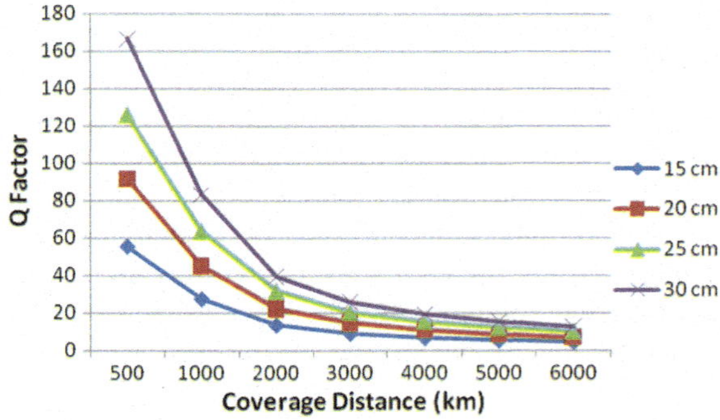

Fig. 5 Graph of Q factor versus antenna aperture versus coverage distance for input power 30 dBm and bit-rate of 30 Gbps

4.4 Consequence of Modulation Format

Modulation format is chosen on the basis of application of the system and transmission range. For long-haul communication, NRZ gives better performance than RZ in terms of bandwidth utilization, whereas RZ offers increased bit-rate and tends to be more robust against distortion. Hence, result concludes that NRZ modulation provides maximum gain (Fig. 6).

Somanath pradhan et al. have achieved the bit-rate of 7.63 Gbps over the distance of 6000 km for the obtained Q factor and BER 6.8 and 5.8e-12 [1]. The proposed paper shows the improvement in the value of Q factor and BER obtained from the designed system using NRZ modulation format for distance of 6000 km are 7.13 and 4.91e-013, respectively, and for RZ modulation format for the distance of 5000 km are 7.98 and 7.087e-016, respectively. It has been seen while simulating the system that for RZ modulation the Q factor obtained for the distance of 6000 km is less than the targeted value of Q factor. Hence, the result concludes NRZ modulation is better for long-haul optical wireless communication.

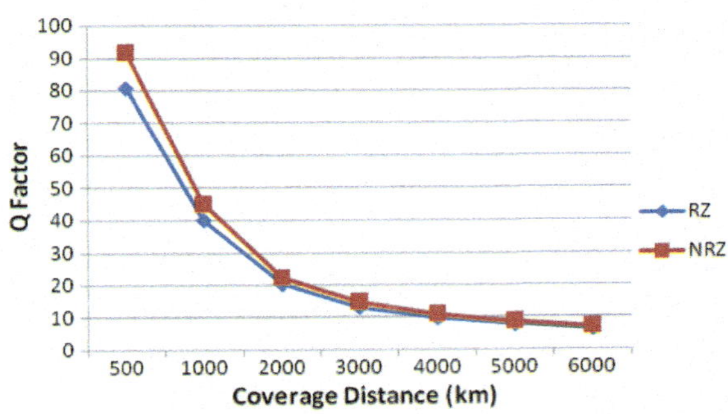

Fig. 6 Graph of Q factor versus coverage distance versus modulation format for the system parameter listed in Table 1

5 Conclusion

Inter-satellite optical wireless communication system is designed and simulated using diversity technique to improve the performance and mitigate the propagation losses. The system is designed to achieve the bit-rate of 30 Gbps for the coverage distance of 6000 km. The effect of diversity has lead to the improvement which is analyzed using results obtained. The different parameters such as transmitter power, bit-rate, and pointing error are varied to reduce the performance deterioration of the system.

References

1. Pradhan S, Sahu PK, Giri RK, Patnaik B (2015) Inter-satellite optical wireless communication system design using diversity techniques
2. Kaushal H, Kaddoum G (2016) Optical communication in space: challenges and mitigation techniques. IEEE
3. Ghassemlooy Z, Arnon S, Uysal M, Xu Z, Cheng J (2005) Emerging optical wireless communication. Adv Challenges
4. Chen CC, Gardner CS (1986) Impact of random pointing and tracking error on the design of coherent and incoherent optical intersatellite communication link, vol 37, no 3
5. Barry JD, Held KJ, Mrecherle GS, Einhoen AJ (1984) Spacecraft system study: a study to define the impact of laser communication systems on their host spacecrafts. Hughes Aircraft Company, Report FR84-75-717, April 1984; prepared for NASA Goddard space flight centre, Greenbelt, MD
6. Held KJ, Barry JD (1986) Precision optical pointing and tracking from spacecraft with vibrational noise. Proc SPIE 616:118–128
7. Germann LM, Nelson R (1986) Pointing, acquisition, and tracking subsystem for space based laser communication. Proc SPIE 616:160–173
8. Hashim AH, Mahad FD, Idrus SM, Supa A (2010) Modeling and performance study of inter-satellite optical wireless communication systems. In: Proceeding of IEEE-international conference on photonics, pp 1–4
9. Sodnik Z, Furch B, Lutz H (2010) Optical intersatellite communication. IEEE J Sel Top Quant Electron 16(5):1051–1057
10. Arnon S (2005) Performance of satellite network with an optical preamplifier. J Opt Soc Am A 22(4):708–715

Resilient Network and Loop-Free Forwarding Using Time Synchronization

Tunisha Varshney and Karan Verma

Abstract Under any routing protocol inconsistency of router create a problem. At the time of link or node failure there is some changes occur in topology which cause some postponement in redesigning of new topology. Router has inconsistent view of network for a few instances. During convergence period, packet may be trapped in forwarding loops. TTL is not adequate when loops are more. It has been watched that even in well oversee systems, failure of connections and router are normal. So as to fulfill the interest for high accessibility if there should arise an occurrence of a failure, quick rebuilding of loop free sending along the ideal ways is significant for a steering plan. The main strength of this paper is that it is not protocol dependent, we can embedded this scheme with any protocol and purpose of this paper is to provide a desirable feature of network, i.e., resilience.

Keywords Resilient · Time-to-live (TTL) · Loop-free forwarding Bidirectional forwarding (BFD) · Breadth-first search (BFS) · Proactive rerouting · Reactive rerouting

1 Introduction

Network resilience is the fascinating feature of any network. Wireless network needs to keep the service running despite a link or node failure. Consequently, recouping from the disappointment themselves by redirecting activity from the fizzled way of the system to the next part of the system, i.e., backup path, known as resilient network. Ample variety of scenarios in wireless where significant number nodes are deployed, they are not directly connected, there are many causes by

T. Varshney (✉) · K. Verma
Department of Computer Science and Engineering,
Central University of Rajasthan, Ajmer, India
e-mail: 2015mtcse018@curaj.ac.in

K. Verma
e-mail: vermakara@gmail.com

V. Janyani et al. (eds.), *Optical and Wireless Technologies*, Lecture Notes in Electrical Engineering 472, https://doi.org/10.1007/978-981-10-7395-3_43

which link and node may be failing, leading to network unavailability, temporarily, or permanently [13]. Survivability of any network impacted by these events like packet loss by failure, attacks, and accidents. Expansive scope of unforeseen occasions are secured by mishaps, some outstanding assaults like DDOS, interruption may harm network possibly and because of failure system did not conveyed expected output. These all events affect the smooth functionality of the system; in this paper, we design the resilient network which is protocol independent. There are two ways to deal with handle connect disappointment protection and restoration. Protection-based plan partitioned into three sections link-based protection, path-based protection, and segment-based protection. Restoration will be sorted in proactive or reactive routing which will portray later. In this paper, we utilized restoration system to deal with connection failure. Where system has to perform some task for keeping system active at the time of fault like (1) detect failure (2) notify to the neighboring node (3) reconfigure the failure so it can continue to operate (backup path calculation), and (4) recovered to failure by rerouting the packet.

In this paper, we proposed a strategy which upgrades the unwavering quality and survivability of the system. In this paper, failure will be detected and notification process is done through time negotiation process by using bidirectional forwarding (BFD) and rerouting is done through the breadth-first search (BFS) algorithm.

1.1 Looping Scenario

As in Fig. 1 shows an example of a looping scenario. There is host '*X*' wants to send its traffic through routers Ra, Rb, and Rc to the host '*Y*'. At the same time, other host wants to send the traffic to the same link which means this traffic also traversed routers Ra, Rb, and Rc. Due to some failure done either with the next router or link failure from the router Rc to further network, overall there is misconfiguration in router Rc. Traffic from host *X* to host *Y* will be forwarded back to the router Rc to router Rb but routing table of Rb is not updated, there is a path to forward the packet only through the router Rc so it further back the packet to router Rc. Therefore, there is problem of looping between two routers Rb and Rc, packet may be loosed due to continuing attempt to use failed component.

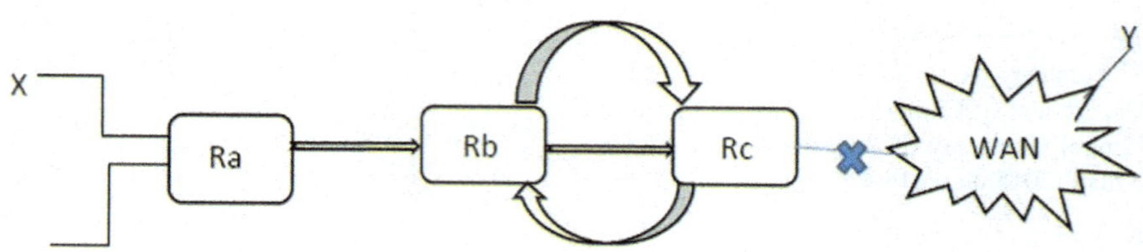

Fig. 1 Formulation of forwarding loops

2 Related Work

In this section, we discussed about existing mechanism that mitigates the effect of forwarding loop—In failure carrying packet (FCP) [1], packet is carried failure information of the network then when packet reached to the failed link then it is reroute and then forwarded to the next link, but it deals with only single failure and here we have to maintain the source map of the network.

In fast convergence fast reroute (FCFR) [12] in this system every router keeps up the two tables before convergence and after convergence. At the season of failure, router trade the after union table in which router have new design of the network. Separation, between before merging and after joining table, is done through one piece which furthermore includes bundle header. This bit will make overhead in packet.

In contrast, our scheme calculates rerouting paths beforehand thus has faster response to link failures. Modify and depend on the current protocol deployment is worth, noting for the fast rerouting scheme. Backup path maintenance is created excessive overhead and in packet header modification increased overhead. Both of the drawbacks the absent in our proposed method and every link failure have only one backup path.

3 Rerouting Scheme

In this section, we first describe how much time a router takes to recover from failure, then gives proposed scheme to make network resilient.

3.1 Overview

To minimize the inimical effect of link and node failure over the real-time application, there is a need of mechanism which provides fast rerouting scheme during convergence.

The concept of the rerouting based on the equation

$$T_{\text{int.Service}} = T_{\text{detect}} + T_{\text{notif}} + T_{\text{recomp.}} + T_{\text{so}}$$

where,

$\mathbf{T_{int.Service}}$ is the aggregate time since the failure happens and gets recognized until packet begins utilizing a backup path and administration is continued. Then again the transmission of parcels is rerouted from an essential way to a reinforcement way. A primary way and the backup path are distinctive (disjoint).
$\mathbf{T_{detect}}$ is the time which is used to detect the failure.

$\mathbf{T_{notif}}$ is the time to notify other neighbor node about the failure that has occured.
$\mathbf{T_{recomp}}$ is the time for recomputing the backup path to the destination, for diverting traffic from the failure path to the backup path.
$\mathbf{T_{so}}$ is the time to switchover from the failure path to the back path.

3.2 Proactive and Reactive Rerouting

In ***Reactive Rerouting, Recomputed, or Reroute*** backup path being calculated after the failure occurs, i.e., T_{recomp}, in this scheme there is need to calculate all four time in calculation of the interruption service time.

In ***Proactive Rerouting, Protection Switching, or Fast Reroute*** backup path is being calculated before the failure occurred, i.e., precompute the backup path. $T_{\mathbf{int.\ Service}}$ time is very less as compare to the *reactive rerouting* because there is no need to calculate T_{recomp} at the time of the interrupt service time calculation.

In this paper, we used proactive rerouting mechanism because fast reroute or proactive. Rerouting provides improved and faster result as compare to the reactive rerouting. It decrease interrupt service time as compare to the non-preplanned schemes.

3.3 Network Model

Detection of Failure and Notify to Neighbors

Bidirectional forwarding detection (BFD): It gives a steady and quick failure discovery component for all the media types like if packet is send through encapsulation, and it can be embedded in any routing protocol or any topologies. Reconvergence time period of BFD is uniform and predictable (Fig. 2).

Working of BFD.

As the Fig. 3 shows the working of BFD. BFD do time negotiation between BFD peers. BFD working is define in the fourth steps

The BFD neighbor session with the BGP neighbor router is torn down.

BFD notifies the local BGP process that the BFD neighbor is no longer reachable.

The local BGP process tears down the OSPF neighbor relationship.

If an alternative path is available the routers will immediately start converging on it.

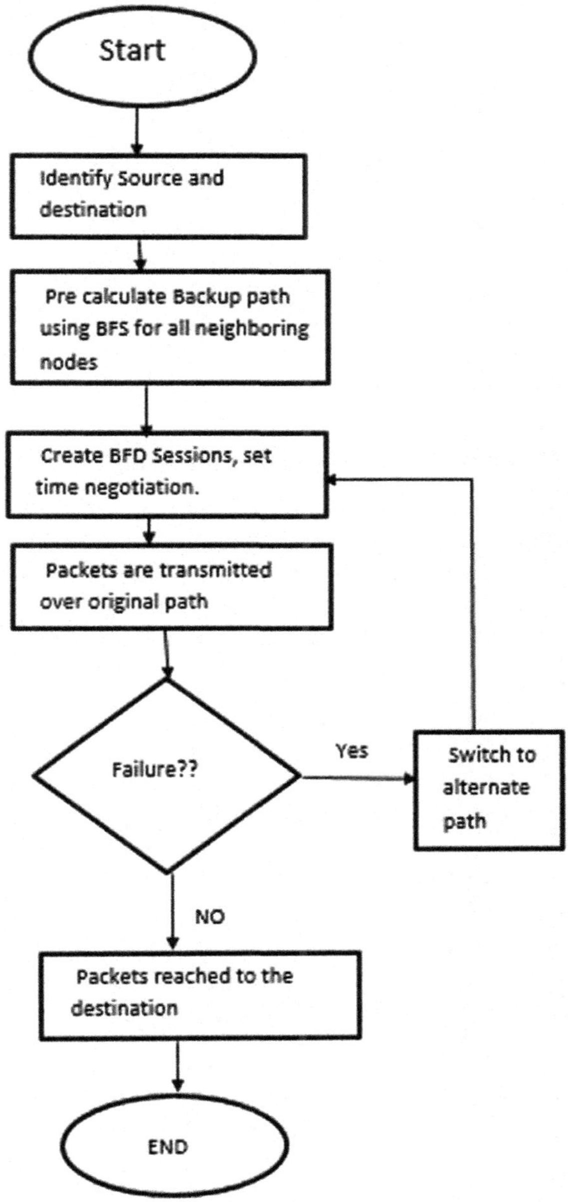

Fig. 2 Proposed network model

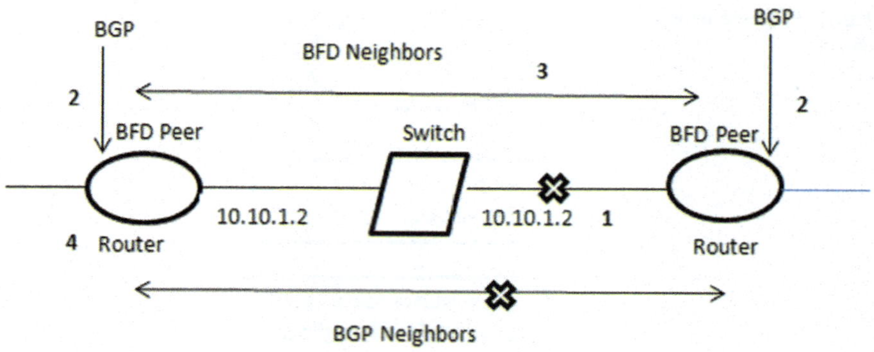

Fig. 3 BFD working

Alternate Path Detection and Switching

BFS algorithm is suitable to find the second path from source to destination. Wireless network is the group of router and wireless links, and we can say it is just like a graph, and BFS is suitable for the graph to find out shortest path in the graph. It takes input as the graph and gives the output as the path from source to the destination. The complexity of the BFS is O (V + E). Before occurrence of the failure, we maintain the backup path table with every node, easily and rapidly forwarding the traffic at the time of the failure.

4 Theoretical Evaluation

4.1 Convergence Time

Although reducing the EIGRP, IS-IS, and OSPF timers can result in minimum detection timer of one to two seconds, BFD can provide failure detection in less than one second in mille seconds, i.e., 50 mm.

4.2 Protocol Independent

This technique is convention autonomous so it can be joined with any directing convention. By utilizing the BFS count of the backup path is less unpredictable O (V + E). In the event that a system is running OSPF and EIGRP over a similar connection to a similar associate, just a single-BFD session will be built up, and BFD will impart session data to both directing conventions.

4.3 Guaranteed Loop Freeness

In this proposed work, looping does not frame in light of the fact that when failure is recognized, it redirects the activity from failure path to the other way then no purpose behind the packet to be in loop. The reason for the looping is that their routing tables do not updated, but in this paper routing table is updated immediately as the failure detected. We divided resiliency is divided in four step which is easy way to recover from the failure without packet loss and improved forwarding.

5 Conclusion and Future Work

In this paper, we have presented a dynamic discovering routing without looping. It tolerates the failure of the small network by dynamically discover the alternate path and divert traffic to his path when node is failed. This mechanism can be attached to the any protocol BGP, OSPF, or other autonomous system. In the future, we plan to experiment with this method, including a simulation and implementation, to evaluate its performance and usability in a real network application.

References

1. Lakshminarayanan K et al (2007) Achieving convergence-free routing using failure-carrying packets. ACM SIGCOMM Comput Commun Rev 37(4):241–252
2. Zhong Z et al (2005) Failure inferencing based fast rerouting for handling transient link and node failures. In: 24th annual joint conference of the IEEE computer and communications societies, vol 4. IEEE
3. Filsfils C et al (2015) The segment routing architecture. In: 2015 IEEE global communications conference (GLOBECOM). IEEE
4. Xia J, Gao L, Fei T (2007) A measurement study of persistent forwarding loops on the Internet. Comput Netw 51(17):4780–4796
5. Raj A, Ibe OC (2007) A survey of IP and multiprotocol label switching fast reroute Schemes. Comput Netw 51(8):1882–1907
6. Kreutz D et al (2015) Software-defined networking: a comprehensive survey, vol 103.1. IEEE, pp 14–76
7. Aldabbagh G et al (2015) Space-efficient and accurate forwarding loop detection method using bloom-filter for fast and reliable internet routing. J Theoret Appl Inf Technol 79(3):456
8. Robertson G, Bedenbaugh J, Nelakuditi S (2010) Fast convergence with fast reroute in IP networks. In: 2010 international conference on high performance switching and routing. IEEE
9. Nelakuditi S et al (2008) Mitigating transient loops through interface-specific forwarding. Comput Netw 52(3):593–609
10. Sundarrajan A, Ramasubramanian S (2015) Fast reroute from single link and single node failures for IP multicast. Comput Netw 82:20–33

Localization Independent Aspects of Topology Control in Wireless Sensor Networks

Tarun Kumar Dubey, Rohit Mathur and Dungar Nath Chouhan

Abstract Topology control (TC) in wireless networks lowers the energy consumption and also minimizes the transmission power. Therefore, energy saving is of supreme importance and is demonstrated by large number of algorithms, methods and protocols proposed for this cause. However, in the context of wireless sensor networks (WSNs), the principle of connected dominating set (CDS) has emerged out as the most popular method for energy efficient TC. This paper presents the performance evaluation of a CDS-based TC algorithm to develop a reduced topology with very low complexity and minimum the need of location information. The simulation results demonstrate the performance of algorithm in terms of low message overhead under reduced topology with a reasonable amount of energy consumption.

Keywords A3 algorithm · Active nodes · Atarraya · Topology control WSNs

1 Introduction

WSNs comprise of lightweight, ad hoc and relatively cheap communication infrastructure for a variety of advanced applications especially in places with limited accessibility to humans including organic compound detection, measurement of ion thrust and photovoltaic levels [1–6]. However, at the same time, limited computation, communication, storage capabilities and energy resources are very

---

T. K. Dubey (✉) · R. Mathur (✉) · D. N. Chouhan
Department of Electronics and Communication Engineering, Manipal University Jaipur, Jaipur, India
e-mail: tarunkumar.dubey@jaipur.manipal.edu

R. Mathur
e-mail: rohit.mathur@jaipur.manipal.edu

D. N. Chouhan
e-mail: dungarnath.chouhan@jaipur.manipal.edu

V. Janyani et al. (eds.), *Optical and Wireless Technologies*, Lecture Notes in Electrical Engineering 472, https://doi.org/10.1007/978-981-10-7395-3_44

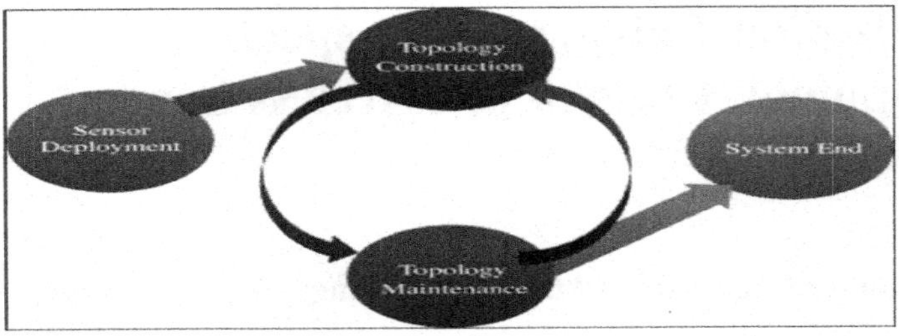

Fig. 1 Topology control cycle of WSNs

well-known major constraints imposed by these networks. This last attribute has the advantage of limiting the lifetime of the network and, hence, has gained a lot of attention by the research community over the last years. The research interventions in designing energy aware protocols, algorithms and mechanisms, with the motive to save energy, have been the topic of many research studies and conventions, all across the globe. One of these mechanisms is TC and is composed of two cyclic mechanisms, topology construction and topology maintenance [7, 8] as shown in Fig. 1. These mechanisms have been used extensively in WSNs which effectively reduce the initial topology of the network to save energy and extend the lifetime of the network [9, 10]. The connectivity and coverage characteristics are preserved with reduced number of active nodes and links. The administrator cannot influence the design of the network after the initial topology is deployed, especially when the location of the nodes is arbitrary.

However, the administrator has command over other parameters of the network like transmission power of the nodes, state of the nodes (active or sleeping) and role of the nodes (cluster head, gateway or regular). By modifying these parameters, the topology of the network can be optimized. However, being active for some time, few nodes will start to run out of energy and would die out. It is a fact that nodes that are in close proximity to the sink spend higher amounts of energy than those away due to the process of message forwarding. Therefore, the network must refurbish the reduced topology in order to preserve connectivity, coverage and energy. There are a wide range of techniques that execute topology construction and can be described into two categories primarily those which reduce the transmission range of the nodes and secondly those which turn the nodes off. The first technique targets the most expensive activity from a node's point of view by reducing the energy needed to transmit. In addition, it also reduces the number of nodes that are able to listen to the messages which in turn reduces energy consumption in the sense by avoiding collisions and reducing interference among the messages. The second technique targets the fact that all the nodes are not necessary to provide coverage or connectivity and even a small group of elite nodes can support the network if the rest go to a state of sleep in which the energy consumption is negligible. The reason behind this technique is that an active idle node wastes energy and produces redundant information; hence, the energy it uses being redundant could be saved in order to replace the node from the small elite set if it fails. Both these techniques have proven to minimize the energy consumption as compared to the case where no topology reduction is applied, but, to our knowledge, it is clear that the performance of none of the existing algorithms has been evaluated by means of simulation to calculate the amount of total energy spent during message broadcast. This paper considers topology reduction in WSNs and evaluates the performance of A3 algorithm in terms of energy consumption without the need of localization information [11–14] since the localization mechanisms introduce additional costs and also increase the energy consumption [15]. The performance of A3 algorithm is evaluated in network having a uniform node distribution with optimal theoretical bounds by means of simulation. The rest of the paper is organized as follows. Section 2 highlights the related work. The description of the A3 algorithm is presented in Sect. 3 followed by its simulation in Sect. 4. Finally, the paper concludes with Sect. 5.

2 Related Work

It is well known that TC can be exercised by dipping down the transmission range of all nodes by the same minimum amount [16]. These techniques are based on the theory that nodes have information about their own positions and the position of their neighbours [17], or they can be furnished with directional antennas that decide the orientation of the nodes within a network [14]. Although both assumptions are convincing, but are costly and complex to implement in practical scenarios. The other methods to control topology, including the method considered in this paper, are based on the CDS paradigms and are well discussed [18]. Here, the idea is not to lessen the transmission range of the nodes but to turn unnecessary nodes to condition off in order to preserve the connectivity and communication. Most of the CDS-based algorithms work in two phases: In phase one, they create a preliminary version of the CDS, and in phase two, they add or remove nodes to obtain a better approximation to the optimal CDS. Two relevant CDS-based algorithms are the Energy Efficient CDS (EECDS) [19] and the CDS-Rule-K [20]. The EECDS algorithm constructs a CDS tree creating maximal independent sets (MIS), which are clusters with non-connected clusterheads, and then picks gateway nodes to attach the clusterheads of the independent sets. The EECDS algorithm proceeds in two phases. The first phase starts with an initiator node that elects itself as a clusterhead and announces it to its neighbourhood. These set of nodes are now "covered". The "covered" nodes will send or pass the message to the uncovered neighbours, two-hop away from the initiator, which start competing to become clusterheads. Once there is a new clusterhead, the process replicates with the four-hop-away nodes from the initiator, until there are no further uncovered nodes.

On the second phase, the covered non-clusterhead nodes compete to become gateways between the clusterheads. The CDS-Rule-K algorithm utilizes the marking algorithm proposed in [20] and the pruning rule included in [21]. The idea is to start from a big set of nodes that achieve a least criterion and prune it according to a specific rule such that a node is active even if there is at least one pair of unconnected neighbours. In the second phase, a node decides to unmark itself if it concludes that all its neighbours are covered by marked nodes with higher priority, as per the level of the node in the CDS tree. The above proposed algorithms are well defined and tested but lack to prove their worth in real scenarios for evaluating the total energy spent in the reduced topology for message transmission in WSNs.

3 Algorithm

A3 algorithm chooses a subgroup of nodes that offer connectivity by establishing a CDS and then turn off all the non-CDS nodes to conserve their energy. The A3 algorithm ensures connectivity and coverage by turning off all redundant nodes in the network, so their energy is conserved for future safeguarding. The main thought behind this is that nodes may conserve energy by going to sleep or low energy-consuming mode, as long as no service request is made, than dropping the power used for transmission purpose.

The procedural steps of the A3 algorithm [22] are summarized as follows:

Step 1 Nodes start with the *Unvisited* state,
stating node starts with the *Active Candidate* state.

Step 2 *Active Candidate* node sends a *HELLO* message to all neighbours,
Assuming the sink node to be the first;
wait for replies from *Unvisited* neighbour nodes

Step 3 All neighbours send a *Parent Recognition* message ID

Step 4 After time *t*, *Active Candidate* stops listening for messages,
If the *Active Candidate* receives at least one answer, it changes its state to *Active Candidate*;
Else it adapts its state to *Sleeping* and turns off

Step 5 Children nodes place themselves in the list;
wait for a time proportional to their position

Step 6 *Sleeping* message is sent,
Active Candidate moves to step 1.

Step 7 On receiving the *Sleeping* message;
State is changed to *Sleeping Candidate*, turns off;
State changes to *Active Candidate*.

This algorithm has been designed to permit each node in the network to finish as an *Active* or *Sleeping* node, besides providing guaranteed connectivity among the nodes in the WSN.

4 Simulation

The computational complexity of A3 algorithm is estimated using Atarraya [23], an event-driven simulator created in Java that presents a new framework for designing and testing of topology control algorithms for WSNs. The evaluation of the A3 algorithm was performed in a topology with 100 nodes and one sink node as shown in Fig. 2, uniformly distributed with unique ID along a square area of side 600 m having a communication range of 100 m and a sensing range of 20 m as shown, respectively, in Figs. 3, 4 and 5. The nodes can handle homogeneous networks in which all nodes have the similar characteristics as well as heterogeneous networks where there is at least one node that is dissimilar from the other nodes. However, the topology

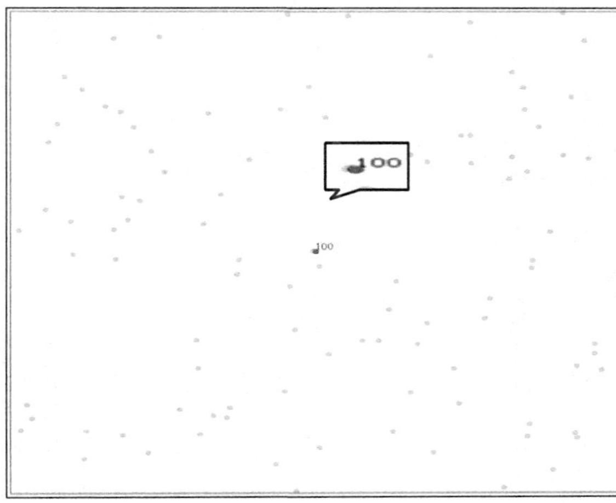

Fig. 2 Deployed topology with sink node

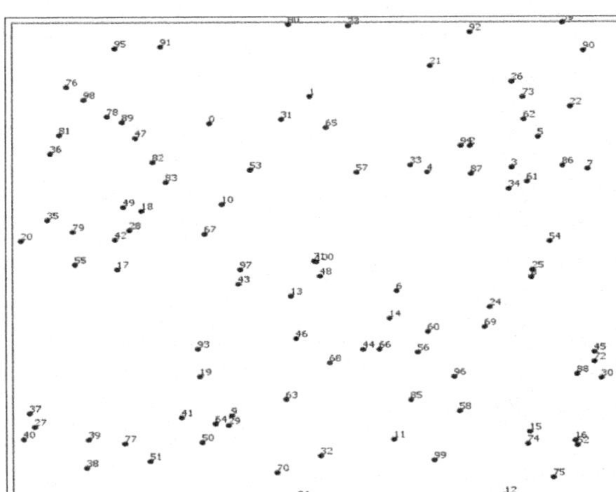

Fig. 3 Nodes with assigned ID

Fig. 4 Sensing range of node

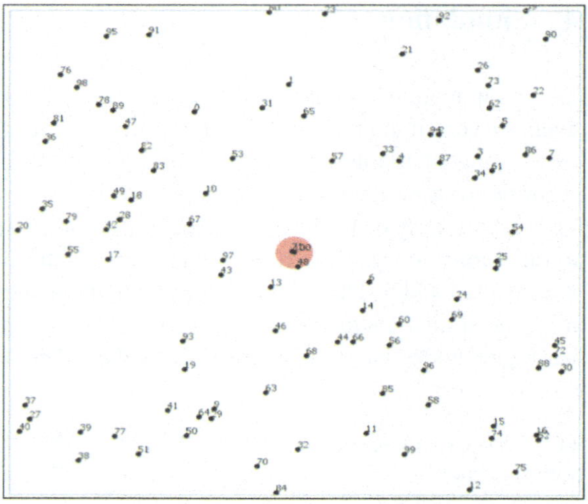

Fig. 5 Communication range of node

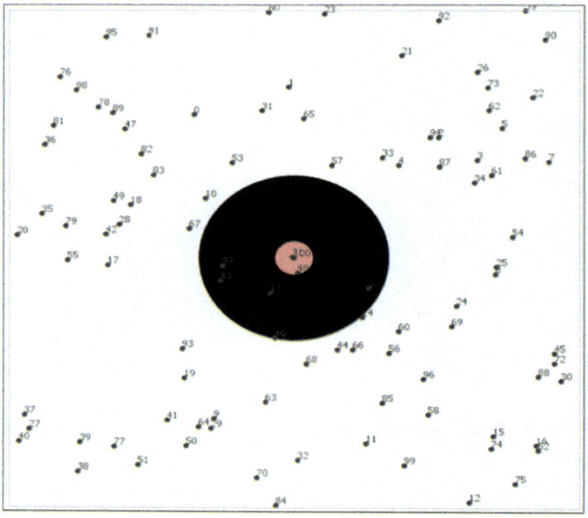

formation is based on subsets of homogeneous nodes and each set outlines a group of nodes that share the same features. The administrator tries to classify a set of weak nodes with small transmission range and low energy and a set of powerful nodes with high transmission range and superior energy. This application allows random distributions of nodes and permits the administrator to create denser zones in the topology if required. Table 1 summarizes the selected parameters for simulation of the A3 algorithm. The visualization of communication topology is shown in Fig. 6.

Figure 7 exhibits the sensing radius of active nodes with the reduced topology, and after the execution of the algorithm, the active nodes determine the

Table 1 Selected parameters for simulation

Parameter	Description
Sink	One
Communication model	Simple
Energy model	Simple
Node energy distribution	Gaussian
Node distribution	Uniform
Topology	Square grid

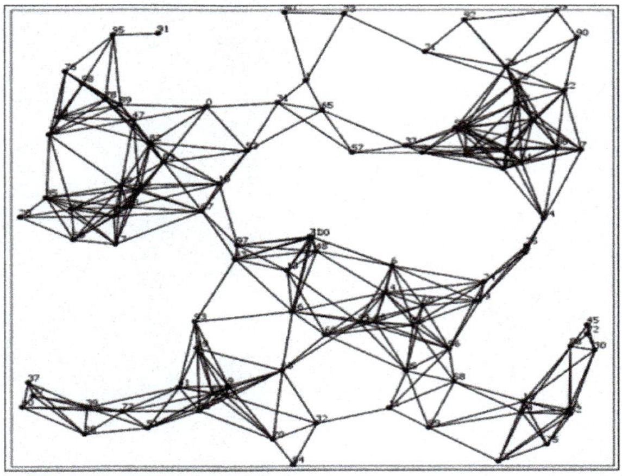

Fig. 6 Communication topology

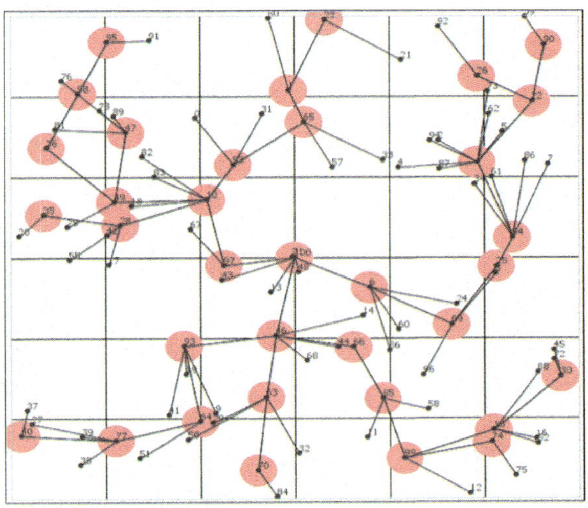

Fig. 7 Sensing radius of active nodes (reduced topology)

communication radius as shown in Fig. 8. This radius covers the maximum possible deployment area. Figure 9 exhibits the average number of neighbours per node.

It is visible from the figures, and based on the results of simulation, we can say that A3 provides wide coverage with the selected active nodes over the same area by using fewer resources. The topology reduction using the A3 algorithm can be well justified even in dense networks with significant performance improvements.

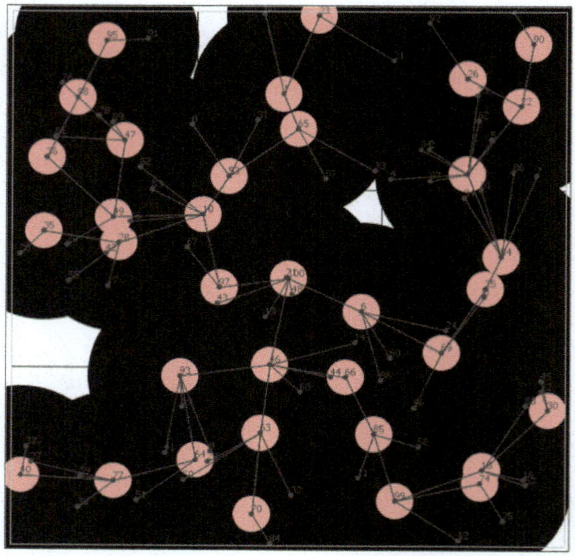

Fig. 8 Communication radius of active node (reduced topology)

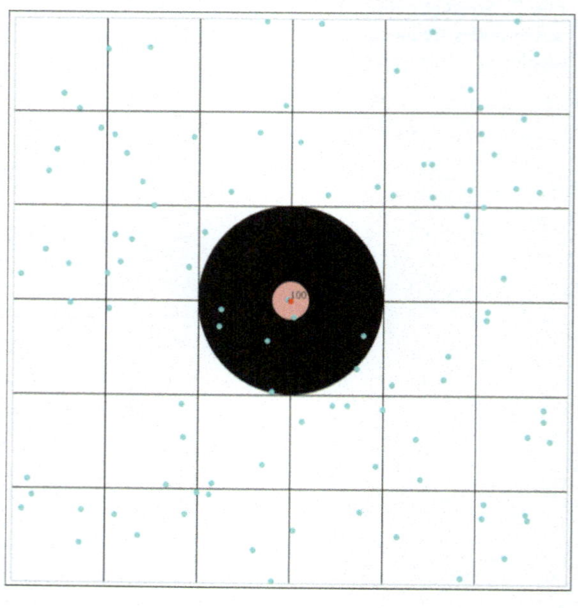

Fig. 9 Average number of neighbours per node

Table 2 Simulation results

Parameter	Value
Number of active nodes	36
Number of messages sent	322
Covered communication area (m)	0.9632
Que size	8
Total energy spent (J)	100821.0
Simulation time (s)	47.012
Average number of neighbours per node	7.866

Table 2 presents the results in terms of values of some key parameters obtained during simulation. It is demonstrated that the results are closely related to the performance evaluation of the A3 algorithm for topology reduction.

To summarize the results from Table 2, it is clear that A3 selected 36 out of 100 nodes as active nodes in order to preserve connectivity in the network. The total message overhead was 322 with an energy usage of 100821.0 J to successfully cover an area of 0.9632 m.

5 Conclusions

The performance evaluation of A3 algorithm is evaluated using simulation tool; the main parameters including energy consumption, covered communication area and average number of neighbours per node in a uniformly deployed WSN are estimated. A3 is a CDS-based algorithm wherein redundant nodes are turned off while maintaining network coverage and connectivity characteristic. Simulation results show the superiority of the A3 algorithm in terms of number of active nodes, messages sent and que size. More significantly, A3 algorithm requires less number of active nodes and exhibits reduced bounded messages and energy usage, compared with non-linear increasing trends. This is the most vital aspect of A3 as TC mechanism where the CDS tree undergoes frequent changes. In addition, the algorithm is computationally inexpensive and simple enough in terms of its message and computation complexity and does not consume excess energy in creating the reduced topology. Further, the algorithm is distributed and does not require any localization information and achieves low message overhead with less computational complexity.

References

1. Spinosi A, Mozumdar MMR, Olivien S, Lavagno L (2011) an extended framework for WSN applications. In: IEEE consumer communications and networking conference. IEEE Press, pp 965–966

2. Culler D, Estrin D, Srivastava M (2004) Overview of sensor networks. J IEEE Comput Netw 37:41–49
3. Akyildiz IF, Su W, Sankarasubramaniam Y, Cayirci E (2002) Wireless sensor networks: a survey. J Comput Netw 38:393–422
4. Pandya HJ, Vaishnav VS (2009) Detection & classification of VOCs using indium tin oxide sensor array and artificial neural network. J Intell Syst Technol Appl 7:72–79
5. Stephen RJ, Rajanna K, Dhar V, Kumar KGK, Nagabushanam S (2004) Thin film strain gauge sensors for ion thrust measurement. J IEEE Sens J 4:373–377
6. Nagaraju J, Murthy MVK (1999) A test facility for photovoltaic refrigerators. J Solar Energy Soc India (SESI) J 9:11–22
7. Labrador MA, Wightman PM (2009) Topology control in wireless sensor networks: with a companion simulation tool for teaching and research. Springer
8. Guan Q, Yu FR, Jiang S, Leung VCM (2011) Capacity-optimized topology control for MANETs with cooperative communications. IEEE Trans Wirel Commun 10(7):2162–2170
9. Rodoplu V, Meng TH (1999) Minimum energy mobile wireless networks. IEEE J Sel Areas Commun 17(8):1333–1344
10. Lee CY, Shiu LC, Lin FT, Yang CS (2013) Distributed topology control algorithm on broadcasting in wireless sensor network. J Netw Comput Appl
11. Tripathi R, Singh YN, Verma N (2012) Two tiered wireless sensor networks-base station optimal positioning case study. IET Wirel Sens Syst 2(4):351–360
12. Sahu OP, Dubey T (2011) A practical solution for location estimation in manually deployed wireless sensor networks. J Wirel Sens Netw 3:117–120
13. Sahu OP, Dubey T (2009) A new approach for self localization of wireless sensor network. J Sci Technol 2(11):1–4
14. Dubey T, Sahu OP (2013) Directional antenna assisted scheme to reduce localization error in wireless sensor networks. J Inf Netw Secur 2(2):183–189
15. Butenko S, Cheng X, Oliveira C, Pardalos PM (2004) A new heuristic for the minimum connected dominating set problem on ad hoc wireless networks. Kluwer Academic, pp 61–73
16. Guha S, Khuller S (1998) Approximation algorithms for connected dominating sets. J Algorithmica 20(4):374–387
17. Kumar V, Arunan T, Balakrishnan N (2002) E-span: enhanced-span with directional antenna. In: IEEE conference on convergent technologies for Asia-Pacific region, vol 2. IEEE Press, pp 675–679
18. Wu J, Cardei M, Dai F, Yang S (2006) Extended dominating set and its applications in ad hoc networks using cooperative communication. IEEE Trans Parallel Distrib Syst 17(8):851–864
19. Wu J, Dai F (2004) An extended localized algorithm for connected dominating set formation in ad hoc wireless networks. IEEE Trans Parallel Distrib Syst 15(10):908–920
20. Wu J, Li H (1999) On calculating connected dominating set for efficient routing in ad hoc wireless networks. In: 3rd ACM international workshop on discrete algorithms and methods for mobile computing and communications, pp 7–14
21. Yuanyuan Z, Jia X, Yanxiang H (2006) Energy efficient distributed connected dominating sets construction in wireless sensor networks. In: IWCMC '06 2006 international wireless communications and mobile computing conference. ACM, New York, pp 797–802
22. Wightman PM, Labrador MA (2008) A3: a topology construction algorithm for wireless sensor networks. In: Global telecommunications conference, 2008. IEEE GLOBECOM. IEEE, pp 1–6
23. Wightman PM, Labrador MA (2009) Atarraya: a simulation tool to teach and research topology control algorithms for wireless sensor networks. In: ICST 2nd international conference on simulation tools and techniques, SIMUTools 2009

Evaluation of Rate Matching of Turbo Code for 3GPP LTE System

Mahmood F. Mosleh and Mais F. Abid

Abstract The huge demand for broadband remote access has promoted the development of the long-term evolution (LTE). The turbo code is one of the most types of forward error correction code that used in Third Generation Partnership Project (3GPP), which standardization works by LTE. One of the major tasks of 3GPP LTE is to increase the speed of radio access to mobile communications. Rate matching (RM) is an important tool used for adaptive the turbo code rate (CR) to increase the throughput of LTE. This research is attempted to evaluate the LTE-system-based turbo code with various CR and compare the data rate for each one with its amount of bit error rate (BER). An RM algorithm based on interlaced and circular buffer applying for 3GPP LTE system has been added to the system in order to reduce redundancy information, thereby increasing the data rate. Using Matlab R2016a package, the proposed algorithm is evaluated and compared with a conventional system in terms of BER and the amount of throughput. The results confirm that RM can increase the CR from 1/3 to which leads to increase the data rate in the same proportion.

Keywords Turbo code · LTE · Rate matching

1 Introduction

The goal of the long term evolution (LTE) is a low latency, optimization packet of radio access technology, and the highest data rate. LTE by the Third Generation Partnership Project (3GPP) showed the standardization works to determine a new

---

M. F. Mosleh (✉) · M. F. Abid (✉)
Electrical Technical Engineering College, Baghdad, Iraq
e-mail: drmahmoodfarhan@gmail.com

M. F. Abid
e-mail: maisfalah70@gmail.com

V. Janyani et al. (eds.), *Optical and Wireless Technologies*, Lecture Notes in Electrical Engineering 472, https://doi.org/10.1007/978-981-10-7395-3_45

method of the high-speed radio access for mobile communications systems. 3GPP LTE, backed by the majority of the operators of telecommunication from all over the world, is generally accepted as the strongest mobile communications system to support the industry of all over the world telecommunication in the following 10 years [1]. To achieve the requirements of the highest data rate LTE, the working group of 3GPP conducted a careful assessment of advanced channel coding candidates (turbo and LDPC codes). Accordingly, it decides to adopt the rate 1/3 WCDMA turbo code with a new contention free internal interleaver according to quadratic permutations polynomial (QPP) in order to facilitate high speed of efficient turbo decoding [2, 3].

Turbo code is an efficient tool, which approaches the limit of Shannon through the use of a convolution codes with a largest constraints length or a block code with the largest blocks length. Turbo codes are used in 3GPP LTE mobile communications. In the design operation for the turbo code of the LTE, the working group of the 3GPP LTE also pays to attract the turbo code performance in jointly with the suggestions of different rate matching (RM) algorithm [4]. RM is an effective role in the field of building the adaptive coding, one of the important features of the standard of modern communications and has a powerful effect on its error-correcting performance [5]. The main function of the RM is to obtain the accurate set of bits to be transmitted within a given transmission time interval (TTI). The RM algorithm punctures or repeats the parity bit to create a desired number of the bits according to a required CR that can be different from the mother CR of the turbo coder [4].

Many researchers attempt with RM to increase the data rate of modern communication. In [4], it has been proposed an efficient RM algorithm based on the circular buffer and interlaced puncturing applying to the 3GPP LTE system. Their results indicate that the algorithm used by such researcher can effectively improve the bit error ratio (BER) performance, particularly at the conditions of the high codes rates, compared with the traditional algorithm in LTE turbo code. Also, the researcher of [5] proposed a method that achieves high throughput by rearranging the orders of the data in the memory in order to improve latency and throughput. They implement the RM in the parallel operation to reduce the latency for the highest data rates applications.

This research is attempted to evaluate the performance of turbo code with and without the RM to clear out the impact of such technique in the term of BER and the amount of throughput. Also, to calculate the value of code rate after applying the RM technique to show the BER and compare with and without existing the RM technique to clear out.

2 Turbo Code for LTE

2.1 Encoder

Turbo code is a type of the high-performance forward error correction (FEC) codes, and that was the first operation codes to approach closely to the capacity of the channel. Turbo codes are used in 3GPP LTE mobile communication. In this research, it has been used a turbo encoder according to LTE standard, which consists of a two parallel convolution code with two 8-state constituent encoders and one convolutional interleaver as shown in Fig. 1. The task of the convolution interleaver is to take the block of the M-bit data and perform a permutation of this block. The performance of turbo code depends on the structure of the interleaver and the permutation sequences [6]. The basic rate of turbo coding according to LTE specifications is 1/3. A block of M-bit data is encoded into a code word with 3M + 12 data bits, where the 12 tail bits are used for the trellis termination. When begin to encode the input bits of the information, the first value of the shift registers of the 8-state constituent encoders shall be all zero. The convolution encoder may be represented as the follows [7]:

$$G_0 = G1 = 1 + D + D2 + D3 + D6 \tag{1}$$

Then convolution encoder basically multiplies the generated polynomials by the input bit, as the follows:

$$A1(k) = G_0 * U(k) = abc\ldots g \tag{2}$$

$$A2(k) = G_1 * U(k) = ABC\ldots G \tag{3}$$

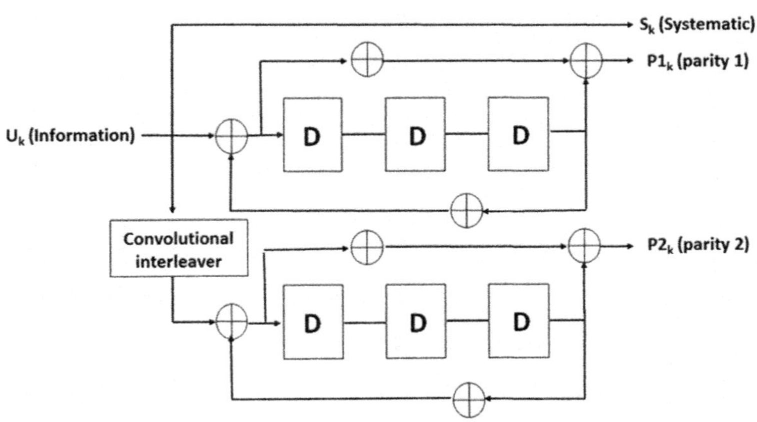

Fig. 1 Structure of the 1/3 CR turbo encoder in LTE [7]

Then, interleaving the outputs from the convolution encoder

$$E(k) = aAbBcC\ldots gG \tag{4}$$

2.2 Decoder

The LTE turbo decoder structure shown in Fig. 2 is composed of two maximum aposteriori decoder (MAP) and interleaver, which installed between these two MAP decoders to make permutations to the sequence of the input. The decoding is an iterative operation, which exchanges the extrinsic information also called (L-posteriori) between MAP decoders. Each iteration of the turbo is split into two half iterations. MAP decoder 1 is enabled through the first half iteration, which receives the soft information of the channel $S1_k$ for the systematic bit, $P1_k$ for the parity bit and the de-interleaved extrinsic information from the MAP decoder 2 L_{e1}. Through the second half iteration, MAP decoder 2 is enabled, and it receives the soft channel information $S2_k$ for the systematic bit after interleaved, $P2_k$ for the parity bit, and the interleaved extrinsic information from the MAP decoder 2 L_{e2}. This iterative operation is repeated until the decoding reached to the maximum number of iterations or has converged.

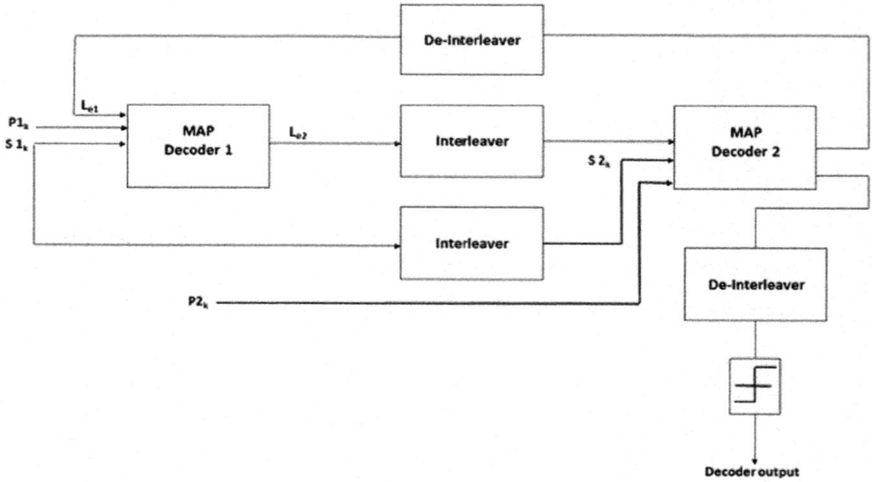

Fig. 2 1/3 rate structure of turbo decoder in LTE

3 Rate Matching for Turbo Codes

RM is used to maximize the data rate transmitted using turbo code. RM process is applied on turbo code after encoding process. The RM for the turbo-coded transport channels is defined per coded blocks and consists of three sub-blocks interleaver to the information bit streams dk_0, dk_1, and dk_2, comes after this the bit collection and bit selection and pruning as shown in Fig. 3.

The stream of the bit is interleaved according to the sub-block interleaver and denoted by d_0^i, d_1^i, and $d_2^i, \ldots, d_{k-1}^i$ where the k is the numbers of bits. The sequence bit of the output from the block interleaver is determined as the follows:

A. Set $C_{subblock} = 32$ to become the columns number of the matrix. And it is numbered from the left to the right as follows: 0, 1, 2, …, $C_{subblock} - 1$.
B. Specify the numbers of the rows of the matrix ($R_{subblock}$), through the finding of the minimum $R_{subblock}$

$$k_\pi = (R_{subblock} * C_{subblock})$$

The rows of the matrix are numbered from the top to the bottom as follows: 0, 1, 2, …, $R_{subblock} - 1$
C. Write the sequence of the input bit, into the ($R_{subblock} \times C_{subblock}$) matrix row by row starting with bit y0 in the column 0 of the row 0 as the following:

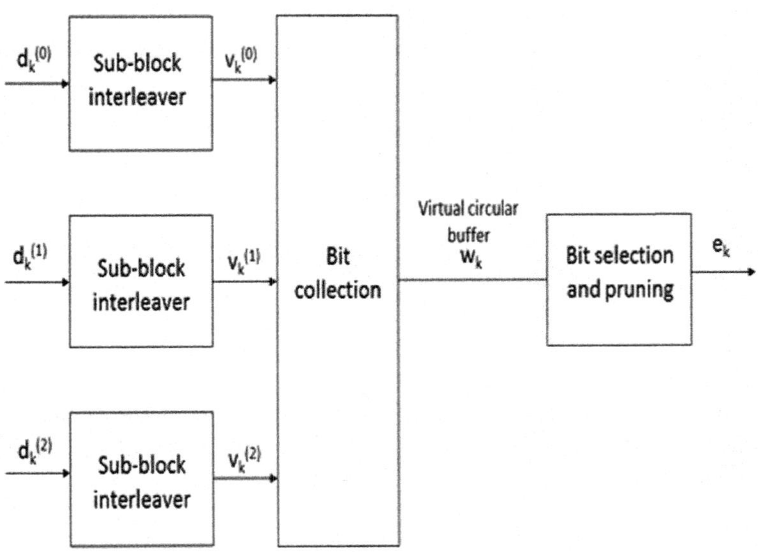

Fig. 3 Rate matching for LTE turbo codes

Table 1 Inter-column permutation pattern

Number of columns $C_{subblock}$	Inter-column permutation pattern [P(0), P(1), …, P($C_{subblock}$—1)]
32	[0, 16, 8, 24, 4, 20, 12, 28, 2, 18, 10, 26, 6, 22, 14, 30, 1, 17, 9, 25, 5, 21, 13, 29, 3, 19, 11, 27, 7, 23, 15, 31]

$$\begin{bmatrix} y_0 & y_1 & \cdots\ y_{C_{subblock}-1} \\ y_{C_{subblock}} & y_{C_{subblock}+1} & y_{2C_{subblock}-1} \\ \vdots & \ddots & \vdots \\ y_{(R_{subblock}-1)*C_{subblock}} & y_{(R_{subblock}-1)*C_{subblock}+1} & \cdots\ y_{(R_{subblock}*C_{subblock}-1)} \end{bmatrix}$$

D. Conduct the permutation to the intercolumn of the matrix according to the P pattern as shown in Table 1, the P(j) is the native columns position of the J-th permutation columns. After the columns permutations, the intercolumns permuted ($R_{subblock}$ × $C_{subblock}$) matrix is equal to this Eq.

$$\begin{bmatrix} y_{P(0)} & y_{P(1)} & \cdots\ y_{P(C_{subblock}-1)} \\ y_{P(0)+C_{subblock}} & y_{P(1)+C_{subblock}} & \cdots y_{(C_{subblock}-1)+C_{subblock}} \\ \vdots & \vdots & \vdots \\ y_{P(0)+(R_{subblock}-1)*C_{subblock}} & y_{P(1)+(R_{subblock}-1)*C_{subblock}} & y_{P(C_{subblock}-1)*C_{subblock}} \end{bmatrix}$$

E. The bit streams $d_k^{(0)}$ and $d_2^{(1)}$ are interleaved according to the sub-block interleaver A and B, respectively. The output of the sub-block interleaver A and B is the bit sequence read out column by column from the intercolumn permuted ($R_{subblock}$ × $C_{subblock}$) matrix. The bits after sub-block interleaving are denoted by $v_0^{(i)}, v_1^{(i)}, v_2^{(i)}, \ldots, v_{k-1}^{(i)}$, The output of the sub-block interleaver C is denoted by $v_0^{(2)}, v_1^{(2)}, \ldots, v_{k-1}^{(2)}$ where $v_k^{(2)} = y_{\pi(\kappa)}$, where

$$\pi(k) = \left(P\left(\lfloor\frac{k}{R_{subblock}}\rfloor\right) + C_{subblock} * (k\,mod R_{subblock}) + 1\right) mod\, k_\pi$$

These output sequences $v_K^{(0)}, v_K^{(1)}$, and $v_K^{(2)}$, across the operation of bit collection and only one output buffer W_k is generated by putting the rearranged systematic bits in the beginning and then followed by the bit by bit interlacing of the two rearranged parity streams [4]. The collected sequence W_k will serve as the circular buffer of length K_w = 3 $K\pi$ and is created as the follows:

$$\begin{aligned} w_k &= v_k^{(0)} && for\ k = 0, 1, \ldots, k_\pi - 1 \\ w_{K\pi} + 2k &= v_k^{(1)} && for\ k = 0, 1, \ldots, k_\pi - 1 \\ w_{K\pi} + 2k + 1 &= v_k^{(2)} && for\ k = 0, 1, \ldots, k_\pi - 1 \end{aligned}$$

The output sequence of the RM is ek, k = 0, 1, …, E − 1 (E is the output sequence length for the r-th coded block of the RM). The dummy bits, which are needed in the operation of LTE turbo code, are assigned by the symbol [NULL]. The step of the bit selection for RM is easily reading the first E bits from the beginning of the buffer. The bits, which have been selected, can be read starting from the k_0-th bit in the buffer determined by the circular buffer size Kw of the transmission [8]. If the buffer end is reached, the reading's operation is continuing through wrapping around the buffer beginning. Therefore, repetition and puncturing are achieved by using unified method [9].

4 System Model

The system model used in this research is shown in Fig. 4. In this section, it will explain the function of each block of such system as follows:

The source is generating bit streams of 2408 bits. The probability of 1's is 0.5. This stream bit is fed to the encoder to apply the encoding process. The encoder block will encode the input data with a CR of 1/3 so that each input bit will be represented by 3 bits. First one is the same input bit called systematic bit, and the second and third are the adaptive parity bits. Then, the encoder output will be 7308 bits after multiplexing the three parallel bits into one series row.

To reduce the redundancy bits, RM is applied after turbo encoder. In this system, the RM block will reduce the redundancy bit (parity bit) that was added by turbo encoder to 4872 bits. That means the RM reduces the bit stream to increase the data rate to 1/2.

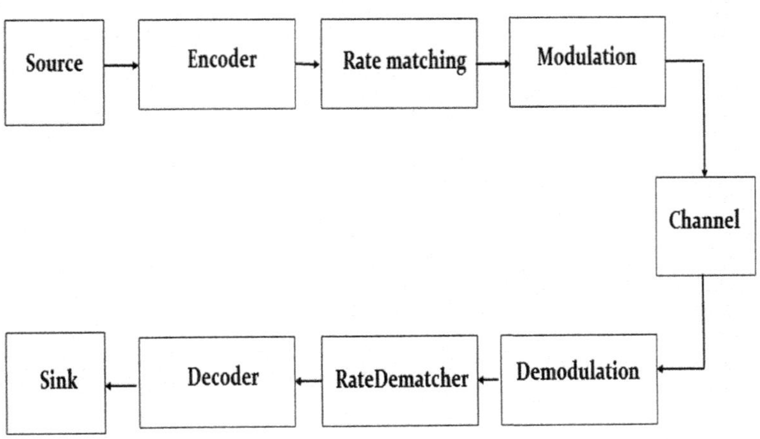

Fig. 4 System model

The modulation block will map the input bits into one symbol. The number of bits in each symbol depended on the level of modulation, and this research has been used QAM, 8QAM, 32QAM, and 64QAM, which maps 2, 3, 5, and 6, respectively. The channel used is AWGN type for the purpose of experimentally testing. The parameters of this channel are Eb/No, number of bits per symbol, power of the input signal, and the period of symbol. The rest blocks form the receiver end, which applies the inverse function of the corresponding transmitted end.

5 Simulation and Result

The results of the simulation are done by using various modulation orders (QAM, 8QAM, 32QAM, and 64QAM) for the system model, which was discussed in Sect. 4. The tool used to measure the performance is by plotting a curve of BER versus SNR. Various CR is used in the simulation 1/4, 1/2, 1/3, 3/5, 2/3, and 3/4. The system is tested for each code individually with all modulation orders. Figure 5 shows all results of each CR corresponded to each modulation level.

The results are arranged to cover the SNR range from 8 to about 26.5 for various CR and a corresponding modulation order. Table 2 is summarizing the results that achieved by simulation at BER 10^{-4}.

Better performance can be achieved using 1/4 CR with QAM modulation order, and it needs only 8.3 dB SNR to achieve 10^{-4} BER. In return, the worst case is at

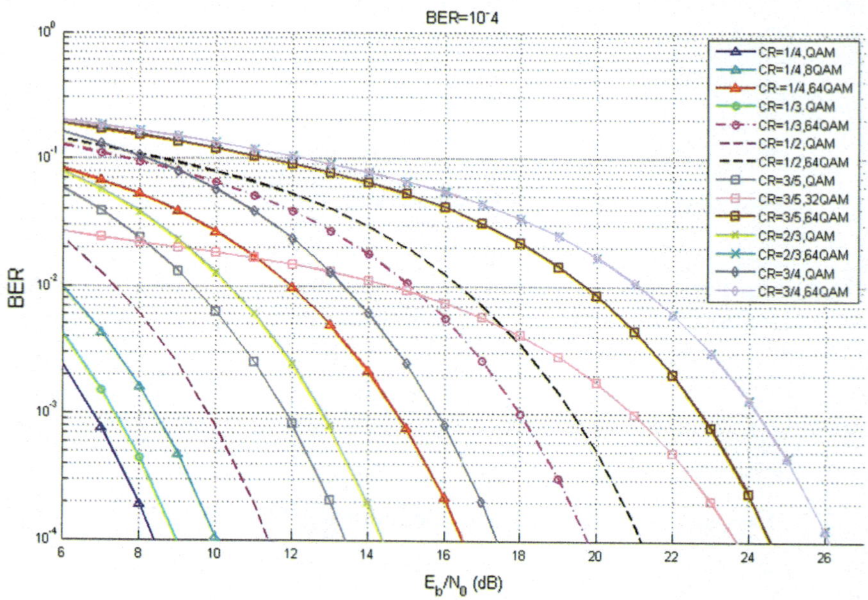

Fig. 5 Results of each CR corresponded to each modulation level

Table 2 Various CR and corresponding modulation order

Code rate (CR)	Modulation order	SNR (dB)
1/4	QAM	8.4
	8QAM	10
	64QAM	16.5
1/3	QAM	9
	64QAM	19.7
1/2	QAM	11.3
	64QAM	21
3/5	QAM	15.5
	32QAM	23.6
	64QAM	24.5
2/3	QAM	14.3
	64QAM	26.1
3/4	QAM	17.3
	64QAM	26.2

3/4 CR with 64QAM; it needs 26.2 dB to achieve 10^{-4} BER. But the data rate in the last one is 3 times more than the first case while the rest vocabularies are distributed between the above two cases in terms of SNR and BER.

To clear out the potential of using RM technique, it has been comparing the performance of such system with and without RM. Figure 6 shows the results in the case of QAM with a CR of 1/3 and 1/2.

Although the inserting of RM technique in the system model causes a degradation in the performance by 0.5 dB at 10^{-4} BER as it clear in Fig. 6, but it improves the turbo code rate from 1/3 to 1/2. Also, the performance of 1/3 CR with RM, which means that the result CR is ½, is achieved 2 dB SNR code gain at 10^{-4}

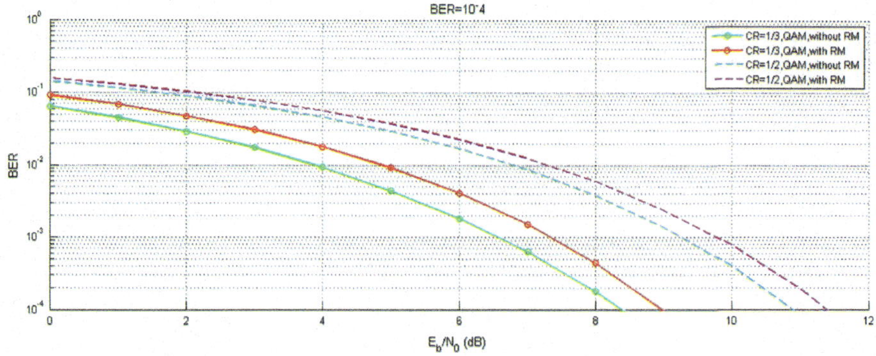

Fig. 6 Results with and without RM code rate

BER as compared with 1/2 CR without RM. While the system is losing 0.5 dB by using RM for each code rate.

6 Conclusions

This paper has adapted turbo code with various values of CR and modulation order; it has been using RM to improve the CR from 1/3 to 1/2. The result shows that the system with RM is achieving 2 dB codes gain while losing 0.5 dB as compared with same code rate. Also, by using various code rare and corresponded multi-modulation order to cover a range of dB from 8.4 to 26.5 dB which is suitable for modern communications.

References

1. 3GPP Technical Specifications 36.212 (2010) Technical specification group radio access network; Evolved Universal Terrestrial Radio Access (E-UTRA); Multiplexing and channel coding (Release 9)
2. 3GPP (2009) 3GPP TS 36.212 v8.7.0 3rd generation partnership project; technical specification group radio access network; evolved universal terrestrial radio access; multiplexing and channel coding (release 8). 3rd generation partnership project, Technical report, May 2009
3. Takeshita OY (2006) On maximum contention-free interleavers and permutation polynomials over integer rings. IEEE Trans Inf Theory 52(3):1249–1253
4. Yu L, Wang X, Liu J (2011) An improved rate matching algorithm for 3GPP LTE turbo code. In: Third international conference on communications and mobile computing
5. Ma C, Lin P (2010) Efficient implementation of rate matching for LTE turbo codes. IEEE
6. Raut RD, Kulat KD (2010) Int J Comput Appl 1(24):0975–8887
7. Sadjadpour HR, Sloane NJA, Salehi M, Nebe G (2001) Interleaver design for turbo codes. IEEE J Sel Areas Commun 19(5):831–837
8. 3GPP Technical Specifications 36.213 (2010) Technical specification group radio access network; evolved universal terrestrial radio access (E-UTRA); Physical layer procedures (Release 9)
9. 3GPP, R1-072604, ericsson, ETRI, freescale, ITRI, LGE, Motorola, Qualcomm, Samsung, ZTE, Way forward on HARQ rate matching for LTE

Copy-Move Detection Using Gray Level Run Length Matrix Features

Saba Mushtaq and Ajaz Hussain Mir

Abstract Copy-move detection is a well-recognized and active area of research owing to great demand of authenticating genuineness of images. Currently, available techniques for copy-move detection fail to accurately locate the tampered region and lack robustness against common post-processing operations like compression, blurring, and brightness changes. This paper proposes a novel technique for the detection and localization of copy-move regions in image using gray level run length matrix (GLRLM) features. In the proposed method, we first divide the forged image into overlapping blocks and GLRLM features are calculated for each block. Features calculated for each block form feature vectors. Feature vectors thus obtained are lexicographically sorted. Blocks with similar features are identified using Euclidean feature distances. Post-processing isolates similar blocks. Results demonstrate the effectiveness of the proposed scheme to locate copy-move forgery and robustness against operations like JPEG compression, blurring, and contrast adjustments.

Keywords Copy-move · Region duplication · Image forgery detection
Image texture · GLRLM

1 Introduction

Over the years, digital images have replaced their analog counterparts in the field of education, media, as evidences almost everywhere. But with immense development of the image manipulation software tools, it has become easy-to-tamper images mostly for malicious reasons; hence, the statement 'seeing is no longer believing' [1] holds true. As a result, various techniques have been developed to tackle the problem of image tampering and authenticate genuineness of images. Presently,

S. Mushtaq (✉) · A. H. Mir
Department of Electronics and Communication Engineering,
National Institute of Technology Srinagar, Srinagar 190006, India
e-mail: sab.mushtaq@gmail.com

V. Janyani et al. (eds.), *Optical and Wireless Technologies*, Lecture Notes
in Electrical Engineering 472, https://doi.org/10.1007/978-981-10-7395-3_46

there are various tampering techniques, namely copy-move, splicing, retouching, re-sampling, compression [2]. Among available image tampering techniques, copy-move detection is the most common one which involves copying a portion of image and pasting it somewhere else in the same image to conceal or multiply a part of the image.

Existing methods for image authentication can be broadly classified into two categories: active authentication and passive authentication [3]. Active authentication includes watermarking and digital signatures [4, 5]. Active authentication requires the availability of original image for the insertion of digital signature or watermarks, while passive authentication requires no prior knowledge about the image and thus gaining enormous attention. Passive image tampering detection is based on the assumption that even though forgery may not be perceived by the naked eye but it does change the underlying statistics of the image. Based on this, a lot of research work is aimed to uncover the forgeries in images. Copy-move is the most common image tampering technique which altogether changes the information conveyed by the image. Moreover, the copied region belongs to the same image so the dynamic range, color, and statistics of copied region remain unchanged making detection of the tampered region even harder [6]. Thus, this paper focuses on the problem of copy-move detection and its localization. Figure 1 shows an example of copy-move forgery with original image and its forged counterpart.

Copy-move forgery detection techniques can be broadly classified into three types—one is the block-based detection, second the key-point-based detection, and third brute force detection [7].

The simplest solution to the problem of copy-move forgery is brute force detection which involves exhaustive search. But the problem with this method is its computational complexity [7]. Block-based techniques give better results than exhaustive search techniques. The forged image is split into blocks of equal size.

Fig. 1 Left is the original image, right is the tampered image

Features are extracted from each block and are compared against other blocks. The result can be an exact match or approximate match. The third one is the key-point-based approach. Key-point-based approaches are faster than block-based approach since the number of comparisons required in key-point-based approach is fewer than block-based approach. In this paper, we propose a block-based copy-move detection technique based on texture of image.

The rest of the paper is organized as follows: Sect. 2 presents the related research in the field of copy-move detection and Sect. 3 introduces the review of texture methods GLRLM. Section 4 explains in detail the proposed forgery detection technique. Section 5 gives account of experimentation and results, and finally conclusion is drawn in Sect. 6.

2 Related Work

A number of methods have been developed for copy-move forgery detection. Among the initial attempts Jessica et al. gave a method which is based on discrete cosine transform, DCT [8]. Exhaustive search method was used by Jessica, followed by block-matching technique based on DCT. This technique is robust against retouching operation but gives no account of robustness against other techniques like JPEG compression.

Popescu and Farid proposed a technique that replaces the feature extraction technique DCT by principal component analysis (PCA) [9] in Jessica's method. This method is considered to be efficient as features used are half that were used by Jessica

Haung et al. proposed a scheme based on DCT features [10]. In this method, the feature vector is reduced by truncating high frequency of coefficients, which results in better performance than Popescu's method [9] in terms of robustness against rotation and in terms of speed in comparison with Jessica's method. This method is highly sensitive only when the copy-pasted areas are not too small.

Muhammad et al. proposed a technique that makes comparison between the approximate LL sub-band and detail HH sub-band from the DyWT technique [11]. This method is not tested for any post-processing operation. All the aforementioned methods are based on a large no. of blocks which result in very high number of feature vectors resulting in high computational complexity. All these methods involve comparison of features using sorting of indices for lexicographical sort. A slight decrease in the number of features in the feature vector results in a considerable improvement in speed and reduction in computational complexity [12].

The advantage with the block-based method is that they give the exact extent and shapes of the copied areas while key-point-based methods only give the location of key points on the copy-pasted regions [13]. Moreover if the forged area exhibits certain structure, it may be entirely missed by key-point-based method [14]. This paper thus focuses on block-based approach. This paper proposes method based on

GLRLM texture features of image. A number of evaluations conducted on realistic image show the effectiveness of our method.

3 GLRLM

In this paper, we intend to use GLRLM to analyze and compare blocks of image so as to identify similar blocks to locate copy-move forgery in image. A brief description of the techniques is given below.

3.1 Gray Level Run Length Matrix (GLRLM)

GLRLM is a pattern of gray intensity pixel in a particular direction from a reference pixel in an image [15]. It is a way of searching the image, across a given direction, for runs of pixels having the same gray level value. The run length is the number of adjacent pixels that have the same gray intensity in a particular direction. GLRLM is defined by intensity of runs (i), length of runs (L), and run direction (θ) from a reference pixel. Mathematically, GLRLM is defined as follows:

$$\text{GLRLM}(\theta) = [S(i, L)|\theta)] \quad (1)$$

S (i, L) is the number of times there is a run of length L having gray level i in direction θ. There are four run length matrices that can be computed for four directions of run (θ = 0°, 45°, 90°, 135°). In this paper, we have considered GLRLM at 0° only since experimentation revealed that considering GLRLM at other angles does not improve the efficacy of the algorithm significantly.

3.2 Features of GLRLM

From the GLRL matrix obtained using Eq. (1), seven features are calculated out of which SRE, LRE, GLN, RLN and RP are given by Galloway [16] and LGRE and HGRE by Chu et al. [17].

4 Proposed Scheme

The proposed scheme can be summed up in Fig. 2 and explained subsequently.

The detailed implementation of the steps is given below.

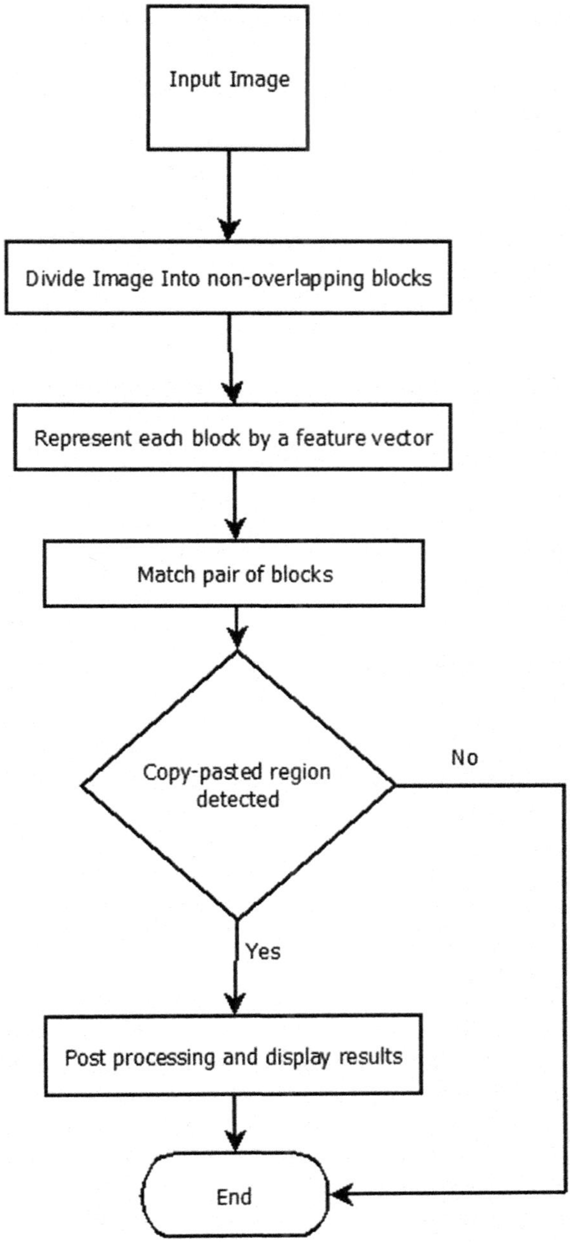

Fig. 2 Proposed algorithm for copy-move detection

4.1 Preprocessing the Input Image

The input color image is transformed into grayscale image using the standard formula

$$Im = 0.299R + 0.587G + 0.114B \tag{2}$$

where R, G, and B represent channels of input color image and Im is the luminance component.

4.2 Dividing Image into Overlapping Blocks

The grayscale image Im is then divided into overlapping blocks of fixed size. Image Im of size $M \times N$ is divided into fixed size square over lapping blocks of size $A \times A$ such that two consecutive blocks differ in a row or a column only. Thus, image is divided into $(M - A + 1)(N - A + 1)$ blocks.

4.3 Extracting Texture Features of the Blocks

Of each overlapping block, GLRLM is calculated and seven GLRLM features are extracted. The features from a block represent that block. An $A \times A$ block is represented by GLRLM feature vector $X = (a_1, a_2, a_3, a_4, a_5, a_6, a_7)$.

4.4 Matching of Texture Features to Locate Similar Blocks

After feature extraction, feature vectors representing each block are arranged in a matrix. The rows of matrix correspond to the feature vectors extracted for each block, and the number of rows in the matrix is equal to the total number of image blocks. Feature vectors are stored in a matrix P of order $(M - A + 1)$ $(N - A + 1) \times 7$. For a large number of blocks, finding similar blocks will be cumbersome if brute force search is used. Thus, lexicographical sorting of feature vectors is proposed by many authors [8, 9]. Lexicographical sorting places similar feature vectors in adjacent rows. If two blocks in the image are similar, their feature vectors will be placed close to each other in sorted matrix. Sorting reduces the comparison time for matching process. The above matrix is lexicographically sorted into P_L. After lexicographical sorting, each row is tested with its neighboring rows for detection of copy-move forgery. In the matrix P_L starting with first row (Xa) its feature distance D is calculated with neighboring rows. Distance D between two rows only when their corresponding block distance in image is more than the distance threshold (T_d). T_d is set according to length of block. Two blocks are similar if $D < T_s$, where T_s is similarity threshold. T_s is predefined and set according to image features. If two blocks are found to be matching, their positions are saved. The same procedure is repeated for all rows of P_L, and matching block positions are saved in matrix α.

4.5 *Post-Processing of Results*

After matching process, the block locations saved in matrix α are those of the copied and pasted blocks. Thus, marking all these locations in the image will isolate the copy-pasted blocks. Proposed algorithm marks all the blocks saved in α to black.

5 Results

In this section, we describe the results obtained with the proposed technique. All experimentation is performed on MATLAB R2009b installed on a computer with i3 2.3 GHz processor and 4 GB memory. Images used were from dataset [18]. The dataset has 200 images in small category of resolution 512 × 512 pixels and stored in png format. The dataset also applies post-processing operations like JPEG compression, blurring, color and contrast adjustments, brightness change to original as well as forged images. The area of copy-move regions varies with images in the database varying from approximately 28 × 14 to about 180 × 180 pixels, i.e., about 15–12% of the image size. The proposed algorithm localizes the forgery, and the performance of the proposed method is evaluated using two measures: true detection accuracy rate '$\dot{T}$' and false detection rate '$\dot{F}$' defined as:

$$\dot{T} = \frac{|c_o \cap c_r| + |P_o \cap P_r|}{|c_o| + |P_o|} \tag{3}$$

$$\dot{F} = \frac{|c_r - c_o| + |P_r - P_o|}{|c_r| + |P_r|} \tag{4}$$

where c_o, P_o are pixels in the copied and pasted regions in the original image to be tested for authenticity and c_r, P_r are the pixels in copied and pasted regions in the result image after detection. || and ∩ present area of the region and intersection of the two regions, respectively, while – means difference of the two regions. $\dot{T}$ indicates the performance of the proposed method to correctly detect pixels of copy-pasted regions in the image and $\dot{F}$ indicates the pixels that were not copied but were indicated to be copied by the proposed method. Closer the value of $\dot{T}$ to 1 and

Fig. 3 Original, forged, and detection result output

$\dot{F}$ to 0, higher is the precision of the system to detect copy-move forgery. In the following experimentation, we selected more than 200 images from CoMoFoD dataset to evaluate the proposed method. The block size was varied between 20×20, 32×32 and 50×50.

Images shown in Fig. 3 give the detection results as obtained with the proposed technique.

Results obtained with the proposed technique were good with all window sizes as long as the window was less than the size of the forged area. Non-regular duplicate regions are also detected as shown in Fig. 3. Table 1 presents the result as obtained on using images with JPEG compression with different quality factors.

The proposed method performs well in comparison with techniques [8, 9, 12] in terms of robustness against JPEG compression. Also the proposed approach has seven features for each block, which is considerably less than for methods based on DCT [8], PCA [9], Gabor magnitude [12]. Since all these techniques divide the

Table 1 True detection accuracy rate and false detection ratio for varying quality factor

Quality factor (Q)	Q = 30	Q = 40	Q = 50	Q = 60	Q = 70	Q = 80	Q = 90
Ṫ	0.62	0.78	0.90	0.92	0.94	0.95	0.96
Ḟ	0.36	0.20	0.17	0.15	0.12	0.08	0.03

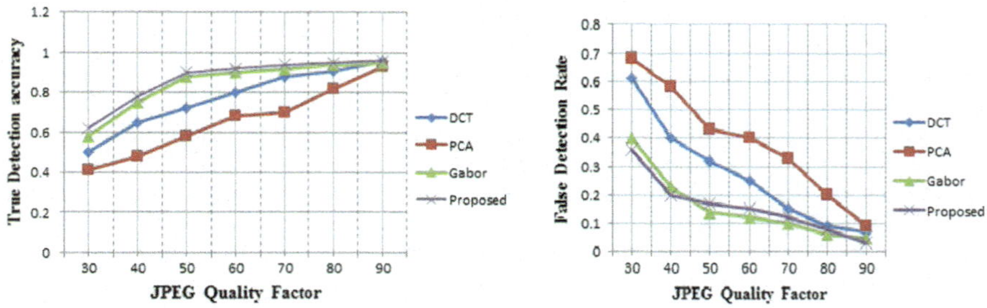

Fig. 4 Comparison of proposed technique with method based on DCT, PCA, Gabor magnitude

image into overlapping blocks and use lexicographical sorting, a reduced number of features representing each block reduce the computation complexity significantly. Since the proposed technique has fewer features in comparison with techniques [8, 9, 12], complexity is reduced. Figure 4 shows a comparison between the proposed technique and techniques given in [8, 9, 12] on the basis of robustness against JPEG compression.

It is clearly evident from the Fig. 4 that the proposed method shows a superior behavior against JPEG compression of varying quality factors. The proposed method outperforms methods [8, 9, 12] in terms of reduced computation complexity and robustness against JPEG compression.

6 Conclusion

Forgeries are carried out to conceal or change information conveyed by the image. Available image-editing software tools make copy-move an even easier task. This paper proposed a robust method for the detection of copy-move forgery based on GLRLM features. The experimental results demonstrate usefulness of proposed method in detecting and isolating the duplicate regions and its robustness against various post-processing operations like JPEG compression. The proposed feature extraction technique has reduced feature dimensionality as compared to most existing techniques making the technique computationally less complex. Thus, this technique can be used in the field of forensics for verifying the authenticity of images against copy-move forgery.

References

1. He Z, Sun W, Lu W, Lu H (2011) Digital image splicing detection based on approximate run length. Pattern Recogn Lett 32(12):1591–1597
2. Al-Qershi OM, Khoo BE (2013) Passive detection of copy-move forgery in digital images: state-of-the-art. Forensic Sci Int 231(1):284–295
3. Mushtaq S, Mir AH (2014) digital image forgeries and passive image authentication techniques: a survey. Int J Adv Sci Technol 73:15–32
4. Katzenbeisser S, Petitcolas F (2000) Information hiding techniques for steganography and digital watermarking. Artech House
5. Cox IJ, Miller ML, Bloom JA, Honsinger C (2002) Digital watermarking. Morgan Kaufmann, San Francisco
6. Kang L, Cheng XP (2010) Copy-move forgery detection in digital image. In: 3rd international congress on image and signal processing (CISP), vol 5, pp 2419–2421
7. Bayram S, Sencar HT, Memon N (2008) A survey of copy-move forgery detection techniques. In: IEEE Western New York image processing workshop. IEEE
8. Jessica F, David S, Jan L (2003) Detection of copy-move forgery in digital images. In: Proceedings of digital forensic research workshop, Cleveland, OH
9. Popescu AC, Farid H (2004) Exposing digital forgeries by detecting duplicated image region. Technical report 2004-515. Hanover, Department of Computer Science, Dartmouth College, USA, p 32
10. Huang Y, Lu W, Sun W, Long D (2011) Improved DCT-based detection of copy-move forgery in images. Forensic Sci Int 206(1):178–184
11. Muhammad G, Hussain M, Bebis G (2012) Passive copy move image forgery detection using undecimated dyadic wavelet transform. Digit Invest 9(1):49–57
12. Lee JC (2015) Copy-move image forgery detection based on Gabor magnitude. J Vis Commun Image Represent 31:320–334
13. Jaberi M, Bebis G, Hussain M, Muhammad G (2013) Improving the detection and localization of duplicated regions in copy-move image forgery. In: 2013 18th international conference on digital signal processing (DSP), pp 1–6
14. Christlein V, Riess C, Jordan J, Riess C, Angelopoulou E (2012) An evaluation of popular copy-move forgery detection approaches. IEEE Trans Inf Forensics Secur 7(6):1841–1854
15. Tang X (1998) Texture information in run-length matrices. IEEE Trans Image Process 7 (11):1602–1609
16. Galloway MM (1975) Texture analysis using gray level run lengths. Comput Graph Image Process 4(2):172–179
17. Chu A, Sehgal CM, Greenleaf JF (1990) Use of gray value distribution of run lengths for texture analysis. Pattern Recogn Lett 11(6):415–419
18. Tralic D, Zupancic I, Grgic S, Grgic M (2013) CoMoFoD—new database for copy-move forgery detection. In: 2013 55th international symposium ELMAR, pp 49–54

A Novel Triple Band Pass Frequency Selective Surface for the Proliferation the Performance of WiMax and WLAN 2.5/3.5/5.5 GHz

Monika Garg, Rekha Chahar and Sanjeev Yadav

Abstract Among the upcoming wireless technologies, WiMax and WLAN provide soaring speed communication with full coverage. However due to assured factors like path loss, interference, and coverage area led to signal degradation. WiMax and WLAN have used different range of bands. But these bands suffer large path loss, so a band-pass frequency selective surface has been projected in this paper which led to escalation in the transmission of WiMax signal to reduce the effect of path loss. In the proposed design, a single-sided FR-4 substrate is inbuilt with the square and three elements that led to triple-band characteristics at 2.5, 3.5, and 5.5 GHz. The dielectric constant of substrate is 4.4, and dimension of substrate is 0.166 λ × 0.166 λ. Further, this paper has been realized with simulated results in CST Microwave Studio.

Keywords Three elements · Triple-band-pass frequency selective surface (FSS) WiMax and WLAN 2.5, 3.5, and 5.5 GHz band

1 Introduction

In the field of electromagnetic engineering for a specific set of frequency band, frequency selective surface (FSS) has been designed to tamper the absorption, transmission, and reflection property of electromagnetic waves. FSS is two-dimensional periodic arrays that consist of one or more metallic patterns in the form of apertures or patches and behave like spatial filters in free space. The frequency response of FSS is described by the unit cell that is the geometry of the

M. Garg (✉) · R. Chahar · S. Yadav
Government Women Engineering College Ajmer, Ajmer, India
e-mail: mgmonikagarg58@gmail.com

R. Chahar
e-mail: raykhachahar2007@gmail.com

S. Yadav
e-mail: sanjeev.mnit@gmail.com

V. Janyani et al. (eds.), *Optical and Wireless Technologies*, Lecture Notes in Electrical Engineering 472, https://doi.org/10.1007/978-981-10-7395-3_47

structure one period. Therefore, a large number of unit cells must be present in order to study the efficient frequency selective behavior. In order to design finite and infinite array, various boundary conditions are applied such as Floquet mode PEC or PMC. In FSS, current flows in-phase led to improved impedance bandwidth, angular stability, and gain [1, 2]. For the analysis of FSS, large number of algorithms can be used such as spectral domain method, periodic method, equivalent circuit method, finite difference time domain method, finite element method [3, 4].

FSS has been studied for wide variety of applications like radar, electromagnetic applications, and satellite applications with resonant unit cells whose dimensions are comparatively half of the wavelength [5, 6]. WiMax and WLAN are the most powerful that provide better communication with full coverage [7].

In the proposed FSS, FR-4 substrate has been used with thickness 1.6 mm, as it is easily available in the market and provides efficient results on the desired frequency for broadband applications. This FSS has worked as band-pass filter at frequency 2.5 GHz, 3.5 GHz, and 5.5 GHz for Bluetooth, WiMax, and WLAN applications, respectively. The structure has realized in CST Microwave Studio in frequency domain solver based on FDTD method. After simulation, various parameters such as insertion, absorption, and return losses are calculated which tell us about how much power has transmitted, absorbed, or reflected from FSS. The various parameters such as inter-element spacing, patch size, and slots affect the results of FSS.

2 Design of the Superstrate

2.1 *Unit Cell*

The designed triple-band-pass FSS is constructed of a single-sided FR-4 superstrate having thickness of 1.6 mm and dielectric constant of 4.4 and that has square shaped having dimension W (mm) × W (mm). On its one-side, a patch is entrenched and included of square and three elements. Due to the combination of square and circle, a 2.5 GHz band is obtained. Moreover, band of 3.5 GHz and 5.5 GHz is obtained due to ellipse and rectangle, respectively. The dimensions are illustrated in Fig. 1 and Table 1.

2.2 *Design of Finite Array*

An array of 8 × 8 element has been constructed based on the proposed FSS unit cell element whose dimensions are Ws (mm) × Ws (mm). Each unit cell element

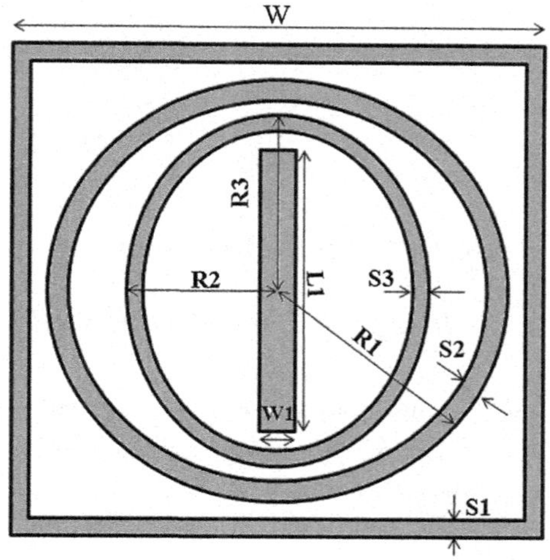

Fig. 1 Front view of the proposed unit cell with dimensions

Table 1 Dimensions of proposed unit cell

Name of dimensions	Symbol	Value in (mm)
Width of substrate	W	20
Width of square	W	20
Radius of circle	R1	9
Major radius of ellipse	R2	6.2
Minor radius of ellipse	R3	7.9
Length of rectangle	L1	12.7
Width of rectangle	W1	1.4
Thickness of substrate	H	1.6

Fig. 2 Front view of the proposed finite array with dimensions

in the finite array is iterated with a periodicity P in both X and Y direction as shown in Fig. 2 and Table 2. Rest of the dimensions are same as that of unit cell element.

Table 2 Dimensions of proposed FSS

Name of dimension	Symbol	Value (in mm)
Width of the substrate	Ws	160
Thickness of the substrate	H	1.6
Periodicity	P	20

3 Results and Discussion

3.1 Insertion Loss of Unit Cell

The simulated results of designed FSS are shown in fig as below; transmission coefficient (S21) is being simulated to know how much power is transmitted from the structure, and it is shown in Fig. 3. The transmission coefficient is −0.27 dB, −0.89 dB, and −0.91 dB at 2.5, 3.5, and 5.5 GHz frequency, respectively.

3.2 Surface Current

At 2.5 GHz, surface current has its effect on the edges of all the three element of patch but it has maximum at the edge of outer element as illustrated in Fig. 4. At 3.5 GHz, surface current has its maximum effect only on the edges of the ellipse as illustrated in Fig. 5. At 5.5 GHz, a maximum effect is seen at the edge of rectangle as shown in Fig. 6.

3.3 Incidence Angle Variation

As illustrated in Fig. 7, to see the behavior of insertion loss and frequency response, incident angles (θ) are varied from 0° to 75°. Maximum signal is transmitted at 0°, and absorption is more with less transmission as we increase the incidence angle.

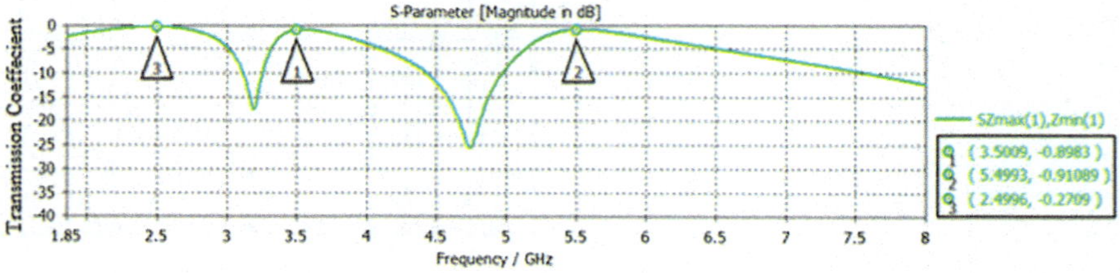

Fig. 3 Simulated transmission coefficient of unit cell

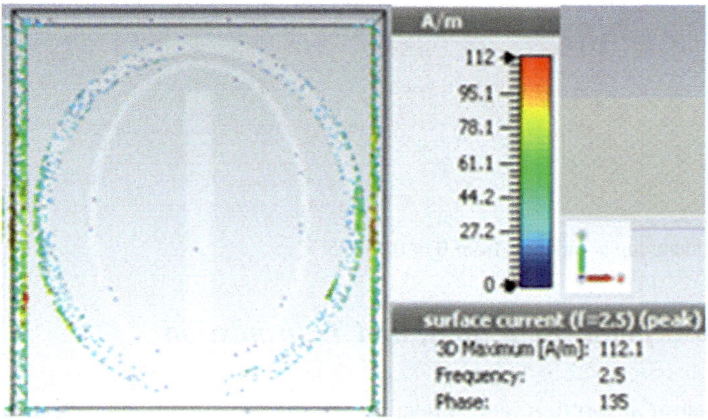

Fig. 4 Surface current at 2.5 GHz

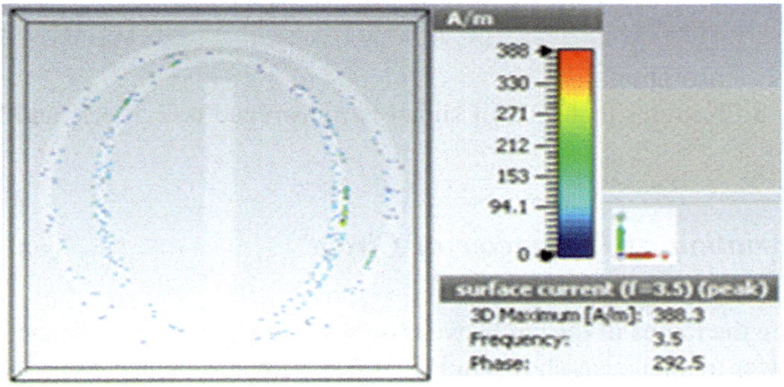

Fig. 5 Surface current at 3.5 GHz

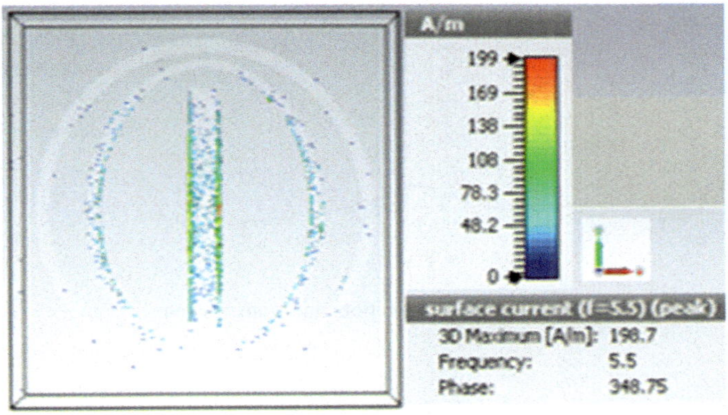

Fig. 6 Surface current at 5.5 GHz

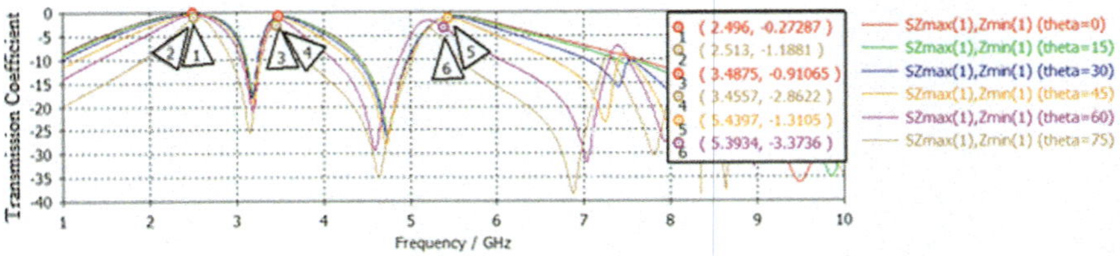

Fig. 7 Incident angle variation from θ = 0° to 75°

3.4 Absorption, Reflection, and Transmission

Transmission, absorption, and reflection are the three main characteristics of FSS. Absorption reveals that the signal is neither reflected nor transmitted. Absorption (A) is calculated using formula,

$$A = 1 - |S11|^2 - |S21|^2$$

where A = Absorption.

Figure 8 illustrates that 92% of signal is transmitted at 2.5, 3.5, and 5.5 GHz.

3.5 Variation of Parameter of Circle

On varying the radius of the circle, we observed a shifting in the resonant frequency toward lower frequency as shown in Fig. 9. Therefore, for better results we consider the radius as 9 mm.

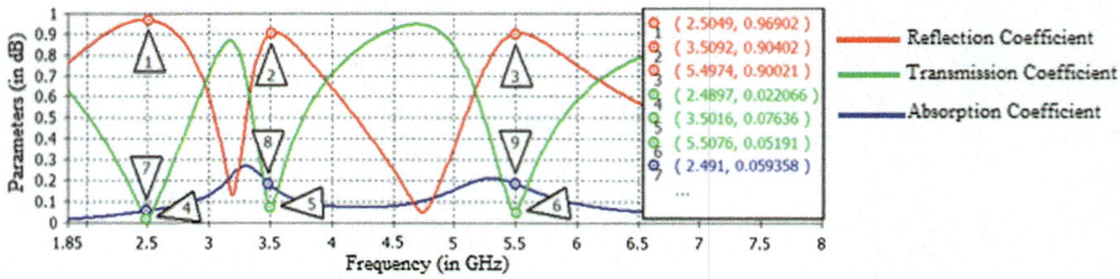

Fig. 8 Simulated results for absorption, reflection, and transmission

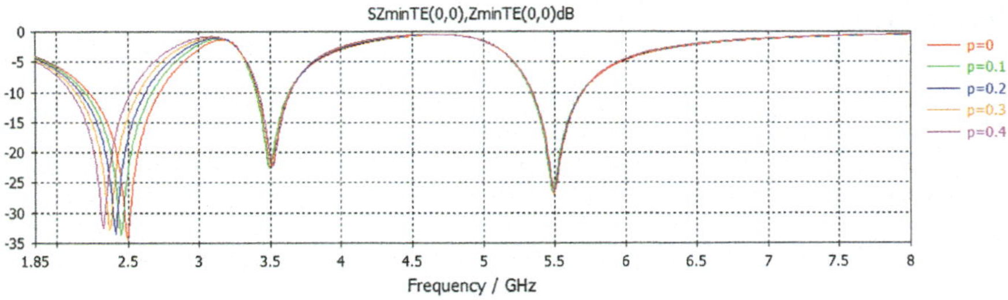

Fig. 9 Variations in parameter of circle

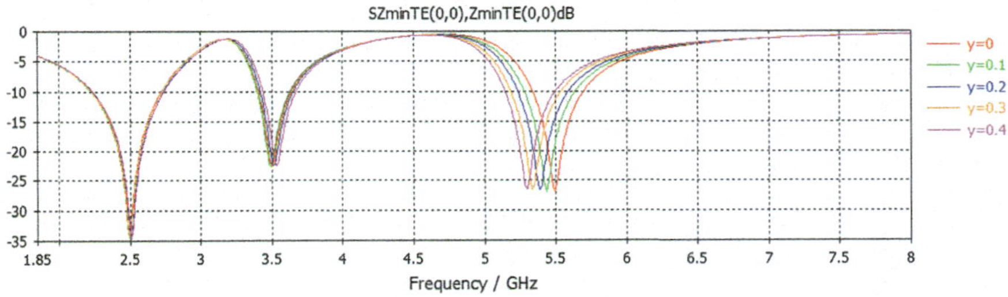

Fig. 10 Variations in parameter of rectangle

3.6 *Variation of Parameter of Rectangle*

On varying the width of the rectangle and keeping the length constant, there is occurred variation in the resonant frequency, i.e., the frequency is shifting toward lower side of the resonant frequency as shown in Fig. 10.

4 Conclusion

A single-sided triple-band-pass frequency selective surface is designed in order to minimize path loss for better transmission. Various parameters are shown in this paper such as incident angle variation, variation in dimensions of the structure which show that 92% of the signal is transmitted at 0° incident angle and rest of it get absorbed which shows enlarged impedance bandwidth, transmission, and gain. In future, it can be extended to other broadband application for frequency selective surface.

References

1. Chen H-Y, Tao Y (2011) Bandwidth enhancement of a U-slot patch antenna using dual-band frequency-selective surface with double rectangular ring elements. MOTL 53(7):1547–1553
2. Yadav S, Peswani B, Sharma MM (2014) A novel band pass double-layered frequency selective superstrate for WLAN applications. In: Confluence, IEEE, pp 447–451
3. Yadav S, Peswani B, Choudhury R, Sharma MM (2014) Miniaturized band pass double layered frequency selective surface superstrate for Wi-Max applications. In: 2014 IEEE symposium, ISWTA
4. Qing A, Lee CK (2010) Differential evolution in electromagnetic. Springer, Heidelberg, Berlin
5. Das S, Sarkar D, Sarkar PP, Chowdhury SK (2003) Experimental investigation on a polarization independent broadband frequency selective surface. In: Proceedings of national conference held in KIIT, Bhubuneswar, Mar 2003
6. Wu TK et al (1992) Multi ring element FSS for multi band applications. Paper presented at the international IEEE AP-S SYMPOSIUM, Chicago
7. Aseri K, Yadav S, Sharma MM (2016) A novel band pass frequency selective surface for the augmentation in the performance of Wi-Max 2.5/3.5 GHz. In: Antenna test and measurement society, 01–03 Feb 2016, Goa, India
8. Chatterjee A, Bisas S, Chanda (Sarkar) D, Sarkar PP (2011) A polarization independent compact multi-band frequency selective surface. In: International conference on current trends in technology, NUiCONE 2011
9. Munk BA (2000) Frequency selective surfaces: theory and design. Wiley-Interscience, New York
10. Wu TK (1995) Frequency selective surface and grid array. A Wiley Interscience Publication, New York, pp 5–7
11. Pozar DM (2005) Microwave engineering, 3rd edn. Wiley; Radisic V, Qian Y, Itoh T (1998) Novel 2-D photonic bandgap structure for microstrip lines. IEEE Microw Guided Wave Lett 8 (2):69–71
12. Li Y, Jiang H, He L, Li H, Zhang Y, Chen H (2006) Multichanneled filter based on a branchy defect in microstrip photonic crystal. Appl Phys Lett 88:081106. Prog Electromagnet Res Lett 31:53
13. Ahn D, Park JS, Kim CS, Kim J, Qian Y, Itoh T (2001) A design of the low-pass filter using the novel microstrip defected ground structure. IEEE Trans Microw Theory Technol 49 (1):86–93
14. Falcone F, Lopetegi T, Laso MAG, Baena JD, Bonache J, Beruete M, Marques R, Martin F, Sorolla M (2004) Babinet principle applied to the design of metasurfaces and metamaterials. Phys Rev Lett 93:197401
15. Kee CS, Jang MY, Park IM, Lim H, Kim JE, Park HY, Lee JI (2002) Photonic band gap formation by microstrip ring: a way to reduce the size of microstrip photonic band gap structures. Appl Phys Lett 80:1520

Compact Triple-Band Stubs-Loaded Rectangular Monopole Antenna for WiMAX/WLAN Applications

Ashok Kumar and Mahendra Mohan Sharma

Abstract A compact triple-band stubs-loaded rectangular monopole antenna for WiMAX/WLAN applications is presented and designed on Rogers RT/Duroid 5880 substrate. The stubs-loaded rectangular monopole antenna is employed a small rectangular patch, 50 Ω microstrip feed line, one T-shaped stub resonator, two asymmetrical inverted L-shaped stubs resonator, and partial ground plane. By loading these stubs, the antenna is resonating at three frequencies at 2.60/3.52/5.45 GHz and obtained impedance bandwidths of 180 MHz (2.50–2.68 GHz), 320 MHz (3.37–3.69 GHz), and 680 MHz (5.15–5.83 GHz), for 2.5/3.5/5.5 GHz WiMAX and 5.2/5.8 GHz WLAN band applications. The antenna provides monopole-like radiation pattern and stable antenna realized gains over the operating bands. In addition, the design approach of the proposed antenna is also discussed.

Keywords L-shaped stub · Rectangular monopole antenna · T-shaped stub WiMAX · WLAN

1 Introduction

The remarkable growth in wireless technology, the demand of compact size, low profile, omnidirectional radiation pattern, and multiband antenna for portable wireless terminal devices is extensively and rapidly increased. The wireless terminal devices, namely worldwide interoperability for microwave access (WiMAX)

A. Kumar (✉)
Department of Electronics and Communication Engineering, Government Women Engineering College, Ajmer 305002, Rajasthan, India
e-mail: kumarashoksaini@gmail.com

M. M. Sharma
Department of Electronics and Communication Engineering, Malaviya National Institute of Technology, Jaipur 302017, Rajasthan, India

V. Janyani et al. (eds.), *Optical and Wireless Technologies*, Lecture Notes in Electrical Engineering 472, https://doi.org/10.1007/978-981-10-7395-3_48

and wireless local area networks (WLAN) systems can integrate with the single system to effectively promote the portability of wireless devices. To meet these requirements, a multiband antenna is a good choice for the reason that different resonance frequencies can be achieved. The numerous types of microstrip-fed/coplanar-fed printed monopole/slot antennas have been reported [1–7]. The various dual-band printed monopole antennas with double-T stubs [1], parasitic inverted L-stubs [2], parasitic T-shaped, and inverted L-shaped stub [3] are reported for WiMAX and WLAN applications. A planar monopole antenna with horizontal H-shaped patch, an L-shaped open-end stub and deformed inverted T-shaped strip [4], a rectangular monopole antenna with rectangular ring and pair of F-shaped stubs [5], microstrip-fed monopole antenna with circular-arc-shaped and L-shaped stub [6], and wide-slot fork-like monopole antenna with dual-ring resonator [7] are demonstrated for triple-band WiMAX/WLAN operation. Unfortunately, those antennas have some disadvantages such as massive size, complex structures, and operation bands. Although all the reported antennas exhibit dual-band [1–3], triple-band performance [4–7] but in comparison, proposed antenna has small size ($30 \times 20 \times 0.79$ mm^3) and simple structure to realize the desired operating bands for WiMAX and WLAN bands.

In this paper, a compact microstrip-fed rectangular monopole antenna with one T-shaped stub resonator and two asymmetrical inverted L-shaped stub resonators is introduced. The stub resonators can be embedded separately on the upper edge of the rectangular patch, and the designed structure is capable of resonating at 2.60/3.52/5.45 GHz frequencies. The proposed antenna exhibits triple-band performance for 2.5/3.5/5.5 GHz WiMAX and 5.2/5.8 GHz WLAN applications. In addition, the stub resonators working independently and individual stub resonator impedance bandwidth can be tuned separately. The detailed antenna configuration discussion and results of the designed antenna are presented in the following Sect. 2 and 3, respectively.

2 Antenna Configuration and Design Approach

To shorten the design of a stubs-loaded rectangular monopole antenna, the step by step evaluation technique is illustrated in Fig. 1a–c, and geometry of the proposed antenna with labeled parameters is depicted in Fig. 1d. All antennas are designed on low-loss Rogers RT/duroid 5880 dielectric substrate with dielectric constant of 2.2, the thickness of 0.79 mm, and loss tangent of 0.0009.

The proposed antenna as shown in Fig. 1d evaluation takes place from the length and width of rectangular patch $Lp \times Wp$; 50 Ω microstrip feed line width of Wf, and partial ground plane $Lg \times W$ and it has a compact size $L \times W$. The rectangular monopole antenna with T-shaped stub resonator shown in Fig. 1a (i.e., antenna I) is resonating at 2.71 GHz that mainly depends on vertical arm length $L1$ and horizontal

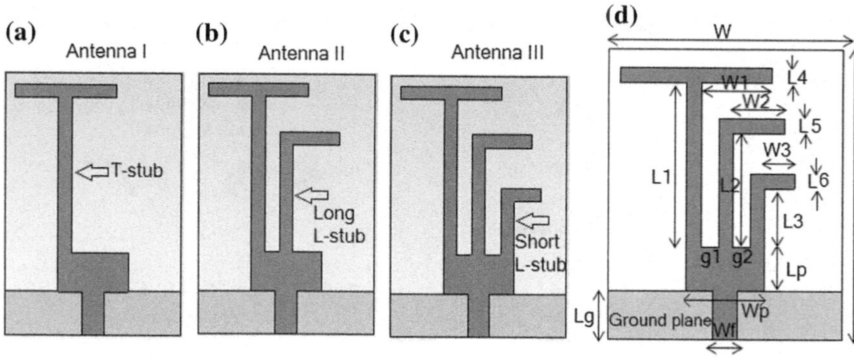

Fig. 1 Step by step evaluation of stubs-loaded monopole antenna **a** antenna I, **b** antenna II, **c** antenna III, and **d** Geometry of proposed antenna with labeled parameters

arm width $W1$ of the T-shaped stub (mentioned in Fig. 1d), and antenna I is called single-band antenna. The inverted long L-shaped stub resonator with vertical arm length $L2$ and horizontal arm width $W2$ is added in antenna I separately with stub gap $g1$ (mentioned in Fig. 1d) as shown in Fig. 1b (i.e., antenna II) gives an additional resonance at 3.54 GHz, and antenna II is called dual-band antenna. In continuation of Fig. 1b, another inverted short L-shaped stub with vertical arm length $L3$ and horizontal arm width $W3$ is added with stub gap $g2$ (mentioned in Fig. 1d) as shown in Fig. 1c (i.e., antenna III) is capable of resonating at 2.60, 3.52, and 5.45 GHz frequencies, and antenna III is called triple-band antenna (proposed antenna). The dimensions of the proposed antenna after optimization on CST Microwave Studio and evaluation are as follows: $L = 30$ mm, $W = 20$ mm, $Lp = 4.5$ mm, $Lg = 5$ mm, $Wf = 2.44$ mm, $Wp = 8$ mm, $L1 = 17.8$ mm, $L2 = 12.3$ mm, $L3 = 6.15$ mm, $L4 = 1.5$ mm, $L5 = 1.5$ mm, $L6 = 1.5$ mm, $W1 = 5$ mm, $W2 = 5.3$ mm, $W3 = 3$ mm, $g1 = 1.7$ mm, $g2 = 1.8$ mm.

The design approach for the excitation of triple resonant mode, the resonant frequency f_{ri}, and effective dielectric constant ε_{eff} can be calculated by Eqs. (1) and (2) as:

$$L_{si} = \frac{\lambda_g}{4} = \frac{c}{4f_{ri}\sqrt{\varepsilon_{eff}}}; \quad i = 1, 2, 3 \tag{1}$$

$$\varepsilon_{eff} \approx \frac{\varepsilon_r + 1}{2} \tag{2}$$

where c = speed of light in a vacuum (3×10^8 m/s), ε_r = dielectric constant of the substrate, λ_g is the guided wavelength and L_{si} is the total length of the stub resonator.

In order to satisfy the triple resonant frequencies of the proposed antenna, the first resonant mode is excited due to the T-shaped stub. The total length of the

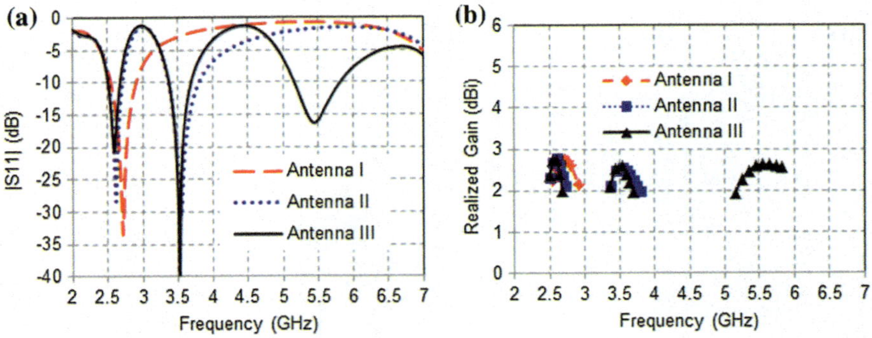

Fig. 2 **a** Reflection coefficient (S11) and **b** realized gains for different antenna configurations of Fig. 1a–c

T-shaped stub can be calculated as $L_{s1} = L1 + W1 = 17.8 + 5 = 22.8$ mm, which is about a quarter of the guided wavelength λ_g and calculated theoretical resonant frequency f_{r1} by (1) and (2) is 2.60 GHz. The second resonant mode is excited due to the inverted long L-shaped stub. The total length of inverted long L-shaped stub can be calculated as $L_{s2} = L2 + W2 = 12.3 + 5.3 = 17.6$ mm, which is about a quarter of guided wavelength λ_g and calculated theoretical resonant frequency f_{r2} by (1) and (2) is 3.37 GHz. The third resonant mode is excited due to the inverted short L-shaped stub. The total length of the inverted short L-shaped stub can be calculated as $L_{s3} = L3 + W3 + L6 = 6.15 + 3 + 1.5 = 10.65$ mm, which is about a quarter of guided wavelength λ_g and calculated theoretical resonant frequency f_{r3} by (1) and (2) is 5.57 GHz. Therefore, we can say that the simulated three resonant frequencies are very close to calculated resonant frequencies. Thus, the proposed antenna structure is verified with the design concept. Figure 2a and b shows the reflection coefficient (|S11|) and realized gain curves for different antenna configurations of Fig. 1a–c, respectively. The comparison of impedance bandwidths ($|S11| \leq -10$ dB) and realized gains is listed in Table 1 for different antenna configurations of Fig. 1a–c.

Table 1 Comparison of bandwidths ($|S11| \leq -10$ dB) and realized gains variations of single-, dual- and triple-band monopole antenna

Antenna configuration	Stub type	Impedance bandwidth (GHz) ($|S11| \leq -10$ dB)	Realized gain variation (dBi)	Antenna response
Antenna I	T-shaped stub	2.55–2.91	2.15–2.75	Single-band
Antenna II	T-shaped stub and inverted long L-shaped stub	2.51–2.70	2.11–2.79	Dual-band
		3.36–3.80	2.0–2.54	
Antenna III	T-shaped stub, inverted long L-shaped stub, and inverted short L-shaped stub	2.50–2.68	2.0–2.72	Triple-band (Proposed antenna)
		3.37–3.69	1.96–2.58	
		5.15–5.83	1.94–2.62	

3 Results and Discussion

The |S11| results of the proposed antenna are shown in Fig. 2a (i.e., antenna III). It is resonating at 2.60/3.52/5.45 GHz frequencies with corresponding impedance bandwidths 180 MHz (2.50–2.68 GHz) for 2.5 GHz WiMAX band, 320 MHz (3.37–3.69 GHz) for 3.5 GHz WiMAX band and 680 MHz (5.15–5.83 GHz) for 5.5 GHz WiMAX and 5.2/5.8 GHz WLAN bands and fractional bandwidths (FBWs) of 6.92, 9.09, and 12.48% with respect to the appropriate resonant frequencies. The surface current distributions at three resonant frequencies 2.6, 3.52, and 5.45 GHz have been illustrated in Fig. 3a–c, respectively, to analyze the performance of the proposed antenna. It can be seen that the current distribution reaches the densest mainly on the T-shaped stub in the same phase as shown in Fig. 3a. It indicates that first resonant mode is excited about at 2.60 GHz due to surface current length L_{s1} of the T-shaped stub. Similarly, it can be seen from Fig. 3b, the surface current distribution reaches the densest on the inverted long L-shaped stub and radiating patch in the same phase. It indicates that second resonant mode is excited at about 3.52 GHz due to surface current length L_{s2} of the inverted long L-shaped stub. The third resonant mode is excited about at 5.45 GHz due to surface current distributed mainly on the inverted short L-shaped stub in the same phase as shown in Fig. 3c. In addition, we can say that the first, second, and third resonant mode is controllable by the total length of stub resonators L_{s1}, L_{s2}, and L_{s3}, respectively. Thus, it may be concluded that the resonant frequency is mainly dependent on the total length of different stub resonators and it has been also verified theoretically in Sect. 2.

The radiation patterns (co-pol and cross-pol) of E- and H-plane at three resonant frequencies 2.60, 3.52, and 5.45 GHz of the proposed antenna are shown in

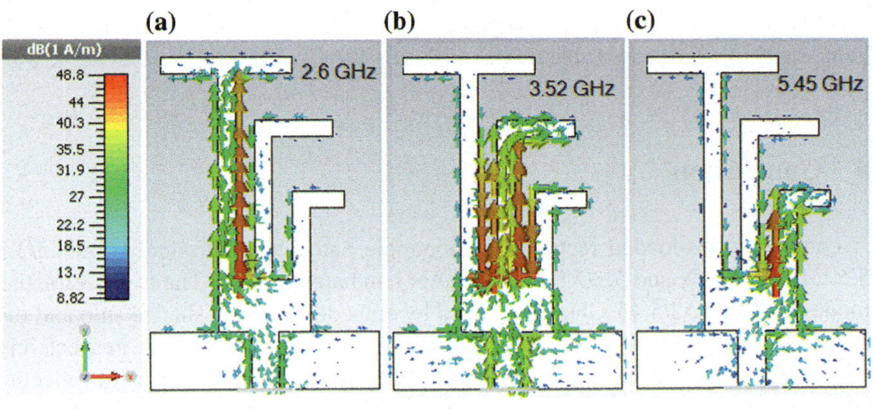

Fig. 3 Surface current distributions of the proposed stubs-loaded rectangular monopole antenna at three resonant frequencies **a** 2.60, **b** 3.52, and **c** 5.45 GHz

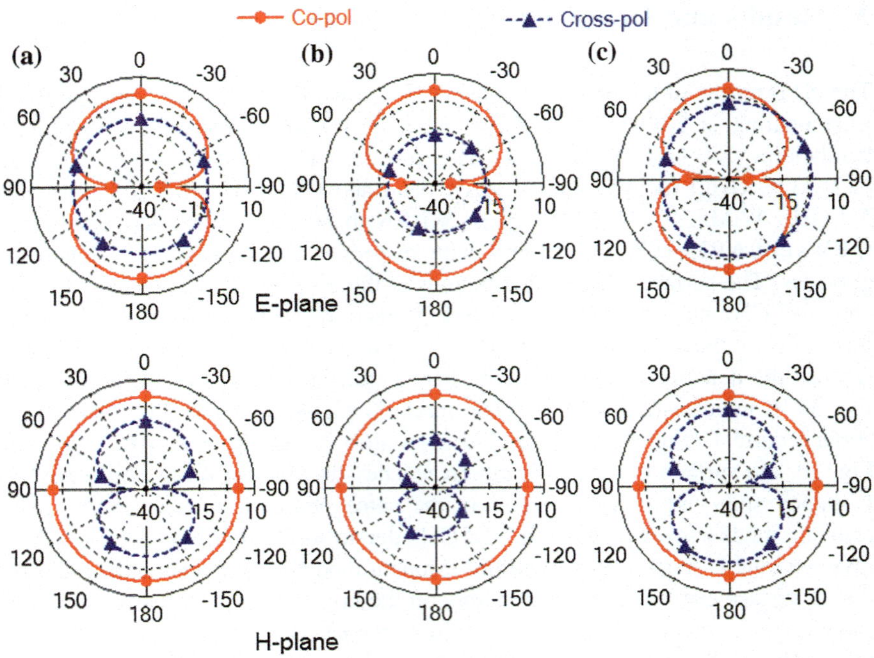

Fig. 4 E- and H-plane radiation patterns for the proposed stubs-loaded rectangular monopole antenna at three resonant frequencies at **a** 2.60, **b** 3.52, and **c** 5.45 GHz

Fig. 4a–c, respectively. It can be seen that the monopole-like pattern in the E-plane and omnidirectional pattern in the H-plane at all resonant frequencies is observed. As we can see that the high level of cross-polarization is observed in both E- and H-plane due to increasing the horizontal components of surface current distributions. The realized gain of the proposed antenna is shown in Fig. 2b and realized gain variations listed in Table 1.

4 Conclusion

A compact stubs-loaded rectangular monopole antenna is introduced for 2.5/3.5/5.5 GHz WiMAX and 5.2/5.8 GHz WLAN band applications. The triple resonance modes at 2.60/3.52/5.45 GHz are excited by embedding one T-shaped stub and two inverted L-shaped stub resonators with FBWs of 6.92/9.09/12.48%, respectively. The design approach of the antenna is explained and studied in terms of reflection coefficient (|S11|), realized gain, surface current distributions, and radiation patterns.

References

1. Kuo Y-L, Wong K-L (2003) Printed double-T monopole antenna for 2.4/5.2 GHz dual-band WLAN operations. IEEE Trans Antennas Propag 51(9):2187–2192
2. Peng L, Ruan C-L (2009) A microstrip fed patch antenna with two parasitic invert L stubs for dual-band WLAN applications. Wireless Pers Commun 57(4):727–734
3. Deegwal JK, Govil MC, Sharma MM (2015) Dual-band printed circular disc antenna with two parasitic stubs for wireless applications. Int J Microw Opt Tech 10(5):314–320
4. Huang SS, Li J, Zhao JZ (2014) Design of a compact triple-band monopole antenna for WLAN/WiMAX applications. Prog Electromagnet Res C 48:29–35
5. Tang Z, Liu K, Yin Y, Lian R (2015) Design of compact triband monopole antenna for WLAN and WiMAX applications. Microw Opt Tech Lett 57(10):2298–2303
6. Shi YW, Xiong L, Chen MG (2015) Compact triple-band monopole antenna for WLAN/WiMAX-Band USB dongle applications. ETRI J 37(1):21–25
7. Liu G, Liu Y, Gong S (2016) Compact tri-band wide-slot monopole antenna with dual-ring resonator for WLAN/WiMAX applications. Microw Opt Tech Lett 58(5):1097–1101

A Triple Band-Reject Frequency Selective Surface for Broadband Applications

Shweta Garg and Sanjeev Yadav

Abstract In this paper, a new FSS design as band-reject filter (BRF) has been proposed by means of periodic structures of frequency selective surfaces (FSS) which has been worked for triple-band frequencies. In the proposed design, a single-sided Teflon (PTFE) lossy substrate is incorporated with three elements that led to triple-band (2.45, 3.5 and 5.5 GHz) characteristics. The substrate of dimension of 0.293 λ × 0.293 λ at center frequency and having a dielectric constant of 2.1 has been used. Three patches are designed of different shapes. Element types FSS is used as a band-reject filter for broadband applications which have much greater bandwidth than its band-pass portion. Further, this paper introduced its simulated results and discussion that is realized in CST Microwave studio. The design is intended at the rejection of triple bands 2.45 GHz, 3.5 GHz, and 5.5 GHz which are for Bluetooth, WiMax, and WLAN, respectively.

Keywords Band-reject filter (BRF) · Bluetooth · Frequency selective surface (FSS) · Triple band · WiMax · WLAN

1 Introduction

Frequency selective surface suppressed the surface wave and reflect plane wave in-phase with high impedance [1]. It has been achieving wide applications in radar system, broadband communication and antenna technology [2]. Frequency selective surface (FSS) has been made up of thin and repetitive surface of planar periodic arrays of metallic slots or patches, and they are designed to absorb, reflect or

---

S. Garg (✉) · S. Yadav
Government Women Engineering College Ajmer, Ajmer, India
e-mail: shwetagarg553@gmail.com

S. Yadav
e-mail: sanjeev.mnit@gmail.com

V. Janyani et al. (eds.), *Optical and Wireless Technologies*, Lecture Notes in Electrical Engineering 472, https://doi.org/10.1007/978-981-10-7395-3_49

transmit electromagnetic fields at desired frequency according to required applications [3–4]. The current applications of FSS are selective shielding of frequencies in airport and military communication, and isolation of unnecessary and harmful radiation in microwave bands in hospitals, schools, domestic environments. FSS has many applications as band pass radome for radio, astronomy and missiles [5–7].

A band stop FSS is reflective for the propagation of electromagnetic waves in a certain frequency. Band stop filters are basically consists of band stop arrays and they are manufactured by using patches [8]. Therefore in this proposed design, three patches are designed of different shapes. Element types FSS is used as a Band reject filter for broadband applications which have much greater bandwidth than its band pass portion [9].

Wireless communication proceeds to benefit from augmentation in wireless internet and wireless home networking areas. Bluetooth, WiMAX and WLAN are wireless broadband technologies that provide inter communication between two or more devices. Bluetooth works on 2.45 GHz frequency [10]. WiMax is most powerful one with high data rate and full coverage. WiMax has different range of band among which WiMax 2.5 GHz and 3.5GHz are mostly used [11–13]. The recent and robust WLANs can provide reliable high speed connectivity and operates in 5–6 GHz band [14].

In this paper, the proposed FSS is used as a band-reject filter (BRF) for broadband applications, which will reject triple band frequencies in the range of S and C band, and the rest of the frequencies will pass. The FSS has been constructed on a single sided square shaped Teflon PTFE (lossy) superstrate [15–16]. This PTFE substrate capable of withstanding high temperature and weather consequences and it is a better insulator. The purpose of using band reject filter is keeping out the signals from undesirable users [17]. The CST MWS has been used for the simulation of proposed design.

2 Design

2.1 Cell Description

Figure 1 and Table 1 illustrated the construction and dimensions of the single unit cell of the proposed FSS. From Fig. 1, the FSS has been printed on a single-sided square-shaped Teflon PTFE (lossy) superstrate with a thickness of 1.6 mm and dielectric constant of 2.1. This PTFE substrate is capable of withstanding heat and weather consequences, and it is a good insulator. It has a dimension of W (mm) W (mm). It consists of three patch elements in which the outer patch is square shaped with dimension L (mm) × L (mm), middle patch is circular in shape with radius R (mm), and the inner patch is square in shape with dimension A (mm) × A (mm) which is deposited on the substrate. The outer patch is embedded for 2.45 GHz band. The middle element is embedded for 3.5 GHz band. The inner

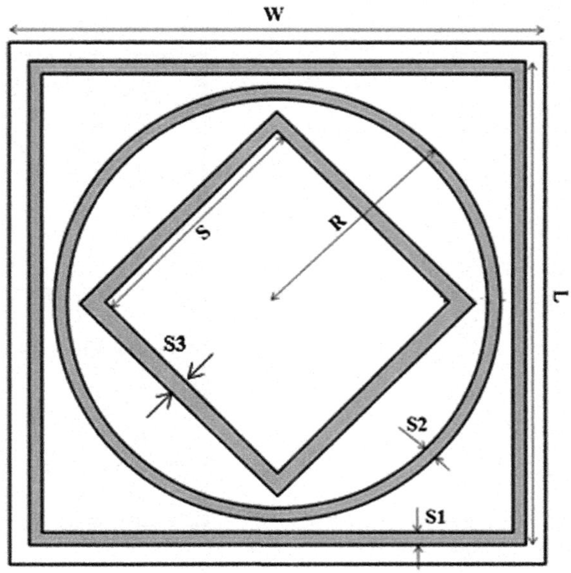

Fig. 1 Design of triple band-stop FSS

Table 1 Dimensions of triple-band FSS

Name	Dimension (in mm)
Dimension of the substrate (W)	22
Substrate thickness (h)	1.6
Length of the rectangular patch (L)	20.7
Width of rectangular patch (S1)	0.25
Radius of the circular patch (R)	19.8
Width of circular patch (S2)	0.2
Side of square patch (A)	13.3
Width of square patch (S3)	0.3

element is embedded for 5.5 GHz band, and all the dimensions are illustrated in Fig. 1 and Table 1

2.2 *Design of Finite Array*

The proposed frequency selective surface has been constructed by cascading an array of 8 × 8 elements whose dimensions are 176 (mm) × 176 (mm) as shown in Fig. 2 and Table 2. With periodicity of 22 mm, the rest of the dimensions are same as that of the single cell.

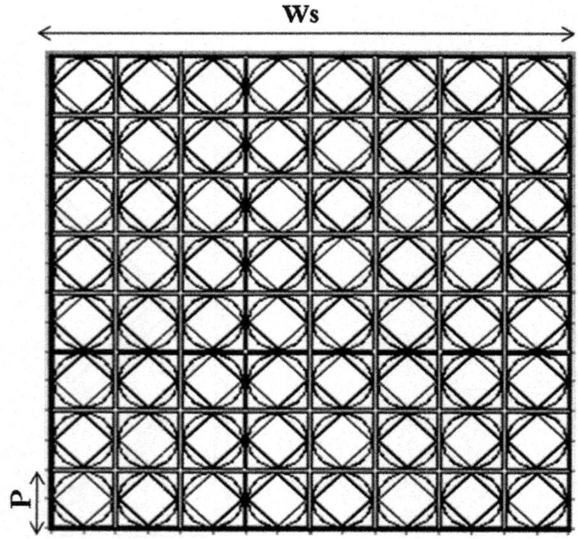

Fig. 2 Front view of the proposed finite array

Table 2 Dimensions of 8 × 8 array FSS

Name of dimension	Symbol	Value (in mm)
Dimension of substrate	Ws	176
Thickness of substrate	H	1.6
Periodicity	P = W	22

3 Results and Discussions

3.1 Transmission, Reflection, and Absorption Coefficient for Single Cell

The triple band-reject FSS resonates at 2.4 GHz, 3.5 GHz, and 5.5 GHz, and the transmitted electric field is −63 db, −49 db, and −69 db, respectively, as shown in Fig. 3. It has a bandwidth 5–6 GHz which resonates at frequency 5.5 GHz. From the absorption graph, it is clear that about 99% of power is reflected and only 0.03% of power is absorbed which is shown in Fig. 5.

3.2 Surface Current

At 2.45 GHz, surface current has its effect on the edges of all the three elements of patch but it has maximum effect at the edge of outer element as illustrated in Fig. 4. At 3.5 GHz, surface current has its maximum effect only on the edges of the

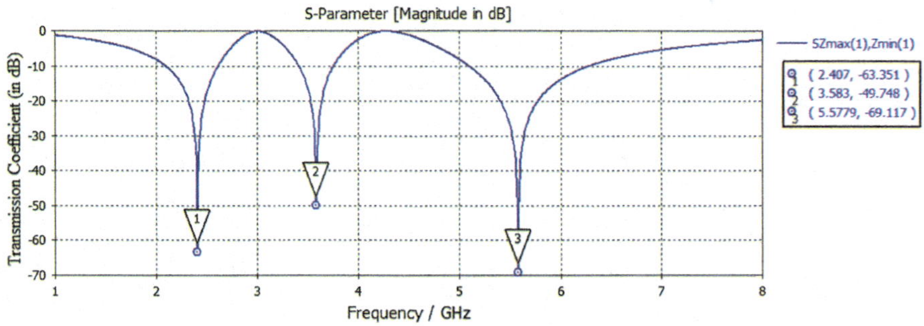

Fig. 3 Simulated transmission coefficient of triple band-stop FSS

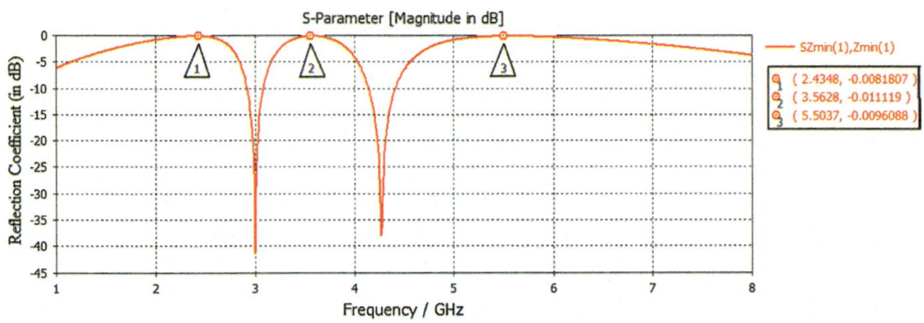

Fig. 4 A simulated reflection coefficient of triple band-stop FSS

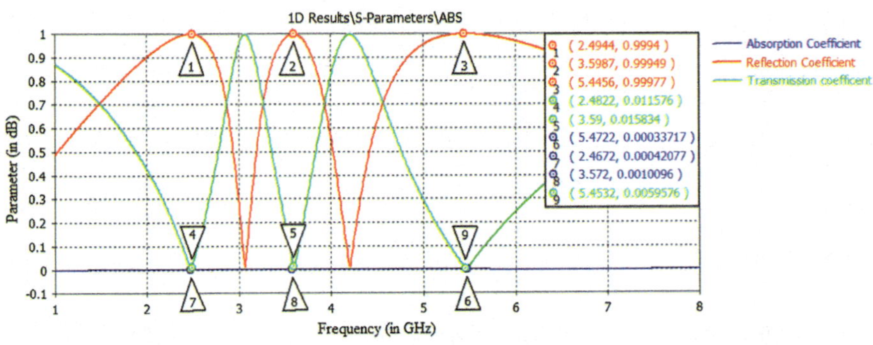

Fig. 5 Simulated transmissions, reflection and absorption coefficient of proposed FSS

circular patch. At 5.5 GHz, surface current has its maximum effect only on the edges of the square patch as illustrated in Figs. 6, 7, and 8, respectively.

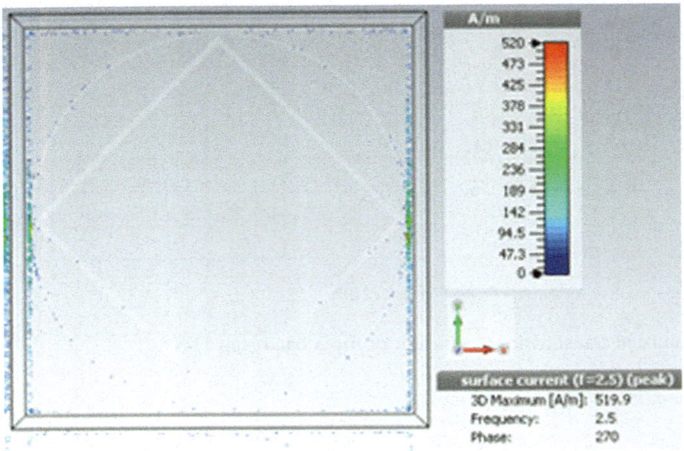

Fig. 6 Surface current at 2.5 GHz

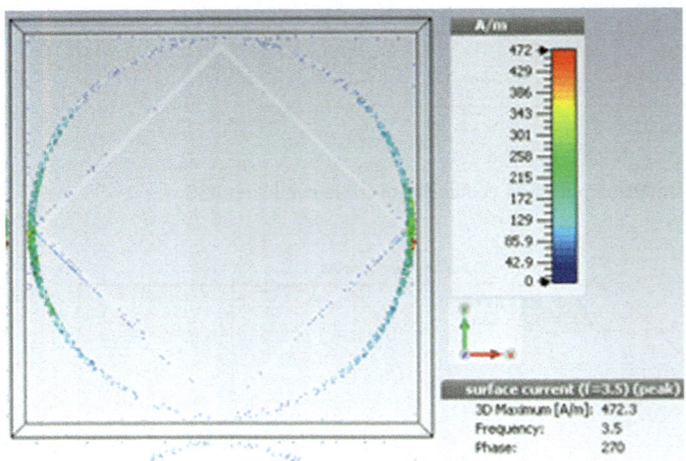

Fig. 7 Surface current at 3.5 GHz

3.3 *Effect on Transmission Coefficient Due to Variation in Parameters*

Further, in this paper, parametric analysis has been shown of different parameters which help in obtaining the requisite frequency and polarization performance at particular dimensions of the proposed design. On varying the width of the outer patch, there occur variations in the resonant frequencies. From Fig. 9, on increasing the width of the patch, the resonant frequency shifts toward lower frequency.

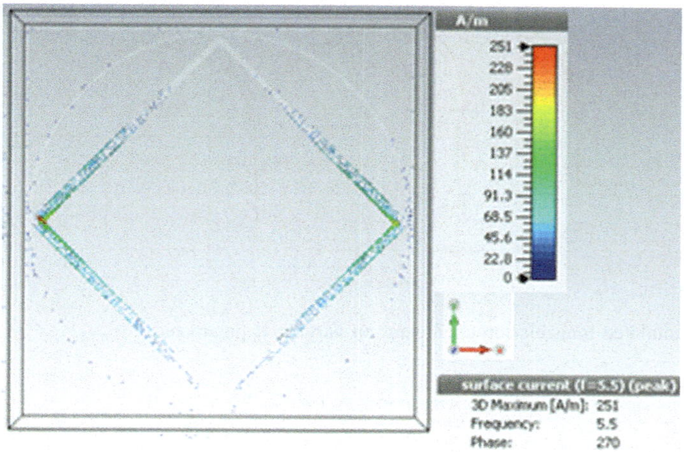

Fig. 8 Surface current at 5.5 GHz

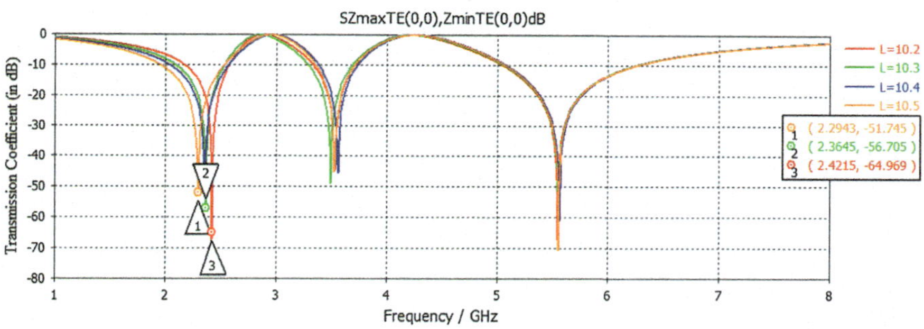

Fig. 9 Simulated transmission coefficient on varying L parameter

Similarly, on varying the width of the middle patch, there occur variations in the resonant frequencies. From Fig. 10, on increasing the width of the patch, the resonant frequency shifts toward lower frequency.

Similarly, on varying the width of the inner patch, there occur variations in the resonant frequencies. From Fig. 11, on increasing the width of the patch, the resonant frequency shifts toward lower frequency.

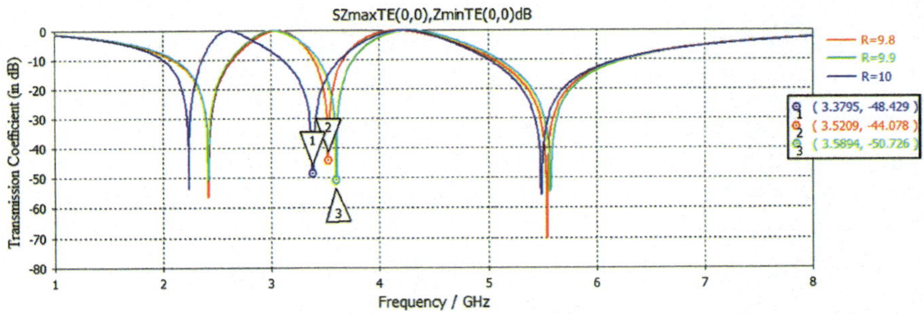

Fig. 10 Simulated transmission coefficient on varying R parameter

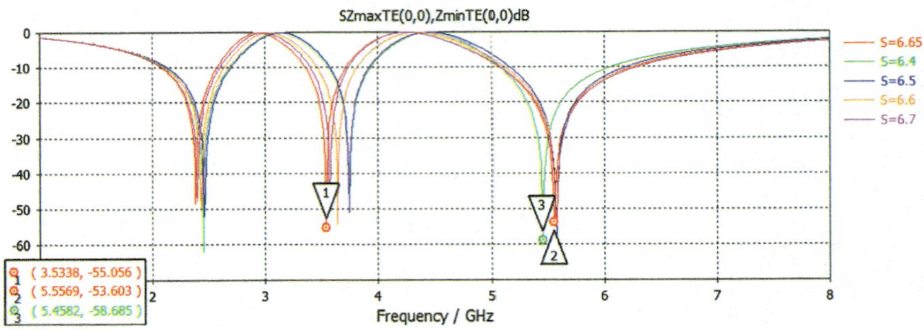

Fig. 11 Simulated transmission coefficient on varying S parameter

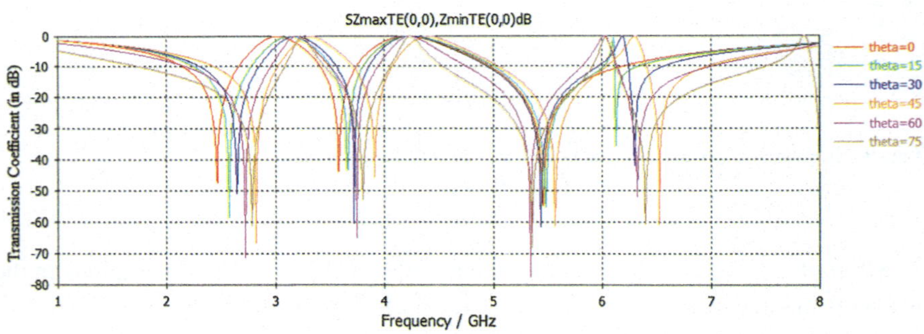

Fig. 12 Simulated transmission coefficient of triple band-reject FSS at different incidence angle

3.4 *Effect on Transmission Coefficient Due to Variation of Incidence Angle on Unit Cell*

From Fig. 12, it is clear that on varying the incidence angle, there has been occurred changes in the resonant frequency, and so the proposed design is polarization dependent.

4 Conclusion

In this paper, a new structure of band-reject filter designed by using frequency selective surfaces is proposed for the triple band. Therefore, three patches are designed of different shapes. Element types FSS is used as a band-reject filter for broadband applications which have much larger bandwidth than its band-pass portion. In the proposed design, a single-sided Teflon PTFE (lossy) substrate which has very good heat and weather resistibility has been used. The outer patch is embedded for 2.45 GHz band for Bluetooth application. The middle element is embedded for 3.5 GHz band for WiMAX application. The inner element is embedded for 5.5 GHz band for WLAN application. In the proposed design, 99% of power is rejected and only 0.03% of power is absorbed.

References

1. Chen H-Y, Tao Y (2011) Bandwidth enhancement of a U-slot patch antenna using dual-band frequency-selective surface with double rectangular ring elements. MOTL 53(7):1547–1553
2. Choudhary S, Mewara HS (2015) A band-reject frequency selective surface with stable response for WLAN applications. In: SPACES-2015, Department of ECE, KL University
3. Munk BA (2000) Frequency selective surfaces: theory and design. Wiley Interscience, New York
4. Aseri K, Yadav S, Sharma MM (2015) A compact frequency selective surface based band-stop filter for WLAN applications. In: 2015 fifth international conference on communication systems and network technologies. IEEE. ISSN 978-1-4799-1797-6/15. https://doi.org/10.1109/CSNT.2015.116
5. Mittra R, Chan CH, Cwik T (1988) Techniques for analyzing frequency selective surfaces—a review. IEEE Proc 76:1593–1615
6. Wu TK (1995) Frequency selective surface and grid array. A Wiley Interscience Publication, pp 5–7
7. Das S, Sarkar D, Sarkar PP, Chowdhury SK (2003) Experimental investigation on a polarization independent broadband frequency selective surface. In: Proceedings of national conference held in KIIT, Bhubuneswar, Mar 2003
8. Wu TK et al (1992) Multi ring element FSS for multi band applications. Paper presented at the international IEEE AP-S symposium, Chicago
9. Aseri K, Yadav S (2015) A novel hexagonal shape-based band-stop frequency-selective Kiran surface with multiband applications. Adv Intell Syst Comput 408
10. Srifi MN, Meloui M, Essaaidi M (2010) Rectangular slotted patch antenna for 5–6 GHz applications. Int J Microw Opt Technol 5(2)
11. Yadav S, Peswani B, Choudhury R, Sharma MM (2014) Miniaturized band pass double layered frequency selective surface superstrate for Wi-Max applications. In: 2014 IEEE symposium, ISWTA
12. Syed IS, Mohd Jenu MZ (2015) Dual-layer frequency selective surface for wide stop-band applications. In: 2015 IEEE international symposium on antennas and propagation and USNC/URSI national radio science meeting, Vancouver, BC, pp 1270–1271
13. Barton JH, Garcia CR, Berry EA, Salas R, Rumpf RC (2015) 3-D printed all-dielectric frequency selective surface with large bandwidth and field of view. IEEE Trans Antennas Propag 63(3):1032–1039

14. Cerveny M, Rigelsford JM (2015) An inductive frequency selective surface for use in secure facilities. In: 2015 Loughborough antennas and propagation conference (LAPC), Loughborough, pp 1–4
15. Abdollahvand M, Encinar JA, Forooraghi K, Atlasbaf Z, Page JE (2016) Tri-band FSS for Ku/Ka bands reflector antennas. In: 2016 10th European conference on antennas and propagation (EuCAP), Davos, pp 1–4
16. Rigelsford JM, Benito SM, Vallecchi A (2014) A tri-band inductive frequency selective surface sub-reflector for satellite communications systems. In: The 8th European conference on antennas and propagation (EuCAP 2014), The Hague, pp 898–900
17. Davenport CJ, Rigelsford JM, Zhang J, Altan H (2013) Periodic comb reflection frequency selective surface for interference reduction. In: 2013 Loughborough antennas and propagation conference (LAPC), Loughborough, pp 615–618

Design of a Spiral-Shaped Slotted Multiband Antenna

Sanjeev Yadav, Santosh Meena and Bhanu Priya Kumawat

Abstract In the proposed paper, a spiral-shaped multiband antenna has been designed for various wireless applications. Antenna can efficiently support four different frequencies as 2.5, 3.8, 4.8, and 6.3 GHz coming under the band of different wireless applications such as GSM, Bluetooth, Wi-Fi, WLAN, WiMAX, and some defense applications. To obtain the multiple bands, multiple spiral-shaped slots are provided with the rectangular patch. Design results of VSWR, surface current, and return loss S11 are shown in this paper. This design is simulated using CST Software.

Keywords Frequency bands · Impedance matching · Multiband antenna Spiral shape

1 Introduction

The application of wireless communication systems has been increased and still continues to increase. Design of an antenna is an important issue in the modern mobile communication technology. An antenna is a device or a system which is used to convert electrical energy into electromagnetic energy at the transmitting end, or electromagnetic energy into electrical energy at the receiving end, and succinctly, it is used to transmit or receive the signal. Because antennas are dependent on frequency bands, they are designed in such a way that they can operate at different frequency bands [1].

S. Yadav (✉) · S. Meena · B. P. Kumawat
Government Women Engineering College Ajmer, Ajmer, India
e-mail: sanjeev.mnit@gmail.com

S. Meena
e-mail: 2012smeena@gmail.com

B. P. Kumawat
e-mail: kumawat.priya7799@gmail.com

V. Janyani et al. (eds.), *Optical and Wireless Technologies*, Lecture Notes in Electrical Engineering 472, https://doi.org/10.1007/978-981-10-7395-3_50

Now mobile systems were designed to operate for one of the frequency bands of fourth-generation long-term evolution (4G LTE) or fifth-generation (5G) system. At present, many mobile communication systems use certain frequency bands such as GSM 900/1800/1900 Global system for mobile communication) bands (890–1900 MHz), UMTS (universal mobile telecommunication systems) bands (1900–2200 MHz and 2500–2700 MHz), Bluetooth band (2400 MHz), WiMAX (world-wide interoperability for microwave access) band (3800 MHz), Wi-Fi (wireless fidelity) bands (2400–2500 MHz), WLAN (wireless local area network) bands (5100–5800 MHz), and (LTE) (long-term evolution) bands (2400–2500 MHz). In most cases, because a single antenna cannot operate at all of these frequency bands of mobile communication, multiband antennas should be used for covering all of these frequency bands. Anyhow, usage of many antennas is usually limited by the volume and cost constraints of the applications. Hence, multiband antennas are essential to provide multifunctional operations for mobile communication [2].

2 Proposed Antenna Design

In many wireless applications, it is significant to design a simple strip line feed antenna for multiband frequencies. Multiple spiral-shaped frameworks have been used to provide both multifrequencies and compact size [3]. However, spiral structures are known to produce large cross-polarization. Therefore, to totally discard the cross-polarization, a four-arm slot spiral is designed. This geometry is symmetric in both x- and y-directions. Hence, as an outcome, cross-polarized components are canceled [4]. In this design, each arm of spiral produces resonant frequency according to spiral length and spiral width (Fig. 1; Table 1).

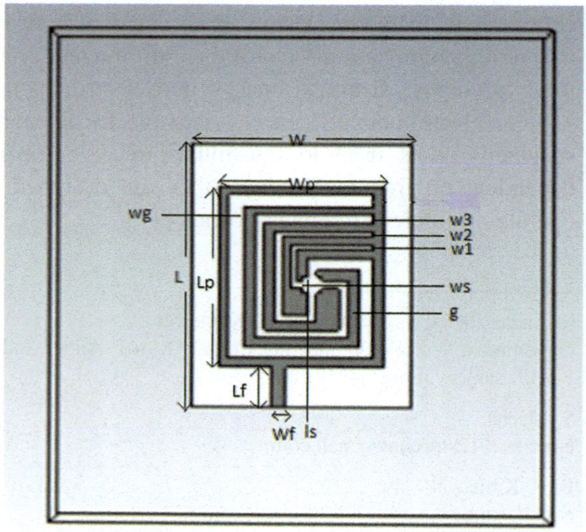

Fig. 1 Spiral-shaped slotted multiband antenna design

Table 1 Parameters of the proposed antenna geometry

Name of the dimension	Symbol	Units (mm)
Length of the substrate	L	32
Width of the substrate	W	22
Length of the patch	Lp	22
Width of the patch	Wp	16.5
Length of the feed	Lf	5
Width of the feed	Wf	1.5
Gap between slots	G	1
Length of smallest arm	Ls	2
Width of smallest arm	Ws	2
Length of largest arm	Lg	20
Width of largest arm	Wg	1.5
Width of slot 1	w1	0.5
Width of slot 2	w2	0.75
Width of slot 3	w3	1

3 Simulation and Measured Results

In this design, four-arm spiral-shaped geometry has been used to obtain the multiband performance. A 32 mm × 22 mm spiral-shaped slotted multiband antenna was designed with 1.6-mm-thick substrate. This proposed design has been fabricated on FR4 substrate having dielectric constant of 4.4 and loss tangent of 0.0024. The overall size of antenna is (L) 0.2930 λ × 0.13094 λ (W) with respect to center frequency. The antenna patch dimensions are 22 mm × 16.5 mm with PEC material. This patch has total 18 slots which have a shape of U slot and L slot, and a rectangular slot with 2 mm × 2 mm dimension. The dimension of the largest slot length is (lg) 20 mm and largest width is (wg) 1.5 mm with a gap g = 1 mm and smallest slot length is (ls) 2 mm and width is (ws) 2 mm with a gap g = 1 mm. In this figure, w1 is the width of the first slot w1 = 0.5 mm, w2 is the width of the second slot w2 = 0.75 mm, and w3 is the width of the third slot w3 = 1 mm. The reduction in the antenna resonant frequency was achieved due to the increase in both electrical and magnetic coupling between the spiral-shaped geometry and radiating antenna [5] (Fig. 2).

Length and width of the substrate in terms of λ are calculated by following formulas:

$$\in_{reff} = \frac{\in_r + 1}{2} + \frac{\in_r - 1}{2}[1 + \frac{12h}{w}]^{-1/2} \tag{1}$$

$$\Delta L = h \times 0.412 \frac{\left(\in_{reff} + 0.3\right)\left(\frac{w}{h} + 0.264\right)}{\left(\in_{reff} - 0.258\right)\left(\frac{w}{h} + 0.8\right)} \tag{2}$$

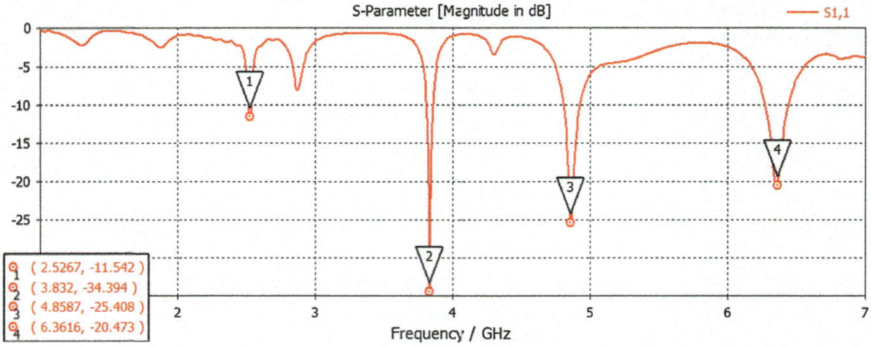

Fig. 2 Simulated and measured results of S-parameter

$$L = \frac{C_0}{2f_r\sqrt{\in_{reff}}} - 2\Delta L \quad (3)$$

$$W = \frac{C_0}{2f_r}\left(\sqrt{\frac{2}{\in_r + 1}}\right) \quad (4)$$

In the measured result of reflection coefficient, four different frequencies have been observed. The bandwidth of the compact multiband antenna can be increased by using thicker substrate. When the resonance frequencies of the slot and the patch are close to each other, then the broad bandwidth could be obtained. Frequency ranges lie from 2.5 to 6.36 GHz with a center frequency 1.92 GHz. Points −11.49, −34.48, −25.36, and −20.29 consist of minimum losses (Figs. 3, 4, 5, 6, and 7).

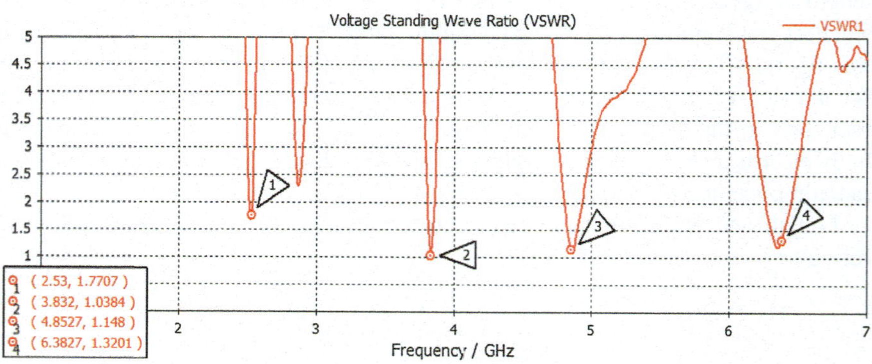

Fig. 3 VSWR results

Fig. 4 Surface current at 2.5 GHz

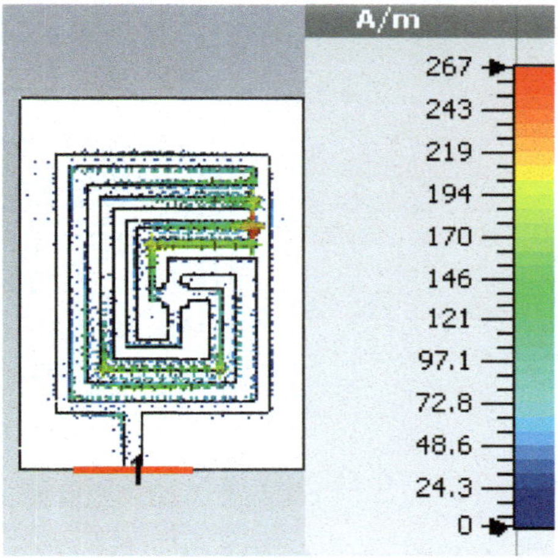

Fig. 5 Surface current at 3.8 GHz

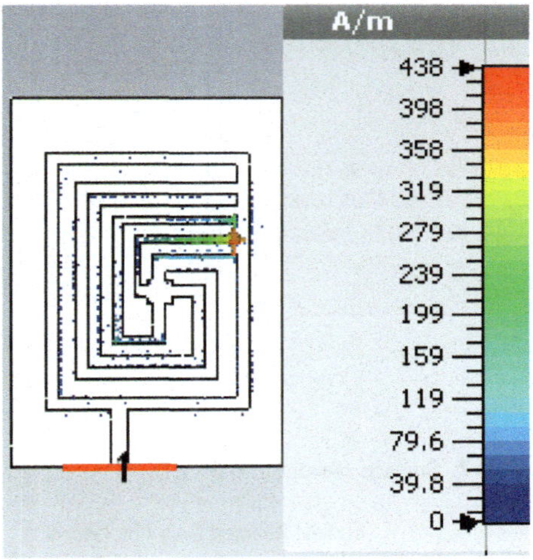

Antenna gain is the combination of directivity and electrical efficiency. Radiation pattern or far-field pattern of an antenna defines variation of the power radiated by an antenna as a function of the direction away from the antenna. Gain and radiation pattern of the slotted multiband antenna at four different frequencies is shown below (Figs. 8, 9, 10, 11).

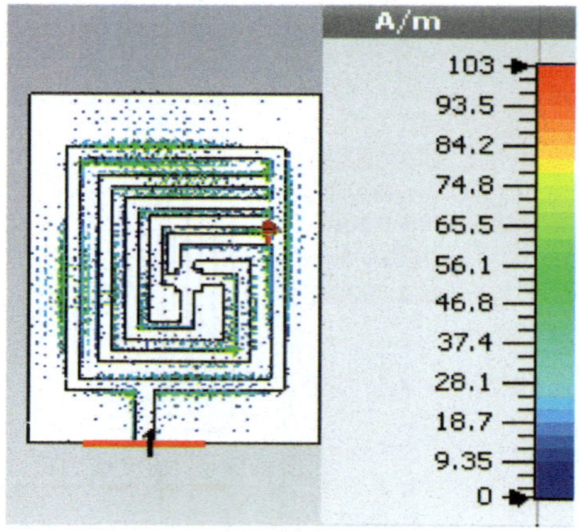

Fig. 6 Surface current at 4.8 GHz

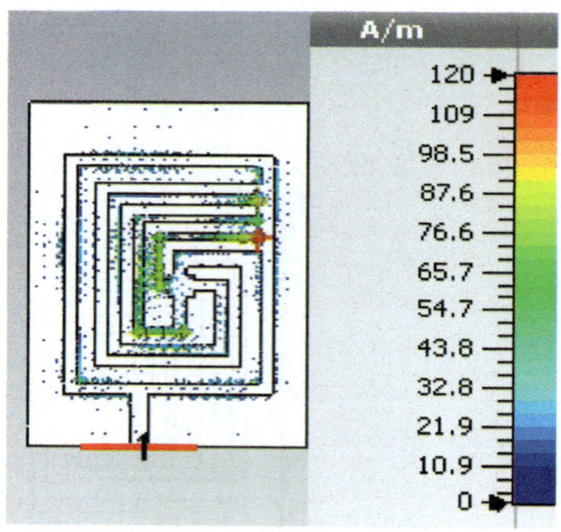

Fig. 7 Surface current at 6.3 GHz

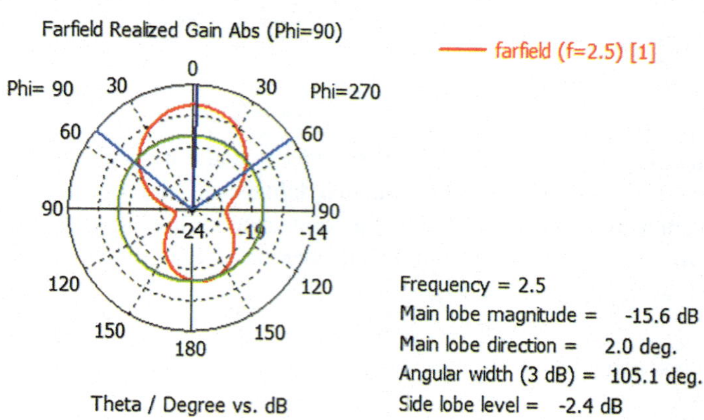

Fig. 8 Gain at 2.5 GHz

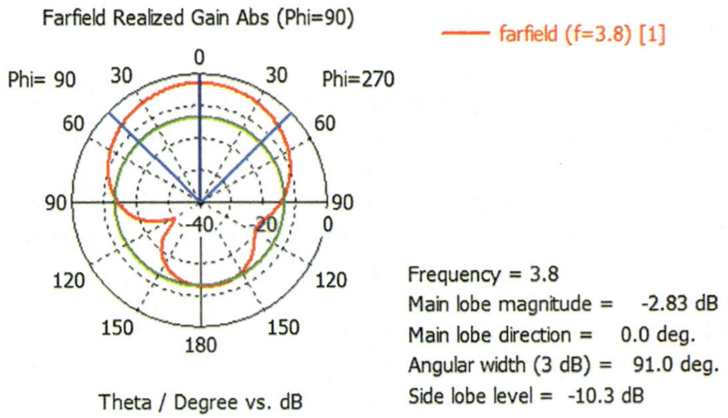

Fig. 9 Gain at 3.8 GHz

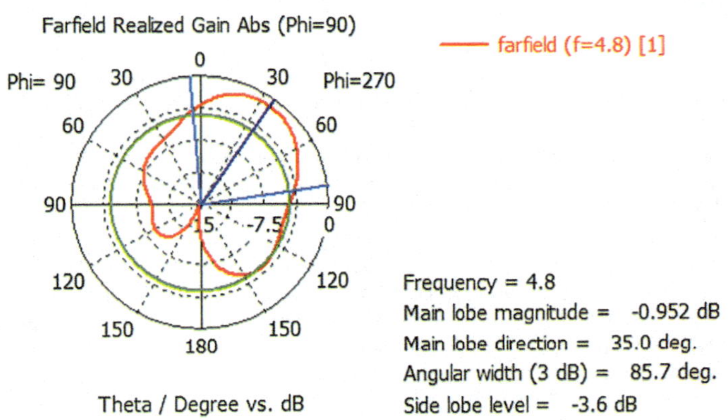

Fig. 10 Gain at 4.8 GHz

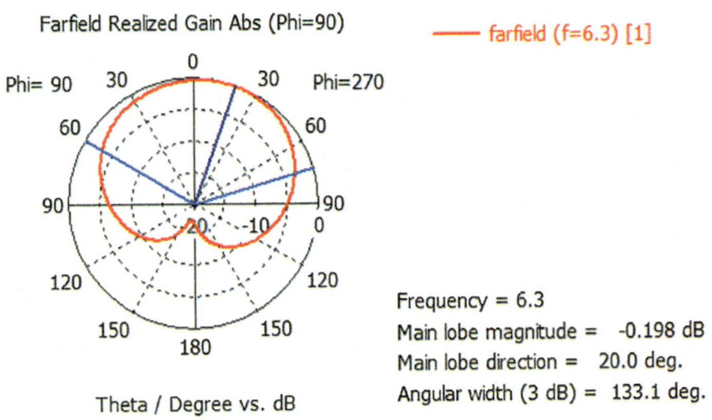

Fig. 11 Gain at 6.3 GHz

4 Conclusion

A new simple strip line feed antenna has been designed with a spiral-shaped multiband antenna that is investigated in this paper. CST Microwave Studio Simulation tool has been used to optimize the design. Obtained result shows that the multi cell spiral shaped geometry generates multi resonant frequencies. The adjustable feeding technique, spiral slots, and the dimension of the antenna made it possible in the accepted frequency bands of 2.5, 3.8, 4.8, and 6.3 GHz in different wireless applications such as Bluetooth, Wi-Fi, WiMAX, and WLAN. Reflection coefficient, current distribution, and gain of the proposed antenna were analyzed. Recently multiband antennas are designed for lower to higher frequency level on multiple bands.

References

1. Ali M, Okoniewski M, Stuchly MA, Stuchly MM (1999) Dual frequency strip-sleeve monopole for laptop computers. IEEE Trans Antennas Propag 47(2):317–323. ISSN 0018-926X
2. Best SR (2008). Electrically small multiband antennas. In: Sanchez-Hernandez DA (ed) Multiband integrated antennas for 4G terminals. Artech House, Boston, USA, pp 1–32. ISBN 978-1-59693-331-6
3. Nashaat D, Elsadek H, Ghali H (2004) Multiband miniaturized PIFA for compact wireless communication applications. Microw Antenna Opt Technol Lett 42(3):230–235
4. Chung Y, Jeon S-S, Kim S, Ahn D, Choi J-I, Itoh T (2004) Multifunctional microstrip transmission lines integrated with defected ground structure for RF front-end application. IEEE Trans Microw Theory Tech 52(5):1425–1432
5. Mandal, MK, Sanyal S (2006) A novel defected ground structure for planar circuits. IEEE Microw Wirel Compon Lett 16(2):93–95

Highly Selective, Closely Spaced Triple-Band Frequency Selective Surface for the Intensification in the Performance of WiMax and WLAN 2.5/3.5/5.5 GHz

Darakshanda Noor, Sandeep Kumar Yadav and Sanjeev Yadav

Abstract With the widespread keenness of the Internet of things in the upcoming future and the sprouting of 5G communication systems, predominantly in residential areas, the number of WLAN and WiMax connected devices may increase exponentially. But due to several factors like path loss, propagation effects and interference led to signal degradation. WiMax and WLAN have used at different ranges of bands. However, these bands suffer from large attenuations, so a band-pass frequency selective surface has been proposed in the paper that led to the intensification in the transmission of WiMax and WLAN signals to reduce the effect of path loss. A single-sided FR-4 substrate has integrated with the square and three decagon elements which results in triple-band (2.5, 3.5 and 5.5 GHz) characteristics. The single-sided substrate dielectric constant is 4.4 with loss tangent 0.025 and thickness 1.6 mm. With respect to different incident angles, offered structure had excellent band-pass response. For evaluating the performance, the CST Microwave computer simulation is used.

Keywords Triple band-pass frequency selective surface (FSS) · Three elements WiMax and WLAN · 2.5, 3.5 and 5.5 GHz band

1 Introduction

WiMax and WLAN are the most powerful as they provide high data rates with much higher speed and coverage area. In earlier days, antennas had used as reflectors but they do not show good results as current flows out of phase from

D. Noor (✉) · S. K. Yadav · S. Yadav
Government Women Engineering College Ajmer, Ajmer, India
e-mail: noorncet@gmail.com

S. K. Yadav
e-mail: sandeep.y9@gmail.com

S. Yadav
e-mail: sanjeev.mnit@gmail.com

V. Janyani et al. (eds.), *Optical and Wireless Technologies*, Lecture Notes in Electrical Engineering 472, https://doi.org/10.1007/978-981-10-7395-3_51

antenna and cancel out. Afterwards Ben A. Munk theory of frequency selective surface (FSS) came into nimble. In this current flows in-phase led to improved impedance bandwidth, angular stability and gain. David Rittenhouse, an American physicist, discovered the basic concept of filtering and diffraction grating used in frequency selective surface while observing his handkerchief and distant flame of a lamp. This gave him an idea of filtering white light into distinct wavelength bands. A notion of frequency selective surface (FSS) has adopted in this paper. High impedance surface that reflects in plane waves and quashing of surface wave are unique features of FSS. Frequency selective surface acts as a filter at any angle of incidence for a plane wave and has band-stop/band-pass frequency-dependent characteristics. Theoretically, FSS consists of 2D periodic array of assured shapes that are either engraved out of conducting layer or are often printed on dielectric substrate for support [1]. Various boundary conditions such as Floquet mode, PEC, PMC, etc., are applied, and insertion and return losses are simulated. The most imperative factor in designing is the spacing between inter-elements, patch and slot types and geometry. Frequency behaviour of an FSS has completely examined by unit cell [2]. It is always advantageous that the operating wavelength should always be greater than proposed unit cell dimensions. Recently, progressive method-based finite element method (FEM), method of moments (MOM), (FDTD), etc., have led to the FSS simulation easy by simulating single unit cell rather than the whole structure. FSS scatters electromagnetic fields that are numerically calculated by different methods as discussed earlier as these methods differed from each other in the way of calculating different FSS configurations [3]. The bandwidth is more often defined in −10 dB for reflection and −0.5 dB for transmission. The structure has realized in CST Microwave Studio software in frequency-domain solver, based on FDTD method.

2 Designing of Superstrate

2.1 Unit Cell

A FR-4 superstrate with thickness 1.6 mm and dielectric constant of 4.4 has used with dimensions W (mm) × W (mm) that is square-shaped. On its one side, a patch has fixed included of square and three elements. Due to the arrangement of square and outer decagon, a 2.5-GHz band has obtained. Two bands, 3.5 and 5.5 GHz, have obtained due to second and third decagons, respectively. The dimensions have illustrated in Fig. 1 and Table 1.

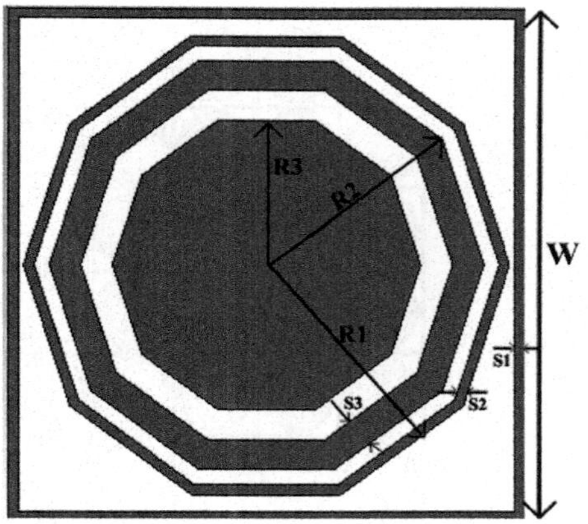

Fig. 1 Front view of the unit cell proposed superstrate with dimensions

Table 1 Dimensions of proposed unit cell

Name of dimensions	Symbol	Value (mm)
Width of substrate	W	20
Width of square	W1	20
Radius of outer decagon	R1	9.4
Radius of middle decagon	R2	8.4
Radius of inner decagon	R3	5.97
Thickness of substrate	H	1.6
Spacing between the square	S1	0.4
Spacing between first decagon	S2	1.2
Spacing between second decagon	S3	0.4

2.2 Design of Finite Array

An array of 8 × 8 element of dimension Ws (mm) × Ws (mm) has been designed. Elements in the array consist of periodicity P and are repeated in both X and Y directions as shown in Fig. 2 and Table 2. Other parameters and the dimensions are same as of unit cell.

3 Results and Discussion

3.1 Insertion Loss of Unit Cell

With the help of PBC (periodic boundary conditions), one can simulate the frequency characteristics of unit cell structure. After assembling the unit cell, the structure works as band pass for three bands 2.5, 3.5 and 5.5 GHz for WiMax and WLAN applications. The transmission response has carried out in anechoic

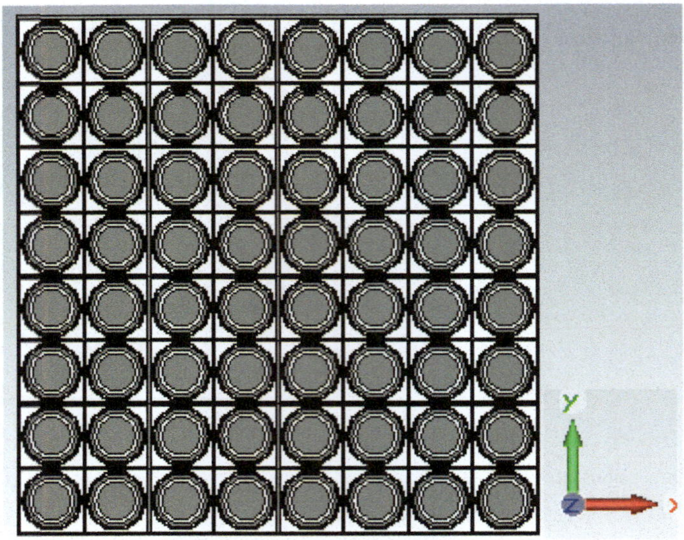

Fig. 2 Front view of the proposed finite array with dimensions

Table 2 Dimensions of proposed finite array

Name of dimension	Symbol	Value (mm)
Substrate width	Ws	80
Substrate height	H	1.6
Periodicity	P	20

chamber by placing the proposed design between two horn antennas, and frequency response of the structure has obtained. Transmission coefficient (S21) has simulated to know how much power has transmitted from the structure. Transmission coefficient at 2.5, 3.5 and 5.5 GHz having insertion loss of −0.32 dB, −1.30 dB and −0.48 dB, respectively, has shown in Fig. 3.

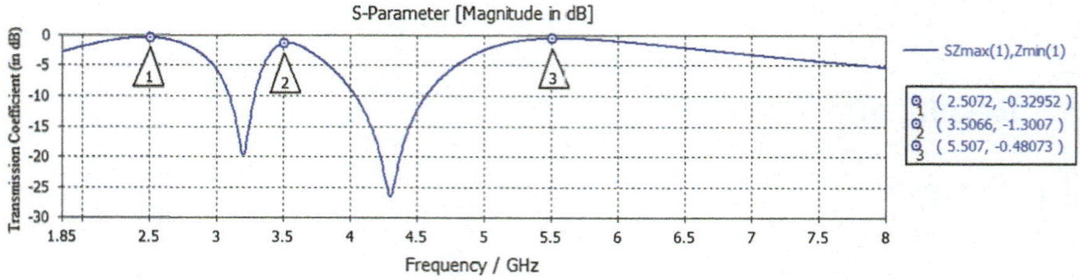

Fig. 3 Simulated transmission coefficient of unit cell

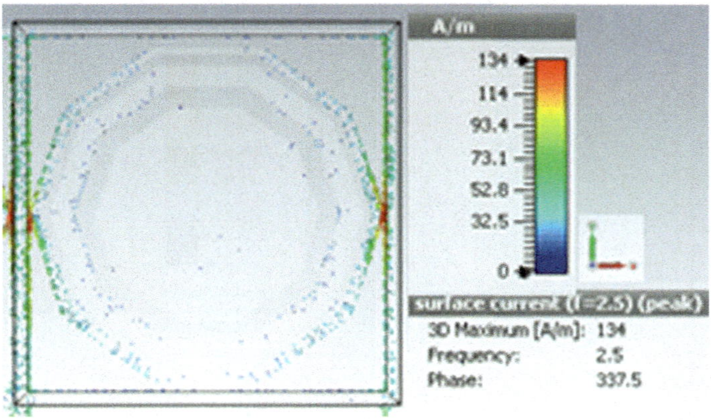

Fig. 4 Surface current at 2.5 GHz

3.2 *Surface Current*

Surface current has an important characteristic, as it decides the frequency response of FSS. Due to mutual effect of inter-elements, surface current has calculated. At 2.5 GHz, surface current has its maximum effect at the edge of outer elements as shown in Fig. 4. At 3.5 GHz, surface current has its maximum effect due to the mutual effect of second and third decagons as illustrated in Fig. 5, and maximum effect has seen at the edge of inner decagon at 5.5 GHz shown in Fig. 6.

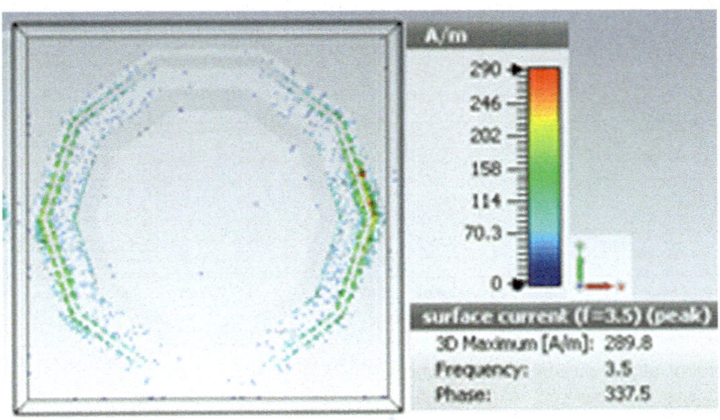

Fig. 5 Surface current at 3.5 GHz

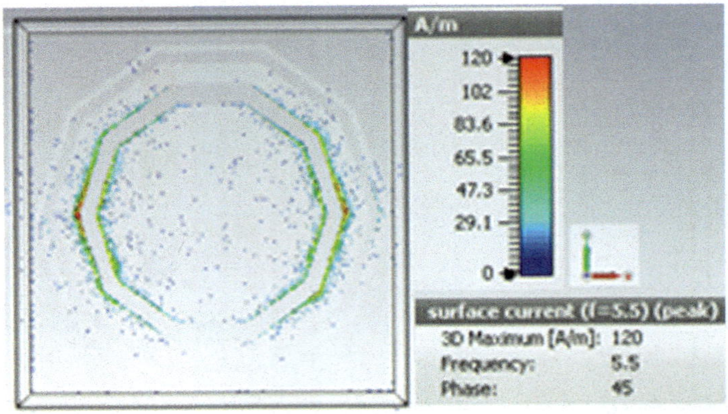

Fig. 6 Surface current at 5.5 GHz

3.3 Incident Angle Variation

While working in space, we exactly do not have any idea of at what angle the electromagnetic will strike the FSS. Therefore, incident angles act as an important parameter as its variation can affect the stability of frequency response of FSS. Important characteristics in band-pass FSS is the insertion loss. Figure 7 shows the variation in incident angle for TE mode from 0° to 75°. With the help of simulation, different resonant frequencies (2.5, 3.5 and 5.5 GHz) for band pass and a high level of stability at different incident angles have achieved. Maximum signal has transmitted at 0°, and absorption is more with less transmission as we increase the incident angle.

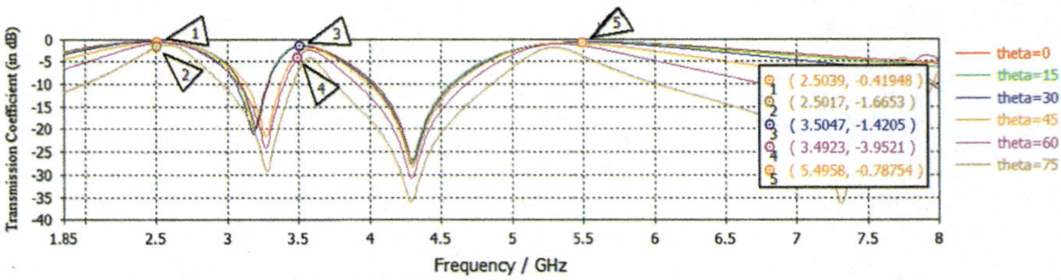

Fig. 7 Incident angle variation from theta = 0°–75°

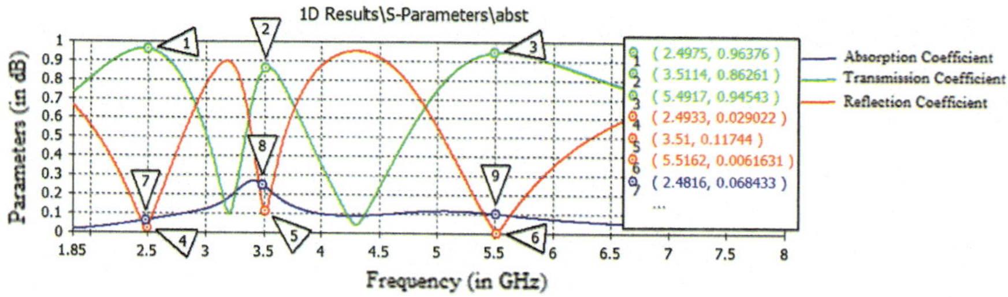

Fig. 8 Simulated results for absorption, reflection and transmission

3.4 *Absorption, Reflection and Transmission*

Transmission, absorption and reflection are the three important characteristics of FSS. Absorption reveals that the signal has neither reflected nor transmitted. Absorption (A) has been calculated, and the equation is as follows:

$$A = 1 - |S11|^2 - |S21|^2$$

where A = Absorption. Figure 8 illustrates that 92% of signal is transmitted at 2.5, 3.5 and 5.5 GHz.

3.5 *Variations in the Outer Decagon*

It has evaluated that on varying the radius of outer decagon, there is shifting in the resonant frequency towards the lower side as shown in Fig. 9. Therefore, for better frequency response at 2.5 GHz, we take the value that has described in Table 1.

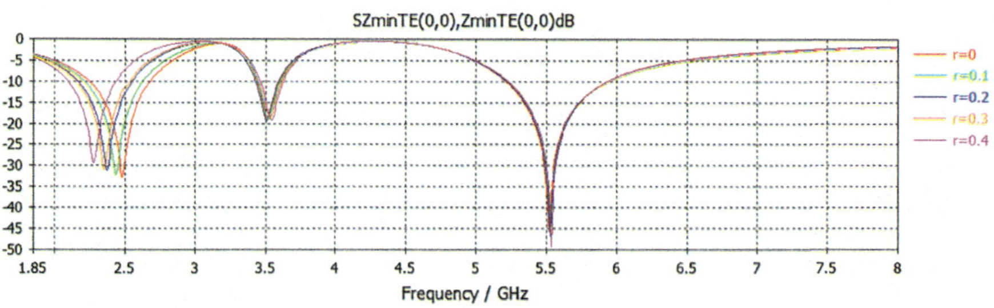

Fig. 9 Variations in the radius of outer decagon

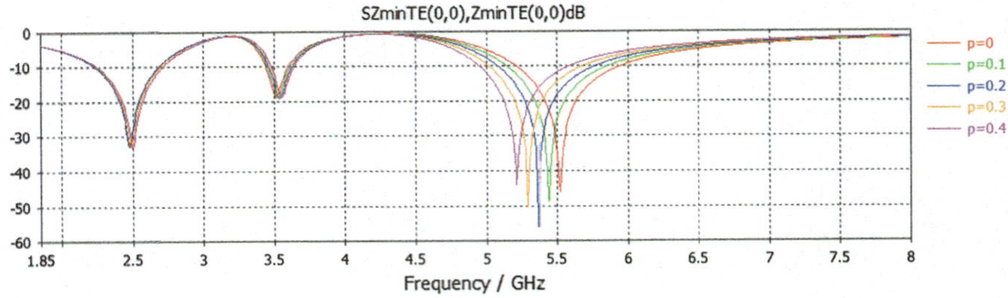

Fig. 10 Variations in the inner decagon

3.6 Variations in the Inner Decagon

On varying the radius of inner decagon, there occurs change in the resonant frequency i.e. it shifts towards left or lower side of the frequency as shown in Fig. 10.

4 Conclusion

With the proposed structure of triple-layered FSS, a band-pass operation for WLAN application has achieved. It has observed that reduction in inter-element spacing in both directions or either direction will lead to increase in bandwidth. FSS had excellent resonance stability, bandwidth and better transmission as path loss is reduced or minimized. In this paper, the insertion loss tells how much power has transmitted whereas return loss gives the idea of how much power has reflected. One can reduce the simulation processing time by simulating only the single unit cell rather than entire FSS structure. Simulation of the results has carried out by CST Microwave Studio in frequency domain using tetrahedral mesh to cover the whole structure. In future, this structure would further extended for multiband WLAN applications.

References

1. Das S, Sarkar D, Sarkar PP, Chowdhury SK (2003) Experimental investigation on a polarization independent broadband frequency selective surface. In: Proceedings of national conference held in KIIT, Bhubaneswar
2. Schoenlinner B, Abbaspour-Tamijani A, Kempel LC, Rebeiz GM (2004) Switchable low-loss RF MEMS Ka-band frequency-selective surface. IEEE Trans Microw Theory Tech 52:2474–2481
3. Qing A, Lee CK (2010) Differential evolution in electromagnetic. Springer, Heidelberg, Berlin

4. Ott R, Kouyoumjian R, Peters L Jr (1967) Scattering by a two dimensional periodic array of narrow plates. Radio Sci 2:1347–1359
5. Yadav S, Peswani B, Sharma MM (2014) A novel band pass double-layered frequency selective superstrate for WLAN applications on fluence. IEEE, pp 447–451
6. Romeu J, Rahmat-Samii Y (2000) Fractal FSS: a novel dual-band frequency selective surface. IEEE Trans Antennas Propag 48:1097–1105
7. Munk BA (2000) Frequency selective surfaces: theory and design. Wiley-Interscience, New York
8. Wu TK (1995) Frequency selective surface and grid array. Wiley Inter Science, pp 5–7
9. O'Nians F, Matson J (1966) Antenna feed system utilizing polarization independent frequency selective intermediate reflector. US patent 3,231,892, 25 Jan 1966
10. Chen C (1970) Scattering by a two-dimensional periodic array of conducting plates. IEEE Trans Antennas Propag AP-18:660–665
11. Huang J, Wu T, Lee S (1994) Tri-band frequency selective surface with circular ring elements. IEEE Trans Antennas Propag 42:166–175
12. Muttra R, Chan CH, Chic T (1988) Techniques for analyzing frequency selective surfaces—a review. Proc IEEE 76:1593–1615
13. Wu TK et al (1992) Multi ring element FSS for multi band applications. In: Paper presented at the international IEEE AP-S symposium, Chicago
14. Romeo J, Rahmat-Samii Y (2000) Fractal FSS: a novel dual-band frequency selective surface. IEEE Trans Antennas Propag 48:1097–1105
15. Parker EA, Savia SB (2001) Active frequency selective surfaces with ferroelectric substrates. Proc Inst Elect Eng Microw Antennas Propag 148:103–108
16. Chen H-Y, Tao Y (2011) Bandwidth enhancement of a U-slot patch antenna using dual-band frequency-selective surface with double rectangular ring elements. MOTL 53(7):1547–1553
17. Munk B, Kouyoumjian R, Peters L Jr (1971) Reflection properties of periodic surfaces of loaded dipoles. IEEE Trans Antennas Propag AP-19:612–617
18. Chatterjee A, Biswas S, Chanda D, Pratim Sarkar P (2003) A polarization independent compact multi-band frequency selective surface. In: International conference on current independent broadband frequency selective surface, proceedings of national conference held in KIIT, Bhubuneswar

Compressed Sensing Based Network Lifetime Enhancement in Wireless Sensor Networks

Prateek Dolas and D. Ghosh

Abstract Increasing network lifetime is the important aspect in wireless sensor network deployed for any application. In this paper, we propose to exploit joint sparsity to remove redundant sensor data to be transmitted to increase network lifetime using compressed sensing. Result indicates that proposed method increases network lifetime significantly.

Keywords Compressed sensing · Network lifetime · Sensor network

1 Introduction

Wireless sensor networks (WSNs) are used these days in various real-time applications such as monitoring of environment, surveillance, wildlife monitoring, military applications, traffic monitoring etc. [1]. The wide range of applicability of WSN is basically due to the availability of small-sized sensor nodes with potential to compute and process data locally. Due to this feature, an ad hoc network of these sensor nodes is used to collect application specific real-time data.

The sensor nodes, besides sensing local environment, participate in communication process as well. The data sensed by any sensor node is forwarded to the central base station or sink via multihop communication in which each sensor node acts as a relay for the data sensed by other nodes. The sensor nodes in WSN are battery powered devices. However, WSNs are usually deployed in unaccessible terrains where it is difficult to maintain or replace the battery. So, the limited availability of energy is a major challenge in sensor network.

Among various processes performed at the sensor node, a major portion of energy is consumed in multihop communication. The major drawback of multihop

P. Dolas (✉) · D. Ghosh
Department of Electronics and Communication Engineering,
Indian Institute of Technology Roorkee, Roorkee 247667, India
e-mail: dolasdec@iitr.ac.in

D. Ghosh
e-mail: ghoshfec@iitr.ac.in

V. Janyani et al. (eds.), *Optical and Wireless Technologies*, Lecture Notes
in Electrical Engineering 472, https://doi.org/10.1007/978-981-10-7395-3_52

communication is the uneven load distribution among the nodes. The nodes in the proximity of sink are heavily loaded and consume the limited available energy more rapidly as compared to the nodes at the periphery of the network. Due to uneven energy consumption, the nodes near to the sink are likely to get their energy drained off soon as a result of which the entire network fails. So, network lifetime maximization or energy preservation is one primary concern in WSN.

Generally, WSNs are deployed in unaccessible terrains. So, due to difficulty in maintenance of the network, low cost and small size of sensor nodes, redundant deployment of sensor nodes is done in WSN. As a result, the data collected in WSN also exhibit redundancy. Consequently, communication of redundant data to the sink causes huge wastage of available energy. So, efficient data gathering in WSN aims at reducing the amount of data to be communicated to the sink by exploiting the redundancy in sensor data in addition to load balancing among all the sensor nodes of the network [7, 8]. This helps to save lot of energy thereby increasing overall network lifetime.

In WSN, readings collected by closely placed sensor nodes in space measure the same phenomenon. So, these readings exhibit spatial (inter) correlation. Similarly, the frequency of sensing the data by any particular sensor node is also high, whereas the surrounding physical phenomenon do not change rapidly. So, the readings recorded by any sensor node exhibit temporal (intra) correlation. Thus, in order to preserve the energy in WSN, redundancy may be removed by exploiting the existing correlation in the sensed data.

Compressive Data Gathering (CDG) [6] based on the phenomenon of Compressed Sensing (CS) [4] reduces the amount of effective data to be transmitted to the sink without excessive computational overhead. CS is based on the principle of sparse signal. It also distributes load equally among all the nodes of the WSN. Thus, CDG helps to enhance overall network lifetime in large-scale WSN by preserving significant amount of energy. Distributed compressive sensing (DCS) [5] is an extension of CS principle to multiple signal measurements which are jointly sparse.

In this paper, we propose to use distributed compressed sensing to achieve high order of data compression for enhancing network lifetime of WSN. Compressed sensing is introduced briefly in Sect. 2. In Sect. 3, joint sparsity models for DCS are discussed. Our proposed scheme to enhance overall network lifetime with simulation results is discussed in Sect. 4. Finally, conclusions are drawn in Sect. 5.

2 Compressed Sensing

Compressed sensing is a sampling paradigm in which sampling and compression occur simultaneously. The data rate at which signal is sampled is far less than the traditional Nyquist rate. Thus, CS can be seen as linear dimensionality reduction technique. This dimensionality reduction of signal requires very less computational overhead which may be achieved locally at sensor nodes with less energy requirement. The tractable recovery of this under-sampled signal at the sink node is nonlinear and complex which may be easily done at the sink as sink nodes are not resource constrained.

Consider an N-dimensional, discrete time, real-valued signal $\boldsymbol{x}$ such that $\boldsymbol{x} \in \mathbb{R}^N$. Using an $N \times N$ orthonormal basis $\boldsymbol{\Psi} \in \mathbb{R}^N$, the signal $\boldsymbol{x}$ may be represented as

$$\boldsymbol{x} = \boldsymbol{\Psi s} \tag{1}$$

where $\boldsymbol{s}$ is an N-dimensional sparse vector. The signal $\boldsymbol{x}$ is said to have sparsity k iff there are k nonzero elements in $\boldsymbol{s}$, where $k \ll N$. Let $\boldsymbol{y}$ be an M-dimensional measurement vector formed by M linear projections of the signal $\boldsymbol{x}$, where $M \ll N$. Thus, $\boldsymbol{y}$ may be represented as

$$\boldsymbol{y} = \boldsymbol{\Phi x} = \boldsymbol{\Phi \Psi s} = \mathbf{D}\boldsymbol{s} \tag{2}$$

where $\boldsymbol{\Phi}$ is an $M \times N$ measurement matrix that is incoherent w.r.t $\boldsymbol{\Psi}$ and satisfies Restricted Isometry Property (RIP), and $\mathbf{D} = \boldsymbol{\Phi\Psi}$ is an $M \times N$ matrix known as Dictionary.

As $M \ll N$, solving the under-determined system of equation of (2) requires additional constraints to obtain a unique solution. A simple way to achieve this is *regularization*. The optimization problem involves objective function $J(\boldsymbol{s})$. This optimization function is defined as

$$\arg\min_{s} J(\boldsymbol{s}) \text{ subject to } \boldsymbol{y} = \boldsymbol{Ds} \tag{3}$$

The commonly used objective functions suggested in literature are the ℓ_0, ℓ_1 or ℓ_p norm. The Euclidean norm ℓ_2 minimization results in unique but non-sparse solution because ℓ_2 norm is a convex function and any convex function promises uniqueness. So, ℓ_2 norm is not an appropriate solution to the CS problem. In some literature, ℓ_0 norm minimization is considered. However, it leads to a non-convex optimization problem which is NP-hard to solve. Accordingly, sub-optimal algorithms have been developed to find an approximate solution. Various sub-optimal sparse signal recovery algorithms found in the literature are Orthogonal Matching Pursuit (OMP) [3], Simultaneous Orthogonal Matching Pursuit (SOMP) [9] etc. The various algorithms based on ℓ_1 norm minimization are Basic Pursuit [2], The FOCul Underdetermined System Solver (FOCUSS), etc.

3 Joint Sparsity Models for DCS

In WSN, each node sends a signal to the sink. Hence, multiple signals are to be recovered at the sink. So DCS is an appropriate solution in this scenario as it is an extension of CS to multiple signal measurement. Besides this, all the signals collected at sink are highly correlated or can be termed as jointly correlated. Thus, they may be represented as sparse signals jointly. In [10], authors have suggested three joint sparsity models (JSMs) as discussed briefly below.

Joint Sparsity Model-I In this model, each signal is composed of two components—*sparse common component* present in all signals and *innovations* unique to every signal, given as

$$\boldsymbol{x}_j = z_c + z_j, \qquad j = 1, 2, 3, \dots J \tag{4}$$

where z_c is the common component to all $\boldsymbol{x}_j$ and has sparsity k_c in basis $\boldsymbol{\Psi}$ and z_j is unique to every $\boldsymbol{x}_j$ having sparsity k_j in the same basis $\boldsymbol{\Psi}$.

Joint Sparsity Model-II In this model, all signals are constructed from the same sparse set of basis vectors, but with different coefficients values. That is,

$$\boldsymbol{x}_j = \boldsymbol{\Psi}\boldsymbol{s}_j, \qquad j = 1, 2, 3, \dots J \tag{5}$$

where each $\boldsymbol{s}_j$ is in general different but has same sparsity k. JSM-II can be considered as a special case of JSM-I with $k_c = 0$ and $k_j = k$.

Joint Sparsity Model-III This model is similar to JSM-I in which the common component is no longer sparse in any basis. That is, it has *non-sparse common component* and *sparse innovation.*

4 Proposed Scheme for Network Lifetime Enhancement

Let $\boldsymbol{X} = [x_{nj}]$ be the $N \times J$ matrix composed of sensor readings collected by J sensor nodes over N time instances, x_{nj} denotes the jth sensor reading at the nth time instance. The rows of $\boldsymbol{X}$ correspond to the samples taken by all the J sensor nodes at the nth time instance. Thus, rows of $\boldsymbol{X}$ exhibit spatial correlation. The columns of $\boldsymbol{X}$ correspond to the samples taken by the jth sensor node at all the N time instances. So, columns of $\boldsymbol{X}$ exhibit temporal correlation. As rows are spatially correlated, there exists a $J \times J$ orthonormal basis $\boldsymbol{\Theta} \in \mathbb{R}^J$ in which all rows of $\boldsymbol{X}$ are sparse with support size of k_s such that $k_s \ll J$. Similarly, columns are temporally correlated so there exists an $N \times N$ orthonormal basis $\boldsymbol{\Psi} \in \mathbb{R}^N$ in which all columns of $\boldsymbol{X}$ are sparse with support size of k_t such that $k_t \ll N$. This scenario is similar to JSM-II described in Sect. 3. Therefore, we may write,

$$\boldsymbol{X} = \boldsymbol{\Psi}\boldsymbol{S}\boldsymbol{\Theta}^T \tag{6}$$

where $\boldsymbol{S}$ is sparse matrix representation of $\boldsymbol{X}$.

Since $\boldsymbol{X}$ is sparse, we choose a measurement matrix $\boldsymbol{\Phi}$ of dimension $M \times N$ where $M \ll J$ and is incoherent with an orthonormal basis $\boldsymbol{\Psi}$ for temporally correlated data. Similarly, we choose a measurement matrix $\boldsymbol{\Omega}$ of dimension $L \times J$ where $L \ll N$ and is incoherent with an orthonormal basis Θ for spatially correlated data. Therefore,

$$\boldsymbol{Y} = \boldsymbol{\Phi X \Omega}^T = \boldsymbol{\Phi \Psi S \Theta}^T \boldsymbol{\Omega}^T = \boldsymbol{D}_t \boldsymbol{S} \boldsymbol{D}_s \tag{7}$$

where $\boldsymbol{Y}$ is a matrix of dimension $M \times L$ formed by linearly projected measurements of $\boldsymbol{X}$. $\boldsymbol{D}_t$ and $\boldsymbol{D}_s^T$ are the overcomplete dictionaries for temporal and spatial correlated data, respectively. As observed, NJ number of samples are compressed to only ML number of samples by applying DCS.

Compression of effective data to be transmitted helps to enhance overall network lifetime as energy consumption is reduced. This can be expressed mathematically as

$$\text{Network Lifetime enhancement} \propto \frac{1}{\text{Energy consumption}} \tag{8}$$

or

$$\text{Network Lifetime enhancement} = 1 - \frac{1}{\text{Compression Ratio}} \tag{9}$$

Simulations were carried out to evaluate the performance of proposed scheme with the help of real data set of sensor readings of environment monitoring deployed at École Polytechnique Fédérale de Lausanne (EPFL), Switzerland. The readings of this data set are spatially and temporally correlated. In our work, we used readings of ambient temperature and relative humidity recorded by $J = 8$ closely located sensor nodes at an approximately constant distance d from the sink at $N = 512$ successive time instances. The readings were spatially and temporally correlated and were sparsely representable in Hadamard basis Ψ and Θ. The temporal sparsity of signal $k_t = 3 \ll N = 512$ and spatial sparsity of signal $k_s = 1 \ll J = 8$.The entries of DCT measurement matrix $\boldsymbol{\Phi}$ are randomly sampled rows of $N \times N$ DCT matrix, and entries of DCT measurement matrix Ω are randomly sampled L rows and J columns of DCT matrix. The entries of $N \times N$ DCT matrix are given by

$$\alpha_{p,q} = \begin{cases} \frac{1}{\sqrt{N}} & p = 0, 0 \le q \le N-1 \\ \sqrt{\frac{2}{N}} \cos \frac{\pi(2q+1)p}{2N} & 1 \le p \le N-1, 0 \le q \le N-1 \end{cases} \tag{10}$$

The number of measurements M is fixed at 4 while L is varied from 2 to 200. By exploiting spatio-temporal correlation simultaneously, the compression ratio ranging from 512:1 to 5.12:1 was achieved. Corresponding to these compression ratios, network lifetime enhancement of 99.8 to 80.4% was achieved. For recovery, we used two different recovery algorithms M-FOCal Underdetermined System Solver (M-FOCUSS) and MMV Sparse Bayesian Learning (MSBL) algorithm. Performance of recovery of original signal with respect to both recovery algorithms was measured in terms of Mean square error (MSE) given by

$$MSE = \frac{1}{NJ}\sum_{n=1}^{N}\sum_{j=1}^{J}\frac{(x_{nj} - \hat{x}_{nj})^2}{x_{nj}^2} \tag{11}$$

where $\hat{x}_{nj}$ is the recovered jth node reading at nth time instance.The process is repeated several times and average MSE is plotted for varying values of number of measurements L. Figure 1a depicts the Root Mean Square Error (RMSE) plot for ambient temperature data while Fig. 1b depicts the RMSE plot for relative humidity data recovered using MSBL algorithm. Figure 2a depicts the RMSE plot for ambient temperature data while Fig. 2b depicts the RMSE plot for relative humidity data recovered using M-FOCUSS algorithm.

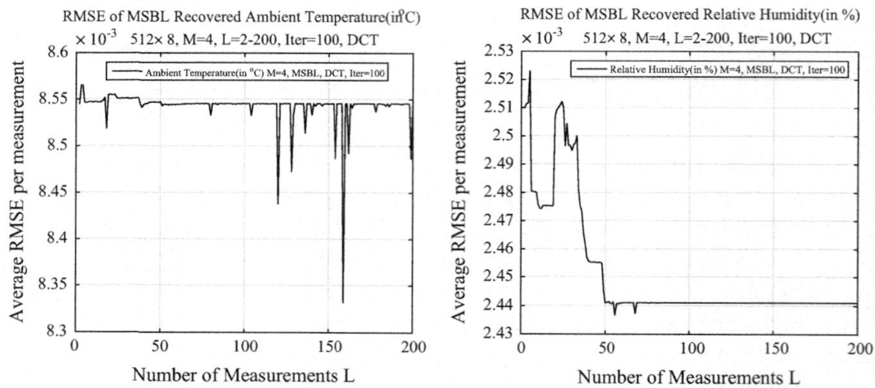

(a) Ambient Temperature (b) Relative Humidity

Fig. 1 Plot of RMSE recovered using MSBL algorithm

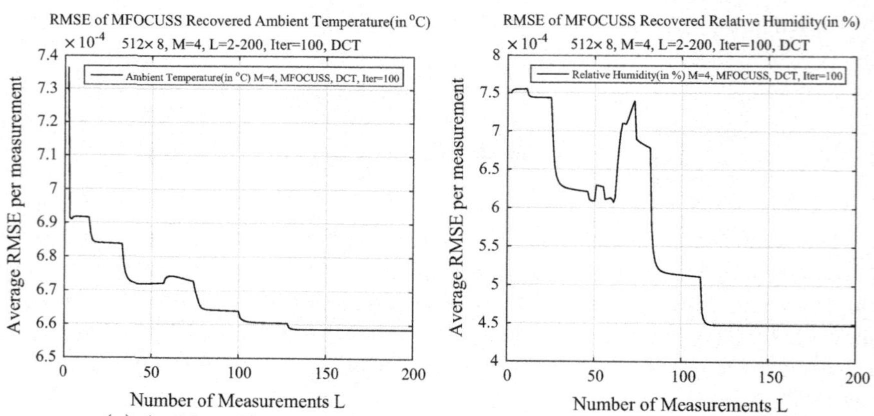

(a) Ambient Temperature (b) Relative Humidity

Fig. 2 Plot of RMSE recovered using M-FOCUSS algorithm

5 Conclusion

Simulation results reveal that a huge compression of sensor data may be achieved by exploiting joint sparsity (spatial and temporal) to remove redundancy in sensor data using DCS in large-scale deployment of WSN. The results compare the RMSE in recovery of original signal at sink using MSBL and M-FOCUSS algorithms. The original data is recovered at with negligible RMSE of the order of 10^{-3} by using MSBL algorithm, and even less, i.e., of order of 10^{-4} in recovery of original data at the sink by using M-FOCUSS algorithm. Thus, proposed scheme enhances the overall network lifetime to great extent. In our simulations, maximum of 99.8% enhancement of network lifetime was achieved with compression ratio of 512:1.

References

1. Akyildiz I, Su W, Sankarasubramaniam Y, Cayirci E (2002) Wireless sensor networks: a survey. Computer Networks 38(4):393–422
2. Chen SS, Donoho DL, Saunders MA (2001) Atomic decomposition by basis pursuit. SIAM Rev 43(1):129–159
3. Davis G, Mallat S, Avellaneda M (1997) Adaptive greedy approximations. Constructive Approximation 13(1):57–98
4. Donoho DL (2006) Compressed sensing. IEEE Transactions on Information Theory 52(4):1289–1306
5. Duarte MF, Sarvotham S, Baron D, Wakin MB, Baraniuk RG (2005) Distributed compressed sensing of jointly sparse signals. Conference Record of the Thirty-Ninth Asilomar Conference onSignals, Systems and Computers 2005:1537–1541
6. Luo C, Wu F, Sun J, Chen CW (2009) Compressive data gathering for large-scale wireless sensor networks. In: Proceedings of the 15th Annual International Conference on Mobile Computing and Networking, ACM, New York, NY, USA, MobiCom '09, pp 145–156
7. So J, Byun H (2016) Load-balanced opportunistic routing for duty-cycled wireless sensor networks. IEEE Transactions on Mobile Computing PP(99):1–1
8. Tan L, Wu M (2016) Data reduction in wireless sensor networks: A hierarchical LMS prediction approach. IEEE Sensors Journal 16(6):1708–1715
9. Tropp JA, Gilbert AC, Strauss MJ (2006) Algorithms for simultaneous sparse approximation: Part i: Greedy pursuit. Signal Process 86(3):572–588
10. Wakin MB, Duarte MF, Sarvotham S, Baron D, Baraniuk RG (2005) Recovery of jointly sparse signals from few random projections. In: Proceedings of the 18th International Conference on Neural Information Processing Systems, MIT Press, Cambridge, MA, USA, NIPS'05, pp 1433–1440

A Small Monopole W-Shaped Antenna for UWB and X-Band Applications with Band Notch for WiMAX

Rajesh Kumar Raj, Roopkishor Sharma, Naresh Pratap Singh Chauhan and Ashish Chaudhary

Abstract This editorial paper presents a small rectangular monopole antenna for UWB and X band with frequency rejection group of WiMAX (3.3–3.7 GHz). An antenna acquires frequency range starting from 2.86 to12.54 GHz. The frequency rejection band is obtained by creating the W-shaped slot in rectangular radiation patch. The antenna is designed on Rogers RO4003 substrate of size 12.0 × 18.0 × 1.6 mm^3. The parametric analysis of VSWR, impedance bandwidth, return loss, gain, input impedance, radiation pattern, etc., of antenna is computed using CST microwave studio.

Keywords Monopole antenna · W-shaped · UWB · Single notch band WiMAX · SMA and FCC

1 Introduction

UWB technology is the one in which transmission of short duration of pulses takes place. There is great attention toward UWB because of various properties such as high data rate, high bandwidth, high speed, low power consumption, and compactness. In addition, it also offers congestion-free hostile receivers and highly secure undetectable and high-accuracy target detection in imaging system and radar to see an object behind the wall and opens it applications for short range and outdoor wireless communication [1]. The FCC allocation of unrestricted UWB band is starting from 3.1 to 10.6 GHz and for X band 8–12 GHz.

Ultra-wide band includes some narrow frequency bands which cause interference such as WiMAX and WLAN operating in 3.3–3.7 GHz range. Thus, the UWB antenna should be in such a way that these interfering frequencies should be

R. K. Raj (✉) · R. Sharma · N. P. S. Chauhan · A. Chaudhary
Government Engineering College, Ajmer, India
e-mail: raj_raj2002@rediffmail.com

A. Chaudhary
e-mail: ashish19n89@gmail.com

V. Janyani et al. (eds.), *Optical and Wireless Technologies*, Lecture Notes in Electrical Engineering 472, https://doi.org/10.1007/978-981-10-7395-3_53

notched; otherwise, there will be interference in the UWB. To avoid this problem, UWB antennas with various antenna configurations have been newly invented [2–6]. In [2] as well as in [3], dissimilar type of slits, a pair of T-shape slot in the square radiation patch, and rod-shaped parasitic element structures on the ground are cut to achieve band notch features, respectively. In [5], by inserting a couple of L-shaped slits on rectangular radiation patch, bandstop function is achieved. In [6], notch frequency function is done by means of U-shaped slots.

It describes a novel UWB printed rectangular monopole aerial with notch band features by inserting the W-shaped slot into the rectangular radiation patch. The projected aerial produces an extensive impedance bandwidth starting from 2.86 to 12.54 GHz with VSWR < 2 requirements for ultra-wide band, which are defined by the Federal Communication Commission, except the narrow band of 3.3–3.7 GHz for IEEE 802.16 (WiMAX system). The antenna gets its design using CST microwave studio. The simulated results of antenna shown in result and discussion section indicate that the projected UWB feeler has good constant omni-directional emission patterns and almost constant gain except the notch band.

2 Antenna Design and Configuration

Practical shape and dimensions of the projected UWB aerial are shown in Fig. 1. It is made of a 1.6 mm thick Rogers RO4003 substrate (permittivity 3.55 and tangent loss 0.0027) with dimension 12.0 × 18.0 mm^2. The basic antenna structure comprises a radiating area with two rectangular slits on the lower edge of the conducting area with a 50 Ω microstrip feed input line and a partial ground plane. The 50 Ω subminiature version A (SMA) connector is associated with the antenna for the purpose of signal communication. Therefore, by etching rectangular-shaped slots on the lower edge of the area as well as by carefully adjusting the length (L_g) of partial ground level and more impedance bandwidth may be reached. In order to mitigate interference by WiMAX, we cut the W-shaped slit in the rectangular radiating patch. In Fig. 1, the W-shaped slit is cut in the middle of the radiating patch and also proportioned with respect to the longitudinal direction.

Table of detail dimensions in millimeter

Symbol	Value	Symbol	Value
L_s	18	W_2	1.5
W_s	12	L_3	6.5
W_p	10	W_3	0.5
L_p	9	W_4	3
L_1	1	L_f	7
W_1	4	W_f	2
L_2	7	L_g	4.5

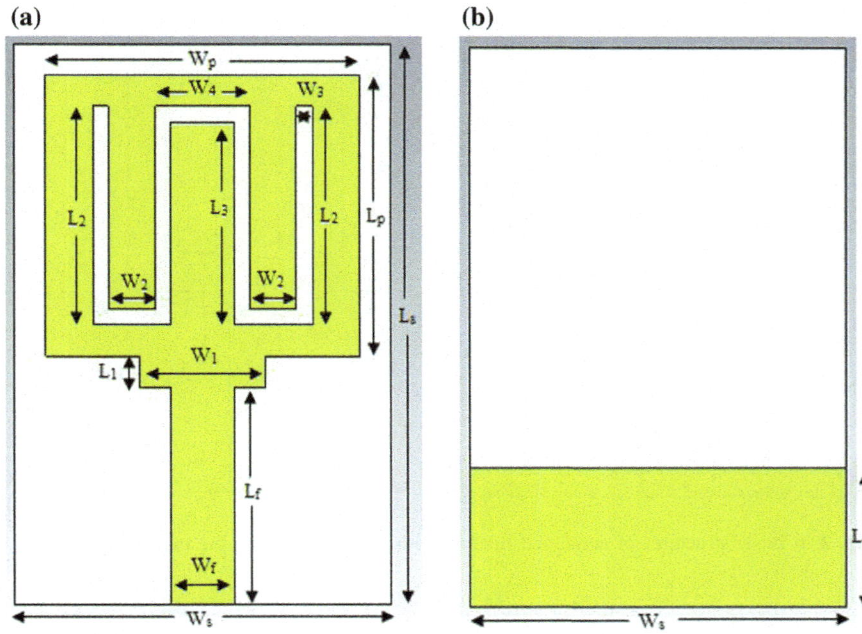

Fig. 1 Practical shape of monopole aerial **a** upper view and **b** rear view

3 Outcomes and Discussion

This unit includes the novel rectangular monopole aerial designed with several factors and computer-generated parametrically for changed prominent factors of the design. Factors of this aerial are calculated by changing only one parameter at a period although keeping others as constant. The computer-generated results are achieved by computer simulation technology (CST).

Figure 2 shows the assembly of aerial used for the purpose of simulation studies. VSWR parameters for rectangular slit aerial with stepping (optimized) (Fig. 2a) and the proposed aerial with W-shaped slot on the radiation patch are shown in Fig. 2b. It is observed that on injecting the W-shaped slot in rectangular radiation patch, bandstop performance is generated. It is also seen that the aerial had an impedance bandwidth (VSWR less than 2) starting from 2.86 to 12.54 GHz. Compared with an ordinary antenna, the proposed UWB antenna with notch band of WiMAX around 3.5 GHz in the frequency range of 3.28–3.7 GHz (Fig. 3).

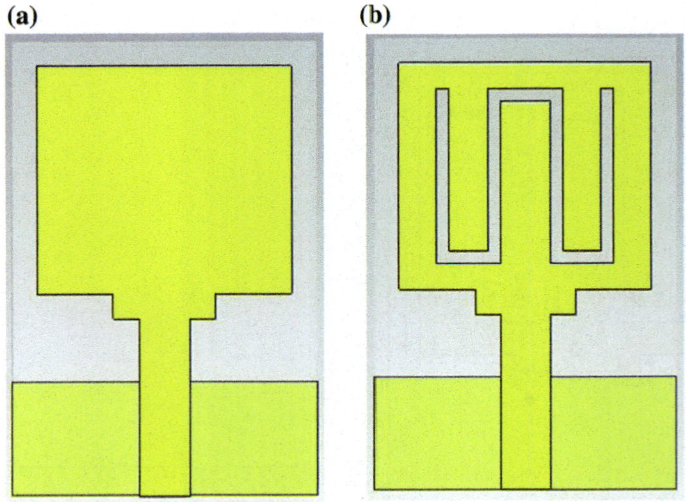

Fig. 2 **a** Basic structures of aerial and **b** aerial with W-shaped slit on the radiation patch

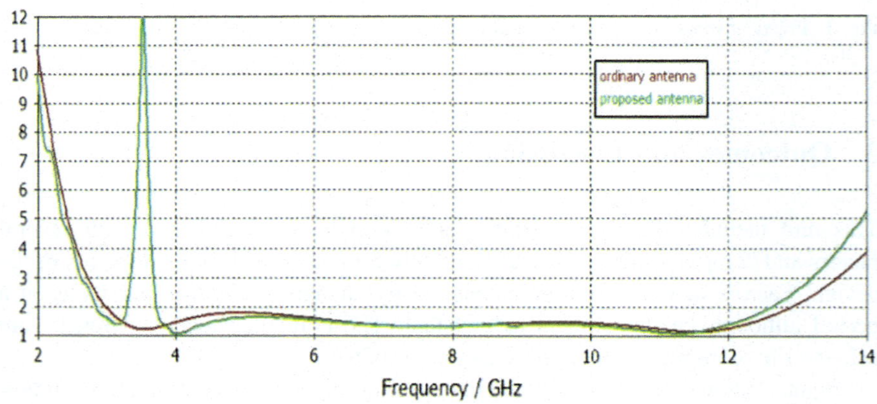

Fig. 3 Simulated VSWR characteristics for antenna structures shown in Fig. 2

Return loss variation with frequency is shown in Fig. 4. Figure 5 shows the virtual surface current 0–276 A/m spreading over the aerial at notch frequency at 3.52 GHz. As detected, there is a compressed current supply along the W-shaped slit at 3.52 GHz. The current path at an inner edge of the slot streams opposite to that at an outer edge, which causes disparaging interference, and antenna becomes nonresponsive to this frequency.

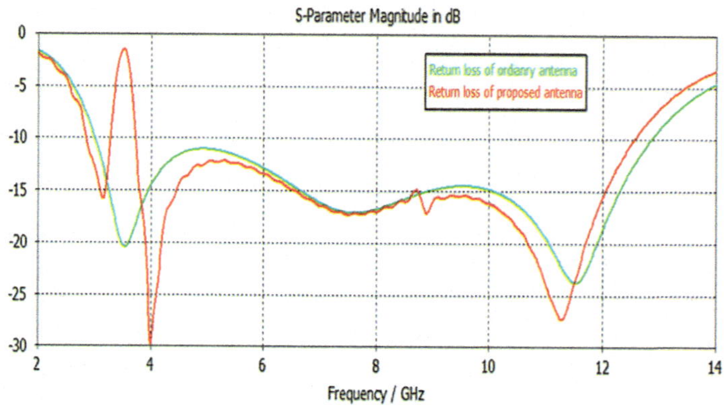

Fig. 4 Comparison of return loss for ordinary antenna and antenna with frequency

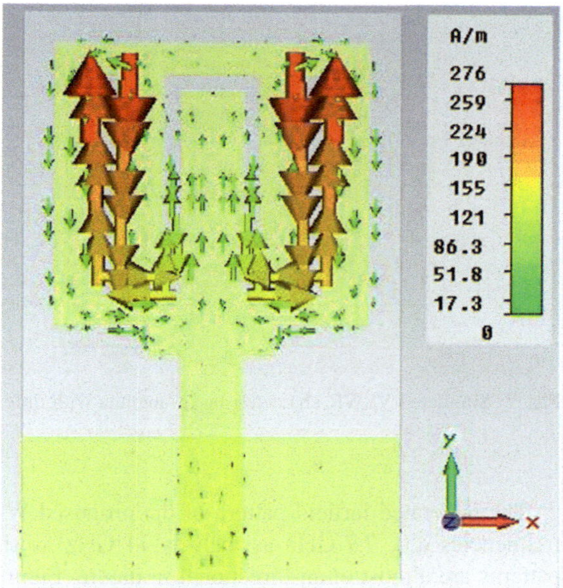

Fig. 5 Simulated surface current distribution at 3.52 GHz

To further investigate the design of proposed UWB antenna, some parametric analyses are carried out. The antenna covers the 2.86–12.54 GHz UWB range by optimizing the length (L_g) of ground plane. The band notch is primarily dogged by the W-shaped slit on the rectangular radiation patch area. Figure 6 as well as Fig. 7 show the simulated bandstop characteristics of the antenna in cases of dissimilar L_2 and W_2, respectively.

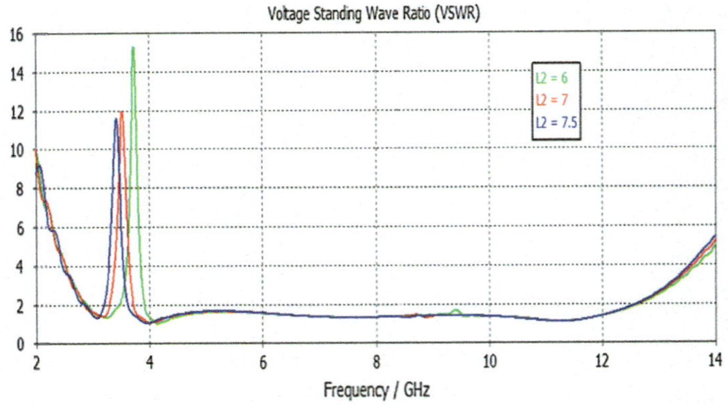

Fig. 6 Simulated VSWR characteristic of antenna with different values of L_2

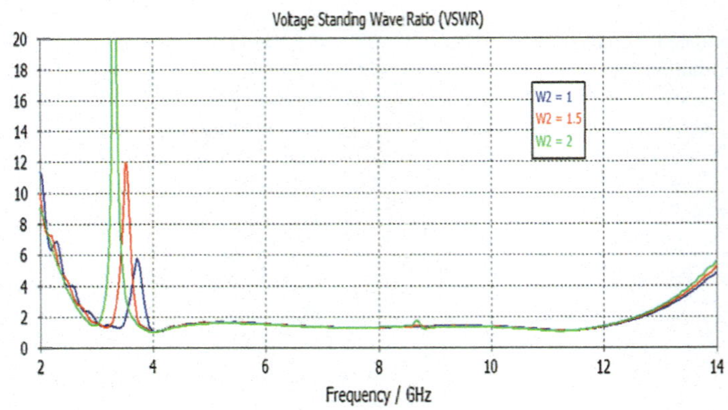

Fig. 7 Simulated VSWR characteristic of antenna with different values of W_2

The generated farfield pattern of the proposed W-shaped notch band antenna at frequencies 4.0, 7.7 GHz as well as 11 GHz as shown in Fig. 8. The radiation patterns are almost omni-directional in nature. Figure 9 shows the constant gain of the antenna except the notch band. Aerial gain with a variation of not more than 4 dB is accomplished outside the notch band.

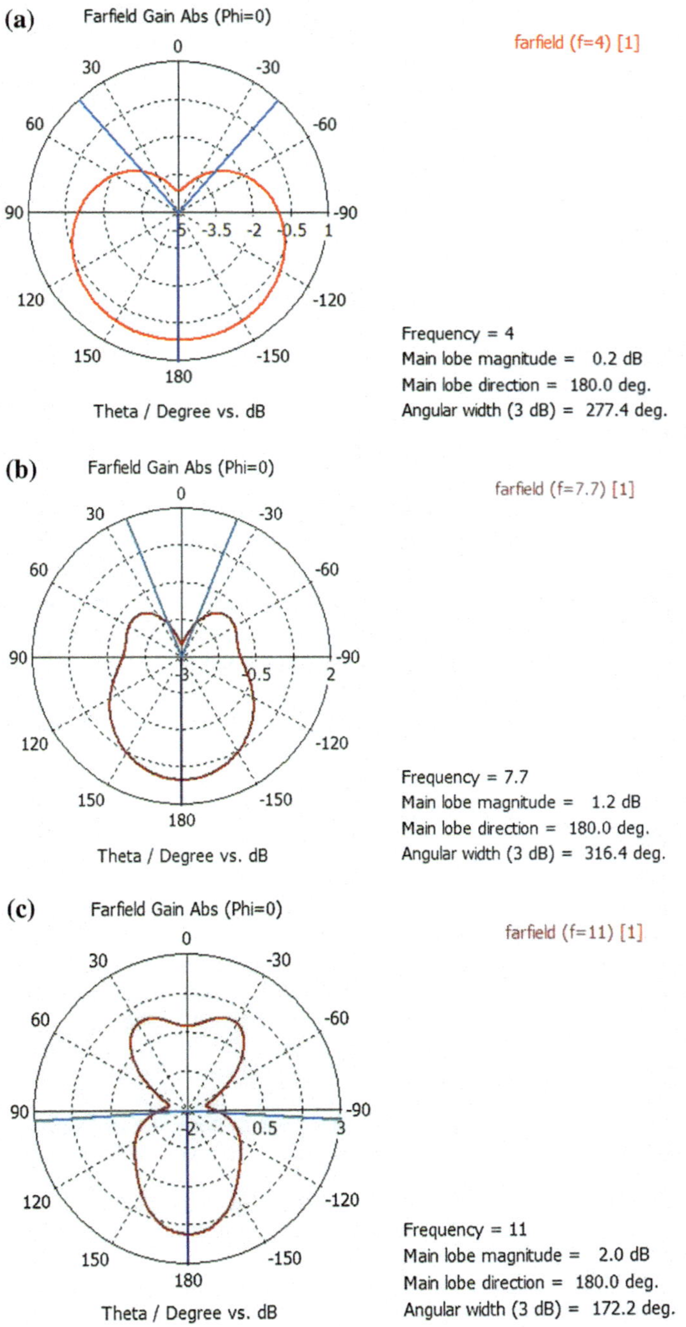

Fig. 8 Simulated farfield radiation pattern of antenna at **a** 4.0 GHz, **b** 7.7 GHz, and **c** 11 GHz

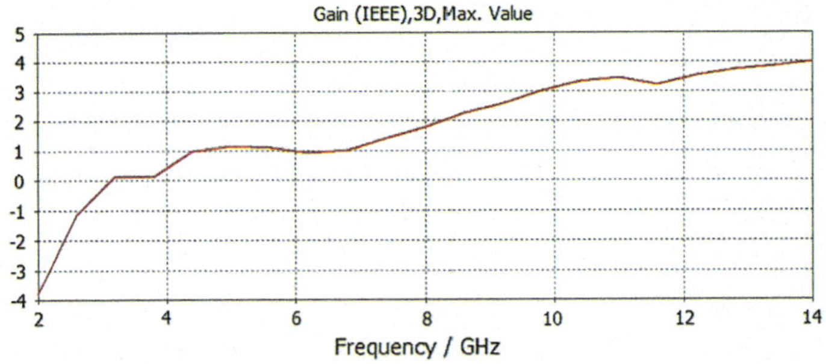

Fig. 9 Simulated gain of antenna

4 Conclusion

A novel W-shaped UWB monopole and X-band antenna with notch band characteristic is shown. By optimizing the length of partial ground plane and cutting the two rectangular slots at the lower edge of radiation patch antenna and stepping achieves the wide bandwidth from 2.86 to 12.54 GHz. By introducing the W-shaped slot in the rectangular radiation patch, bandstop for WiMAX is created. The antenna gives wide impedance bandwidth, VSWR, good radiation characteristics, and return loss characteristics. Also, an aerial size of 12.0 × 18.0 mm^2 is very small, and the gain is higher than 2 dB for X band with simple ground.

References

1. First report and order, revision of part 15 of the commission's rules regarding ultra-wideband transmission systems FCC. FCC (2002)
2. Rostamzadeh M, Mohamadi S, Nourinia J, Ghobadi C, Ojaroudi M (2012) Square monopole antenna for UWB applications with novel rod-shaped parasitic structures and novel v-shaped slots in the ground plane. IEEE Antennas Wirel Propag Lett 11:446–449
3. Ojaroudi M, Yazdanifard S, Ojaroudi N, Sadeghzadeh RA (2011) Band-notched small square-ring antenna with a pair of t-shaped strips protruded inside the square ring for UWB applications. IEEE Antennas Wirel Propag Lett 10:227–230
4. Jalil YE, Chakrabarty CK, Kasi B (2012) A compact ultra wideband antenna with WLAN (IEEE 802.11a) band rejection. In: 1st IEEE international symposium on telecommunication technologies, pp 1–5
5. Singh A, Raj RK (2014) Dual band notched small monopole antenna with a novel t-shaped slot for UWB applications. In: IEEE international conference on advances in engineering & technology research (ICAETR), pp 1–4
6. Mehranpour M, Nourinia J, Ghobadi C, Ojaroudi M (2012) Dual band-notched square monopole antenna for ultrawideband applications. IEEE Antennas Wirel Propag Lett 11:172–175
7. Computer simulation technology (2014)

Vehicle-to-Vehicle Communication—A Vertical Handover Algorithm Based on Vehicle Speed

S. Meghana and P. C. Jain

Abstract Vehicular communication is a part of intelligent transport system (ITS) which provides an intelligent way of transport to avoid accidents. Vehicle-to-Vehicle communication (V2V) is carried out through Vehicular Ad hoc Networks (VANET). This paper discusses about the MAC (Media Access Control) layer protocol, features of IEEE802.11p standard, and the handover protocol problems it is facing. Due to vehicle high speed and the non-homogeneous structure of the network infrastructure, support of connectivity by VANET is still a challenging task. In this paper, a vertical handover algorithm has been analyzed using vehicle speed, and the results have been supported by simulation.

Keywords VANET · ITS · DSRC · WAVE · 802.11p · Handover

1 Introduction

There are many accidents occurring on roads due to negligence and lack of proper intelligent transport system (ITS). Studies proved that 60% of accidents can be avoided if a warning message is sent to that vehicle at least half a second before accidents. National Highway Traffic Safety Administration (NHTSA) has decided and is committed to improve the safety and mobility of transport by designing an ITS. The vehicle-to-vehicle (V2V) communication is to avoid accidents by sending position and speed data of vehicle in transit to another vehicle over an ad hoc mesh network. It provides 360° awareness of surrounding threats [1]. It supersedes techniques like blind spot detection, rear parking sonar, and backup camera. The main purpose of vehicle-to-vehicle communication is to provide an intelligent

S. Meghana · P. C. Jain (✉)
EE Department, School of Engineering, Shiv Nadar University, G. Noida 201314, UP, India
e-mail: premchand.jain@snu.edu.in

S. Meghana
e-mail: sm867@snu.edu.in

V. Janyani et al. (eds.), *Optical and Wireless Technologies*, Lecture Notes in Electrical Engineering 472, https://doi.org/10.1007/978-981-10-7395-3_54

means of transport service. This is introduced in order to avoid accidents between vehicles by sending warning messages to each other. These warning messages consist of information regarding vehicles speed, emergency stopping, brake status, etc. V2V communication is like an additional step for warning the drivers. V2V communication uses dedicated short-range radio communication (DSRC) devices to transmit messages. This paper examines all the elements of technology used, and emphasis is given mainly on communication protocols and handover problem that occur while communicating with vehicle and a roadside unit (RSU). Our main aim is to increase throughput and to decrease the latency for effective transmission of an EWM (Emergency Warning Message). In this paper, Sect. 2 discusses V2V communication, Sect. 3 discusses handover protocol and its simulation, and finally Sect. 4 concludes the paper.

2 Vehicle-to-Vehicle Communication

Vehicular ad hoc network (VANET) is a self-organized network in which the vehicles get connected. VANET provides efficient route information to other vehicles and informs drivers about traffic conditions, accidents, road conditions. It is similar to mobile ad hoc networks (MANET). In VANET, all nodes are vehicles which move at high speed compared to the speed of MANET nodes. VANET sets up and maintains a communication network among a group of vehicles along with the help of roadside base station or a controller. VANET turns every participating vehicle into a wireless router or node, allowing vehicles approximately 100—300 m of each other to connect, and in turn creates a network with a wide range. Some vehicles can fall out of the signal range and are dropped out of the network while other vehicles within the range can join the network. This results in creating a mobile Internet. VANET topology changes frequently as the position of vehicle changes due to high mobility and the random speed of the vehicles.

2.1 Inter-Vehicle Communication

Inter-vehicle communication uses multi-hop multicast/broadcast to communicate between each vehicle. Collision warning messages broadcast from V2V across multi-hops. This is suitable for short range. A vehicle communicates with another vehicle by using different protocols. In such cases, the receiver takes appropriate decision on the basis of emergency messages received and accordingly takes appropriate action. To avoid collision in V2V communication, location-based multicast and broadcasting are used [2]. Multi-hop communication propagates the message in the absence of RSU infrastructure. However, in low-density vehicular network, V2V communication is not very good due to large distances.

There are two types of messages forwarding in inter-vehicle communications: Naive broadcasting and Intelligent broadcasting. In Naive broadcasting, vehicles broadcast messages periodically and at regular intervals. Messages are ignored by a vehicle if received from a vehicle behind it. However, if message comes from vehicle in front of it, the vehicle sends its own broadcast message to vehicles behind to it. In this way, all enabled vehicles moving in the forward direction get all broadcast messages. This method generates a large number of messages, and thus, data is not utilized properly. In Intelligent broadcasting, if broadcast message comes from a vehicle behind it, as an acknowledgment, it ceases broadcasting because this vehicle now knows that the vehicle behind it has received its message and will also assume that vehicle behind it will also broadcast the message to the rest of the vehicles. If the vehicle behind it receives messages from a number of sources, then it will entertain only the first message and discard the others.

2.2 *Vehicle-to-Infrastructure Communication*

Vehicle-to-infrastructure (V2I) provides solution for longer range vehicular networks by using wireless cellular networks like 3G, 4G-LTE and supports V2I protocol. The communication between vehicles and RSU is supported by V2R protocol. In V2R, warning messages are sent first to RSU, and then RSU broadcasts it to all vehicles in the range. The roadside infrastructure involves additional installation costs. V2I provides large BW link between vehicles and RSU. The RSU can be deployed after every km to obtain high data rates required during heavy traffic.

2.3 *V2V/V2R Communication Protocol*

When a vehicle has mechanical failure or detects road hazards, the vehicle generates an EVM (Emergency warning message) and keeps one copy with him for retransmission, if required. The vehicle broadcasts to neighboring vehicles, and it transmits periodically to RSU also through transceivers with different frequency bands till it receives the message with the same ID from the vehicle behind and RSU, respectively. When RSU receives EWM from source vehicle, it replaces with own ID and forwards to all vehicles within the range [3].

2.4 *Dedicated Short Range Communication*

Dedicated short range communication (DSRC) combines WiFi and GPS positioning to give a 360° awareness of all vehicles around them. It helps to provide the driver with a warning to avoid collision using on-board computer. The cars share their positions very rapidly even though two cars could be around a corner. DSRC allows cooperative collision avoidance (CCA) in which cars warn each other about changing conditions to significantly improve safety. DSRC is a two-way short-to-medium range wireless communication that permits very high data transmission in V2V communication. The main reason for using DSRC in V2V communication is to detect hazards in vehicle's path even the driver in such a position is not able to see.

DSRC is lower overhead version of IEEE 802.11a WLAN (Wireless Local Area Network) [4] and is named as IEEE802.11p standard and is suitable for high-speed vehicle communication. DSRC PHY (Physical) layer is targeted to operate in a licensed spectrum around 5.9 GHz band with 75 MHz BW as opposed to IEEE 802.11a that allows only the unlicensed frequency band 5 GHz. In MAC (Media Access Control) layer, the DSRC band plan consists of seven channels each with 10 MHz BW and that include one control channel and 6 service channels. It can support a large family of vehicular safety and non-safety applications. Some requirements of MAC are low latency and high reliability. The PHY layer provides better radio propagation with respect to multi-path reflection delay, and Doppler effects occur due to high-speed vehicles and road environment. The end-to-end delay is low and packet delivery ratio is high compared to 802.11a [5].

The 802.11p standard adds wireless access to vehicular networks and implements OSI layers stack. Upper layers of OSI follow IEEE1609 family standard as shown in Fig. 1.Wireless Protocol works at licensed band of 5.9 GHz with 1 km range and data rate of 6—27 Mbps. DSRC PHY layer adopts same OFDM as 802.11a. Figure 1 shows DSRC Protocol.

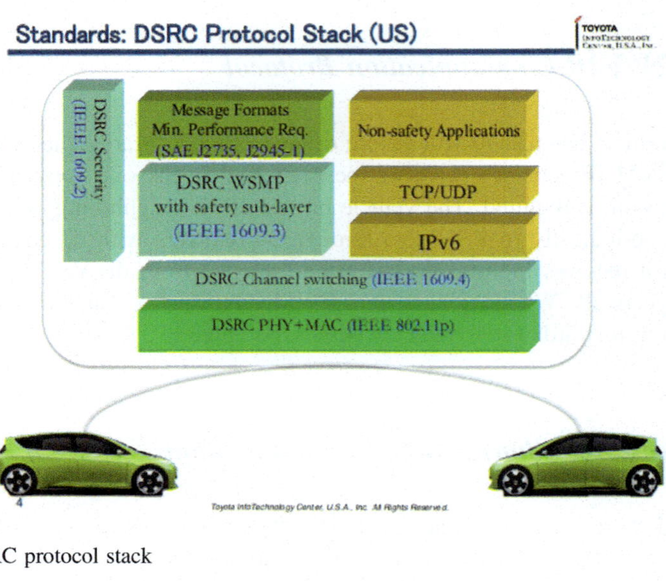

Fig. 1 DSRC protocol stack

2.5 *Wireless Access for Vehicular Environment*

Wireless access for vehicular environment (WAVE) is referred as IEEE 802.11p standard. It is designed for wireless VANET to support safety, quality, and reliable data communication [5]. The 5.9 GHz licensed band with 75 MHz BW (5.85–5.925 GHz) is used to set up communication between V2V and RSU. WAVE provides the real-time communication and takes care of transport safety and minimizes traffic congestion. It also benefits for the transport sustainability. WAVE ensures the traffic information collection and immediate transmission with stability and keeps the information security. US FCC channel format reserves seven channels each with 10 MHz BW. Service channel (SCH) provides two-way communications while control channel (CCH) provides management and safety information. WAVE uses the CCH for communication safety and control information while SCH is used for IP-based communication services. Consequently, each communication zone must utilize CCH as a control channel used for safety messages, and then it may utilize one or more SCH from available service channels. Vehicles initiating connections are discouraged to choose same SCH used by immediate neighbors.

3 Handover Protocol

Emergency warning message is sent from vehicle-to-RSU and from V2V. While sending the information, there could be handover problem between vehicles due to which sending EWM becomes a problem. The handover problem occurs when the received signal decreases and falls below the minimum signal strength required for the communication [6]. When a vehicle moves from one cell to another that belongs to the same network, it is called horizontal or homogenous handover. Vertical handover (VHO) handover occurs when vehicles overlap two heterogeneous wireless networks during moving. In such scenarios, frequent switching not always necessary to take place from one network to another. The speed-based VHO (S-VHO) algorithm outperforms in terms of throughput and delay [7]. VANET deals with the wireless problems such as high-speed vehicles, packet loss due to obstacles, often change in topology, network fragmentation, channel capacity, and coverage. A great deal of effort is required and has to be worked upon to offer new optimization techniques and to design efficient, powerful data transfer protocols. In such networks, handover is a difficult task, because no central authority exists for finding the data paths among different nodes.

3.1 Speed-Based Vertical Handover Protocol

Vehicle speed is used as assessment criteria for vertical handover [8]. It aims to minimize VHO frequency, number of VHO occurrences, and maximize throughput measured at the vehicle. Handover decision takes place on the basis of time taken by a vehicle to cross a cell and the estimation of data rate. Let us consider two networks: one serving network (SN) when vehicle enters in the cell and the other candidate network (CN) when vehicle leaves that cell. The switching mechanism from SN to CN is called vertical handover which is driven by vehicle speed or by infrastructure using RSU and received signal power.

We assume that the vehicle is driven at a constant speed v, where v is the vehicle speed vector whose direction is indicated with an arrow in Fig. 2. The vehicle location (P) is tracked and updated continuously with the help of GPS (Global Positioning System) receiver in the vehicle. Let t_{in} be the time instant and Pin is the position at SN (serving network) when a vehicle enters in a wireless cell, t_{out} and Pout be the time instant and position at CN (candidate network), respectively, when vehicle leaves wireless cell, Δx be the distance traveled inside the wireless cell during the cell crossing time interval $T = (t_{out} - t_{in})$, R be the radius of the wireless cell, and φ be the angle between the vehicle's line of sight with the RSU and the direction of the vehicle, then

$$\Delta T = \Delta x/v = 2R\cos\varphi/|v| \quad (1)$$

Throughput Θ(t) evaluated for ΔT as product of BW and the time interval.

$$\Theta(\Delta T)_{SN} = B_{SN} \cdot \Delta T \quad (2)$$

where B_{SN} is the bandwidth (bps) of the service network (SN).

$$\Theta(\Delta T)_{SN} = B_{SN}[2R\cos\varphi/|v|] \quad (3)$$

Now, consider a vehicle connected to a SN and entering in a CN, and let L be the handover latency, and δ be the hysteresis factor. Hysteresis factor avoids vertical handover occurrence when two competing networks have negligible BW difference

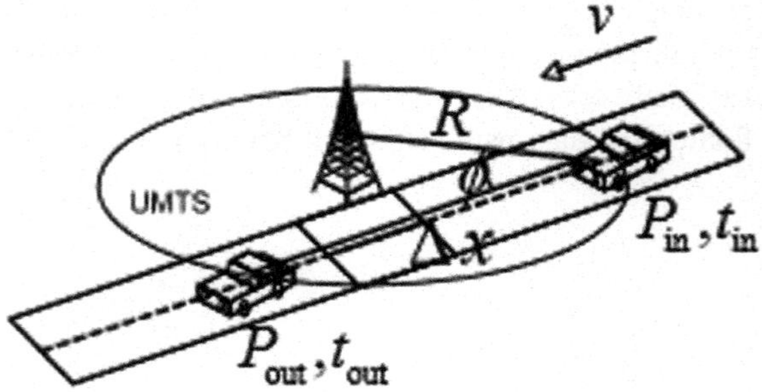

Fig. 2 Vehicle mobility mode

$(B_{CN} - B_{SN})$. Handover latency L between two wireless networks is the time interval during which vehicle does not receive any data due to control plane (socket switching) signaling messages exchange between two networks. Valid handover results in throughput increase when a vehicle moves over two wireless networks. To maximize the throughput at CN when a system initiates a handover from the SN to the CN, the throughput at CN is

$$\Theta_{CN} \geq \Theta_{SN} \tag{4}$$

Taking into consideration of handover latency (L) and hysteresis factor (δ) and using Eq. (2),

$$B_{CN}(\Delta T - L) \geq (B_{SN} + \delta)(\Delta T) \tag{5}$$

$$B_{CN} \geq [B_{SN} + \delta]/[1 - (L/\Delta T)] \tag{6}$$

The above Eq. (6) states that handover latency and vehicle speed play together a crucial role in VHO decision because ΔT is indirectly related to velocity of vehicle (Eq. 1). The throughput at CN depends not only on BW but also on crossing time (ΔT), hysteresis factor (δ), and handover latency (L). If the new handover rate overcomes the old rate by at least a value equal to the hysteresis threshold, the crossing time evaluates whether it makes sense to proceed with a handover or not. The vertical handover threshold T_{VHO} can be obtained by using Eq. (1) and (5)

$$T_{VHO} = (2R\cos\varphi\, B_{CN}L)/\{\Delta X[B_{CN} - (B_{SN} + \delta)]\} \tag{7}$$

S-VHO algorithm takes three criteria into consideration: first determines data rate estimation, then cross time estimation, and thereafter handover decision takes place. The input to the algorithm is vehicle speed, time of vehicle entering into wireless cell, and location of vehicle at SN. If the throughput of CN is greater than throughput of SN (Eq. 5), then stay at SN otherwise find crossing time estimation. If the crossing time ΔT is higher than or equal to T_{VHO}, then only handover switches to CN otherwise stay at SN.

Simulation has been carried out for a vertical handover in heterogeneous vehicular ad hoc networks using MATLAB/SIMULINK. Simulation results show nonlinear behavior between latency and speed of the vehicle (Fig. 3). As the speed of the vehicle increases, latency decreases which implies that the rate of handover decreases. This is because by the time the handover occurs, the vehicle crosses the cell indicating no further need of handover. As the speed of vehicle increases, the time spent by the vehicle within the wireless cell decreases.

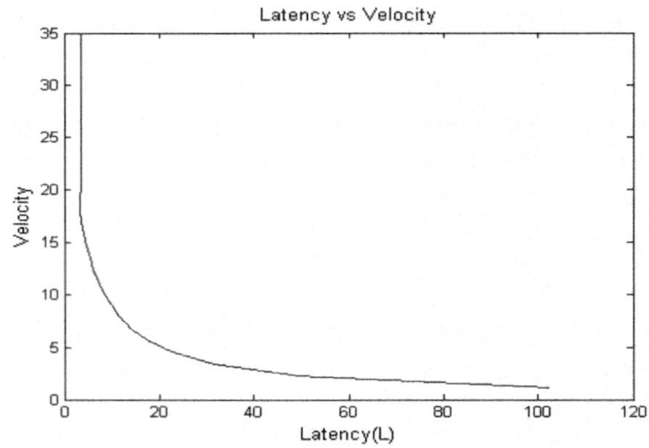

Fig. 3 Velocity (m/s) versus latency (ms)

4 Conclusions

An analytical model and simulation for a vertical handover in heterogeneous vehicular ad hoc networks has been carried out in this paper. A network switching does not improve throughput when a vehicle approaches candidate network (CN) with high data rate. S-VHO uses handover hysteresis factor to improve throughput and delay and is driven by the vehicle's speed. To maintain acceptable levels of throughput, jitter, and delay, the vehicle should maintain its given speed limit. The throughput is dependent on BW, cell crossing time, vehicle speed, and overhead in protocol of control plane adopted.

Acknowledgments The authors are thankful to Prof. Dinkar Prasad, Head, EE Dept., and Associate Director, School of Engineering., Shiv Nadar University, G. Noida (UP) for providing necessary resources and infrastructure to conduct this project work and for his encouragement and permission to publish this paper.

References

1. Molisch AF, Tufvesson F, Karedal J, Mecklenbräuker CF (2009) A survey on vehicle-to-vehicle propagation channels. Proc IEEE Wirel Commun 16(6):12–22
2. Biswas S, Tatchikou R, Dion F (2006) Vehicle-to-vehicle wireless communication protocols for enhancing highway traffic safety. IEEE Commun Mag 44(1):74–82
3. Hu B, Gharavi H (2011) A joint vehicle-vehicle and vehicle-roadside communication for highway traffic safety. Int J Veh Technol, Hindawi Publishing Corporation
4. IEEE standard for information technology-telecommunications and information exchange between systems—local and metropolitan area networks-specific requirements-part 11: wireless LAN medium access control (MAC) and physical layer (PHY) specifications. IEEE Std. 802.11-2012 (2012)
5. Li YJ (2015) An overview of the DSRC/WAVE technology. NICTA, NSW, Australia, pp 1–15

6. Kwak D, Mo J, Kang M (2009) Investigation of handoffs for IEEE 802.11 networks in vehicular environment. In: Proceedings on first internatinal conference on ubiquitous and future networks (ICUFN-2009). Hongkong, pp 89–94
7. Yan Z, Zhou H, Zhang H, Zhang S (2008) Speed based probability-driven seamless handover scheme between WLAN and UMTS. In: Proceedings on 4th international conference on Mobile Ad-hoc and Sensor Networks, Los Alamitos. CA, USA, pp 110–115
8. Vegni AM, Esposite F (2010) A speed based vertical handover algorithm for VANET. In: Proceedings of 7th international workshop on intelligent transportation. Humburg

Design a Slotted Microstrip Patch Antenna at 60 GHz for Millimeter Wave Mobile Communication

Jyoti Saini and S. K. Agarwal

Abstract This proposed paper presents a design of slotted microstrip patch antenna at 60 GHz. This proposed antenna is suitable for millimeter wave mobile communication. The proposed antenna is of size 10 × 10 mm^2 and printed on a material Roger RT5880. The substrates have a relative permittivity 2.2, loss tangent 0.0009, and a thickness 1.6 mm. This antenna is simulated by using computer simulation technology software. The design consists of a two U slot and two rhombic shapes radiating on the patch and 50 ohm microstrip line feeding method used. The simulated antenna results show the return loss −31.45 dB and gain 6.03 dB.

Keywords Slotted microstrip patch antenna · Millimeter wave
CST software

1 Introduction

Fast advancement in the field of telecommunication requires a great and drastic change in systems and techniques used for communication. In this step to develop new communication techniques, there are many people who are researching in the direction to increase the number of users as well as demand of bandwidth for high speed. In this direction, the most important rising millimeter wave technologies are used in last few years. Although this technology is well known since many years, but in last 5–6 years due to commercial perspective, there is a great advancement in silicon technologies and results having low cost. These papers mainly target on 60 GHz of wireless technologies that are enabled for many new applications which are feasible at higher value of carrier frequencies [4, 5].

J. Saini (✉) · S. K. Agarwal
Government Women Engineering College Ajmer, Ajmer, Rajasthan, India
e-mail: 001sainijyoti@gmail.com

S. K. Agarwal
e-mail: skagarwal5@rediffmail.com

V. Janyani et al. (eds.), *Optical and Wireless Technologies*, Lecture Notes in Electrical Engineering 472, https://doi.org/10.1007/978-981-10-7395-3_55

When a concept of large bandwidth comes in mind, we have to move towards 60 GHz this band being allocated the unlicensed bandwidth. This large unlicensed band of 60 GHz technology helps in increasing the capacity and flexibility, which is irresistible for the wireless technologies. Due to slotted dimensions of 60 GHz radio wave, at the user terminal it also permits multiple antenna solutions that are very difficult, or we can say not impossible at lower frequencies [1].

2 Structure of Slotted Antenna and Dimensions

In the basis of several parameter studies, the structure of slotted microstrip patch antenna is presented in Fig. 1. The proposed antenna design of 10 × 10 mm^2 built on a 1.6-mm-thick Roger RT5880 substrate has dielectric material $\varepsilon r = 2.2$ with the loss tangent $\tan\delta = 0.0009$. Patch of the length and width is 2.187 and 3.950 mm, respectively. The feed line length is 3.09 mm, and width is 0.9 mm. Optimized parameters are used in slotted microstrip patch antenna. For simulation purpose, electromagnetic software CST microwave studio has been used.

The width of the patch antenna is calculated by

$$W = \frac{v_0}{2f_r}\sqrt{\frac{2}{\in_r + 1}}$$

The length of the patch antenna is calculated by

$$L = L_{eff} - 2\Delta L$$

where L_{eff} is the effective length of patch

ΔL is the extended length of patch

L_{eff} and ΔL are calculated by

$$L_{eff} = \frac{v_0}{2f_r\sqrt{\in_{reff}}}$$

$$\Delta L = h \times 0.412\frac{\left(\in_{reff} + 0.3\right)\left(\frac{w}{h} + 0.264\right)}{\left(\in_{reff} - 0.258\right)\left(\frac{w}{h} + 0.8\right)}$$

where ε_{reff} is the effective dielectric constant of microstrip patch antennas on dielectric substrate and is calculated by

$$\in_{reff} = \frac{\in_r + 1}{2} + \frac{\in_r - 1}{2}\left[1 + \frac{12h}{w}\right]^{-1/2}$$

A dimension of the ground plane is 10 mm x 10 mm, and the thickness of 0.035 mm has been used. The two U slots and two rhombic shapes cut on the patch.

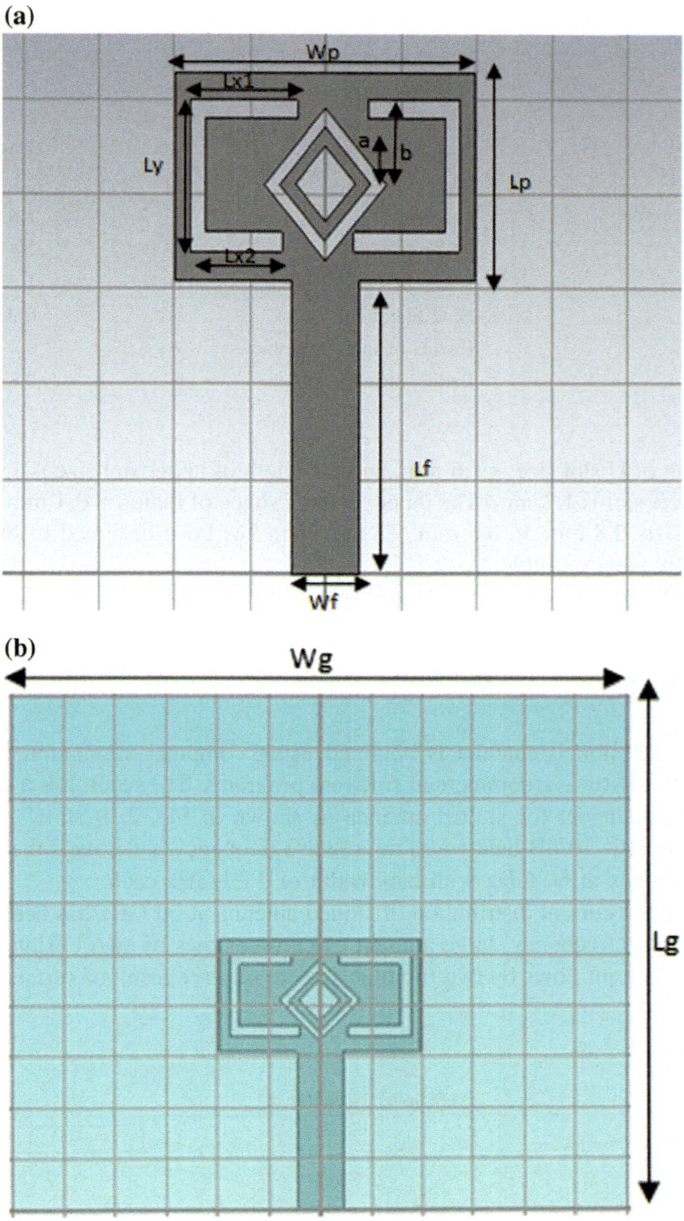

Fig. 1 Structure of slotted patch antenna: **a** Patch view. **b** Ground view

Table 1 Proposed antenna dimensions

Name of dimensions	Symbol	Value (mm)
Length of ground	Lg	10
Width of ground	Wg	10
Length of patch	Lp	2.187
Width of patch	Wp	3.950
Feed line length	Lf	3.09
Length of slot	Ly	1.6
Width of slot 1	Lx1	1.4
Width of slot 2	Lx2	1.2
Feed line width	Wf	0.9
Radius of inner rhombic	a	0.4
Radius of outer rhombic	b	0.8

The length of U slot size is 1.6 mm, and the width of upper slot size is 1.4 mm and of the lower slot is 1.2 mm. The inner rhombic shape of radius is 0.4 mm, and outer slot radius is 0.8 mm × 0.6 mm. The antenna has been designed to the work at 60 GHz frequency (Table 1).

3 Results and Comparisons

The result of slotted antenna is obtained using computer simulation technology (CST) and microwave commercial software programs. The result has a return loss/ reflection coefficient for slotted antenna as shown in Fig. 2. It shows the slotted antenna has its −10 dB bandwidth microwave frequency and covers the millimeter wave frequency at 60 GHz with bandwidth of 3.29 GHz each.

The surface current distribution of slotted antenna at 60 GHz has been shown in Fig. 3. At this frequency, large amount of current flows by two U-slots and small amount of current flows by two rhombic slots to observe result of surface current at

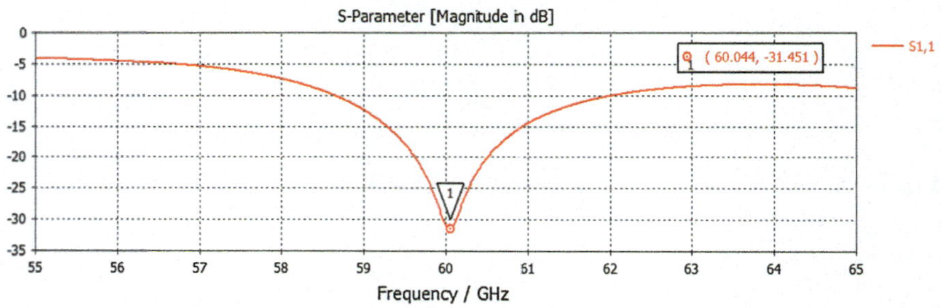

Fig. 2 Reflection coefficient |S11| versus frequency of proposed antenna

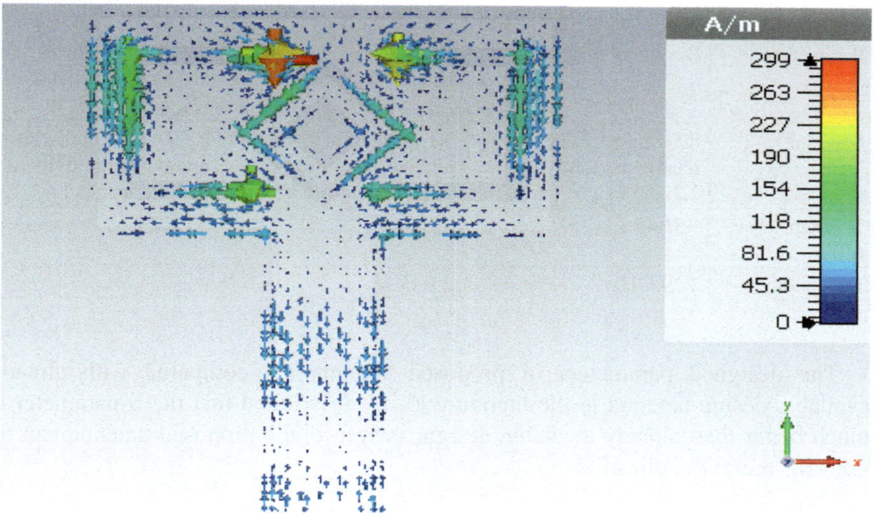

Fig. 3 Surface current distribution of the slotted antenna at the resonant frequency

60 GHz frequency. The surface waves having some losses that are cause for both slots coupled with patch.

The three-dimensional radiation pattern of the slotted antenna represents gain at 60 GHz frequency with 6.03 dB, and radiation efficiency −1.738 dB is shown in Fig. 4 (Table 2).

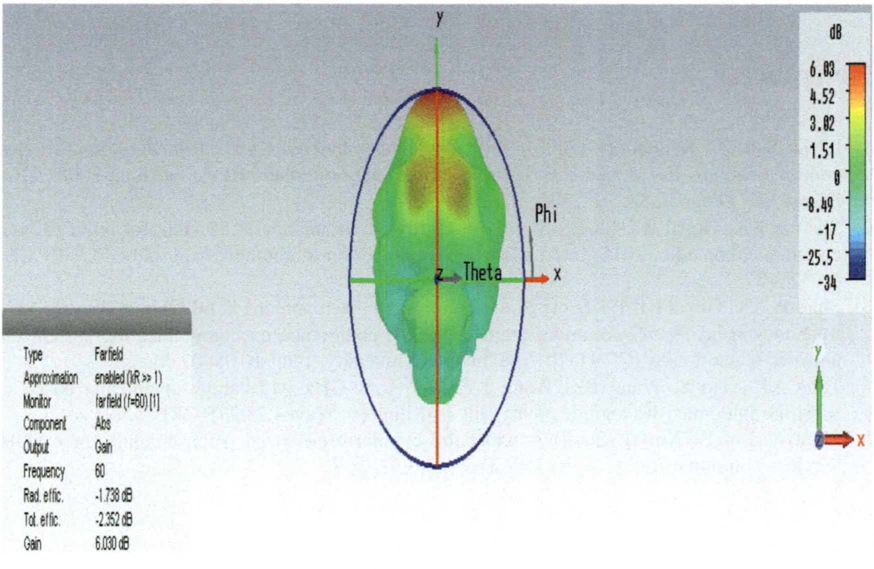

Fig. 4 Three-dimensional radiation patterns at 60 GHz

Table 2 Comparison of designed antenna with already designed antenna in the literature

Parameters	Designed antenna	Already designed antenna in literature [2]	Already designed antenna in literature [3]
Material	RT Duriod 5880 with relative permittivity 2.2 and height 1.6	RT Duriod5880 with relative permittivity 2.2 and height 1.6	RT Duriod5880 with relative permittivity 2.2 and height 1.6
S-parameter (dB)	−31.45	−26.69	−13.258
Bandwidth	3.29 GHz	2.6 GHz	–

The designed parameters of proposed antenna are compared with already available design antenna in the literature [3, 2]. It is found that the S-parameter is much better than already available design; therefore, the proposed antenna can be used for the 5G applications.

4 Conclusion

This proposed paper, slotted microstip patch antenna for the 5G mobile communication, is presented. This proposed design results show the center frequency at 60.048 GHz and the bandwidth at 3.27 GHz, ranging from 58.649 GHz to 61.924 GHz, respectively. At a time, slotted antenna is very thin and uses low dielectric material. The proposed microstrip patch antennas have gain upto 6.03 dB.

References

1. Li K, Sato T, Kajitani N (2008) Wideband planar antennas for millimeter-wave wireless communications. In: Antennas and propagation society international symposium, 2008, AP-S. Digest San Diego, USA, no. 505-2, July 5–12
2. Sharma KK, Goyal RK (2015) Slotted microstip patch antenna at 60 GHz for point to point communication. In: 2015 international conference on communication networks (ICCN), pp 371–373
3. Sharma KK, Goyal RK (2016) H-Slotted microstrip patch antenna at 60 GHz millimeter wave frequency band for 5G communication. In: IEEE conference on communication system and network technologies (CSNT-2016). Chitkara University, Punjab, 05–07 March
4. Tang XJ, Xiao S, Wang BZ, Wang J (2007) A 60-GHz wideband slot antenna based on substrate integrated waveguide cavity. Int J Millimeter Waves 28:275–281
5. Zhou R, Liu D, Xin H (2012) A wideband circularly polarized patch antenna for 60 GHz wireless communications. Wirel Eng Technol 3:97–105

Study of Electromagnetic Radiation Effects on Human Body and Reduction Techniques

Rupesh Acharya, Durgesh Kumar and Garima Mathur

Abstract It is 2017 and mobile communication technology has completely engraved itself into our lives. Mobile communication technology usage has been quickly develop internationally in the last 10 years, which has ensued in concern of public about the harmful health effect of electromagnetic (EM) radiation that are discharged by cell phones. The exposure of electromagnetic radiation (EMR) has received an equitable share of participation in literature; however, this paper is one to represent the effects of the electromagnetic fields induced by cell phones on human body, plans related to electromagnetic vulnerability in mobile communication and present feasible ways of reducing the effects.

Keywords Epidemiological · EM radiation · Specific Absorption Rate (SAR) Mobile communication system

1 Introduction

In this modern era of technology, with the rapid development of information technology, more and more electronic devices/gadgets are invented, and the electromagnetic radiation of electronic devices/gadgets is getting more and more aid. Nowadays, the number of cell phone usage has been quickly expanded internationally in the last ten years, which has ensued in concern of public about the bad effects on health of electromagnetic radiation that are discharged by cell phones. Mobile phone technology has revolutionized the Indian telecommunication scenario; more than 100 crore people are using mobile phones. All in all, a major

R. Acharya (✉) · D. Kumar · G. Mathur
Poornima College of Engineering, Rajasthan Technical University, Jaipur, India
e-mail: 2014pceecrupesh@poornima.org

D. Kumar
e-mail: durgesh.kumar@poornima.org

G. Mathur
e-mail: drg.mathur@poornima.org

V. Janyani et al. (eds.), *Optical and Wireless Technologies*, Lecture Notes in Electrical Engineering 472, https://doi.org/10.1007/978-981-10-7395-3_56

problem is the growing concern of public about the harmful effects on health of the interaction between EM radiation and the human body [1, 2]. A higher brain cancer peril showed by epidemiological studies in human being who had used handset for more than 10 years [3]. There is abundant experimental proof of alteration in brain activity of human brain caused by the mobile phone exposure [4]. Concerning the endocrine system, the hypothalamus-hypophysis-thyroid axis plays a staple function in profiles of human hormone which impacts the body emergence, progress, metabolism, and action of nervous system. It has been entrenched that even a petite alteration in thyroid hormone stratums flowing in the blood is ample to interpolate function of brain. It is (thyroid gland) one of the most vulnerable and vital organs and can be a target for any sort of EM radiation [5]. The Specific Absorption Rate (SAR) used as an indicator of EM radiation in the study of microwave biologic. The SAR caused by the human vulnerability to EM fields, and it is the standard criteria to measure or calculate the amount of EM radiation energy engrossed by the body's unit mass tissue. The expression Specific absorption rate (SAR) is defined as

$$\mathrm{SAR} = \frac{d}{dt}\left(\frac{dw}{dm}\right) = \frac{d}{dt}\left(\frac{dw}{\rho dv}\right) \tag{1}$$

In the actual measurement, the radiation power is not easy to obtain; therefore, the expression can be deformed as

$$\mathrm{SAR} = \frac{\sigma E^2}{\rho} \tag{2}$$

The electrical conductivity and density are known, so that if we know the electric field intensity the value of Specific Absorption Rate (SAR) will be measured or calculated. Specific Absorption Rate (SAR) is measured or calculated individually for different parts of body due to their pace of preoccupation. Thus, SAR metrics is further subdivided into

(1) Whole-Body Averaged Specific Absorption Rate (wbaSAR): It is elucidated as the ratio of power (total) enwrapped in the body and the mass of body. For regulative conformity, the wbaSAR is averaged over duration of 6 min. The Whole-Body Averaged SAR is expressed as

$$\text{Whole Body Averaged SAR} = \frac{1}{M}\int_R \sigma |E|^2 \delta v \tag{3}$$

where

M => mass (complete) of body
R => area of the body
V => volume of complete tissue of the body

(2) Organ-Specific Averaged Specific Absorption Rate (osaSAR): It is elucidated as mass average of the SAR in a specific tissue or organ in the body. Organ-Specific Averaged SAR is expressed as

$$\text{Organ Specific Averaged SAR} = \frac{1}{M_{organ}} \int_{organ} \sigma |E|^2 \delta v \tag{4}$$

3) Peak Spatial-Averaged Specific Absorption Rate (psaSAR): The utmost local Specific Absorption Rate mean over a particular tissue's mass, customarily 10 gm, above duration of 6 min is known as peak spatial-averaged SAR [6].

Specific absorption rate (SAR) measure unit is watt/kilogram and as stated by Telecommunication department of India, SAR value must be made available to the consumers at the point of sale.

2 Effects of Em Radiation on Human Body

When a human body comes in contact with EMR, it engrosses radiation, as human body is made up of 70% liquid. It is alike to that of cookery in the oven where the water in the food substance is made hot first. Microwave engrossment consequence is much more substantial by the parts of human body which comprise more liquids (blood, water, etc.), like brain which comprise of anent 90% water. Consequence is easily noticeable where the motion of the liquid is less, for instance, heart, eyes, brain, and etcetera. The higher molecular substances, water molecules, and proteins are the main corporeal to engrossment the EM radiation in the human body. When the surface of skin comes in the contact with EM radiation energy, the oil, water molecules, and other substances in skin start riveting the radiation energy. With the depth of the radiation energy in the skin, energy crumble in a very short distance, so the SAR in the skin layer minify is visible. The skull has low water content and mainly composed of minerals (calcium and phosphorus), so the absorption of EM radiation in the skull layer is very low. The amount of water in the brain tissue is comparatively high, and there is more molecular material in tissue of brain, which leads to the brain tissue absorption of EM radiation energy are obvious. After the EM radiation preoccupancy of the skin and skull, the radiation energy has certain attenuation, the SAR value in the layer of brain tissue was significantly lower than the layer of skin. The SAR value in the layer of brain tissue attenuation speed is greater than skin layer because the brain tissue electrical parameters are higher or greater than skin layer. In human head, the EM radiation engrossment is mainly concentrated on the skin and brain tissue, the engrossment rate on skull is very low. The water level in the human body is gradually loss, with the increase of the age, which leads to the electrical parameters decreases [7]. The electrical parameter of children and adolescents is higher than adult. Engrossment of EM radiation by an adult, 10-year-old and a 5-year-old child from mobile phone is shown in the Fig. 1. The yellow area at the base is the spot of mobile phone by the ear.

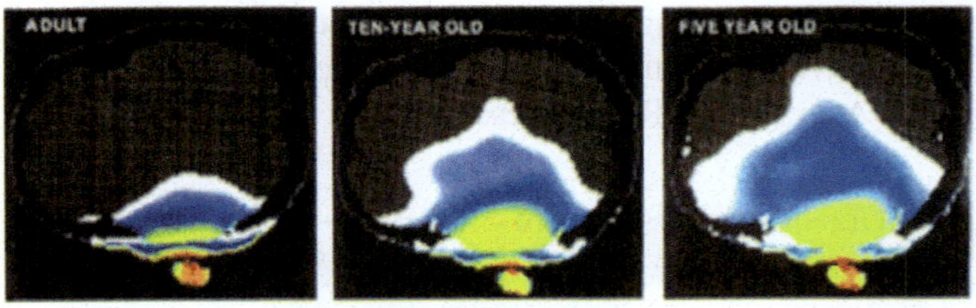

Fig. 1 EM radiation absorption from mobile phone based on age

2.1 Effect on Male Fertility

Electromagnetic radiation that is emitted by mobile phones can affect the man fertility. According to the American Society for Reproductive Medicine (2006) reported that utilization of mobiles by men is related with diminution in sperm count, quality of semen, viability, motility, and customary morphology and is associated to the time span of mobile use. Studies have established 30% sperm minify in intense users of mobile, in inclusion to harm of sperms. Average count of sperm was established to be at 59 million sperm per ml of seminal liquid compared to 83 million for men not regularly influenced to handset radiation [8]. Alike, the study established that motility—sperm's capability to swim—was influenced by handset transmission. Adult male who create calls for long duration had not many speedily motile sperm, 36.3% likened with 51.3% for adult male who create no calls. Cell phone also has an effect on sperm quality of men, when merely having handset switched ON in a pant as cell phones sporadically channelize information to mobile towers to establish contact. The radiation which is emitted by handset can also create DNA shatter in cells of sperm that can change in form or nature and cause cancer. Harm to DNA of sperm step-up the peril further and can move on the genetic changes to succeeding generations.

2.2 Effect on Ear

Tinnitus, popularly known as "Ringxiety"—is the psychological disorder of hearing illusory sound and sense of handset ring and it has been described amid millions of users of handset in the globe. Human beings with severe tinnitus may have problem hearing, working, or even sleeping. Radiation which is discharged by handset may harm the intricate operations of inner ear, and long-term and intense handset utilization for more than 4–5 years and for prolonged durations more than 30 min per day are at a greater peril of growing hearing loss, which cannot be transposed.

2.3 *Effect on Pacemaker*

EM radiation exposure from mobile phone can affect the pacemaker, impulse generators and Implantable Cardiovascular Defibrillators carrying patients. Mobile phone signals cause EM hindrance with the devices and interpose with its proper working. Signals generated by the functions as ringing, turning on, talk and hanging up, hold low frequencies components that can step in with the planted pacemakers causing them to begin to be arrhythmical which in harmful state may put the patient to death. Because of these reason, it is advisable do not keep the handset squarely above the implanted pacemakers (shirt's pocket). Handset should not be used with the left ear, if the pacemaker is planted in the left side. It should be used with right ear.

3 EMR Exposure Guidelines and Limits

For mobile telephony, the Global System for Mobile (GSM), 3G and 4G networks operate at the frequency range of 900–2600 MHz. The parceling of unalike frequencies is rigorously headed by International Telecommunication Union (ITU) and nationally by the pertinent regulatory authorities. EMR is categorized into ionizing and non-ionizing radiation as given in Fig. 2. EMR with enough energy to abolish densely bound electrons from the trajectory of an atom, hence making the atom to become ionized is called as ionizing radiation. EM radiation which do not have sufficient energy to ionize the atoms is called as non-ionizing radiation.

According to the 2011 report by the International Agency for Research on Cancer (IARC) of the World Health Organization (WHO) arrived at a judgment that

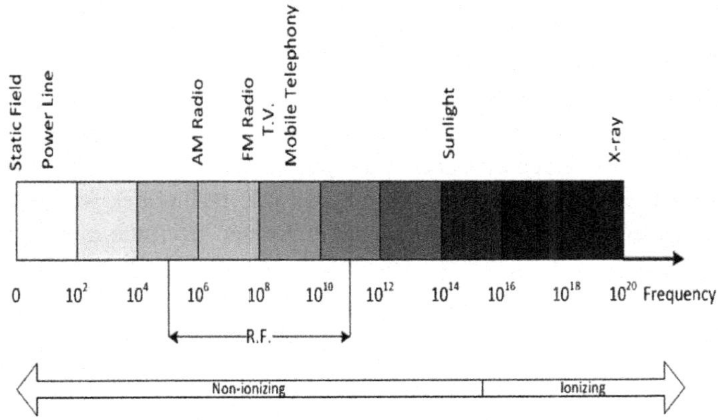

Fig. 2 Electromagnetic spectrum

electromagnetic radiation (EMR) is mayhap carcinogenic to people and categorized it as Group 2B, a class used when there is confined proof of carcinogenicity in people [9]. The electromagnetic radiation (EMR) vulnerability principles and limits have been established by several international and national agencies and adopted by the relevant regulatory authorities. The International Commission on Non-Ionizing Radiation Protection (ICNIRP) gave the principles and limits in their report [10] and mobile phones in the market should not exceed this limit. These principles apply to both public and occupational exposure and are based on epidemiological studies and laboratory. Adults who are trained and having knowledge of the possible effects of EM exposure and take necessary safeguard are referred as occupational. And public vulnerability refers to the people of all ages. ICNIRP sets the SAR value of 1.6 w/kg which is actually for 10 g tissue over a period of 6 min [10]. Many people are not aware about the short-term and long-term possible effects of EM exposure on the human body, so it is required to make the people aware about this.

ICNIRP SAR exposure limits

	Public	Occupational
Whole-Body (w/kg)	0.08	0.4
Head (w/kg)	2	10

4 Em Radiation Reduction Techniques

For increasing the spectral efficiency and peak data rate, the IV back-to-back generations of handset have been designed. It is anticipated that other standard such as energy efficiency, latency, and electromagnetic radiation vulnerability will have an indispensable role in the design of handset in the fifth generation (5G). With increasing usage of cell phones by people over a longer duration of times, perturb anent the harmful health effects of electromagnetic radiation which are emitted by the mobile phones are rapidly growing.

As there is head-on connection between the transmit power, specific absorption rate, and electromagnetic radiation vulnerability. We present some techniques that have been shown to reduce the specific absorption rate or transmit power of mobile communication system. Following are the possible techniques which are used to reduce the electromagnetic radiation which is emitted from the mobile phones.

4.1 Specific Absorption Rate Shielding

It is a method of lessening the specific absorption rate with the minimal abasement of performance of antenna. This technique involves the use of ferrite material or metamaterial attachment between handset and head. Metamaterial is based on a special resonant characteristic of single cell which generates negative permittivity or negative permeability. It has been shown in [11] that the use of metamaterial reduces the SAR by 27–52% at 900 MHz and 1800 MHz, respectively, over 1 g of tissue with performance of antenna being less influenced. Ferrite material has very small conductivity which leads to smaller induced currents when uncovered to electromagnetic wave [12] and they abandon magnetic field part of an incident electromagnetic waves [13].

4.2 Control of Power

It is merely mentioned attuning the output transfer power level of base station or handset to maximized the accepted power of wanted signals, meliorate system potential, reduce interference, and in addition power consumption. Control of power focuses at lessening interference and transmits power level; it can also diminish in the action, electromagnetic vulnerability stratum.

4.3 Beam Forming

Typically, omnidirectional antennas have been used at base station of mobile communication system because antenna has no information regarding location of handset's utilizer and the environment's state. In order to arrive at the wanted mobile phone location omnidirectional antennas transmit maximum power. Beam forming is referred as process of using an array antenna in a specific direction. By aiming antenna rays in wanted directions, the interferences in the system and electromagnetic radiation to the not meant directions reduce with the help of beam forming. Nearly all of signal is aimed in assured direction, thus beam forming helps in ameliorating the SINR and reduction in transmit power [14]. Nowadays, adaptive/smart antenna technology is used. This antenna technique utilizes antenna's array that builds a ray for transference/acceptance toward each wanted handset utilizer in mobile communication system [14–16]. In beam forming, each handset utilizer's signal is multiplied by composite weights which conforms the magnitude and phase of the signal toward wanted destination (Fig. 3).

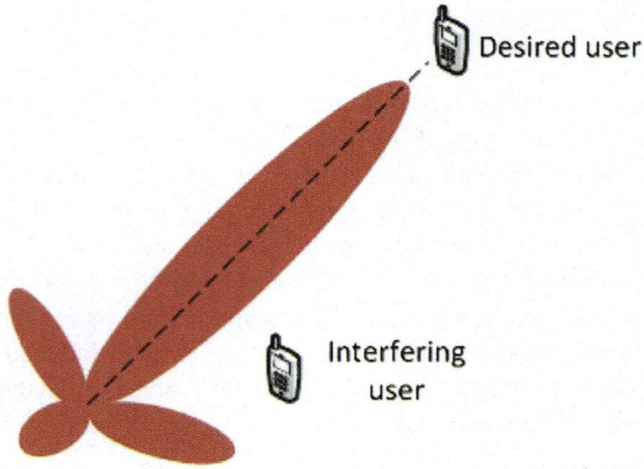

Fig. 3 Adaptive antenna system beam formation

4.4 *Massive MIMO*

MIMO stands for multi-input multi-output. Massive MIMO is an abstract idea by which a large number of antennas are positioned at the base station of mobile communication to treat a much lesser amount of handset customer. This leads to meliorate the energy productivity, spectral productivity, and lesser transfer power, which could lead to reduce electromagnetic radiation vulnerability stratums in system.

5 Conclusion

A comprehensive study of EM radiation effects on human body from handset has been done. We showed the effects of EMR on human body, EMR vulnerability guidelines and limits, reduction techniques in this paper.

According to my knowledge of the current scenario, there is serious perturb anent the harmful health effects of EMR of mobile phone in the surrounding. Even though the research work on harmful/bad health effects of EMR from handset is indecisive, there is small possibility of it being carcinogenic. The amount EMR vulnerability to handset's user rests on the mobile phone technology, distance of user from antenna, and duration of call or use of mobile.

References

1. Mohan Kumar D (2014) Mobile phones radiation 2010. Accessed 15 Aug 2014 [Online]. Available: http://www.electroschematics.com/5200/mobile-phone-radiation/

2. Dutta N (Jan 2013) Can electromagnetic radiation from mobile towers harm you? Accessed 15 Aug 2014 [Online]. Available: http://www.thehealthsite.com/diseases-conditions/can-electromagnetic-radiationfrom-mobile-towers-harm-you/
3. Hardell L, Carlberg M, Soderqvist F, Mild KH (2008) Meta-analysis of long-term mobile phone use and the association with brain tumours. Int J Oncol 32:1097–1103
4. Luria R, Eliyashu I, Hareuveny R, Margaliot M, Meiran N (2009) Cognitive effects of radiation emitted by cellular phone: the exposure side and time. Bio Electromagnet 30:198–204
5. Lu M, X-Y Wu (2009) Study of specific absorption rate (SAR) induced in the human endocrine glands for using mobile phones. In: 7th Asia Pacific international symposium on electromagnetic compatibility, IEEE 2016 by cellular phone: the exposure side and time. Bio Electromagnetics, vol 30, pp 198–204
6. Sambo YA, Heliot F, Imran MA (2015) A survey and tutorial of electromagnetic radiation and reduction in mobile communication systems. IEEE Commun Surv Tutorials 17(2)
7. Zhu C, Wu D, Ping L, Liang W (2015) The effects of head parameters on the specific absorption rate in the human head. IEEE 2015 8th International Conference on Biomedical Engineering and Informatics (BMEI)
8. Kumar G (December 2010) Report on cell tower radiation
9. IARC classifies radiofrequency electromagnetic fields as possibly carcinogenic to humans. Lyon, France, Press Release No. 208 (May 2011)
10. Guidelines for limiting exposure to time-varying electric, magnetic, and electromagnetic fields (up to 300 Ghz). Health Phys 4(74):494–522 (Apr 1998)
11. Hwang J, Chen F Reduction of the peak SAR in the human head with metamaterials. IEEE Trans Antennas Propag 54(12):3763–3770
12. Ragha L, Bhatia M (2010) Evaluation of SAR reduction for mobile phones using RF shields. Int J Comput Appl 1(13):80–85
13. Pretorius J (Nov 2004) Design and manufacture of a ferromagnetic wave absorber for cellular phone radiations. In: Proceedings of 12th International Symposium on EDMO. Berg-en-Dal, South Africa, pp 119–123
14. Stevanovic I, Skrivervik A, Mosig JR (Jan 2003) Smart antenna systems for mobile communications. Ecole Polytechnique Federale De Lausanne, Technical Report. Lausanne, Switzerland
15. Jain RK, Katiyar S, Agrawal NK (2011) Smart antenna for cellular mobile communication. VSRD Int J Electr Electron Commun Eng 1(9):530–541
16. Nowicki D, Roumeliotos J (April 1995) Smart antenna strategies. Mobile Communications International. London, UK

Challenges and Issues in Implementation of Underwater Wireless Sensor Networks

Rubal Bansal, Saurabh Maheshwari and Payal Awwal

Abstract Underwater wireless sensor networks (UWSNs) is an emerging technology. Undersea, it is not possible for us to frequently change the batteries of sensors deployed in network so some strategies are taken under consideration to increase the node capabilities in UWSNs. Being the least explored, issues and challenges in underwater networks oceans always remain one of the tempting topics for researchers. In this paper, we have compiled constituents of UWSN, issues and challenges. The challenges related to routing strategies require essential exposure to the researchers. Recent articles have been analyzed, discussed further working directions have been described and major network design issues at different layers have also been discussed. The paper has importance for the researchers looking to work in this area and looking for future problems.

Keywords UWSNs challenges · Layer issues · Applications
Overview UWSNs · Routing strategies

1 Introduction

Underwater sensor networks are another conventional approach utilized for submerged, i.e., beneath the water. Researchers are working on some of the major issues to improve the output and techniques used in this zone. This is to a great extent an unexplored zone and as of late it has entranced people to investigate it. Common or man-made debacles that have occurred in the course of the last few years have stimulated huge enthusiasm for observing maritime situations for

R. Bansal (✉) · S. Maheshwari · P. Awwal
Computer Science, Gweca, Ajmer, India
e-mail: ruby0071@gmail.com

S. Maheshwari
e-mail: saurabh@gweca.ac.in

P. Awwal
e-mail: ershweta@gmail.com

V. Janyani et al. (eds.), *Optical and Wireless Technologies*, Lecture Notes in Electrical Engineering 472, https://doi.org/10.1007/978-981-10-7395-3_57

logical, ecological, business, well-being, country security, and military needs. Wireless sensor networks are the systems which are associated by various gadgets and share their data through remote hubs. More details on different types of WSNs [1].Architecture of sensor nodes in details is discussed in [2]. Group of network devices that can communicate the various underwater applications through a technique is called an underwater wireless sensor network. The particular destination of the underwater environment introduce new challenges for the research community, they share the same principles as their terrestrial counterparts. The objective of this work is to contribute to this world of the IoT by modeling, analyzing, and studying the performance of UWSNs. Routing protocol plays an important role in data transmission and practical application. In future prospective, the ocean will supply many important resources to industrial and human needs [3]. Various industries will work on seabeds and collect valuable items and used then to renew energy sources which will be a beneficial sort. So many different requirements like fixed or mobile, short or long lived, best effort or death are used in UWSNs; these requirements can result in different design [3]. When we talk about underwater wireless sensor networks comparison between terrestrial and underwater sensor networks are always discussed [4]. There are many applications of underwater environment for monitoring the health of river and marine environments. As it is really costly and difficult for humans to operate these areas, so drivers and sensor networks are developed and deployed for this purpose [5]. The applications divided in two categories below:

Long-term non-time critical applications: In this category, the aquatic monitoring applications, deploying of sensor nodes to cover a spatial monitor applications. *Short-term time critical acoustic exploration*: The applications which deals with short duration and short part come in this category. The application which uses less time in responding they are being followed below in this category [6]. Routing as we all know plays a vital role in technologies used for wireless networks; it is one of the major challenges as with the increase in wireless sensor networks, it is not possible to have a suitable protocol for all applications. UWSNs have been studied widely in past years [7].

2 Issues for UWSNs

In this section, we will highlight various emerging UWSNs issues with respect to different layers, i.e., physical, data link, network, transport, and application layers. In light of work talked about in this way, plainly many issues are still cleared out. These issues could be further dealing of researchers based on different layers of networks [8].

Physical Layer Issues—Underwater deals with different frequency waves like frequency, optical, or acoustic waves. Objectives of UWC are always to improve the bandwidth and efficiency. To achieve this, various modulation schemes are applied like frequency shift keying (FSK), in which bus carry information generally to select carrier frequencies of signal. Frequency hopped (FH), it was introduced as FSK, was unable to wait the channel clearing so this improves bandwidth rate. Code division multiple access (CDMA) one of the important underwater technology which permits random, overlapping access to shared channel. Modulation cab also be done spread spectrum, i.e., can be coherent or not coherent. Estimation and tracking are needed for coherent. DSSS, different spreading codes are needed for non-coherent ones. The concept of multicarrier modulation for available bandwidth to divide is important. Due to this concept, OFDM was introduced, and MIMO, i.e., Multiple transmitter multiple receiver are also introduced. Some of the issues are necessity of developing less expensive modems for transmitting and receiving for underwater. For convergence, low-complexity sub-optimal filters must be designed. With appropriate energy consumption, real time could be obtained in underwater communication.

Data Link Layer Issues—This layer is normally implemented in software as a network card driver. Mac data communication protocol sub-layer is part of DL layer. As physical layer, provides means for transmitter bits and network layer, responses for end to end packet delivery. Mac Protocols are radio-based, that deals with fact propagation delay is legible but in underwater scenario, propagation is five times so these are always an issue. Frequency division multiple accesses (FDMAs) are channelized MAC protocol, which divides available bandwidth, but due to different bandwidth this was not holding more. Time division multiple access (TDMA)—it is technology of accessing medium based on time slots. It is vitality sparing and adaptable. It is hard to accomplish time synchronization in submerged situation because of spread postponement. Code division numerous get to it is one of the most encouraging innovation for generally arranges. ALOHA, it is class of MAC protocols conventions that don't attempt to forestall parcel impact. A version of ALOHA, i.e., slotted Aloha is provided by researchers. Basically, aloha increase the propagation delay and give other time slot division. Some of the issues are high transport relationship and low desert connection properties are looming taken care. Ideal data packet length is expected to most extreme productivity. Complexity should be decreased so that coders and encoders are taken care off. Feasibility and energy efficiency should be noted by researchers. When battery is reaching its depleting point then distributed protocols should be taken care of.

Network Layer Issues—The course from source to goal with different trademark is the principle capacity of system layer. In system layer, there are two techniques for directing, i.e., virtual circuits steering, which is better in UWAN and other is bundle switch directing which are ordered into proactive steering and responsive steering conventions. Some of the issues like simulation models and tools need to be improved. Robust algorithm can be made because quality of link cannot be attained. Redevelopment of good algorithm and protocols need to be taken into consideration to provide strict latency, to detect failures or battery depletion. Data transmission dynamics are to be studied, i.e., accurate modeling is needed. AUVs need to be integrated and mechanism to enable communication between sensors n AUVs. Efficient submerged point area disclosure modules are to be created in the event of topographical steering conventions.

Transport Layer Issues—In underwater sensor networks, this layer is the least explored till date. In wireless networks end points and short link delays are taken into consideration but in underwater due to propagation delay it is not possible. Existing protocols are not available for underwater because RTT is needed which is not possible in UWSN. In cross-layer module, this layer can be tightly integrated with data link layer. For underwater environment, there are some principles to complete this layer, i.e., shadow zones, minimum energy consumption, rate-based transmission of packets, out of sequence packet forwarding, cross-layer interaction-based protocol operation, reliability, SACK-based loss recovery. Some of the issues are new compelling systems custom fitted to the submerged acoustic channel should be produced. New occasion transport dependability metric definitions need to be proposed, in light of the occasion demonstrate and on the submerged acoustic channel show. For dealing sensor detailing, rate some strategies are required, to counteract blockage and boost the system throughout proficiency and additionally the vehicle dependability in data transfer capacity restricted submerged systems. To engage setback excusing applications tweaked to the specific submerged necessities. Diverse functionalities at the information affiliation and transport layer, for example, channel get to, dependability and stream control, ought to be generally organized and considered.

Application layer Issues—The capacity of utilization layer is to give a system administration convention that makes equipment and programming detail of the lower layers straightforward to administration applications. Communication between sources and resources are to be determined, synchronization of communication are some of functionalities which takes place for assigning text and event data. Different protocols are assigned in application layer for terrestrial network like FTP, SMTP, SRB a client server middleware. Very little exertion has been made to address the particular needs of submerged. Rather than planning a totally new arrangement of conventions, one can alter existing conventions of ground-based remote systems to meet the submerged acoustic system (UAN) needs. Along these lines, it is important to comprehend the application territories and the correspondence issues for UANs, and to apply its uniqueness into the current application convention. Through application layer you can access the devices, modifications, or fault diagnosis of system which is attached to the underwater networks. The work targets to conquer the weaknesses of a layered approach that needs data sharing crosswise over convention layer [8].

3 Challenges for UWSNs

Before coming to underwater sensor networks, we first study about the terrestrial networks and a very genuine question comes to mind that can we apply the methods of terrestrial to underwater but answer is due to some basic differences between both discussed above they cannot be directly used in underwater environment. Attenuation and absorption of acoustic waves, propagation delay, bandwidth, bit error rate, node movement, architecture 3D, expensive GPS devices and location algorithms, devices more complex and expensive though could be listed in challenges of UWSN. Since so many researches have been carried out but still given below are some of the open issues still can be considered for doing research further.

Propagation Delay—When a signal received then the amount of time required traveling it between the sender and receiver is known as propagation delay. The computation is done on ratio of link length and speed for a medium. It is due to order for magnitude higher than radio frequency, i.e., longer latencies. This is one of the major challenges researchers always have. In [9], PDTCA protocol is introduced which offers higher throughput than the protocols. In [10], propagation delay aware opportunities are introduced which uses positively obtained local information to increase the concurrent transmission.

Bandwidth—The bit rate of available information capacity expressed in multiple of bits per second. It is supported by network interface. Radio frequency communications are not effective in underwater communication due to medium. Potential UWSN application such as measuring the amount of pollution from fish's farm at seabed requires transmission lot of data. Transmission data rates need to be increased so that it is possible for exploring high frequency transmission of energy. The range and frequency both are used to communicate acoustic signals. The baud (bandwidth), quality of audio waves, bit error rate are some of factors which has to be always considered than radio waves on terrestrial basis increasing the accuracy of these factors. The investigation has to be done which in return becomes a challenge [5].

Energy Consumption—One of the significant difficulties in UWSNs is said to be energy. Wireless sensor nodes in underwater network have restrictive source of energy (batteries) and backup of their batteries is absolutely difficult and costly [11]. Hence, energy efficiency has always remained as a major concern in UWSNs. Moreover, the variation in energy consumption of nodes reduces the network lifetime and creates network holes. Many protocols and work had been done but still no accurate and reliable energy model defined and researchers keep working on it and keep proposing new idea to save energy and making the node alive for longer. Depths of nodes are taken under consideration. In [11], author proposed multilink, scalable, robust, and energy-efficient model. In [12], 3D routing guideline which overcomes the suspension of unstable 3D holes is introduced. Hydro-cast transfers the data packets to the surface by using the measured pressure levels. In [12], energy-efficient hybrid clustering protocol (EEHC) which works by deploying low energy consuming sensor nodes and balancing of energy. Many more protocols such as energy-efficient hybrid routing protocol (EEHR) [13], multipath routing (MPR) [3], energy-efficient depth-based routing (EEDBR) [14] are proposed. Thus, energy saving and increasing the battery life is always a major criterion of research.

Mobility—Because of water currents, underwater nodes move, so it is an challenge though, mobility models, i.e., networks models which deals with the movement of water with the help of their structure and mechanisms. Data mulling, data is transported to different physical location and interaction between mobile agent and static agent takes place for uploading and downloading. Submerged hubs are in persistent movement due to water streams; hence, finding hubs submerged turns out to be significantly more urgent; thus, due to break down of nodes, the network topology makes information unreliable. In [15], a model which captures the actual movement of nodes in large water bodies like oceans provides a clear picture regarding the problems.

Localization—Information about the location of autonomous vehicles is always important as it is a useful routing strategy due to the fact of expensive and complex nature of GPS. Therefore, how to get the information through their place where it is being used and details about it always an essential and decisive open issue in underwater sensor nodes [5]. In [16], a DNR positioning is proposed for localization which is based on DNR beacons which gets the position, i.e., xy axis while floating and then they dive into water. There are two types of localization, i.e., range-free and range-based localization. In localizing of AUV is somehow solved with higher localization ratio with lower cases [17]. Routing, the above feature is important, as the fact discussed earlier also due to its complex and expensive issues. So it is still an open issue.

Attenuation—Geometrical spreading, transmission reflection, absorption, scattering are some of factors which corresponds to weakening of signals which in return reduces the amplitude of electric signals. Exhaustion of sound in seabed for predicting transmission loss in shallow is always an important discussion [18]. A technique, i.e., time wrapping signal processing is used to enables resolution of nodes. Due to high radio frequency there can be performance affection and high variation in knowledge of power receiving. Electromagnetic wave attenuation is being used in short-range sensors which return gives good, accurate sensor model. OFDM, multiplexing technique which drives multicar communication for UW environment.

Cost, hardware protection, reliability-cost, as we all know are always a major concern and the topic we are discussing, i.e., beneath water, a thought really different and must be needing special techniques arrangements which will effect cost. The devices or material used are no more supplied by common vendors of sensors rather research-oriented activities can provide so it automatically affects the cost. Sending data back to surface is more challenging than to collect data. Failure in underwater sensors is one of the issues which is taken under consideration because of environment constraints.

Intelligent Algorithms—Working on underwater systems is always a sensitive topic for researchers. Since just couple of impossible to miss calculations have been given in UASN directing conventions, in this way, how to utilize these excellent calculations and to tackle the debate that exist in submerged acoustic sensor systems has been a constantly open issue in crisp years [5].

4 Conclusion

In this paper, we surveyed about the major challenges which come across while working with underwater wireless sensor networks. One of the major steps is to deal with these major challenges and then working with our provider to make the best deals which will facilitate a successful environment for underwater. To give researchers a clear vision the above said challenges should be kept in mind for developing further. According to the survey, different challenges are discussed. Still there is a need to have an optimal approach that combines the benefits of both approaches and to eliminate the drawbacks of each.

References

1. Smith RN, Schwager M, Smith SL, Jones BH, Rus D, Sukhatme GS (2011) Persistent ocean monitoring with underwater gliders: adapting sampling resolution. J Field Robot 28(5):714–741
2. Kasture A, Raut A, Thool S (2014) Visualization of wireless sensor network by a java framework for security in defense surveillance. In: 2014 International conference on electronic systems, signal processing and computing technologies. Nagpur, pp 256–261. https://doi.org/10.1109/ICESC.2014.49
3. Heidemann J, Stojanovic M, Zorzi M (2012) Underwater sensor networks: applications, advances and challenges. Phil Trans R Soc A 370(1958):158–175
4. Gautam KK, Gautam NK, Agrawal PC Memory required for wireless sensor nodes on the basis of characteristics and behaviour when using tinyOS
5. Li N, Martínez JF, Meneses Chaus JM, Eckert M (2016) A survey on underwater acoustic sensor network routing protocols. Sensors 16(3):414
6. Climent S, Capella JV, Meratnia N, Serrano JJ (2012) Underwater sensor networks: a new energy efficient and robust architecture. Sensors 12(1):704–731
7. Ahmed M, Salleh M, Channa MI (2016) Routing protocols based on node mobility for underwater wireless sensor network (UWSN): a survey. J Netw Comput Appl
8. Manjula RB, Manvi SS (2011) Issues in underwater acoustic sensor networks. Int J Comput Electr Eng 3(1):101
9. Guo X, Frater MR, Ryan MJ (2006) A propagation-delay-tolerant collision avoidance protocol for underwater acoustic sensor networks. In: OCEANS 2006—Asia Pacific. Singapore, pp 1–6. https://doi.org/10.1109/OCEANSAP.2006.4393849
10. Noh Y, Lee U, Han S, Wang P, Torres D, Kim J, Gerla M (2014) DOTS: a propagation delay-aware opportunistic MAC protocol for mobile underwater networks. IEEE Trans Mob Comput 13(4):766–782
11. Edvinoe Christina DPS, Jothi Chitra R (2011) Energy efficient secure routing in wireless sensor networks. In: 2011 international conference on emerging trends in electrical and computer technology. Tamil Nadu, pp 982–986. https://doi.org/10.1109/ICETECT.2011.5760262
12. Chen YS, Lin YW (2014) A mobicast routing protocol with carry-and-forward in vehicular ad hoc networks. Int J Commun Syst 27(10):1416–1440
13. Javaid N, Hussain S, Hafeez T, Maqsood H, Zarar S (Nov 2015). EEHR: energy efficient hybrid routing protocol for underwater WSNs. In: 2015 10th International Conference on Broadband and wireless computing, communication and applications (BWCCA). IEEE, pp 20–26

14. Wahid A, Lee S, Jeong HJ, Kim D (2011). Eedbr: energy-efficient depth-based routing protocol for underwater wireless sensor networks. In: Advanced computer science and information technology. Springer, Berlin, Heidelberg, pp 223–234
15. Caruso A, Paparella F, Vieira LFM, Erol M, Gerla M (2008) The meandering current mobility model and its impact on underwater mobile sensor networks. In: INFOCOM 2008—the 27th conference on computer communications. IEEE, Phoenix, AZ. https://doi.org/10.1109/INFOCOM.2008.53
16. Erol M, Vieira LFM, Gerla M (2007) Localization with Dive'N'Rise (DNR) beacons for underwater acoustic sensor networks. In: Proceedings of the second workshop on underwater networks (WuWNet '07). ACM, New York, NY, USA, 97–100. http://dx.doi.org/10.1145/1287812.1287833
17. Nicolaou N, See A, Xie P, Cui JH, Maggiorini D (2007) Improving the robustness of location-based routing for underwater sensor networks. In: OCEANS 2007. Europe, Aberdeen, pp 1–6. https://doi.org/10.1109/OCEANSE.2007.4302470
18. Tan H-P, Diamant R, Seah WKG, Waldmeyer M (Oct 2011) A survey of techniques and challenges in underwater localization, Ocean Eng 38(14–15):1663–1676, ISSN 0029-8018. http://dx.doi.org/10.1016/j.oceaneng

Microstrip Line Geometric Variation Consequences for Linear Parameters of Microwave Amplifiers

Mahavirsingh Rajpurohit, Kishor G. Sawarkar, K. B. Pramod, Kushal Tuckley and Kumarswamy

Abstract This article analyzes and traces the design arrangement of the broadband low noise amplifier (LNA) with geometrical varieties in microstrip line structure divulge circa over the linear parameters like gain, noise figure (NF), return loss (RL), and stability of LNA. Firstly, design of LNA utilizing linear microstrip lines after that LNA design with mitering of microstrip lines with same input and output matching networks had been done and give essential comparisons of parameters. All design configuration, measurements, and estimations are conveyed, carried in applied wave research (AWR) microwave office tool. Likewise, the simulations result shows that with mitered (U-bend) microstrip line in band 7.5–10 GHz parameters, input return loss (IRL) is enhanced by critical amount and NF, gain, output return loss (ORL) are improved by minor fraction but stability degrades by small fraction correspondingly. The proposed LNA with mitered microstrip lines will be favorable for the designer and engineers to analyze dissect effect and as investigate the impact from claiming geometric variation about microstrip lines. Purpose of this study is to analyze effect of geometric variation on broadband low noise amplifier instead of components values variation.

Keywords Microwave LNA · Microstrip line · Linear measurements and hybrid microwave integrated circuits

---

M. Rajpurohit · K. G. Sawarkar · K. B. Pramod (✉)
MCT's RGIT, Mumbai, India
e-mail: pramod63putta@gmail.com

K. B. Pramod
JAIN University, Bangalore, India

K. Tuckley
R&D, AGV System, Mumbai, India

Kumarswamy
RVCE, Bangalore, India

V. Janyani et al. (eds.), *Optical and Wireless Technologies*, Lecture Notes in Electrical Engineering 472, https://doi.org/10.1007/978-981-10-7395-3_58

1 Introduction

A low noise amplifier (LNA) is a device utilized as a part of communication systems which increases exceptionally feeble signal caught by the antenna. Practically, in any specialized communication device, the LNA is found near to the receiving antenna; indeed, the primary part after the antenna is the low noise amplifier. LNA to help coveted and boost the interested signal power while including as meager noise and distortion signal mutilation as conceivable [1]. An LNA circuit design is the combination of the measurements like low noise figure (NF), high gain, and stability over the complete range of frequencies in the interested band. The common topology of the LNA comprises of three blocks: input matching network (IMN) (passive network), the amplifier itself (transistor), and the output matching network (OMN) (passive network) [2, 3].

Hybrid Microwave Integrated Circuit (HMIC) based amplifiers are by all record by all account not the only course to LNA. Over the span of the latest two decades, there have also been advances in the change of HMIC LNAs. These amplifiers incorporate the greater part of the transistors, vendor lumped components, transmission lines, and matching networks onto an individual circuit board; this mix gives the HMIC a couple of central focuses and its judicious as well [4]. HMIC are utilized because of a few points of interest. Each segment can be intended for ideal execution; least loss microwave components can be made by picking the ideal microstrip substrate; trim modification is conceivable and has high-influence ability since the high-influence creating components can be ideally heat-sinked. High electron mobility transistor (HEMT) is innovation of decision for high-power microwave hardware as it has vast band crevice, high immersion transporter speed, great warm conductivity, and low noise figure [5]. The fundamental component of pHEMT innovation is signal-to-noise ratio for higher-frequency range making them a preferred device as compared to its counterparts [6].

2 Research Background

This fragment will portray measure generally works of the LNA plan which is submitted for the considerable execution [7]. In this work, new strategy to diminish the measure of matching networks of microwave amplifiers utilizing the slow-wave qualities of the microstrip with DGS has been proposed. With a specific end goal to abbreviate the size, while keeping the same electrical length, a few DGS examples were received in matching networks. Those resultant extents of the microstrip line with DGS are just 54 and 56% in input and output matching networks, individually, contrasted with the essential amplifier, and significantly, those enhancer exhibitions would have upheld. Lim et al. [8] to their article, demonstrated that the expanded SWF could make connected. Previously, lessening those sizes about microwave circuits, an amplifier need has been diminished utilizing the VPDGS. Due to the increased electrical length by VPDGS, the physical size of the original amplifier might have been diminished without incredulous expense for exhibition. The resultant lengths of the microstrip lines with VPDGS were only 38.5 and 44.4% of the unique ones to input and output matching networks [9]. For their paper, an LNA need has been planned using multilayer on-chip inductor to improve the LNA execution. Furthermore, exhibit that prescribed strategy enhances the performance of noise figure and provides a better matching network qualities at the input and output side of the LNA. Additionally, it gives a decent power gain. Increasing the on-chip inductor metal layers decreases the spiral resistance and enhances the noise figure.

3 Design Blocks of LNA

Designed LNA is divided into three segments: input matching network, intermediate matching network, and output matching network as shown in Fig. 1. Figure 2 shows input matching network comprises of π-type network utilizing of L-C-L combination having values $L_1 = 1$ nH, $L_2 = 1.5$ nH, and $C_1 = 0.3$ pf. This quality got in the wake of setting microstrip line for connection between components which

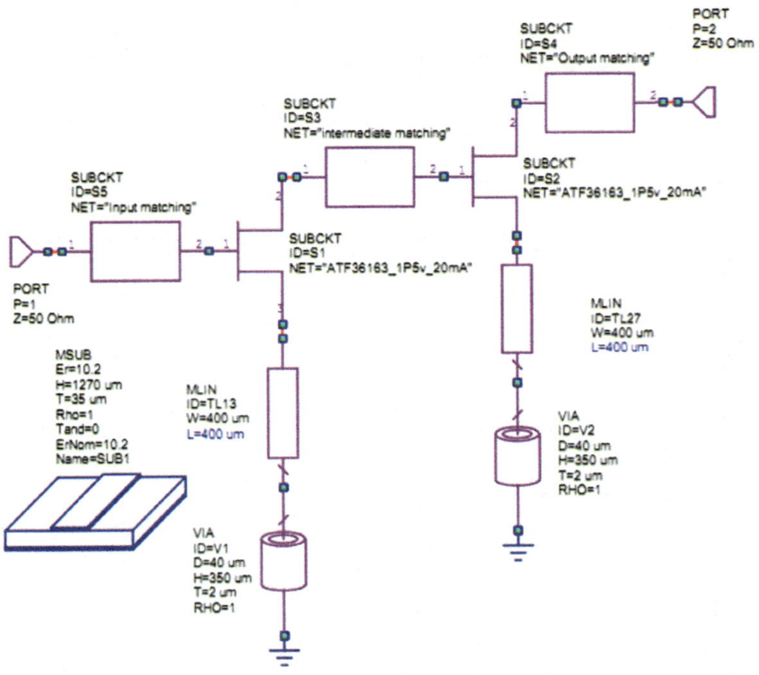

Fig. 1 Shows block diagram of LNA design

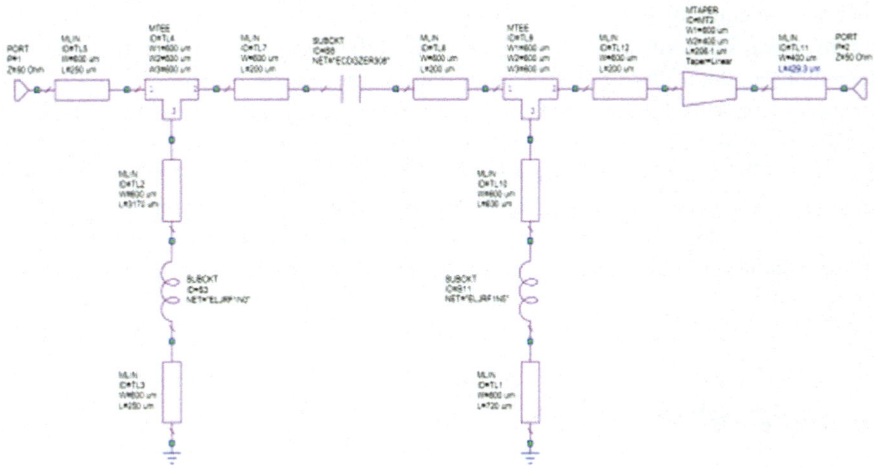

Fig. 2 Shows input matching network of LNA design

are responsible for variation in values of lumped component due to having their own particular impedances.

Figure 3 shows intermediate matching network comprises of L-type network utilizing L-C combination having values L1 = 0.5 nH and C1 = 4 pf. Complex conjugate method is used for this network. Figure 4 shows output matching network comprises of L-type network utilizing L-C combination having values L1 = 2.2 nH and C1 = 0.5 pf. All the matching networks designed using Smith chart analysis method.

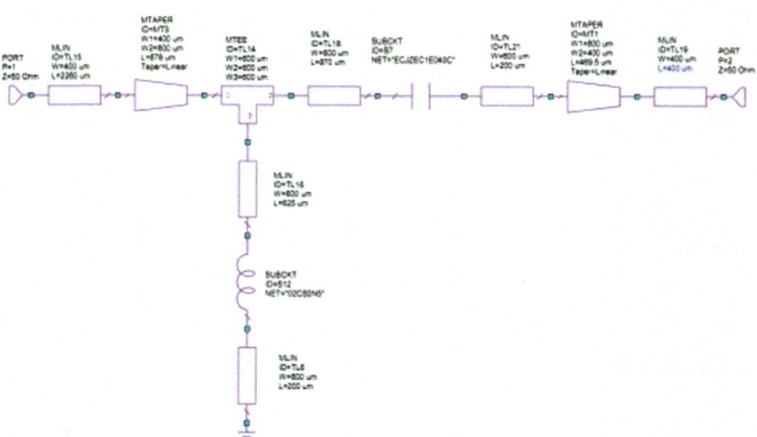

Fig. 3 Shows intermediate matching network of LNA design

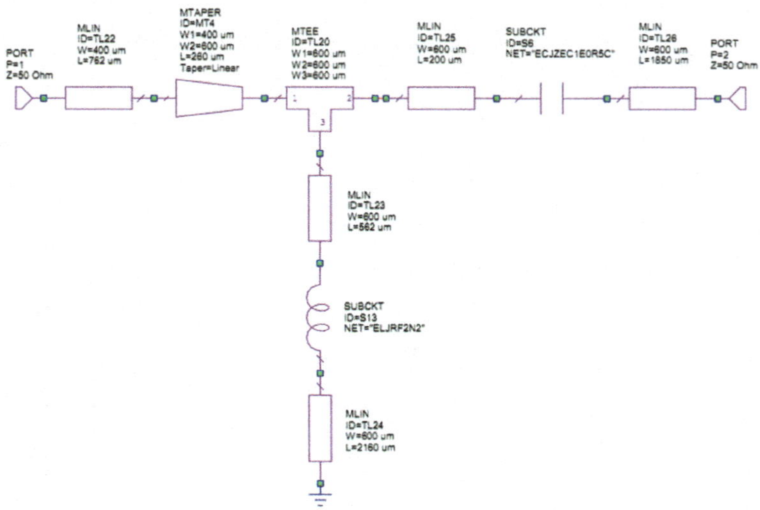

Fig. 4 Shows output matching network of LNA design

3.1 Case 1: Design with Linear Microstrip Line

Figure 5 shows linear microstrip line prototype from AWR tool having length 1600 um, width 200 um, and port 1 and port 2 are input and output port of microstrip line Fig. 6.

Layout Design set up by taken care of certain convention. MLIN is used to connect components (inductor, capacitor, transistor); MTEE is used to connect three-point intersection of components, and MTAPER is used for smooth tapering between ports by utilizing library of AWR Microwave Office Simulation Tool [10]. For transmission power PT from the source and the reflected power PR, the return loss in dB as shown in Fig. 8 is given by [4] (Fig. 7):

Fig. 5 Microstrip line (MLIN)

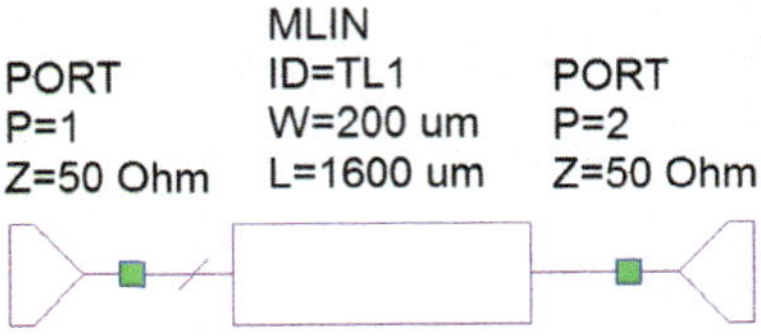

Fig. 6 Layout of microstrip line (MLIN)

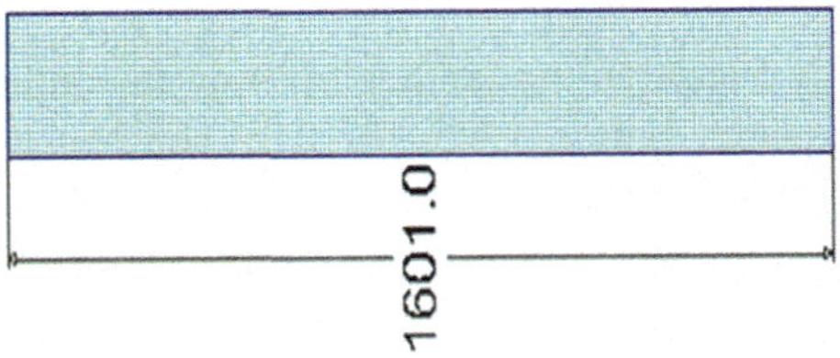

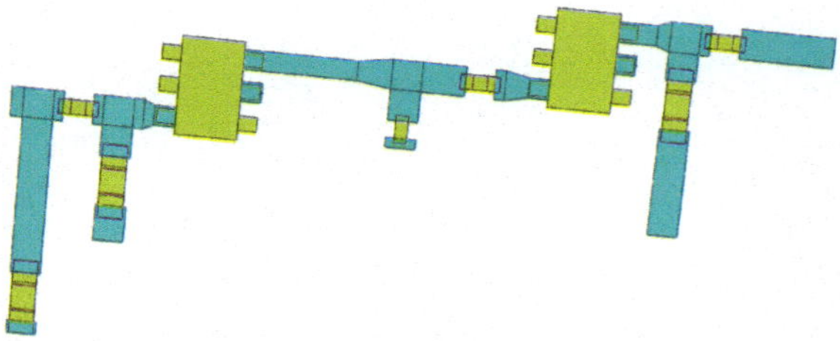

Fig. 7 Shows layout of the LNA design with linear microstrip line

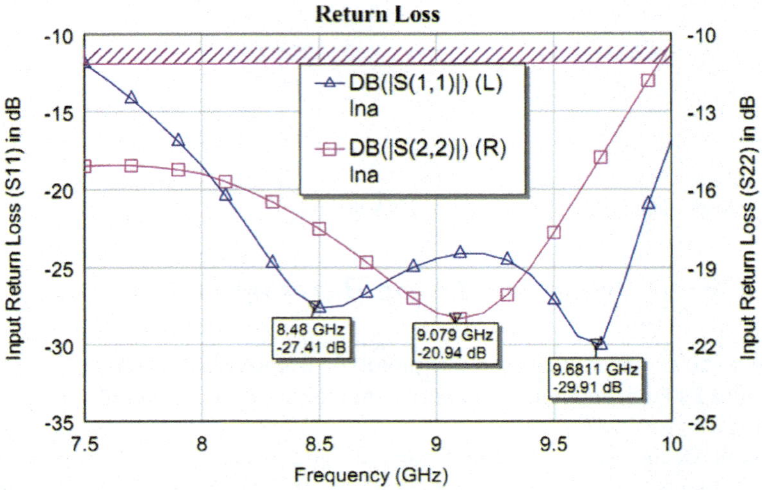

Fig. 8 Shows return loss of the LNA design with linear microstrip line

$$\mathrm{RL(dB)} = 10log_{10}\left(\frac{P_T}{P_R}\right) \quad (1)$$

For unconditional stability [11]:

$$\mathrm{K} = \frac{1 - |S_{11}|^2 - |S_{22}|^2 + |\Delta|^2}{2|S_{12}S_{21}|} > 1 \quad (2)$$

$$|\Delta| = |S_{11}S_{22} - S_{12}S_{21}| < 1 \quad (3)$$

Figure 9 shows that LNA is unconditionally stable over entire range of frequency between 7.5–10 GHz. The Noise Figure for cascaded LNA will be given by, and shown in Fig. 11 (Fig. 10).

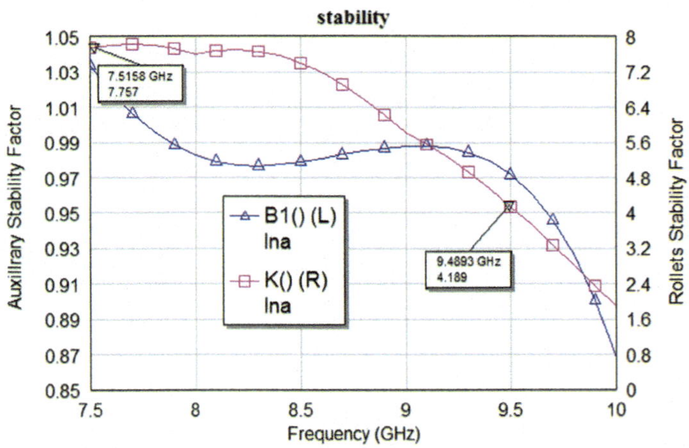

Fig. 9 Shows rollets factors K and auxiliary factor B1 of LNA design with linear microstrip line

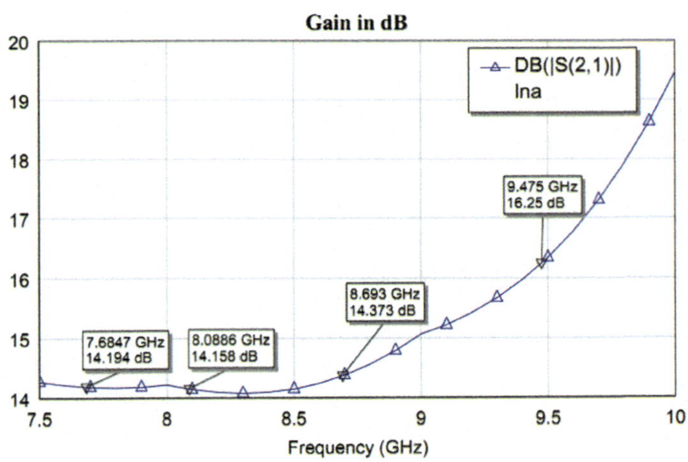

Fig. 10 Shows gain of the LNA design with linear microstrip line

$$F_{total} = F_1 + \frac{F_2 - 1}{G_1} + \frac{F_3 - 1}{G_1 G_2} + \frac{F_4 - 1}{G_1 G_2 G_3} + \ldots \quad (4)$$

where Fn is the noise factor, and Gn is available power gain, individually of the nth phase. Note that both magnitudes need aid communicated as ratios, not clinched alongside decibels [4]. Designed LNA with linear microstrip line is unconditionally stable over entire range of frequency between 7.5–10 GHz having input return loss in range of −11.87 to −30.02 dB (S11), output return loss −15.08 to −20.97 dB (S22) with gain in range of 14.08–19.455 dB and noise figure <1.4 dB.

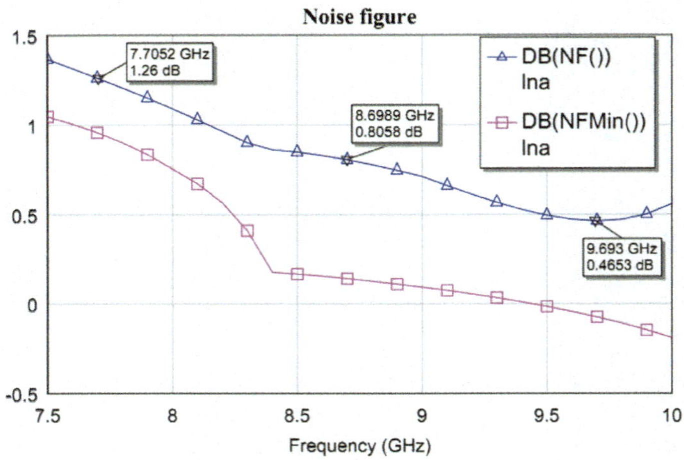

Fig. 11 Shows noise figure of the LNA design with linear microstrip line

3.2 *Case 2: Design with Mitered (U-Bend) Microstrip Line*

For mitering microstrip line, meander line 2 closed form (MTRACE2) is used instead of microstrip line (MLIN) in AWR Microwave Office with miter length to width ratio = 1. Whole design remains same with same matching networks (Figs. 12, 13, and 14).

Figure 15a shows centerline approach, and the equivalent length (L_{eq}) is equal to length **ABC** and will be given by [12]:

$$L_{eq} = 2L_2 + W \tag{5}$$

Figure 15b shows modified centerline approach, and the equivalent length (L_{eq}) is equal to length **DEFG** which shows actual current flow will be given by [12]:

$$L_{eq} = 2L_2 + \frac{\sqrt{2}}{2}W \tag{6}$$

According to Fig. 16a, due to deviation in current density that is high current density at edges and low current density at center of microstrip line, current path

Fig. 12 Meander line 2 closed form (MTRACE2)

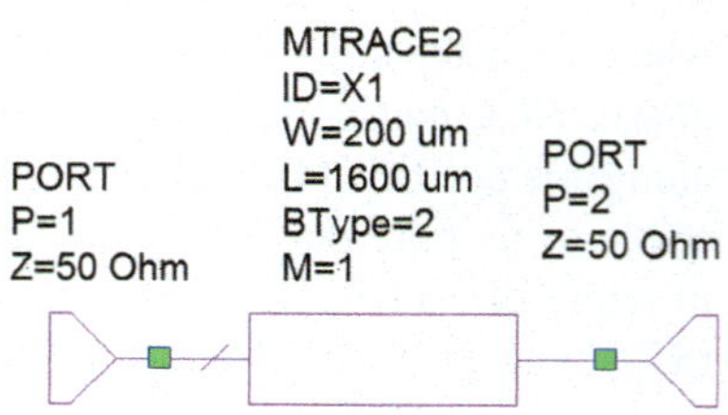

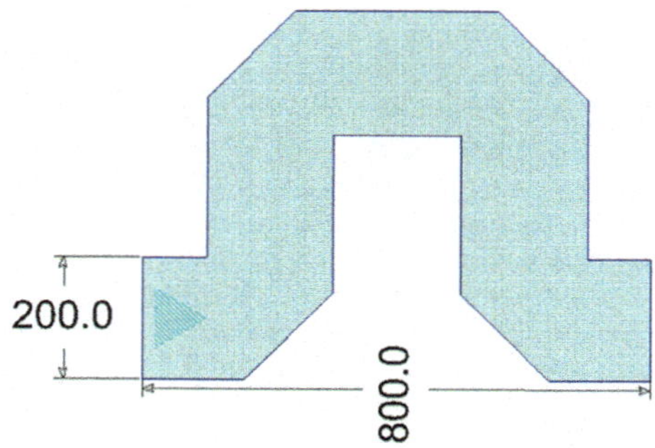

Fig. 13 Layout of meander line 2 closed form (MTRACE2)

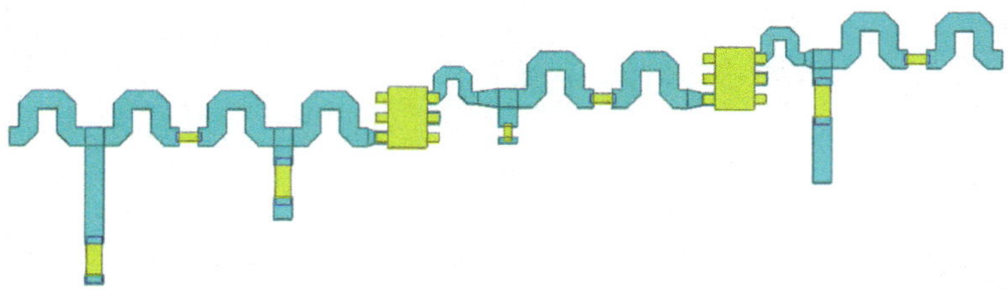

Fig. 14 Shows layout of the LNA design with mitered (U-bend)

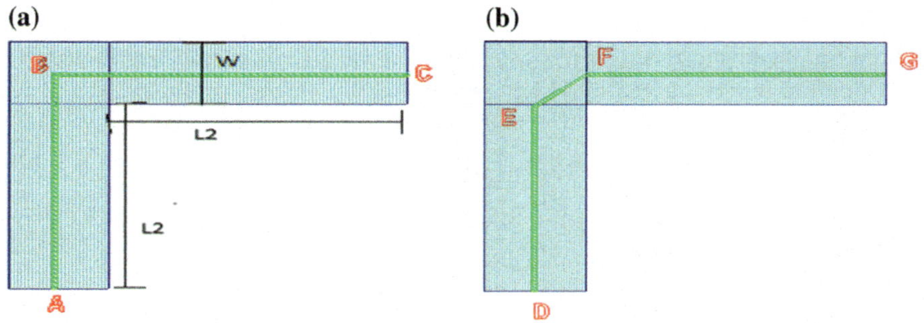

Fig. 15 Determination of equivalent length for unmetered right-angled bend. **a** centerline approach, **b** modified centerline approach

will not follow the centerline path (**ABC**), but deviates toward the shortest path (**MNO**). For the unmitered right-angled bend, the corrected current path by modifying the centerline path, Eq. (5), into a path following the inner edge more closely (**DEFG**) is given by Eq. (6). The shortest path length, L_{shrt} (**MNO**), follows from Fig. 16a is given by:

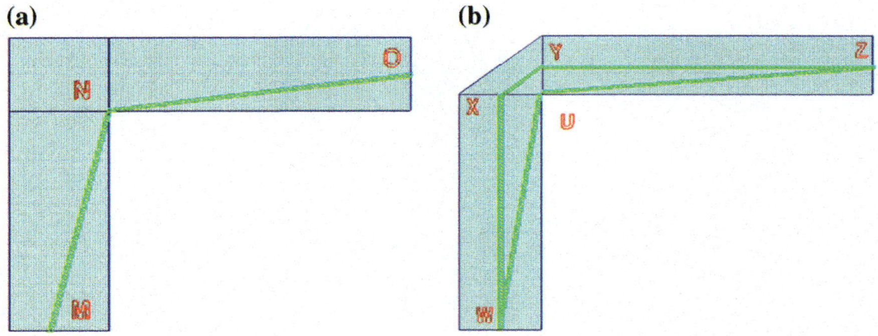

Fig. 16 Current path approximations. **a** Shortest path. **b** Shortest path and modified centerline path

$$L_{short} = \sqrt{\left(\left(\frac{W}{2}\right)^2 + (L_2)^2\right)} \tag{7}$$

The equivalent length of the 50% mitered right-angled microstrip bend, L_{eqmit2} (**WXYZ**), is now calculated as [12]:

$$L_{eqmit2} = \sqrt{L_{Short}.L_{eq2}} \tag{8}$$

where L_{eq2} is given by Eq. (5).

Optimum miter is given by [13]:

$$\frac{X}{D} = 0.52 + \left(0.65 * e^{\left(-1.35*\left(\frac{W}{H}\right)\right)}\right) \tag{9}$$

Range of usage:

$$0.5 \leq \frac{W}{H} \leq 2.75 \text{ and } 2.5 \leq \varepsilon_r \leq 25$$

where ε_r = dielectric constant (from associated substrate), H = substrate thickness (from associated substrate), W = conductor width, in specified units W $\geq$ 0 for layout

The equivalent circuit of a microstrip corner is shown in Fig. 17a. The values of the components are as follows [14];

$$C[pF] = W.\left[(10.35 * \varepsilon_r + 2.5) * \left(\frac{W}{h}\right) + (2.6 * \varepsilon_r + 5.64)\right] \tag{10}$$

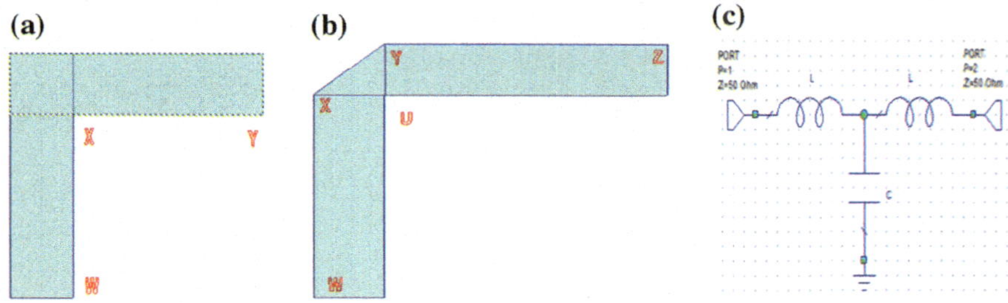

Fig. 17 **a** Microstrip corner, **b** mitered corner, **c** equivalent circuit

$$\mathrm{L[nH]} = 220 * h * \left[1 - 1.35 * \exp\left(-0.18 * \left(\frac{W}{h}\right)^{1.39}\right)\right] \tag{11}$$

The values for a 50% mitered bend are as follows [14]:

$$\mathrm{C[pF]} = \mathrm{W.}\left[(3.93 * \varepsilon_r + 0.62) * \left(\frac{W}{h}\right) + (7.6 * \varepsilon_r + 3.80)\right] \tag{12}$$

$$\mathrm{L[nH]} = 440 * h * \left[1 - 1.062 * \exp\left(-0.177 * \left(\frac{W}{h}\right)^{0.947}\right)\right] \tag{13}$$

With W constantly width of the Microstrip accordance Furthermore h stature the height of the substrate. The Z-parameters to those provided for proportional little sign out might make composed Similarly as underneath mathematical statement What's more would simple to change over with to scattering parameters [14].

$$Z = \begin{bmatrix} j\omega L + \left(\frac{1}{j\omega C}\right) & \frac{1}{j\omega C} \\ \frac{1}{j\omega C} & j\omega L + \left(\frac{1}{j\omega C}\right) \end{bmatrix} \tag{14}$$

Figure 18 clearly indicates input return loss (S11) at left side axis and output return loss (S22) in the right side of the axis of LNA design with mitering (U-bend) and describes the significant reduction in return loss. Figure 19 shows that LNA is unconditionally stable over entire range of frequency between 7.5 and 10 GHz. For compensation of Return Loss various techniques have been implemented on basis of transmission line bends such that Capacitive effect due to storage of charges at corners, bends and Inductive effect due to flow of charges along transmission line will reduce (Figs. 20 and 21).

A standout amongst them is transmission line curves might make mitered What's more rounded, the place the miter technobabble removes metal the place there is no present flow, what's more that lessens those capacitances What's more inductance. Finishing along these lines enhances the VSWR and diminishes the

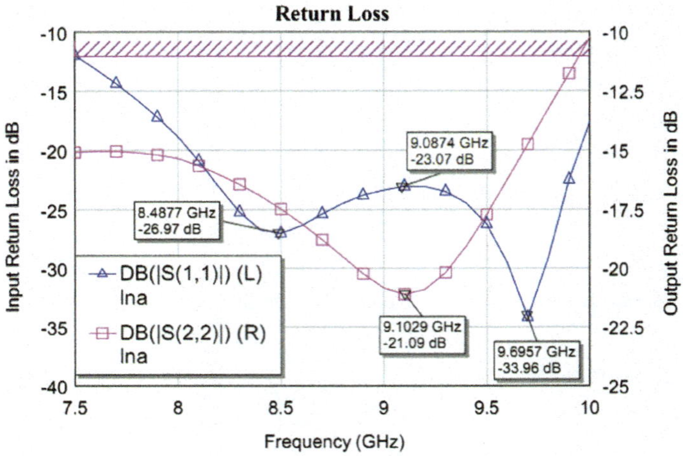

Fig. 18 Shows return loss of LNA design with mitered (U-bend)

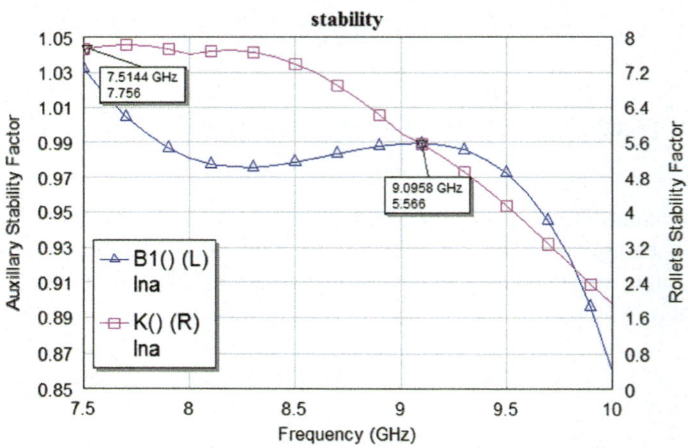

Fig. 19 Shows rollets factors K and auxiliary factor B1 of LNA design with mitered (U-bend)

give back the return loss [15]. Designed LNA with mitered (U-bend) microstrip line is unconditionally stable over entire range of frequency between 7.5 and 10 GHz having input return loss in range of −11.94 to −34.03 dB (S11), output return loss in range of −15.07 to −21.08 dB (S22) with gain in range of 14.08–19.461 dB and noise figure <1.4 dB (Table 1).

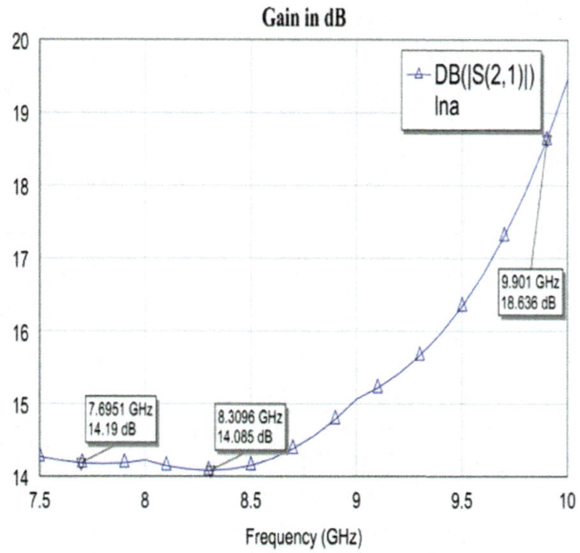

Fig. 20 Shows gain of the LNA design with mitered (U-bend)

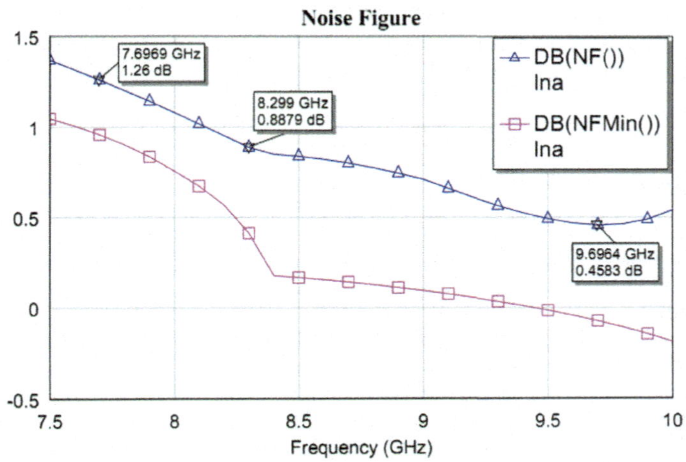

Fig. 21 Indicates NF of LNA circuit designed using with mitered (U-bend)

Table 1 Comparision of fundamental parameters of LNA

S. No.	Parameters	Gain (dB)	Noise figure (dB)	Return loss	
				S11(dB)	S22(dB)
1.	Linear microstrip line	14.08–19.455	0.4655–1.368	−11.87 to −30.02	−15.08 to 20.97
2.	With mitered (U-bend)	14.08–19.461	0.459–1.368	−11.94 to −34.03	−15.07 to −21.08

4 Conclusion

In this paper, two-stage LNA circuit has been designed, composed, and simulated for the frequency range of 7.5–10 GHz, and moreover, examination of impact mitering of microstrip line is analyzed. The paper demonstrates that for two-stage LNA with U-bend mitering improvement in input return loss (S11) and minor improvement in output return loss (S22), noise figure and gain but minor reduction in stability occur. These works have been illustrated, demonstrated, analyzed, and simulated for performance of the LNA design using industrial standardise AWR microwave office tool. This proposed work can be further upgraded by utilizing diverse mitering approaches and geometries. One of the advantages of this design is simplicity and easy to implementable, and hindrance is augmentation in size of LNA because of bigger width of microstrip line.

References

1. You L, Wang C, Liu F (2010) A 3–5 GHz ultra-wideband low noise amplifier employing noise cancellation. In: 2010 IEEE international conference on ultra-wideband (ICUWB), vol 1. IEEE, pp 1–4
2. Yelten MB, Gard KG (2009) A novel design procedure for tunable low noise amplifiers. In: 2009 IEEE 10th annual wireless and microwave technology conference WAMICON'09, Clearwater, FL, pp 1–5. https://doi.org/10.1109/WAMICON.2009.5207313
3. Jha CK, Gupta N (2012) Design of a front end low noise amplifier for wireless devices. In: 2012 students conference on engineering and systems (SCES), Allahabad, Uttar Pradesh, pp 1–4. https://doi.org/10.1109/SCES.2012.6199043
4. Pramod KB, Kumaraswamy HV (2016) The design, optimization and characterization of 7 GHz ultra low noise figure amplifier using hybrid MIC technique for satellite mobile applications. Commun Appl Electron (CAE) 6(4)
5. Mishra UK, Parikh P, Wu Y-F (2002) AlGaN/GaN HEMTS-an overview of device operation and application. Proc IEEE 90:1022–1031
6. Arsalan M, Amir F, Khan T (2014) pHEMT LNA design and characterization for 4G applications. In: 2014 IEEE 17th international multi-topic conference (INMIC), Karachi, pp 61–66
7. Lim JS, Park JS, Lee YT, Ahn D, Nam S 2002) Application of defected ground structure in reducing the size of amplifiers. IEEE Microwave Wirel Compon Lett 12(7):261–263

8. Lim JS, Lee YT, Kim CS, Ahn D, Nam S (2002) A vertically periodic defected ground structure and its application in reducing the size of microwave circuits. IEEE Microwave Wirel Compon Lett 12(12):479–481
9. Hasaneen ESAM, Okely N (2012) On-chip inductor technique for improving LNA performance operating at 15 GHz. Sci Res Circuits Syst 3:334–341
10. Sawarkar KG, Pramod KB, Tuckley K, Kumaraswamy HV (2015) The frequency, time design analysis of noise figure optimization of a wideband PHEMT hybrid LNA with flat gain for WiMAX application. In: International conference on communication technology (ICCT 2015)
11. Pozar DM (2012). Microwave engineering, 4th edn
12. Visser HJ (2007) Equivalent length design equations for right-angled microstrip bends. In: The second European conference on antennas and propagation, EuCAP 2007, Edinburgh, pp 1–6
13. http://cp.literature.agilent.com/litweb/pdf/ads2008/ccdist/ads2008/MBEND3_(90degree_Microstrip_Bend_(Optimally_Mitered)).html
14. Kirschning M, Jansen RH, Koster NHL (1983) Measurement and computer-aided modeling of microstrip discontinuities by an improved resonator method. IEEE MTT-S international microwave symposium digest, pp 495–497
15. Margomenos A, Rida A (2009) System and method for improving performance of coplanar waveguide bends at mm-wave frequencies. US patent, US20100182107 A1
16. Avago Technology. ATF-36163 datasheet

A Novel Image Steganography and Steganalysis Technique Based on Pattern Searching

Surya Srivastava, Prateek Thakral, Vani Bansal and Vivek Shandil

Abstract Image steganography is a method, which hides a message into a cover image and generates a stego-image. A secret message is embedded into the image as noise, which is nearly impossible to be detected by human eyes. The stego-image is then sent to the receiver who can extract the message with or without stego-key, which depends on the hidden scheme. Most of the hiding techniques hide the message straight away into the significant bits of the image pixels therefore disturbing the resolution, which degrades the quality of the image and makes the image prone to various types of attacks. A new approach that can make the message more secure and enhance the image quality is proposed here. This approach uses LSB substitution method along with the proposed algorithm. The suggested method hides the secret message based on searching the pattern for {00, 01, 10, 11} and matching the above pattern with last two bits of each pixel in RGB format. The secret message along with two RGB images is then taken as an input where the message is converted into binary form first, and the complete algorithm is applied to hide the secret message within the images. Inversely, using the concept of a dictionary file and applying the algorithm accordingly can retrieve the secret message. The proposed method hides and retrieves the message based on a combination of the two images. The two images work as lock and key and both are necessary for proper retrieval of the message. The quality of the image from which the original message is extracted is not affected at all.

Keywords Image · Steganography · Pixel matching · LSB
PSNR ratio · MSE

S. Srivastava · P. Thakral (✉) · V. Bansal · V. Shandil
National Institute of Technology, Kurukshetra, Haryana, India
e-mail: uneekprateek@gmail.com

V. Janyani et al. (eds.), *Optical and Wireless Technologies*, Lecture Notes in Electrical Engineering 472, https://doi.org/10.1007/978-981-10-7395-3_59

1 Introduction

The word steganography is taken from the Greek words "steganos" which means "covered or secret" and "graphia" means "writing or drawing." So the literal meaning of steganography is "covered writing." Cryptography provides confidentiality by altering the message; steganography on the other hand does not alter the message structure but hides it inside a cover object [1]. Thus, cryptography is a way of shielding the message alone, steganography is used to conceal the fact that communication is being made, as well as hiding the contents of the message. Steganography has been used since the ancient times for hidden communication. One of the early examples of steganography is from Chinese history. The secret text was written on fine silk or paper, which was then rolled or enclosed with wax. The person who had to deliver the message then swallowed the ball or would hide it in his inner parts. Another example of steganography is from the Vietnam War where the imprisoned US armed forces flaunted hand signals during a photograph session to transfer some military message. Image steganography is a process that hides the message into cover image and generates a stego-image, which contains the hidden message [2]. Secret message is embedded into the image as noise, which is nearly impossible to be detected by human eyes. Stego-image is then sent to the receiver without anyone else knowing that it contains the hidden message. The receiver can extract the message with or without stego-key, which depends on the hiding scheme. The rest of the paper is organized as follows: Sect. 2 discuses the related work. The proposed scheme and algorithm are explained in detail in Sect. 3 of the paper. The results after performing various experiments on various tools are shown in the form of graphs in Sect. 4. The complete paper is concluded lastly in Sect. 5.

2 Related Work

Al-Shatnawi [3] analyzes that his method hides the secret message based on searching about the identical bits between the secret messages and image pixels values on comparing it with the LSB benchmarking method for hiding the secret message, which hides the secret message directly in the least two significant bits of the image pixels. The results of proposed and the LSB hiding methods were discussed and analyzed based on the ratio between the number of the identical and the non-identical bits between the pixel color values and the secret message values. The results obtained by the proposed method and the LSB hiding method in terms of ratio of precision in improving the image quality were 83% and 43%, respectively.

Kumar and Vemali [4] proposed a method that converts the secret message into image format. The secret message is then hidden based on searching about the identical pixel between the secret image pixels and the cover image pixels values. This piece of work discusses a tool that will not only work on 24-bit indexed color images but also is able to hide secret text messages inside the selected segment of the image. The embedding capacity of this approach is very high.

Tulasidasu et al. [5] describe block division technique in which there is a secret image and a cover image, which is divided into square grids using block division handling. It increases the security and image quality levels. Without stego-key, no one could abstract the information about the stego-image, which helps in confidential communication. It helps to enhance the visual quality of the image and the chances of losing the information are reduced.

Thangadurai and Sudha Devi [7] describe an LSB-based image steganography using two types, grayscale image and color image. The technique used for grayscale images is simple LSB method of hiding the data in LSB of each pixel. To store the data in color images, two arrays are used to store pixel data and message data. The data is then embedded in each pixel. The hiding capacity of this method is very high.

3 Proposed Work

Most of the hiding techniques hide the message straight away into last two significant bits of the image pixels therefore disturbing the resolution, which degrades the quality of the image and makes the image prone to attacks. A new approach that can make the message more protected and enhance the image quality is proposed. The suggested method hides the secret message based on searching the pattern for {00, 01, 10, 11} and matching the above pattern with last two bits of each pixel in RGB format. The proposed algorithm is explained followed by Fig. 1.

Encoding:
Inputs: Two RGB images, secret message.
Output: Stego-image.

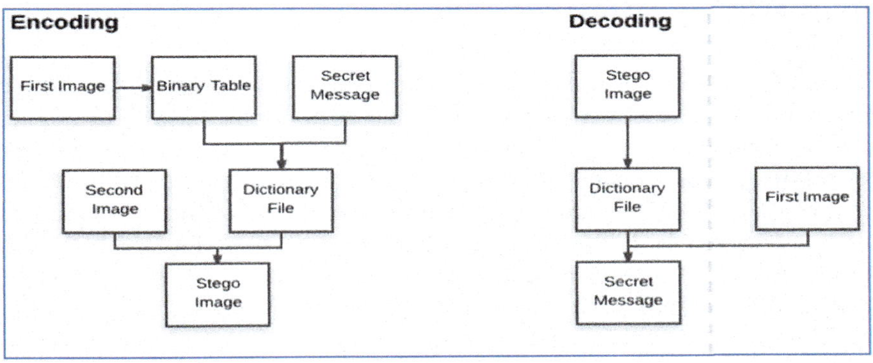

Fig. 1 Steganography system using the proposed algorithm

1. Begin: scan the image row by row and convert it into binary format.
2. Convert the secret message into binary format.
3. Start sub-iteration 1: Choose one pixel from image randomly. Take the Red part, i.e., the first 8 bits representing the value of R (red) in RGB format. Check the last two bits of R (red) value and check if the pattern is "00"; if it is "00," then store the location of the pixel in a binary table else repeat step 3.
4. End sub-iteration 1.
5. Repeat step 3 for every pattern, i.e., "01," "10," and "11" (repeat until pixel locations are found for every pattern).
6. Divide the binary message into blocks of two bits.
7. Take the first block and match it with the pattern 00, 01, 10, and 11. Store the corresponding pixel location in the dictionary file as per the binary table and the matched pattern.
8. Repeat step 7 for all blocks of message.
9. Convert dictionary file into binary format.
10. Save dictionary file in the second image using LSB method.
11. End.

Decoding:

1. Begin: Extract dictionary file from the second image in which it was stored using LSB algorithm in step 10.
2. Start sub-iteration 1: Take the first pixel location from the dictionary file. Retrieve the last two bits of the message from the red part of the pixel, i.e., first 8 bits of the pixel.
3. End sub-iteration 1.
4. Repeat step 2 until all the blocks of the message are retrieved.
5. Convert the produced binary message into human-readable format.
6. End.

4 Results

The proposed hiding algorithm has been implemented using VB.net and C# programing language on Microsoft Visual Studio 2010 on Intel i3 2.0 GHZ system running windows 10. The algorithm is used to hide the secret message **"the quick brown fox jumps over the lazy dog"** using two images in Fig. 2a and b. The first image is used for pixel mapping and converting the message into pixel locations using a binary table. In this example, we got pixel locations 160 and 141 for the pixel combination 00. Similarly, we got pixel location for the other patterns 11, 10, and 01. After converting the message in the form of pixel locations using the binary table, we get our dictionary file. Now, the dictionary file is stored in the second image using LSB method.

Fig. 2 **a** Image for pixel mapping, **b** Image to store dictionary file

Fig. 3 Two images before and after storing the dictionary file. **a** Original image, **b** Stego-image

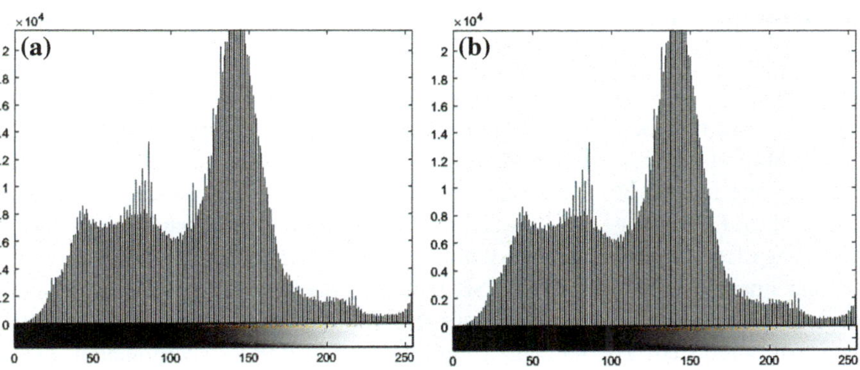

Fig. 4 **a** Original image , **b** Stego-image

The two images before and after storing the dictionary file are shown in Fig. 3. The results thus obtained from the original image and the stego-image is shown after that in the form of graphs.

The results for the above example are mapped into the graphical form where comparison of our proposed algorithm is done with the previous existing steganography technique. The below two different graphs are plotted against our considered example. First graph represents the original image whereas second graph represents image after applying steganography process. Two important parameters that are taken into consideration is PSNR ratio as well as mean square error (MSE).

Figure 4a and b show the color graphs for original and stego-image, respectively.

PSNR is defined using the mean squared error (*MSE*).

PSNR (in dB) is calculated as:

$PSNR = 20 \times \log10\ (255 \div \sqrt{MSE})$

PSNR for first image = infinite

Since there is no change in pixel data of first image, value of MSE is zero.

PSNR for second image = 58.0487

To calculate the average of PSNR of two images, since the value of PSNR of first image is infinite, we take its value as 100.

$PSNR = (100 + 58.0487)/2 = 79.02435$

The high value of PSNR indicates that the difference between these images is very low, i.e., MSE is very low as it is the denominator in calculating value of PSNR. Thus, the proposed method is more effective than LSB method, which has a PSNR value in the range of 40–60.

5 Conclusion

In this paper, a new algorithm of steganography was proposed and implemented. The proposed method hides the secret message using a combination of two images. The quality of the first image remains unchanged as it was just used for converting the message into pixel locations using dictionary file. Since no data of pixel was changed and the image pixel was used just for referencing and forming corresponding dictionary file, the quality of image remains unaffected. The quality of the second image is changed as per LSB but since the dictionary file only contained pixel locations, the file was smaller in size and thus change in the quality of the image was reduced. The message cannot be retrieved unless combination of both the images is used. Thus, the proposed method acts as a lock and key using a combination of two images thus robustness of this method is very high as compared to simple LSB method. This method is very much suitable for covert communication.

References

1. Morkel T, Eloff JHP, Olivier MS (2005) An overview of image steganography. In: Proceedings of the fifth annual information security south Africa conference (ISSA2005), June/July
2. Kaur S, Bansal S, Bansal RK (2014) Steganography and classification of image steganography techniques. In: International conference on computing for sustainable global development
3. Al-Shatnawi AM (2012) A new method in image steganography with improved image quality. Appl Math Sci 6(79):3907–3915
4. Kumar JNVRS, Vemali C (2014) A novel secure image hiding on indexed images using pixel-matching technique, India 7(3):358–362. ISSN: 0974-6471
5. Tulasidasu M, Lakshmisirisha B, Rasool Reddy K (2015) Steganography based secret image sharing using block division technique. In: IEEE international conference on computation intelligence and communication network IEEE. https://doi.org/10.1109/CICN.2015.227
6. Alaa A, Altaay J, Sahib SB, Zamani M (2012) An introduction to image steganography techniques. In: International conference on advanced computer science applications and technologies
7. Thangadurai K, Sudha Devi G (2014) An analysis of LSB based image steganography techniques. In: 2014 International conference on computer communication and informatics (ICCCI-2014), 03–05 Jan
8. Eloff JHP, Olivier MS, Morkel T (2005) An overview of image steganography. In: Proceedings of the fifth annual information security South Africa conference (ISSA 2005), June/July
9. Nagar P, Mohammad N, Kumar N, Sharma N (2014) A secure implementation and analysis of image steganographic method for data security in clouds. Int J Adv Res Comput Sci Softw Eng. 4(6). ISSN: 2277 128x

DOA Estimation of Coherent Sources Using QPSO in WSN

M. Shree Prasad and Trilochan Panigrahi

Abstract Here we formulate the problem of direction of arrival (DOA) estimation of correlated sources when single snapshot is available in wireless sensor network (WSN). A computationally inexpensive fitness function is used instead of maximum likelihood for DOA estimation of correlated signals. In order to achieve faster convergence, the quantum particle swarm optimization (QPSO) algorithm is proposed to optimize the cost function. The performance of QPSO is analyzed and compared with particle swarm optimization (PSO). Simulation results indicate that the QPSO algorithm provides better estimation accuracy and faster convergence rate compared to PSO algorithm.

Keywords Wireless sensor network · Quantum particle swarm optimization
DOA estimation · Correlated source localization

1 Introduction

High-resolution direction of arrival (DOA) estimation is an important signal processing task in source of WSN [1]. Subspace- and maximum likelihood (ML)-based technique are used for estimating DOA of superimposed signal impinging on a sensor network [2], but subspace-based approach requires preprocessing techniques for DOA estimation of correlated signals [3–5]. The ML-based DOA estimation techniques are computationally intensive, but provide good performance even for highly correlated signals and under low snapshot and low SNR scenarios.

In many applications [6], the number of available snapshots is limited and the source signals are correlated either due to multipath propagation or intentional

M. Shree Prasad (✉) · T. Panigrahi
Department of E.C.E, National Institute of Technology Goa, Farmagudi, Goa, India
e-mail: shreeprasadm@gmail.com

V. Janyani et al. (eds.), *Optical and Wireless Technologies*, Lecture Notes in Electrical Engineering 472, https://doi.org/10.1007/978-981-10-7395-3_60

jamming. The single snapshot DOA estimation based on compressive sensing approach proposed in [7] suffers from off-grid effects. In [8, 9], a computationally efficient methodology compared to ML estimation technique is described for joint DOA and amplitude estimation of multiple emitter signals with single snapshot. Metaheuristic optimization algorithms are analyzed, and their performances are compared.

The evolutionary algorithms are used for ML-DOA estimation of uncorrelated signals in WSN [10, 11]. The PSO is one of the best-suited algorithms for DOA estimation for single snapshot as well [8]. Here we consider the fitness function provided in [8] for the DOA estimation of correlated signals in WSN. Since the quantum particle swarm optimization (QPSO) algorithm proposed in [12] has faster convergence rate compared to PSO algorithm, therefore the fitness function is optimized by QPSO algorithm for estimating the DOA of the correlated signals. Other advantage of QPSO algorithm is that it requires tuning for only one control parameter. The simulation results show the performance of proposed QPSO algorithms better than the PSO.

2 Problem Formulation

2.1 Single Snapshot Signal Model

Consider a two-dimensional sensor network (SN) of arbitrary geometry with P isotropic sensors. Let K narrowband superimposed signals with DOAs, $\theta = \theta_1 \ldots \theta_K$ impinge the SN with center frequency ω_c. The vector of signal received across the SN for single snapshot is represented as [8, 11]

$$\boldsymbol{s} = \boldsymbol{A}\boldsymbol{m} + \boldsymbol{n} \tag{1}$$

where $\boldsymbol{s} \in \mathrm{C}^{P \times 1}$ is the received vector at the SN, $\boldsymbol{m} \in \mathrm{C}^{K \times 1}$ is the vector of K emitter signals, and $\boldsymbol{n} \in \mathrm{C}^{P \times 1}$ is the spatially white noise with variance σ^2. The steering matrix, $\boldsymbol{A} \in \mathrm{C}^{P \times K}$ for the SN is defined as

$$\begin{aligned} \boldsymbol{A}(p,k) &= \exp\left(j\tfrac{2\pi}{\lambda}\left[x_p \sin(\theta_k) + y_p \cos(\theta_k)\right]\right) \\ p &= 1 \ldots P; k = 1 \ldots K \end{aligned} \tag{2}$$

Here, λ is the wavelength, (x_p, y_p) is the position of the pth sensor, and m_k is the source signal arriving in θ_k direction.

The source signal covariance matrix for two correlated source signals of equal power is defined as

$$C = E[mm^H] = \begin{bmatrix} 1 & \rho \\ \rho & 1 \end{bmatrix} \tag{3}$$

where $E[\cdot]$ is the expectation operator, H is the Hermitian operator, ρ is the coefficient of correlation between source signals 1 and 2. The covariance matrix $\boldsymbol{C}$ is non-diagonal and singular if the source signals are correlated [2].

2.2 Fitness Function for DOA Estimation

The fitness function $f(\theta)$ takes the value zero if the candidate solution that is K DOAs estimated by the swarm intelligence algorithm matches the actual DOAs of the K superimposed signals in the absence of the noise at the sensor. The fitness function $f(\theta)$ correlates the actual snapshot with the synthetic snapshot generated from the candidate solution [8].

$$f(\theta): R^K \rightarrow R: \theta = (\theta_1 \ \ldots \ \theta_K) \rightarrow [0, 1] \in R \tag{4}$$

The expression for $f(\theta)$ is given as

$$f(\theta) = 1 - \frac{\boldsymbol{s}^H \boldsymbol{d}}{\|\boldsymbol{s}\| \times \|\boldsymbol{d}\|} \tag{5}$$

where $\boldsymbol{d} \in \mathrm{C}^{P \times 1}$ is the synthetic snapshot and is defined as per [8].

3 DOA Estimation of Correlated Sources with Single Snapshot Using QPSO Algorithm

3.1 Quantum Particle Swarm Optimization

Quantum PSO (QPSO) algorithm is a variant of classical PSO and has particles with quantum behavior [12]. In QPSO, a swarm of S particles search for optimum solution in a K-dimensional solution space. The position vector of the jth particle in the iteration i is denoted as $\boldsymbol{x}_j^i = \left[x_{j1}^i, x_{j2}^i, \ \ldots \ , x_{jK}^i\right]$. The swarm in QPSO is associated with personal best and global best of the particle. The personal best of the jth particle is denoted as $\boldsymbol{b}_j^i = \left[b_{j1}^i, b_{j2}^i, \ldots, b_{jK}^i\right]$, and global best position is represented as $\boldsymbol{g}^i = \left[g_1^i, g_2^i, \ldots, g_K^i\right]$. During the search process for optimal solution, position of the particles in the swarm at the central node is updated according to the following equations [12].

$$\boldsymbol{x}_j^{i+1} = r^i \mp \operatorname{sgn}(u - 0.5) \cdot \beta^i \cdot abs\left(\boldsymbol{mbest}^i - \boldsymbol{x}_j^i\right) \circ \ln\left(\frac{1}{\boldsymbol{u}}\right) \tag{6}$$

where

$$\boldsymbol{r}^i = \frac{\phi_1 \circ \boldsymbol{b}_j^i + \phi_2 \circ \boldsymbol{g}^i}{\phi_1 + \phi_2} \tag{7}$$

$$\begin{aligned} \boldsymbol{mbest}^i &= \frac{1}{S} \sum_{j=1}^{S} \boldsymbol{b}_j^i \\ &= \left[\frac{1}{S} \sum_{j=1}^{S} b_{j1}^i, \frac{1}{S} \sum_{j=1}^{S} b_{j2}^i, \ldots, \frac{1}{S} \sum_{j=1}^{S} b_{jK}^i\right] \end{aligned} \tag{8}$$

In the above equations, $\circ$ indicates element-wise multiplication, i represents iteration number, S represents population size, and β^i is the contraction-expansion coefficient, the only parameter to be tuned in the QPSO algorithm. sgn(•) represents signum function, and u is the uniform random variable distributed in the interval *0–1*. ϕ_1, ϕ_2, and $\boldsymbol{u}$ are K-dimensional random vectors whose components are independent random variables uniformly distributed in the interval *0* and *1*.

Based on [13], β_i is varied from *1.778* to *0.5* over the iteration and is represented as

$$\beta_i = \frac{(0.5 - 1.778)}{Maximum\_Iteration} \cdot i + 1.778 \tag{9}$$

3.2 DOA Estimation of Correlated Sources Using QPSO Algorithm

This section describes DOA estimation of correlated sources using single snapshot available at the SN by optimizing the $f(\theta)$ using QPSO algorithm. The detail of the algorithm is described in Algorithm 1.

Algorithm 1: QPSO algorithm for single snapshot DOA estimation of coherent sources

```
Initialization of WSN and QPSO algorithm
– Initialize the locations of the sensors in the wireless sensor network.
– Initialize the particles in the population with random position in QPSO algorithm.

QPSO algorithm for single snapshot DOA estimation of correlated
  sources
for i ← 1 to Maximum_Iteration do
    for j ← 1 to S do
        Map the positions of the particles of the population to the solution
        vector in the solution space.
        From the solution vectors, evaluate the fitness function in (5)
        According to the value of the fitness function update the particles
        personnel best location b_j^i and group best position g^i
        Compute the mbest^i according to the (8).
        Compute the vector r^i defined in the (7).
        Update the particle position according to the (6).
        if particle position out of the solution boundary then
            Adjust the particle position
        end
    end
    Check for the termination criteria.
end
```

4 Simulation Results

In this section, simulation results are presented to demonstrate the performance of PSO and QPSO algorithms for DOA estimation of correlated signals for different correlation values with just a single snapshot in WSN. The overall simulation results shows that the proposed QPSO algorithm with new fitness function is giving better performance over the PSO and ML-DOA estimation techniques [11].

The performance of all algorithms is compared in terms of root-mean-square error (*RMSE*) and probability of resolution (*PR*). The convergence rate is also compared. The RMSE is a measure of accuracy of estimated DOA and is computed as [11]

$$RMSE = \sqrt{\frac{1}{KN_{run}} \sum_{i=1}^{N_{run}} \sum_{k=1}^{K} \left(\widehat{\theta}_k(i) - \theta_k\right)^2} \tag{10}$$

where K is the number of sources, N_{run} is the number of Monte Carlo simulations, $\widehat{\theta}_k(i)$ is the estimate of the kth DOA in the ith run, and θ_k is the true DOA of the kth source. The PR is a measure of the ability to resolve two closely spaced sources. Two sources are said to be resolved in given simulation run if both $\left|\widehat{\theta}_1 - \theta_1\right|$ and $\left|\widehat{\theta}_2 - \theta_2\right|$ are less than $\frac{|\theta_1 - \theta_2|}{2}$.

The number of K multiple superimposed emitter signals which can detect and distinguish depends on the number of sensor nodes in the WSN. Hence, in the

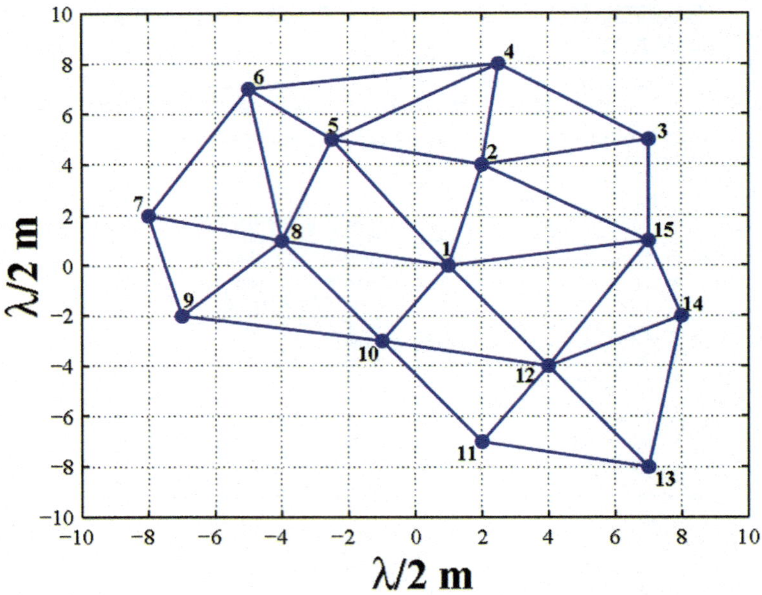

Fig. 1 WSN topology used in simulation

simulations a static WSN with $P = 15$ number of sensor nodes is considered and its topology is shown in Fig. 1. In the simulation, Node 1 is taken as reference node. It is assumed that $K = 2$ correlated signals with true DOA [112°, 122°] is impinging on the sensor network. The simulation results comparing the performance of PSO and QPSO algorithms for correlation factor $\rho = 0.99$ is discussed below. In the simulation, the SNR range is considered from *−20* to *25* dB. The average over *100* Monte Carlo simulations is plotted in each figure.

4.1 Convergence Performance

The convergence comparison of PSO and QPSO algorithms for new fitness function (5) and ML for DOA estimation [11] is depicted in Fig. 2. In the simulation, the value of *SNR* considered as *22* dB, and it is assumed that the emitter signals from two sources are correlated with $\rho = 0.99$. It can be observed from Fig. 2 that the DOA estimation using $f(\theta)$ has better estimation accuracy compared to ML-DOA estimation [11]. From Fig. 2a, it can be observed that RMSE attains a steady-state value at *40*th iteration for QPSO-based DOA estimation using fitness function $f(\theta)$. Further, from Fig. 2b, QPSO-based DOA estimation using $f(\theta)$ achieves *100%* PR at *16*th iteration.

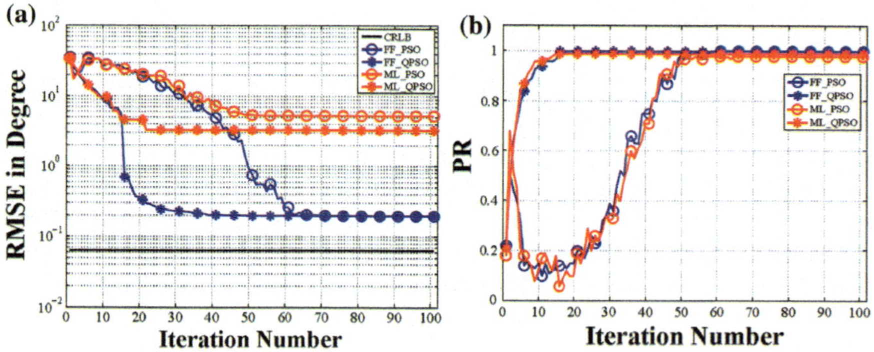

Fig. 2 **a** Plot of *RMSE* versus iteration number for *SNR* = *22* dB and $\rho = 0.99$, **b** plot of *PR* versus iteration number for *SNR* = *22* dB and $\rho = 0.99$

4.2 *Performance Comparison for Different Correlation Coefficient*

The performance of PSO and QPSO algorithms for DOA estimation using $f(\theta)$ in terms of *RMSE* and *PR* is compared for $\rho = 0$ and *0.9999* with swarm size = *60*. The swarm intelligence algorithms are run for *100* iterations for comparison of *RMSE* at steady state across all *SNR* values. The steady-state performance for both the algorithms is comparable. But QPSO has faster convergence than PSO algorithm. Further, from Fig. 3b, it is observed that due to the availability of just a single snapshot the performance for uncorrelated sources improves after *SNR* = *5* dB compared to correlated sources.

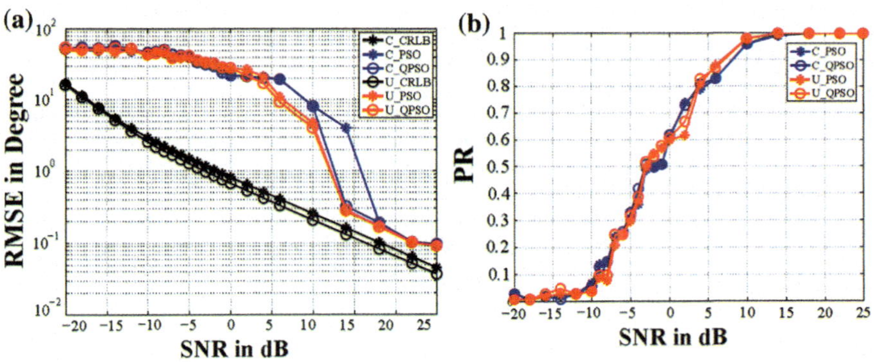

Fig. 3 **a** Plot of *RMSE* versus *SNR* for $\rho = 0$ and *0.9999*, **b** plot of *PR* versus *SNR* for $\rho = 0$ and *0.9999*

5 Conclusion

In this paper, the performance of PSO and QPSO is compared for DOA estimation of correlated sources with a single snapshot for a WSN with arbitrary array geometry. Based on the simulation results, the DOA estimation based on the fitness function (5) yields better estimation accuracy compared to the computationally intensive ML technique. Analyzing the simulated results, it can be inferred that at high *SNR*s, QPSO algorithm is able to distinguish correlated targets and yields better estimation accuracy with fewer iterations compared to PSO algorithm.

References

1. Rawat P, Singh KD, Chauchi H, Bonnin JM (2014) Wireless sensor networks: a survey on recent developments and potential synergies. J Supercomput 68(1):1–48
2. Krim H, Viberg M (1996) Two decades of array signal processing research: the parametric approach. IEEE Signal Process Mag 13(4):67–94
3. Shan TJ, Wax M, Kailath T (1985) On spatial smoothing for direction-of-arrival estimation of coherent signals. IEEE Trans Acoust Speech Signal Process 33(4):806–811
4. Gan L, Luo X (2013) Direction-of-arrival estimation for uncorrelated and coherent signals in the presence of multipath propagation. IET Microwaves Antennas Propag 7(9):746–753
5. Chen H, Hou CP, Wang Q, Huang L, Yan WQ (2014) Cumulants-based Toeplitz matrices reconstruction method for 2-D coherent DOA estimation. IEEE Sens J 14(8):2824–2832
6. Hacker P, Yang B (2010) Single snapshot DOA estimation. Adv Radio Sci 8:251–256
7. Fortunati S, Grasso R, Gini F, Greco MS, LePage K (2014) Single-snapshot DOA estimation by using compressed sensing. EURASIP J Adv Signal Process 2014(1):120
8. Errasti-Alcala B, Fernandez-Recio R (2012) On the use of particle swarm optimization for single snapshot DOA and frequency estimation. In: 6th European conference on antennas and propagation (EUCAP), pp 3635–3639
9. Errasti-Alcala B, Fernandez-Recio R (2013) Performance analysis of metaheuristic approaches for single-snapshot DOA estimation. IEEE Antennas Wirel Propag Lett 12:166–169
10. Panigrahi T, Roula S, Gantayat H (2016) Application of comprehensive learning particle swarm optimization algorithm for maximum likelihood DOA estimation in wireless sensor networks. Int J Swarm Intel 2
11. Panigrahi T, Panda G, Majhi B (2013) Maximum likelihood source localization in wireless sensor network using particle swarm optimization. Int J Signal Imag Syst Eng 6(2):83–90
12. Sun J, Xu W, Feng B (2004) A global search strategy of quantum-behaved particle swarm optimization. In: IEEE conference on cybernetics and intelligent systems 1:111–116
13. Sun J, Xu W, Fang W (2006) Quantum-behaved particle swarm optimization algorithm with controlled diversity. Springer, Berlin, Heidelberg

A Silver-Coated Scheme for Detection and Prevention Against Vampire Attack in Wireless Sensor Network

Richa Kumari and Pankaj Kumar Sharma

Abstract An ad hoc wireless sensor network is used to monitor the environment of a specific geographical area in which they gather the information from physical environment and transfer the information to the other node. Nowadays, the main problem that occurs in WSN is the diffusion of battery power at individual sensor device. Battery-power is one of the most critical aspect while deals with sensor devices because every nodes operated by battery and battery is limited in each node. A different variety of intrusion called vampire attack occurs at routing layer and medium access control layer, which are disastrous, arduous to recognize and very simple to implement by applying one nasty node sends random protocol-independent packets so that it drains the energy form sensory device. A single vampire can raise the network power consumption by an aid of O (N) in the poor case, where N is the number of nodes in WSN. We develop a scheme to reduce the impact of this type of attack for AODV routing protocol. The objective of this paper is detection and prevention of directional antenna vampire attack on stateful protocol. The suggested scheme restricts the consumption of energy from those nodes which are not the part of honest path, so trouble of directional antenna vampire attack can be minimized to a certain range.

Keywords Ad hoc wireless sensor network · Denial-of-service
Routing protocol · Vampire attack

R. Kumari (✉) · P. K. Sharma
Department of Computer Science and Engineering,
Government Mahila Engineering College, Ajmer, Rajasthan, India
e-mail: richa.kumari7171@gmail.com

P. K. Sharma
e-mail: pankaj.gmeca@gmail.com

V. Janyani et al. (eds.), *Optical and Wireless Technologies*, Lecture Notes in Electrical Engineering 472, https://doi.org/10.1007/978-981-10-7395-3_61

1 Introduction

Ad hoc WSN is infrastructure-less wireless network made for a particular application or task. Sometime when infrastructure is not available and impractical to implement then we can use ad hoc WSN. A wireless sensor network includes numerous sensing terminals known as sensor node which is distributed over a specific area to observe environmental status, such as pressure, temperature and sound. They coordinately transfer this information over the network [1]. All are small in size, featherweight and portable. They can function separately in rigid places where a human presence is hazardous or even impossible. Lifetime of the sensor node is dependent on their batteries. Replacing or recharging their batteries is impossible in rigid places, so it is needed to asset energy-efficient routing protocols. Every sensor node has a transducer, microcomputer, transceiver and battery for power source. Transceiver consumes energy much more than the energy consumed at microcontroller, so sensor devices are aided to do in-network data processing as much as possible.

2 Assumption and Routing Protocol

2.1 Assumption

- At first time, when the legitimate node is attacked by the vampire that node cannot be active at the time of route discovery. Initially, the status of all sensor nodes is set to be normal.
- Sensor node will never send RREQ to the abnormal node by checking its status.

2.2 Ad Hoc On-demand Distance Vector Routing Protocol (AODV) [2]

It is reactive in nature, which means that it establishes a path between the two nodes when the communication is required. In AODV, three types of packets are used: route request packet (RREQ): it helps to begin the pathfinding procedure, route reply packet (RREP): it helps to finalize the path and route error packet (RERR): it helps to inform the network of a path failure.

3 Vampire Attack

It is a type of denial-of-service attack which drains the power of the sensor node [3].

3.1 Vampire Attack on Stateless Protocol

It is also categorized into two attacks.

Carousel attack: Intruder builds a fake packet that includes series of loops of certain route, so that equal nodes will traverse by the one packet many times as shown in Fig. 1.

Stretch attack: Intruder builds up a fake lengthy route so that packets traverse through a huge amount of node rather than the honest number of nodes as shown in Fig. 2.

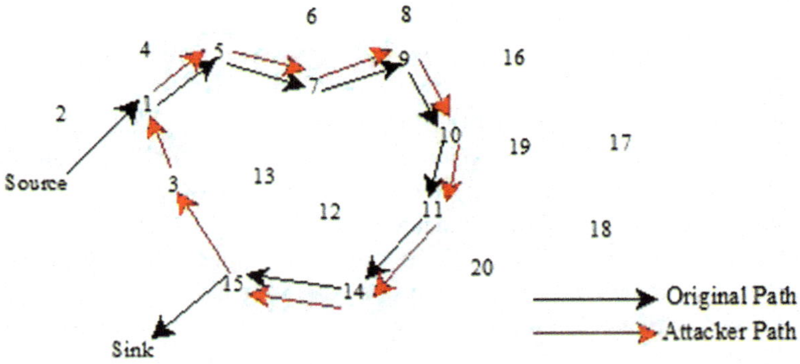

Fig. 1 Carousel attack

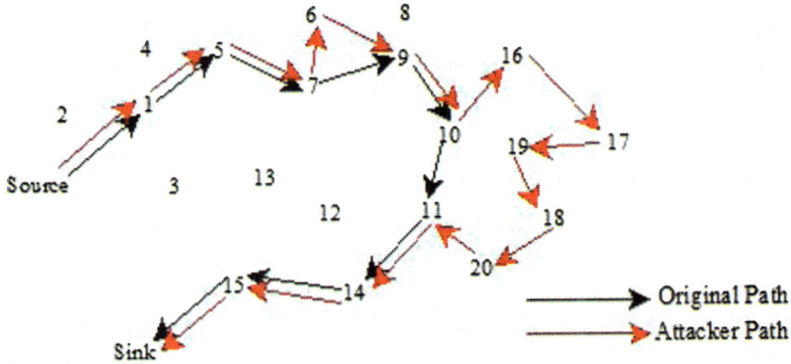

Fig. 2 Stretch attack

3.2 Vampire Attacks on Stateful Protocol

In this protocol, network topology and network states are known in advance by each node, so forwarding decisions at each node are done by using stored state or situation.

Directional antenna attack: In this attack, intruder has small control over the packet because of stateful protocol, but by using directional antenna they can still waste energy. Attacker redirects a packet in any section of the WSN, so packet traverses unnecessarily into network and consumes energy. This attack is also called half wormhole attack [4].

Malicious discovery attacks: DSR and AODV are susceptible to vampire because the path discovery may begin by nodes at any time, not just during the topology change. There are various ways for attacker to activate topology change; it may falsely claim that a new link between two fake nodes or a link is a fail between two non-existent nodes.

4 Related Work

Vasserman et al. [5]**,** proposed PLGP method against this attack. PLGP is vulnerable to vampire attack, so they have developed PLGPa (PLGP with attestation).

Vanitha and Dhivya [6]**,** the authors proposed a new routing against these attacks in sensor network named valuable secure protocol.

Umakanth and Damodhar [7]**,** the authors proposed the energy-weighted monitoring algorithm (EWMA) to detect the vampire attack.

Patil and Gaikwad [8]**,** the authors proposed a method for this attack by using EWMA and found corresponding trust value of each node.

Chumble and Ghonge [9]**,** the authors proposed a new routing protocol against vampire known as vampire attack removal protocol (VARP).

Patel and Soni [10]**,** the authors proposed a method for detection of vampire attack for AODV.

5 Proposed Work

This paper proposed a silver-coated scheme algorithm (SCSA) for early recognition and restrain of the directional antenna vampire attack. We have already discussed directional antenna vampire attack on stateful protocol [3]. This scheme is named as silver-coated scheme because the silver coating is used for identification and restraining of the real vampire. Here, every intermediate node performs some mathematical function to protect itself that works as silver coat. This approach has mainly three phases through which every packet goes and performs some operations on incoming data packet. In the first phase, each middle node will check whether it is in active path or not. In the second phase, if the node is in active path then it validates its previous node (it detects looping and jumping of data packet). In the third phase, each node checks its next node status (normal/abnormal) before forwarding a data packet for secure transmission against vampire.

5.1 Implementation Steps of Proposed Method

Step. 1: Establish a wireless sensor network. Each node maintains its routing table with the following fields: dest_addr, next_hop, hop_count, dest_seq# and life_time of the entry.

Step. 2: Whenever a node wants to send the information to other node, it will check whether it has a valid route (VR) in routing table to destination or not.

```
if(VR == 1)
        {Select that route for routing;}
else
        {It will discover the route using AODV routing protocol;}
```

Step. 3: After selecting a path to destination, the source node generates data packet with one extra field U_id of current node (CN) and forwards it to the next node (NN).

Step. 4: After receiving the data packet, each middle node checks that is it really in active route from source (S) to destination (D)? We do this because in directional antenna attack, packet may be redirected using directional antenna through vampire node from a part of the network to a node which is not a part of honest route.

Step. 5: A node is in active route if and only if it have been received before RREQ and RREP control packet from the source and destination, respectively. In other words, if node is in active path then it should have entry for source address and destination address in its routing table field "dest_addr".

```
if(datapkt→S_addr == rtable→dest_addr && datapkt→D_addr ==
rtable→dest_addr)
        {Move to step 6;}
else
{Previous node is abnormal. Current node will inform to IPS (Intrusion detection
and hindrance System)by sending U_id of previous node and goto step 2;}
```

Step. 6: When a node is in active path then the current node will validate the previous node because data packet may be redirected to a node which lies around in honest or active path from particular source to destination to make a loop or jump in that path.

Step. 7: Previous node is valid only when the data packet receive from same node that the RREQ packet has received from.

Here, $dest\_addr_{S\_addr}$ = entry of dest_addr field in routing table corresponding to the S_addr in the packet.

$$if\left(datapkt \rightarrow U_{id} == rtable \rightarrow dest_{addr_{S_{addr}}} \rightarrow next_{hope}\right)$$

$\{Move\ to\ step\ 8;\}$

$else$

$\{Previous\ node\ is\ abnormal.\ Current\ node\ will\ inform\ to\ IPS\ (Intrusion\ detection\ and\ hindrance\ System)\ by\ sending\ U\_id\ of\ previous\ node\ and\ goto\ step\ 2;\}$

Step. 8: At last, the current node (CN) will check the status of next node.

$if(status == normal)$

$\{Forward\ the\ data\ packet\ to\ the\ next\ node;\}$

$else$

$\{Inform\ source\ node\ for\ this\ abnormality\ and\ goto\ step\ 2\ to\ select\ another\ path;\}$

Step. 9: End the process of data packet routing (Fig. 3).

6 Simulation Result

The proposed scheme is implemented by using NS2.35. Here, we simulate to evaluate the performance of the SCSA as compared to the AODV in presence of vampire attack. About 15 nodes are taken in this model. Energy consumption and throughput of the SCSA and existing AODV protocol in the presence of vampire are compared in the following graph (Table 1; Figs. 4 and 5).

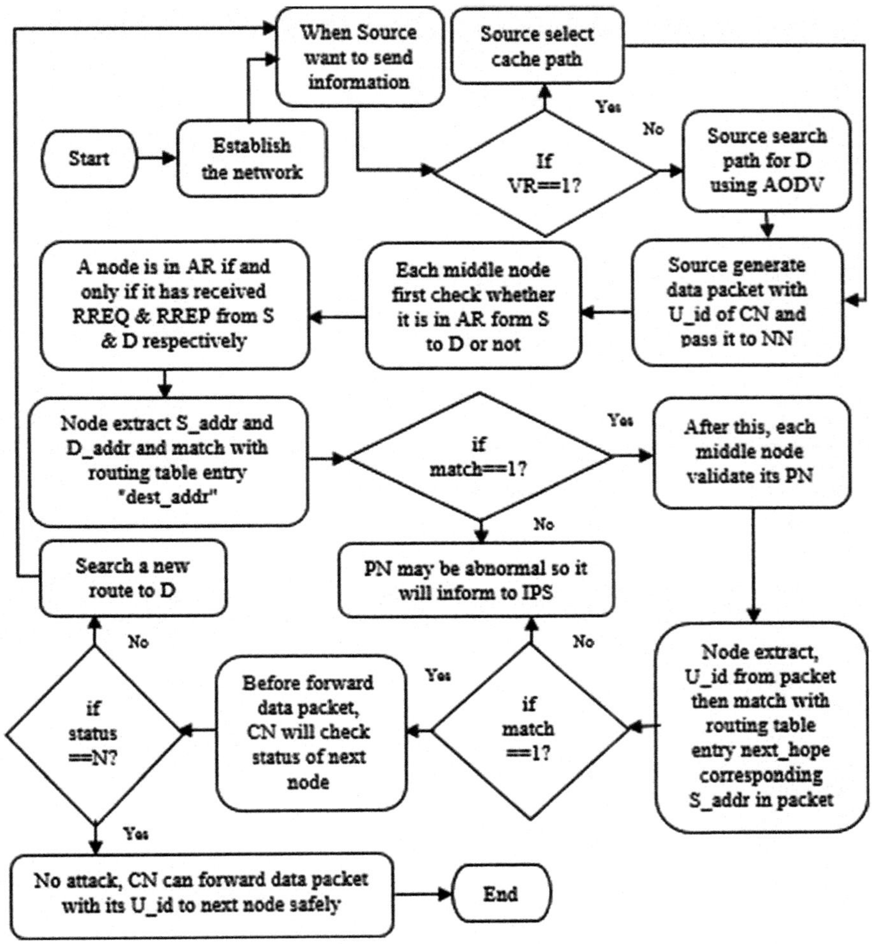

Fig. 3 Flowchart of Silver-Coated Scheme against Directional Antenna Vampire Attack

Table 1 Comparison table of energy consumption and throughput

Simulation time	Energy consumption (Joules)		Throughput (Kbit/s)	
	AODV in presence of vampire	After using proposed SCSA	Presence of vampire in AODV	After using proposed SCSA
10	175	170	200	250
15	270	250	280	360
20	370	345	350	480
25	460	425	420	600
30	555	555	550	550

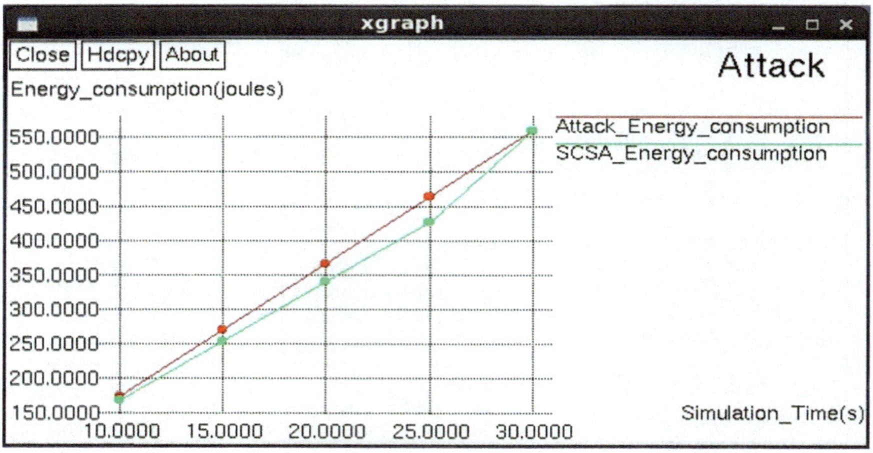

Fig. 4 Energy consumption at different simulation time

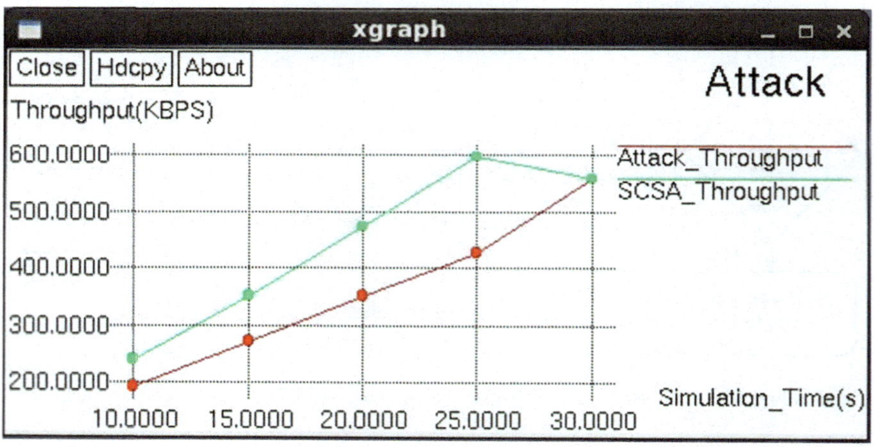

Fig. 5 Throughput at different simulation time

7 Conclusion

Several novel routing protocols have invented to mitigate vampire attack which is mentioned in related work. But there are no work done in AODV routing protocol for identification and restrain of directional antenna vampire attack.

This paper provides a different approach for identifying and preventing the directional antenna vampire attack for AODV by adding one extra variable U_id of current node in data packet then performing some computation on incoming data packet and node's routing table to detect this attack. For prevention, each intermediate node checks the status of the next node before forwarding the data packet.

This paper represents the performance of the existing AODV protocol and proposed SCSA protocol which are evaluated by using small number of vampire node in random generated topology of 15 sensor node. The simulation result shows energy consumption and throughput of the SCSA which are compared with the existing AODV routing protocol. Using this scheme, we can control the unnecessary network energy usage to a certain extent.

References

1. Al-Karaki JN, Kamal AE (2004) Routing techniques in wireless sensor networks—a survey. Wireless communications, IEEE 11.6
2. Naqishbandi TA, Sheriff I (2014) A resilient strategy against energy attacks in WSN and future IoT. Int J Adv Res Comput Sci Softw Eng 4(2):766–773. ISSN: 2277 128X
3. Kumari R, Sharma P (2016) A literature survey on detection and prevention against vampire attack in WSN. In: International conference on computer, communication and computational sciences [IC4S 2016], pp 122–126, Ajmer (Raj.), August 2016
4. Hu Y-C, Johnson DB, Perrig A (2003) Packet leashes: a defense against wormhole attacks in wireless ad hoc networks. In: Proceedings of the IEEE INFOCOM, pp 8–13
5. Vasserman EY, Hopper N (2013) Vampire attacks: draining life from wireless ad hoc sensor network. IEEE Trans Mobile Comput 12(2):318–332
6. Vanitha K, Dhivya V (2014) A valuable secure protocol to prevent vampire attacks in wireless ad hoc sensor networks. In: IEEE international conference on innovations in engineering and technology (ICIET'14), vol 3, special issue 3, March 2014
7. Umakant, Damodhar J (2014) Resource consumption attacks in wireless ad hoc sensor networks. Int J Eng Res 3(2):107–111. ISSN: 2319-6890(online), 2347-5013(print)
8. Patil A, Gaikwad R (2015) Preventing vampire attack in wireless sensor network by using trust. Int J Eng Res Technol 4(06). ISSN: 2278-0181
9. Chumble SC, Ghonge MM (2015) Simulation of mitigation of vampire attack in wireless ad-hoc sensor network. Int J Advent Res Comput Electr (IJARCE) 2(9). E-ISSN: 2348-5523
10. Patel AA, Soni SJ (2015) A novel proposal for defending against vampire attack in WSN. In: International conference on communication systems and network technologies. https://doi.org/10.1109/CSNT

Design of Five-Band Microstrip Antenna for Radar and Satellite Communications

Indra Bhooshan Sharma, Fateh Lal Lohar, Ravi Kumar Maddila, Abhinav Deshpande and M. M. Sharma

Abstract A single-feed microstrip antenna for five-band satellite communications and RADAR applications is presented in this paper. The proposed antenna is basically rectangular microstrip patch antenna with overall dimension of about $30 \times 36.7 \times 1.6$ mm^3 with partial and defective ground. The patch has circular cuts at the three corners and a slot in the middle. The antenna resonates at frequencies f1 = 3.88 GHz (2.55–4 GHz), f2 = 6.96 GHz (4.5–7.94 GHz), f3 = 10.57 GHz (9.50–12 GHz), f4 = 16.14 GHz (14.74–17.10 GHz), and f5 = 18.82 GHz (18.23–19.61 GHz) covering standard microwave frequency bands, namely S (2–4 GHz), C (4–8 GHz), X (8–12 GHz), Ku (12–18 GHz), and K (18–26 GHz), respectively [1]. Extensive simulations are accomplished using CST microwave studio [2]. The antenna has minimum return loss in range −37 to −48 dB for all the five bands. The antenna shows better impedance matching results, i.e., VSWR close to unity. Also, it has better directivity and efficiency for all the bands.

Keywords Five bands · Circular cuts · Defective ground · S-band C-band · X-band · Ku-band · K-band

---

I. B. Sharma (✉) · F. L. Lohar · R. K. Maddila · A. Deshpande · M. M. Sharma
Department of Electronics and Communication Engineering, Malviya National Institute of Technology, Jaipur 302017, India
e-mail: ibsharma1@gmail.com

F. L. Lohar
e-mail: 2015pec5100@mnit.ac.in

R. K. Maddila
e-mail: rkmaddila.ece@mnit.ac.in

A. Deshpande
e-mail: adeshpande93@gmail.com

M. M. Sharma
e-mail: mmsjpr@gmail.com

V. Janyani et al. (eds.), *Optical and Wireless Technologies*, Lecture Notes in Electrical Engineering 472, https://doi.org/10.1007/978-981-10-7395-3_62

1 Introduction

A rapid development of wireless communication systems is seen in the past decade. This fast development is brings about a wave of new wireless devices and systems to fulfill the demand of multimedia applications [3]. For this, the antennas are the most important devices for the establishment of the communication links over a long distance [4].

Sometimes, the need arises to operate the device at different frequency bands which require the corresponding antenna to have multiband characteristics. Also, a broadband characteristic is required for faster data transfer rate [1, 5]. Different frequency bands have different areas of applications [4].

The above characteristic is well accomplished by microstrip patch antenna. Microstrip patch antenna is gaining popularity because of its low profile, light in weight, planar configuration, least fabrication cost, and capability to integrate with microwave circuits [6]. But these antennas suffer problems of having narrowband characteristics. Many researchers are working over this demerit. Different methods have been devised to design the antenna for broadband and multiband operations [7–10]. Some of these techniques are stack and multilayers, shorting walls, V-shaped configuration, defective ground, etc. [11], out of which the most famous technique is defective ground technique. Keeping all these requirements into consideration, the microstrip patch antenna is designed which works effectively in five bands, namely S (2–4 GHz), C (4–8 GHz), X (8–12 GHz), Ku (12–18 GHz), and K (18–26 GHz) [1]. The stepwise procedure has been illustrated in the result section.

Firstly, the bandwidth is enhanced using partial grounding techniques and corner cuts. After that, the slots are cut in both the ground plane and the patch which is mainly the cause of the creation of the five bands. The proposed antenna covers the whole standard microwave bands, eliminating any wastage in the bandwidth. So any system connected to this antenna can work effectively in these microwave bands. The antenna shows minimum return loss and unity VSWR due to perfect impedance matching between feed line and the patch. Also, the antenna shows good efficiency in radiating all the five bands.

Future implementation of this antenna can include frequency reconfigurability characteristics, making the antenna to switch between the bands using switches like optical switches, PIN diode.

2 Antenna Design

Fig. 1 shows the schematic diagram of the proposed antenna with (a) top view showing the design of the patch and (b) bottom view showing the ground plane. The antenna is designed on an FR-4 substrate having dielectric constant of 4.3 and loss tangent of about 0.02. The patch has a circular cut in the three corners with

Fig. 1 An arrangement of the microstrip patch antenna

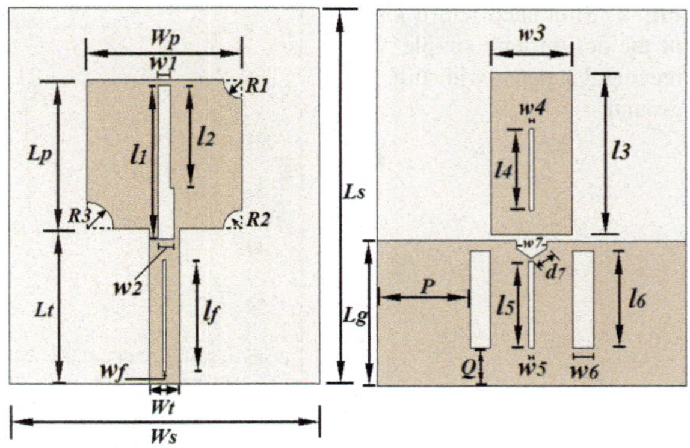

(a) Top View (b) Bottom View

Table 1 Parameters of the proposed antenna

Parameter	Ws	Ls	Wp	Lp	Lt	Wt	R1
Value (mm)	30	36.7	15	14.5	15.2	2.95	1.8
Parameter	R2	R3	l1	w1	l2	w2	lf
Value (mm)	1.8	2.6	15	1.2	10	1.6	10.9
Parameter	wf	l3	w3	l4	w4	l5	w5
Value (mm)	0.3	15.8	7.8	8	0.4	7.8	0.5
Parameter	l6	w6	d7	w7	Lg	P	Q
Value (mm)	9.5	2	12	1.7	14.2	9	6.85

different radii. Also, it has a slot at the center. At the back, two symmetric slots are cut at the bottom part of the ground plane and one slot in the upper part of the ground plane which is the cause of the multiband behavior of the antenna. The patch is given microstrip line feed. However, other feeding techniques can also be used, but impedance matching process becomes a tedious task in those cases. Here, with the help of microstrip line impedance can be matched by varying width of the patch. The line also has a slot to provide perfect impedance matching. Different notations have been illustrated in the diagram whose values are shown in Table 1.

3 Results and Discussion

The above antenna has been simulated using CST Studio Suit [2]. Initially, a simple rectangular patch antenna is designed on an FR-4 substrate with full ground. The return loss is shown in Fig. 2. It is observed that the antenna highly resonates at a particular frequency, and in addition to this, a minor band is also formed.

But for multiband operation, it is required to enhance the bandwidth. One of the famous methods to do this is to use the partial ground technique. By making use of

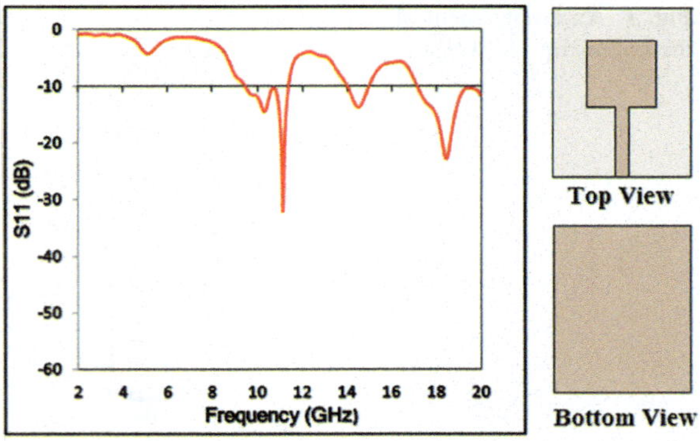

Fig. 2 Simulated return loss of the design with simple rectangular patch with full ground

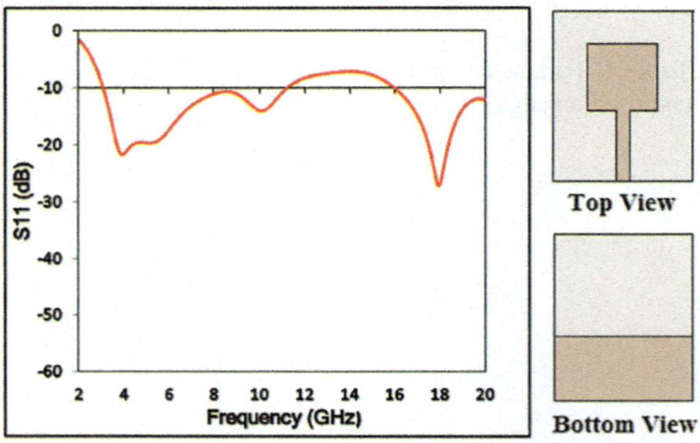

Fig. 3 Simulated return loss of the design of simple rectangular patch with partial ground

this, we obtained the response as shown in Fig. 3. We can see that the bandwidth is enhanced, forming two bands with some return loss.

Our main aim is to design a multiband antenna which is obtained by cutting the three corners of the patch circularly with different radius and also introducing a second ground plate above the initial ground plate, as shown in Fig. 4. Here, we can see that the second band is not properly formed and antenna exhibits high return loss for each band. The bandwidth is also low for each band. So for the improvement of the bandwidth and for minimum return losses, we introduced slots at the patch and two symmetrical slots in the ground plane. The result is shown in Fig. 5. It is observed that exact five bands are created with good bandwidth. But the impedance matching is still not perfect for each band resulting in VSWR not close to unity. So, for the improved impedance matching for each band, we used two extra slots of length lf, l4, l5 (according to Fig. 1). Thus, we get the best results as shown in Fig. 6.

The slots mentioned above improve the matching performance of the antenna which can be observed from the Smith chart as shown in Fig. 7. The chart shows the input impedance at the frequencies ranging from 2 to 20 GHz.

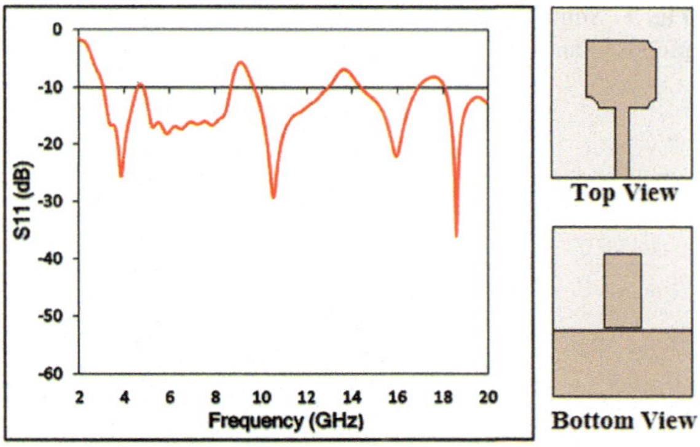

Fig. 4 Simulated return loss of the design with circular cuts and second ground plate

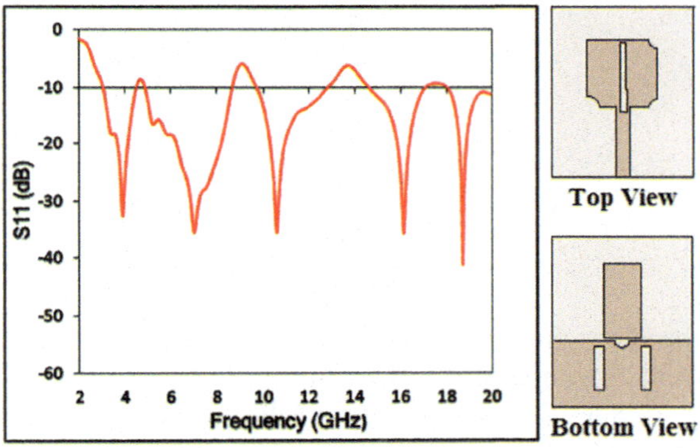

Fig. 5 Simulated return loss of the design by inclusion of the slots

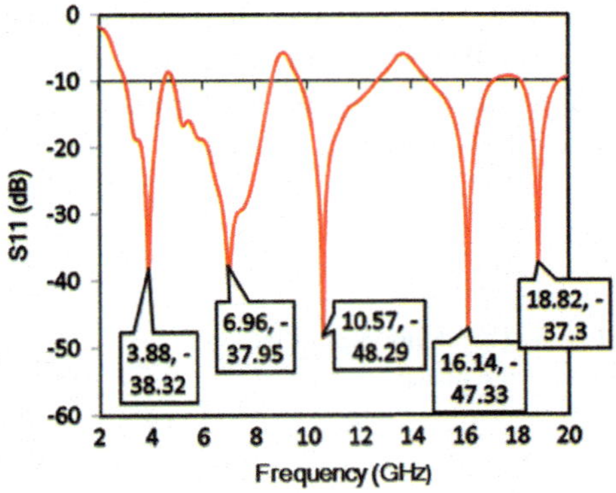

Fig. 6 Final return loss of the design with geometry as shown in Fig. 1

The input impedance of antenna is at f1 = 3.88 GHz (50.37 − j1.04) Ω, f2 = 6.96 GHz (48.73 – j0.73), f3 = 10.57 GHz (49.41 – j0.15), f4 = 16.14 GHz (50.16 + j0.12), and f5 = 18.82 GHz (48.53 – j0.58). It can be seen that the real part of the input impedance is approximately 50 Ω and the imaginary part is negligible making perfect matching with the input 50 Ω feed line. So VSWR

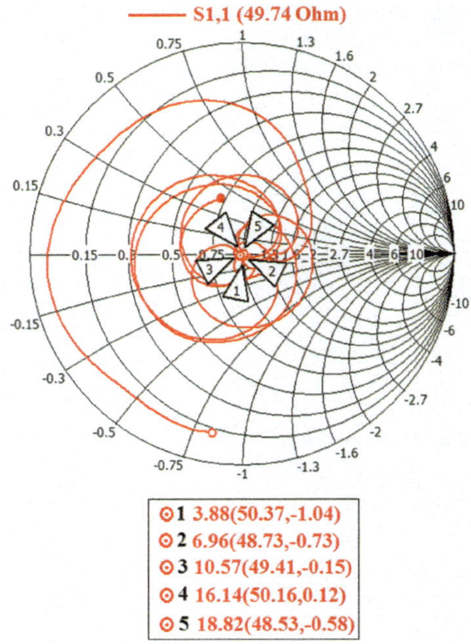

Fig. 7 Smith chart of proposed antenna

becomes close to unity as desired which means that the complete power is transmitted without any reflections.

Figure 8 shows the plot of directivity in dB for frequencies ranging from 2 to 20 GHz. It can be seen that the directivity is at its maximum at the desired band of frequencies as compared to the other frequencies. This shows that the antenna has more directives in the desired frequency band. The total percentage efficiency of the antenna throughout the frequency range 2–20 GHz is shown in Fig. 9. It can be observed from the plot that the total percentage efficiency is around 80–90% for the five bands, which is desired.

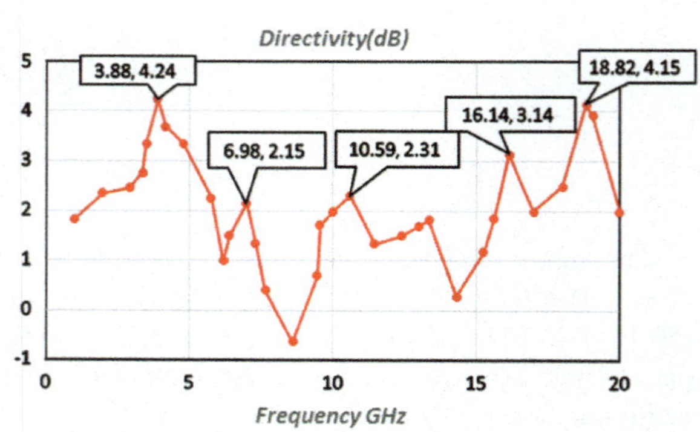

Fig. 8 Directivity plot of the designed antenna

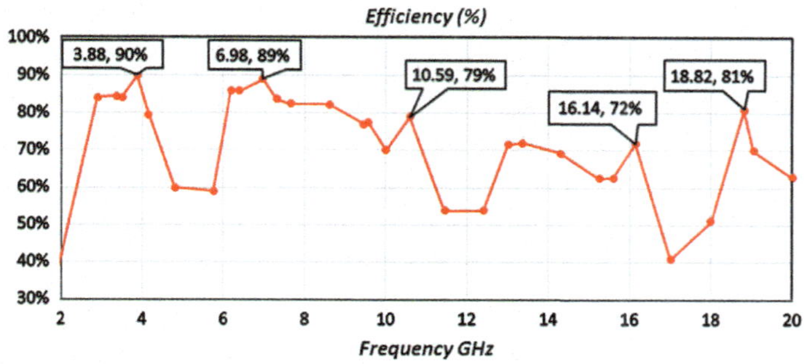

Fig. 9 Percentage efficiency plot of the proposed antenna

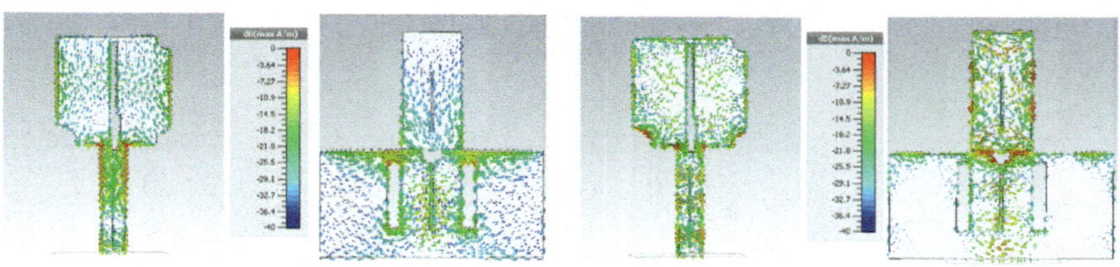

At 6.19 GHz Frequency

At 18.803 GHz Frequency

Fig. 10 Surface current distribution of the antenna at two resonant frequencies

The performance of the antenna can be well understood by the study of the surface current distribution. Current distribution for the two different frequencies is shown in Fig. 10.

The surface current is more at all the five resonant frequencies as compared to other frequencies. From the current distribution, we can see that in the patch area,

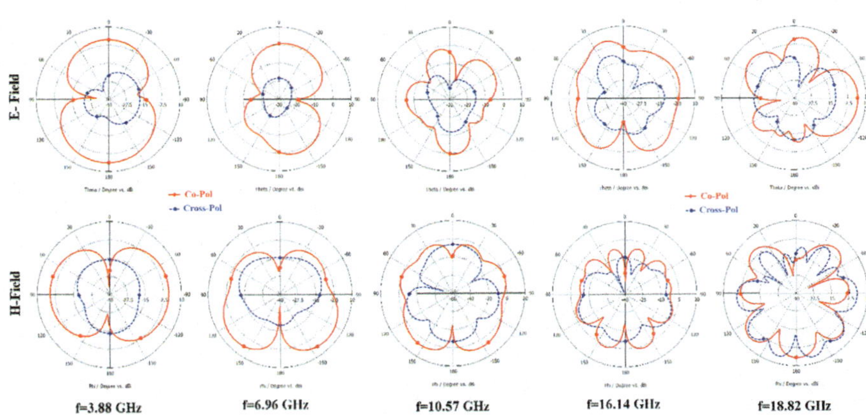

Fig. 11 Simulated co- and cross-polarization radiation plot of E-plane and H-plane of the five-band MSA antenna

Table 2 Deep analysis of designed antenna performance

Parameter	f1 = 3.88 GHz	f2 = 6.96 GHz	f3 = 10.57 GHz	f4 = 16.14 GHz	f5 = 18.82 GHz
Return loss (dB)	−38.32	−37.95	−48.29	−47.33	−37.33
VSWR	1.02	1.02	1.0	1.0	1.02
Frequency band (GHz)	2.55–4	4.5–.94	9.5–12	14.74–17.10	18.23–19.61
B.W (GHz)	1.45	3.14	2.50	2.36	1.38
B.W (%)	44.27	50.48	23.2	14.82	7.29
Efficiency (%)	90	89	79	72	81
Directivity (dB)	4.24	2.15	2.31	3.14	4.15

the currents are more confined to the slots rather than complete area of the patch. Also, the current is denser over the corner cuts. On the back side, at the ground plane, the current density is more around the slot in comparison with complete ground area, and this is the main cause of the formation of bands having lower return loss.

Figure 11 shows the co- and cross-polarization for E-field and H-field separately. The plot is shown for all the five bands. When the antenna is in its transmitting mode, the polarization is defined. The pair of orthogonal components of and Φ directions over a sphere is usually evaluated at each band. These components can later be called as co- and cross-polarization. The value of the cross-polarization is less as compared to the co-polarization which is desired for an effective antenna. And the designed antenna performances are presented in Table 2.

If the desired polarization component is vertical polarization, then this component is co-polarization component, and on the other hand, horizontal component is cross-polarization component. The performance of the antenna can be well judged from the performance table. From the given data, it can be concluded that the antenna has lower return loss, VSWR closer to unity showing better matching, considerable bandwidth for each band, better efficiency, and directivity properties.

4 Conclusion

The five-band microstrip patch antenna for RADAR and satellite communication has been designed and simulated. The antenna works effectively in all the four bands, i.e., f1 = 3.88 GHz, f2 = 6.96 GHz, f3 = 10.57 GHz, f4 = 16.14 GHz, and f5 = 18.82 GHz covering standard microwave frequency bands, namely S (2–4 GHz), C (4–8 GHz), X (8–12 GHz), Ku (12–18 GHz), and K (18–26 GHz) [1], respectively. The antenna has minimum return loss, better efficiency, and directivity properties.

References

1. Sharma IB, Lohar FL, Maddila RK, Deshpande A, Sharma MM (2017) Tri band reconfigurable microstrip patch antenna using shorted strip technique. In: 12th international conference on microwaves, antenna, propogation and remote sensing, ICMARS, February, 2017
2. CST Microwave Studio (2013) User's manual. www.cst.com
3. Noori O, Chebil J, Rafiqul Islam Md, Khan S (2011) Design of a triple-band h slot patch antenna. In: IEEE international RF and microwave conference (RFM 2011), pp 235–238
4. Kayalvizhi A, Prasanna G (2015) Multi band operating antenna for Wi-Fi, WiMAX and radar application. In: IEEE sponsored second international conference on electronics and communication systems (ICECS '2015'), pp 1017–1020
5. Pathak N, Chandrasekaran N (2015) Multi-band antenna using defective ground structure. In: IEEE ICCSP 2015 conference, pp 1152–1155

6. Mehedi Hasan Md, Raqibull Hasan Md, Tabassum R (2015) Investigation of multiband antenna for wireless communication. In: 2nd international conference on electrical engineering and information and communication technology (ICEEICT), pp 1–5
7. Mondal K, Sarkar PP (2015) Multiband patch antenna with modified ground plane for wireless communication systems. In: IEEE conference on research in computational intelligence and communication networks (ICRCICN), pp 377–380
8. Kumar A, Sharma IB, Sharma MM (2016) Reconfigurable circular disc monopole UWB antenna with switchable two notched stop bands. In: International IEEE India conference, INDICON, December 2016
9. Kumar A, Sharma IB, Saraswat RK, Sharma MM (2016) Dual band-notched circular disc monopole UWB antenna with switchable five notched stop bands. In: Proceedings of the Asia-Pasific microwave conference
10. Sharma IB, Lohar FL, Maddila RK, Deshpande A, Sharma MM (2017) Switchable five frequency band microstrip antenna using shorted parasitic elements. In: 12th international conference on microwaves, antenna, propagation and remote sensing, ICMARS, February 2017
11. Abutarboush HF, Nilavalan R, Cheung SW, Nasr KM (2012) Compact printed multiband antenna with independent setting suitable for fixed and reconfigurable wireless communication systems. IEEE Trans Antennas Propag 60(8):3867–3874

Tri-Band Microstrip Patch Antenna for C, X, and Ku Band Applications

Indra Bhooshan Sharma, Fateh Lal Lohar, Ravi Kumar Maddila, Abhinav Deshpande and M. M. Sharma

Abstract In this chapter, a tri-band microstrip patch antenna for C, X, and Ku band applications has been designed. The antenna having the dimensions of 30 × 35 1.6 mm^3 is simulated using CST Microwave Studio (Sharma et al, 12th international conference on microwaves, antenna, propagation and remote sensing, ICMARS, 2017) [1]. The microwave signal is supplied by microstrip transmission line. The minimum return loss (−40 to −52 dB) for each band is achieved by cuts at the corner and defective ground resulting in better VSWR. The antenna operates in five band f1 = 4.85 GHz (4.05–7.10 GHz), f2 = 10.04 GHz (9.05–12 GHz), f3 = 15.09 GHz (13.77–16 GHz) covering approximately entire C band (4–8 GHz), X band (8–12 GHz), and Ku band (12–18 GHz) (Sharma et al, 12th international conference on microwaves, antenna, propagation and remote sensing, ICMARS, 2017) [1] coming under microwave applications.

Keywords Quad band · Cross-slot · Defective ground · X band Ku band · K band

I. B. Sharma (✉) · F. L. Lohar · R. K. Maddila · A. Deshpande · M. M. Sharma
Department of Electronics and Communication Engineering,
Malviya National Institute of Technology, Jaipur 302017, India
e-mail: ibsharma1@gmail.com

F. L. Lohar
e-mail: 2015pec5100@mnit.ac.in

R. K. Maddila
e-mail: rkmaddila.ece@mnit.ac.in

A. Deshpande
e-mail: adeshpande93@gmail.com

M. M. Sharma
e-mail: mmsjpr@gmail.com

V. Janyani et al. (eds.), *Optical and Wireless Technologies*, Lecture Notes in Electrical Engineering 472, https://doi.org/10.1007/978-981-10-7395-3_63

1 Introduction

In the past decade, the wireless communication has been developed drastically [2]. The rapid growth of satellite and radar communication system and also the enormous use of radars require a new type of antennas such as small antennas, broadband antennas, and multifrequency antennas. The communication systems require broadband antennas for a fast data transfer rate [3].

Microstrip patch antenna satisfies such requirement [4]. These types of antenna are low profile, relatively easier in construction, low weight, low cost, comfortable to planar and non-planar surfaces, simple design, and manufactured using printed circuit board making them inexpensive [5]. Sometimes, systems like RADAR, GPS, and satellite too far from each other are required to operate at different frequencies. Microstrip patch antenna can be used to fulfill these requirements without using different antennas for different frequencies. Thus, there is always a challenging task for the designer to design antenna having multiple frequencies and broadband characteristics.

However, patch antennas suffer from narrowband problems. But in the past decade, researchers have devised various techniques for the enhancement of the bandwidth for multiband behavior, one out of which is a partial grounding technique which is used in this proposed antenna. Keeping all these requirements in mind, we have devised an antenna operating in three different bands [6–9].

The paper presents microstrip patch antenna with partial ground having two slots. The proposed antenna covers entire C band (4–8 GHz), X band (8–12 GHz), and Ku band (12–18 GHz) [1] with good bandwidth. The corner cuts and slots in the partial ground together are responsible for multiband operation having better bandwidth for each band. The antenna also has a very good return loss plot having return losses as low as −40 to −52 dB, showing perfect impedance matching between the patch and the feed line. Also, there is a need for the reconfigurable antennas which make use of switching techniques to switch from one band to other. Thus, the antenna can be further modified by reconfiguring in such a way that each band (one out of the three) is activated at a time by making use of switches.

2 Antenna Design

The design specification has been discussed below. The antenna is simulated in Microwave CST Studio [10]. The antenna has been designed over the FR-4 substrate ($\epsilon r = 4.3$, $\tan\delta = 0.02$). The overall design and notations are shown in Fig. 1. The antenna consists of a simple rectangular patch with circular cuts at the corners, having different radii.

The feeding technique used is microstrip line feeding technique. We can use other feeding techniques, but impedance matching in those cases is not easily obtained. Impedance matching in microstrip line feeding can be easily controlled by varying the width of the line only. At the back, there is a partial ground which has

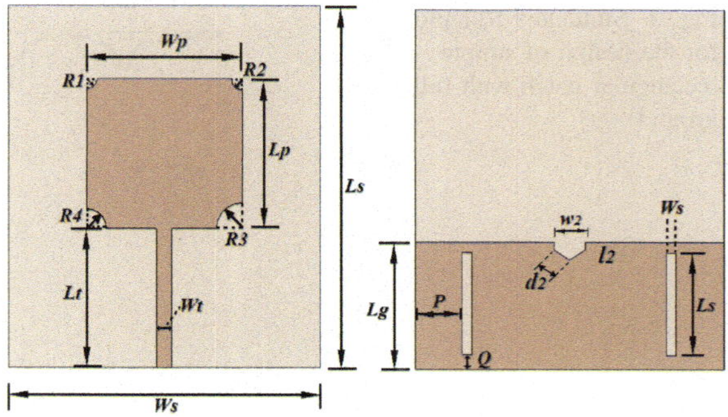

(a) Top View (b) Bottom View

Fig. 1 An arrangement of the microstrip patch antenna

Table 1 Parameters of the proposed antenna

Parameters	Ws	Ls	Wp	Lp	R1	R2
Values (mm)	30	35	15	14.5	1	1.1
Parameters	R2	R4	Lt	wt	Lg	P
Values (mm)	2.5	1.9	13.5	1.45	12.5	4.5
Parameters	Q	w2	l2	d2	Ls	Ws
Values (mm)	1	3	0.75	1.8	10	1

two symmetrical slots. These slots are the main reason by which we are getting multiband characteristics. The rest of the details regarding the dimensions and parameters have been illustrated in Table 1.

3 Results and Discussion

The antenna has been designed and simulated using Microwave CST Studio [10]. Initially, a simple rectangular microstrip antenna with complete ground is designed whose result is shown in Fig. 2. It was observed that the single band with high return loss is formed. So it was required to improve the return loss and enhance the bandwidth. Thus, we used the popular method to improve the bandwidth, i.e., by using partial ground at the back. Figure 3 shows the plot for this design.

In this Fig. 3, we can see that though we used the partial ground there is not any appreciable enhancement in bandwidth and also not much improvement in the return loss.

After this, the rectangular patch is circularly cut at the four corners with the radius as illustrated in Table 1. Figure 4 shows the return loss for this arrangement. It is seen that the performance of the antenna is improved a lot. The bandwidth has been enhanced, and also, the three bands start forming. This is because as the truncating corner has been introduced, the length of the patch gets reduced. This allows more density of surface current on the patch. However, the original length is

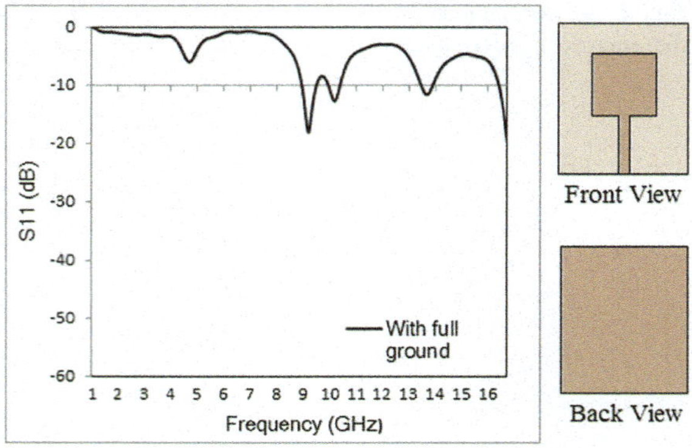

Fig. 2 Simulated S11 plot for the design of simple rectangular patch with full ground

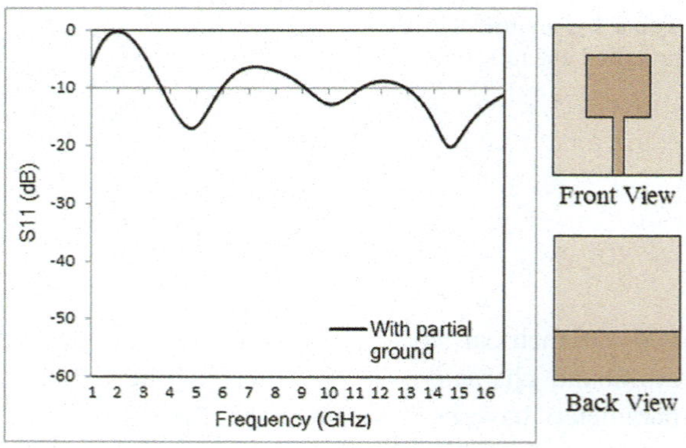

Fig. 3 Simulated S11 plot for the design of simple rectangular patch with partial ground

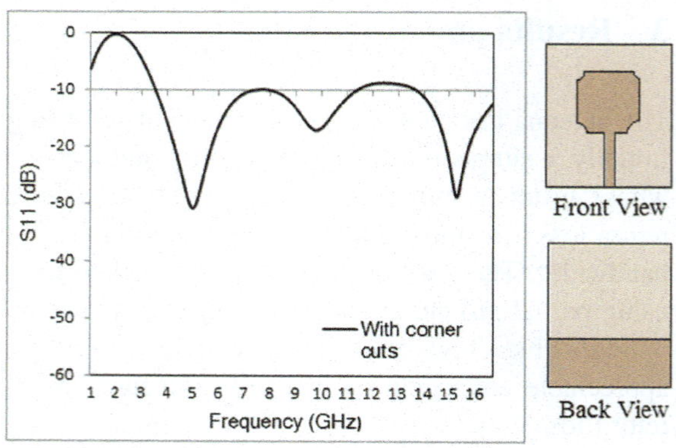

Fig. 4 Simulated return loss of the design with circular cuts

responsible for desired resonant frequencies, so it is unaltered. But the bands have some amount of return loss which is not desired. And also the bands are not distinguishable. We still require lower return loss for each band.

Finally, the two symmetrical slots have been cut in the ground side as shown. This reduces the loss of the antenna which gives lower return losses, and we get the result as shown in Fig. 5. We can see that the three bands are created which are distinguishable, and a minimum return loss of the order of −40 to −52 dB is obtained.

Figure 6 shows the plot of both directivity (in blue) and total efficiency (in red). The directivity and efficiency obtained for band f1 = 4.85 GHz is 2.99 dB, 91%; f2 = 10.04 GHz is 1.21 dB, 84%; and f3 = 15.09 GHz is 1.27 dB, 72%, respectively. So we can observe from the above data that as the frequency increases the value of efficiency slightly decreases, but the desirable values of directivity and total efficiency is achieved for each band.

The antenna's working principle is better understood by studying the surface current distribution. The surface current distribution of the antenna is shown in Fig. 7. The current is given for the frequencies of 4.8 and 10.04 GHz. As at these frequencies the return loss is minimum, the current density is more as compared to other frequencies. As it can be observed that the current is confined more at the corner cuts rather than the whole area of the patch. Also, at the ground plane, the currents are distributed more around the slots instead of the whole ground area. This is the main cause of the multiband behavior of the antenna.

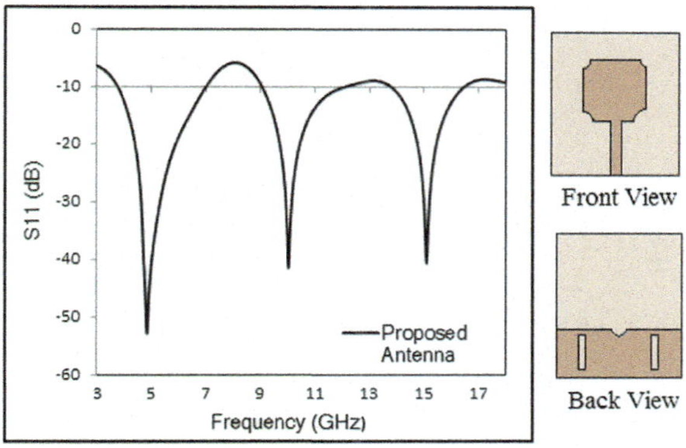

Fig. 5 Final S11 plot of the antenna

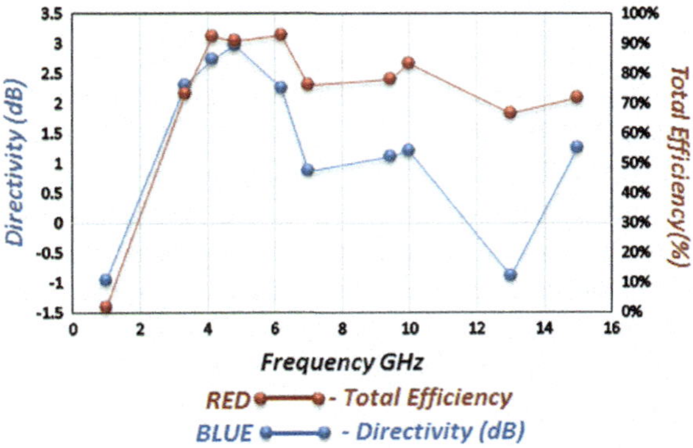

Fig. 6 Combined plot of directivity and total efficiency

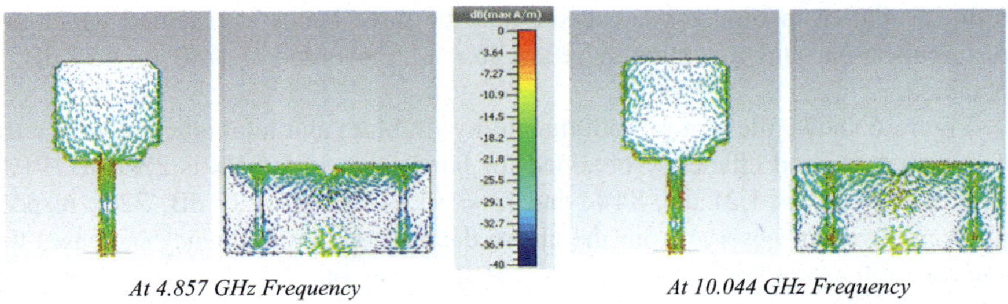

Fig. 7 Surface current distribution of the antenna at two different resonant frequencies

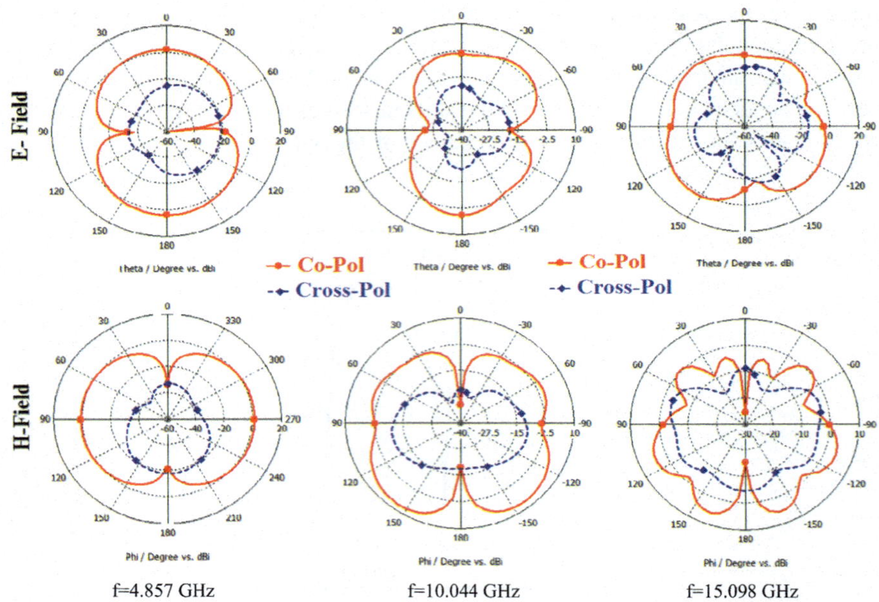

Fig. 8 Simulated co- and cross-polarization radiation plot of E-plane and H-plane the tri-band MSA antenna

Fig. 8 shows the co- and cross-polarizations for E-field and H-field for each band. Polarization of the antenna is defined in its transmitting mode for a given direction. The polarization is usually resolved into a pair of orthogonal component at each band over the sphere, along the and Φ directions. These components are known as co-polarization and cross-polarization.

In our case, the magnitude of the cross-polarization is lower as compared to that of the co-polarization. At higher frequencies, the value of cross-polarization little bit increases. The overall performance of the antenna is presented in Table 2.

Table 2 Antenna performance comparing with useful parameter for three different frequencies

Parameter	f1 = 4.85 GHz	f2 = 10.04 GHz	f3 = 15.09 GHz
Return loss (dB)	−52.89	−42.57	−40.75
Frequency band (GHz)	4.05–7.10	9.05–12	13.77–16.45
B.W (GHz)	3.05	2.95	2.72
B.W (%)	54.54	28.07	18.15
Total efficiency (%)	91	84	72
Directivity (dB)	2.99	1.21	1.27

4 Conclusion

The tri-band microstrip patch antenna has been successfully designed and implemented using CST Microwave Studio [10]. It is observed that the three bands come under the standard microwave bands, i.e., C band (4–8 GHz), X (8–12 GHz) and Ku band (12–18 GHz) [1]. As shown in Table 2, the antenna has a minimum return loss ranging from −40 to −52 dB, better directivity and efficiency, i.e., at f = 4.857 GHz (2.99 dB, 91%), f = 10.04 GHz (1.21 dB, 84%), f = 15.09 GHz (1.27 dB, 72%).

References

1. Sharma IB, Lohar FL, Maddila RK, Deshpande A, Sharma MM (2017) Tri band reconfigurable microstrip patch antenna using shorted strip technique. In: 12th international conference on microwaves, antenna, propagation and remote sensing, ICMARS, February 2017
2. Kumar A, Sharma IB, Sharma MM (2016) Reconfigurable circular disc monopole UWB antenna with switchable two notched stop bands. In: International IEEE India conference, INDICON, December 2016
3. Ojaroudi N, Ghadimi N, Ojaroudi Y, Ojaroudi S (2014) Frequency reconfigurable monopole antenna for multimode wireless communications. ACES J 29(8)
4. Noori O, Chebil J, Rafiqul Islam Md, Khan S (2011) Design of a triple-band h slot patch antenna. In: IEEE international RF and microwave conference (RFM 2011), pp 235–238
5. Abutarboush HF, Nilavalan R, Cheung SW, Nasr KM (2012) Compact printed multiband antenna with independent setting suitable for fixed and reconfigurable wireless communication systems. IEEE Trans Antennas Propag 60(8):3867–3874
6. Kayalvizhi A, Prasanna G (2015) Multi band operating antenna for Wi-Fi, WiMAX and radar application. In: IEEE sponsored second international conference on electronics and communication systems (ICECS '2015'), pp 1017–1020
7. Kumar A, Sharma IB, Saraswat RK, Sharma MM (2016) Dual band-notched circular disc monopole UWB antenna with switchable five notched stop bands. In: Proceedings of the Asia-Pasific microwave conference, December 2016
8. Sharma IB, Lohar FL, Maddila RK, Deshpande A, Sharma MM (2017) Switchable five frequency band microstrip antenna using shorted parasitic elements. In: 12th international conference on microwaves, antenna, propagation and remote sensing, ICMARS, February 2017

9. Dong YD, Hong W, Kuai ZQ, Yu C, Zhang Y, Zhou JY, Chen J-X (2008) Development of ultrawideband antenna with multiple band-notched characteristics using half mode substrate integrated waveguide cavity technology. In: IEEE Trans Antennas Propag 56(9):2894–2902
10. CST Microwave Studio (2013) User's manual. www.cst.com

A Design of Compact Planar Active-Integrated Inverted-F Antenna (AI-PIFA) for Mobile Handsets

Ashok Gundumalla, Sachin Agrawal and Manoj Singh Parihar

Abstract A class-E power amplifier integrated with a rectangular ring planar inverted-F antenna (PIFA) is presented in this paper. The proposed PIFA is very compact and occupying less volume 16 × 10 × 5.5 mm^3. The measured result shows that the proposed PIFA offers an impedance bandwidth of 250 MHz from 2.3–2.55 GHz, which is useful for LTE band 40 and WLAN (2.4–2.48 GHz). The maximum power-added efficiency (PAE) of class-E power amplifier is found to be 70% at 2.3 GHz. Further, the proposed PIFA is integrated with the class-E power amplifier to design an active-integrated antenna. The simulated results of the active-integrated antenna are presented. It is found that the gain of the active-integrated antenna system is increased from 2.4 dBi (passive) to 17.2 dBi (active).

Keywords Active-integrated antenna (AIA) · Class-E power amplifier (PA) Planar inverted-F antenna (PIFA)

1 Introduction

With recent development in wireless communication, active antenna attracted huge attention from the industry and research community for streamlining the appearance of the mobile terminals. An antenna integrated with active devices, such as bipolar junction transistor (BJT) or field effect transistor (FET), is generally referred to as active-integrated antenna (AIA). It offers several advantages such as compact size,

A. Gundumalla (✉) · S. Agrawal · M. S. Parihar
Discipline of Electronics & Communication Engineering, Indian Institute of Information Technology, Design & Manufacturing Jabalpur, Jabalpur, India
e-mail: gundumallaashok@iiitdmj.ac.in

S. Agrawal
e-mail: sachin.agrawal@iiitdmj.ac.in

M. S. Parihar
e-mail: mparihar@iiitdmj.ac.in

V. Janyani et al. (eds.), *Optical and Wireless Technologies*, Lecture Notes in Electrical Engineering 472, https://doi.org/10.1007/978-981-10-7395-3_64

increased bandwidth, and high gain. Depending on configuration and applications, AIA can be categorized in five groups, power-amplifying active antennas [1], mixing active antennas [2], oscillating active antennas [3], low-noise active antennas [4], and non-Foster active antennas. In this paper, a miniaturized PIFA integrated with the low supply class-E power amplifier is presented.

In AIA, a high-efficiency power amplifier is needed to increase the overall efficiency. Indeed, power amplifier consumes most of the dc power in the transmitter section due to its low efficiency. In recent year, a huge attention is paid to increase the efficiency of power amplifiers [5–9]. In this work, a class-E power amplifier is proposed using an enhanced PHEMT ATF-54143. It offers a high power-added efficiency (PAE) at very low input power supply as compared to the earlier reported works [4, 10]. For instance, a class-E amplifier using the MwT-8 GaAs MESFET is proposed in [10] achieves the maximum PAE of 64% for an input power of 10.4 dBm at VDC = 4 V, whereas the proposed design obtained the maximum PAE of 70% for an input power of 8 dBm and Vdd = 3 V. In addition, the proposed amplifier design uses the enhanced PHEMT devices, which tender the use of single-polarity supply voltage.

Furthermore, an increasing demand for various applications on a single handset requires integrating several components on a single chip. Therefore, a compact antenna is required to accommodate in small size mobile handset. In addition, the chosen antenna should possess the high efficiency, moderate bandwidth, low cost, and low profile that can be fulfilled by PIFA [11–13]. In this paper, a compact PIFA is proposed for LTE and WLAN. The PIFA consists of a ground plane, an elevated radiating patch shorted to ground using shorting pin which results in 50% size reduction as compared to other conventional antennas. In [4], a patch antenna of size 40×40 mm^2 and gain −8 dBi is co-designed with PA which increases the overall size of AIA, whereas the proposed PIFA achieves a 2.4 dBi gain using the dimension of $16 \times 10 \times 5.5$ mm^3. The paper is organized as follows: Section 1 describes the working of the proposed PA followed by simulated results. Section 2 discusses the working principle of the proposed antenna and the measured results, whereas Sect. 3 demonstrates the integration of proposed PA with the proposed PIFA.

2 Class-E Power Amplifier Design

The class-E PA is designed and simulated in Agilent's ADS-2014. The proposed structure uses the enhanced PHEMT ATF-54143. The typical circuit topology for the class-E PA is presented in [8, 9]. In this paper, class-E using shunt capacitor and shunt filter topology [8] is used as shown in Fig. 1. First the S-parameters of the device are measured by using the input and output 50 Ω lines and bias Tee for the DC biasing. From the measured S-parameters, input and output reflection coefficients are measured. By using input matching, the device is matched to standard

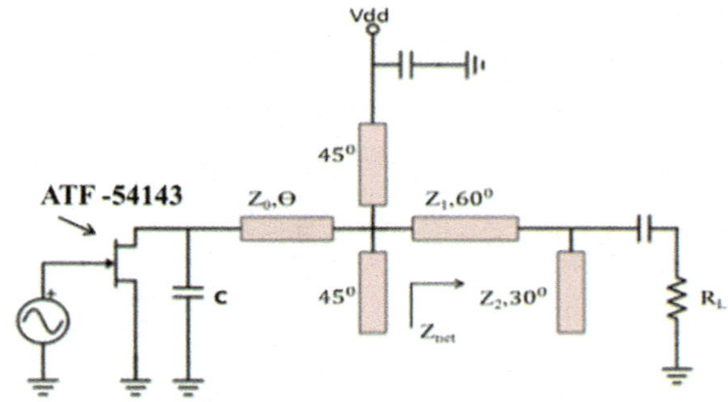

Fig. 1 Circuit diagram of the class-E power amplifier

source load of 50 Ω. The output reflection coefficient is calculated by using the Eqs. (1).

$$\Gamma_{\text{out}} = S_{22} + \frac{S_{12}S_{21}\Gamma_s}{1 - S_{11}\Gamma_s} \tag{1}$$

In Fig. 1, the L-type impedance transformer is used to match the load to input impedance of the antenna, while the output-matching network is used for the impedance transformation and the harmonic suppressors. Two λ/8 transmission lines (45° transmission line) are used to short-circuit the second harmonic, whereas the 30° transmission line for third harmonic. The simulated schematic of transmission line class-E PA is shown in Fig. 2. Here, Z_1 and Z_2 are calculated from the design as per given equations in [8].

Drain capacitance (C_{DS}) of the device is 0.08 pF, which is very less value for switching action of the transistor; therefore, the external capacitance (C_s) used in design is 1 pF. The amplifier is designed for optimum impedances which are obtained from the load-pull and source-pull techniques. The simulated S-parameters of the proposed PA are illustrated in Fig. 3. As can be seen, PA achieved the maximum impedance bandwidth of 400 MHz from 2.1–2.5 GHz. Moreover, the

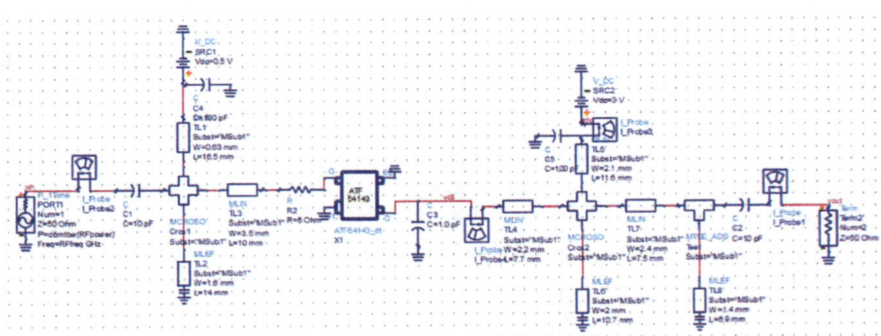

Fig. 2 Simulated circuit schematic of class-E PA with transmission lines

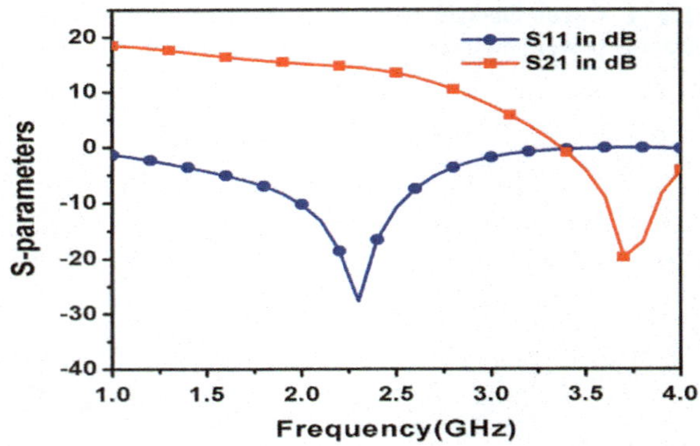

Fig. 3 S-parameter of the power amplifier

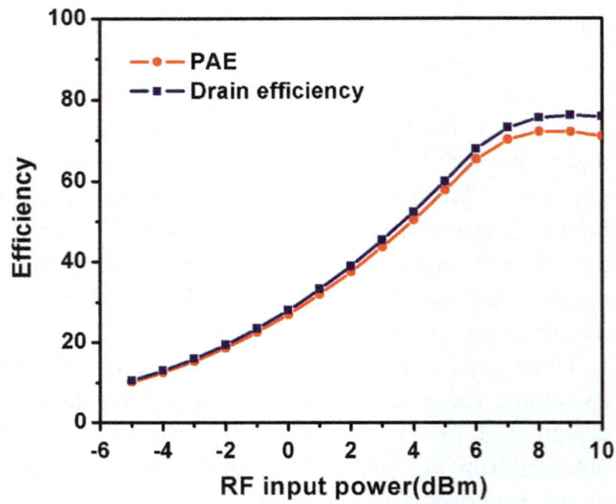

Fig. 4 Efficiencies of the power amplifier

gain is 15.5 dB over the range of 1–2.5 GHz. As depicted in Fig. 4, the PAE and drain efficiency of the proposed PA are 70% and 74%, respectively.

In the proposed PA, the gate is biased at cutoff voltage of 0.55 V and drain supply voltage is fixed at 3 V. The simulated output power as a function of input power is shown in Fig. 5. As seen, the PA exhibited an output power of 0.15 W for an input power of 8 dBm. The drain voltage and current are depicted in Fig. 6. It can be seen that the current and voltage waveform follows the switching action phenomena, which promises high efficiency due to low power loss. Load network consists of two quarter wavelength open stubs. TL6 and TL8 are used to suppress the second and third harmonics. Whereas, the series high-impedance line TL7 is used to transform the optimum load of 20–50 Ω load impedance. Input matching is designed at optimum load to obtain the maximum PAE. The bias network is

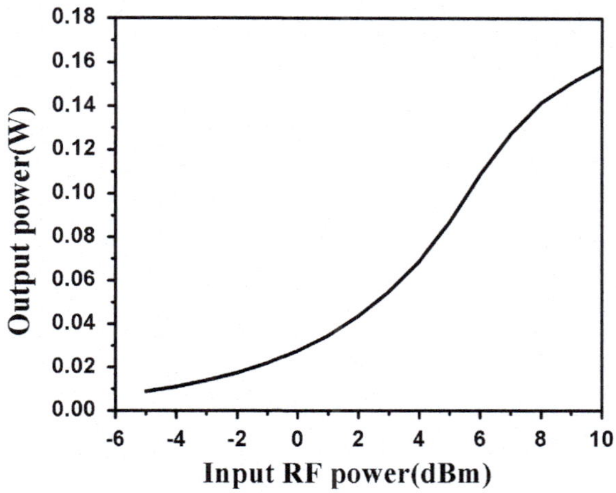

Fig. 5 Output power versus input power

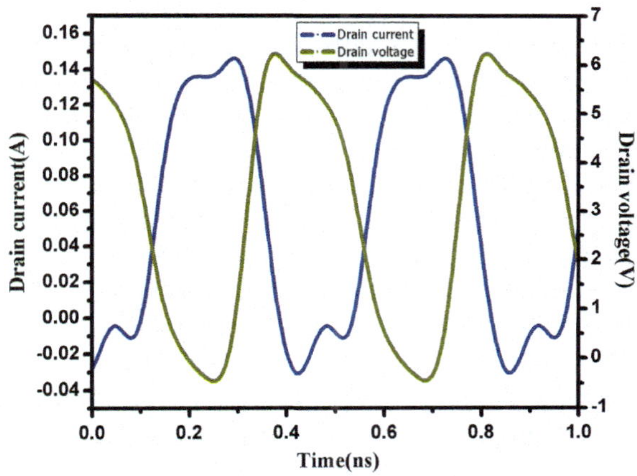

Fig. 6 Drain current and voltage waveform at 2.3 GHz

designed by using quarter wavelength transmission lines TL1 and TL5, respectively.

The transmission line TL4 is used to avoid the short-circuit at the drain terminal. The output voltage and current waveforms are illustrated in Fig. 7(a) and Fig. 7(b), respectively. As seen, the voltage and current waveform are purely sinusoidal and free from harmonic. The input power of 8 dBm ensures that the PA is in saturation mode. To obtain the stability of the device, a series resistance (R) of 5 Ω is inserted between input matching and the gate terminal of the PHEMT, which eliminates the unwanted oscillations.

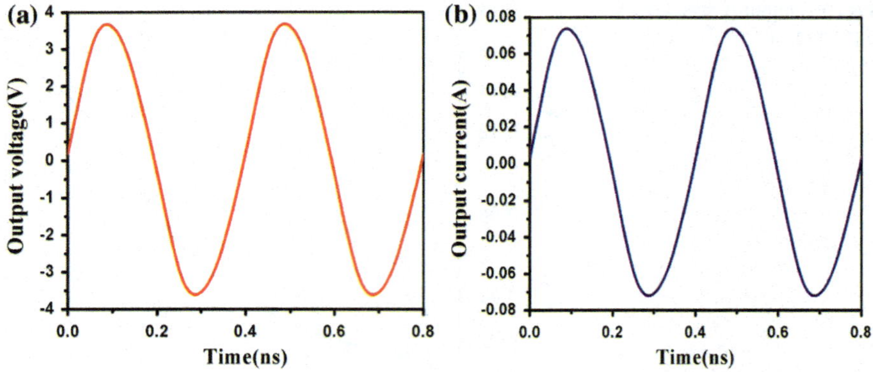

Fig. 7 **a** Voltage and **b** current waveform at 2.3 GHz

3 Rectangular Ring-Shaped PIFA

This section contains a design of rectangular ring-shaped PIFA which is proposed to integrate with the class-E power amplifier. The 3D view of the proposed PIFA is shown in the Fig. 8. Addition of shorting plate to the top plate made this antenna as a λ/4 resonator. It benefits in 50% size reduction as compared to the conventional patch antennas. The proposed PIFA is placed on the Rogers RO4725JXR of height (H) 0.77 mm and dielectric constant of 2.55. In designing the PIFA, meandering structure is used for compactness.

A rectangular metal sheet of dimension 30 × 60 mm^2 is used as a ground plane. Rectangular-shaped PIFA having the thickness of 0.1 mm is used in the simulation. The slot is created in both rectangular-shaped patches, which increases the surface current propagation length that resulted in antenna miniaturization. The one end of the antenna is supported by a strip of height (H_1) mounted on the substrate, whereas the other end is supported by the shorting plate of height H_2. The height of the antenna is 5.5 mm. The PIFA is feed at the edge of rectangular ring using the metal plate, which is connected to a microstrip line printed on the substrate. The length and width of the PIFA are first estimated using the equation given below.

Fig. 8 3D view of the proposed PIFA

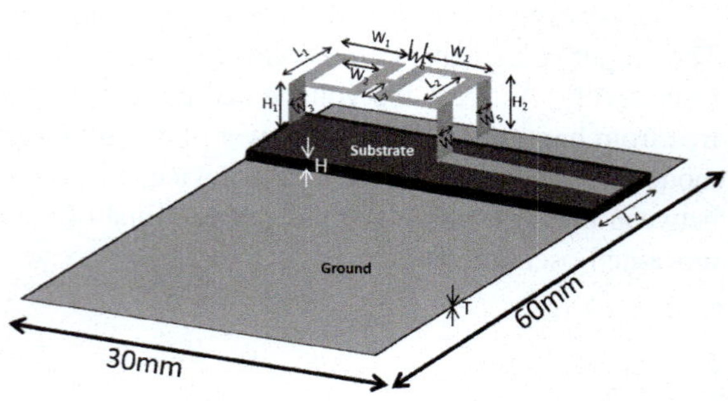

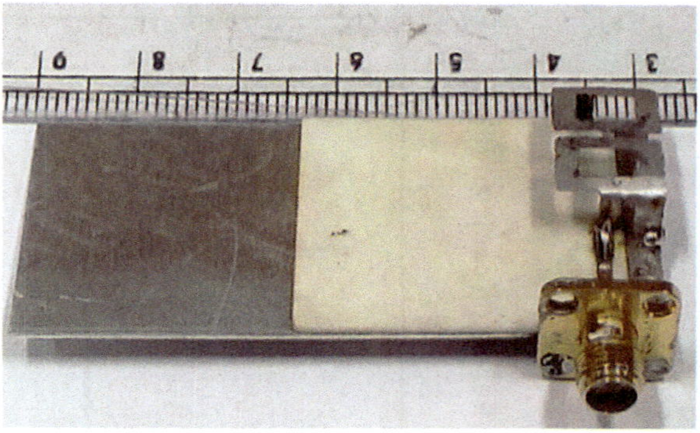

Fig. 9 Photograph of the fabricated PIFA

$$f_0 = \frac{c}{4(W+L)}$$

whereas L = 16 mm, W = 10 mm. PIFA is resonating at a natural frequency of 2.9 GHz. Meandering structure of the proposed PIFA is reason for the decrease of natural resonating frequency and used for the compactness. Now the width of the resonating PIFA is decreased but the length of PIFA is increased due to meandering structure. Now the length is increased to L = 30 mm and width is decreased to W = 2 mm. This dimensions gives the resonating frequency of 2.3 GHz.

Later, the antenna height is optimized to miniaturize the size, which facilitates to accommodate more number of components on the substrate. The height of PIFA plays major role in the impedance bandwidth so the height is optimized to get an impedance bandwidth which covers entire LTE band-40 which is 100 MHz. The optimized dimensions of the antenna are as follows: $L_1 = 10$, $L_2 = 6$, $L_3 = 4.5$, $L_4 = 30$, $W_1 = 7.5$, $W_2 = 4.5$, $W_3 = 4$, $W_4 = 1.6$, $W_5 = 2.5$, $W_6 = 1$, $H_1 = 5$, $H_2 = 5.5$ (all in mm). The photograph of the fabricated PIFA is shown in Fig. 9. The simulated and measured S-parameter of the proposed antenna is shown in Fig. 10. The measured result shows reasonable agreement with the simulated one; slight difference can be accounted for fabrication imperfections.

As seen, the measured S-parameter covers the impedance bandwidth of 250 MHz from 2.3 to 2.55 GHz. The simulated gain and total efficiency of the PIFA are illustrated in Fig. 11. It can be seen that PIFA achieved maximum gain and efficiency of 2.4 dBi and 93% at 2.35 GHz, respectively. The simulated radiation patterns in *xz*- and *yz*-plane at two different frequencies 2.3 and 2.4 GHz are depicted in Fig. 12a, b. It can be seen that in *yz*-plane, the radiation pattern is bi-directional, whereas in *xz*-plane, it is omni-directional.

Active-Integrated Antenna (AIA)

Integration of power amplifier and antenna is carried through co-design approach where the output impedance of the power amplifier is matched to the input impedance of the antenna. The AIA is simulated by using the CST and ADS co-simulation. The S-parameter model of the proposed class-E power amplifier is

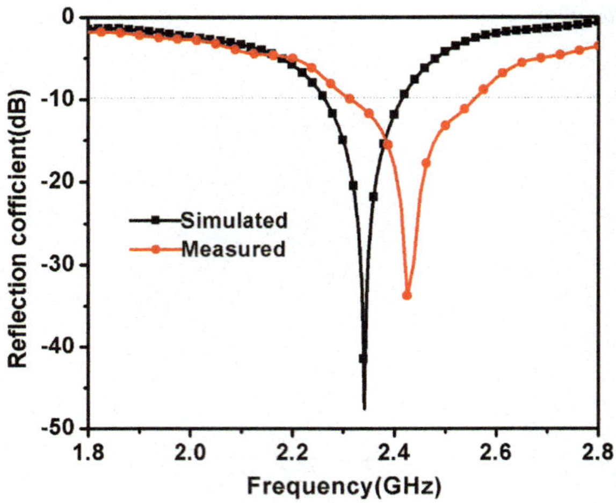

Fig. 10 Measured and simulated reflection coefficient

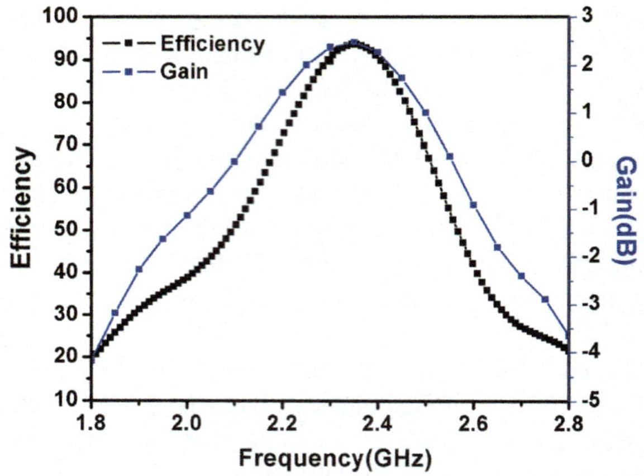

Fig. 11 Simulated gain and efficiency of the PIFA

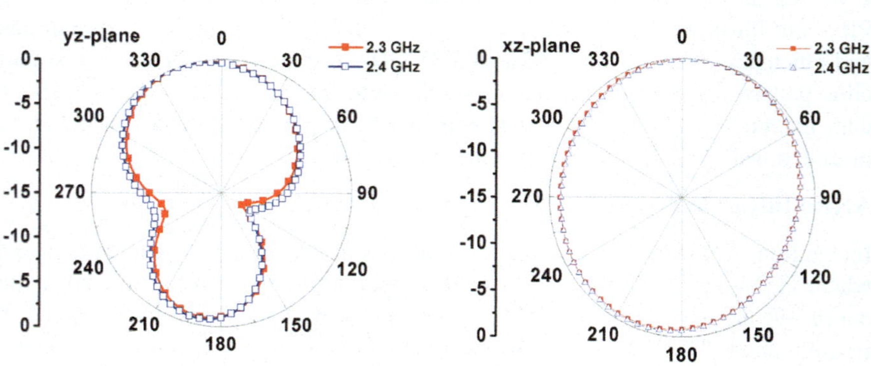

Fig. 12 Radiation patterns of the proposed PIFA

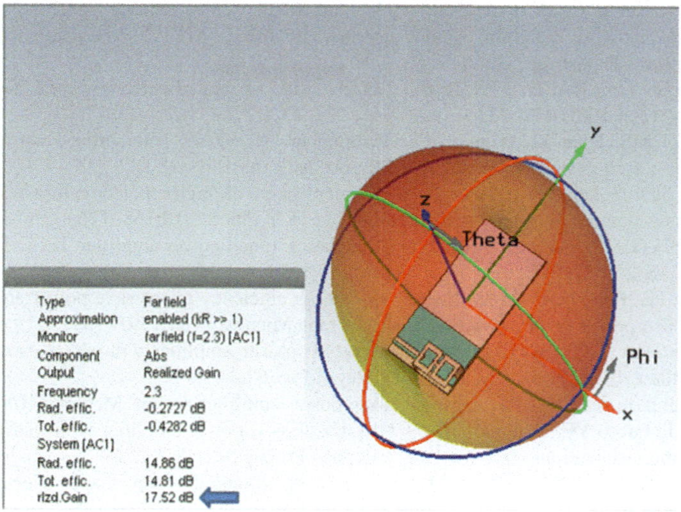

Fig. 13 Simulated radiation pattern and gain of AIA at 2.3 GHz

integrated with the antenna in CST. The radiation pattern of the proposed AIA system is shown in Fig. 13. As noticed, the realized gain of the AIA is 17.52 dBi at 2.3 GHz, which was 2.4 dBi before PA integration. Therefore, a large improvement in gain is achieved due to PA.

4 Conclusion

This paper presents a power amplifier-based active-integrated antenna. The proposed power amplifier achieves the maximum PAE of 70% at 2.3 GHz. The output power is 0.15 W for an input power of 8 dBm. Furthermore, the proposed PIFA is fabricated and the measured result is compared with simulated one. It found that the proposed antenna offers the impedance bandwidth of 250 MHz from 2.3 to 2.55 GHz. The proposed miniaturized active PIFA may provide more area to accommodate more components in the manufacturing of the mobile handsets. The IA-PIFA system provides the realized gain of 17.52 dBi.

References

1. Qin Y, Gao S, Sambell A (2006) Broadband high-efficiency circularly polarized active antenna and array for RF front-end application. IEEE Trans Microw Theory Tech 54(7), 2910–2916

2. Choi W, Cheon C, Kwon Y (2006) A V-band MMIC self oscillating mixer active integrated antenna using a push-pull patch antenna. In: IEEE MTT-S International Microwave Symposium Digest, pp 630–633
3. Chang K, York RA, Hall PS, Itoh T (2002) Active integrated antennas. IEEE Trans Microw Theory Tech 50(3):937–944
4. Sharawi MS, Dhar SK, Hammi O, Ghannouchi FM (2016) Miniaturised active integrated antennas: a co-design approach. IET Microw Antennas Propag 10(8):871–879
5. Sokal NO, Sokal AD (1975) Class-E a new class of high-efficiency tuned single-ended switching power amplifiers. IEEE J Solid-State Circuits SC-10:168–176
6. Raab FH (1977) Idealized operation of the class E tuned power amplifier. IEEE Trans Circuit Syst CAS-24:725–735
7. Mader TB, Popovic Z (1998) Switch mode high efficiency microwave power amplifiers in a free space power combiner array. IEEE J Trans Microw Tech 1391–1397
8. Grebennikov A (2016) High-efficiency class-E power amplifier with shunt capacitance and shunt filter. IEEE Trans Circuits Syst I Regul Pap 63(1):12–22
9. Grebennikov A (2004) RF and microwave power amplifier design. McGraw-Hill, New York
10. Mury T, Fusco VF, Cantu H (2007) 2.4 GHz class-E power amplifier with transmission-line harmonic terminations. IET Microw Antennas Propag 1(2):267–272
11. Yun G (2007) Planar inverted-F antenna with rectangular cap shape. Electron Lett 43 (12):656–657
12. Chen HM, Lin YF, Cheng PS, Lin HH, Song CTP, Hall PS (2005) Parametric study on the characteristics of planar inverted-F antenna. IEE Proc Microw Antennas Propag 152(6):534–538
13. Lin CH, Saito K, Takahashi M, Ito K (2012) A compact planar inverted-F antenna for 2.45 GHz on-body communications. IEEE Trans Antennas Propag 60(9):4422–4426

GPS-Aided AODV Routing Protocol for MANET

Bhupendra M. Parmar and Kishor G. Maradia

Abstract This paper describes suitability of Global Positioning System (GPS) with Ad hoc On-Demand Distance Vector (AODV) to enhance routing mechanisms in Mobile Ad hoc NETworks (MANETs). AODV is modified where GPS-equipped nodes uses location information to set-up path from source to destination. Route REQuest (RREQ) and Route REPly (RREP) packets are modified to include GPS location of source and destination. Routing table of node is also modified to include GPS information of nodes in a network and is updated from time to time during initialization of network and packet sharing. GPS information in routing table of node is used to set-up path whenever node has data to send. Position information decreases number of nodes involved during path set-up process which ultimately decreases large control overhead during source to destination path formation. Whereas conventional AODV uses global flooding of RREQ during route discovery which results in large control overhead for higher node density networks. In this technique, relative speed is also taken into account while choosing node during route discovery process. If relative speed of next node (to be selected as next hope during route discovery) is higher than limit may result in premature route break and therefore such nodes avoided as next hope. Performance of proposed mechanism is measured using NS3.25 simulations with different scenarios and varying load in a network. Comparative analysis of simulation results shows better performance of GPS-aided AODV in terms of reduction in control overhead, end-to-end delay, route discovery time and increase in throughput.

Keywords MANETs · GPS · AODV · Routing · Normalized routing load
Packet delivery ratio · End-to-end delay · Throughput · Jitter
Mobility · Packet lost

B. M. Parmar (✉)
GTU, Ahmedabad, Gujarat, India
e-mail: parmarbhupendra@gmail.com

K. G. Maradia
EC Department, GEC, Gandhinagar, Gujarat, India
e-mail: kgmaradia@yahoo.com

V. Janyani et al. (eds.), *Optical and Wireless Technologies*, Lecture Notes in Electrical Engineering 472, https://doi.org/10.1007/978-981-10-7395-3_65

1 Introduction

During recent years, world has seen great technological advancement in the area of wireless communication networks. MANTEs are such communication network which are formed temporarily on ad hoc basis and operates dynamically without any central administration. Communication among participating nodes took place hope to hope or node to node. This means intermediate nodes are used to send and receive packet from source to destination. Since networks are highly dynamic and temporary in nature, quality of service is always an issue. To achieve high-quality performance of MANETs, various issues needed to be addressed like accessibility, routing, power consumption, security, scalability.

Routing in MANETs is very important issue and is different from wired networks, because here the networks are dynamic due to mobility of nodes and varied geographical conditions. Each node in a network may not know and in direct connection with all other nodes in a network. It is challengeable to find path from source to destination and to manage it until the communication is over. Many routing protocols are proposed for MANETs, most of which are inherited from wired networks. Due to difference in nature of wired and wireless networks, these protocols need improvements. In these routing protocols, major concern is about process of path formation (route discovery and optimization) and route maintenance (for broken paths). Comparative studies of various routing protocols show AODV is prime candidate as routing protocol [1, 2].

AODV is an on-demand routing protocol, that is, path-finding process initiated whenever node has packet to transmit. It employs special sequence number to identify fresh routes. In AODV, source node and intermediate nodes have memory to store next hope information corresponding to each flow of packet. Whenever node has data to transmit, it floods Route Request (RREQ) packet through network. It employs optimization mechanism to find best path among multiple paths available. Destination node responds to source node with Route Reply (RREP) packet, and then both parties exchange the data packets [3]. In AODV, one of the issues which need attention is that it uses global flooding during path set-up between source and destination. It causes large overhead and high number of duplicate packets, which ultimately increases congestion in a network and decreases efficient use of network resources like bandwidth.

GPS is a very accurate, low-cost positioning system available worldwide [4]. GPS can be used to get position information of nodes to form path between various nodes in a network. Nodes can exchange own GPS information with each other, use appropriate algorithms to find and manage broken paths.

2 Related Works

For position-based routing, use of GPS information to locate nodes and routing in MANET's seems attractive option because of availability of low-cost GPS receivers. Therefore, many GPS-aided routing protocols and techniques are proposed to improve overall performance of MANET's. Ko and Vaidya [5] proposed location-aided routing (LAR) protocol reduces control overhead by limiting the network area into small request zones and estimated zones. Parkinson et al. [4] and Thongthavorn et al. [6] shows GPS can be used to obtain location of node, and the location information is exchanged when node wanted to transmit the packet. Deb et al. [7] proposed location-aided cluster-based energy-efficient routing (LACBER) which provides routing for GPS scare areas where few nodes are GPS enabled and smart antennas by measuring RSSI of signal and angle of arrival. Srivastava and Sudarshan [8] proposed zone-based energy-efficient routing protocol (ZEEP) for mobile sensor network using zone which uses concept of dynamic forwarding to save nodes from excessive computation. Anbalagan and Julian [9] proposed routing method called Handover Count-Based Location-Aided Routing for Cluster Networks (HCBLR). In HCBLR, the expected region is calculated based on the handover count value of the mobile node, which forwards the flooding messages only to the cluster heads that are in the expected region. Faizul Islam et al. [10] proposed Localization algorithm based on a force directed method that allows estimating the approximate location of each node based on network topology information from a local OLSR database with enhancement from known landmarks as reference points. A new approach proposed by Mukherjee et al. [11] which is based on concept of "encounter" in which a trust-based routing protocol named AER-AODV protocol which evaluates direct trust with average encounter rate (AER) and successful cooperation frequency. Indirect trust is evaluated using the revised D-S evidence theory. Asenov and Hnatyshin [12] proposed GPS-enhanced AODV routing called GeoAODV limits the route discovery process to the search region that is likely to contain the path to destination.

3 GPS-Aided Routing Method

In conventional AODV, source node uses global flooding to establish routes to destination; these routes may not be shortest. While in proposed method, routing is position-based where source node selects shortest path to destination using GPS information of destination node as discussed in next section. Here, RREQ and RREP packets are modified to include GPS location of source and destination nodes. RREQ packet also made include hope counts; if hope counts exceeds prescribed limit, RREQ expires and dropped. Nodes are equipped with memory to store GPS information. Node acquires GPS locations of all nodes with unique ID while network initialization phase using hello packets and stores it temporarily until next update.

3.1 Route Discovery Process

Whenever node (called source node) has data to send it looks for GPS location of destination into its routing table. If GPS location of destination is available, then its follow procedure depicted using flowchart in Fig. 3. Source node transmits RREQ to its neighbour node having GPS location most nearer to destination node and satisfies mobility condition (discussed in next section). If neighbour node is not destination node, then it transmits RREQ to next neighbour having GPS location most nearer to destination node. Destination nodes reply with RREP on transverse path upon receiving RREQ after that exchange of data may takes place. This procedure continues until destination node found or hope counts expires. If this method fails or GPS location of destination is not available at source node, then it follows conventional method with mobility information to establish route between source and destination node. Example scenario in Fig. 1 shows how path is set-up between source node and destination node. In this example, values of *x* and *y* axis are assumed equivalent to GPS longitudes and latitudes while height kept zero (ground level). *N7* is assumed to be source node and *N1* is destination node. It is also assumed that *N7* has GPS information of *N1*. Now *N7* initiated route discovery to *N1*. The steps are,

1. Source node *N7(3, 7)* compares destination node N1(8, 2) with neighbour nodes *N3(5, 8), N5(2, 5)* and *N4(5, 5).*
2. *N4* is winner node in step—1 *so N7* sends RREQ to *N4.*

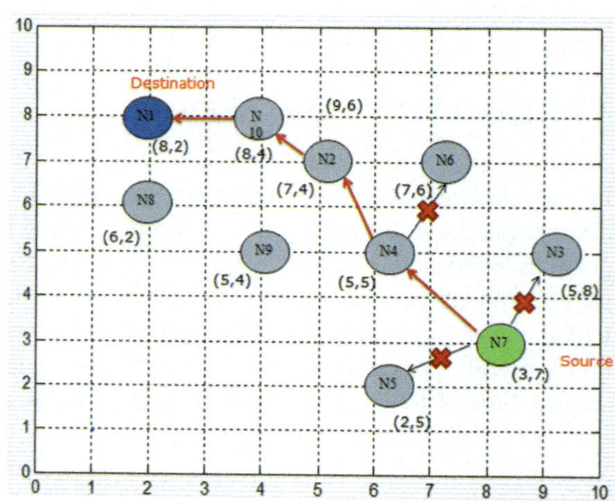

Fig. 1 Path set-up process

Fig. 2 Route maintenance process

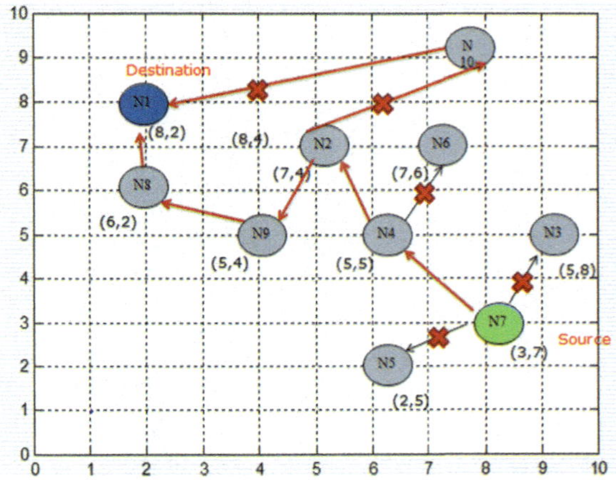

3. *N4*, *N2* and *N10* repeat same procedure to forward RREQ.
4. Destination node *N1* replies with RREP intended to destination node *N7* on reverse path traversed by RREQ.

3.2 Relative Speed

Relative speed (RS) and direction of neighbour node are taken into account while forwarding RREQ packet. If neighbour node which is having most nearer location to destination but relative speed large and direction is opposite than forwarding RREQ is avoided to that node. If nodes are moving in opposite direction than route formed between them is highly prone to be broken. Whenever source node (intermediate nodes too) forwards RREQ packet to neighbour it just ensures that both are moving in similar direction. Relative speed of two nodes (node *N1* and *N2* in Fig. 1) is calculated from GPS speed information using following formula.

$$RS(t) = |V_{N1}(t) - V_{N2}(t)| \tag{1}$$

where $V_{N1}(t)$ = speed of node *N1* and $V_{N1}(t)$ = speed of node *N2*

3.3 Route Maintenance

Route maintenance is done at node level. If node in a route finds route breaks, then it uses GPS location of next possible hope using its path cache and routing table. Flowchart of route maintenance process is shown in Fig. 4. Figure 2 shows example scenario how broken path is repaired at node level. In this, *N10* move

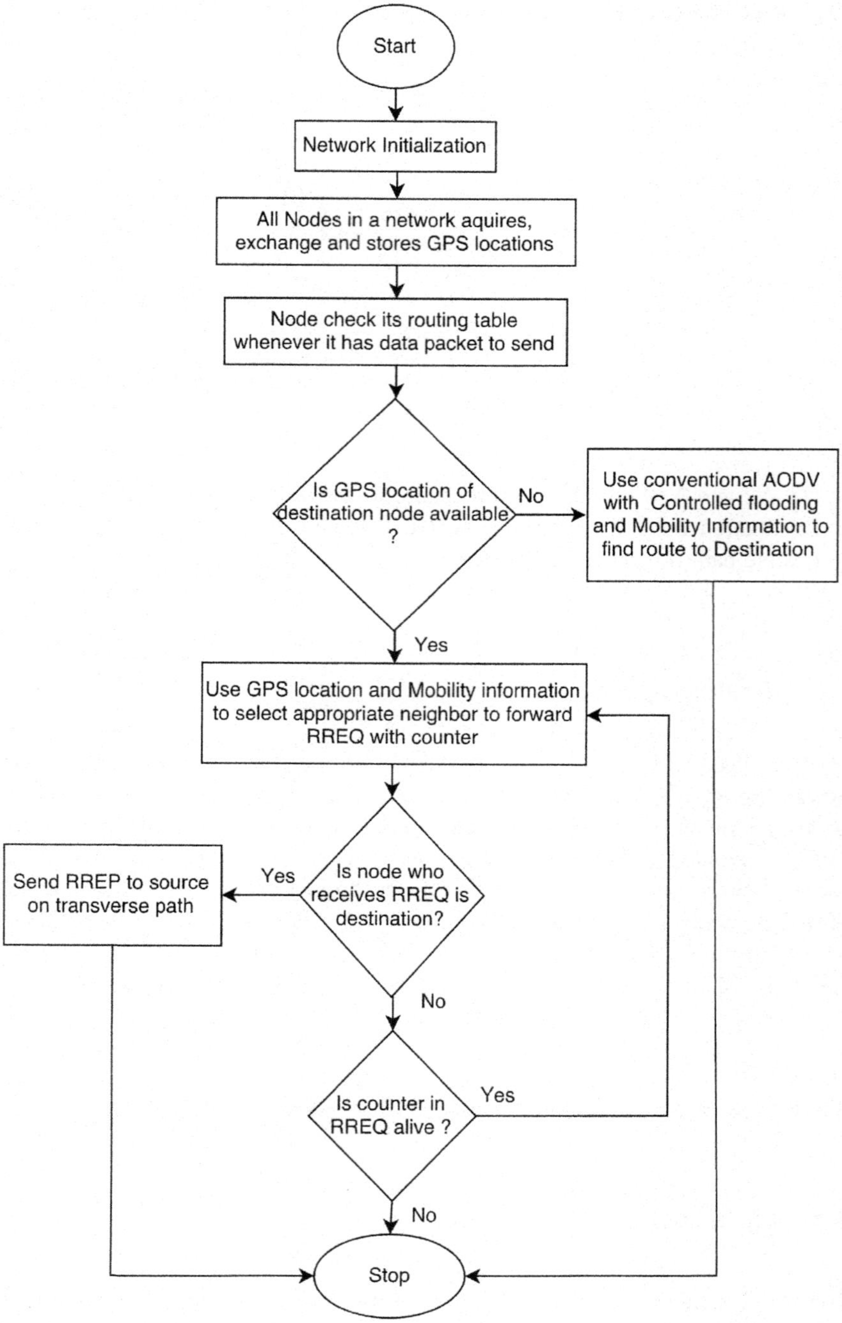

Fig. 3 Process flow—route discovery

away by certain GPS locations. Therefore, link between *N2* and *N1* breaks down. As intermediate node *N2* finds alternate route to *N10* through *N9* and *N8*. Most of times whenever destination or intermediate node moves by predetermined units of

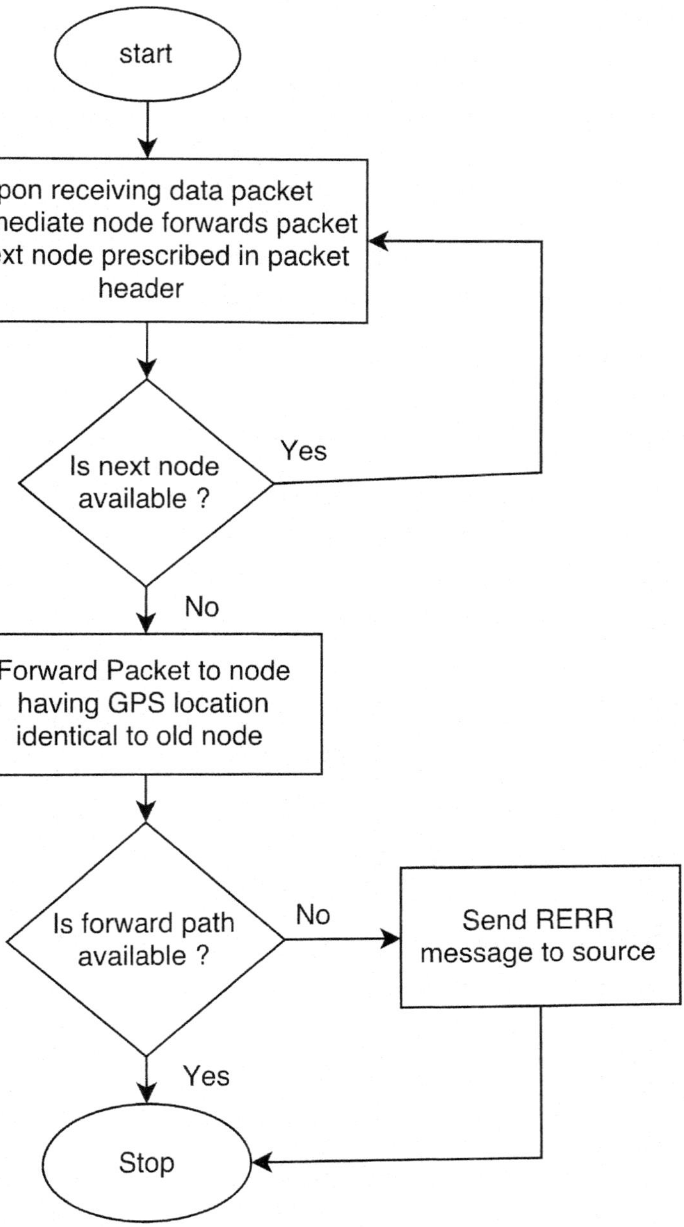

Fig. 4 Process flow—route maintenance

GPS location it gets modified to inform this to all nodes. If intermediate node fails, it sends route error message to source node (Fig. 3).

4 AODV Versus GPS-Aided AODV

AODV routing table modified to include GPS information and path caches. RREQ and RREP packets are modified to include GPS information as depicted in Fig. 5 and Fig. 6, respectively. Source node finds route to destination using GPS information available in routing table and mobility information.

5 Simulation and Performance Analysis

5.1 Simulation Model

Performance of proposed scheme is evaluated by means of simulation using NS-3.25. *Geographic positions class* of NS-3 is used to obtain GPS information of nodes in a network. Here, we have simulated conventional AODV protocol and GPS-aided AODV for different scenarios and parameters as given in Table 1. Sink is related to NS-3tracing system where trace sources are generators of events and trace sinks are consumers. Trace file and flow-monitor output of NS3 are analysed to calculate and measure various performance metrics.

<table>
<tr><td>0</td><td>1</td><td>2</td><td>3</td><td>4</td><td>5</td><td>6</td><td>7</td><td>0</td><td>1</td><td>2</td><td>3</td><td>4</td><td>5</td><td>6</td><td>7</td><td>0</td><td>1</td><td>2</td><td>3</td><td>4</td><td>5</td><td>6</td><td>7</td><td>0</td><td>1</td><td>2</td><td>3</td><td>4</td><td>5</td><td>6</td><td>7</td></tr>
<tr><td colspan="8">Type</td><td colspan="5">Flags</td><td colspan="11">Reserved</td><td colspan="8">Hope Count</td></tr>
<tr><td colspan="32">RREQ ID</td></tr>
<tr><td colspan="32">Destination IP Address</td></tr>
<tr><td colspan="32">Destination GPS Location</td></tr>
<tr><td colspan="32">Destination Sequence Number</td></tr>
<tr><td colspan="32">Source IP Address</td></tr>
<tr><td colspan="32">Source GPS Location</td></tr>
<tr><td colspan="32">Source Sequence Number</td></tr>
</table>

Fig. 5 RREQ packet format

<table>
<tr><td>0</td><td>1</td><td>2</td><td>3</td><td>4</td><td>5</td><td>6</td><td>7</td><td>0</td><td>1</td><td>2</td><td>3</td><td>4</td><td>5</td><td>6</td><td>7</td><td>0</td><td>1</td><td>2</td><td>3</td><td>4</td><td>5</td><td>6</td><td>7</td><td>0</td><td>1</td><td>2</td><td>3</td><td>4</td><td>5</td><td>6</td><td>7</td></tr>
<tr><td colspan="8">Type</td><td>R</td><td>A</td><td colspan="9">Reserved</td><td colspan="5">Prefix Sz</td><td colspan="8">Hope Count</td></tr>
<tr><td colspan="32">Destination IP Address</td></tr>
<tr><td colspan="32">Destination GPS Location</td></tr>
<tr><td colspan="32">Destination Sequence Number</td></tr>
<tr><td colspan="32">Source IP Address</td></tr>
<tr><td colspan="32">Source GPS Location</td></tr>
<tr><td colspan="32">Lifetime</td></tr>
</table>

Fig. 6 RREP packet format

Table 1 Simulation parameters

Parameter	Value/Type
Number of nodes (and sink nodes)	10(5), 20(15), 30(15), 40(20), 50(25), 100(50)
Total simulation time	100 s
Data packet length	64 bytes
IPV4 packet length	92 bytes
Data rate	2048 bps
Network area	300 × 1500 m
Propagation delay model	Constant speed
Propagation loss model	Friis prop. loss model
Mobility model	Random waypoint
Physical standard	IEEE 802.11b
Transmit power	8 dBm
Node speed	100 ms
Pause time	0 ms

5.2 *Performance Metrics*

To measure suitability of proposed technique, different packet level metrics are used. Performance metrics used are Throughput, Packet Delivery Ratio, End-to-End Delay, Normalized Routing Load, Packet Loss and Jitter which is the time variation in arrival of consecutive packets at receiver. Jitter is the result of variable path lengths, traffic route availability, etc. Flow-monitor output of NS3 calculates "*jitterSum*" automatically. Average jitter calculations are based on analysis of flow-monitor output file.

Throughput (TP) is average successful transmission rate in kbps [13]. TP is ratio of total bytes received (*tbr*) to simulation time difference between last packet received (*lpr*) and first packet transmitted (*fpt*).

$$TP(\text{kbps}) = tbr/(lpr - fpt)*(8/1000) \tag{2}$$

Packet Delivery Ratio (PDR) gives number of transmission attempts per packet received [13]. PDR is the ratio between the received packets by the destination and the generated packets by the source.

$$PDR = pt/pr \tag{3}$$

End-to-End Delay (EED) is average time taken by packets to reach destination in seconds [13]. EED is simulation time difference between transmission (*tt*) and reception (*rt*) of particular packet.

$$EED(\mathrm{s}) = tt - rt \tag{4}$$

Normalized Routing Load (NRL) is defined as the number of routing (control) packets transmitted per data packet sent to the destination [13]. NRL is the ratio of routing packets (*rp*) and data packets transmitted.

$$NRL = rp/pt \tag{5}$$

Packet Loss (PL) is measure of number of data packets actually received at receiver [13]. PL is difference of total data packets transmitted (*pt*) and total packets received (*pr*). Data packet may be lost due to link failure, buffer overflow, looping, retransmissions, buffer overflow, etc.

$$PL = pt - pr \tag{6}$$

5.3 Result Analysis

Set of simulations are performed for increasing number of nodes with fixed transmission power and constant speed of nodes. Numbers of nodes varied are 10, 20, 30, 40, 50 and 100 with number of sink nodes as 5, 10, 15, 20, 25 and 50, respectively. The simulation time is 100 s for each set of parameters.

From Fig. 7, it is observed that throughput is low for nodes less than 40. This is because with area of 300 × 1500 m, nodes are spread across transmission and discovery range of each other. For nodes greater than 40–100, throughput is stable in the range of 165 kbps. Here, GPS-aided AODV performs better due to fact that the path formed is shortest with mobility considerations. In case of GPS-aided AODV, routes are less prone to break due to mobility considerations.

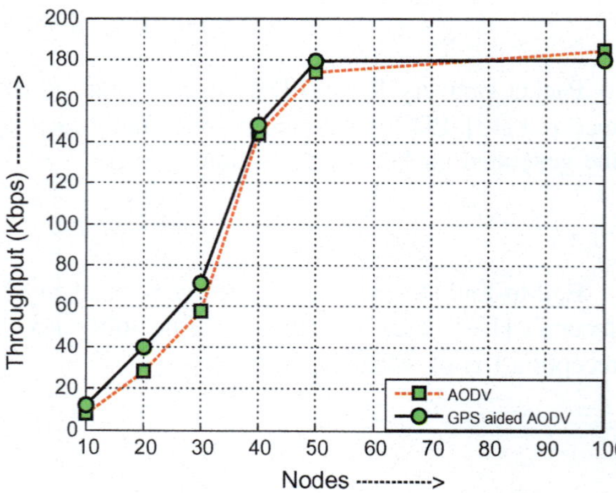

Fig. 7 Increasing nodes V/S throughput

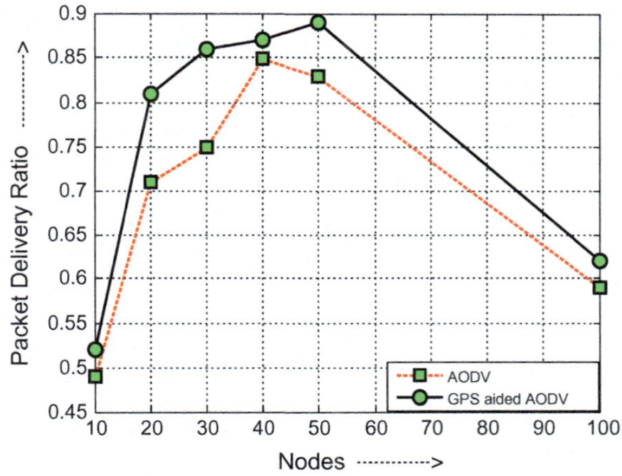

Fig. 8 Increasing nodes V/S PDR

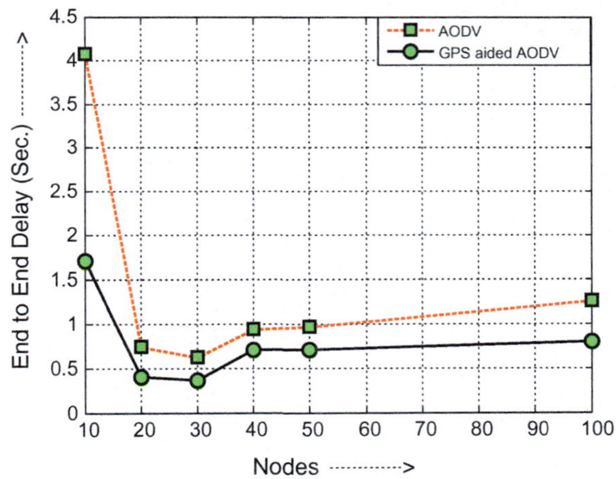

Fig. 9 Increasing nodes V/S end-to-end delay

From Fig. 8, it is observed that PDR is less for nodes up to 20 and then increases around 0.8. This shows that initially less number of nodes per unit area are available to form route. This also results in weak routes formations which result in packet loss. Here, GPS-aided AODV performs better compared to AODV because nodes forming routes are less relative speed. Route maintenance mechanism is at node level (where route breaks) so route maintenance is fast, and less number of packets are dropped during route breaks.

From Fig. 9, we can see that end-to-end delay in the case of AODV is less compared to AODV because end-to-end routes formed are shortest.

From Fig. 10, we see that average time variation in packet arrival at receiver for GPS-aided AODV is marginally better because of better route selection mechanism. In GPS-aided AODV long live strong paths are provided for end-to-end connection

Fig. 10 Increasing nodes V/S jitter

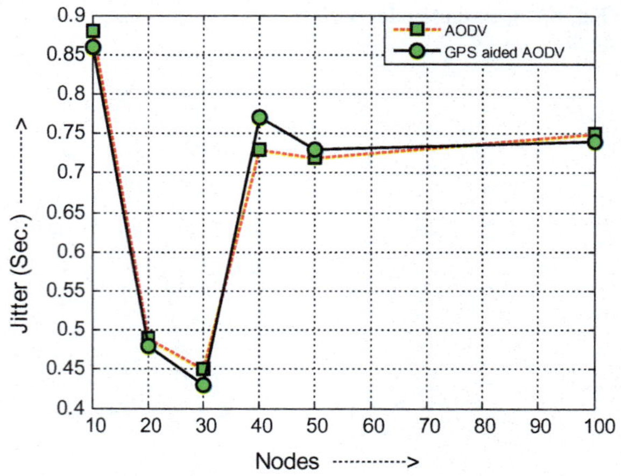

Fig. 11 Increasing nodes V/S NRL

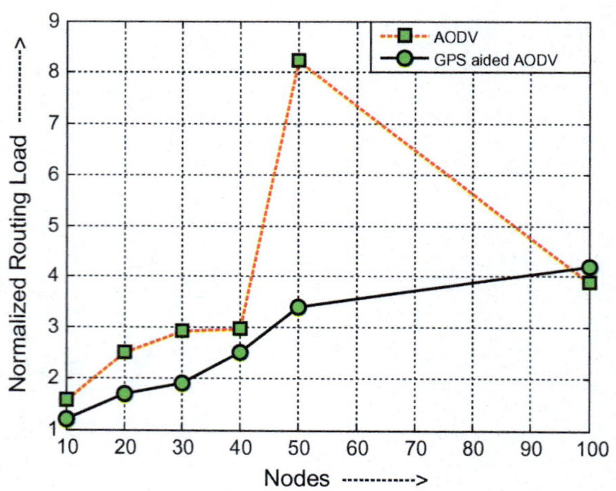

which results in smooth flow of packets from source to destination, and therefore the packets arrived are at less time variations.

From Fig. 11, it is observed that GPS-aided AODV outperforms conventional AODV because control packets generated are far less in number. Since conventional AODV uses global flooding during route discovery process, whereas GPS-aided AODV uses control flooding. It is also because route maintenance is at node level and rate of route breaks is less due to strong route selection and mobility considerations.

Figure 12 shows packet lost with increasing number of nodes where GPS-aided AODV performs better because of strong and short routes available. Number of packet lost for both conventional and GPS-aided AODV is large beyond 50 nodes because of increase in the traffic and therefore congestion.

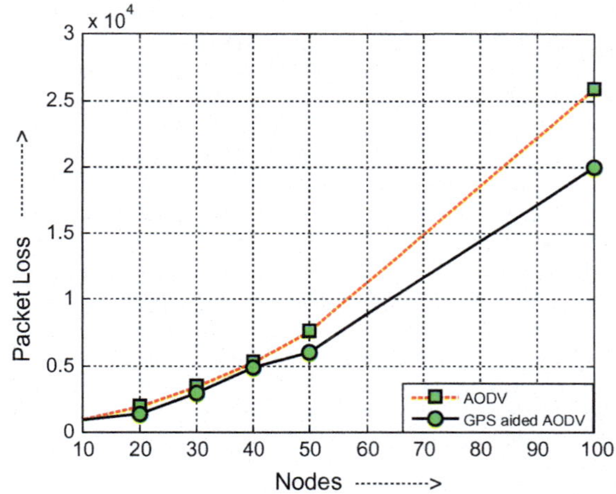

Fig. 12 Increasing nodes V/S packet loss

6 Conclusion and Future Work

These results show that performance of GPS-aided routing is better than conventional AODV. This is because in our method, controlled flooding is used during route discovery process. Selected routes are shortest and strong due to mobility considerations. Route maintenance process is handled at node level. The point of advantage in GPS-aided AODV is number of control packets generated are far less compared to conventional AODV. GPS-aided technique can be extended to more sophisticated and efficient routing protocol in future. More mobility parameters are to be used to enhance the performance in future.

References

1. Bilandi N, Verma HK, Kumar N (2013) Comparative analysis of ad hoc routing protocols based on users point of view. In: 3rd IEEE international conference IACC
2. Parmar B, Maradia KG (2015) Comparative analysis of routing protocols for MANET using packet level diagnostic metrics. IJETEE 11(4). ISSN: 2320-9569
3. Perkins CE, Royer EM (1999) Ad hoc on demand distance vector routing. In: Proceedings of IEEE workshop on mobile computing systems and applications 1999, pp 90–100, February 1999
4. Parkinson B et al (1996) Global positioning system: theory and application. In: Progress in astronautics and aeronautics, vol 163
5. Ko Y-B, Vaidya NH (1998) Location-aided routing (LAR) in mobile ad hoc networks. In: Proceedings of the 4th annual ACM/IEEE international conference on mobile computing and networking, pp 66–75, 25–30 October 1998
6. Thongthavorn T, Narongkhachavana W, Prabhavat S (2014) A study on overhead reduction for GPS-assisted mobile ad-hoc networks. In: TENCON 2014—2014 IEEE region 10 conference

7. Deb D et al (2009) LACBER: a new location aided routing protocol for GPS scare MANET. Int J Wirel Mobile Netw (IJWMN) 1
8. Srivastava JR, Sudarshan TB (2013) ZEEP: zone based energy efficient routing protocol for mobile sensor networks. In: ICACCI conference
9. Anbalagan R, Julian A (2014) Handover count based location aided routing for cluster networks (HCBLR). In: 2014 IEEE international conference on advanced communication control and computing technologies (ICACCCT)
10. Faizul Islam ZM et al (2012) OLSR-based coarse localization in tactical MANET situational awareness systems. In: 2012 IEEE international conference on communications (ICC)
11. Mukherjee S et al (2015) A novel encounter based trust evaluation for AODV routing in MANET. In: Applications and innovations in mobile computing (AIMoC), pp 141–145. https://doi.org/10.1109/AIMOC.2015.7083843
12. Asenov H, Hnatyshin V (2009) GPS-enhanced AODV routing. In: International conference on wireless networks (ICWN), Las Vegas USA, July'09
13. Subbarao MW (1999) Ad hoc networking critical features and performance metrics. In: Wireless communications technology group, NIST, 7 October 1999

Dual-Band Rectangular Microstrip Patch Antenna Design for RF Energy Harvesting

Jyoti Jain and Abha Sharma

Abstract A rectangular probe feed microstrip patch antenna is simulated and presented. The designed antenna operates at the frequency band 2.45 GHz (Wi-Fi) and 4.9 GHz (WLAN). This paper presents a rectangular patch of 28.28 mm × 28.28 mm which is placed on the substrate of 50 mm × 50 mm. The position of the probe is optimized to achieve dual-band characteristics. This paper also presents a method to design an energy harvesting device depending on the type of energy available (Wi-Fi and WLAN) for low consumption electrical devices.

Keywords Rectangular microstrip patch antenna · Wi-Fi · WLAN
RF energy harvesting · Dual band

1 Introduction

The probe feed and microstrip line feed rectangular and circular microstrip patch antennas are well explored in [1]. The effect of the probe in polygonal patch antenna on antenna performance is explained in [2]; it also suggests that the position of the probe plays important role in tuning the resonant frequency and matching impedance. Further, this technique can be used to explore rectangular patch antenna and finding the optimized feed point.

Probe feed pentagonal and hexagonal shapes with polygonal slot are used to improve the gain of antenna [3]; it also presents the surface current for pentagonal and hexagonal patch antenna for 18, 19, and 20 GHz. The perturbation of the edges of polygon affects the scattering parameter as discussed in [4]. Performance analysis of various polygonal shaped patch antennas such as circle, triangle, rectangle,

J. Jain (✉) · A. Sharma
Department of Electronics & Communication Engineering, Rajasthan College of Engineering for Women, Rajasthan, India
e-mail: er.jyotijain04@gmail.com

A. Sharma
e-mail: sharmaabha0307@gmail.com

V. Janyani et al. (eds.), *Optical and Wireless Technologies*, Lecture Notes in Electrical Engineering 472, https://doi.org/10.1007/978-981-10-7395-3_66

pentagon, and hexagon is presented [5]. These techniques to analyze the polygon can be implemented to analyze rectangular patch antenna.

The microstrip patch antenna design can be used for RF energy harvester for low power consumption electrical and electronic devices were introduced by [6]. A circular microstrip antenna array for ambient RF energy harvesting has been proposed for making dual band and extending the frequency bandwidth from 1.85 to 1.93 GHz and from 2.0 to 2.1 GHz which can cover GSM-1800 and UMTS-2100 bands [7]. A three-band GSM850 (850 MHz), PCS1900 (1.9 GHz), and Wi-Fi (2.4 GHz) RF-DC converting system to harvest wireless energy is presented [8]. The design and measurement results of an RF energy harvester aimed to power sensor nodes like temperature, humidity, radiation in an indoor environment is presented that operates at the 2.42 GHz Wi-Fi frequency band [9]. A dual-band monopole antenna for GSM band applications is presented, which is used for energy harvesting system at resonance frequencies of 915 MHz and 1800 MHz with a bandwidth of 124.2 MHz and 196.9 MHz, respectively [10]. The hexagonal patch antenna can achieve a bandwidth of 400 MHz in C-band centered at 5.5 GHz (WLAN) is presented [11].

This paper presents a rectangular microstrip patch antenna with dual-band characteristics for the RF harvesting application. The rectangular patch antenna is simulated using CST software, and simulated results are studied and analyzed for different positions of the feed. The feed point is optimized to achieve dual band with 50Ω impedance. Finally, the observations are concluded in the last section of the paper.

2 Antenna Design

The schematic of rectangular patch antenna is designed in CST software as shown in Fig. 1. The proposed antenna structure uses a FR-4 Substrate of relative permittivity of 4.3. The substrate dimensions are 50 mm × 50 mm × 1.6 mm. A proposed rectangular patch has a dimension of 28.28 mm × 28.28 mm. The ground dimension with dimension 40 mm × 40 mm is used for simulation. A probe feed of radius 0.62 mm is used to excite the rectangular patch antenna. The probe position is chosen by varying the position of the probe along x-axis such that it radiates at the desired frequencies.

3 Simulation, Parametric Study, and Results

CST software is used to design and simulate the rectangular patch antenna. The effect of probe position (Feedx) on reflection coefficient (S_{11}), impedance (Z_{11}), surface current, and far-field directivity (in dB) is studied and analyzed. 'Feedx' is the distance of feed point from the origin along x-axis.

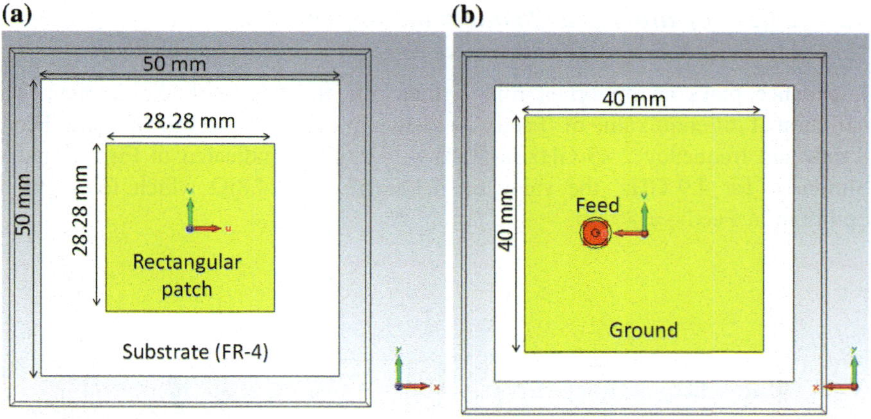

Fig. 1 Schematic of the proposed rectangular shaped patch antenna with square ground **a** top view **b** bottom view

3.1 *Effect of the Feed Position on Reflection Coefficient*

The rectangular patch antenna is sensitive to feed position as shown in Fig. 2. The effect of feed over reflection coefficient S_{11} can be observed by changing the value of 'Feedx' from 0 to 15 mm along the length of the rectangle. It is observed that the minimum value of the reflection coefficient for 2.45 GHz is at 8 mm with a value −24.79 dB as shown in Fig. 2. The band at 2.45 GHz below −10 dB ranges from 2.38 to 2.52 GHz. The value of S_{11} for 4.9 GHz is −22.41 dB and range below −10 dB is 4.79 to 5 GHz.

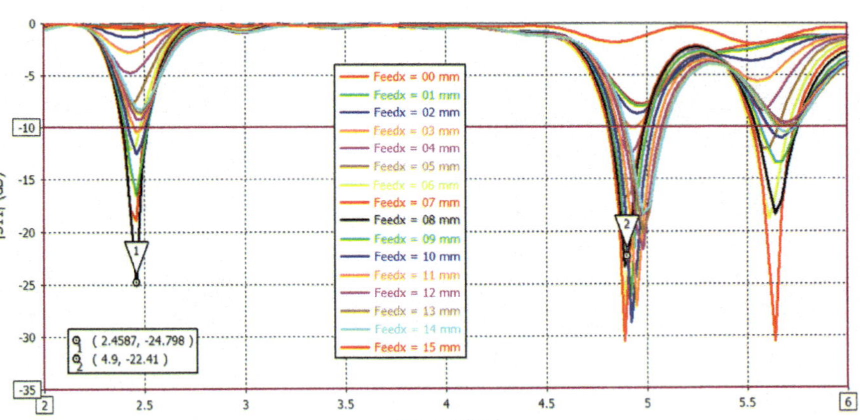

Fig. 2 Simulated reflection coefficient (in dB) for rectangular patch antenna with different values of feed positions (Feedx)

3.2 Effect of the Feed Position on Impedance

Impedance plays an important role in matching the feed with the antenna. The variation at different value of 'Feedx' is shown in Fig. 3. The value of Z_{11} at Feedx 8 mm and frequency 2.45 GHz is 50.58 − 4.21jΩ as indicated in Figs. 3 and 4. Similarly, for 4.9 GHz, the value of Z_{11} is 53.67 − 4.8jΩ which found to be optimum at Feedx = 8 mm.

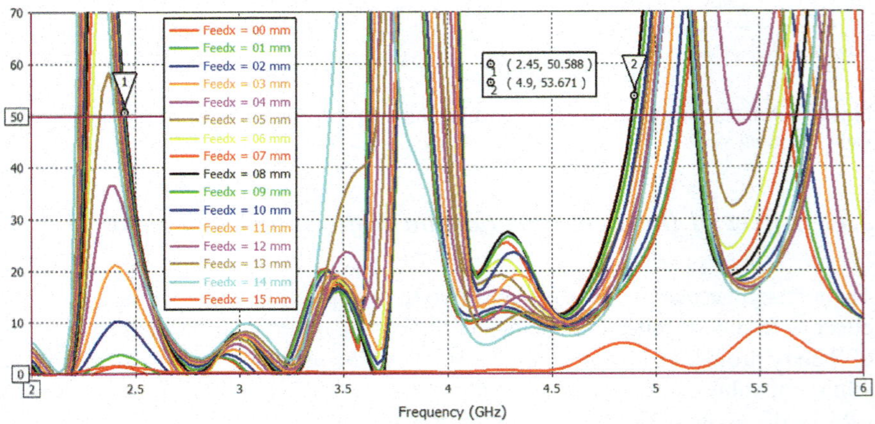

Fig. 3 Simulated impedance real part (in Ω) versus frequency matrix for rectangular patch antenna with different values of feed positions (Feedx)

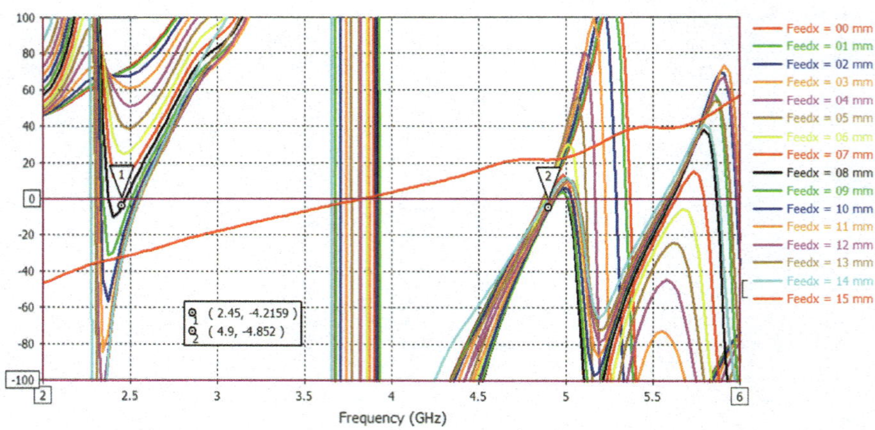

Fig. 4 Simulated impedance imaginary part (in Ω) versus frequency matrix for rectangular patch antenna with different values of feed positions (Feedx)

3.3 *Effect of the Feed on Surface Current*

The surface current in the rectangular patch antenna at frequencies 2.45 and 4.9 GHz is generated for the optimum value of Feedx = 8 mm as shown in Fig. 5a, b.

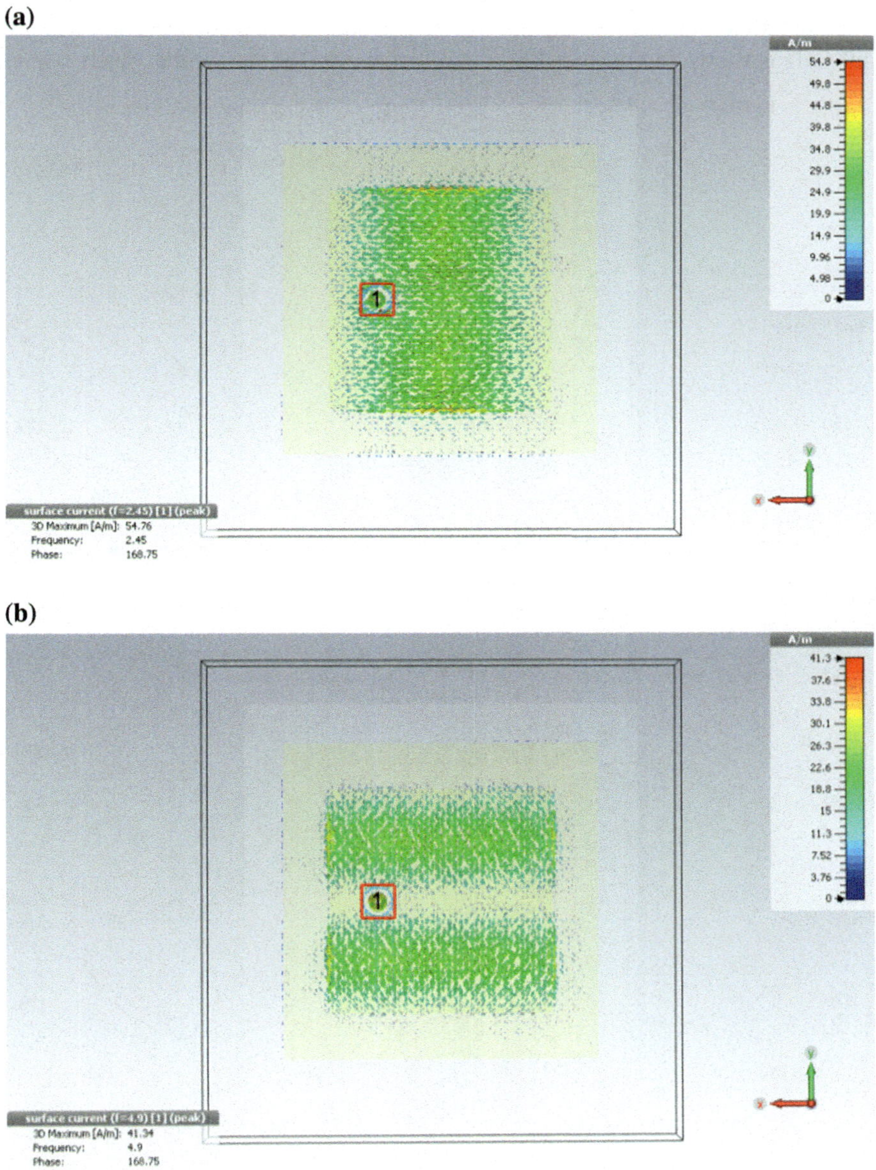

Fig. 5 Simulated surface current distribution of rectangular patch antenna for feedx = 8 mm for the frequency **a** 2.45 GHz **b** 4.9 GHz

The surface current suggests that the antenna is resonating at TM_{10} mode at 2.45 GHz and TM_{02} at 4.9 GHz.

3.4 *Effect of the Feed on Far-field Directivity*

The directivity of the rectangular patch antenna at frequencies 2.45 GHz and 4.9 GHz is shown in Fig. 6 and Fig. 7, respectively. The rectangular patch antenna

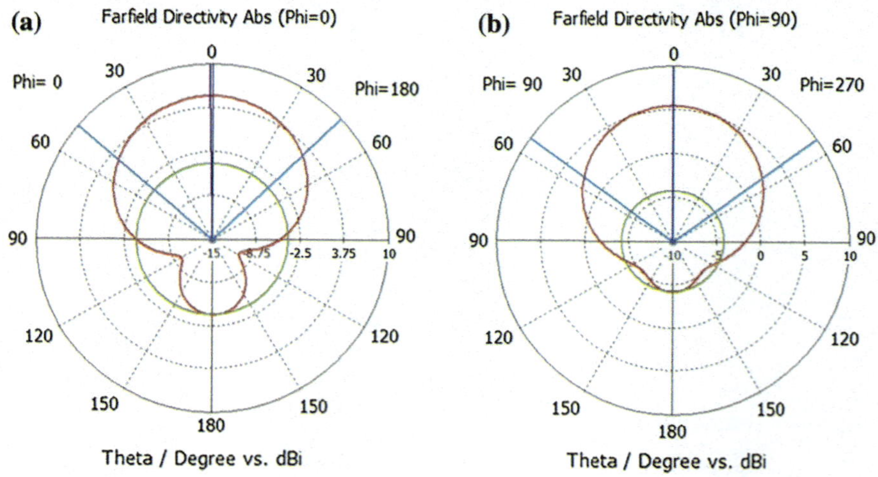

Fig. 6 Simulated far-field directivity at frequencies 2.45 GHz for **a** E-plane (phi = 0°) **b** H-plane (phi = 90°)

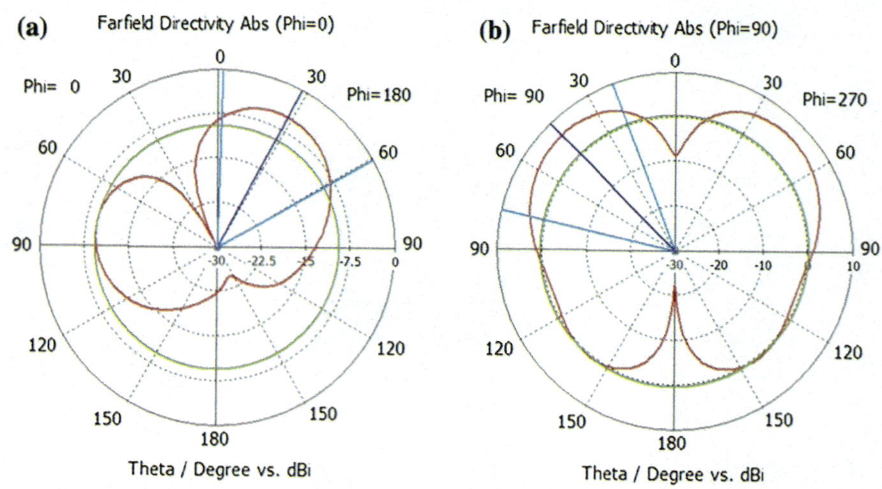

Fig. 7 Simulated far-field directivity at frequencies 4.9 GHz for **a** E-plane (phi = 0°) **b** H-plane (phi = 90°)

is highly directive at 2.45 GHz with value 5.44 dBi at 0° for both the E- and H-plane, while at 4.9 GHz, it is observed to the main lobe shifted to 60° with 6.4 dBi value in H-plane.

4 Conclusion

A rectangular microstrip patch antenna with dual-band characteristics has been obtained and the results are simulated. The results suggest that the proposed antenna structure is suitable for RF energy harvesting applications. The rectangular patch antenna resulted in a bandwidth of 140 MHz at 2.45 GHz (Wi-Fi) and 200 MHz at 4.9 GHz (WLAN) with the substrate thickness of 1.6 mm. This suggests that the antenna can be used for RF energy harvesting.

References

1. Balanis CA (2008) Antenna theory, analysis and design. Wiley, New York. ISBN: 0-471-66782-X
2. Joshi A, Pranathi P, Singhal R (2015) Probe feed effect in polygonal patch antennas. In: 2015 IEEE 4th Asia-Pacific conference on antennas and propagation (APCAP), Kuta, pp 126–127
3. Joshi A, Singhal R (2015) Gain improvement in polygonal patch antennas. In: 2015 annual IEEE India conference (INDICON), New Delhi, pp 1–5
4. Joshi A, Singhal R (2015) Edge perturbation effect in polygonal patch antennas. In: 2015 international conference on signal processing, computing and control (ISPCC), Waknaghat, pp 97–101 (2015)
5. Joshi A, Singhal R (2016) Performance comparison of probe-fed polygonal patch antennas for L-band applications. In: 2016 IEEE region 10 conference (TENCON), Singapore, pp 1307–1311
6. Ramesh GP, Rajan A (2014) Microstrip antenna designs for RF energy harvesting. In: 2014 international conference on communication and signal processing, Melmaruvathur, pp 1653–1657
7. Xu LJ, Huang B, Bai X, Mao HP (2016) A dualband and broadband antenna array for ambient RF energy harvesting. In: 2016 IEEE international conference on ubiquitous wireless broadband (ICUWB), Nanjing, pp 1–3
8. Zhou Y, Huerta C, Hinojosa J (2016) Three-band ambient wireless energy harvesting system. In: 2016 IEEE international symposium on antennas and propagation (APSURSI), Fajardo, pp 613–614
9. Alneyadi F, Alkaabi M, Alketbi S, Hajraf S, Ramzan R (2014) 2.4 GHz WLAN RF energy harvester for passive indoor sensor nodes. In: 2014 IEEE international conference on semiconductor electronics (ICSE2014), Kuala Lumpur, pp 471–474
10. Zakaria Z, Zainuddin NA, Abd Aziz MZA, Husain MN, Mutalib MA (2013) Dual-band monopole antenna for energy harvesting system. In: 2013 IEEE symposium on wireless technology & applications (ISWTA), Kuching, pp 225–229
11. Joshi A, Singhal R (2016) Probe-fed regular hexagonal narrow-slot antenna with reduced ground plane for WLAN applications. In: 2016 IEEE region 10 conference (TENCON), Singapore, pp 1312–1316

A Compact Dual-Band Frequency Selective Surface for Gain Enhancement of a Dual-Band Antenna

R. Adeline Mellita, D. S. Chandu and S. S. Karthikeyan

Abstract A compact, single-layered dual-band frequency selective surface (FSS) is proposed in this paper. It is used as a reflector to enhance the gain of a dual-band printed monopole antenna. The antenna operates at 2.4 and 5 GHz WLAN bands. The FSS has two concentric rectangular open loop structures which resonate, respectively, at the designed frequencies. It is angularly stable and polarization independent. By using the FSS as a reflector, the gain of the antenna has been improved by 50%. The proposed FSS and printed antenna have been simulated using full-wave simulator and their prototypes have been fabricated and measured. Good concordance between the simulated and measured results has been obtained.

Keywords Angular stability · Frequency selective surface (FSS)
Incident wave · Printed monopole antenna

1 Introduction

With an increase in the number of people using communication services, wireless communication technology has gained prominence as it supports more users, increases portability, and reduces cost. The advancement of wireless technology brought the need for miniaturization which resulted in designing of multiband and wideband antennas for communication. As cost always remains the prime controlling factor in any industry, the printed antennas with their inexpensive manufacturing and ease of design have attained importance. But, it has certain

R. Adeline Mellita (✉) · D. S. Chandu · S. S. Karthikeyan
Department of Electronics Engineering, Indian Institute of Information Technology, Design and Manufacturing, Kancheepuram, Chennai 600127, India
e-mail: cds15m005@iiitdm.ac.in

D. S. Chandu
e-mail: edm14d002@iiitdm.ac.in

S. S. Karthikeyan
e-mail: ssk@iiitdm.ac.in

V. Janyani et al. (eds.), *Optical and Wireless Technologies*, Lecture Notes in Electrical Engineering 472, https://doi.org/10.1007/978-981-10-7395-3_67

disadvantages like low gain, low impedance bandwidth and unwanted radiations from its feed and junctions and solutions to these problems are still a challenge to the researchers. One way to enhance the gain while retaining the antenna properties is to use a frequency selective surface (FSS) as a reflector. In this paper, a dual-band FSS is proposed and its effect on the radiation gain of a dual-band antenna is studied for various configurations.

FSS [1] is one of the primitive technologies which was kept hidden for many decades as its primary use was only in the military for manufacturing of RADARS, spy-crafts, etc. The applications of FSS in the fields of microwave, antenna, and satellite communication have surfaced the technology to the industrial world. Few of its general applications are as spatial filters [2], RCS reducers in radomes [3], beam splitters [3], absorbers [4], dichroic sub-reflectors [5], or common antenna reflectors, etc. In [6], it was shown that an FSS can be designed using slots or patches in the form of strips, apertures, loops or, selective combinations which can act as filtering elements. The FSS was simple and easy to manufacture but its operation was limited to single band. The size, shape, geometry, and periodicity of the unit cell determine the filtering characteristics and dual-band FSS can be realized by proper design of these geometrical parameters. As long-distance communication seeks for high-gain antennas, the use of FSSs as gain enhancers is being researched extensively. In [7, 8] a dual-band single-layered FSS is proposed. It is embedded within the antenna and the gain was enhanced in the two bands by 46% and 30%, respectively. In [9, 10], the FSS is placed at an optimized distance from the antenna and a gain improvement of 54 and 35% is obtained. However, the FSSs in [7–8] have higher cell size and do not account for angular stability or polarization independence. A compact FSS is designed in [10] with electrically small dimensions of the unit cell. But the design does not provide significant gain enhancement. A compact and efficient design of a FSS is in the needs of the market to enhance the gain and thereby the coverage area of a dual-band antenna. This paper presents the design of an angularly stable and polarization-independent FSS resonating at the IEEE 802.11a/b WLAN frequencies. An antenna radiating at the 2.4 GHz (say, f_1) and 5 GHz (say, f_2) WLAN band is designed and the gain performance is analyzed for various configurations of the FSS and explained in the following sections.

2 FSS and Antenna Design

2.1 Proposed Unit Cell Description

The unit cell of the proposed FSS has two concentric open square loops. The perimeter and area of the loops are adjusted to tune the resonance such that the WLAN bands corresponding to 2.4 and 5 GHz fall within the band-reject region. The outer loop responds at the 2.4 GHz band and the inner loop resonates at the 5 GHz band. The loop dimensions are optimized using the inverse proportionality relationship between the physical length and frequency. The perimeter of the outer open loop is 124.9 mm, which is one wavelength long at the lower operating frequency of 2.4 GHz. The inner open loop has length of 71.6 mm which is the value of 1.2 λ, at a higher operating frequency of 5 GHz. The length of the inner loop is chosen more than one wavelength only to account for the radiation interference caused by the close proximity of the outer loop. The dimensions (in mm) of the proposed FSS structure are shown in Fig. 1a. The proposed structure is etched on an FR4 board of thickness 0.8 mm having relative permittivity (ε_r) of 4.4 and a loss tangent 0.02.

2.2 *Working Principle of the Proposed FSS*

Floquet's principle is used in the analysis of FSS, according to which every unit cell responds analogously to an incident plane wave. Thus, a unit cell is designed and analyzed to understand the response of a sheet containing many such cells. Since the number of unit cells used does not affect the overall response of the sheet, the FSS is considered to be robust. The length of the loop and the separation between the elements determines the resonant frequencies. But, in case of a closed loop, only the outer perimeter is considered as the resonant length. Thus, an open loop structure aids miniaturization and the proposed FSS has a compact structure. The two loops together provide a wide dual-band response at the frequency bands of 2.1–2.7 and 4.5–5.8 GHz. The transmission and reflection coefficient of the proposed FSS are shown in Fig. 1b shows that the compact unit cell gives angular stability to the FSS. The symmetric design imparts polarization independence to the FSS. These properties can be verified from Fig. 2a, b. It can be observed that there is a negligible difference in S_{21} with change in incident angle in both TE and TM modes. Therefore, the proposed FSS is angularly stable and polarization independent.

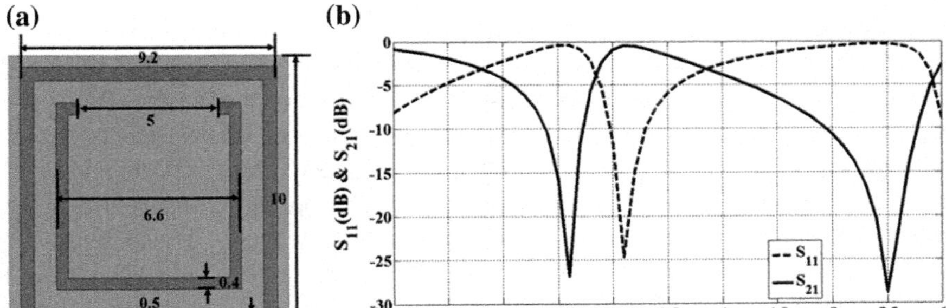

Fig. 1 **a** Unit cell structure (all dimensions are in mm) and **b** transmission and reflection characteristics of the FSS

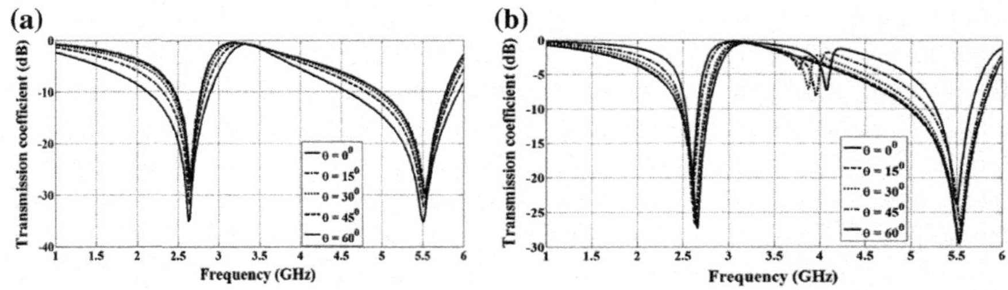

Fig. 2 **a** TM mode response and **b** TE mode response for oblique incident angles

2.3 *Dual-Band Antenna Design*

A printed antenna resonating at WLAN frequencies of 2.4 and 5 GHz is designed. It is etched on one side of a 1.6 mm thick FR4, the other side has a partial ground plane with a height of 11 mm. The proposed monopole antenna is loaded with an open loop and it is capacitively coupled to an inverted U-shaped patch with extra extended arms to make an open loop. At lower frequency band, the monopole and the loaded open loop show inductive behavior, whereas the gap between the loops gives capacitance. The electric fields are tightly coupled between the two open loops at the higher frequency band and the antenna exhibits dual-band characteristics which can be verified from the current distribution plots shown in Fig. 3. The length of the top-loaded loop (124.9 mm) together with the length of the monopole (15 mm) determines the lower cutoff frequency of the antenna (2.1 GHz) in the lower operating band. The perimeter of the inverted U-shaped loop is 1.2 λ at f_2 which corresponds to 73.4 mm. The radiating monopole and ground dimensions (in mm) are shown in Fig. 4a.

3 Results and Discussions

Initially, the dual-band antenna is simulated and its performance is analyzed. Later, the FSS is placed below the antenna and the antenna's performance is re-analyzed. The radiation characteristics of the antenna remain same with and without FSS and it can be verified from Fig. 4b. To enhance the gain, the distance between the antenna and FSS is to be chosen such that the reflected wave from the FSS interferes constructively with the antenna's radiation. The optimum distance for maximum gain enhancement at both the bands is 0.25 λ, where λ considered as the wavelength at lower resonant frequency, i.e., 2.4 GHz. Let this be called as case 1. But, it can be seen from Fig. 5b that the gain of the antenna has improved only by 28% in both the bands. An interesting observation made during this process is that the gain of the respective band increases as the dual-band FSS layer approaches closer to the 0.25 λ of the corresponding resonant frequency. To justify this

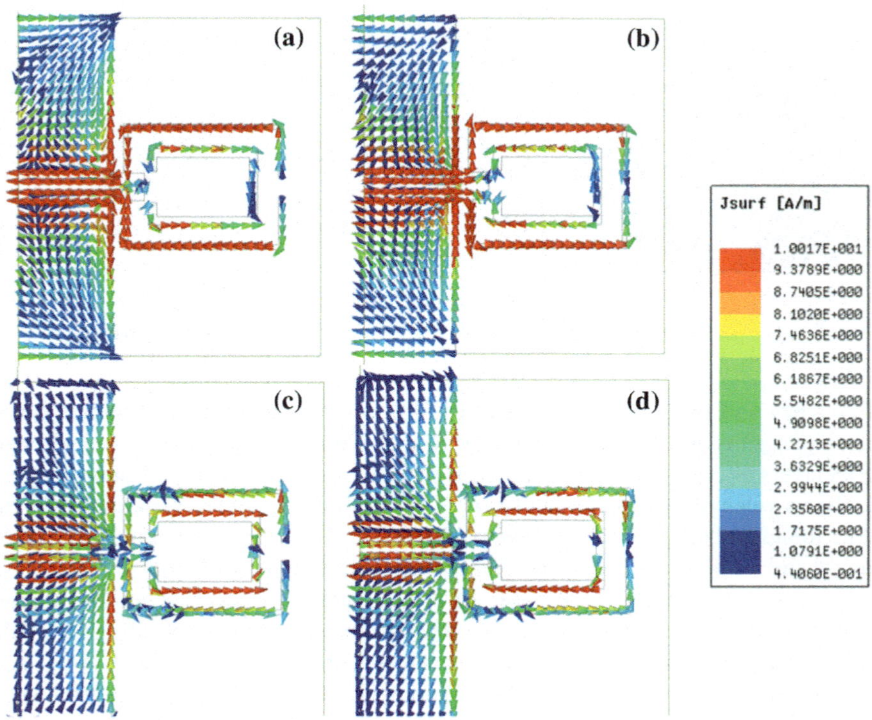

Fig. 3 Current distribution **a** at ωt = 0 at f_1, **b** at ωt = T/4 at f_1, **c** at ωt = 0 at f_2, and **d** at ωt = T/4 at f_2

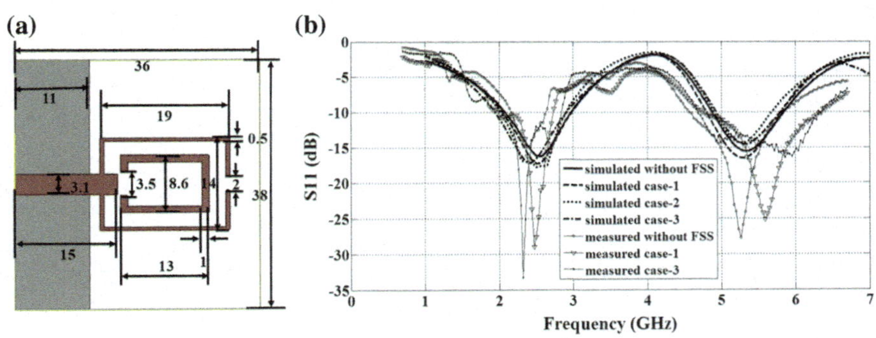

Fig. 4 **a** Antenna structure (top-view) **b** simulated and measured reflection characteristics of antenna with the FSS

argument, in case 2, the proposed dual-band FSS was modified by etching the outer loop patch on an FR4 layer and the inner loop patch on another FR4 layer with the same thickness. Each layer was placed at 0.25 λ of its corresponding resonant

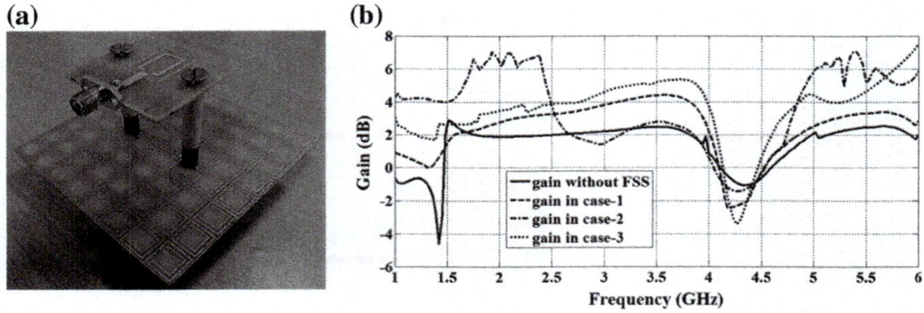

Fig. 5 **a** Perspective view of antenna with the FSS **b** gain of antenna with and without FSS

frequency and the gain improvement of 60% is observed from simulations. Since the paper aims at compact design of FSS, the solution was to stack two dual-band FSS layers as case 3. The stacked FSS is placed at 0.25 λ from the antenna, where λ is the wavelength corresponding to $(f_1 + f_2)/2$. In this case, the gain improvement was found to be 50%. This interesting behavior is due to the combination of two dual-band FSS layers and it can be explained by the frequency response of the combined dual-band FSS. It is observed that response generated is almost similar to a wideband FSS, but with negligible discrepancies in the simulated results. The fabricated prototype of the dual-band antenna with FSS is shown in Fig. 5a, and spacers are provided to adjust the separation between them. The gain performance for all the cases is shown in Fig. 5b. The simulated radiation patterns of the dual-band antenna are plotted for the *xz*-plane at 2.4 and 5 GHz for all three configurations in Fig. 6a, b, respectively. The dual-band antenna radiates more unidirectional patterns in the upper hemisphere when the FSS is installed. The proposed FSS provides significant gain improvement of the designed printed dual-band antenna and its performance is compared in Table 1 with the other existing dual-band FSS designs in the literature. λ_g is the guided wavelength at f_1.

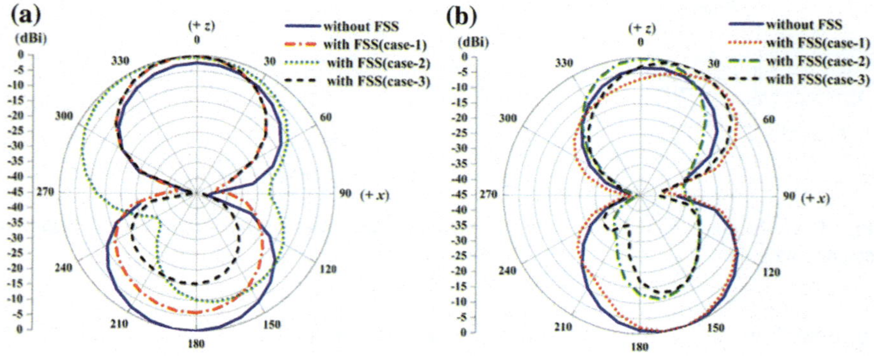

Fig. 6 Simulated radiation patterns in the xz-plane of the dual-band antenna with and without the FSS **a** at f_1 **b** at f_2

Table 1 Comparison of proposed structure with existing gain enhancing FSS

Refs.	FSS resonant frequency (GHz)	Cell dimension	% Gain enhanced (%)
[7]	2.45, 5.8	$0.40\lambda_g \times 0.35\lambda_g \times 0.059\lambda_g$	46
[9]	2.45, 3.5	$0.39\lambda_g \times 0.39\lambda_g \times 0.021\lambda_g$	54
[8]	2.45, 5.8	$0.43\lambda_g \times 0.43\lambda_g \times 0.057\lambda_g$	30
[10]	2.4, 5.4	$0.16\lambda_g \times 0.16\lambda_g \times 0.013\lambda_g$	35
Proposed FSS	2.4, 5.4	$0.13\lambda_g \times 0.13\lambda_g \times 0.011\lambda_g$	50

4 Conclusion

A compact, cost-efficient, angularly stable, and polarization-independent single-layered dual-band FSS has been designed and used to enhance the gain of a printed dual-band antenna operating at 2.4 and 5 GHz WLAN bands. Three different configurations of the antenna with FSS are studied and critically analyzed for gain enhancement and peak gain of the antenna is improved by 28%, 60%, and 50%, respectively. It is also observed that the radiation patterns are stable and the proposed FSS contributes to significant reduction in back radiation of the designed antenna.

References

1. Munk BA (2011) Frequency selective surface: theory and design. Wiley-Interscience, New York
2. Franchi PR, Mailloux RJ (1983) Theoretical and experimental study of meta l grid angular filters for sidelobe suppression. IEEE Trans Antennas Propag AP-31 3:445–450
3. Pelton E, Munk B (1976) A streamlined metallic radome. IEEE Trans Antennas Propag AP-22 6:799–803
4. Saleh AAM, Semplak R (1976) A quasi optical polarization independent diplexer for use in the beam feed system of millimeter wave antennas. IEEE Trans Antennas Propag AP-24 6:780–785
5. Agarwal VD, Imbriale WA (1979) Design of a dichroic cassegrain sub-reflector. IEEE Trans Antennas Propag AP-27 4:466–473
6. Rashid AK, Li B, Shen Z (2014) An overview of three-dimensional frequency-selective structures. IEEE Antennas Propag Mag 56(3):43–67
7. Chen HY, Tao Y (2010) Antenna gain and bandwidth enhancement using frequency selective surface with double rectangular ring elements. In: Proceedings of antennas propagation EM theory (ISAPE), pp 271–274

8. Chen HY, Tao Y (2011) Performance improvement of a U-slot patch antenna using a dual-band frequency selective surface with modified Jerusalem cross elements. IEEE Trans Antennas Propag 59(9):3482–3486
9. Pengthaisong K, Krachodnok P, Wongsan R (2013) Design of a dual band antenna using a patch and frequency selective surface for WLAN and WiMax. In: Proceedings of electrical engineering/electronics, computer, telecommunications and information technology (ECTI-CON), vol 41, pp 1–4
10. Liu Y, Ye QW, Xiao B, Yang H (2012) A novel application of frequency selective surface in dual-band WLAN antenna. In: Proceedings of antennas propagation EM theory (ISAPE), pp 512–514

Techniques to Detect Clickjacking Vulnerability in Web Pages

Priya Jyotiyana and Saurabh Maheshwari

Abstract Clickjacking attack attracts clients towards clicking on items unsuspectingly to set in motion unpleasant web pages that may be punctual to perform undesirable operations on genuine web pages without any notifying the clients. Specifically, victims can be misled through clicking on items from different websites, for example, online social networking sites (Facebook, Twitter), shopping websites (Amazon), etc. This paper discusses in particular the efficient defence techniques, which comprise a tedious process of preventing clickjacking attacks. The study of different types of malicious activities, vulnerabilities and their weaknesses in web program framework is a challenging task. We study different approaches and defence techniques with the purpose of providing a valuable reference to clickjacking attacks in order to enable a safe and secure web environment. In addition, this paper provides a review of injecting different types of clickjacking attacks in a webpage constrained environment. To fight against clickjacking attacks, it is important to see how these attacks happen in the present reality along with the comparative study of the state of the art solutions.

Keywords Clickjacking attacks · Likjacking attacks · Cursorjacking attacks
Clickjacking defenses

1 Introduction

Clickjacking refers to the phenomenon through which an individual's personal information can be hijacked when he/she clicks on a web link or url activating vital activities. In 2008, Jeremiah Grossman and Robert Hansen "rediscovered" the issue

P. Jyotiyana (✉) · S. Maheshwari
Department of Computer Science and Engineering,
Government Women Engineering College, Ajmer, India
e-mail: priya7653@gmail.com

S. Maheshwari
e-mail: dr.msaurabh@gmail.com

V. Janyani et al. (eds.), *Optical and Wireless Technologies*, Lecture Notes in Electrical Engineering 472, https://doi.org/10.1007/978-981-10-7395-3_68

and uncovered an attack in which a user could click on a safe button, without knowing that this safe button could be camouflaged by the fake button and lose vital personal information [1]. The trap behind this scenario ot the attack is to put the focus on the site inside an iframe and overlay it with a finished diverse page. Through this overlay the client could click on specific button or connection. The bright universe of data security has aggregated a somewhat vast cluster of related attack strategies—double clickjacking, likejacking, cursorjacking [2]. The initial report of clickjacking attack was introduced in web program in 2008 [3].

Objectives of Attacker

- Attacker can distract the user through the attractiveness of web page.
- Attacker can mislead them to visit the malicious web page to click on undetectable target on button.
- Attacker insert the source content in some tags in web script, in iframe tag, diminish the frame opacity to zero and point them to click on above apparently certified button click.
- Without the user's knowledge or warning, the user click has been captured and made the attacker capable of utilizing the user click for noxious purposes.

Two steps are required for clickjacking attack. **First,** the malicious web page designed by an attacker that creates an object like a Twitter button, etc., genuine web page can overlap the invisible object with the other visible object. **Second,** when any user clicks the invisible overlap with visible genuine message or button, the genuine message is framed under the invisible target element so that if any user clicked on the message or button, the unwanted action happens on the target which is overlapped with the invisible iframe element. Attacker site exhibits a user interface of their application and conceals the sensitive component by setting the opacity to zero with cascading stylesheet. The attacker, as a rule, overlays the malicious page over an ordinary page, which requires a click. These days, research attempts are being made to recognize and overcome the clickjacking attacks [4].

1.1 Motivation and Scope

The power of Google Chrome as well as Firefox lies in the fact that they can store, control a substantial number of extensions, in which some extensions are poisonous, or may turn into malicious extensions a while later. The target of over-protecting a program attacker have controlled the extension in somehow, favoured script successively inside the program. By this strategy, an attacker can get access to the program code inside position, as well as have an attractive component to allow the increase of speed of the surprise attack to complete exchange to the client device [5]. While program growth is based on continuous running on a privileged advantage phase, external of conventional program, and safety measures courses of action as well as examine and control various different effective concentration of attacker. In 2011 to the first quarter of 2012, 757 clickjacking related spaces were enlisted in the United States. Indeed, even Facebook gives a common stage to empower the sharing, yet at the same time cannot ascertain the dangers—whether the message is a clickjack or not. Current arrangements about fighting clickjacking (particularly on Facebook) are to confirm Web script from client, and convincing clients to prepare the module of program is a major test. Clickjacking is just watching the malicious scripts at the client sides, and can't have enough flexibility toward the scripts' variation and confusion [6]. This paper structured as follows: Sect. 2 gives a well-known theory of examples of clickjacking attacks. Section 3 clarifies the different possible surviving attacks and defence techniques. In Sect. 4 it describes the detection of clickjacking attacks, which gives the details of previous implementation at the client side, server side and proxy level detection mechanism. We conclude the paper in Sect. 5.

2 Background Details

2.1 Examples of Clickjacking Attack

Clickjacking has various sorts of assaults. Among them, the most prevalent are cursorjacking and Likejacking.

- **Cursorjacking attack:** Cursorjacking is the idea presented by Eddy Bordi. In this, clients signify the utilization of a usual cursor representation. Accordingly they showed cursor gets moved to the other side from the real mouse position. With perceptive situating of page components, the attacker guides client clicks to required components [3]. One of the benefits of this type of attack is that objective connection might be unmistakable as well as positioned at its real place. The client's cursor is supplanted with a false one through an attacker. Fake cursor ought to be moved moderately to its typical location, so it will give fake input of cursor area to the client (Fig. 1).

Fig. 1 Cursorjacking attack scenario

Fig. 2 Likejacking attack scenario

- **Likejacking attack:** It is a form of clickjacking in which harmful coding is connected with a Facebook Like button. In the likejacking attack, an attacker builds a website that has its own control of button click and it says that click here to get the free ipod and at the top of web page, the attacker has stacked an inmail account and arranged precisely whatever other messages like—delete all messages button specifically on the top of the free ipod button. User tries to click on that free ipod click button. However, in reality, the user has clicked on the 'delete all messages button click'. Through this, the attacker got access to use the client's information. The client can be deceived into clicking 'like' button, on an attacker web page. By doing this, the 'like' button overlaps with the hidden button [7] (Fig. 2).

3 Surviving Attacks and Defence Techniques

3.1 Surviving Attack Techniques

Clickjacking attacks can be classified into three types: compromising display integrity; compromising temporal integrity; compromising pointer integrity; wack a mole attack; cursor spoofing attack; double clickjacking attack.

- **Compromising display integrity:** Attacker wishes that the user click on the fake button so the attacker has wrapped the target button with an opaque frame. The user maybe confused as the attacker can be ambiguous by using the opaque target button atop a PayPal button. The button is wrapped with the fake frame and the user without realising it has clicked on the pay button but in reality the real click is made on the PayPal button [3, 8]. The attacker can draw a distraction under the objective component by utilizing CSS property and z-index property. Attacker may totally cover the objective component with a foggy distraction, yet make the fake 'unclickable' by setting the CSS property pointer actions to 'none'.

- **Compromising pointer integrity:** The assailant may abuse pointer respectability by showing a fake cursor symbol far from the pointer. This is known as cursorjacking. Utilizing the CSS cursor property, an assailant can without much problem, cover the default cursor and automatically draw a fake cursor somewhere else or on the other hand, set a custom mouse cursor symbol to a misleading picture that has a cursor symbol moved a few pixels off the first position. Another variation of cursor control includes the blinking cursor which demonstrates console centre. The blinking cursor may confuse the victim into feeling that they are writing content or text into the assailant's input field, though they are really connecting with the objective component [8].
- **Compromising temporal integrity:** An aggressor tricks the client in the little interim of instance so as to manipulate the User Interface component. Client chooses towards click on User Interface component. The focus on User Interface was covered by a catch component and the real snap performs on the covered button/component [3]. Prior assaults controlled visual setting to trap the client or casualty into sending contribution to the wrong User Interface component. A free method for accomplishing a similar objective is before genuine click happens.
- **Cursor spoofing attack:** This attacker threatens the mouse cursor's 'honesty'. Attacker conceals the webcam consent discourse box inside a commercial and trades off the accuracy of the cursor. The false cursor gives fake input of pointer area to the client. A hilarious video advertisement fool around naturally that constrains the client to click on skip promotion connects. The client clicks on the connection as well as the genuine cursor gets authorization of client's webcam from adobe streak player webcam setting [7].
- **Double clickjacking attack:** The double tap assault depends on compromising temporal integrity. The program doesn't ensure against Framebusting. The double click attack does attraction as well as switch so as to make the clients do a double tap directly later than the principal click. Aggressor changes client centre towards the Google pop-up window below the privilege earlier than the second click [3].
- **Wack on mole attack:** It is the combination of pointer and temporal integrity attack. The aggressor traps the client to click and all of a sudden switch Facebook 'like' catch and gets the profile of the client. The assailant requests that the client play a whack-a-mole diversion and advise her to score high and gain remuneration by tapping on click appearing on screen area. All through the diversion, assailant utilizes a fake cursor to control where the client attention ought to be. At a later point in the amusement game, switch in a Facebook 'like' catch at the genuine cursor's area, deceiving client to tap on it [3].

3.2 Surviving Defence Techniques

- **User confirmation:** At the point when the casualty clicks, he is provoked for an affirmation. Facebook at present uses this instrument for the 'like' catch, requesting client affirmation at whatever point asked for originates from blacklisted domain [9]. This system corrupts client experience and it is additionally defenceless against double tap assaults.
- **User interface randomization:** The component's User Interface format is randomized. For instance, as opposed to having Facebook 'like' button at a settled area, it ought to have been placed in an irregular area. This method has its own weakness as the attacker may request that the casualty clicks at various circumstances on various areas to figure the genuine position of the 'like' button [9].
- **Framebusting**: Framebusting is the procedure that does not permit the objective component from rendering in an iframe. This can be accomplished by utilizing JavaScript code for XFRAME choices. But this procedure confronts similarity issues. The Framebusting arrangement is authorized on all caches, connections, and structures on a solitary website page as well as consequently, should not be utilized if web page is permitted to be installed in an outsider web page. Web applications need to implement distinctive approaches on various components on a solitary site page [10].
- **Opaque overlays policy:** This strategy is utilized as a part of Gazelle web program. All cross starting point casings are compelled to be rendered indistinctly. Once more, this procedure is additionally not exceptionally powerful [11].
- **Visibility detection on click:** In this strategy the straightforward edges are permitted to be rendered, however mouse clicks are blocked if the program identifies that the clicked cross root outline is mostly unmistakable or undetectable. For instance Adobe uses such assurance to glimmer player webcam get to exchange in reaction to webcam clickjacking assaults [11].

4 Detection of Clickjacking Attacks

4.1 At the Client's Side

Despite the fact that the same-beginning approach should shield doubting websites from interfering with each other, it neglects to prevent any of the clickjacking assaults. There is no well-known strategy to piece all unnoticeable iframes.

- **User Confirmation:** Huang et al. An answer for avoidance clickjacking at the client side is to produce an affirmation discussion for end clients to comprehend a possibility outside the area of relevance of the click. Facebook at present sends this approach for the 'like' click, requesting affirmation at whatever point demands originate from blacklisted domain. Sadly, this approach degrades client encounter, particularly on single-click. Moreover, this approach may just concentrate on particular protests as opposed to all [4].
- **UI Randomization:** Huang et al. Another system is to ensure an objective component by randomizing the User Interface format of a true genuine web page. Subsequently, an attacker would neglect to effectively find the position of the objective component [4].
- **Blocking of Mouse Click**: Huang et al. Rather than totally threatening confining, an optional approach is to permit the rendering of straightforward ends yet blocking mouse clicks, if a program identifies that the clicked cross-origin frame border is not completely unambiguous. Adobe has added such assurance to Flash Player's Webcam to exchange because of webcam clickjacking assaults. This guard just secures the exchange and is not accessible for other web content [9].
- **ClickIds**: Balduzzi et al. It cautions clients when the clicked component covers with other interactive components. Shockingly, ClickIds can't identify attack in view of fractional overlay or editing, as well as it demonstrates false positive cautioning. Specifically, obvious ad standards, and element drop down menus can produce interactive components that may be covered by previous interactive components on the major page [12].
- **NoScript:** Heiderich et al. The NoScript is a Firefox program augmentation that can permit the execution of JavaScript code from client specific web destinations. It secures against clickjacking assault in the program. NoScript consents to framebusting code in light of Frame Break imitating. Along these lines, if an encircled page contains a 'casing busting' script, it consequently replaces the confined page as the highest record, free of the setting of the program to empower or impair the JavaScript code [13].

4.2 At the Server Side

We discuss two well-known server side defence techniques:

- **HTTP Header:** Rydstedt et al. The header can have three conceivable qualities: deny (keeps any space from surrounding the substance of a URL), same-source (just permits confining of pages from the present website), and allow from (an arrangement of whitelisted areas that can outline a given URL). Notwithstanding, this procedure requires sending HTTP header from each of the pages that need insurance against clickjacking assaults. Another trouble is the absence of support at the program side to comprehend the header and act fittingly (e.g., more seasoned renditions of IE and Firefox will overlook the header) [10].

- **HEAD Element:** Huang et al. A workaround for managing absence of program support for HTTP header-based arrangement towards affect a unique HTML HEAD component contain an exceptional style id with no show for an encircled page. The web page is shown just on the off chance that it is situated as the top window [9].

4.3 At the Proxy-Level (ProClick)

Shahriar et al. It is an intermediary-based method that can capture and break down demands (issued from programs) and response pages (got from remote Web servers) to test the assault side effects early. ProClick which has four phases: ask for examination, sending demand to website, reaction page examination, and sending [5] (Table 1).

Table 1 Comparitive study of different Clickjacking Defences

	Types	Tools and techniques	Description	Limitations
1	Client side defence [4]	Framebusting	At the time of development it can easily include the framebusting code	For double framing, framebusting code is not efficient
2	Server side defence [10]	X-frame option HTTP response	Mainly the resourceful technique used for website HTTP header	Not used widely because overhead of coding is so high
3	Framework [5]	ProClick	It is not a detectable the web page on the basis of rating instead it can check the symptoms of attack	Still developing source code not accessible for free
4	Browser plug in [13]	NoScript	It is still powerful with clearclick technology	Only available for Mozilla web browser
5	Browser plug in [12]	ClickIDs	It detects the mouse click events and verifies the features of web page	It relates only with clickable basics
6	Click analysis technique [4]	Adaptive UI Randomization	On the basis of sample clicks it can detect statistically	Poor User Interface because it provides randomization

5 Summary and Discussion

Clicks are created by human activity and the whole web demand procedure is finished, which implies that the site page is stacked, as well as the pictures, Flash, JavaScript code, and so forth. A human may have action on the website page, for example, mouse clicks, key strokes and scroll bar exercises and the site hit time is no less than a few seconds. It represents the danger for security and privacy of the computer user. The greater part of the detection advertising businesses enhances movement utilizing its own double worldview into justifiable click and invalid click using their own particular measurements. Be that as it may, except for a little rate of obvious sincerity to justifiable click and robot movement, by far, most of click can't be resolved to be valid or invalid. Or maybe, they are of a specific quality which relies upon client expectation.

5.1 Our Contribution

- The power of Chrome as well as Firefox lies in the fact that they can store and control a substantial number of extensions, in which some extensions are poisonous, or may turn into malicious extensions a while later.
- For the right usage, web engineers must have awareness on avoidance strategies of the countermeasures. After reviewing all the clickjacking attacks and countermeasures, we give the optimized proposed strategy for avoidance of clickjacking attack.
- We try to attempt the challenge and detect all the vulnerabilities by using optimized proposed strategy, avoid the defenceless elements that feature on web pages as well as send the notification to the user.
- We have additionally tried that numerous well known browser extension components and features are combined together to test clickjacking attack, in this manner, we can enhance the work in the proposed way.

6 Conclusions and Future Work

The aim of this paper is to survey all the clickjacking attacks and defences by the utilization of Iframe attack and many other features on web pages. Alongside this, the review is intended to exhibit the impact of utilizing iframes attack as a part of web pages and how this could be utilized as defencelessness in favour of the user. We could discuss the victim's mouse pointer and track his mouse button to make him click on a covered iframe layer, set on top of the real site content. So, we could also trade off the victim's web-based social networking profile by likejacking i.e. making the victim to like a page on Facebook unintentionally. The finished review show a few ways to deal with attacks and protects both client and server sides from such attacks. Clickjacking is the present trendy expression, it is a program issue. The programmers abuse different bugs and introduces them in the working of the web programs and also uses new tool effective hacking methods by using clickjacking. Along these lines clickjacking is a genuine attack and we have to discover such powerful and versatile arrangements that can obstruct this in all regards.

From the point of future perspective, clickjacking attack is extremely unpredictable. These clickjacking attacks framework that require concern include the web pages security management, excessively careless security management, making reliable vulnerabilities of internet web pages. To prevent these attacks we can develop the security extension and provide it a safe and secure environment for all users and offer better results as compared with the previous extension.

References

1. Simpkins L, Yuan X, Kim J (2014) A course module on clickjacking. In: Proceedings of the 2014 information security curriculum development conference. ACM
2. Takamatsu Y, Kono K (2014) Clickjuggler: checking for incomplete defenses against clickjacking. In: 2014 twelfth annual international conference on privacy, security and trust (PST). IEEE
3. Akhawe D, He W, Li Z, Moazzezi R, Song D (2014) Clickjacking revisited: a perceptual view of UI security. In: 8th USENIX workshop on offensive technologies (WOOT 14)
4. Huang L-S, Moshchuk A, Wang HJ, Schecter S, Jackson C (2012) Presented as part of the 21st USENIX security symposium (USENIX security 12)
5. Shahriar H, Devendran VK, Haddad H (2013) ProClick: a framework for testing clickjacking attacks in web applications. In: Proceedings of the 6th international conference on security of information and networks. ACM
6. Sood AK, Enbody RJ (2011) Frametrapping the framebusting defence. Netw Secur
7. Faghani MR, Nguyen UT (2014) A study of clickjacking worm propagation in online social networks. In: 2014 IEEE 15th international conference on information reuse and integration (IRI). IEEE
8. Shamsi JA, Hameed S, Rahman W, Zuberi F, Altaf K, Amjad A (2014) Clicksafe: providing security against clickjacking attacks. In: 2014 IEEE 15th international symposium on high-assurance systems engineering. IEEE
9. Van Goethem T, Chen P, Nikiforakis N, Desmet L, Joosen W (2014) Large-scale security analysis of the web: challenges and findings. In: International conference on trust and trustworthy computing. Springer International Publishing
10. Rydstedt G, Bursztein E, Boneh D, Jackson C (2010) Busting frame busting: a study of clickjacking vulnerabilities at popular sites. In: IEEE Oakland Web 2
11. Selim H, Tayeb S, Kim Y, Zhan J, Pirouz M (2016) Vulnerability analysis of Iframe attacks on website. In: Proceedings of the 3rd multidisciplinary international social networks conference on social informatics. ACM
12. Balduzzi M, Egele M, Kirda E, Balzarotti D, Kruegel C (2010) A solution for the automated detection of clickjacking attacks. In: Proceedings of the 5th ACM symposium on information, computer and communications security. ACM
13. Heiderich M, Niemietz M, Schuster F, Holz T, Schwenk J (2012) Scriptless attacks: stealing the pie without touching the sill. In: Proceedings of the 2012 ACM conference on computer and communications security. ACM

Location-Based Services: Current State of The Art and Future Prospects

K. Rajalakshmi and Mukta Goyal

Abstract As technology evolves, the use of smartphones for personal work, shopping, learning, entertainment is growing day by day. One of the major applications of mobile devices is in the area called location-based services (LBS). LBS is a software-oriented service provided by smartphones that perform control operations using location-based data. The objective of this paper is twofold—first, to study deeper about existing LBS methodologies and next to analyze the applications of LBS in various fields. Further, this work focuses on comparative analysis of LBS based upon different attributes like interoperability, extendibility, scalability, and flexibility. For the purpose of comparison, many research papers are considered and studied analytically. Finally, in order to emphasis our analysis, for methods of location-based services and their applications, a novel application GoClues is developed for tourism industry.

Keywords GoClues · Location-based service · Android · Web applications Smart shopping · Autonomous car system

1 Introduction

Today's state of the art in mobile computing is that data can be accessed anywhere and anytime through smartphones and personal digital assistants. This trend opens up wider research in mobile computing with integration of enhancing computing methods and wireless communication technology. It is to be noted that, on one hand, the mobile communication has to handle much more complex data communication, other than simple digital voice communication. On the other hand,

K. Rajalakshmi (✉) · M. Goyal
Department of Computer Science and Information Technology,
Jaypee Institute of Information Technology, Noida, India
e-mail: k.rajalakshmi@jiit.ac.in

M. Goyal
e-mail: mukta.goyal@jiit.ac.in

V. Janyani et al. (eds.), *Optical and Wireless Technologies*, Lecture Notes in Electrical Engineering 472, https://doi.org/10.1007/978-981-10-7395-3_69

enhanced mobile computing technologies should be able to provide the spectrum of software level services to the mobile users. Once such on demanding service, that is provided, by advanced mobile computing platform for users is location-based services (LBS).

LBS is a software-oriented service that provides virtually and also physical control of operations using location-based contextual data. This paper discusses two aspects of LBS. The first discussion is on technologies used in LBS and next is on applications of LBS. To begin, the current method to provide LBS to the users is surveyed. It mainly focuses on three technological platforms of LBS namely (i) Web application [1], (ii) Android platform [2], (iii) Open Geospatial Consortium [3] as given in Fig. 1a. Secondly, the application of LBS is approached as (i) standalone LBS applications and as (ii) integrated LBS applications as given in Fig. 1a. The standalone LBS application does not need any other technologies to be integrated and just LBS is sufficed to completely control the functionalities of such applications. For example, the autonomous car system and facial recognition [4] the car can be more customized and less prone to robbery. Another example is health care centers and indoor taxi cab app for positioning and routing on outdoor environments using [5, 6].

On the other hand, integrated LBS technologies are incorporated with other technologies like cloud computing, Bluetooth technologies. For example, in [7], application to find the location of the nearest school [7] with the user's location is stated. The proposed approach exhibits integration of several computation-based technologies such as location services, vehicular systems incorporating mobile network. Next, the LBS is used to secure the computation offloading of location device in cloud computing applications [8–10]. These applications are created using Android Studio, Java, and NetBeans. And next, the application for community and local societies has been developed for providing data called InfoPoint [11]. The InfoPoint detects and attempts to transfer multimedia content to nearby mobile devices using Bluetooth connectivity. In this, an adaptive inquiry access code algorithm for Bluetooth is proposed. This algorithm mainly optimizes the time taken to page nearby devices in InfoPoint application. This work, Sect. 2 provides deeper insight into various platforms of LBS. Section 3 discusses LBS applications

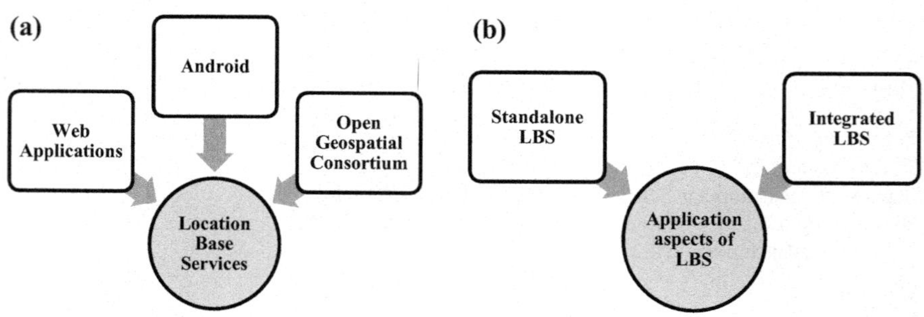

Fig. 1 **a** Technological platforms of LBS, **b** application aspect of LBS

and challenges in each of these applications. Next, in Sect. 4, a creative LBS application named GoClues is discussed. Finally, Sect. 5 concludes and summarizes the importance of LBS.

2 Various Platforms of Location-Based Services

Location-Based Services using Web Applications: TUGIS [1] is one of the major platforms used as Open Geospatial Consortium (OGC)-based services. It is a flexible platform with a high emphasis on extendibility, scalability for larger database management system. It provides geographic information services (GIS) and location-based services (LBS) dynamically. Earlier used Web applications services in LBS had stateless servlet. Once deployed, the servlet can accept as many map generation requests possible in its current state. However, the demerit of this proposed approach was that the current state of users needed to be tracked at every instance. In fact, the tracking process was a complex and time-consuming task. This demerit was taken care by using JSP pages using MapBean. The MapBean component of Java keeps track of current state. This limitation is overcomed by TUGIS, implementing dynamic Web mapping user interfaces. The clients access certain map segments on a Web portal for which they receive traffic intensity notifications by using specialized TUGIS publishing module.

Location-Based Services using Android platform: Android is a powerful wave operating system used to design a huge number of LBS-based applications. It offers Open Development Environment based on Linux-based Kernel [2]. Hardware access is available to a number of applications using API's. Google Maps which give full access to maps programmatically and Geocoding support lets one create a large variety of map applications using Android API's and Android rich Graphics Library. The various components such as location manager manage all other components need in LBS system. Location provider is used to determining physical location of the user. Geocoder is divided into two parts such as reverse geocoding and forward geocoding. Reverse geocoding converts latitude and longitude into the street address, and forward geocoding converts street address to geographical coordinates.

Location-Based Services using Open Geospatial Consortium (OGC) Map Services: Open Geospatial Consortium (OGC) servers have been used to implement many interoperable LBS applications which are independent of device or data providers [3]. OGC open location service, Web map services, Web feature services (WFS) are some useful Web services used for development. Replacing the earlier wireless browsers, mobile devices now offers HTML, XHTML, Cascading Style Sheets, JavaScript, which is increasingly gaining support as well as third-party browsers. The shortcomings of proposed framework are weak network

Table 1 Comparison analysis of various technologies used for providing location-based services

Methods	Interoperability	Extendibility	Scalability	Flexibility	Remark
TUGIS—location-based services in Web app (2008)	High	Possible	Possible	Yes	Support various languages
Location-based services using android (2013)	High	Yes	Yes	Yes	Easy to run on android virtual machine
A framework to develop location-based services (2015)	No	Yes	Moderate	No	Limited language support and fixed set of commands to operate

connectivity, location detection, and location-based services. In the present scenario, numerous research works are performed to evaluate LBS under OGC with different WMS and WFS implementations. A comparative analysis is shown in Table 1.

3 Applications and Challenges of Location-Based Services in Various Domains

Tourism and Community [3]: In today's information era, the exponential use of mobiles and personal digital assistants (PDAs) has opened up further advanced ways of interactive location-based services. Of special interest, LBS is used into construct ad hoc personal area network for tourism industry. Major challenges faced were like low latency transactions, high capacity, sufficient data transfer bandwidth.

Smart Shopping [12, 13]: Smart shopping can be done using location-based services which attract both online and offline users. The main services of online to offline are delivery, location-based coupon, and mobile easy payment system. These applications have numerous functions such as scan the barcode of products through smartphone camera and then receive the products information such as list of products in ascending or descending order of price. Online stores like Amazon, eBay are connected to these applications. Since beacon has small error range and small amount of electricity consumption, more accurate information can be provided.

Autonomous Car System [4]: Here, the aim is to reduce the problem of stolen vehicles used for crimes and accidents. In this method, the image is broken into number of blocks of gray pixels and given to one of the best face recognition systems, Weber Descriptor (WLD) where WLD is calculated in different neighborhood. All the WLD histograms obtained from blocks of image are concatenated to produce face from final images. Next, the extracted feature set of face image is compared with a feature set of database. Then, on real time, user identification is done. If the unauthorized entry is detected, the unauthorized image is sent to the user via MMS module and seeing the image. The image processing of the data is done by compressing the image using discrete cosine transform (DCT). The system of the car is turned on only after the verification of the user.

Health Services [6]: In medical community, the location-based services are useful to identify specific medical institutions by providing location of medical specialty hospitals. In this case, the LBS works based on the customers' health care specifications and user profile. In this application, user's location is obtained using probabilistic-based multilateral method. It computes the wireless signal strength based on the antennas whose distances are in the range of the requested mobile device. Then, the physical distance between the mobile device and the required location is calculated based on the signal strength by wireless receiver. After finding the required locations, the information related to offered medical services is retrieved semantically based on the ontology method. A prior location information is set as input parameter for each semantic searching. Once the suitable locations are calibrated, the indicators with different colors are used. These indicators have range of colors according to the distance between the device and medical centers. Further, these indictors provide various information like the name, the address, and the Euclidean distance to each medical center.

IoT-Based Smart Museum [14]: In recent trend, the museums and art galleries provide information about historical facts for the visitors through conventional techniques like paper-based brochures, pamphlets and through digital system like audio guides. Major disadvantages of these techniques are not smart or easiness enough. To overcome, these naive methods, wearable devices along with smart computing can be incorporated. For this purpose, wearable devices would detect the current location of user. These location data would be communicated and processed at centralized system. On completion of the computation process, the user will be provided with assistance, based on user's preferences, in different formats like text, pictures, audio, or video contents. To enhance the performance of LBS, in this application, the multimedia contents requested by user would be offloaded in shared environments like cloud or distributed systems.

4 Experimental Results

Based on android platform and using Dijkstra algorithm, an application called GoClues is developed. The GoClues is developed using android platform for the tourists to make their travel a memorable experience. It covers three cities with ten best locations to be visited known as Geo_Sites and ten Geo_Points in each location to cater your desire to travel and to make user's journey worth.

Any Geo_Site is supposed to be a large area where many points, viz., Geo_-Points with its own name, altitude, latitude, and longitude. For example, Qutab Minar in New Delhi is heritage complex where many points in addition to Qutab Minar itself are of interest, e.g., Aulia Minar, Iron Pillar, Masjid. If the direction goes wrong, the application throws an alert about it. At any Geo_Point, all the nearby Geo_Points of that location are shown along with their directions and shortest paths. The algorithm used for connecting various geopoints is Dijkstra algorithm. Data is collected for various locations through application as soon as a user clicks the application; it saves the latitude and longitude of the current location in the database. SqLite and PHP administration is used to store the database, which stores the location description and name of the location as shown in Fig. 2. Next task given was to do user analysis for observing the kind of users visiting the monuments or any travel sites. For this, data is crawled from many traveling Web sites and applied classification and segmentation algorithms for the users. Cluster analysis is being done so that the user can be understood better, and better application as per the user's requirement is developed. In order to fulfill this perspective, we went to Qutab Minar and collected data on our own and found many problems, which as a tourist can face.

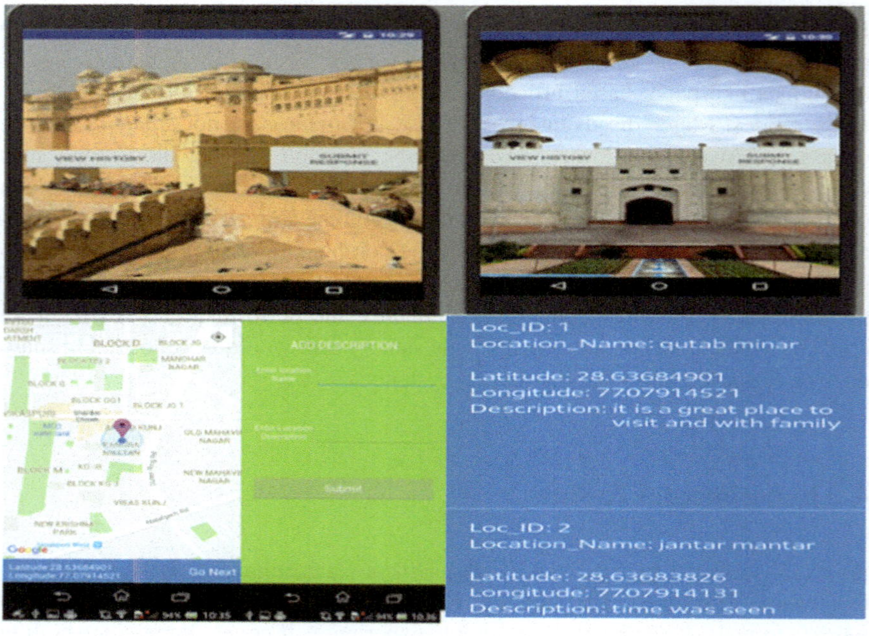

Fig. 2 Location-based service for tourism application based on Dijkstra algorithm

5 Conclusion

This paper discusses the technology used for location-based service. It also discusses the challenges to implementing the location-based service in various applications. Next, LBS-based application GoClues is developed using the android platform. The future work is to launch the GoClues on the cloud platform and to analyze performance enhancement of this application.

References

1. Muresan O, Gavrea B, Gorgan D (2006) TUGIS-location based services in web applications. In: IEEE international conference on automation, quality and testing, robotics, vol 1. IEEE, pp 436–441
2. Kumar S, Qadeer MA, Gupta A (2009) Location based services using android. In: 3rd IEEE international conference on Internet multimedia services architecture and applications, vol 9, pp 335–339
3. Fernando N, Dias D, Wijesekara S (2010) A framework to develop location based services applications using OGC map services. In: Fifth international conference on information and automation for sustainability. IEEE, pp 521–526
4. Faisal M, Thakur A (2016) Autonomous car system using facial recognition and geo location services. In: 2016 6th international conference cloud system and big data engineering (confluence). IEEE, pp 417–420
5. Kilinc C, Mostafa SAM, Islam RU, Shahzad K, Andersson K (2014) Indoor taxi-cab: real-time indoor positioning and location-based services with ekahau and android OS, pp 223–228
6. Martinez ML, Vazquez DA, Maya D, Olvera X, Guzmán G, Torres M, Moreno M (2014) Geospatial recommender system for the location of health services. In: 14th international conference on computational science and its applications
7. Setiyono A, Sedo R, Pramono SH (2015) Application search school location method using location based services (LBS) based on J2ME. In: Intelligent technology and its applications (ISITIA). IEEE, pp 319–322
8. Neto JLD, Macedo DF, Nogueira JMS (2016) Location aware decision engine to offload mobile computation to the cloud. In: 2016 IEEE/IFIP network operations and management symposium (NOMS). IEEE, pp 543–549
9. Cao N, Wang C, Li M, Ren K, Lou W (2014) Privacy-preserving multi-keyword ranked search over encrypted cloud data. IEEE Trans Parallel Distrib Syst 25(1):222–233
10. Wang C, Chow SS, Wang Q, Ren K, Lou W (2013) Privacy-preserving public auditing for secure cloud storage. IEEE Trans Comput 62(2):362–375
11. Mahmood FM, Salam ZABA (2013) A conceptual framework for personalized location-based Services (LBS) tourism mobile application leveraging semantic web to enhance tourism experience. In: 2013 IEEE 3rd international advance computing conference (IACC), pp 287–291
12. Ros M, D'Souza M, Postula A, MacColl I (2011) Location based services with personal area network for community and tourism applications. In: IET international communication conference on wireless mobile and computing (CCWMC 2011), pp 432–437, IET

13. Zhu YQ, Ou DX, Shen T (2014) Online adaptive positioning algorithm for public location services in indoor places. In: 17th International IEEE conference on intelligent transportation systems (ITSC). IEEE, pp 1515–1520
14. Anju S, Joseph J (2015) Location based service applications to secure locations with dual encryption. In: International Conference on Innovations in information, embedded and communication systems (ICIIECS). IEEE, pp 1–4

A Reconfigurable Microstrip Antenna with Defected Ground Structure for Radio Applications

Harshal Nigam, Shubhi Jain and Ruchika Jain

Abstract A design of circular patch antenna having elliptical slot with defected ground substrate is being presented. It is designed for radio application. The antenna is reconfigurable, and we can switch between the C-band and X-band applications depending on the switching of diodes. Broadband applications in C-band and RADAR applications in the X-band can be switched. It is observed through analysis that dimensions of the antenna and slots are key factors for improving bandwidth of the geometry, here the antenna is fabricated on FR4 substrate with an elliptical slot and defected ground structure whose dimensions are optimized to get the best results and further the antenna is made reconfigurable.

Keywords Circular antenna · Elliptical slot · Broad bandwidth
Defected ground

1 Introduction

Antennas are having tremendous applications during last few years such examples are radio application, wireless communication [1]. Since antennas having many limitations like broad bandwidth polarization, larger size for better improvement nowadays antenna need compact size, high VSWR, high gain for fabrication, etc. Nowadays radiation pattern, VSWR are having important aspects in modern communication system. The antenna application will be planar and compact structures which can directly put inside the hand set. It will provide broadband performances which are all limitation of conventional microstrip antenna [2]. Many techniques have been proposed such as stacked patches. A new technique will be developed, i.e., defected ground structure to enhance the requirement of radio application [3].

H. Nigam (✉) · S. Jain · R. Jain
Swami Keshvanand Institute of Technology, Management & Gramothan, Jaipur, India
e-mail: hrshlnigam@gmail.com

V. Janyani et al. (eds.), *Optical and Wireless Technologies*, Lecture Notes in Electrical Engineering 472, https://doi.org/10.1007/978-981-10-7395-3_70

2 Antenna Design

Thus, antenna is designed with circular patch with defected ground structure. The circular patch is as shown in Fig. 1 which is designed on FR4 substrate, the top surface of substrate (30 × 30 mm) has a circular patch. The bottom of substrate has defected from (30 × 13 mm). There is an elliptical slot in the patch as shown in Fig. 2. The defected ground structure is as shown in Fig. 3. Then the next step is to make the antenna reconfigurable by using two diodes, i.e., D1, D2 on the slot, D1 horizontal diode and D2 vertical diode as shown in Figs. 4 and 5. These diodes can be switched ON and OFF and vary the return loss for both the bands.

Fig. 1 Front view of circular geometry

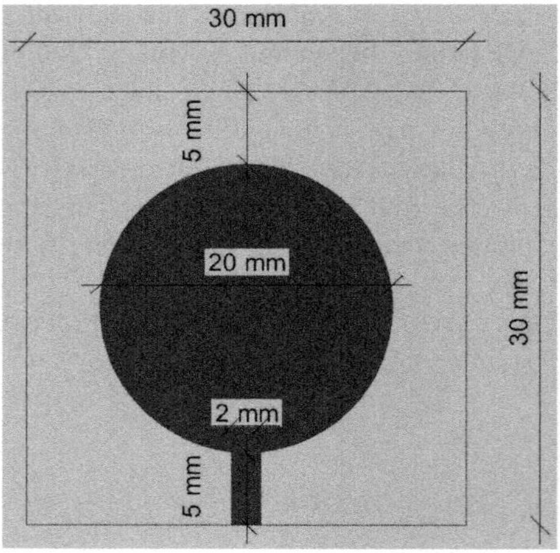

Fig. 2 Front view of circular geometry

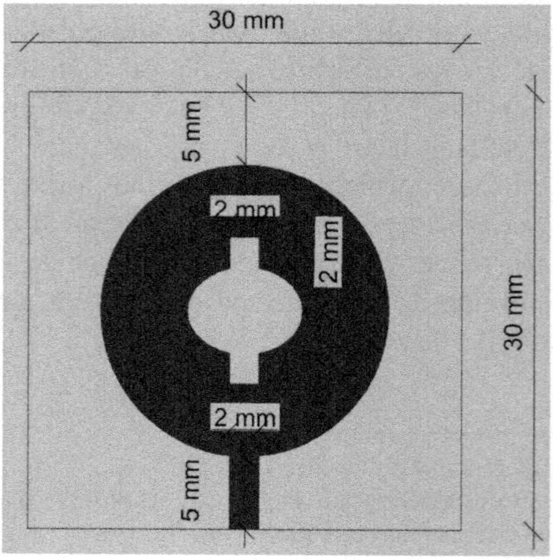

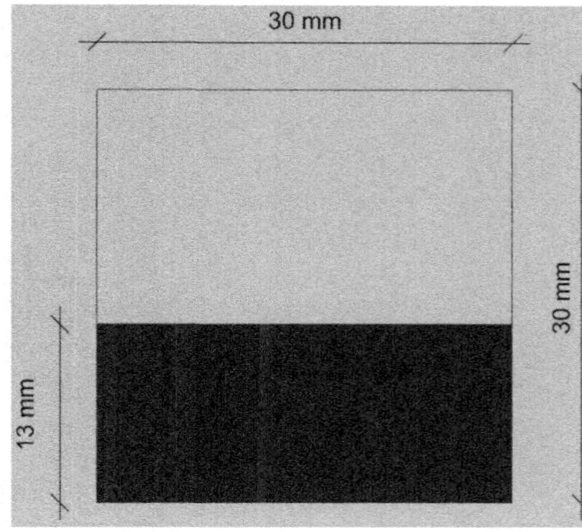

Fig. 3 Back view ground design of truncated antenna

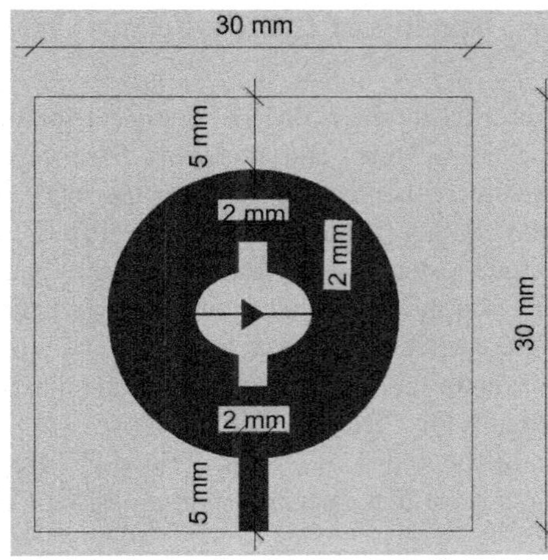

Fig. 4 Front view of elliptical patch with diode in horizontal plane (D1)

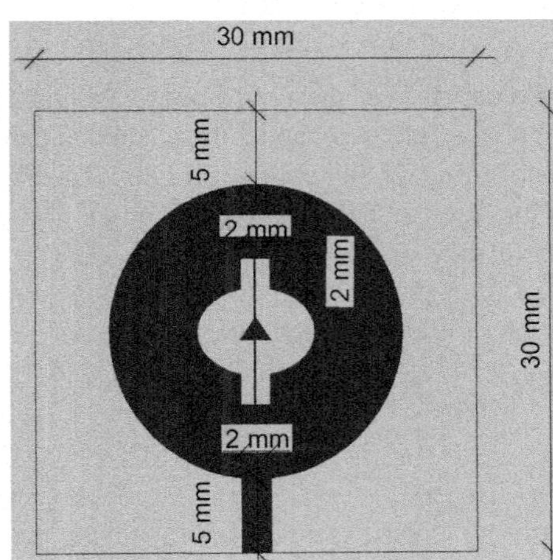

Fig. 5 Front view of elliptical patch with diode in vertical plane (D2)

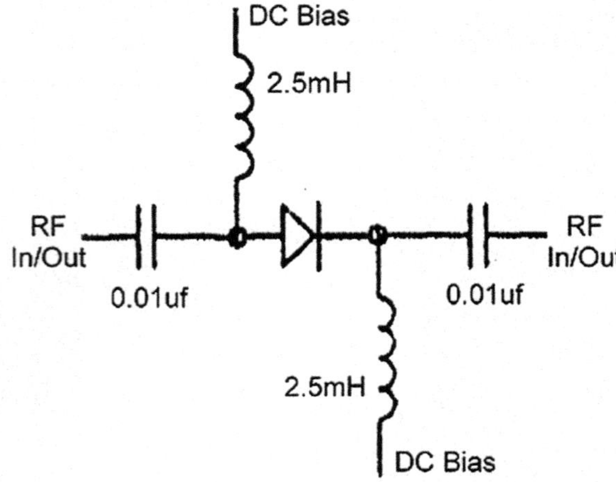

Fig. 6 Biasing circuit for PIN diode

3 Biasing of the Diode

The PIN diode switch is popular in microwave circuit applications due to its fast switching times and relatively high current handling capabilities. Conventional electromechanical RF switches are inherently speed limited devices due to inertial and contact potential effects. The PIN diode can operate at speeds orders of magnitude faster than mechanical switches and can be placed in packaging measuring a fraction the size of mechanical RF switches. The PIN diode along with other solid state switches utilize a semiconductor junction as the RF control element which accounts for the increase in switch speed and reduction in package size. The bias circuit for PIN diode is as shown in Fig. 6. The capacitors bloc the DC and inductors block AC, Thus, RF and DC are separated. The diode can be forward bias or reverse bias depending on polarity of DC at the DC bias terminals.

4 Antenna Results

We can switch the two diodes ON and OFF and vary S11 for the two bands as shown in the two bands; thus, antenna can work in both the bands or a single band depending on switching of diodes. The S parameter results can also be seen for the four cases as below figures (Fig. 7). The polar pattern is also shown in Fig. 8 at a frequency of 8.8 GHz.

Table of reconfigurable diodes

Case	D1 (results)	D2 (results)	C-band (7–7.5 Ghz) first band	X-band (8.5–9 Ghz) second band
I	On	On	−21 (Return loss)	−17 (Return loss)
II	On	Off	−30	−50
III	Off	On	−19	−13
IV	Off	Off	−28	−26

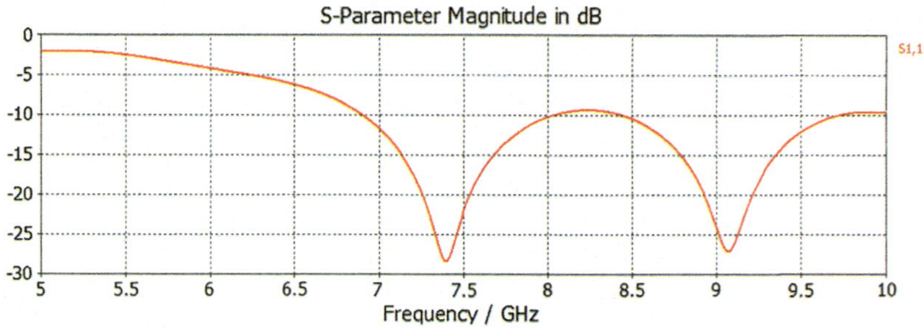

Return loss verses Frequency when both the diodes off.

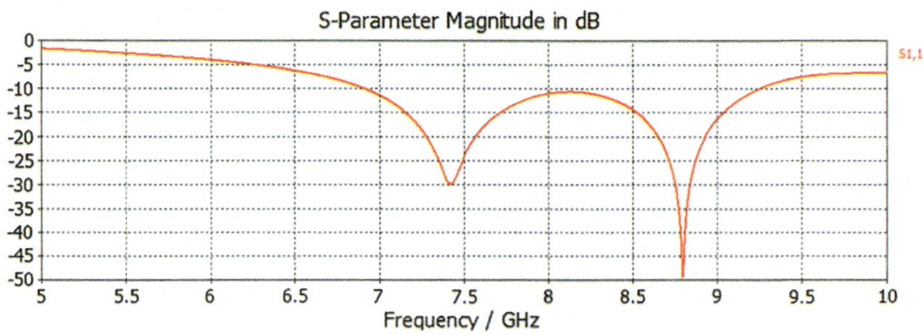

Variations of elliptical slot on return loss when First diode is ON.

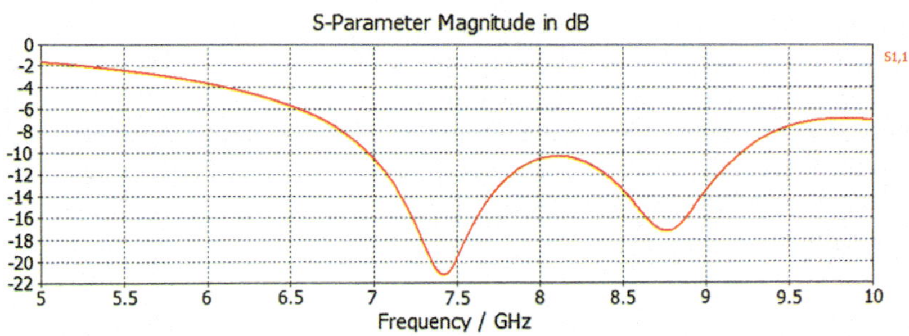

Results of S-parameters when both the diodes is ON.

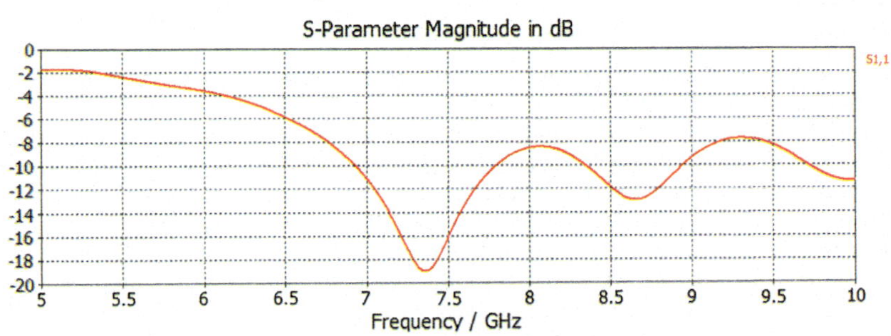

Results of S-parameter when second diode is ON

Fig. 7 S-parameter results

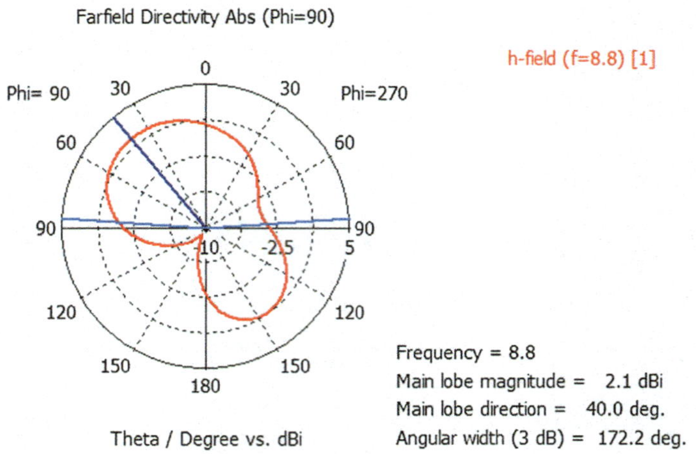

Fig. 8 Front view of polar pattern

5 Conclusion

As defected structure of circular patch design was presented. We optimize various results impedance matching, S-parameters, etc., with the help of defected ground structure. These antennas are reconfigurable and designed on C-band and X-band, i.e., (7–7.5 Ghz) and (8.5–9 Ghz), and it will be applicable for broadband applications. We can observe that change in first band is not much but the second band can be made reconfigurable S11 can vary from −13 dB (very high) to −50 dB (very low) attenuation; thus, we can control this band by changing the diodes ON and OFF condition.

References

1. Garg R, Bhartia P, Behl I, Ittipiboon A (2001) Microstrip antenna design handbook. Artech House Inc., Norwood, MA
2. Chen ZN, Chia MYW (2006) Broadband planar antennas, design and applications. Wiley
3. Kandwal A, Sharma R, Khah SK (2013) Bandwidth enhancement using z-shaped defected ground structure for a microstrip antenna. Microw Opt Technol Lett 55(10):2251–2254
4. Guha D, Biswas S, Biswas M, Siddiqui JY et al (2006) Concentric ring-shaped defected ground structures for microstrip applications. IEEE Antennas Wirel Propag Lett 5:402–405
5. Bansal R, Jain A, Kumar M, Meena RS (2012) Compact dual band microstrip loop antenna using defective ground plane. In: Proceedings of international conference on communication systems and network technologies, pp 20–22
6. Geng JP, Li JJ, Jin RH, Ye S et al (2009) The development of curved microstrip antenna with defected ground structure. Prog Electromagnet Res, 53–73

Compact Olympic Ring-Shaped Planar Antenna for K Band Applications

Ushaben Keshwala, Sanyog Rawat and Kanad Ray

Abstract A single-feed linearly polarized microstrip antenna for K band satellite communication is presented in this paper. The geometry consists of a rectangular strip coupled with five rings which appears like Olympic ring-shaped antenna with overall dimension of $12 \times 12 \times 1.6\ \text{mm}^3$. By considering partial ground, tremendous improvement in the radiation performance of the proposed antenna is observed. The printed antenna has been simulated using full-wave simulator, and it shows impedance bandwidth of 42% in the K band ranging from 17.2 to 26.7 GHz. In addition to radiation characteristics, the design approach of the proposed antenna is also discussed.

Keywords Olympic ring shape · Planar antenna · K band

1 Introduction

In recent time, the demand for compact size antennas for wireless communication has increased tremendously and so the extensive researches on compact microstrip antenna design. The emerging segments of communication in K band include maritime communications, cellular backhaul, and other mobile service. Recently, the microstrip antennas have been paid much attention by the researcher due to their

---

U. Keshwala (✉)
Electronics and Communication Engineering Department, Amity School of Engineering, Amity University, Noida, Uttar Pradesh, India
e-mail: usha_keshwala30@yahoo.com; ukeshwala@amity.edu

S. Rawat
Electronics and Communication Engineering Department, Manipal University, Jaipur, India
e-mail: sanyog.rawat@jaipur.manipal.edu

K. Ray
Amity School of Applied Science, Amity University, Noida, Rajasthan, India
e-mail: kanadray00@gmail.com

V. Janyani et al. (eds.), *Optical and Wireless Technologies*, Lecture Notes in Electrical Engineering 472, https://doi.org/10.1007/978-981-10-7395-3_71

interesting characteristics and outstanding features like low cost, light weight, compact size, integration with microwave integrated circuits, and ease of fabrication [1–3].

The performance of these antennas is dependent on their physical dimensions. As the size of the antenna decreases, the gain, bandwidth, and efficiency decreases, and hence, the full K band spectrum cannot be utilized [4–6]. Many techniques have been proposed to overcome the limitations of gain and bandwidth which includes changing the dielectric constant with notched microstrip antenna which increases the bandwidth slightly at the cost of quality factor [7], stacking the microstrip antenna and slotting the ground which in turn increases the bandwidth and gain [8, 9]. Many dual band antennas have been proposed for Ku and K band applications [10–13]. The techniques which are used to enhance the bandwidth also make the antenna design complex.

In this paper, we have simulated compact Olympic ring-shaped with single microstrip line feed. In sects. 2 and 3, we have summarized the modeling of antenna and discussed the results, respectively.

2 Proposed Antenna Design and Geometry

2.1 Compact Olympic Ring-Shaped Antenna

The proposed antenna is designed on FR-4 dielectric substrate with single feed. For linear polarization, the single feed is sufficient. The antenna is placed in x-y plane, and normal axis is parallel to z axis. The partial ground plane is printed on the bottom of the substrate. The antenna is designed based on cavity model for thin substrate ($h << \lambda_o$) and has a very compact size of 12×12 mm^2 [14]. The patch is designed with five rings of 0.5 mm width arranged in Olympic shape with the outer radius of ring is $R = 1.5$ mm and inner radius of $r = 1$ mm. The geometry is simulated using finite element method-based computer simulation technology (CST) microwave studio. The initial design (front and back view) of antenna and concerned results are presented in Fig. 1a, b, c, and d. The geometry shows multiband behavior in the present state, and VSWR < 2 is obtained for the resonant frequencies (i.e., at 13.4, 21.1, 24.6, and 28.3 GHz), but some modification needs to be done in order to achieve broadband performance.

2.2 Parametric Analysis and Variation of Ring Width in Olympic Shape Antenna

The initial geometrical parameters of the antenna are altered for the best optimum results in K band applications. The parametric analysis for the ground size variation

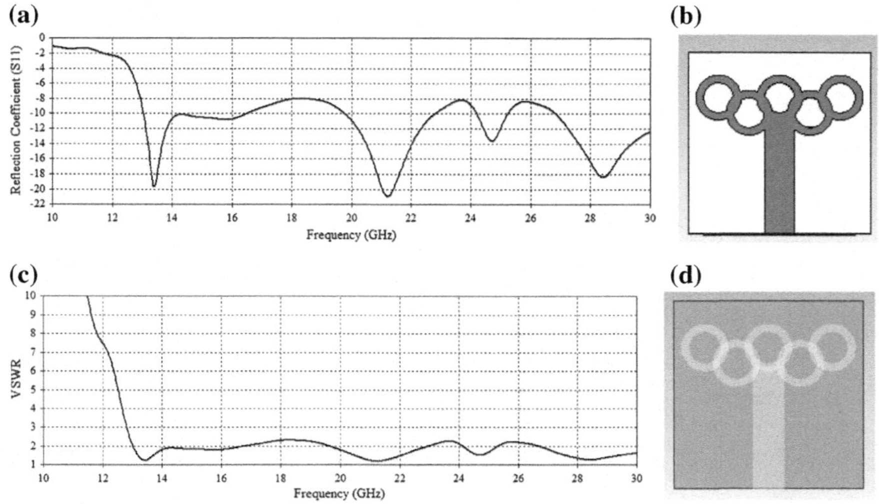

Fig. 1 **a** Variation of return loss with frequency for initial design, **b** front view of initial design, **c** variation of VSWR with frequency for initial design, **d** back view of initial design

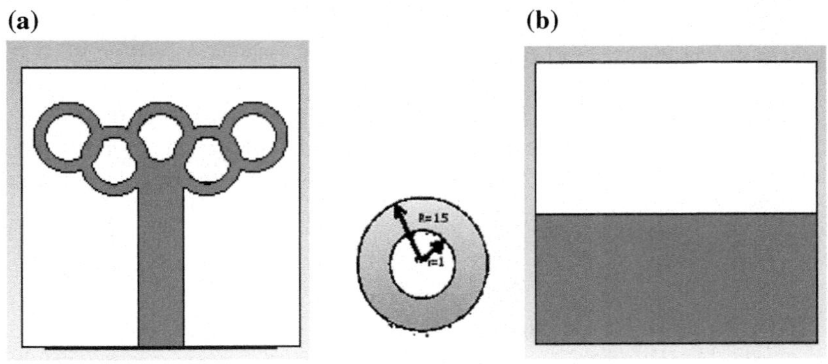

Fig. 2 **a** Front view of proposed antenna design, **b** back view of proposed antenna design

and ring width variation is carried out, and the final antenna prototype of the proposed antenna is selected. The final geometry parameters and prototype antenna are shown in Fig. 2a and b. The ring width is 0.5 mm; ground size is 5.5 mm, and all other parameters are same as in the initial design of the antenna.

The antenna characteristic variation can be studied as a function of ring width while keeping the radius of the outer ring constant (to keep the antenna size compact, i.e., 12 × 12 mm^2), and optimally best suited width can be selected for the wide bandwidth and optimum performance of the antenna. The results for the

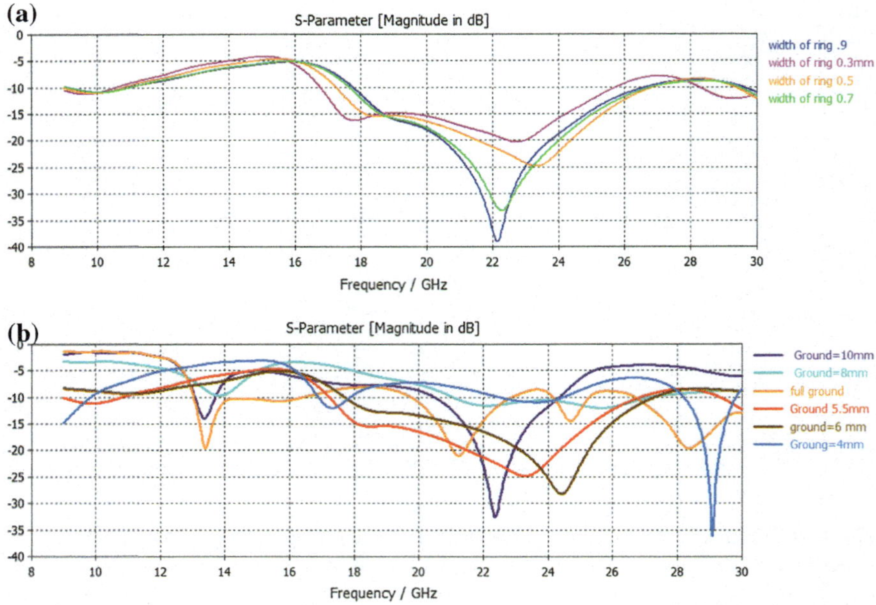

Fig. 3 **a** Effect of varying width of the ring on reflection coefficient as a function of frequency, **b** effect of varying ground size on reflection coefficient as a function of frequency

same are presented in Fig. 3. The width of the ring is varied from 0.3 to 0.9 mm. The best results are obtained for the 0.5 mm width of the ring for the 5.5 mm ground size, and the antenna characteristics are presented for the selected width of the ring in Fig. 3a and b.

3 Results and Discussion

The various radiation performance results of the proposed antenna are presented in the succeeding section. The return loss characteristics and VSWR for the concerned antenna are shown in Figs. 4 and 5, respectively. It can be depicted from the graph that impedance bandwidth of 9.2 GHz (17.2–26.7 GHz) has been obtained in the K band. The VSWR curve for the same is also shown in Fig. 4. The VSWR < 2 is obtained in the impedance bandwidth range. The proposed antenna resonates at 23.08 GHz which makes it suitable for the K band applications.

The gain variation with respect to frequency and variation of radiation efficiency with frequency for the proposed antenna is given in Fig. 6a and b, respectively. The stable positive gain is obtained in the frequency range 17.2–26.7 GHz range. The

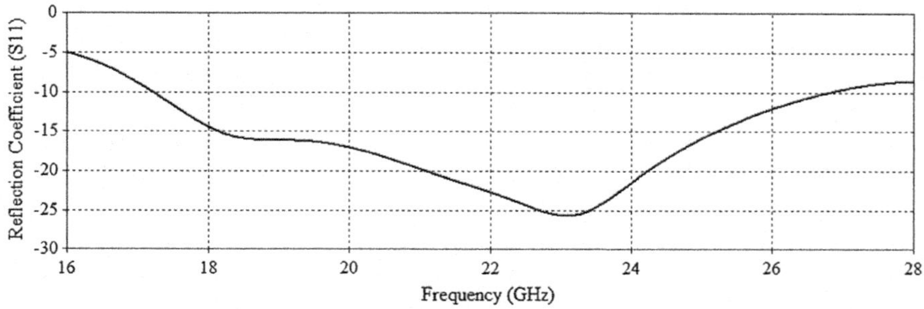

Fig. 4 Variation of reflection coefficient (S11) with frequency for proposed design

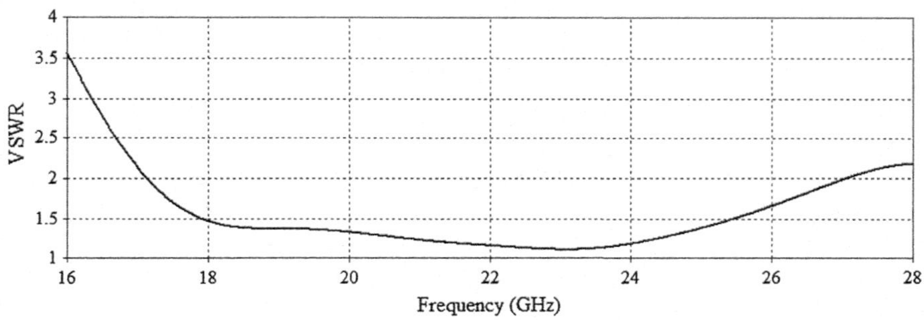

Fig. 5 Variation of VSWR with frequency for proposed design

maximum gain of 2.7 dBi is achieved at frequency of 25.5 GHz. The efficiency is almost above 70% in the obtained bandwidth. The axial ratio for the proposed antenna is 40 dBi constant in the given frequency range.

The surface current distribution for the antenna at resonant frequency is displayed in Fig. 7. The current distribution is more in the middle of ground plane and lower edges of the patch.

The E- and H-plane radiation patterns at resonant frequency of 23.08 GHz are shown in Fig. 8. The radiation pattern shows the bidirectional characteristic of antenna in E-field pattern with main lobe at 100° and angular width of 111.6°. In H-field radiation pattern, the main lobe appears at 125° and minor side lobes are also present. This bidirectional radiation pattern makes the antenna suitable candidate for the K band applications.

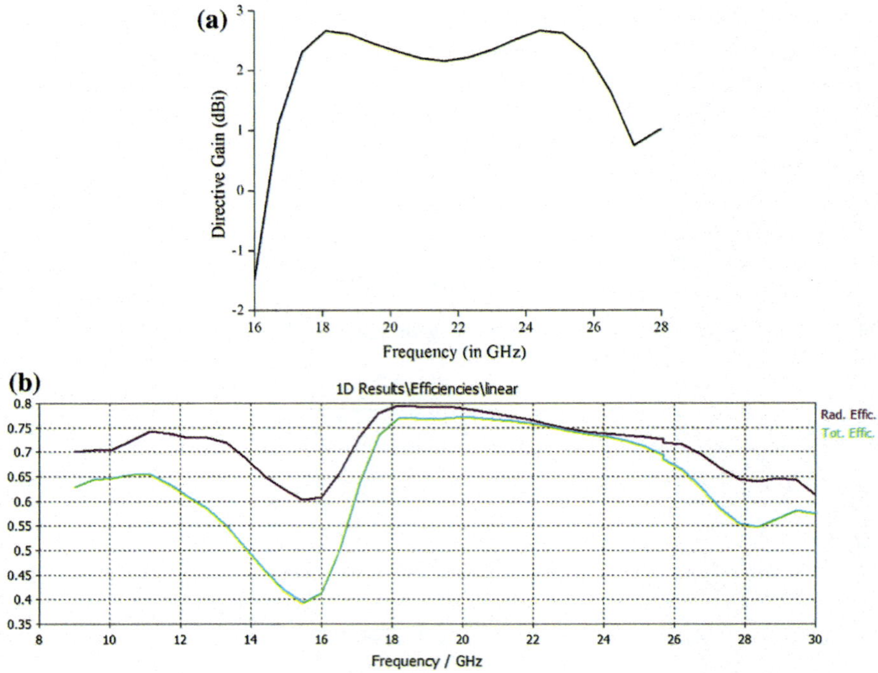

Fig. 6 **a** Variation of gain with frequency for proposed design. **b** variation of efficiency with frequency

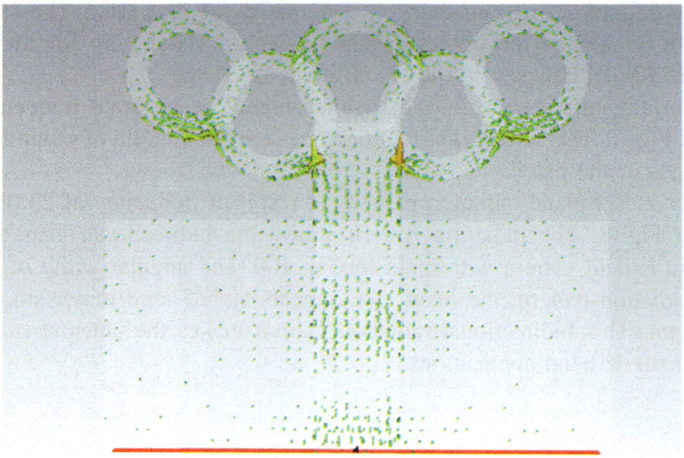

Fig. 7 Surface current distribution at resonant frequency

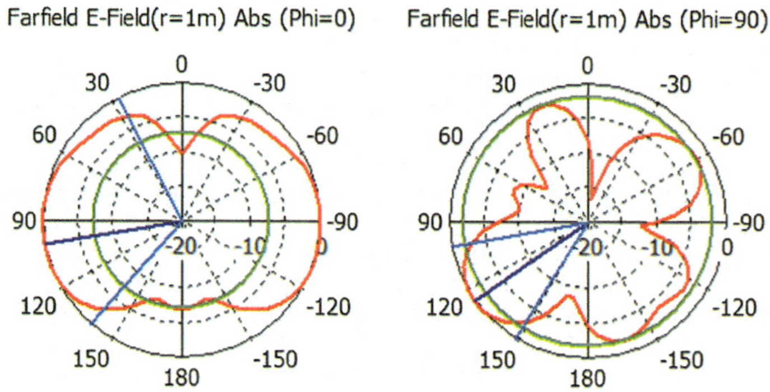

Fig. 8 E-field and H-field radiation pattern at resonant frequency

4 Conclusion

The compact Olympic ring-shaped linearly polarized planar antenna is presented for the K band applications. The design approach of the antenna is explained and studied in terms of return loss, realized gain, surface current distributions, and radiation patterns. The simulated results have been analyzed and discussed for the K band applications. The antenna provides impedance bandwidth of 9.5 GHz (17.2–26.7 GHz) and positive gain over the complete range of bandwidth. The compact size, wide bandwidth, and positive gain make the proposed antenna suitable for the K band applications.

References

1. Wong KL, Chen LY (2014) Small-size LTE/WWAN tablet device antenna with two hybrids feeds. IEEE Trans Antennas Propag 62(6):2926–2934
2. Bath A, Thakur A, Sharma J, Prasad B (2014) Design of a rectangular patch antenna. Int J Electr Electron Eng (IJEEE) 1(1):1–6
3. Rawat S, Sharma KK (2015) Annular ring microstrip patch antenna with finite ground plane for ultra-wideband applications. Int J Microw Wirel Technol 7(2):179–184
4. Balanis CA (2005) Antenna theory, analysis and design, 3rd edn. JohnSons, Inc.
5. Pozar MD (1992) Microstrip antennas. In: Proceedings of IEEE, 1
6. Garg R, Bhartia P, Bahl I, Ittipiboon A (2001) Handbook. Artech House, Inc.
7. Zhang L, Zhang Q, Hu C (2010) The influence of dielectric constant on bandwidth of U-notch microstrip patch antenna. In: Proceedings of IEEE international conference on ultra-wideband (ICUWB), vol 1, pp 1–4
8. Sharma MM, Yadav S, Kumar A, Ranga Y, Deepak B (2011) Compact elliptical microstrip patch antenna with slotted Ku-band applications. In: Applied electromagnetics conference. IEEE, pp 1–3
9. Parikh H, Pandey S, Modh K (2012) Wideband and high gain microstrip antenna for Ku-band application. In: International conference on engineering (NUiCONE). Nirma University, pp 1–5

10. Azim R, Islam MT, Misran N (2011) Dual polarized microstrip patch antenna for Ku-band application. Informacije MIDEM, 41(2):114–117
11. Ullah MH, Islam MT, Mandeep JS, Misran N (2013) Design and analysis of a multi band electrically small antenna using ceramic material substrate. Przegląd Elektrotechniczny 89 (1):271–274
12. Dubey SK, Pathak SK, Modh KK (2011) High gain multiple resonance Ku-band microstrip patch antenna. In: Proceedings of the IEEE applied electromagnetics conference. Kolkata, India, pp 1–3
13. Misran N, Islam MT, Yusob NM, Mobashsher AT (2009) Design of a compact dual band microstrip antenna for ku-band application. In: Proceedings of the international conference on electrical engineering and informatics (ICEEI '09). Selangor, Malaysia, pp 699–702
14. Peng C, Zheqiang W, Guangli Y (2016) A wideband WLAN 2.4/5.2/5.8 GHz MIMO antenna based on cavity mode in full metal cover tablet computer. In: Progress in electromagnetic research symposium (PIERS), vol 64, pp 169–177

A Monopole Antenna with Reconfigurable Notched Characteristics from WLAN-band Notched UWB to ITU-band Notched UWB Antenna

Dinesh Yadav, Mahesh P. Abegaonkar, Shiban K. Koul, Vivekanand Tiwari and Deepak Bhatnagar

Abstract This paper presents a monopole ultra wideband (UWB) antenna with reconfigurable notched characteristic. The proposed antenna is capable to switch its resonance from 5.2 GHz centre frequency band-notched UWB to 8.2 GHz centre frequency band-notched UWB applications. It consists of H-shaped reconfigurable parasitic resonator in ground plane to obtain the band notched at 5.2 and 8.2 GHz centre frequency bands by the mean of two p-i-n diodes. The impedance bandwidth enhancement of the antenna is achieved by inserting a slot in the ground plane. The proposed antenna is optimized and simulated using CST—microwave studio software.

Keywords Band notched · Parasitic resonator · Reconfigurable · Switchable notch · Ultra wideband

---

D. Yadav (✉) · V. Tiwari
Department of Electronics and Communication Engineering,
Manipal University Jaipur, Jaipur, Rajasthan, India
e-mail: dinesh.yadav@jaipur.manipal.edu

V. Tiwari
e-mail: vivekanand.tiwari@jaipur.manipal.edu

M. P. Abegaonkar · S. K. Koul
Centre for Applied Research in Electronics (CARE),
Indian Institute of Technology (IIT) Delhi, Hauz Khas, New Delhi, India
e-mail: mpjosh@care.iitd.ac.in

S. K. Koul
e-mail: s.k.koul@ieee.org

D. Bhatnagar
Microwave Lab, Department of Physics, University of Rajasthan, Jaipur,
Rajasthan, India
e-mail: dbhatnagar_2000@rediffmail.com

V. Janyani et al. (eds.), *Optical and Wireless Technologies*, Lecture Notes in Electrical Engineering 472, https://doi.org/10.1007/978-981-10-7395-3_72

1 Introduction

The monopole ultra wideband (UWB) antenna attracts more attention among the various antenna configurations, because it can be easily matched over the entire UWB bandwidth starting from 3.1 to 10.6 GHz [1]. These antennas are capable to utilize whole spectrum of UWB, when no interfering signal appears in close vicinity. Moreover, such a wide frequency range of UWB system may affect the existing narrowband systems, such as WLAN (5.15–5.35 GHz and 5.725–5.825 GHz bands), WiMAX (3.3–3.6 GHz band), HIPERLAN (5.45–5.725 GHz band) and ITU (8.02–8.4 GHz band). Therefore, to reduce the signal interference of UWB system with narrowband systems and vice-versa, UWB antenna with band-notched characteristics is required. Practically, the physical size of the antenna does not allow to design separate antennas for each application. In recent years, researchers are more interested in frequency reconfigurable antennas, which can switch resonance from one operating band to another operating band, without changing the physical size of the single fed antenna by using diode switches. Therefore, various single fed antenna configurations have been proposed that can be switched in one and more operating bands [2–5]. In [4], a reconfigurable monopole antenna with controllable band-notched characteristics using two diodes is demonstrated experimentally. Several printed monopole antennas with reconfigurable band-notch characteristics have been investigated, to minimize the signal interference due to presence of narrowband signals between 3.1 and 10.6 GHz UWB system [6–9]. In [6], a single band-notched (5.03–5.94 GHz) reconfigurable slot antenna with a switchable characteristic over the UWB spectrum and notched-band ultra wideband antenna with p-i-n diode switch have been demonstrated. While in [7], RF microelectromechnical system (MEMS) switches are incorporated to achieve the reconfigurable band-notched characteristics. By inserting two slots on the radiating patch to embed two p-i-n diodes for achieving switchable single and dual band notch performance by controlling the bias voltage of the p-i-n diodes is reported in [9].

In this paper, a monopole UWB antenna with H-shaped parasitic switchable resonator in the ground plane is presented by using diode switches. The proposed antenna is proficient to switch its resonance from 5.2 GHz band notch to 8.2 GHz band notch with in the UWB applications. The biasing circuitry comprises of two p-i-n diodes and three inductors.

2 Antenna Designing and Analysis

The front and back view of the proposed circular monopole frequency reconfigurable antenna with all dimensions in mm is shown in Fig. 1. The overall size of antenna is $30 \times 22\,\text{mm}^2$, which is designed on 0.76 mm (h) thick RO3003 substrate (loss tan-

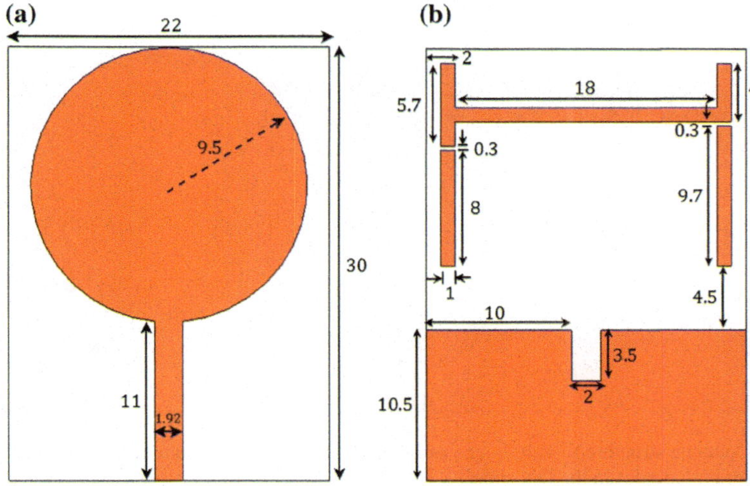

Fig. 1 Monopole UWB antenna structure **a** front view and **b** back view (all dimensions in mm)

gent $\tan\delta = 0.0013$ and dielectric constant $\varepsilon_r = 3$). The width of the 50 Ω microstrip line is 1.92 mm, which is calculated using [10] for the lower cutoff frequency $f_r =$ 3.1 GHz. The optimized radius of the circular radiator to achieve wide impedance bandwidth is 9.5 mm, which leads to fix the length of the fed line at 11 mm. The back side of the proposed design consists of defective ground plane and H-shaped parasitic resonator. The size of metallic ground is $10.5 \times 22\,\text{mm}^2$, which has a rectangular slot of the size $3.5 \times 2\,\text{mm}^2$ at the centre to the upper edge of metallic ground. The slot is inserted in metallic ground plane to enhance the impedance bandwidth 3.1–10.6 GHz. The H-shaped parasitic resonator is placed 4.5 mm above to the ground plane and symmetric to the radiating edge of the circular patch. The orientation of the parasitic resonator is such a way that the vertical stubs of the H-shaped resonator is tangential to the radius of the upper circular patch. Two slots of the size 0.3 mm are inserted at a distance of 8 and 9.7 mm from the lower edge of H-shaped parasitic resonator to place two p-i-n diodes in between the slots. The antenna structure with two p-i-n diodes (D_1 and D_2), three inductors (L_1, L_2 and L_3) and three biasing pads of the size $2 \times 1\,\text{mm}^2$ is shown in Fig. 2a. In design, diodes are used to connect and disconnect the stubs of the H-shaped parasitic resonator, and inductors are used as RF chokes in bias circuitry. At zero bias and forward bias voltage, the antenna structure is shown in Fig. 2b and Fig. 2c, respectively.

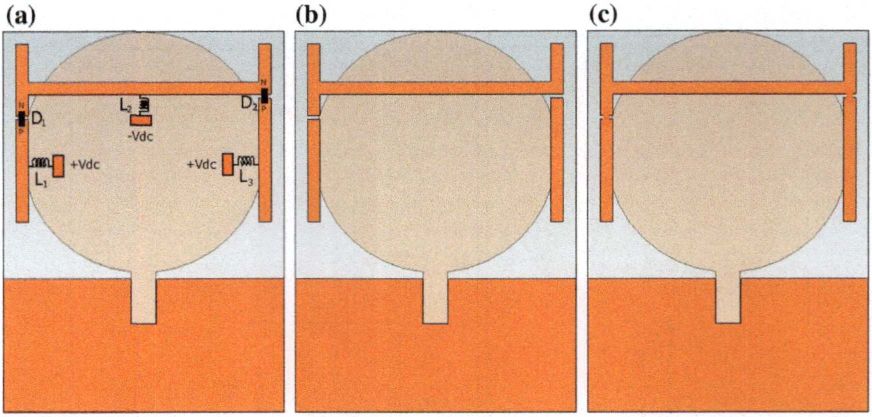

Fig. 2 Antenna structure **a** with diodes and biased circuitry **b** with both diodes in the OFF-state and **c** with both diodes in the ON-state

3 Results and Discussions

The full wave simulation of the proposed antenna structure is being done using computer simulation technology (CST)—microwave studio software [11]. The comparison of the magnitude of return loss at different physical states of the antenna is shown in Fig. 3. It can be observed that the basic optimized UWB antenna with circular radiating patch and the ground plane in the absence of slot and without parasitic resonator has potential to achieve the wide impedance bandwidth from 3.2 to 9.5 GHz. To maximize the UWB impedance bandwidth, specifically at higher operating band, a rectangular slot is inserted at the metallic ground plane, and as a result, enhancement in impedance bandwidth from 3.1 to 13 GHz is achieved. Now in next step, the designing of band-notched UWB antenna with switchable notch is achieved using H-shaped parasitic resonator consist of biased circuitry. In CST software, the ON- and OFF-state of the diodes could be done by the connected metal and the disconnected metal. When both diodes (D_1 and D_2) are in the OFF-state with the zero bias voltage, then H-shaped parasitic resonators vertical stubs are isolated with upper structure of parasitic resonator. As a result, the upper separated parasitic structure produces the band notched at 8.2 GHz centre frequency ITU band. Similarly, when both diodes (D_1 and D_2) are in the ON-state with the forward bias voltage, then H-shaped parasitic resonator shows continuity for the resonance and provides the band notched at 5.2 GHz centre frequency WLAN band. Table 1 depicts the performance of antenna in terms of operating and notched band in response to state of the diodes and shapes of the ground plane.

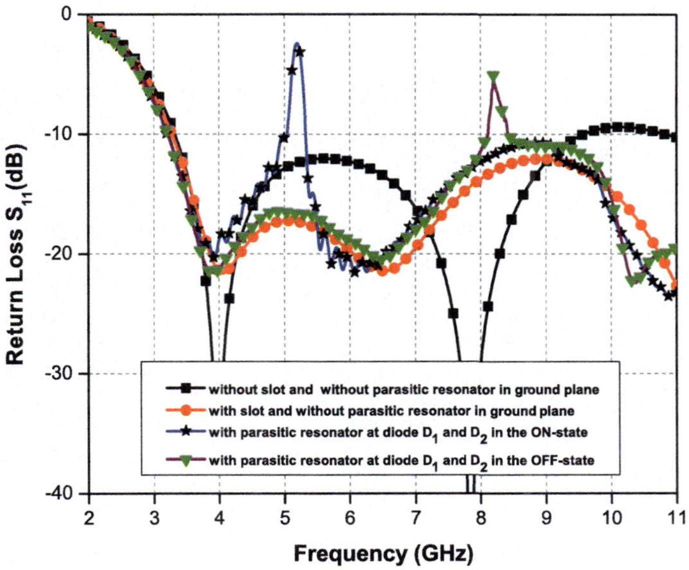

Fig. 3 Comparison of return loss for the proposed antenna at different antenna configurations

Table 1 Simulated frequency bands at different antenna configurations

Antenna configurations	Frequency bands (GHz)	
	Operating bands	Notched bands
Without slot and without parasitic resonator in ground plane	3.2–9.5	–
With slot and without parasitic resonator in ground plane	3.1–13	–
With parasitic resonator at diode D_1 and D_2 in the ON-state	3.1–5 and 5.35–13	5–5.35
With parasitic resonator at diode D_1 and D_2 in the OFF-state	3.1–8 and 8.45–13	8–8.45

3.1 Surface Current Distribution

The surface current distribution of the proposed antenna at notched centre frequencies (5.2 and 8.2 GHz) with diode bias states is shown in Fig. 4. It can be seen that, at 5.2 GHz with diodes (D_1 and D_2) in the ON-state, the surface currents are strongly concentrated around the H-shaped parasitic resonator as shown in Fig. 4a. At 8.2 GHz, as diodes (D_1 and D_2) goes into the OFF-state, the H-shaped parasitic resonator disintegrates into three parts, and the surface current has major concentration

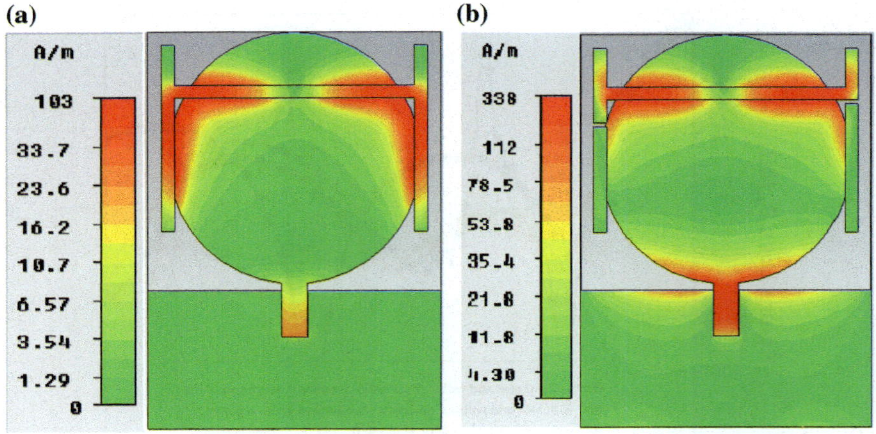

Fig. 4 Surface current distribution at notched frequencies **a** 5.2 GHz with both diodes in the ON-state and **b** 8.2 GHz with both diodes in the OFF-state

on the upper section as shown in Fig. 4b. The hindrance offered to the radiation due to concentration of the surface current along the parasitic resonators causes band notched at 5.2 GHz (both diodes in the ON-state) and 8.2 GHz (both diodes in the OFF-state).

3.2 *Radiation Pattern and Realized Gain*

Figure 5 shows the E- and H-plane radiation pattern of the proposed antenna at the resonant frequencies of the operating band at 4, 7 and 9 GHz. The state of the diodes does not have any significant effect on the radiation pattern at the resonant frequencies of the operating band. It can be observed that E- and H-plane patterns at three operating frequency are almost omnidirectional. The comparison between simulated

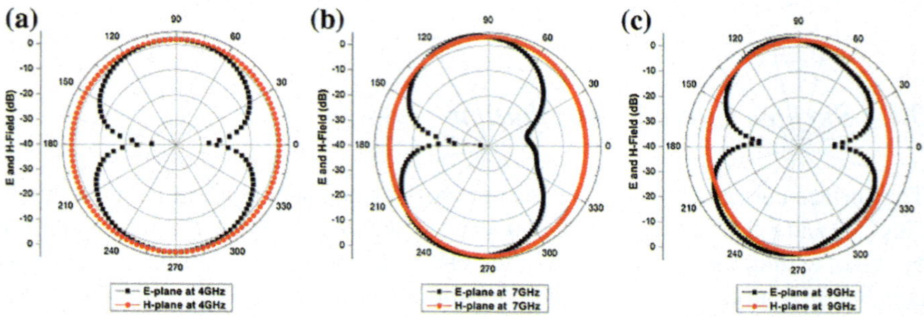

Fig. 5 The computed E-plane and H-plane radiation patterns at **a** 4 GHz **b** 7 GHz and **c** 9 GHz

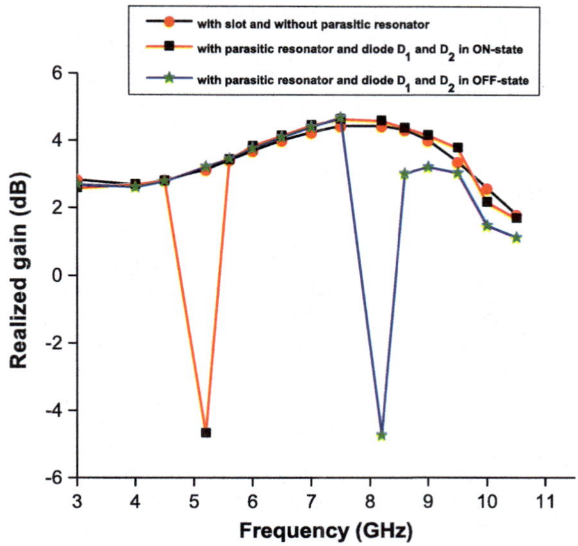

Fig. 6 Comparison between realized peak gains at different antenna configurations

peak realized gains of the proposed antenna at different configurable of the proposed geometry is illustrated in Fig. 6. In first configuration, a monopole circular antenna with slot in ground plane and without parasitic resonator, it is observed that antenna exhibits good gain over the entire operating band, and the maximum gain of 4.5 dB is achieved at 7.5 GHz. However, in second configuration, H-shaped parasitic resonator is added with biased circuitry (diodes D_1 and D_2 in the ON-state), a drastic decrement in the realized gain at 5.2 GHz centre band frequency is observed. Similarly, in third configuration, with parasitic resonator and with biased circuitry (diodes D_1 and D_2 in the OFF-state) approximately same gain as first antenna configuration is achieved over the entire band, except appreciable drop is observed at 8.2 GHz centre band-notched frequency. In all three antenna configuration, simulated realized gains are stable and uniform at resonating frequencies of the operating band accept notched band. It is also observed that in second and third physical state, due to the loading of the passive components (diodes and RF–choke inductors), at higher frequencies, the achieved realized gain is less than 2 dB.

4 Conclusion

A monopole circular UWB antenna with reconfigurable H-shaped parasitic resonator has been studied. It is found that the antenna is capable of utilizing the entire UWB bandwidth with switchable band-notched characteristics, when interfering narrowband signal (WLAN/ITU) is appearing in close proximity. The switching of the notched band can be controlled by the bias voltage of p-i-n diodes, and single antenna can be exploited for the two operating modes without changing the physical dimensions of the structure.

References

1. US Federal Communications Commission et al. Fcc revision of part 15 of the commission's rules regarding ultra-wideband transmission systems: first report and order. Technical report, technical report. Feb 2002
2. Tariq A, Ghafouri-Shiraz H (2012) Frequency-reconfigurable monopole antennas. IEEE Trans Antennas Propag 60(1):44–50
3. Wei F, Qin PY, Jay Guo Y (2015) A wideband-to-narrowband tunable antenna using a reconfigurable filter. IEEE Trans Antennas Propag 63(5):2282–2285
4. Ojaroudi N, Amiri S, Geran F (2012) Reconfigurable monopole antenna with controllable band-notched performance for UWB communications. In 2012 20th telecommunications forum (TELFOR). IEEE, pp 1176–1178
5. Al-Zayed AS, Kourah MA, Mahmoud SF (2014) Frequency-reconfigurable single-and dual-band designs of a multi-mode microstrip antenna. IET Microwaves, Antennas Propag 8(13):1105–1112
6. Valizade A, Ghobadi C, Nourinia J, Ojaroudi M (2012) A novel design of reconfigurable slot antenna with switchable band notch and multiresonance functions for UWB applications. IEEE Antennas Wirel Propag Lett 11:1166–1169
7. Nikolaou S, Kingsley ND, Ponchak GE, Papapolymerou J, Tentzeris MM (2009) UWB elliptical monopoles with a reconfigurable band notch using MEMS switches actuated without bias lines. IEEE Trans Antennas Propag 57(8):2242–2251
8. Loizeau S, Sibille A (2014) Reconfigurable ultra-wide band monopole antenna with a continuously tunable band notch. IET Microwaves Antennas Propag 8(5):346–350
9. Badamchi B, Nourinia J, Ghobadi C, Shahmirzadi AV (2014) Design of compact reconfigurable ultra-wideband slot antenna with switchable single/dual band notch functions. IET Microwaves Antennas Propag 8(8):541–548
10. Kumar G, Ray KP (2003) Broadband microstrip antennas. Artech House
11. CST Microwave Studio—Computer simulation technology (2013)

Enhancing QoS for Multimedia Services Using Mobility-Aware Bandwidth Estimation Algorithm in MANETs

Lokesh Sharma, Chhagan Lal, Devi Prasad Sharma and Pallavi Kaliyar

Abstract Over the years, a large array of multimedia services started using mobile ad hoc networks (MANETs) due to its attractive features such as low-cost quick deployment, node mobility, and distributed networking. However, multimedia applications also require provisioning of adequate quality of service (QoS) to enhance quality of experience (QoE) of end users. In order to provide QoS in a network, the accurate knowledge of the remaining network resources at any time is the key. But, due to the inherent characteristics of MANETs such as dynamic topology, high bit-error-rate communication channels, and decentralized control, estimating remaining network resources is a challenging task. In this paper, we propose a novel remaining bandwidth estimation algorithm which considers the node (or link) mobility into consideration while performing dynamic bandwidth estimation on a node. The estimated link bandwidth value is then used during the route discovery process to select routes with adequate bandwidth and thus increases the QoS of the data flow. The proposed algorithm is verified using real-time video traffic (i.e., H.264/SVC encoded video traces) using a industry standard simulator called Qualnet. The simulation results show that the proposed approach increases the decision accuracy of the soft admission control process and thus leads to the enhancement of the quality of the received video traffic.

Keywords Mobility · QoS · Bandwidth estimation · AODV

L. Sharma (✉) · D. P. Sharma
Manipal University Jaipur, Jaipur, India
e-mail: lokesh.sharma@jaipur.manipal.edu

D. P. Sharma
e-mail: deviprasad.sharma@jaipur.manipal.edu

C. Lal
University of Padova, Padova, Italy
e-mail: chhagan@math.unipd.it

P. Kaliyar
SKIT, College of Engineering, Jaipur, India
e-mail: pallavi.kaliyar@gmail.com

V. Janyani et al. (eds.), *Optical and Wireless Technologies*, Lecture Notes in Electrical Engineering 472, https://doi.org/10.1007/978-981-10-7395-3_73

1 Introduction

The IEEE 802.11-based networks provide QoS level using the service differentiation stated in the IEEE 802.11e amendment. However, still there is no solution defined as standard for estimation and allocation of the network resources. Bandwidth is one of the most critical network resources for real-time flows. Due to the absence of precise estimation of available bandwidth, it is difficult to provide safe admission control. The estimated available bandwidth can be defined as maximum throughput achieved between two adjacent nodes while keeping the existing ongoing ows in the network. Thus, the available bandwidth is influenced by the current ows in the network. Consequently, if the admission of ows whose bandwidth needs is more outstanding than the available bandwidth in network, leads to false admission control. Therefore, accuracy in the estimation of the usable bandwidth is a decisive and vital issue.

To provide true admission control without disturbing the existing flows, we proposed an improved approach for bandwidth estimation by considering the mobility of the nodes. Nodes mobility causes the frequent link failures and re-transmissions. The effects of the mobility are not considered in the existing solution for network resource estimations. Further, on the basis of true available bandwidth estimations, a reactive route discovery process proposed that discovering end-to-end path from source to destination node covering all the intermediate nodes satisfies the bandwidth requirement requested by the application to support desired QoS.

The remaining part of the paper is organized as follows: The next part is the review of the existing solutions and methodologies suggested for bandwidth estimations in the literature. Then, we present our proposed methodology. It is followed by the simulation results of the comparison of proposed new protocol with the existing solutions like BAND-MR [1], which is one of the closest QoS-aware reactive routing solutions suggested, and AODV routing protocol [2]. In the last section, we provide the conclusion and future scope of the proposed work.

2 Related Work

This section presents the existing solutions for bandwidth estimation and admission control to support QoS-enabled delivery of multimedia traffic in MANETs. Hanzo and Tafazolli [3] in their paper presented a survey of the main contributions to the group of QoS routing solutions for MANETs available in the period 1997–2006. They also presented the process and their communications with the medium access control (MAC) protocol. In another paper, Prasad et al. [4] explained the state-of-the-art bandwidth estimation techniques, metrics, and practices used, and tools that instrument with them.

Sarr et al. [5], in another paper, broadly categorized the bandwidth estimation approaches in two streams: (1) Active approaches: used for end-to-end estimation of the available bandwidth by sending the sequence of the probe packets along the route. (2) Passive approaches: uses the local information to estimate the bandwidth utilization. The estimation is based on the channel sensing (ideal and busy time of channel). In another paper, the authors [6] proposed new solutions for enhancing the performance of QOS-aware routing and admission control protocols in the presence of mobility, shadowing, and fluctuating link signal-to-interference-plus-noise ratio (SINR).

In another literature, Belbachir et al. [7] focused on the QoS enabling using resources allocation and estimation. They considered bandwidth as a key resource, and in order to estimate it, they proposed a new approach named Improved Bandwidth Estimations through Mobility incorporation (IBEM) MANETs. In another paper, Chaudhari and Biradar [8] proposed a scheme of estimating available bandwidth (ABW) in MANETs using the actual channel utilization and collision rate. They modified the existing collision probability model and dynamically computed separate Lagrange interpolation polynomial at each node according to node behavior.

3 Proposed Methodology

For efficient admission control in MANET, information about the available bandwidth at each node as well as its neighbors is required. The available bandwidth of any node can be calculated at its physical layer by passively monitoring the network activities. In [9], the authors suggested a method to measure available bandwidth of a node based on the busy and ideal time of the channel. When a node transmits and receives information from the neighbor node, the channel will be considered as busy period (denoted as Pr_{busy}). Let the busy/ideal period of the medium get updated in every T time known as the computational/measurement time period. Then, the ideal time of the channel can be expressed by $Pr_{idle} = (1 - Pr_{busy})$. The available bandwidth at any node on the basis of ideal time can be defined as

$$\mathbf{BW} = \frac{Pr_{idle}}{T} \cdot C \tag{1}$$

where C is the channel capacity. However, the expresses mentioned in Eq. 1 are not correct as different nodes have different contention ratio and sensing range. Let us assume two nodes V_1 and V_2 create a wireless link as shown in Fig. 1. The available bandwidth at the node V_1 and V_2 will be different. Therefore, the available bandwidth of the link between V_1 and V_2 is defined as

$$\mathbf{BW}_{\{V_1,V_2\}} = min\{\mathbf{BW}_{V_1}, \mathbf{BW}_{V_2}\} \tag{2}$$

Fig. 1 Wireless link connecting 2 nodes

Therefore, the available bandwidth of the link will be the minimum of the bandwidths available at two nodes connecting through the wireless link. However, as per the authors [5], the expression written in Eq. 2 is only valid when the computation period T of nodes V_1 and V_2 does not overlap, and hence, ideal period of channel in both communicating nodes is synchronized. This requires strict clock synchronization of the communication nodes. If the ideal period of channel in the nodes is not synchronized, it is suggested to use the average of the overlap times to evaluate the available bandwidth [2]. So, on single wireless link between nodes V_1 and V_2, the available bandwidth is newly defined as being different. Therefore, the available bandwidth of the link between V_1 and V_2 is defined as

$$\mathbf{BW}_{\{V_1,V_2\}} = \left(\frac{Pr_{idle}^{V_1}}{T}\right) \cdot \left(\frac{Pr_{idle}^{V_2}}{T}\right) \cdot C \tag{3}$$

where $Pr_{idle}^{V_1}$ and $Pr_{idle}^{V_2}$ are the ideal periods of channel at nodes V_1 and V_2, respectively, and $(\frac{Pr_{idle}^{V_1}}{T}) \cdot (\frac{Pr_{idle}^{V_2}}{T})$ is the average of the channel availability on the entire link.

For more reliability and accuracy in the measurement of bandwidth, we need to consider more network features. Mobility is the vital issue in MANETs. Mobility leads to frequent topology change and link failures because of the change in the distance between the communicating nodes. Let us take an example for mobility as shown in Fig. 2. At the time instance t_1, the two nodes V_1 and V_2 are not in the communication range of each other. That is, the distance d between the two nodes is larger than communication distance $D_{comm}(d > D_{comm})$. Thereafter, at time instance

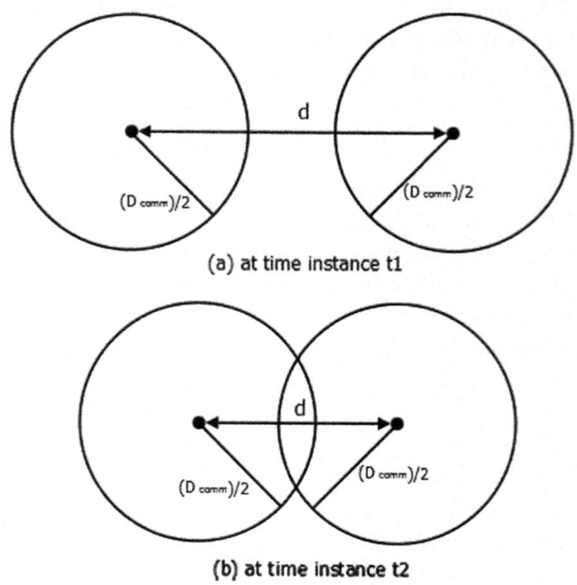

Fig. 2 Example of mobility of two modes

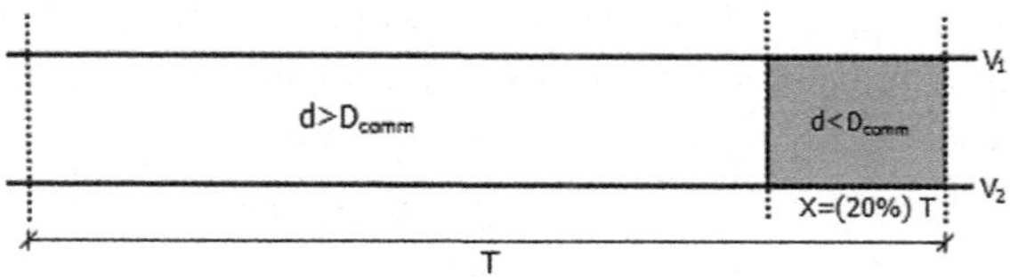

Fig. 3 Link status during T measurement period

t_2 both the nodes come into the communication range of each other. The distance d between the two nodes is smaller than communication distance $D_{comm}(d < D_{comm})$.

Let us take an example shown in Fig. 3, the wireless link state between nodes V_1 and V_2 during the measurement period T. Here, it is mentioned that this link exists only during $X = 20\%$ of T (X value denotes the percentage of the link presence during the period T). We intend to compute the available bandwidth on the link available between the nodes V_1 and V_2 (the scenario depicted in Fig. 3) with the measurement method defined in Eq. 3. As per the author [main paper ref], this method largely overestimates the available bandwidth. This is due to the negligence of the mobility and leads to false admission control and QoS violation. Now, the expression to evaluate the bandwidth can be defined as including the effect of mobility

$$\overbrace{\mathbf{BW}}_{\{V_1,V_2\}} = \mathbf{BW}_{\{V_1,V_2\}} \cdot MC \tag{4}$$

where $\mathbf{BW}_{\{V_1,V_2\}}$ is mentioned in Eq. 3 and MC is the mobility criteria. With the predicted value of MC, let us consider the scenario depicted in Fig. 4. It illustrates a link between nodes V_1 and V_2 during consecutive measurement period T, with two [(a) and (b)] alike activities scenarios (the same bandwidth consumption), but in different states:

- *Stable Link* (Fig. 4a): the communication distance D_{comm} between the two nodes of this link is greater than the distance d of at any time; we note: $\forall t_i / i \in [0, \ldots, n]$,

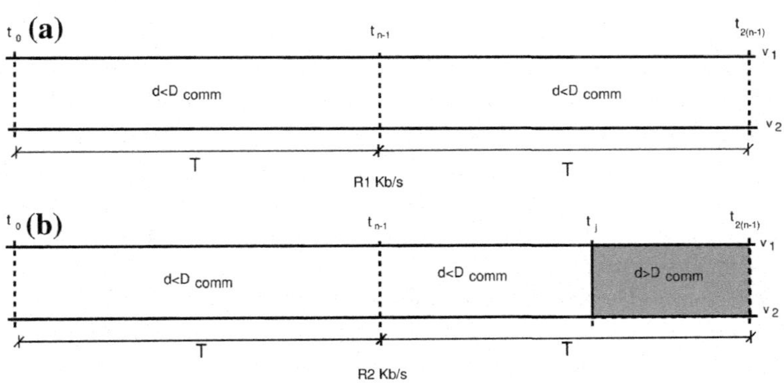

Fig. 4 Link expiration during bandwidth estimations. **a** Stable link. **b** Unstable link

$D_{comm} > d$ (where t_i is time unit of calculus in T period, and $n \in N^*$). This type of links in wireless networks is considered as stable. Most of the authors till now measure the available bandwidth on the stable links.

- *Unstable Link* (Fig. 4b): When we consider a link between two nodes is dynamic or unstable, at some instance, the communication distance D_{comm} between the two nodes of this link is less than the distance d, $D_{comm} < d$ (which can be observed at t_j instant in Fig. 4b), and we note: $\forall t_i / i \in [0, \ldots, n]$.

Algorithm 1 Mobility-aware QoS route discovery process

// Variables used in algorithms
N = Node, S = Source Node, D = Destination Node, I = Intermediate Node
BW_{req} = Bandwidth required for admission control, BW_{est} = Estimated Bandwidth
if Node S has data flow for Node D **then**
    S estimate the BW_{est} using the Eq. 1
    if S(BW_{est}) > (BW_{req}) **then**
        Accept the data flow and generate RREQ
    else
        discard the data flow
    end if
end if
if Node I receives a non-duplicate *RREQ* message **then**
    I calculates BW_{est} using Eq. 7
    if I(BW_{est}) > (BW_{req}) **then**
        Node I update the info in RREQ Packet and
        broadcast further RREQ packet
    else
        generate RERR packet and send to Node S
    end if
end if
if D receives a non-duplicate RREQ message and BW_{req} satisfied **then**
    Node D sent a unicast RREP to Node S
end if

As the scenario depicted in Fig. 4a, b, in the first measurement period T (from t_0 to t_{n-1}), the estimated bandwidths R_1(Kb/s) and R_2(Kb/s) between nodes V_1 and V_2 will be equal as there is no link breakage due to the nodes mobility ($D_{comm} > d$). But in the next measurement period T (from t_{n-1} to $t2_{n-1}$), the estimated bandwidths R_1(Kb/s) and R_2(Kb/s) between nodes V_1 and V_2 will be different as link breakage occurs because of the node mobility ($D_{comm} < d$) at time instance t_j, where $t_{n-1} < t_j < t2_{n-1}$. Therefore, in second scenario (Fig. 4b), the link is only available from t_{n-1} to t_j time instance of T. The estimated bandwidth will depend on the probability of link availability. Therefore, to estimate the available bandwidth R_2(Kb/s) scenario (Fig. 4b), we have to take into account the fragment (Y) and the mobility criteria M (defined as link existence probability) during the measurement period T

$$MC = \frac{t_j - t_{n-1}}{T} \tag{5}$$

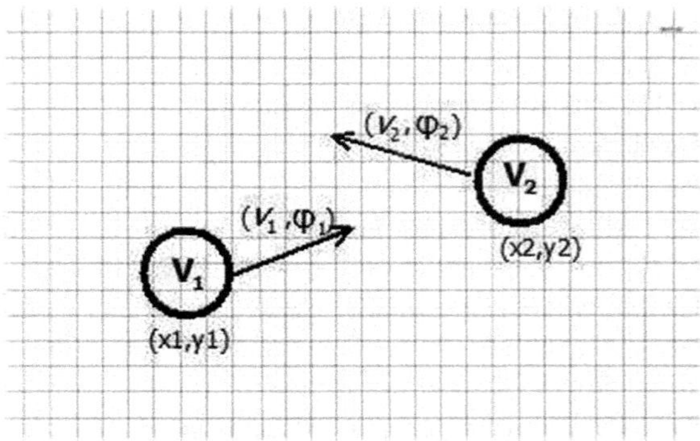

Fig. 5 Mobility of the nodes (coordinates, velocity, and direction)

t_j is the time instance ($t_{n-1} < t_j < t2_{n-1}$) when the $D_{comm} < d$ and link failure occurs.

To identify the value of t_j, authors [10] suggested fairly comprehensive approach based on the motion velocity (v) and the motion direction (φ). According to [11], each node has its geographical coordinates provided with global positioning system (GPS). Consequently, as shown in Fig. 5, each mobile node in network is able to identify the coordinates of its position, the initial coordinates, and the destination coordinates of its motion. Each node informs its neighbor nodes about its coordinates and motion parameters (direction and velocity). Consider that node V_1 (respectively V_2) moves with velocity v_1 (respectively the velocity of node V_2 (v_2)) and follows direction φ_1 (respectively the direction of node V_2 (φ_2)). t_j value is calculated as (e.g., at node V_1):

$$t_j = \begin{cases} \infty & if(v_1 = v_2) \wedge (\varphi_1 = \varphi_2) \\ \frac{-(pq+rs)+\sqrt{(p^2+q^2)d^2-(ps-rq)^2}}{(p^2+r^2)} & \text{otherwise} \end{cases} \tag{6}$$

where $p = v_1 \cdot \cos(\varphi_1) - v_2 \cdot \cos(\varphi_2)$, $q = (x_1 - x_2)$, $r = v_1 \cdot \sin(\varphi_1) - v_2 \cdot \sin(\varphi_2)$, $s = (y_1 - y_2)$.

The objective of the back-off process is to reduce the probability of channel access for some time period. During the back-off time period, channel may be ideal, but it will be considered as interfering (busy). Because of this, a portion of bandwidth is lost during the back-off time period. Hence, we also consider the collision probability of the measurement to be more realistic, and final accurate bandwidth is given in Eq. 7

$$\overbrace{\mathbf{BW}}_{\{V_1,V_2\}} = \left(\frac{Pr_{idle}^{V_1}}{T}\right) \cdot \left(\frac{Pr_{idle}^{V_2}}{T}\right) \cdot C \cdot MC \cdot (1 - Pr_c) \cdot (1 - backoff) \tag{7}$$

4 Algorithms Design

AODV HELLO packets are being used for exchanging the mobility parameters between the neighbor nodes. Modified HELLO packets carry the movement velocity and the coordinates of the beginning and ending of its motion. Using these beginning and ending coordinates, nodes calculate the direction of the motion respectively using Eq. 1. When network layer receives data from upper layers that to be delivered at destination node, it estimates the available bandwidth using the equation. If the available bandwidth is greater than the bandwidth requisite of the application to support QoS delivery of the traffic, the source node broadcasts the RREQ packet to initiate the route discovery process. We have modified the RREQ packets as shown in figure to carry the requisite bandwidth and the medium availability rate.

When a RREQ packet reaches an intermediate node, the node estimates the available bandwidth of the link using Eq. 7. The bandwidth requirement is being validated against the estimated available bandwidth as per the admission control constraints and QOS support. If the validation is true, the node replaces the information passed by the predecessor node in RREQ packet by its own information and broadcasts it further; else, the node discards the RREQ packet.

When the RREQ packet reaches the destination node and the admission control requirements found acceptable, the node generates the RREP packet and acknowledges it back to the source node on the reserve route. When a node finds that the available bandwidth is not sufficient to support the requisite QOS and admission control requirements or link failure occurs due to nodes mobility, then it generates a RERR packet and forwards it to the source node. On receiving this REER packet, the source node pauses the transmission of the data and initiates new route discovery process.

5 Simulation and Result Analysis

In this section, we present the details of the simulation setup used to test the effectiveness of our proposed approach and analysis of the results obtained. We use the Qualnet simulator to create the target MANET scenario which consists of the following main parameters: (i) Simulation time is set to 500 s; (ii) random waypoint mobility model with constant node mobility of 10 m/s is configured; (iii) routing protocols used are AODV [2], BAND-MR [1], and MBA-AODV (i.e., our proposed); (iv) the network consists of 60 nodes deployed randomly in an area of $1000 \times 1000\,\text{m}^2$; and (v) video trace file generated from a real-time video stream with variable bit rates per second (i.e., from 33 bytes to 37,318 bytes) are used to model the source nodes.

5.1 Result Evaluation

The network is stressed with the increase in network load in each simulation round by introducing two new data sessions in the network. The source–destination pairs are selected randomly in the network. Each simulation result is the average of ten simulation runs done on different seed values.

As shown in Fig. 6, the ACR (i.e., the ratio of the total admitted to the total requested data sessions) of the AODV is 100% as no admission control scheme is running on it, while the ACR of BAND-MR and MBA-AODV starts dropping as the number of sessions increases beyond four. This is because as the number of sessions increases the network is not able to accommodate new data sessions due to the lack of bandwidth on discovered routes. However, the accurate and mobility-aware nature of our proposed MBA-AODV is able to accommodate more number of data sessions in the network and thus increases the network throughput without decreasing the network performance.

Figure 7 shows the QSR (i.e., percentage for which an admitted data session gets the requested bandwidth) for all the comparing protocols. As shown in Fig. 7, the AODV performs worse due to the un-availability of any admission control method, while the BAND-MR protocol's QSR decreases with increase in the number of data flows in the network. This is because the BAND-MR is unable to accurately estimate the remaining bandwidth of links during its route discovery process as it does not consider the node mobility which is addressed in MBA-AODV. This metric also shows that the MBA-AODV protocol is able to provide QoS guarantees for the whole duration of the data flow and thus increases the QoE for the end user.

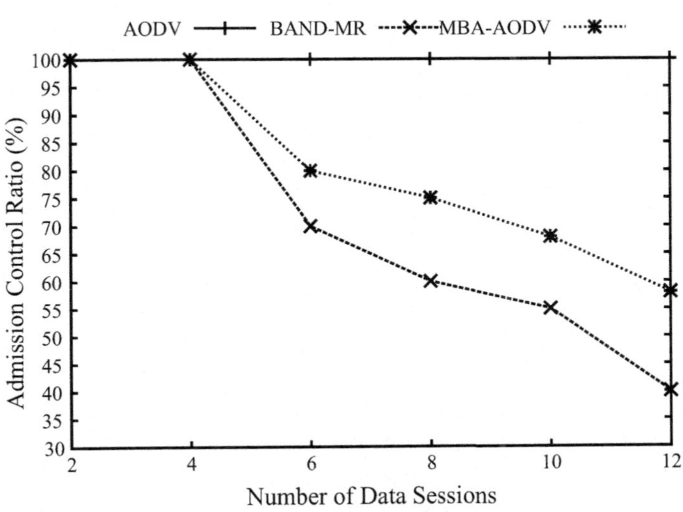

Fig. 6 Admission control ratio with increased network load

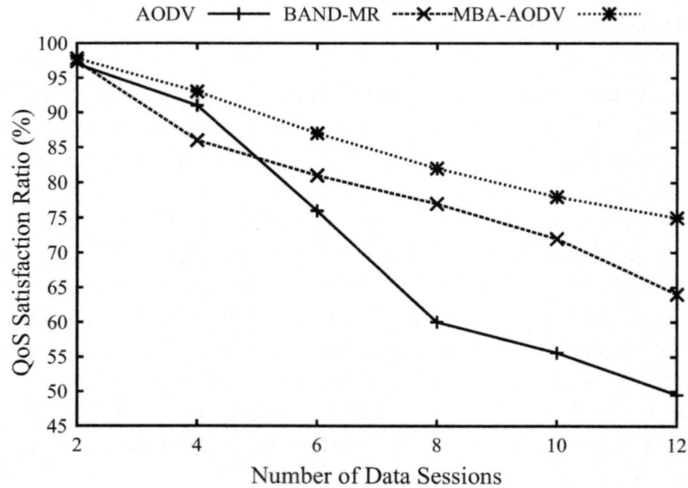

Fig. 7 QoS satisfaction ratio with increased network load

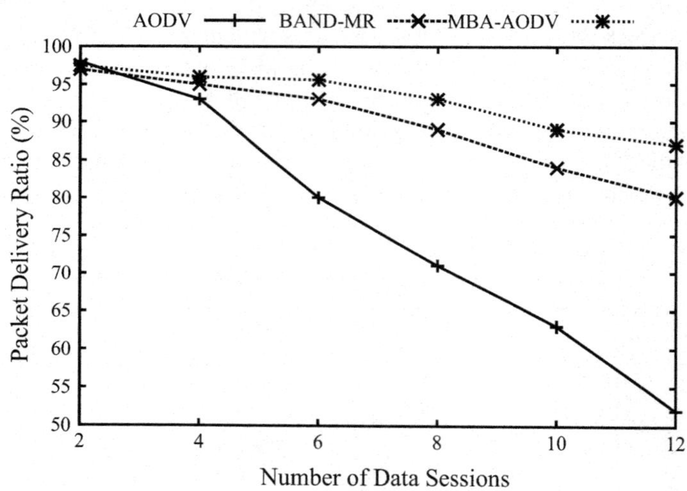

Fig. 8 Packet delivery ratio with increased network load

Figures 8 and 9 show the effect on PDR and EED of the increase in network load for all the three comparing protocols. As shown in Fig. 8, the PDR of all the protocols decreases with increase in network load because the contention and intermediate routes queue overflows. However, the PDR is high for MBA-AODV as it includes only the data sessions for which the required bandwidth is available on the discovered routes, and the accuracy in the estimation method is the reason behind its higher PDR when compared with the BAND-MR protocol. Similarly, the EED of

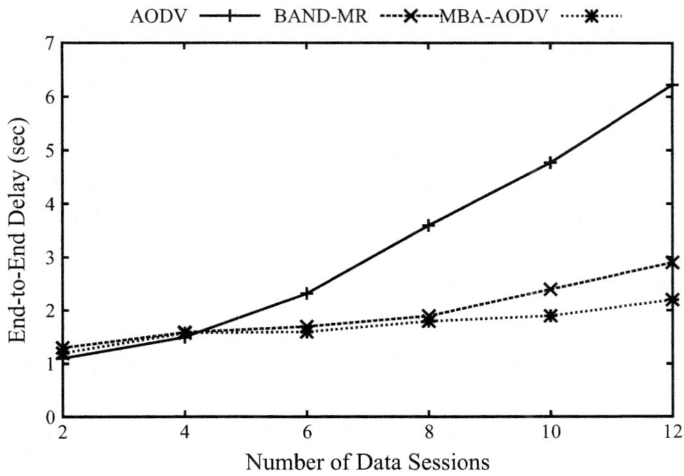

Fig. 9 End-to-end delay with increased network load

MBA-AODV is lower than the BAND-MR because the number of re-routing is lower in case of MBA-AODV due to the selection of high-bandwidth routes for data flow transmission. Also, the lower contention in MBA-AODV leads to the lower number of MAC layer re-transmissions, which leads to the lower EED.

6 Conclusion

In this paper, we presented a bandwidth-aware routing protocol that uses node mobility as an additional metric during its route discovery process. The simulation results clearly show the effectiveness of proposed approach along with the admission control ratio and QoS satisfaction ratio. It can be concluded from the theoretical and simulation analysis that using the mobility and admission control techniques into account during the route discovery greatly enhances the QoE of the received video traffic for the end user as its increases the packet delivery ratio and decreases the end-to-end delay for the received packets. In future, we are planning to emulate our proposed approach with real-time video streaming to evaluate its effectiveness in real-world scenarios.

References

1. Lal C, Laxmi V, Gaur MS, Ko SB (2015) Bandwidth-aware routing and admission control for efficient video streaming over MANETs. Wirel Netw 21(1):95–114
2. Perkins C, Belding-Royer E, Das S (2003) Ad hoc on-demand distance vector (AODV) routing. Technical report

3. Hanzo L, Tafazolli R (2007) A survey of QoS routing solutions for mobile ad hoc networks. IEEE Commun Surv Tutorials 9(2):50–70
4. Prasad R, Dovrolis C, Murray M, Claffy K (Nov 2003) Bandwidth estimation: metrics, measurement techniques, and tools. IEEE Netw 17(6):27–35
5. Sarr C, Chaudet C, Chelius G, Lassous IG (2008) Bandwidth estimation for IEEE 802.11-based ad hoc networks. IEEE Trans Mobile Comput 7(10):1228–1241
6. Hanzo II L, Tafazolli R (2011) QoS-aware routing and admission control in shadow-fading environments for multirate manets. IEEE Trans Mobile Comput 10(5):622–637
7. Belbachir R, Maaza ZM, Kies A (2013) The mobility issue in admission controls and available bandwidth measures in MANETs. Wirel Pers Commun 70(2):743–757
8. Chaudhari SS, Biradar RC (Feb 2014) Collision probability based available bandwidth estimation in mobile ad hoc networks. In: The Fifth international conference on the applications of digital information and web technologies (ICADIWT'2014), pp 244–249
9. Chen L, Heinzelman WB (March 2005) QoS-aware routing based on bandwidth estimation for mobile ad hoc networks. IEEE J Sel Areas Commun 23(3):561–572
10. Guimarães R, Cerdà L, Barceló JM, García J, Voorhaen M, Blondia C (2009) Quality of service through bandwidth reservation on multirate ad hoc wireless networks. Ad Hoc Netw 7(2):388–400
11. Kaplan E, Hegarty C (2005) Understanding GPS: principles and applications. Artech house

Magnetic Field Line-Based Node Deployment Strategy for Wireless Sensor Networks

Vipin Pal, Anju Yadav, Vinay Kanungo, Yogita and R. P. Yadav

Abstract Node deployment is the fundamental design issue of wireless sensor networks because the node location in the region of interest affects the fulfilment of application's requirement as well as the other performance metrics. Work of this paper presents framework of a new deterministic node deployment strategy that finds the location of nodes in RoI for critical section with the resemblance of magnetic field lines. Experimental results signify that the proposed solution has coverage of nearly complete area.

Keywords Node deployment · Coverage area · Magnetic lines
Wireless sensor networks · Critical section

1 Introduction

Wireless sensor network consists of various spatially distributed sensor nodes and not less than one sink or base station [1]. Spatially distributed nodes work in collaboration and communicate the gathered information to the base station directly

V. Pal (✉) · A. Yadav
Manipal University, Jaipur, India
e-mail: vipinrwr@gmail.com

A. Yadav
e-mail: anju.anju.yadav@gmail.com

V. Kanungo
SKIT, Jaipur, India
e-mail: kanungo.vinay@gmail.com

Yogita
UIET, Panjab University, Chandigarh, India
e-mail: thakranyogita@gmail.com

R. P. Yadav
MNIT, Jaipur, India
e-mail: rpyadav.ece@mnit.ac.in

V. Janyani et al. (eds.), *Optical and Wireless Technologies*, Lecture Notes in Electrical Engineering 472, https://doi.org/10.1007/978-981-10-7395-3_74

or multi-hop. Low cost and small in size requirements for sensor nodes restrict the resources and features that can be imposed. Limited battery power and difficulties to recharge/replace the battery unit, because of nearly inaccessible location of sensor nodes, make efficient energy consumption of nodes as the prime design issue for wireless sensor networks. Various energy efficient approaches are proposed in the literature, like clustering, sleep–wake scheduling, data aggregation\fusion, node deployment, to protract the lifetime of network by consuming energy in economical way [2]. For most of applications of wireless sensor networks, node deployment is the fundamental design issue and sits above all the other design issue because the location of nodes affects the fulfilment of the application's requirement.

Node deployment is the first step towards the operation of wireless sensor network. Node deployment is classified as random deployment and deterministic deployment [3]. Depending on the requirements and RoI of wireless sensor networks strategy of node deployment changes—wildlife monitoring, hostile area monitoring applications are possible with random deployment while military surveillance, critical area monitoring require deterministic deployment. A poor node deployment approach diminishes the network connectivity and data processing that results in reduction of performance.

Work of this paper presents a deterministic node deployment strategy for critical area monitoring. The main aim of proposed strategy is monitor the predetermined critical sections of RoI for a prolonged time along with covering nearly complete RoI. In the proposed node deployment strategy, nodes are placed around the critical section as resembles of magnetic field lines. The critical section is considered as a magnetic bar, and nodes are deployed over the magnetic field lines of that magnetic bar. Magnetic field is always stronger near the magnets and decreases with the increase in distance from the magnet. Concept of magnetic field lines is imposed for node deployment, as more nodes are placed near the critical area of RoI which results in better quality of monitoring and the less number of nodes are placed far away from critical region. Performance of proposed solution is examined by calculating coverage accomplished by proposed scheme and experimental results affirm that the proposed node deployment strategy covers nearly complete RoI.

Rest of paper is organized as Sect. 2 which describes the various node deployment strategies proposed in the literature; operations of proposed node deployment strategy are elaborated in Sect. 3, and analysis of proposed solution through experimental results is demonstrated in Sect. 4, Sect. 5, which concludes the proposed work.

2 Related Work

Node deployment is one of prime design issue for wireless sensor networks, as the fulfilment of the network application and various network performance metrics depend on the nodes location in RoI. Strategies for deployment of nodes in the RoI can be classified as random and deterministic. Selection of node deployment strategy depends on the requirements of application, type of sensor, environment conditions of RoI. Various deployment strategies are proposed in the literature keeping in mind the requirements of applications and network performance parameters.

Authors in [4] proposed a solution of energy hole problem of uniform node distribution by considering a regular hexagonal cell architecture CRH [5] that offers the largest coverage area. To minimize the energy hole, node deployment is completed in two phases. First phase ensures the connectivity and coverage by placing few nodes at some strategic locations. In the second phase, nodes are placed at some predetermined prioritized locations. To achieve the energy balancing among nodes, more nodes are placed near to sink area while number of deployed nodes decreases in proportion to the distance from the sink. Simulation results affirm the claims of load balancing and extended network lifetime.

Authors in [6] proposed a variable acceleration straight-line movement-based non-uniform random deployment strategy in contrast to uniform node distribution to achieve better network load balancing. In the proposed approach, the thrower deploys nodes at a fix time interval over a straight line with a constant rate varying acceleration. The acceleration decreases as the thrower progresses towards the centre and increase the acceleration by passing the centre. RoI is considered as a circle and thrower moves over the diameter. Experiments' results ensure the load balancing possessed by the proposed solution.

Authors in [7] proposed a node deployment strategy for mobile nodes that is based on Orthogonal Teaching Learning-Based Optimization (OTLBO) in order to improve network coverage. Authors in [8] proposed a node deployment strategy based on optimization theory to achieve intelligent monitoring of agricultural field with minimum number of nodes. Authors in [9] also use hexagonal cell architecture for node deployment that minimizes the cost per unit area. Furthermore, to monitor a larger area, a multi-sink hexagonal cell architecture is proposed. The numerical analysis shows that cost of deployed sensor nodes is significantly reduced while having better load balance network.

Work of this paper proposed a deterministic node deployment strategy for monitoring of critical sections in RoI.

3 Proposed Solution—Magnetic Field Lines-Based Node Deployment

Node deployment in the region of interest for wireless sensor networks is prime issue for successful operations as total coverage area and connectivity as well as energy efficiency, load balancing, network longevity depend on the node deployment. Monitoring of critical section in RoI requires special effort for node

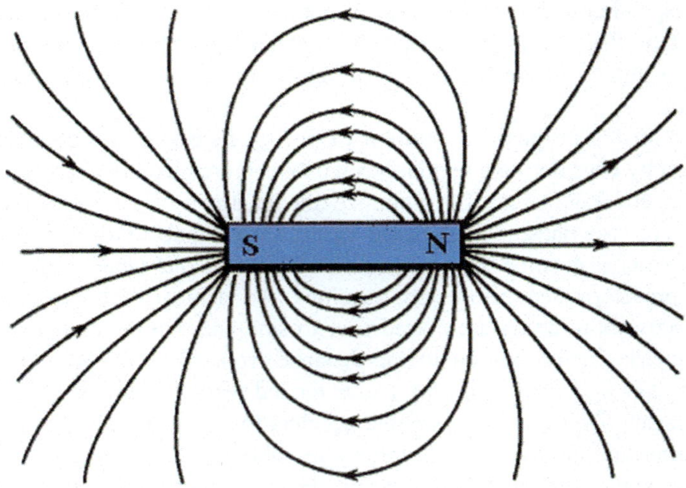

Fig. 1 Magnetic field lines of a dipole [10]

deployment than the other sections because critical sections should be monitored more precisely, quality of data should be more reliable, and monitoring should be longer than the other sections.

Work of this paper proposed a node deployment strategy for wireless sensor networks that deploy nodes following the magnetic field lines of a dipole. Magnetic lines start from North Pole and end at South Pole of the dipole hence are always in closed loop as shown in Fig. 1 [10]. Strength of magnetic is high near the dipole and decreases with the increase in distance from the dipole. Further, the magnetic lines force out the other lines hence have low intensity as we move from the dipole [10].

In the proposed solution, critical sections in the RoI are considered as a dipole. Node density is higher near the critical section as in the magnetic field of dipole and density decreases with the increase in distance from the critical section. Node deployment follows the magnetic lines of dipole and determines the node location in the RoI. Density of node is changed at a constant varying speed as in [6]. Rest of the RoI has random node deployment. Node deployment of proposed solution for 100 nodes over 100 × 100 m^2 area is shown in Fig. 2. There are four critical sections in the RoI which are marked as red '+' in the figure while sensor nodes are demonstrated as blue '*'. More nodes are placed near the critical section while the normal sections are deployed with random distribution of nodes.

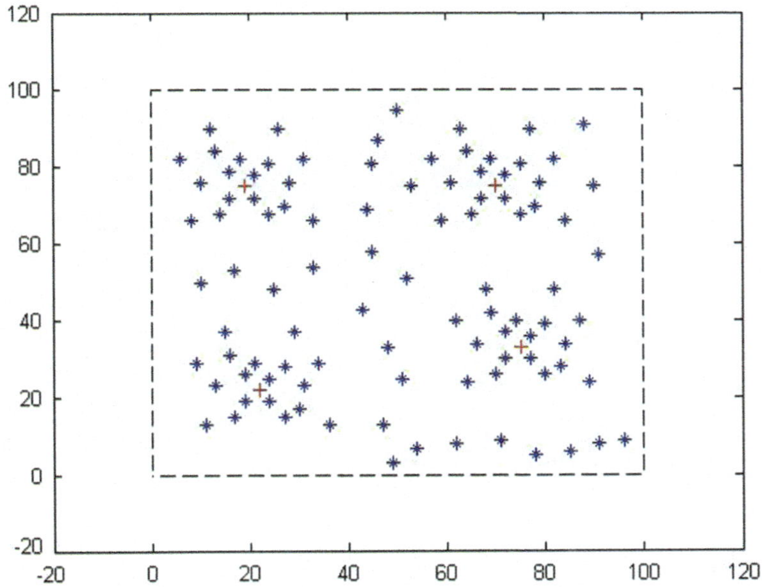

Fig. 2 Node deployment for proposed solution

4 Simulation Results

Wireless sensor network of 100 nodes deployed over 100 × 100 m^2 area. Sensing range of sensor nodes is circulator of radius 10 m, and communication range is higher than sensing range. All nodes are homogeneous in nature and are fixed after deployment. Critical section in the RoI is identified in prior, and for the simulation experiments, there are four critical sections. Total coverage area is considered as the performance metrics for proposed solution that is collective coverage of all deployed nodes [11].

Table 1 shows the comparison of total area covered by the deployed nodes in proposed solution with random node deployment. Figures 3 and 4 demonstrate the covered area in case of random deployment and proposed solution, respectively. In case of random deployment, total covered are higher than the complete area of RoI, as few nodes are sensing outside the RoI while in case of proposed solution all nodes are placed well inside the RoI and have less probability to cover outside area.

Table 1 Comparison of total covered for random and proposed node deployment strategy

Node deployment strategy	Total area of RoI (m^2)	Covered area (m^2)
Random	10000	10800
Proposed	10000	9870

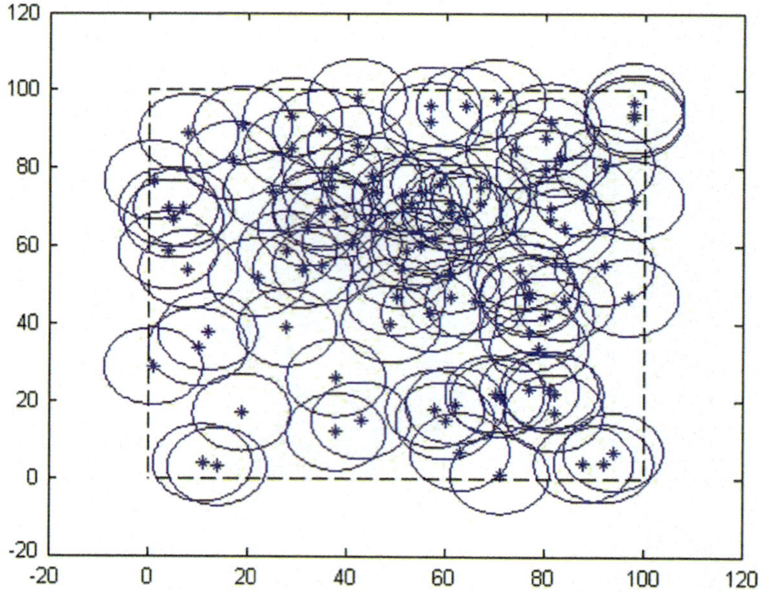

Fig. 3 Covered area of random node deployment

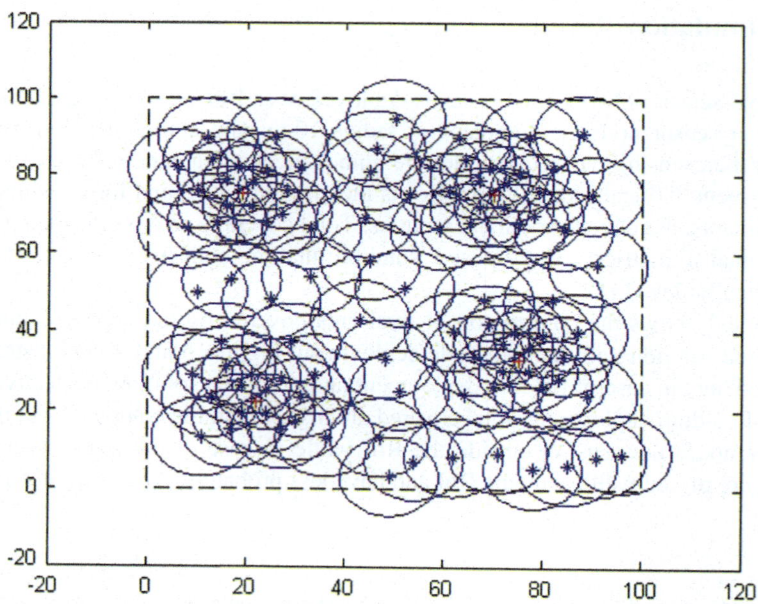

Fig. 4 Covered area of proposed node deployment strategy

Further, it is observed from the Figs. 3 and 4 that almost complete area is covered in both solution but critical section is well monitored in case of proposed solution while critical section is not well monitored in random deployment.

5 Conclusion and Future Scope

Node deployment is prime design issue for wireless sensor networks. Work of this paper presented a deterministic node deployment strategy for critical section monitoring in RoI of wireless sensor networks that is based on magnetic field lines of dipole which have node density higher near the critical sections. Simulation results confirmed that the proposed solution covered the almost all area. In future, we will examine the performance of proposed solution for network longevity and load balancing.

References

1. Zheng J, Jamalipour A (2009) Wireless sensor networks a networking perspective. Wiley
2. Anastasi G, Conti M, Di Francesco M, Passarella A (2009) Energy conservation in wireless sensor networks: a survey. Ad Hoc Netw 7(3):537–568
3. Abdollahzadeh S, Navimipour NJ (2016) Deployment strategies in the wireless sensor network: a comprehensive review. Comput Commun 91–92:1–16
4. Halder S, Ghosal A, Bit SD (2011) A pre-determined node deployment strategy to prolong network lifetime in wireless sensor networks. Comput Commun 34:1294–1306
5. Halder S, Ghosal A, Sur S, Dan A, DasBit S (2009) A lifetime enhancing node deployment strategy in WSN. FGIT, LNCS 5899:296–308
6. Chao SHA, Ru-Chaun W (2013) Energy-efficient node deployment strategy for WSNs. J China Univ Posts Telecommun 20(1):54–57
7. Prakash A, Yadav RK, Gupta D (2015) Sensor node deployment based on OTLBO in WSNs. In: 3rd international conference on recent trends in computing, pp 988– 995
8. Zou S, Yang F (2016) Optimized algorithm of sensor node deployment for intelligent agriculture monitoring. Comput Electr Agric 127:76–86
9. Tiegang F, Guifa T (2013) Deployment strategy of WSN based on minimizing cost per unit area. Comput Commun 38:26–35
10. Griffiths DJ (2013) Introduction to electrodynamics. Pearson 4th edn.
11. Senouci MR, Mellouk A (2016) Deploying wireless sensor networks: theory and practice. Elsevier

Application of Analog Electronic Circuits in Secure Communication: A Review

Manish Kumar Thukral, Karma Sonam Sherpa and Kumkum Garg

Abstract In recent times, secure communication has been one of the most exciting areas of research, and many analog electronics circuits have been found to be useful in this area. In this paper, different types of analog circuits exhibiting nonlinear dynamics have been reviewed. It has been shown that such circuits can generate signals which have frequency spectrum characteristics similar to noise. This leads to an important conclusion that a low-cost spread spectrum-based secure communication model can be developed using analog circuits. A few of the secure communication models developed using these analog circuits have also been reviewed. It is expected that this paper would enable readers to explore more analog circuits which can be used in secure communication in a better way.

Keywords Nonlinear dynamics · Chaotic communication · Spread spectrum Chua's circuit · Coherent receiver · Non-coherent receiver

1 Introduction

Analog chaotic communication (ACC) is a technique used presently in secure communication. Secure communication employing analog chaotic circuits for hiding messages is known as (ACC). The main essence of ACC lies in the fact that this type of secure communication can be implemented by using simple analog circuits [1, 2].

Another important advantage is that this technique provides an alternative to tradition spread spectrum communication in terms of cost-effectiveness [3, 4].

There are three main components of ACC-based secure communication system, namely (i) transmitter (ii) receiver, and (iii) communication channel. In the transmitter

M. K. Thukral (✉) · K. Garg
Manipal University Jaipur, Dahmi Kalan, Jaipur, Rajasthan, India
e-mail: manishkumarthukral1984@gmail.com

K. S. Sherpa
Sikkim Manipal Institute of Technology, Majhitar, Sikkim, India

V. Janyani et al. (eds.), *Optical and Wireless Technologies*, Lecture Notes in Electrical Engineering 472, https://doi.org/10.1007/978-981-10-7395-3_75

side, the message signal to be transmitted is encrypted using signals generated by analog chaotic circuits.

The encryption of message in ACC is done using either chaotic masking or chaotic modulation technique [5, 6]. In chaotic modulation, the bifurcation parameter of the analog chaotic circuit is modulated by the message signal [7–11]. Bifurcation parameter is nothing but the parameter of the circuit which when varied changes the qualitative behavior of the circuit. In chaotic masking, the message signal is encrypted by simply adding it, with chaotic signal. The frequency spectrum characteristic of message signal after encryption by either of these two techniques is same as that of noise signal. In this way, the message is disguised as a noise signal to the eavesdropper.

At the receiver, the message is decrypted using suitable methodology. There are different two types of receivers named as coherent [12] and non-coherent receivers [13] based on techniques used to decrypt the message. In coherent receiver-based ACC, identical chaotic signal generators are employed at both transmitter and receiver side and are synchronized.

In non-coherent approach, message extraction by the receiver is done without synchronization.

To implement ACC, various types of analog chaotic circuits have been used like Chua's circuit [14], particle-in-a-box circuit [15], Lorentz Circuit [16], Duffing Holmes Oscillator [17].

One interesting thing to note is that most of the ACC-based secure communication has been coherent receiver based. But there are major drawbacks of coherent receiver-based ACC which are as follows:

- To implement two identical chaotic systems at receiver and transmitter end with exact parameters matching is impossible [18].
- If the power level of message signal is significantly above the chaotic signal then only synchronization is possible [18].
- Coherent receiver-based chaotic communication cannot be used in radio communication as it is highly sensitive to channel noise [18].

Non-coherent receiver-based chaotic communication does not suffer from these drawbacks but it exists in digital communication only. From literature survey [10, 11], it is evident that non-coherent receiver-based chaotic communication is possible if the discrete time map of the chaotic signal generator is available.

In recent times, various authors have published work related to nonlinear dynamics phenomenon exhibited by power electronics-based converters [19, 20]. It is interesting to note that signals generated by power electronics-based DC-DC converters in chaotic condition possess spread spectrum characteristics [21].

The prime aim of this paper is to introduce the readers with potential of analog circuits in area of secure communication.

The paper is organized as follows: in Sect. 2, various analog electronic circuits have been discussed which exhibit nonlinear dynamics and in Sect. 3, various secure communication models which have been developed based on analog chaotic circuits have been discussed.

2 Nonlinear Dynamics in Analog Electronics Circuits

In simple terms, one can define nonlinear dynamical systems as those systems which can be mathematically modeled in the form of nonlinear differential equations or difference equations. Various types of analog electronics circuits exhibiting nonlinear dynamics have been reported in the literature and are discussed below.

2.1 *Chua's Circuit*

Chua's circuit has been one of the first analog circuits found, which exhibits nonlinear dynamics. It was developed by Leon O. Chua and gained popularity as a model for chaos [14, 22]. Figure 1 shows circuit diagram of Chua's circuit.

The Chua's circuit contains an inductor L, a linear resistor R, two capacitors C_1 and C_2, and a single voltage controlled nonlinear resistor N_R. The nonlinear resistor N_R can be realized in the form of an operational amplifier [23], diodes [24], transistors [25], and operational trans-conductance amplifiers [26]. A mathematical model of Chua's circuit, developed using basic Kirchhoff's voltage and current law [11], is given below.

$$\left.\begin{aligned} \frac{dV_2}{dt} &= \frac{I}{C_2} - \frac{G}{C_2} * (V_2 - V_1) \\ \frac{dV_1}{dt} &= \frac{G}{C_1} * (V_2 - V_1) - \frac{1}{C_1} * f(V_1) \\ \frac{dI}{dt} &= -V_2/L \end{aligned}\right\} \quad (1)$$

In [27], nonlinear dynamics analysis of Chua's circuit has been done by varying parameter $f(V_1)$.

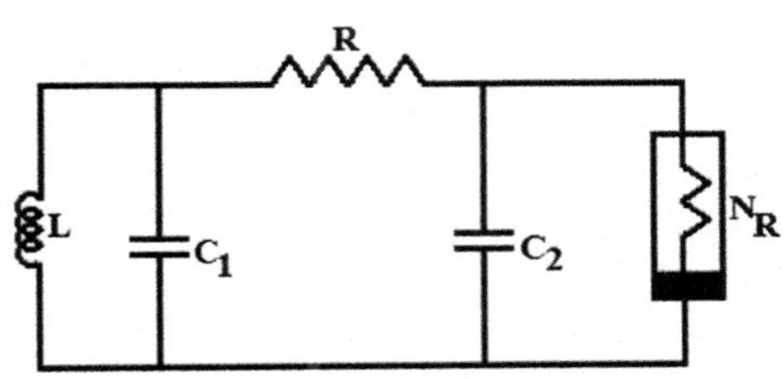

Fig. 1 Chua's circuit diagram [22]

2.2 Particle-in-a-Box Circuit

Particle-in-a-box is a mechanical model. It basically has a particle having a mass *m* in a box which can move along with the x-axis through distance L. Its model as derived in [15] is shown in Fig. 2.

The mathematical model in the form of nonlinear differential equations as reported in [15] is given as

$$\frac{md^2x}{dt^2} + \frac{\mu dx}{dt} + f(x) = mgsin\alpha \tag{2}$$

where,

$m=$ Mass of the particle, $\mu=$ Viscous Coefficient, $f(x)=$
Force applied by the spring, $g=$ Gravitational Acceleration, and $\alpha=$ Variable angle to make box oscillate

Its model in form of electrical parameters has been developed in [15] as

$$(R_1C_1C_2) * \frac{d^2V_2}{dt^2} + \left(\frac{R_2C_2}{R_1}\right) * \frac{dV_2}{dt} + I_D(V_2) = -V_E/R_E \tag{3}$$

Equation (3) is a nonlinear differential equation which can be split into sets of two differential equations as

$$\dot{V}_1 = -\frac{V_1}{R_1C_1} - \frac{I_D(V_2)}{C_1} - \frac{V_E}{C_1R_E} \tag{4}$$

$$\dot{V}_2 = \frac{V_1}{R_2C_2} \tag{5}$$

Figure 3 shows equivalent analog electronic circuit of particle-in-a-box model.

In the circuit, shown in Fig. 3, following parameter values are set to observe nonlinear dynamics

$$R_1 = 47\,\text{K}\Omega,\ R = R_E = R_2 = 10\,\text{K}\Omega,\ R_L = 0.51\,\text{K}\Omega,\ C_1 = C_2 = 12\,\text{nF}$$

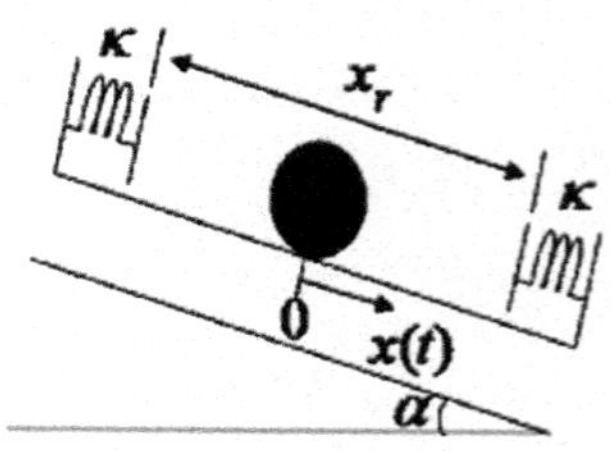

Fig. 2 Particle-in-a-box model [15]

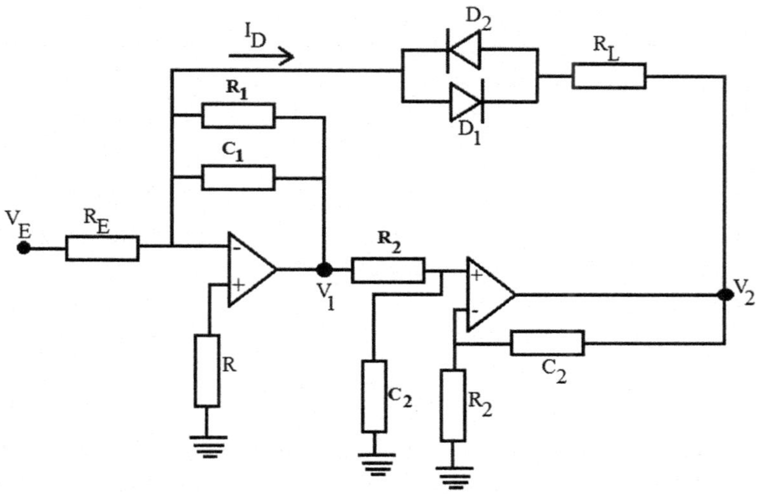

Fig. 3 Particle-in-a-box electronic circuit [15]

Also, parameter $V_E = V_{max} \sin(\omega t)$ is taken as bifurcation parameter. At frequency $\omega = 12208\,\text{rad/s}$, parameter V_E is varied, and chaos is reported in [15] at $V_{max} = 2.93\,\text{V}$.

2.3 Current Mode Controlled DC-DC Boost Converter

Current mode controlled boost converter is a power electronics-based circuit. It is basically a DC voltage regulator. The name is boost because the output voltage of the circuit is stepped up version of the input voltage. It has been reported in various literature that DC-DC converters operating in closed loop exhibit nonlinear dynamics [28–30]. This converter is one of them operating in closed loop. Figure 4 shows circuit diagram of current mode controlled boost converter as discussed in

Fig. 4 Current mode controlled boost converter [19]

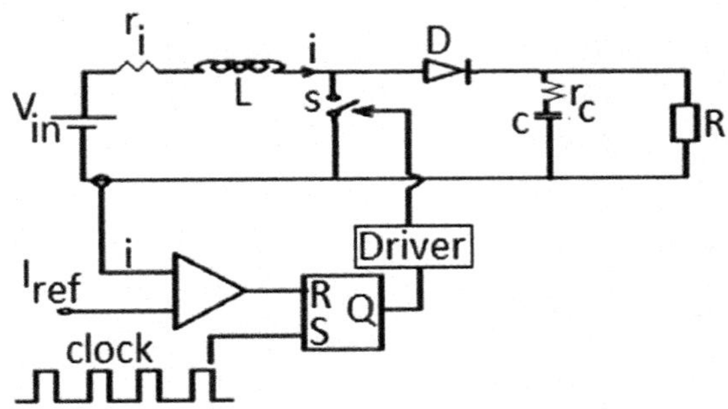

[19] and [31]. The mathematical model of the circuit has been developed in the form of discrete time map in [19]. The circuit parameters considered in [19] are as follows:

Load Resistance $R = 20\ \Omega$, Capacitor C Value $= 120\ \mu F$, Inductor L Value $= 27\ \mu H$, Inductor parasitic resistance value $r_l = 1.2\ \Omega$, and Capacitor parasitic resistance value $r_c = 0.1\ \Omega$.

The nonlinear dynamics in the circuit is analyzed considering input voltage v_{in} as bifurcation parameter where period-II, period-IV bifurcation, and chaos has been reported in [19] at 34.9 V, 24.4 V, and 21.8 V, respectively.

2.4 Lorentz Circuit

Mathematical meteorologist, E. N. Lorenz, in 1963, introduced a system of three nonlinear differential equations which showed chaotic dynamics.

The electronic circuit equivalent equations of Lorentz as derived in [16] are given as follows

$$\left.\begin{array}{l} \dot{u} = \left(\frac{1}{R_5 C_1}\right) * \left[\left(\frac{R_4}{R_1}\right) v - \frac{R_3}{R_2 + R_3} * \left(1 + \frac{R_4}{R_1}\right) u\right] \\ \dot{v} = \left(\frac{1}{R_{15} C_2}\right) * \left[\frac{R_{11}}{R_{10} + R_{11}} * \left(1 + \frac{R_{12}}{R_8} + \frac{R_{12}}{R_9}\right) * \left(1 + \frac{R_7}{R_6}\right) u - \frac{R_{12}}{R_8} * v - \frac{R_{12}}{R_9} * uw\right] \\ \dot{w} = \frac{1}{R_{20} C_3} * \left[\frac{R_{19}}{R_{16}} * uv - \frac{R_{18}}{R_{17} + R_{18}} * \left(1 + \frac{R_{19}}{R_{16}}\right) w\right] \end{array}\right\} \quad (6)$$

The corresponding circuit implementation of Lorentz equations, as done in [16], is shown in Fig. 5.

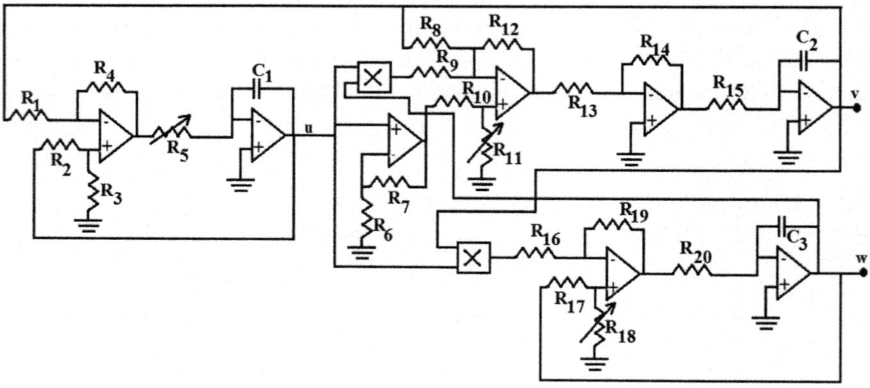

Fig. 5 Lorentz circuit [16]

The chaotic behavior of the Lorenz system has been reported in [16] at following values of the circuit parameters

$$R_1, R_2, R_3, R_4, R_6, R_7, R_{13}, R_{14}, R_{16}, R_{17}, R_{19} = 100\,\text{K}\Omega,\ R_5, R_{10} = 49.9\,\text{K}\Omega,\ R_8 = 200\,\text{K}\Omega,$$
$$R_9, R_{12} = 10\,\text{K}\Omega,\ R_{11} = 63.4\,\text{K}\Omega,\ R_{15} = 40.2\,\text{K}\Omega,\ R_{18} = 66.5\,\text{K}\Omega,\ R_{20} = 158\,\text{K}\Omega,\ C_1, C_2, C_3 = 500\,\text{pF}$$

3 Application of Analog Chaotic Circuits in Secure Communication

In various literature [32–36], it has been shown that different types of analog circuits exhibiting nonlinear dynamics generate signals having noise-like characteristics. This resulted in development of new area of secure communication where analog chaotic circuits found their application in secure communication [16, 37, 38] in the form of ACC.

3.1 Review of Coherent Receiver-Based ACC Model

In [12], Lorentz circuit has been used as a chaotic signal generator on transmitter side. Based on synchronization technique proposed in [39], receiver is designed in [12]. The proposed coherent receiver-based ACC is analyzed with speech signal "He has the bluest eyes" which is encrypted using chaotic masking technique.

Another type of coherent receiver-based ACC has been proposed in [18], where a Duffing oscillator has been used as a chaotic signal generator at the transmitter end. The message is encrypted using chaotic modulation technique. The message is decrypted by estimating the bifurcation parameter at the receiver end, based on synchronization process [40].

In [41], Chua's circuit has been used as analog electronic circuit for chaotic signal generation. The receiver designed is an inverse system of the Chua's circuit in a sense that what is input to the Chau's circuit in transmitter end will be output at receiver end.

New form of ACC based on coherent receiver has been proposed in [42] where message is encrypted using chaos shift keying technique (CSK) [38, 43, 44]. Here, the message signal which is in the form of +1 or −1 is encrypted as a strange attractor of Chua's circuit. The signals generated by Chua's circuit are transmitted, and the receiver estimates the Chua's attractor and hence the message. The attractor is estimated using synchronization technique.

3.2 Review of Non-Coherent Receiver-Based Chaotic Communication Model

The main backbone of non-coherent receiver is discrete time model of dynamical system or statistical properties of signals generated by chaotic circuits. It means those dynamical systems which have discrete time modeling can be used to develop non-coherent receiver-based chaotic communication.

Few of the non-coherent receiver-based chaotic communication models developed are reviewed as follows:

In [10], the encryption of the message signal is done by embedding it in the bifurcation parameter of logistic map using chaotic modulation technique. On the receiver side, the bifurcation parameter of the logistic map is estimated using recursive least square (RLS) method. Basically, the discrete time map model of the logistic map at receiver side is employed whose bifurcation parameter is estimated using RLS technique.

Again, logistic map has been used in [11] to encrypt the message using chaotic modulation technique. The receiver is non-coherent type where radial basis function (RBF)-based neural network is trained to estimate the bifurcation parameter.

In [45], a non-coherent receiver-based chaotic communication is developed where three types of discrete time systems, namely logistic map, skew tent map, cubic maps, etc., are used to transmit the message. The signals generated by three discrete time systems are used to encrypt the message. At the receiver end, based on the statistical properties of signals generated by these dynamical systems, the message is decrypted.

In a recent development, power electronics-based analog circuits are also being used to design non-coherent receiver-based secure communication model. In [31], current mode controlled boost converter has been used to design a secure communication model.

4 Conclusion

This paper reviews a few of the analog electronics circuits which exhibit nonlinear dynamics and are used in chaotic communication. Through literature review, it has been shown how by using simple analog electronics circuits, a secure communication model can be developed. Various types of chaotic communication models that can be developed using analog circuits have been discussed. The main aim of the paper is to make the reader aware of chaotic communication, which is an emerging topic in the area of secure communication.

References

1. Li S, Alvarez G, Li Z, Halang WA (2007) Analog chaos-based secure communications and cryptanalysis: a brief survey. In: 3rd international IEEE scientific conference on physics and control
2. Li Z, Li K, Wen C, Soh YC (2003) A new chaotic secure communication system. IEEE Trans Commun 51:1306–1312
3. Eisencraft M, Fanganiello RD et al (2012) Chaos based communication systems in non-ideal channels. J Commun Nonlinear Sci Numer Simulat 17:4707–4718
4. Kennedy M, Rovatti R, Setti G (2000) Chaotic electronics in telecommunications. CRC Press, Inc., Boca Raton, FL, USA
5. Murali K, Leung H, Yu H (2003) Design of noncoherent receiver for analog spread-spectrum communication based on chaotic masking. IEEE Trans Circuits Syst I FundamTheory Appl 50:432–441
6. Jiang ZP (2002) A note on chaotic secure communication systems. IEEE Trans Circuits Syst I Fundam Theory Appl 49:92–96
7. Chen YY (1996) Masking messages in chaos. Europhys Lett 34:245–250
8. Milanovic V, Zaghloul ME (1996) Improved masking algorithm for chaotic communication system. Electron Lett 32:11–12
9. Dmitriev AS, Panas AI, Starkov SO, Kuzmin LV (1997) Experiments on RF band communications using chaos. Int J Bifurcation Chaos 7:2511–2527
10. Leung H, Lam J (1997) Design of demodulator for the chaotic modulation communication system. IEEE Trans Circuits Syst I Fundam Theory Appl 44:262–267
11. Chow TWS, Feng JC, Ng KT (2000) An adaptive demodulator for the chaotic modulation communication system with rbf neural network. IEEE Trans Circuits Syst I Fundam Theory Appl 47:902–909
12. Cuomo KM (1993) Circuit implementation of synchronized chaos with applications to communication. Phys Rev Lett 71:65–68
13. Kennedy MP, Kolumbán, G (2000) Special issue on noncoherent chaotic communications. IEEE Trans Circuits Syst I Fundam Theory Appl 47:1661–1662
14. Hu Z, Chen X (2006) Synchronization of chaotic cryptosystems based on Chua's circuits with key functions. Dynam Contin Discrete Impulsive Syst A Math Anal 13:489–49
15. Pizolato JC, Romero MA, Neto LG (2008) Chaotic communication based on the particle-in-a-box electronic circuit. IEEE Trans Circuits Syst I Fundam Theory Appl 55:1108–1115
16. Cuomo KM, Oppenheim AV, Strogetz SH (1993) Synchronization of Lorenz-based chaotic circuits with application to communications. IEEE Trans Circuits Syst II Analog Digit Process 40:626–633
17. Lindberg E, Bumeliene S (2008) Analog electrical circuit for simulation of the duffing-holmes equation. Nonlinear Anal Model Control 13:241–252
18. Kis G, Jákó Z, Kennedy MP, Kolumbán G (1998) Chaotic communications without synchronization. In: Proceedings 6th IEE conference telecommunication, Edinburgh, UK, pp 49–53
19. Banerjee S, Chakrabarty K (1998) Nonlinear modeling and bifurcation in boost converter. IEEE Trans Power Electron 13:252–260
20. Cheng KWE, Liu M, Wu J, Cheung NC (2001) Study of bifurcation and chaos in the current-mode controlled buck-boost dc-dc converter. In: IECON'01: the 27th annual conference of the IEEE industrial electronics society, pp 838–843
21. Banerjee S, Baranovski AL, Marrero JLR, Woywode O (2002) Minimizing electromagnetic interference problems with chaos. IEICE Trans 85:1–11
22. Kennedy MP (1993) Three steps to chaos-part-I: evolution. IEEE Trans Circuits Syst I Fundam Theory Appl 40:640–656
23. Kennnedy MP (1992) Robust OPAMP realization of chua's circuit. Int J Frequen 46:66–80

24. Zhong GQ, Ayrom F (1985) Experimental confirmation of chaos from chua's circuit. Int J Circuit Theory Appl 13:93–98
25. Matsumoto T, Chua L, Tokumasu K (1986) Double Scroll via a two-transistor circuit. IEEE Trans Circuits Syst 33:828–835
26. Cruz JM, Chua LO (1992) A CMOS IC nonlinear resistor for chua's circuit. IEEE Trans Circuits Syst I Fundam Theory Appl 39:985–995
27. Kennedy MP (1993) Three steps to chaos-part-II: evolution. IEEE Trans Circuits Syst. I Fundam Theory Appl 40:657–674
28. Deane JHB, Hamill DC (1990) Instability, sub-harmonics, and chaos in power electronics circuits. IEEE Trans Power Electron 5:260–268
29. Banerjee S, Poddar G, Chakrabarty K (1996) Bifurcation behavior of buck converter. IEEE Trans Power Electron 11:437–439
30. Hamill DC, Deane JHB, Jefferies DJ (1992) Modeling of chaotic dC–dC converters by iterated nonlinear mappings. IEEE Trans Power Electron 7:25–36
31. Thukral MK, Sherpa KS (2013) Chaotic modulation-based spread spectrum communication using complex dynamics of chaotic dC–dC current mode controlled boost converter. Int J Secur Commun Netw 6:1053–1063
32. Stavroulakis P (2005) Chaos applications in telecommunication. CRC Press
33. Rohde GK, Nichols JM, Bucholtz F. Chaotic signal detection and estimation based on attractor sets: applications to secure communications. Int J Chaos 18
34. Cummings FE (1977) The particle in a box is not simple. Amer J Phys 45:158–160
35. Eisencraft M, Kato DM, Monteiro LHA (2010) Spectral properties of chaotic signals generated by the skew tent map. J Signal Processing 90:385–390
36. Galleani L, Biey M, Gilli M, Presti LL (1999) Analysis of chaotic signals in the time-frequency plane. In: Conference proceedings of the ieee-eurasip workshop on nonlinear signal and image processing (nsip'99). Antalya, Turkey, June 20–23
37. Fodjouong GJ, Fotsin HB, Woafo P (2007) Synchronization modified van der pol-duffing oscillators with offset terms using observer design: application to secure communications. Phys Scrip 75:638–644
38. Kocarev L, Halle K, Eckert K, Chua LO, Parlitz U (1992) Experimental demonstration of secure communications via chaotic synchronization. Int J Bifurc Chaos 2:709–713
39. Pecora LM, Caroll TL (1990) Synchronization in chaotic systems. Phys Rev Lett 64:821–825
40. Sobiski DJ, Thorp JS (1998) PDMA-1: chaotic communication via the extended kalman filter. IEEE Trans Circuits Syst I Fundam Theory Appl 45
41. Feldmann U, Hasler M, Schwarz A (1996) Communication by chaotic signals: the inverse system approach. Inte J Circuit Theory Appl 24:551–579
42. Dedieu H, Kennedy MP, Hasler M (1993) Chaos shift keying: modulation and demodulation of a chaotic carrier using self-synchronizing chua' s circuits. IEEE Trans Circuits Syst II Analog Digit Signal Process 40
43. Hayes S, Grebogi C, Ott E (1993) Communicating with chaos. Phys Rev Lett 70:3031–3034
44. Kolumb´an G, Kennedy MP, Chua LO (1998) The role of synchronization in digital communications using chaos—part II: chaotic modulation and chaotic synchronization. IEEE Trans Circuits Syst I Fundam Theory Appl 45
45. Tse CK, Lau FCM (2003) A return map regression approach for non-coherent detection in chaotic digital communications. Int J Bifurc Chaos 13:685–690
46. Thukral MK, Sherpa KS, Garg K (2016) Design of a robust receiver for chaotic switching maps of dc-dc power electronics transmitter converters for secure communication. J Secur Commun Netw 9:4404–4415

A Comparative Investigation of Microstrip Patch Antenna Using FR-4/RT-Duroid/RO3003 Materials for Bluetooth and WLAN Applications

Pourush Bara and Dinesh Yadav

Abstract The purpose of this paper is to design, compare, and analyze the radiation properties of the rectangular microstrip patch antennas for three different materials at resonant frequencies of 5.2 and 2.4 GHz. The proposed rectangular patch antennas have been designed using the materials, glass epoxy substrate with dielectric constant ($\varepsilon_r = 4.3$) and loss tangent (tan $\delta = 0.025$), RT/Duroid (RT/5880) with dielectric constant ($\varepsilon_r = 2.2$) and loss tangent (tan $\delta = 0.0009$), and Rogers (RO3003) with dielectric constant ($\varepsilon_r = 3$) and loss tangent (tan $\delta = 0.0013$). The proposed rectangular patch antennas are excited using 50 Ω microstrip line. The reflection coefficient curve for the return loss less than −10 dB for both operating frequency with three materials is compared. Various parameters like the radiation pattern, gain, return loss, and VSWR are obtained using CST Microwave Studio software.

Keywords Antenna parameters · Bluetooth · Material properties WLAN

1 Introduction

A simple rectangular microstrip patch antenna consists of a dielectric substrate having fixed dielectric constant sandwiched between the radiating patch and a ground plane [1]. The dielectric constant of substrate plays a very important role in determining the size and effective gain of an antenna. There are many dielectric substrates that can be used for designing microstrip patch antenna. The dielectric constant usually ranges from 2.1 (Teflon) to 16.9 (diamond) [2]. The dimensions of

P. Bara (✉) · D. Yadav
Electronics and Communication Engineering, Manipal University Jaipur, Jaipur, Rajasthan, India
e-mail: bpourush@gmail.com

D. Yadav
e-mail: Dinesh.yadav@jaipur.manipal.edu

V. Janyani et al. (eds.), *Optical and Wireless Technologies*, Lecture Notes in Electrical Engineering 472, https://doi.org/10.1007/978-981-10-7395-3_76

the patch correspond to the resonant frequency of antenna but the selection of right substrate material is also very important for designing an efficient antenna [3]. Microstrip patch antennas are mostly used nowadays and may take any geometrical shapes like rectangular, triangular, circular, helical, ring, and elliptical as per requirement due to its various advantages such as light weight, small size, low cost, robust, compatibility with integrated circuits, high performance, and easy to install [4]. Microstrip patch antennas can operate in both dual band. Here, we have done our investigation on frequencies 2.4 and 5.2 GHz which correspond to Bluetooth and IEEE 802.11a WLAN, respectively [5, 6]. Microstrip patch antennas are used for communication purposes and are widely used in military, civil, medical, and commercial applications. Therefore, it is required to select the best substrate material for desired applications [7].

2 Designing and Calculations

The design parameters to be calculated are as follows:

Width of patch, Length of patch, Input impedance, Inset feed depth, Length of feed line, and Width of feed line. The parameters are calculated from the formulas given below.

2.1 Width (W)

The width of the patch is calculated using the following equation.

$$W = \frac{c}{2f_r}\sqrt{\frac{2}{\varepsilon r + 1}} \tag{1}$$

where

W width of the patch
C speed of light
εr value of the dielectric substrate

2.2 Effective Refractive Index (εr_{eff})

The radiations that travel from the patch toward the ground pass through air and some of them pass through the substrate called as fringing field. Since both the air and the substrates have different dielectric values, so we find the value of effective dielectric constant. The value of the effective dielectric constant (εr) is calculated using the following equation.

$$\varepsilon r_{eff} = \frac{\varepsilon r + 1}{2} + \frac{\varepsilon r - 1}{2}\left[1 + 12\frac{h}{w}\right]^{-1/2} \quad Where \quad W/h > 1 \tag{2}$$

2.3 Length (L)

Due to fringing fields, actual size of the antenna is increased by an amount of (ΔL). Therefore, the increase in length (ΔL) of the patch is calculated using the following equation.

$$\frac{\Delta L}{h} = 0.412\frac{(\varepsilon r_{eff} + 0.3)\left(\frac{w}{h} + 0.264\right)}{(\varepsilon r_{eff} - 0.258)\left(\frac{w}{h} + 0.8\right)} \tag{3}$$

where 'h' = height of the substrate

The length of the patch is now calculated using the equations mentioned below.

$$L_{eff} = \frac{c}{2f_0\sqrt{\varepsilon r_{eff}}} \tag{4}$$

$$L = L_{eff} - 2\Delta L \tag{5}$$

2.4 Input Impedance (Z_{in})

It is very important parameter to be determined because after knowing the input impedance of the patch we can correctly match it with the input feed line for effective transmission of power.

The input impedance of the patch is calculated using the below-mentioned equation.

$$Z_{in} = 90\frac{Er^2}{(Er - 1)}\left(\frac{L}{W}\right)^2 \tag{6}$$

2.5 Inset Feed Depth (Y_o)

Based on the input impedance that we calculated, we now calculate the inset feed depth the point on the patch at which the impedance is 50 Ω so as to get good matching and low VSWR. Minimum reflection coefficient indicates minimum return loss.

The inset feed depth is calculated using the following equation.

$$Z_{in}(y=y_0)=Z_{in}(y=0)\cos^2\left(\frac{\pi}{L}y_0\right) \tag{7}$$

2.6 *Width of Feed Line* (W_f)

The width of feed line is calculated using the following equation.

$$\frac{Wf}{h}=\frac{2}{\pi}\left[\beta-1-ln(2\beta-1)+\frac{\varepsilon r-1}{2\varepsilon r}\left(ln(\beta-1)+0.39-\frac{0.61}{\varepsilon r}\right)\right] \tag{8}$$

For feeding the microstrip patch antenna, there are different methods, for example, feed line method and coaxial probe feeding method. But mostly coaxial probe method is used [8].

2.7 *Length of Feed Line* (L_f)

The length of feed line is calculated using the following equation.

$$L_f=\frac{\lambda}{4\sqrt{\varepsilon r_{eff}}} \tag{9}$$

3 Results and Discussions

Using the equations mentioned above, three square microstrip patch antennas with different substrate material are designed to work at resonant frequencies of 5.2 and 2.4 GHz. For feeding the microstrip patch antenna, feed line method is used. The simulation is carried out in CST Microwave Studio software (Fig. 1).

3.1 *For 5.2 GHz Operating Frequency*

The dimensions of the rectangular patch antennas are calculated using all the three materials for 5.2 GHz shown in the table (Table 1).

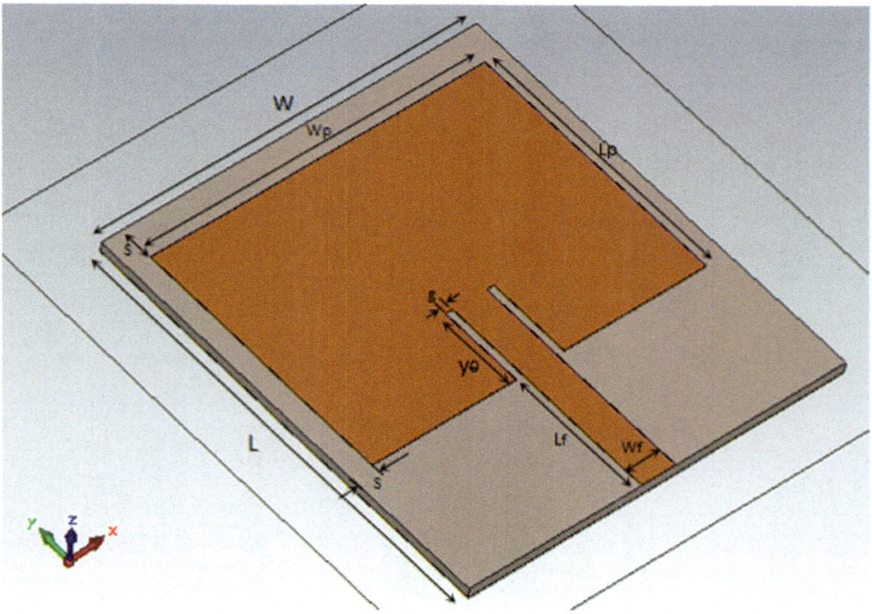

Fig. 1 Design parameters of patch antenna (s = 2.5, g = 0.5)

Table 1 Rectangular microstrip patch antenna dimensions

Substrate material	RT5880	FR-4	RO3003	Units
Length of substrate (L)	31.5	23.2	27.45	mm
Width of substrate (W)	29	22.7	25.4	mm
Length of patch (L_p)	19	13.34	16.35	mm
Width of patch (W_p)	24	17.7	20.4	mm
Antenna size (W × L)	913.5	526.64	697.23	mm^2
Patch size ($W_p \times L_p$)	456	236.12	333.54	mm^2
Feed line width (W_f)	2.45	3.1	1.9	mm
Inset feed depth (Y_0)	6.8	4.84	5.8	mm
Feed line length(L_f)	10	7.36	8.6	mm

3.1.1 Radiation Pattern (3D)

The direction of the maximum gain of the antenna is above the patch while minor lobes are on the opposite side. The maximum antenna gain is observed by the material RT-Duroid which is 7.27 dB (Fig. 2).

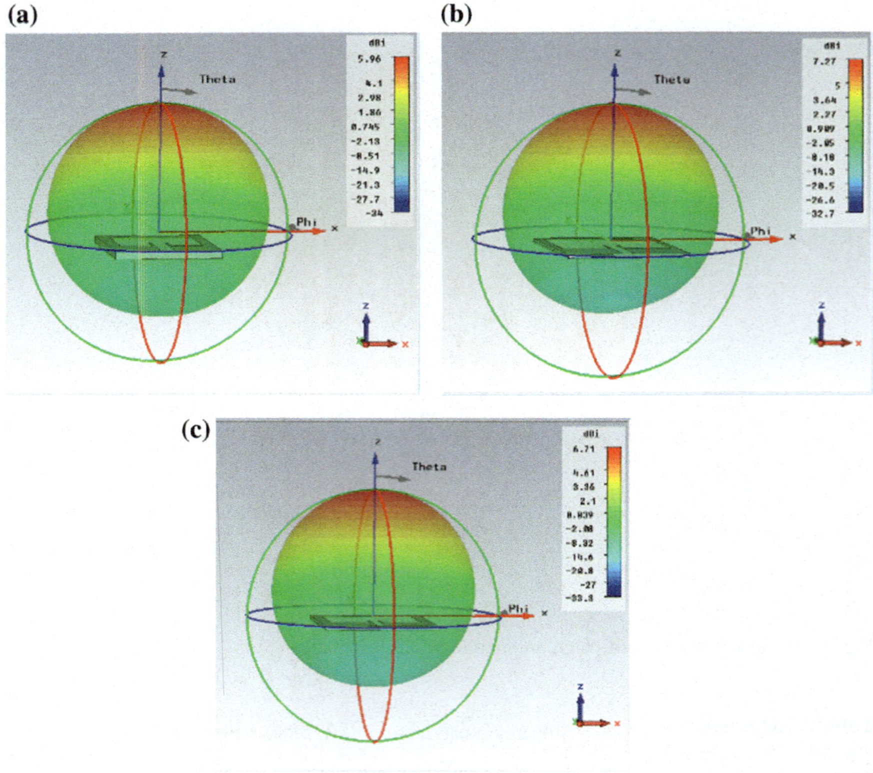

Fig. 2 3-D radiation pattern of the designed antennas. **a** FR-4, **b** RO3003, **c** RT5880

3.1.2 Radiation Pattern (2D)

See Fig. 3.

3.1.3 Return Loss (S11)

The minimum return loss is observed from material FR-4 which is −27.027 dB at 5.1882 GHz (Fig. 4).

3.1.4 VSWR

At 5.188 GHz, the minimum VSWR is observed from material FR-4 is 1.0932 (Fig. 5 and Table 2).

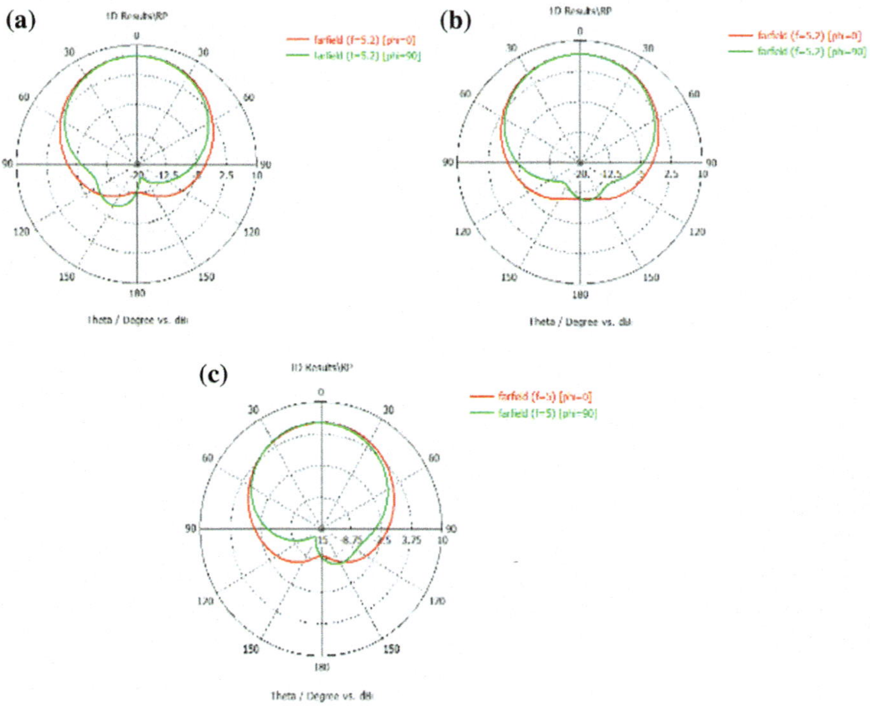

Fig. 3 2-D (E-field and H-field) radiation pattern of the designed antennas. **a** RT5880, **b** RO3003, **c** FR-4

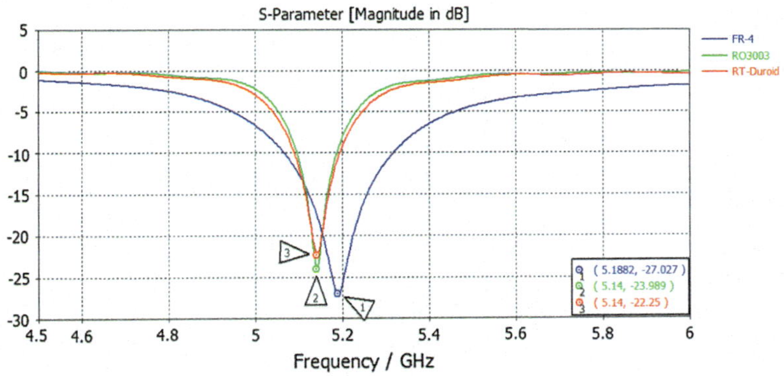

Fig. 4 S-parameter (magnitude in dB) for the three antennas

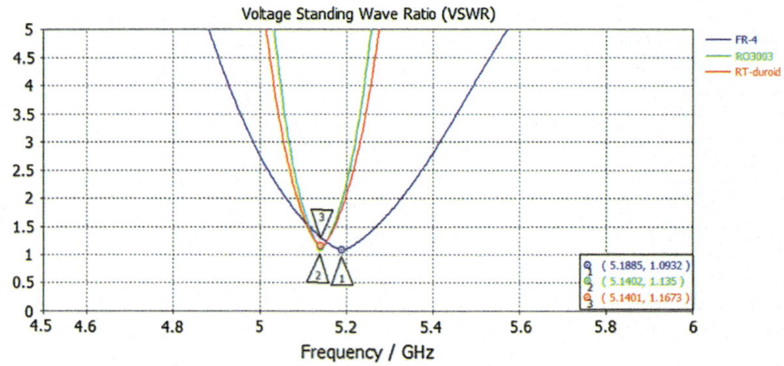

Fig. 5 Voltage standing wave ratio (VSWR) versus frequency graph

Table 2 Rectangular microstrip patch antenna parameter comparison

Substrate material	RT5880	FR-4	RO3003	Units
Dielectric constant (ε_r)	2.2	4.3	3	–
Center frequency (f_c)	5.14	5.19	5.14	GHz
Frequency band (BW)	5.09–5.19	5.068–5.317	5.1–5.18	GHz
Gain (G)	7.27	5.96	6.71	dB
VSWR (at f_c)	1.167	1.1	1.13	–
Input impedance (Z_{in})	270	286.4	260	ohm

3.2 For 2.4 GHz Operating Frequency

The dimensions of the rectangular patch antennas are calculated using all the three materials for 2.4 GHz shown in the table below (Table 3).

Table 3 Rectangular microstrip patch antenna specifications

Substrate material	RT5880	FR-4	RO3003	Units
Length of antenna (L)	65.6	47.9	56.65	mm
Width of antenna (W)	54.4	43.4	49.2	mm
Length of patch (Lp)	41.8	29.8	35.85	mm
Width of patch (Wp)	49.4	38.4	44.3	mm
Antenna size (W × L)	3538.6	2078.8	2787.2	mm^2
Patch size (Wp × Lp)	2065	1142.8	1584.6	mm^2
Feed line width (W_f)	2.43	3.1	1.9	mm
Inset feed depth (Y_0)	14.86	10.9	12.8	mm
Feed line length (L_f)	21.3	15.6	18.3	mm

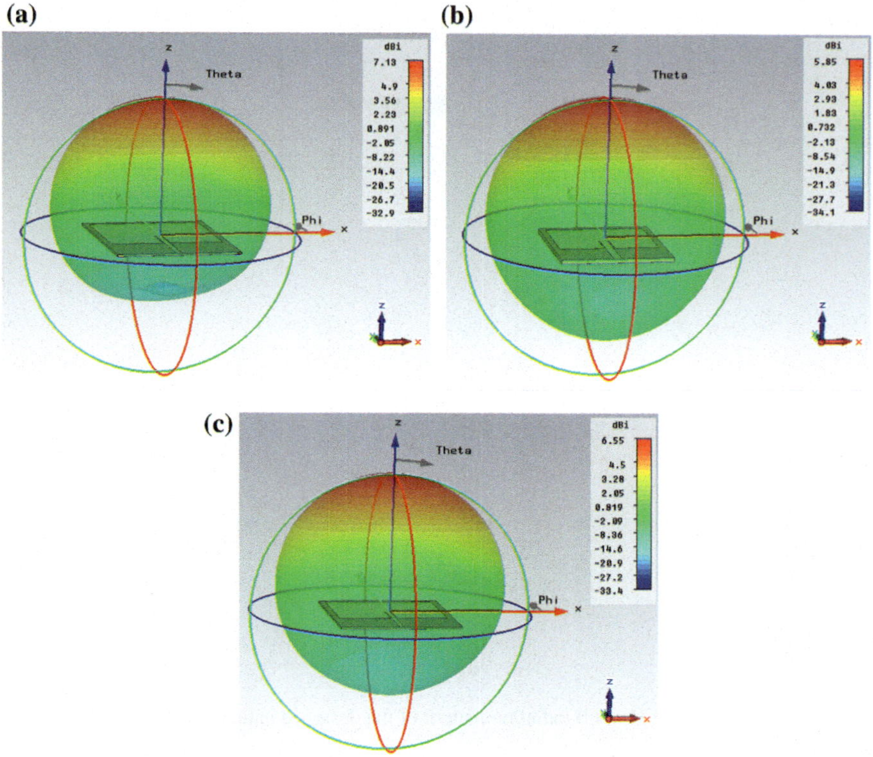

Fig. 6 Shows the 3-D view of radiation patterns of the three antenna. **a** RT5880, **b** FR-4, **c** RO3003

3.2.1 Radiation Pattern (3D)

The maximum antenna gain is observed by the material RT-Duroid which is 7.13 dB (Fig. 6).

3.2.2 Radiation Pattern (2D)

See Fig. 7.

3.2.3 Return Loss (S11)

The minimum return loss is observed from material RO3003 which is −34.937 dB at 2.377 GHz (Fig. 8).

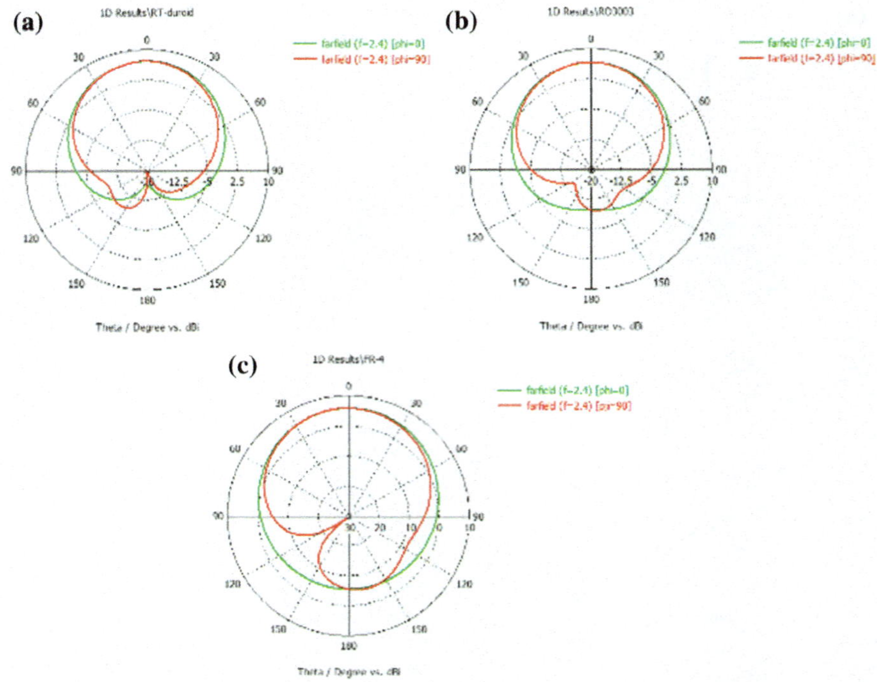

Fig. 7 2-D (E-field and H-field) radiation pattern of the designed antennas **a** RT5880, **b** RO3003, **c** FR-4

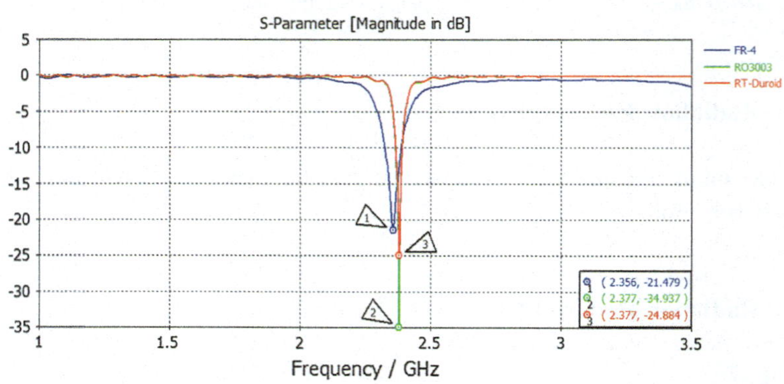

Fig. 8 S-parameter (magnitude in dB) for the three antennas

3.2.4 VSWR

At 2.377 GHz, the minimum VSWR is observed from material RO3003 is 1.0365 (Fig. 9 and Table 4).

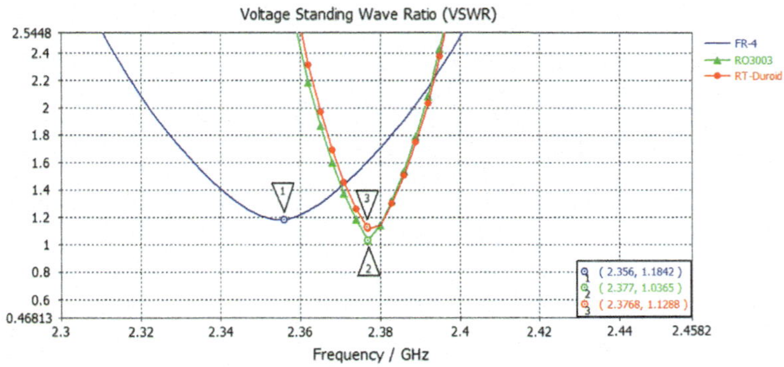

Fig. 9 Voltage standing wave ratio (VSWR) versus frequency graph

Table 4 Rectangular microstrip patch antenna parameter comparison

Substrate material	RT5880	FR-4	RO3003	Units
Dielectric constant (ε_r)	2.2	4.3	3	–
Center frequency (f_c)	2.41	2.35	2.38	GHz
Frequency band (BW)	2.39–2.42	2.32–2.38	2.36–2.39	GHz
Gain (G)	7.13	5.85	6.55	dB
VSWR (at f_c)	1.06	1.17	1	–
Input impedance (Z_{in})	260	302.87	266.4	ohm

4 Conclusion and Future Scope

After observing the comparison results and evaluating different parameters, it can be seen that the antenna with substrate material (RT5880) is achieving better gain compared to the other two materials. The gain achieved by (RT5880) is 7.13 and 7.27 dB for 2.4 GHz and 5.2 GHz, respectively, and antenna with substrate material (FR-4) is giving least gain. So for an application like satellite where we require high-gain RT-duroid is better material for antenna designing as its loss tangent is low. It is also observed that the antenna with substrate material (FR-4) is smallest in size as compared to the other two materials as it has the higher dielectric constant. So for applications requiring small antenna size like military (FR-4) is better material to select for antenna designing. In future, the antennas with more materials can be investigated and compared with the experimental results.

References

1. Balanis CA (2005) Antenna theory analysis and design, 3rd edn. Wiley (2005)
2. Nawale PA et al (2014) Int J Eng Res Appl 4(6)(Version 3):123–129. ISSN: 2248-9622

3. Stutzman WL, Thiele GA (1998) Antenna theory and design, 2nd edn. Wiley, New York (1998)
4. Mandal D, Kar R, Bhattacharjee AK (2008) Progress Electromagn Res C, 1:191–198
5. Ghosh CK, Parui SK (2010) Design, analysis and optimization of a slotted microstrip patch antenna array at frequency 5.25 GHz for WLAN-SDMA system. Int J Electr Eng Inf 2(2)
6. (2014) vol 03 Issue 05. https://www.ijret.org
7. https://www.en.wikipedia.org/wiki/Microstrip_antenna
8. Chakraborty U, Chatterjee S, Chowdhury SK, Sarkar PP (2011) A Compact microstrip patch antenna for wireless communication. Prog Electromagn Res C 18:211–220

Author Index

V. Janyani et al. (eds.), *Optical and Wireless Technologies*, Lecture Notes in Electrical Engineering 472, https://doi.org/10.1007/978-981-10-7395-3

The manufacturer's authorised representative in the EU is Springer Nature Customer Service Centre GmbH, Europaplatz 3, 69115 Heidelberg, Germany. If you have any concerns regarding our products, please contact ProductSafety@springernature.com

Printed and bound by CPI Group (UK) Ltd, Croydon, CR0 4YY
01/10/2026
02225871-0020